GOLDEN'S DIAGNOSTIC RADIOLOGY

SECTION 6:
Roentgen Diagnosis of Diseases of Bone
THIRD EDITION **VOLUME TWO**

VOLUMES OF
Golden's Diagnostic Radiology Series

Section 1: **Diagnostic Neuroradiology** (Taveras & Wood)
Section 2: **Radiology of the Nose, Paranasal Sinuses and Nasopharynx** (Dodd & Jing)
Section 3: **Radiology of the Chest** (Rabin & Baron)
Section 4: **Radiology of the Heart and Great Vessels** (Cooley & Schreiber)
Section 6: **Roentgen Diagnosis of Diseases of Bone** (Edeiken)
Section 7: **Clinical Lymphography** (Clouse et al.)
Section 8: **Urologic Radiology** (Sussman & Newman)
Section 10: **Radiologic, Ultrasonic, and Nuclear Diagnostic Methods in Obstetrics** (Campbell et al.)
Section 17: **Tomography—Physical Principles and Clinical Applications** (Littleton)
Section 20: **Diagnostic Nuclear Medicine** (Gottschalk & Potchen et al.)
Section 21: **Radiology of the Colon** (Dreyfuss & Janower)
Section 22: **Radiology of the Gallbladder and Bile Ducts** (Hatfield & Wise)

Roentgen Diagnosis of Diseases of Bone

THIRD EDITION VOLUME TWO

JACK EDEIKEN, M.D.

Professor and Chairman, Department of Radiology; Jefferson Medical College, Thomas Jefferson University; Philadelphia, Pennsylvania; Consultant in Radiology; M.D. Anderson Hospital, Houston, Texas

SECTION 6
GOLDEN'S DIAGNOSTIC RADIOLOGY
John H. Harris, Jr., M.D., Series Editor

WILLIAMS & WILKINS
Baltimore/London

Copyright ©, 1981
Williams & Wilkins
428 E. Preston Street
Baltimore, Md. 21202, U.S.A.

All rights reserved. This book is protected by copyright. No part of this book may be reproduced in any form or by any means, including photocopying, or utilized by any information storage and retrieval system without written permission from the copyright owner.

Made in the United States of America

Second Edition 1973
    Reprinted 1978, 1979
First Edition 1967
    Reprinted 1969, 1970, 1975

Library of Congress Cataloging in Publication Data

Edeiken, Jack.
    Roentgen diagnosis of diseases of bone.

    (Golden's diagnostic radiology; section 6)
    Includes bibliographies and index.
    1. Bones—Diseases—Diagnosis. 2. Bones—Radiography. I. Title. II. Series.
[DNLM: 1. Bone and bones—Radiography. WN200 G181 sect.6]
RC78.G6 sect.6, 1981   [RC930.5]   616.07′572s
ISBN 0-683-02744-1 (set) [616.7′1′07572] 80-12566

Composed and printed at the
Waverly Press, Inc.
Mt. Royal and Guilford Aves.
Baltimore, Md. 21202, U.S.A.

Dedication

To my family:
   Yale Frederick Edeiken
     Beth Sue Edeiken
       Matt Monroe
         Wendy Edeiken Feins
         Herb Feins
           Nanette Edeiken Cooperman
            Harry Cooperman
             Louis M. Edeiken
              Douglas Monroe
               Brett Monroe
                 Pamela Feins
                  Jason Feins
                   Anneliese

Preface to the Third Edition

The *Roentgen Diagnosis of Diseases of Bone* was begun in 1962 and the first edition was published five years later. Since advances and changes had taken place over the five years, revisions for a second edition began almost immediately, and after six years of preparation this updated edition appeared in 1973, this time in two volumes. Immediately after publication, work began on the third edition, to include 35 new subjects, 500 new photographs, and updating of all sections.

From the beginning this work was designed to provide those interested in osteoarticular abnormalities with an understanding of their radiologic diagnosis. In most of the chapters, there is an approach before the discussion of the specific diseases. Hard-core radiologic information is presented against the background of characteristic pathologic and clinical findings. The goal is to serve as a basis for greater skill and accuracy in skeletal radiology, and the radiographic reproductions, many of which appeared in the first two editions, have been carefully selected toward this end. Some of the diagnoses have been modified in light of added current information, but, of course, it is the interpretation of the condition, and not its appearance, which changes as medicine moves forward.

The third edition goes beyond the first two and covers conditions overlooked or unknown before, including a review of current information on skeletal maturation, with charts. The dysplasias and mucopolysaccharides have been put at the end of Volume II, since they act more as reference material than as a reading text.

Because of my known interest in skeletal radiology, I have been privileged to see innumerable examples of less common and rare diseases sent from the four corners of the earth. It is impossible to list everyone whose correspondence has broadened my knowledge, but with this Preface gratitude is extended. I am deeply indebted to them for allowing me to share in their experiences which have enriched this text and widened its scope.

Indebtedness has been incurred to Gilbert and Ring for their glossy prints of unusual fidelity, for their photographic expertise, and for their sympathetic involvement in our problems.

Maureen Curran, Barbara Schilling, Betty Rehmann, Raelea Foxwell, and Blanche Edeiken Kramer have spent endless hours editing and typing this work, and the author thanks them for their patience and their generous devotion to the task.

Acknowledgment is given for the help and the cooperation of the publishers, The Williams & Wilkins Company. Their interest has lightened the load. Particular thanks is given to Ruby Richardson, Alice Reid, Diana Welch and Wayne Hubbel for their personal supervision of the manuscript.

With a work that requires so much added time, friends, colleagues and family are sure to suffer. My personal thanks go to my very close friends, Jerome and Andrea Wiot, Gerald and Glenn Dodd, Harold and Ruth Jacobson, Cathy and John Harris, Jr., Ray and Connie Mandell, John and Sally Fenlin, Ruth and Herbert Weiman, Richard Rothman, Barbara and Harold Levick, Gertrude and Hugh Weiss, and Arlene and Robert Kaskey, who supported me in my times of need.

My colleagues at Thomas Jefferson University Hospital, Doctors Richard Brennan, John Curtis, Stephen Feig, Gham Hur, Harvey Koolpe, K. Francis Lee, Esmond

Mapp, Gerald Nissenbaum, A. Edward O'Hara, Vijay M. Rao, Gary Shaber, Robert M. Steiner, Bruce Stratt, Noble Thompson, David Weiss, Barry B. Goldberg, Catherine Cole-Beuglet, Alfred Kurtz, Leonard Ellenbogen, Mortimer B. Hermel, Morton G. Murdock, Stephen Pripstein, and Irvin Wexlar, unstintingly accepted my absences from patient service and cooperated in covering for me so frequently.

Finally, my family, scattered all over the country, understood my infrequent visits and supported me by their love.

JACK EDEIKEN, M.D.

Contents

VOLUME ONE

	Preface to the Third Edition	vii
Chapter 1	Basic considerations of bone and cartilage	1
Chapter 2	General radiologic approach to bone lesions	8
Chapter 3	New bone production and periosteal reaction	11
Chapter 4	Bone tumors and tumorlike conditions	30
Chapter 5	Arthritides	414
Chapter 6	Bone ischemia and osteochondroses	679
Chapter 7	Osteomyelitis	727
	Author Index	xi
	Subject Index	xxxvii

VOLUME TWO

	Preface to the Third Edition	vii
Chapter 8	Metabolic and dystrophic bone disease	829
Chapter 9	The anemias	1047
Chapter 10	Congenital hip dislocation	1084
Chapter 11	Spondylolisthesis	1089
Chapter 12	Calcinosis, calcifications, and myositis ossificans	1095
Chapter 13	Osseous manifestations of metal poisoning	1120
Chapter 14	Skeletal maturation	1148
Chapter 15	Dysplasias	1179
Chapter 16	Mucopolysaccharidoses	1465
Chapter 17	Chromosomal abnormalities	1487
Chapter 18	Lipidoses	1520
	Author Index	xi
	Subject Index	xxxvii

8
Metabolic and Dystrophic Bone Disease

Metabolic bone disease is bone dystrophy and includes endocrine imbalance, disturbances due to vitamin deficiency or excess, osteoporosis, and osteomalacia.

The term dystrophy, meaning disturbance of nutrition, is applicable to metabolic and endocrine bone diseases, since these affect the bone by depriving it of normal mineral or organic components. Dystrophy, however, should be distinguished from dysplasia, which means a disturbance of bone growth; the two terms are not interchangeable and should not be confused.

The terms osteomalacia and osteoporosis are used to describe specific bone dystrophies. Osteomalacia is a failure of deposition of calcium salts in bone matrix, and osteoporosis, a deficiency of the organic matrix of bone. Osteoporosis is sometimes used nonspecifically to describe decalcification or bone atrophy, but, when it is used specifically of a bone dystrophy, the above definition applies.

Radiographic evaluation of bone formation and resorption is difficult. Variations in technique may preclude the accurate evaluation of density, which varies with the individual, and considerable loss of bone must occur before it is apparent roentgenographically.[1] Large areas of bone destruction may not be roentgenographically demonstrable.[2,3] The clinical terms, osteoporosis, deossification, demineralization, osteosclerosis, osteolysis, etc., are used almost interchangeably and compound the confusion. It would be preferable to avoid such terms, referring to opacity changes merely as either decreased or increased density unless the underlying condition is known. Decreased density may be due to lack of mineralization, lack of bony matrix, or increased rate of destruction. The factors influencing bone metabolism are complicated and incompletely understood. A simplified overview of normal bone metabolic mechanisms is as follows.

Calcium is the principal skeletal mineral; 99% of the body calcium is concentrated in bone. Serum calcium occurs as protein-bound and ionic in equal amounts. Approximately 3% of serum calcium is citrate and phosphate complex.

Ionized calcium is the most important. Physiologically, parathormone responds to the amount of ionized calcium, not the total serum calcium levels. The protein-bound calcium, mostly albumin, is pH dependent and decreases with acidosis.

Calcium is absorbed from the gastrointestinal tract by active transport and simulated diffusion, which is influenced by vitamin D. Calcium binding to protein occurs in the intestinal mucosa, controlled by vitamin D. The percentage of calcium absorption decreases with increased intake; a low calcium diet will increase the efficiency of absorption, provided there is adequate vitamin D. The urinary excretion of calcium is directly related to dietary intake, and excretion greater than 500 mg per 24 hr indicates hypercalciuria.

Phosphorus is absorbed by active transport across the intestinal lumen and requires the presence of sodium. Aluminum hydroxide gel in the gut binds phosphorus and decreases absorption, a mechanism used in treating chronic renal disease.

Phosphate excretion is increased by estrogens and parathormone, and decreased by growth hormone, vitamin D, and glucocorticoids. Normally 85–95% of phosphorus is resorbed by the tubules, and a decrease in total resorption of phosphorus is a diagnostic test for hyperparathyroidism. The serum phosphate level has no direct effect on the secretion of parathormone or calcitonin.

Vitamin D sources are dietary absorption, and the conversion of inactive precursors in the skin to active metabolites via sunlight. Vitamin D is responsible for adequate calcium absorption from the gastrointestinal tract, and necessary for the synthesis of the calcium binding protein in the small intestinal mucosa.

Vitamin D is required for the parathormonal effects on bone, and stimulates osteoclastic and osteocytic resorption of bone.

Parathyroid hormone (parathormone) increases calcium release from bone, elevating the serum calcium level. It also promotes bone remodeling by stimulating mesenchymal proliferation and osteoclast induction, and accelerates the lytic activity of osteocytes. The principal effect is an increase in osteoclasts, but the initial increase in serum calcium is rapid and is believed to be secondary to osteocytic osteolysis and perhaps osteoclast stimulation. Vitamin D in pharmacologic concentration is required for the action of parathormone, and stimulates bone resorption via osteocytes and osteoclasts.

Parathyroid hormone also acts on the kidneys and thereby regulates tubular resorption of phosphorus.

Thyrocalcitonin is a polypeptide produced by the parafollicular cells of the thyroid gland; it inhibits bone resorption. Its action is rapid and probably mediated through osteocytes, and occurs in the absence of thyroid hormone, although it is more obvious during high levels of resorption. Calcitonin may be responsible for the osteosclerosis of renal osteodystrophy, and it causes persistance of bone sclerosis after the parathyroid removal.

Thyroid hormone influences metabolism generally, and has no specific daily regulation. Hyperthyroidism causes increased calcium release from bone, resulting in osteoid seams. Hypothyroidism results in a decreased modeling rate and bone blood flow.

Growth hormone is a strong stimulus to bone formation, causing increased skeletal mass and normal trabecular and cortical bone formation.

Gonadal hormones (androgens, estrogens, and synthetic steroids) are anabolic agents that depress bone resorption without influencing bone formation.

Adrenocortical steroids: The skeletal response to steroids is species specific; in man, it increases bone resorption and decreases bone formation.

Reifenstein[4] provides a logical scheme for the analysis, physiology, and chemical changes of metabolic disease:

I. Too little calcified bone
   A. Too little bone formation
      1. Too little formation of matrix: osteoporosis
      2. Too little calcification of matrix: rickets or osteomalacia
   B. Too much bone resorption
      1. Too much resorption of matrix and mineral: osteitis fibrosa generalisata (hyperparathyroidism)
      2. Paget disease
II. Too much calcified bone
   A. Too much bone formation
      1. Too much formation of matrix: hyperosteogenesis, e.g., excessive stress, chemical poisoning
      2. Too much calcification of matrix: apparently nonexistent
   B. Too little bone resorption
      1. Too little resorption of matrix and calcium: osteosclerosis
         a. Congenital osteoblastic defect: osteopetrosis
         b. Deficient parathyroid hormone
            (1) Hypoparathyroidism
            (2) Pseudohypoparathyroidism.

OSTEOPOROSIS

Osteoporosis is caused by a deficiency of bone matrix, but there is normal mineralization of the remaining bone. Accelerated bone resorption is the usual cause,[5] but decreased bone formation is found in Cushing syndrome, with steroid administration, and disuse or immobilization osteoporosis.

Roentgenographic evaluation of bone density is difficult, because 30% of bone must be lost before it is apparent with routine studies. Cortical thickness measurements of the metacarpals, radius, and humerus[6] have proved useful. The standards used must be of the same age and sex, and probably from the same geographic region. The normal bone mass (density) changes with age, increasing from infancy to age 35–40, and then progressively decreasing at the rate of 8% per decade in women, and 3% in men.[7]

Roentgenographic quantitative bone densitometry is not accurate without careful control of filtration (almost monochromatic) beam, film quality, and darkroom techniques.[8] Iodine-125 radiodensitometric techniques have proved successful.[9]

A physiologic classification follows[1, 10]:

I. Defects in the osteoblasts
   A. Congenital (osteogenesis imperfecta)
   B. Lack of stress and strain (disuse atrophy)
   C. Deficient estrogen (primary ovarian agenesis and postmenopausal osteoporosis)
II. Defects in the matrix
   A. Deficient androgen (eunuchoidism, senile state in males)
   B. Deficient protein (malnutrition, hyper-

thyroidism, uncontrolled diabetes, Cushing syndrome, prolonged stress, and scurvy)

III. Excessive utilization of calcium.

Osteoporosis results from decreased production or increased resorption of bone matrix, usually lack of osteoblast activity or from protein deficiency. Osteoblastic inactivity is often accompanied by low serum alkaline phosphatase, although normal levels may be present with severe osteoporosis.

Deficiency of bone matrix precludes mineralization, and decreased density results. The blood chemical changes are normal unless there is rapid osteoporosis, and may remain normal even during rapid demineralization. A classification of osteoporosis according to cause follows:

Endocrine:
1. Cushing syndrome
2. Hyperthyroidism
3. Hypothyroidism
4. Hypogonadism
   A. Turner syndrome
   B. Secondary postmenopausal
5. Acromegaly

Deficiency diseases:
1. Scurvy
2. Malnutrition
   A. Anorexia nervosa
   B. Kwashiorkor
   C. Concentration camp starvation

Idiopathic:
1. Juvenile
2. Adult

Congenital:
1. Homocystinuria
2. Osteogenesis imperfecta

Induced via drugs:
1. Heparin
2. Cortisone
3. Vitamin A

Neoplastic:
1. Multiple myeloma
2. Metastatic
3. Liver tumors in children

Miscellaneous:
1. Rheumatoid arthritis
2. Senility
3. Immobilization
4. Regional migratory osteoporosis
5. Liver disease
6. Pregnancy
7. Lactation.

Senile Osteoporosis

Senile osteoporosis may result from several factors: (1) osteoblast inactivity; (2) lack of gonadal hormones, necessary for osteoblast activity; (3) deficiency of protein; and (4) inadequate diet.

Postmenopausal Osteoporosis

The postmenopausal background is common in osteoporosis, due to gonadal hormone deficiency and decreased osteoblast activity.[11] (Eunuchoidism also causes osteoblast inactivity.) Saville[12] studied 80 women with spine fractures due to osteoporosis and compared them to a controlled group of women matched for age.

The osteoporotics were shorter, lighter, and less often of Italian and Negro stock. Their milk-drinking habits were similar to controls, and menopause occurred at a similar age. They sustained more long bone fractures than controls and were especially prone to repeat fractures of the hip, humerus, and wrist. Neither long bone fractures nor spine fractures were more common in osteoporotics who did not drink milk than in those who drank more than one glass a day, but the latter had slightly thicker radial cortices. Nearly half of of the osteoporotics had fractured T12, none had fracture above T3, and none developed neurologic sequellae as a result of vertebral fracture. Nine patients sustained spine fractures without obvious trauma, but more than one-third were lifting objects when the spine was in a slightly flexed position. Eighteen percent sustained recurrent spine fractures while under observation.[12]

Disuse Osteoporosis

Disuse atrophy results from lack of stress and strain on the bone, as when parts of the skeleton are immobilized, removing normal stress and leading to osteoporosis. Ordinarily, disuse osteoporosis is relieved when the affected part is mobilized, but prolonged immobilization produces irreparable bone damage, especially in adults.

Acute immobilization osteoporosis is frequently caused by paralysis or body cast, especially in young patients.[13] Without the stress on bone, osteoblasts are inactive and older bone is not replaced. Calcium withdrawal may be so rapid that hypercalcemia results.[14] Excessive urinary excretion may lead to renal calculi. Administration of large doses of vitamin D and calcium to young patients in casts is harmful, since it exaggerates the condition created by immobilization. The reduction of calcium intake and increased fluid consumption relieves the hypercalcemia and hypercalciuria. With mobilization, the osteoblasts resume normal activity, and calcium again is laid down in newly formed matrix.

Osteoporosis Due to Protein Deficiency

Protein deficiency, or abnormal protein metabolism, may result in osteoporosis because matrix is not produced[10] as in malnutrition, nephrosis, diabetes mellitus, Cushing's syndrome, and hyperthyroidism.[15]

The bone changes of infantile scurvy are a form of osteoporosis occurring during the period of endochondral bone growth and will be discussed in detail elsewhere.

Heparin Osteoporosis

Griffith et al.[15] reported on the relationship between long term use of large amounts of heparin and symptomatic osteoporosis. Patients treated with 10,000 units daily for 1-15 years did not develop symptoms of osteoporosis. However, of 10 patients given 15,000-30,000 units for 6 months or longer, 6 developed spontaneous vertebral or rib fractures and pain. Bone biopsies of 2 patients revealed a soft bone matrix easily cut by the pathologist's knife. The bone marrow revealed a slight plasmocytosis. Renal and parathyroid functions were normal in a few patients studied. Back pain improves dramatically after withdrawal of heparin therapy.

The mechanism that causes the osteoporosis is unknown, but it is postulated that heparin has a direct local stimulating effect on bone resorption.[16] This is supported by the fact that heparin stimulates bone resorption in tissue culture.[17]

Marfan and Hurler syndromes may represent naturally occurring hyperheparin states; elevated systemic levels of heparin have been found.[18] Bone resorption in mast cell disease[19] may also be the result of a hyperheparin state.

Roentgenographic Features of Osteoporosis

The roentgenographic appearance takes two forms, adult and infantile, the differences depending on bone maturation. The infantile form, epitomized by the changes of scurvy and osteogenesis imperfecta, will be discussed subsequently.

The main defect in osteoporosis is deficiency of bone matrix; it has the same roentgenographic appearance in all conditions producing it. Its diagnosis depends on both the clinician and roentgenologist. First, there must be no evidence, either roentgenologic or clinical, of osteomalacia or hyperparathyroidism. Second, there should be evidence of a condition, such as senility, the postmenopausal state, immobilization, or disturbance of protein metabolism, to explain osteoporosis.

The roentgenologic changes of osteoporosis are often discovered when well advanced. The changes are subtle, but identifiable. The most striking is cortical thinning of long bones with irregularity of endosteal surfaces (Figs. 923-928). The thin cortex maintains normal mineral content and appears dense, in contrast with that of deossification. The result is generalized bone density decrease, with thin but usually dense cortices. Spongy bone loses some of the trabeculae; those remaining are in line of stress, and increase in density and width (Fig. 929).

Identical changes occur in vertebrae, and vertebral plates become thin but dense in contrast with deossified bone (Figs. 930 and 931). The endosteal surface of the vertebral plate is irregular which distinguishes it from senile osteoporosis (Fig. 932). The trabecular pattern is accentuated because of loss of horizontal trabecular structure and maintenance of vertical trabeculae (lines of stress) (Figs. 933 and 934).

There is a notable absence of osteophyte formation and ossification of the anterior and lateral spinal ligaments. Fractures occur in the brittle bones of osteoporosis, especially the femoral neck and centra of dorsal vertebrae. Anterior wedging is the most dramatic manifestation. All vertebral bodies may compress, and pressure of disks causes biconcavity (Fig. 934).

Severe osteoporosis occurs in Cushing syndrome, and, if the condition remains untreated, multiple fractures of vertebral centra occur. The vertebral plates become thick and fuzzy during the active phase of Cushing disease (or with corticosteroid excess), a distinguishing feature and an excellent means of differentiation. With healing, cortical plates become thin and appear dense. Fat deposition may cause mediastinal widening and loss of clavicular companion shadow (Fig. 935).

Although osteoporosis is usually discovered in the final stage, it is possible to observe the progressive changes, especially when an extremity is immobilized. Immobilization leads to disuse atro-

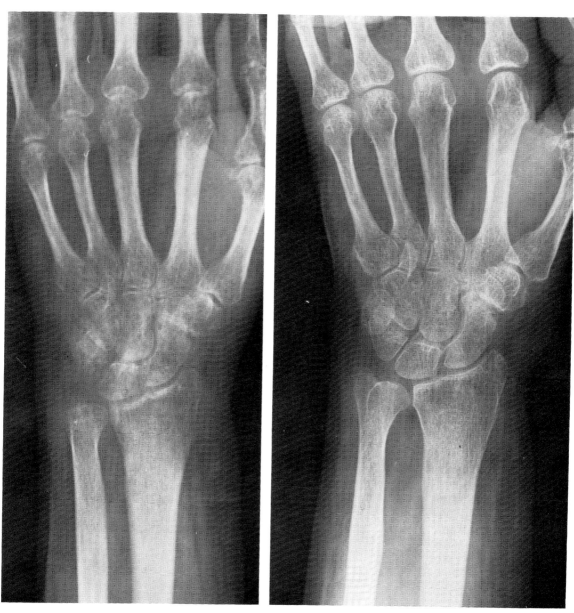

Fig. 923. Sudeck Atrophy

(*Left*) Acute deossification. The ends of the bones and the carpal bones reveal marked loss of mineral content. The cortices are maintained in density and thickness. This patient had severe pain for 10 days after trivial injury. (*Right*) Same patient, 6 months later, reveals cortical thinning and irregularity of the endosteal surface. The minerals from the cortices have been mobilized and, although the spongy portions of bones are as deossified as in the first photograph, they do not stand out because of deossification in the cortices. At this time the patient was asymptomatic.

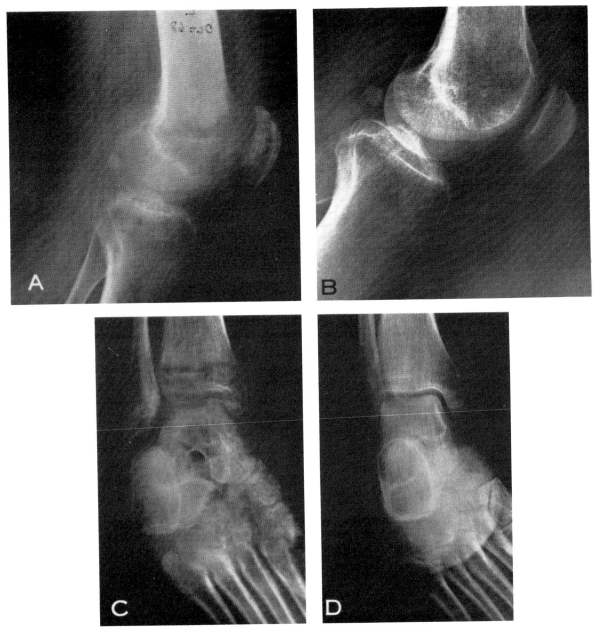

FIG. 924. REGIONAL MIGRATORY OSTEOPOROSIS

A young man with regional pain and swelling of rapid onset that persisted for 6 months and then gradually improved. The knee pain began in October 1968 and in the ankle in December 1968. (*A*) 1968: Diffuse deossification adjacent to the joint. (*B*) 1970: Reossification. (*C*) December 1968: Marked deossification. (*D*) 1970: Reossification. (Courtesy of Dr. Robert Steiner, Flemington, N.J.)

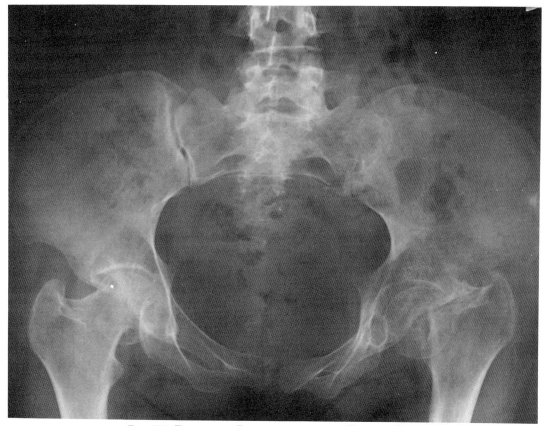

Fig. 925. Transitory Demineralization of Femoral Head
This patient developed demineralization of the iliac bone and femoral head and neck during pregnancy. Subcapital fracture occurred.

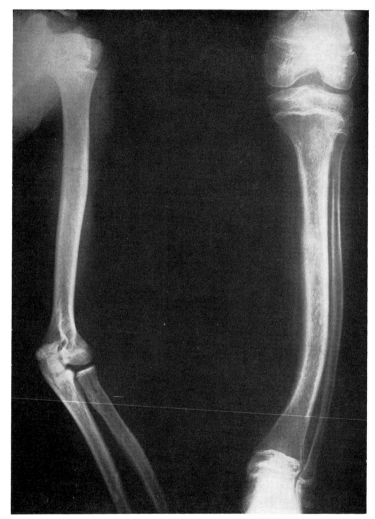

Fig. 926. Osteoporosis Due to Disuse

Lower extremities were immobilized after osteotomies to correct deformities due to rickets. The tibia reveals deossification with radiolucency of the spongy portion of the bone and irregularity of the cortices. The upper extremity bones do not show deossification.

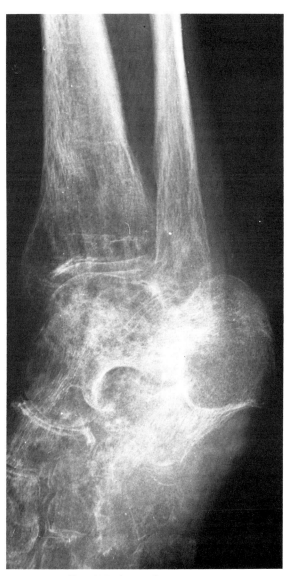

FIG. 927. ACUTE OSTEOPOROSIS
Patchy decrease in density in the tarsal bones and distal ends of the tibia and fibula. Thinning of the cortex of the tibia and irregularity of the endosteal surface. The thin cortex maintains normal mineral content and appears dense in contrast with the remainder of the deossification.

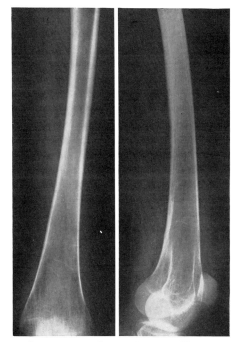

FIG. 928. OSTEOPOROSIS DUE TO PRIMARY HYPOGONADISM

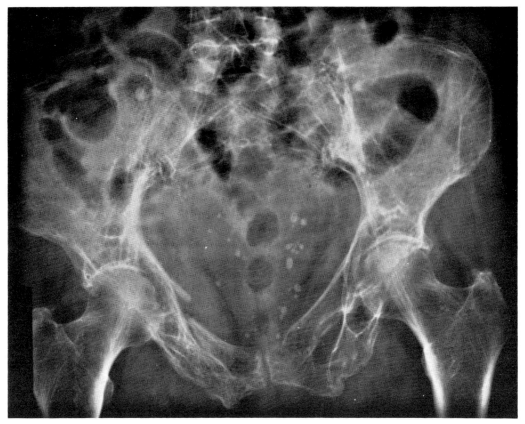

FIG. 929. OSTEOPOROSIS

Spongy bone has lost most of the trabeculae. Those remaining are the lines of stress and increase in density and width. This is particularly prominent in the necks of the femurs and the iliac bones. Multiple fractures are present in the rami of the pelvis. (Courtesy of Herbert M. Stauffer, Temple University Hospital, Philadelphia, Pa.)

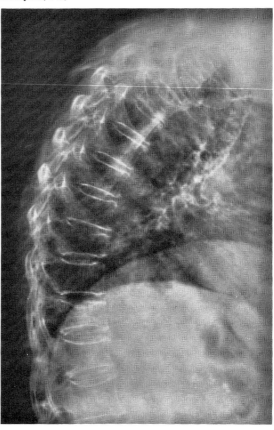

FIG. 930. OSTEOPOROSIS

Vertebral plates are thin but dense in contrast with appearance seen in the deossification of bone. Multiple anterior compression fractures. (Courtesy of Herbert M. Stauffer, Temple University Hospital, Philadelphia, Pa.)

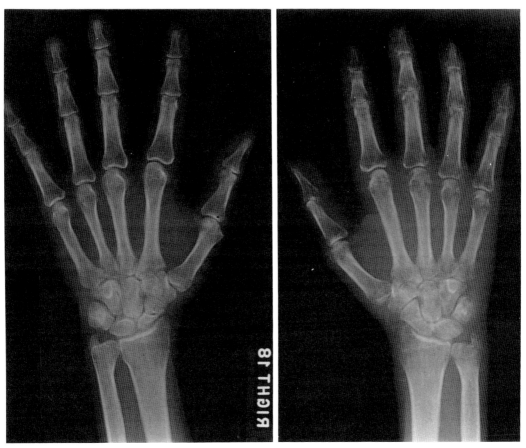

Fig. 931. Sudeck Atrophy of the Left Hand

Clinical Sudeck atrophy: The right hand is normal; the left shows periarticular deossification. The cortices remain normal in thickness, indicating acute disease. Deossification in the spongy bone occurs before cortical bone resorption. Note the resorption of the tufts which is also spongy bone. This roentgenogram was made 2 months after injury to the humerus.

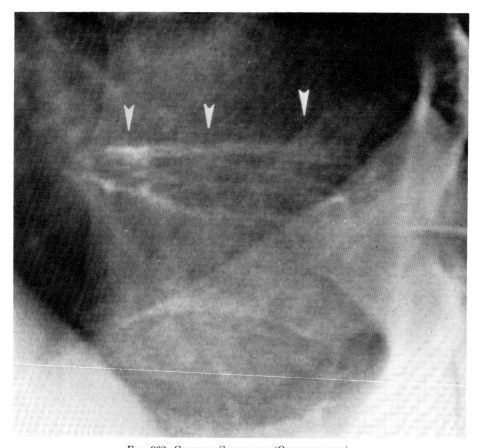

FIG. 932. CUSHING SYNDROME (OSTEOPOROSIS)
Compression of one vertebra and increased density of the vertebral plates. The endosteal surface is irregular (*arrows*) indicating an active osteoporotic phase rather than physiologic osteoporosis.

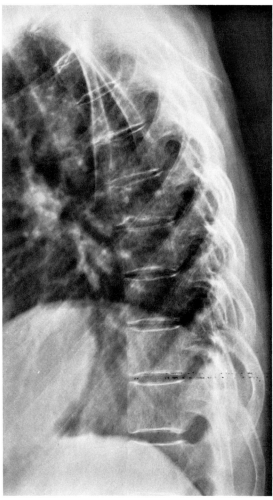

Fig. 933. Osteoporosis Due to Cushing Disease
Vertebral plates are thin and appear dense. The vertical trabecular pattern in the lower dorsal spine is accentuated. (Courtesy of Herbert M. Stauffer, Temple University Hospital, Philadelphia, Pa.)

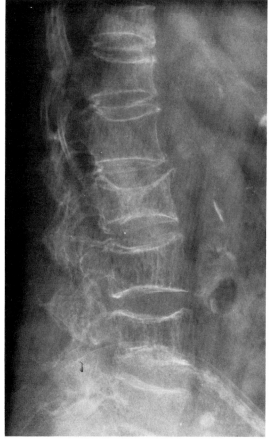

Fig. 934. Osteoporosis (Cause Unknown)
Deossification of the vertebrae and loss of horizontal trabecular structures. The vertical trabeculae are in the lines of stress and are widened and prominent. Pressure of disk causes biconcavity of some of the vertebrae. (Courtesy of Herbert M. Stauffer, Temple University Hospital, Philadelphia, Pa.)

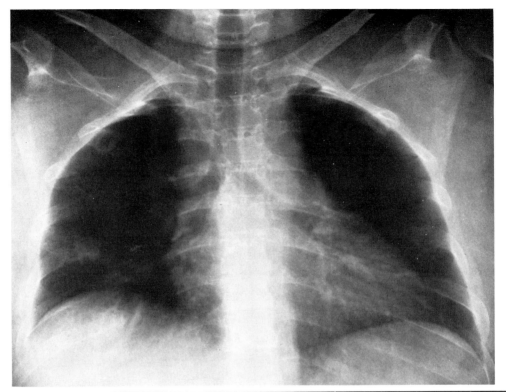
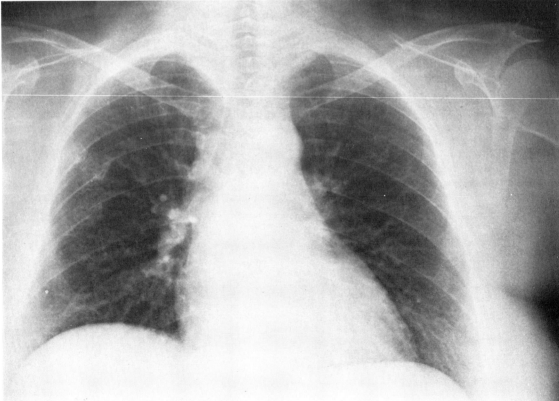

Fig. 935. Cushing Syndrome

Mediastinal widening and loss of the clavicular companion shadow due to fat deposition. Multiple healed rib fractures. The bones are deossified. (*Top*) Supine: (*bottom*) erect.

phy, the early phase of osteoporosis. First, density decreases in cancellous portions of bone, because minerals are more easily mobilized here than in compact bone (Fig. 936). The contrast between deossified bone and destroyed cortical bone is striking. Then the cortex thins, and a patchy decrease in density occurs throughout (Fig. 937). As osteoporosis progresses, the bone loses this patchiness, and changes proceed to the final appearance as described above.

There is no cure for adult osteoporosis, and, once the condition is established, full restitution of bone will not occur. Certain forms, such as disuse osteoporosis or Cushing syndrome, show improvement after removal of the cause. The roentgenographic appearance of osteoporosis in children is reversible, which may be due to the child's ability to reconstitute bone rather than to the actual healing of osteoporotic bone.

Regional Migratory Osteoporosis (Transitory Osteoporosis)

This is an unusual syndrome described by Duncan et al.[1] It is more common in men and is usually seen in the 4th and 5th decades of life. There is regional pain and swelling with a rapid onset of osteoporosis localized in the painful areas (Fig. 924). Spontaneous involvement of other regions may occur concomitantly or later. Occasionally it is accompanied by periosteal new bone formation.

The rapid onset of painful transient osteoporosis may occur in single or multiple joints, either successively or concurrently. Other conditions which are probably the same or closely related include migratory osteolysis of the lower extremity, transient osteoporosis of the hip, transient osteoporosis of foot and knee, peculiar artropatia rarefacente dell'anca, sympathetic dystrophy of lower limbs, and reflex sympathetic dystrophy of the foot.[2-9]

Osteoporosis may be partial, in that one or two fingers may be involved; or zonal, such as a portion of the femoral head. Patients with these localized types tend to recover in 6 months. Patients with partial transitory osteoporosis may recover completely, after which a migratory type may ensue, which also may be zonal. There are cases of the distal femur or one tibial plateau being involved.

The partial transitory osteoporosis occurs mostly in men and usually proceeds to a complete osteoporosis within 3 months. It then subsides within a period of 6-7 months.[10]

Lequesne described three types or grades of osteoporosis: (1) loss of the transverse trabeculation, (2) that on the tension side of the bones, and (3) compression trabeculae. This pattern is best observed in the hip and the os calcis. The same type of osteoporosis occurs in Paget disease and therefore the pattern of trabecular loss is: (a) loss of transverse trabeculae, (b) thickening of the tension, and (c) compression trabeculae.[10]

Transitory Demineralization of Femoral Head

Rosen[1] reported 3 patients with transitory demineralization of the femoral head and believed the entity was related to regional migratory os-

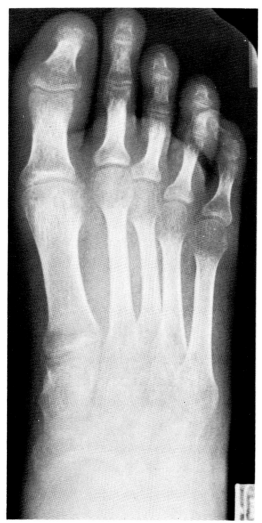

FIG. 936. ACUTE OSTEOPOROSIS (MINOR TRAUMA)
Deossification in the distal ends of the metatarsals and phalanges is striking. The cortical thickening is maintained and density along the shaft maintained. This is because there is easy mobilization of minerals in the spongy portion of bone, whereas the cortical bone maintains minerals for quite some time.

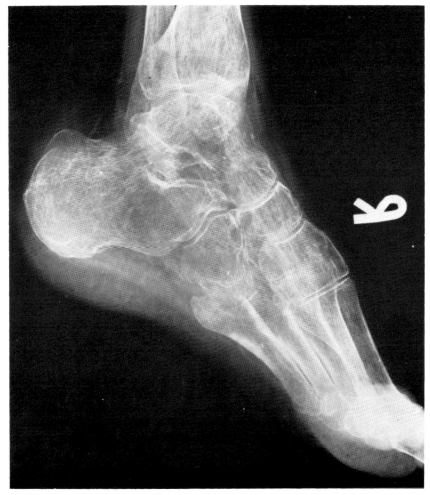

FIG. 937. OSTEOPOROSIS (LONG STANDING)
Compare with Figure 927. Homogeneous deossification. After some time the patchy deossification is lost and the bone is totally demineralized.

teoporosis. One of his cases and some in the literature have had multiple bone involvement. It occurs frequently in pregnancy, and a mechanical compression syndrome is the suggested cause.

Pathologic fracture is a serious complication (Fig. 925).

Jaundice Osteoporosis

Atkinson et al.[1] described osteomalacia and osteoporosis coexisting in patients with prolonged jaundice. Most have osteomalacia but some have osteoporosis.

Sudeck Atrophy[1]

This condition, also called posttraumatic reflex atrophy of bone, has long been difficult to explain. It often follows rather trivial injury, but sometimes occurs after major trauma. The affected part suffers partial loss of motor function, mild to severe vasomotor and trophic changes, demineralization of the bones near the site of trauma, and mild to severe aching pain.[2,3] In all instances the degree of disability is out of proportion to the trauma. The aching pain is often severe and is not relieved by immobilization, as are most injuries.

Roentgenologic Findings. The hands and feet are the most common sites. The osteoporosis involves all of the bones of the affected extremity, but the changes are greatest in the periarticular portions. In the acute form there is mottled irregular rarefaction of the bones. The mottling is lost as the disease progresses and the osteoporosis becomes diffuse, with thinning not only of the trabeculae of the spongiosa but also of the cortex. Since the changes are greatest near the joint, the articular cortex of the bone may become so thin as to be difficult to see, but under close observation it will be found intact (Fig. 923). This observation is important to distinguish Sudeck's atrophy from arthritis, especially tuberculous arthritis.

During healing the bones recalcify, but in many instances never regain normal density, although the symptoms may be relieved completely.

The cause of the osteoporosis in Sudeck atrophy is not known. Perhaps the immobilization resulting from the intense pain is a factor; however, it is also likely that the vasomotor changes are involved with the bone decalcification.

Idiopathic Juvenile Osteoporosis

Idiopathic juvenile osteoporosis occurs just prior to the onset of puberty, runs a limited course, and usually ends in healing before or during puberty without residual deformity.[1-4]

Clinical Features. A previously healthy child complains of spinal pain and/or pain in the ankles and knees about 2 years before puberty. The pain is due to metaphyseal infractions and is easily confused with arthritis when the radiologic abnormalities are not discovered.

Pathologic Features. There is an increase in bone resorption, probably due to increase in osteoblastic absorption.[5] A low positive to markedly negative calcium balance is the only consistent abnormality.[1]

Roentgenographic Features. The dorsal and lumbar spine are almost invariably affected[1] with one or several vertebral bodies collapsed. The number of bone trabeculae is decreased in the vertebral bodies and the vertebral end plates are increased in clarity. The degree of involvement usually is proportional to the severity of change in the appendicular skeleton. Anterior wedging of the vertebral bodies occurs earliest in the dorsal spine, and kyphosis and scoliosis may result depending on the severity of the disease. A "codfish" appearance occurs later in the lumbar vertebral bodies due to the marked expansion of the intervertebral disks, which causes a biconcave impression on the vertebral bodies. Displacement of the apophyseal rings may cause steplike impressions of the lower and upper anterior surfaces of the vertebral plates.[1]

Reconstitution of the shape of the vertebral bodies occurs when remission begins during puberty; it takes 1–4 years, and a normal outline eventually occurs in almost all patients. Kyphosis and scoliosis tend to correct spontaneously. Restoration of the patient's height is rarely complete.[1, 6] Growth lines (ghost vertebrae) may persist.[1]

There is osteoporosis of the pelvis in all cases with severe disease, and protrusio acetabuli, triradiate pelvic deformity, and medial slip of the femoral heads occur, as may metaphyseal fracture of the femoral neck.[1]

The skull is not seriously affected although mild vault and sella turcica demineralization may occur.

The ribs are osteoporotic and fractures are frequently seen in severe cases.

The bones of the extremities are usually osteoporotic, as evidenced by cortical thinning. Metaphyseal fractures may be symmetrical and begin as incomplete cortical fractures which progress across the bone. The bilateral symmetrical features are similar to the Milkman syndrome of osteomalacia. Fractures usually affect the weight-bearing bones, particularly near the ankles and knees, but non-weight-bearing bones are also affected. The femoral condyles may be enlarged and deformed. Healing of fractures is usually normal, but delayed union and pseudoarthrosis may occur.[1]

The hands and feet usually show marked demineralization with fractures in severe cases.[6] Osteogenesis imperfecta tarda may simulate idiopathic juvenile arthritis. The following features of this condition include: blue sclera, marked pelvic deformity, deafness, wormian bones, abnormal teeth, and increased hydroxyproline,[5] which aid in distinguishing it from idiopathic juvenile osteoporosis.

Infantile Scurvy (Barlow Disease)

Although many historical accounts of scurvy in adults have been recorded, there is little mention historically of infantile scurvy. This is probably because, up to the present time, breast feeding was almost universal, and human milk has a much higher concentration of ascorbic acid than cow's milk. With non-human milk feeding, the infant has become more dependent on exogenous sources of ascorbic acid, being unable to synthesize vitamin C as do other mammals.[1]

Etiology. A deficiency of ascorbic acid causes abnormal function of the osteoblast and defective osteogenesis resulting in osteoporosis. The ascorbic acid is a strong reducing agent, and also influences connective tissue metabolism,[2] folic acid metabolism,[3] iron metabolism,[4, 5] blood coagulation,[6] and vasoconstrictive response to epinephrine.[7]

Clinical Features. The clinical symptoms and signs of infantile scurvy, in order of frequency, are irritability, tenderness and weakness of the lower extremities, a scorbutic rosary of the ribs, legs drawn up and widely spread, pseudoparalysis, bleeding of the gums (usually where teeth have erupted), and fever.[1,8] Most symptoms appear between 6 and 9 months of age,[8,9] and although it was previously believed not to develop so early, several cases have been documented at less than 3 months, probably due to prenatal deprivation.[10]

Laboratory Features. The normal plasma level of ascorbic acid is 0.6 mg per 100 ml, so a laboratory value of at least this level excludes scurvy. A level of 0.2 mg per 100 ml, or less, indicates ascorbic acid deficiency. However, not even a 0 level assures clinical scurvy, since the deficiency must be maintained for 3–6 months before symptoms occur. A more sensitive indicator of diminishing ascorbic acid level is the buffy layer (white blood cell and platelet layer) content, but this test is not available in most laboratories.[1,8]

Pathologic Features. The morphologic changes of scurvy are due to interference with the formation of intercellular substance, i.e., collagen, osteoid, dentin, and vascular endothelium. Both chondroblastic and osteoblastic activity is disordered,[11,12] the bone changes being most prevalent where the growth is normally most rapid: at the sternal end of the ribs, the distal end of the femur, the proximal end of the humerus, both ends of the tibia and fibula, and the distal end of the radius and ulna. Development and calcification of new cartilage cells occur normally, but the removal of the calcified cartilage matrix is interfered with. Normal bony trabeculae are not formed and, where bone is already formed, resorption occurs, due to immobilization and failure of osteogenesis under normal or accelerated osteoclastic activity.[11,13]

Roentgen Features. Although several months may elapse between the clinical onset of the disease and the development of typical roentgen signs ("latent radiographic"),[1,14] the roentgenogram is probably the easiest and most practical method of confirming the diagnosis as well as assessing treatment. The roentgen changes help determine the severity as well as the extent of the disease, especially in children under 2 years of age. Mild cases may reveal no abnormal roentgen features.[14]

No particular feature is alone specific to the disease, and a combination of them is needed for the diagnosis.[1,13] The following characteristic roentgen changes, in order of frequency, are found in multiple films with bilateral involvement.[1,8,10,11,13,14]

1. DENSE METAPHYSEAL LINE. This is due to an intensification of the zone of preparatory calcification, resulting from the matrix formation failing, but not the cartilage calcification. It has been referred to as the white line of Frankel. It is nonspecific, however, since it may also be seen in lead or phosphorus poisoning, in the treatment of syphilis with bismuth, and in healing rickets (Figs. 938–940).

2. "GROUND GLASS" OSTEOPOROSIS. This appears at the end of the shaft, with blurring or disappearance of trabecular markings, and is very characteristic of scurvy (Figs. 938 and 940).

3. "HALO" OSSIFICATION CENTER. The same process that produces the Frankel line will affect the epiphyseal ossification center, since it is surrounded by bone-producing periosteum. The entire interior of the ossification center will be radiolucent, in contrast to the dense periphery. This has been referred to as "Wimberger ring" (Fig. 938).

4. CORTICAL THINNING. This is best noted by comparing the diameter of both cortices to the diameter of the medullary cavity; in the normal subject they are equal, or nearly so (Fig. 938).

5. CORNER SIGN. This is due to a subepiphyseal infraction, or separation of the epiphysis from the metaphysis. Comminution occurs and results in mushrooming of the epiphysis (Figs. 940–942). A similar change may be seen in syphilitic osteochondritis. When the comminution is greater in the center of the end of the bone, cupping is produced. This is distinguished from rachitic cupping by the zone of preparatory calcification, which is concave on the cartilage side and is irregular in rickets but stands out sharply in scurvy.

6. LATERAL SPURS. These metaphyseal spurs project at right angles to the axis of the shaft (Figs. 940 and 941). They may be due to mushrooming of the epiphysis on the metaphysis, as described under the corner sign, or they may represent the earliest calcification of periosteum elevated by a subperiosteal hemorrhage. They have been referred to as "Pelkan spurs."

7. SUBPERIOSTEAL HEMATOMAS. Characteristically these occur in the ends of long bones. They may not be visible until 2 weeks or more after the onset of clinical symptoms. It is not the subperiosteal hemorrhage that calcifies but the

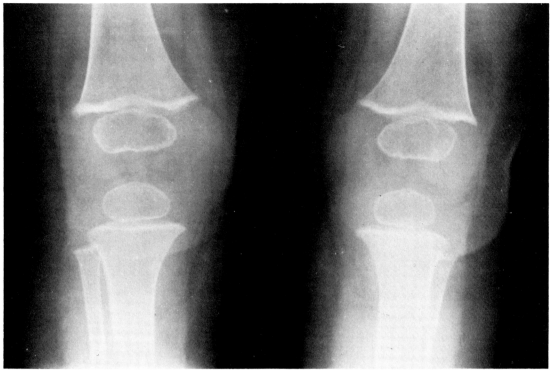

FIG. 938. SCURVY
Dense line at the zone of preparatory calcification (white line of Frankel) and a dense margin around the epiphyses (Wimberger ring). The bone has a ground glass appearance with disappearance of the fine trabecular markings. The cortices are thinner than normal.

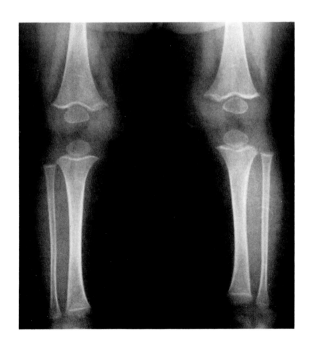

FIG. 939. SCURVY
Zone of temporary calcification is widened and dense. The epiphyses are outlined by a dense line.

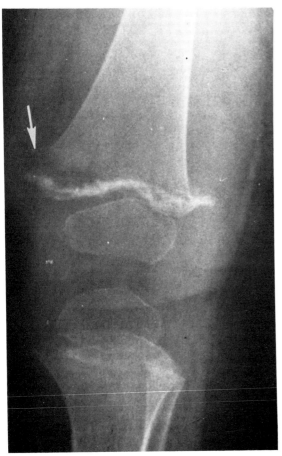

Fig. 940. Scurvy (Early)
The "corner sign" at the medial aspect of the femur just below the epiphyseal plate (*arrow*). Irregularity of the epiphysis due to fractures and a spur has formed on the lateral margin. Deossification produces a ground glass appearance of the bony structures. The epiphyses are sharply outlined with rarefied centers. (Courtesy of George T. Wohl, Philadelphia General Hospital, Philadelphia, Pa.)

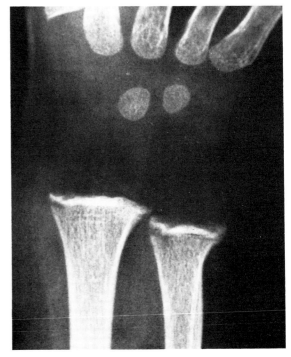

Fig. 941. Scurvy
The "white line of scurvy" is prominent. Incomplete transverse radiolucent zone below this line. Irregularity of the epiphysis at the corners due to infractions (corner sign).

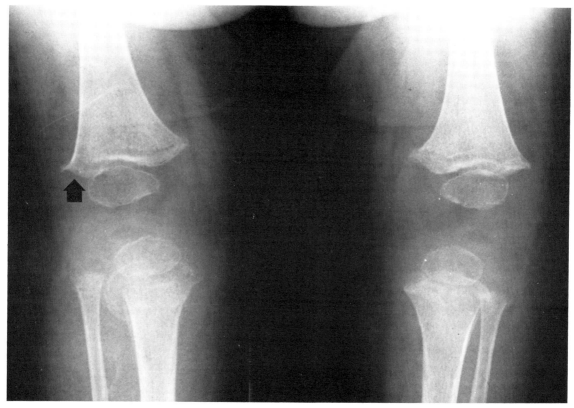

FIG. 942. SCURVY (HEALING)

The "white line of scurvy" is still present but not as marked. Spur (*arrow*) at the the lateral margin of the right femur resulting from fracture through the scurvy zone. (Courtesy of George T. Wohl, Philadelphia General Hospital, Philadelphia, Pa.)

elevated periosteum, secondary to resumption of bone formation. The subperiosteal hemorrhage is gradually resorbed. Calcification of the elevated periosteum is believed to be the surest radiographic sign of healing (Figs. 943, 945, and 946).

8. METAPHYSEAL FRACTURES. These are usually subperiosteal comminuted fractures at the ends of the long bones, extending only partially through the width of the bone. They are quite rare and probably reflect trauma on weakened bone.

9. ATROPHIC SCURVY LINE. This is a radiolucent zone on the shaft side of the Frankel white line, marking where the matrix is not converted to bone. Subepiphyseal infraction occurs here. It has been referred to as the "Trummerfeld zone" (Fig. 944).

10. SOFT TISSUE EDEMA. This relatively rare finding may cause gross deformity of the soft tissues.

11. SPINAL CHANGES. Radiographic changes in the spine are extremely rare in infants under 12 months, because they depend on the degree and duration of the osteoporosis as well as the amount of weight-bearing, and most cases of infantile scurvy are diagnosed and treated before these factors become significant. The collapsed biconcave vertebral bodies typical of conditions causing osteoporosis may be seen in patients over 1 year old.[15]

With healing, cortices become thicker and trabeculae of the spongiosa become clearly defined[4] (Fig. 945). The epiphyseal plates become less densely calcified and broader, and the scurvy zone disappears. As bone grows, the thickened epiphyseal plate is buried in the shaft and becomes a transverse line of increased density in the substance (Fig. 946). Subperiosteal hematomas resorb. Subepiphyseal separations are reduced, and the shaft is realigned to the epiphyseal center. The centers of ossification return to normal density, and permanent deformity almost never occurs.

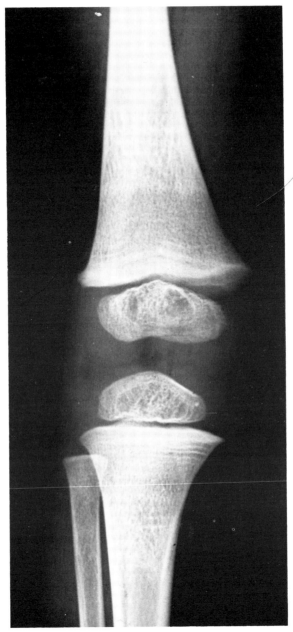

Fig. 946. Scurvy (Healed)

Thickened epiphyseal plates have become buried in the shaft and are now transverse lines of increased density in the substance. The rarefied areas in the center of the epiphysis have been maintained. Subperiosteal hematomas have resorbed. (Courtesy of George T. Wohl, Philadelphia General Hospital, Philadelphia, Pa.)

OSTEOMALACIA

Osteomalacia results from insufficient mineralization of osteoid (bone matrix). Normally there is a balance between osteoid formation and mineralization, and osteomalacia may be considered a result of either excessive osteoid or insufficient mineral.

Frost's[1] concept of high and low remodeling rates is important in understanding the condition. A high remodeling rate indicates excessive osteoid formation and a normal or decreased mineralization rate. A low remodeling rate indicates normal osteoid production and diminished mineralization. With the high remodeling rate there is rapid resorption of underlying bone and progressive osteoporosis with the osteomalacia. With the low remodeling rate, free bone surfaces are covered with osteoid, preventing resorption. Thus, less mineral is available to the blood, and it is necessary to resorb bone to release the necessary ions. When osteoclasts finally do reach portions of the bone, there is rapid resorption, forming small radiolucent osteoid seams (Looser zones). They are pathognomonic of osteomalacia, and most indicate the low remodeling rate type.

Blood and urine chemical changes are not constant but usually serum calcium and phosphorus are low, and the solubility product of serum calcium and phosphorus ions below normal. Serum alkaline phosphatase level is elevated because of the osteoblastic hyperactivity.[2] Insufficient absorption of calcium from the gastrointestinal tract, or excessive excretion of calcium or phosphates, ultimately leads to calcium deficiency and osteomalacia.

Vitamin D is necessary for proper calcium assimilation through the gastrointestinal tract.[2, 3] A frequent cause of osteomalacia or rickets is a condition in which vitamin D is not utilized, thought to be due to resistance to the action at end organs (resistant rickets). Vitamin D deficiency sufficient to cause osteomalacia in adults is rare in this country. Rickets is osteomalacia in early life during endochondral growth. Since its manifestations are somewhat different from those of adult osteomalacia, it is discussed separately.[4]

Most cases of vitamin D-resistant or persistent rickets (rachitis tarda) probably represent osteomalacia caused by renal tubular insufficiency (Fanconi syndrome), a primary defect in renal tubular phosphate resorption.

The histologic diagnostic criteria are not standardized, but the number of osteoid seams per unit area of undecalcified rib cortex is a sensitive measurement.[1] The width of the osteoid seams is another measurement.

Tetracycline deposition occurs in sites of active bone formation, and may be studied by fluorescein microscopy in undecalcified bone sections.[5] Tetracycline administration and successive bone biopsies are used to measure a mean linear rate, or first approximation for the amount of osteoid calcified between biopsies; this is called the appositional rate. The histologic diagnosis of osteomalacia is based on an excess of osteoid seams and decreased appositional rate.

According to Arnstein et al.[6] this separates osteomalacia of low modeling rate disorders (e.g., pseudohypoparathyroidism) from osteoporosis. If the appositional rate is normal, even though the seam count is high, a mineralization defect is unlikely. Diseases such as Paget, hypoparathyroidism, and thyrotoxicosis may have high seam counts, but the modeling rates are increased, ruling out osteomalacia.

The following classification has been modified from Arnstein, Frame, and Frost.[6]

  I. Primary vitamin D deficiency—rickets
 II. Gastrointestinal malabsorption
     A. Partial gastrectomy
     B. Small intestinal disease
        1. Glutenin-sensitive enteropathy and regional enteritis
     C. Hepatobiliary disease
        1. Chronic biliary obstruction and biliary cirrhosis
     D. Pancreatic disease; chronic pancreatitis
III. Primary hypophosphatemia; vitamin D deficiency rickets
 IV. Renal disease
     A. Chronic renal insufficiency
     B. Renal tubular disorders
        1. Renal tubular acidosis
     C. Multiple renal defects
  V. Hypophosphatasia and pseudohypophosphatasia
 VI. Fibrogenesis imperfecta osseum
VII. Axial osteomalacia
VIII. Miscellaneous diseases
     A. Hypoparathyroidism
     B. Hyperparathyroidism
     C. Thyrotoxicosis
     D. Osteoporosis
     E. Paget disease of bone
     F. Fluoride ingestion
     G. Ureterosigmoidostomy
     H. Neurofibromatosis
     I. Osteopetrosis
     J. Macroglobulinemia
     K. Malignancy.

Roentgenologic Features of Adult Osteomalacia

Osteomalacia may occur without roentgenologic evidence. Changes are often nonspecific, and deossification may be the only sign (Figs. 947–950). This deossification is uniform, without trabecular detail. The cortices of the long bones

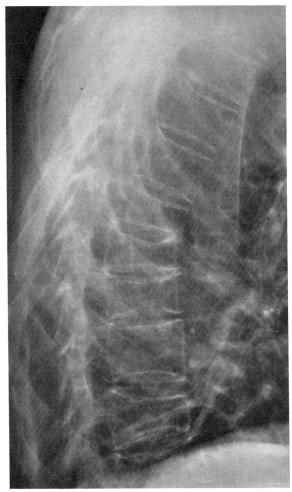

FIG. 947. OSTEOMALACIA (CAUSE UNKNOWN; HISTOLOGICALLY PROVED)

Deossification of the bones of the thorax with multiple compression fractures of the dorsal vertebrae. (Courtesy of Herbert M. Stauffer, Temple University Hospital, Philadelphia, Pa.)

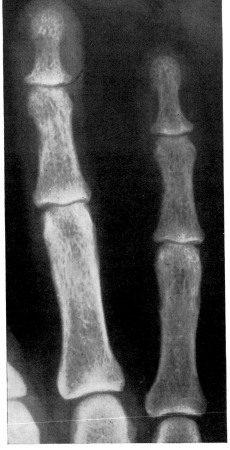

FIG. 948. OSTEOMALACIA DUE TO SPRUE

Decalcification of the spongiosa and cortex but there is no evidence of the lacelike periosteal resorption of bone which occurs in hyperparathyroidism.

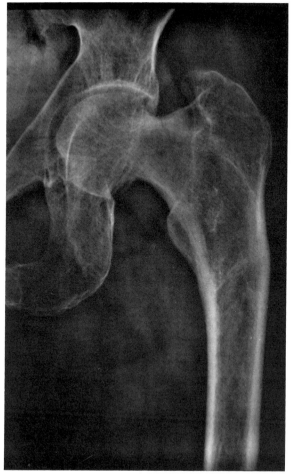

FIG. 949. OSTEOMALACIA (CAUSE UNKNOWN)
Deossification and thinning of the cortex. Most of the trabeculae are indistinct. (Courtesy of Herbert M. Stauffer, Temple University Hospital, Philadelphia, Pa.)

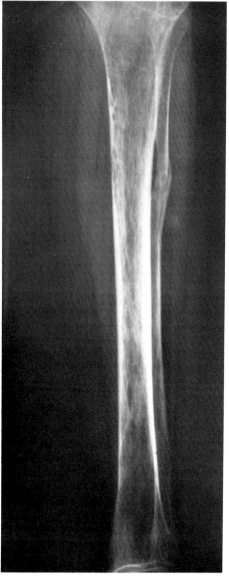

FIG. 950. OSTEOMALACIA (CAUSE UNKNOWN)
Deossification of the bone and thinning of the cortex which has produced mottled radiolucent areas. Healing fracture of the fibula. (Courtesy of Herbert M. Stauffer, Temple University Hospital, Philadelphia, Pa.)

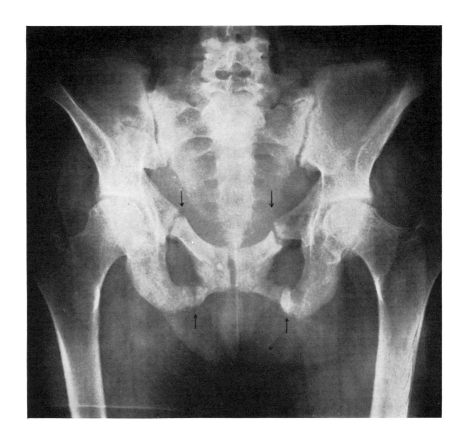
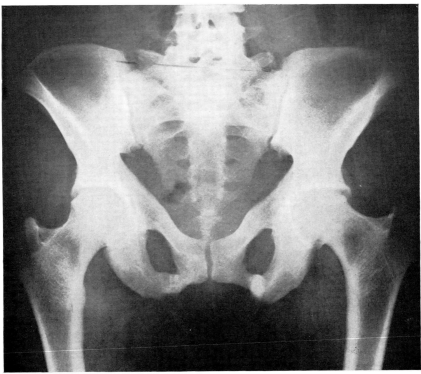

FIG. 951. OSTEOMALACIA DUE TO SPRUE

(*Top*) Bilateral symmetrical fractures which develop first as pseudofractures. (*Bottom*) The fractures have healed as the result of treatment consisting of a high intake of vitamin D, liver extract, and calcium.

METABOLIC AND DYSTROPHIC DISEASE

are thin, and the skull may be mottled and resemble that of hyperparathyroidism.

Pseudofractures are frequently found; these are ribbonlike zones of decalcification extending into bones at approximately right angles to the margin.[7-10] Pseudofractures are infractions with attempted healing; osteoid is formed in the defect, but, because of mineral deficiency, healing is incomplete (Figs. 951–953). They may be present when the bone is otherwise normal roentgenologically. They are often bilateral and symmetrical, and occur at the axillary margin of the scapula, femoral neck, pubic and ischial rami, and ribs (Figs. 954 and 955).

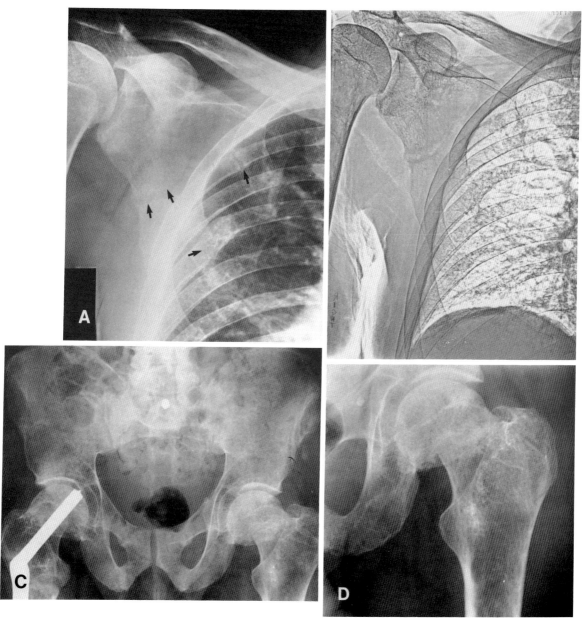

FIG. 952. OSTEOMALACIA

Multiple fractures in the scapula, ribs, femurs and pubic ramus. (A) Chest: Fractures of the scapula and 4th rib (*vertical arrows*), and a healed fracture of the 6th rib (*horizontal arrow*). (B) Xerography of (A) revealing fractures to better advantage. (C) Pelvis: Fracture of the body of the right iliac crest, the right pubic ramus, and a subcapital fracture of the left hip. A fracture of the right hip has been fixed with a nail and plate. All trabecular patterns are disturbed. (D) Subcapital fracture of the left hip; the trabecular pattern is abnormal.

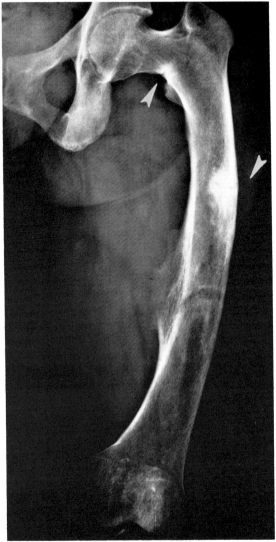

FIG. 953. OSTEOMALACIA (CAUSE UNKNOWN)
Deossification of the femur with bowing and coxa valga. Pseudofractures are present in the neck and shaft of the femur (*arrows*). (Courtesy of Herbert M. Stauffer, Temple University Hospital, Philadelphia, Pa.)

The pseudofractures of osteomalacia cannot be distinguished roentgenologically from the similar lesions of Paget disease, fibrous dysplasia, and osteogenesis imperfecta, but in the latter case considerable bone abnormality is associated. Multiple symmetrical fractures may also occur in patients without evidence of osteomalacia or osteogenesis imperfecta. This has been related to serum pyrophosphate which is an inhibitor of calcification.[11, 12] The elevated serum pyrophosphate may be seen in other conditions such as osteogenesis imperfecta, but when there is no evidence of other disease it is possible that there is a defect in the phosphate metabolism.

Rickets or Vitamin D Deficiency

Rickets is a systemic disease resulting from deficiency of vitamin D in the diet or from lack of exposure to ultraviolet rays.[1, 2] It is osteomalacia, but, occurring during enchondral bone growth, it presents features different from adult osteomalacia.

The human skin contains provitamin D_3 (7-dehydrocholesterol) which is activated by ultraviolet rays. This vitamin is not active, but is transported to the liver, where it is hydroxylated, and from there to the kidney, where it is hydroxylated again and becomes dihydroxy-vitamin D_3, the most active form of the vitamin. It acts as a hormone. Liver or kidney disease may cause rickets if the hydroxylation cannot take place. Infants in colder areas of the world may not receive enough sunlight to activate provitamin in the skin, and rickets will develop unless vitamin D is supplemented. Premature infants are prone to rickets.[4, 5] It rarely develops before the age of 6 months and is most frequent at the end of the first year.

Vitamin D-resistant rickets may be due to hypophosphatasia, unrecognized renal rickets secondary to tubular insufficiency, or end organ resistance to vitamin D.

Pathologic Features. NORMAL EPIPHYSIS. The epiphysis consists of four zones. From the epiphyseal ossification center toward the shaft they are: (1) zone of resting cartilage, (2) zone of proliferating cartilage, (3) zone of maturing cartilage, and (4) zone of degenerating cartilage. The last named is the zone of preparatory calcification and appears as a dense transverse line on the roentgenograms, where osteoblast and vessel invasion takes place, to lay down osteoid on the calcified cartilage bridge.

DEVELOPMENT OF BONE IN RICKETS. Rickets is characterized by defect of bone growth due to lack of normal mineralization. Cartilage cells of the epiphyseal plate grow and reproduce normally but fail to calcify and degenerate.[6] Continuously produced, these cells begin to heap up and disturb the columnar configuration. This causes a patchy widening of the epiphyseal plate, seen roentgenographically as a radiolucent area, so that the ends of the bones are frayed and the epiphyseal line irregular. Mineralization of the osseous and cartilage matrix fails, the zone of preparatory calcification does not form, and a

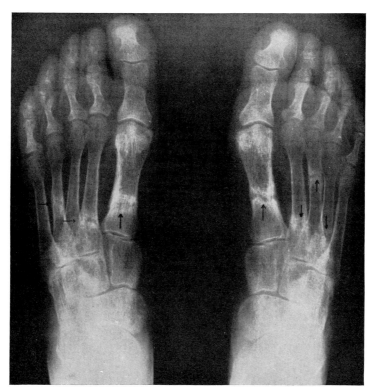

FIG. 954. OSTEOMALACIA DUE TO SPRUE
Multiple pseudofractures of the metatarsal.

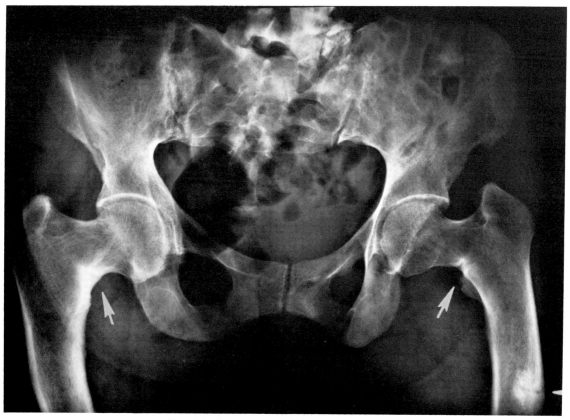

FIG. 955. OSTEOMALACIA (MILKMAN SYNDROME)
Deossification of the bones and symmetrical pseudofractures of the neck of the femora (*arrows*). Pseudofracture of the shaft of the left femur (*arrow*). (Courtesy of Herbert M. Stauffer, Temple University Hospital, Philadelphia, Pa.)

malleable nonrigid tissue is produced instead. It becomes compressed, cupped, and flared.

Changes in bone shafts are caused by failure of osteoid mineralization, and a shell of subperiosteal osteoid tissue surrounds the shaft. The osteoid continues to heap; underlying cortical bone is resorbed and replaced by unmineralized osteoid. Thus the shaft loses rigidity, and molding and fractures occur. Without mineralization, osteoblasts cannot break down bone, and overproduction of osteoid elevates the periosteum.

HEALING. When vitamin D is replaced, cartilage calcifies, and the zone of preparatory calcification becomes identifiable. Cartilage cells then

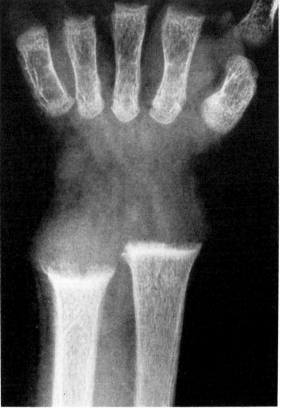

FIG. 957. RICKETS

Irregularity of the epiphyseal plates and a thin irregular temporary zone of calcification. Cupping of the ulnar metaphysis. The trabeculations are atrophied and the remaining trabeculations are prominent. Longitudinal resorption lines (vertical striations) are present in the cortices of the radius and ulna. The trabecular pattern is not ground glass as with scurvy.

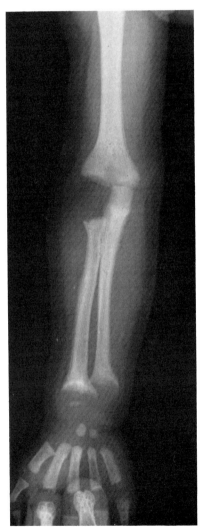

FIG. 956. RICKETS

This is early rickets. Irregularity of the epiphyseal plates, the thin and irregular temporary zone of calcification, and the concavity of the metaphyses. (Courtesy of George T. Wohl, Philadelphia General Hospital, Philadelphia, Pa.)

degenerate normally, and invasion of blood vessels and osteoblasts allows normal epiphyseal growth. The skeleton is rapidly mineralized. The larger subperiosteal osteoid collections are mineralized and resorbed, and bone outline returns.

CHEMICAL PATHOLOGY. Serum calcium usually is normal or reduced. Reduction of the serum phosphorus is an important feature[7]; alkaline phosphatase is elevated.

Clinical Manifestations. Early recognition of rickets is difficult, but it should be suspected in a patient with a history of inadequate vitamin D. Craniotabes may be the first sign, manifested by softening of the posterior parietal bones. Also, enlargement of the costochondral junction of the ribs produces the characteristic "rachitic rosary." The wrists and ankles swell.

In advanced rickets, calcarial sutures are open

and soft; frontal and parietal bossing gives the head a boxlike appearance. Dentition is delayed, and caries extensive.[8] The sides of the thorax are flattened, and longitudinal grooves develop behind the rachitic rosary. Pigeon breast deformity also occurs. There is swelling of the joints and curvature of the long bones.

Roentgenographic Features. Early roentgenographic features are seen most readily in the fastest growing portion of bones,[9, 10] such as the sternal ends of the ribs, proximal ends of tibia and humerus, and distal ends of radius and ulna. Irregularity of the epiphyseal plates is the earliest change. The zone of temporary calcification becomes thin and indistinct. The metaphyses become concave (Figs. 956-958).

Later, the zone of temporary calcification disappears. The concavity of the metaphysis becomes more marked ("cupping") in weight-bearing bones. Fraying of the ends of the shaft causes threadlike shadows of calcification which extend into the transparent epiphyseal cartilage. Cortical spurs project at right angles to the metaphysis and enclose the epiphyseal cartilage where the

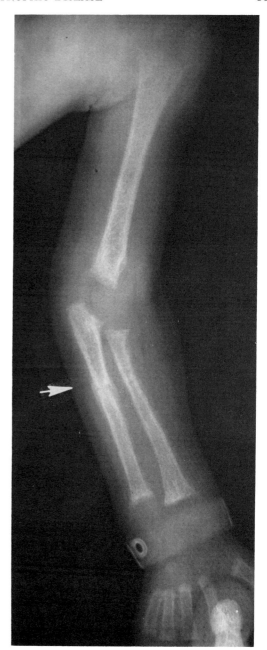

FIG. 959. RICKETS
Temporary zone of calcification at the distal end of the radius and ulna has almost completely disappeared. Cupping. The epiphyses are indistinct. A healing fracture in the shaft of the ulna (*arrow*). (Courtesy of George T. Wohl, Philadelphia General Hospital, Philadelphia, Pa.)

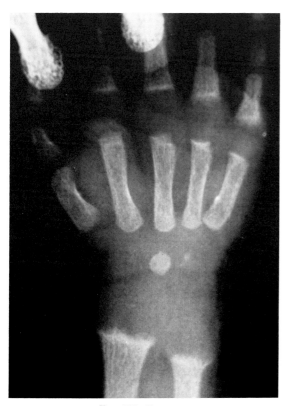

FIG. 958. RICKETS
Irregularity of the epiphyseal plates and prominent trabeculations. The bones are not ground glass and there is no dense ring around the epiphysis as in scurvy.

endochondral ossification is arrested (Figs. 958 and 959).

As the disease progresses, the cartilage of the epiphyseal plate proliferates, but new bone cannot form, and the distance from the end of the shaft to the epiphyseal center increases. Cupping and fraying become pronounced. The trabeculation of the shaft is coarse, because of the loss of fine trabeculae and the accentuation of the remaining trabeculae by cortical thinning. Sometimes the cortex is widened by periosteal deposition of poorly calcified osteoid. Pseudofractures seldom occur, but green-stick fractures are frequent (Fig. 959). The epiphyseal ossification centers show blurred margins—a distinction from scurvy, in which they are sharply outlined. Delayed appearance of ossification centers is due to failure of calcification (Figs. 960 and 961).

The zone of temporary calcification is the first to calcify with healing, and it widens as healing progresses. Outlines of centers of ossification become more distinct, and delayed centers appear (Figs. 962 and 963).

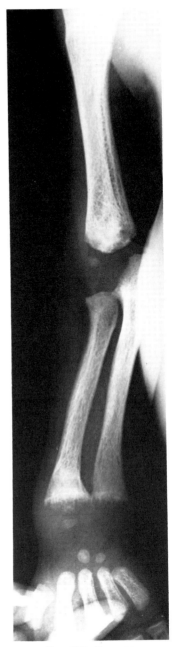

FIG. 961. RICKETS

Temporary zones of calcification of both the radius and ulna are indistinct and frayed. Perpendicular trabeculae extending into the epiphyseal area. The epiphyses are indistinct. (Courtesy of George T. Wohl, Philadelphia General Hospital, Philadelphia, Pa.)

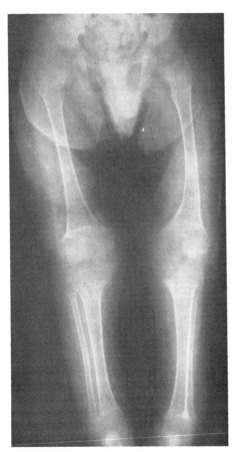

FIG. 960. RICKETS

Metaphyses are indistinct and concave. Cupping and bending of the rachitic intermediate zone of the lower part of the left femur. The epiphyseal ossification centers are indistinct and hardly visible.

Deformities are common and sometimes permanent (Figs. 962 and 964), resulting from conditions during the active stage: (1) molding of the epiphysis, (2) bowing of soft diaphysis, and (3) green-stick fractures. When rickets is severe,

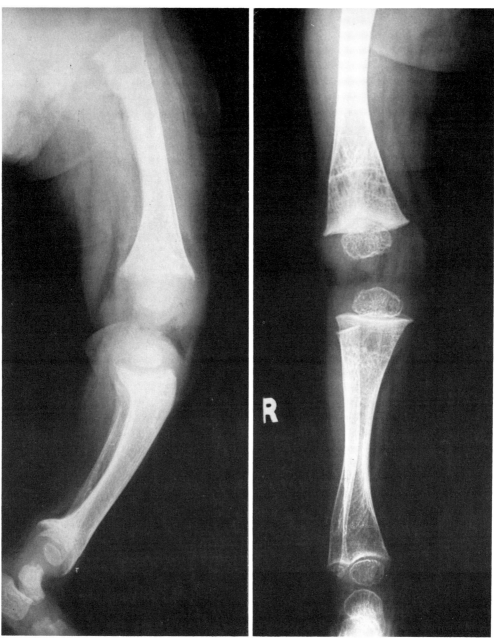

FIG. 962. RICKETS

(*Left*) Active untreated rickets. The temporary zone of calcification is irregular, frayed, and indistinct. The epiphyses are indistinct. Beginning cupping at the metaphyseal portions of bone. (*Right*) Six months later, after treatment with vitamin D. The zones of temporary calcification have reappeared. The epiphyses are well outlined. Trabecular distortion in the shaft and metaphyseal areas. (Courtesy of George T. Wohl, Philadelphia General Hospital, Philadelphia, Pa.)

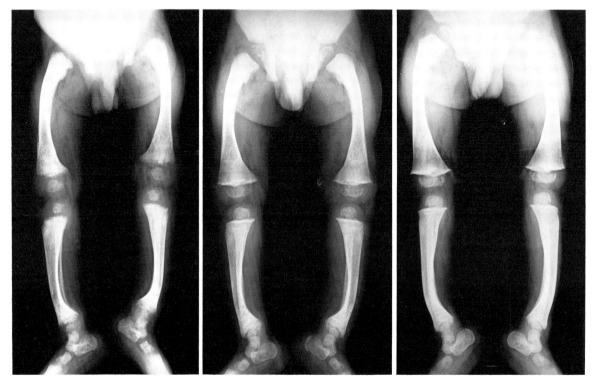

Fig. 963. Rickets

(*Left*) The zones of temporary calcification are thin and indistinct and the trabeculae are frayed. The metaphyses are concave at the distal ends of the tibia. The epiphyses are irregular and indistinct. Bowing of the tibias. (*Center*) Three weeks later, after treatment. The zones of temporary calcification are beginning to calcify but are still irregular. The outlines of the ossification centers are more distinct. (*Right*) Two weeks later. The trabeculations have become more distinct. The zone of temporary calcification is wider and the epiphyses show irregularity due to the previous disease but are well mineralized. Bowing of the distal end of the tibias persists. (Courtesy of George T. Wohl, Philadelphia General Hospital, Philadelphia, Pa.)

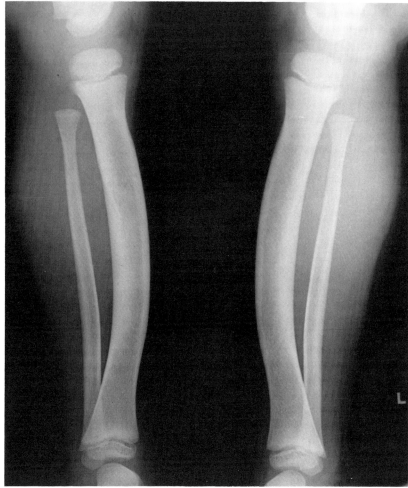

FIG. 964. RICKETS

This is healed rickets. The deformity of the tibias is the only residual of previous severe disease. (Courtesy of George T. Wohl, Philadelphia General Hospital, Philadelphia, Pa.)

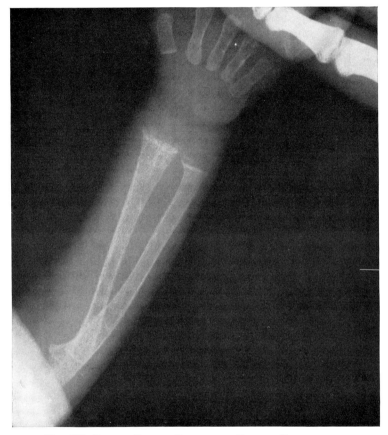

Fig. 965. Rickets Due to Congenital Biliary Obstruction
Appearance of the bones is no different than that in vitamin D intake insufficiency.

with prolonged multiple fractures, the roentgenographic features are similar to osteomalacia. The widened epiphyseal plates are also present and hyperparathyroidism may eventually ensue (Fig. 966).

When the children's form of osteomalacia called rickets is due to conditions other than vitamin D deficiency, such as celiac disease and hereditary osteomalacia,[11] the roentgenographic manifestations are similar (Fig. 965).

Steatorrhea

Steatorrhea may lead to osteomalacia. Fats are not assimilated, and fat-soluble vitamins D, A, K, and E are not utilized.[1] The inability to utilize vitamin D prevents gastrointestinal absorption of calcium, which combines with fats to form insoluble soap and rapidly passes through the gastrointestinal tract. Inability to absorb vitamin K leads to hemorrhagic diatheses.

In this country, steatorrhea is the most common cause of adult osteomalacia. The serum calcium may be extremely low (as low as 3.5–4.0 mg per 100 ml), and latent tetany may be present.[2] The roentgenologic examination of the small intestine may show disordered motor function.

Fibrogenesis Imperfecta

Fibrogenesis imperfecta is an idiopathic disease of older individuals, probably the result of an acquired vitamin D resistance.[1]

Clinical Features. Only a few cases have been reported,[1-4] and both men and women are affected. The onset of bone pain begins in the 6th or 7th decade, usually in the limbs and trunk, and becomes progressively severe over a period of 1–3 years. The pain is related to weight-bearing and improves with rest. If untreated, numerous fractures occur.

Elevated alkaline phosphatase is the only abnormal blood chemistry.[1-4] An excess of urinary and fecal calcium excretion causes negative calcium balance, and there is also a negative phosphorus balance.

Treatment with vitamin D, although it produces an increase in bone pain, eventually brings improvement.[1] If resistance to vitamin D treatment ensues, improvement may again be achieved with dihydrotachysterol[1] after initial pain exacerbations.

Pathologic Features. Grossly, the bones are brittle but also soft.[4] Histologically, there is a widespread defect in the formation of collagen fibers of newly laid down bone matrix, easily demonstrated by the use of a polarizing microscope[1,4] or stained by the Gomori reticulum method.[1]

Radiographic Features. In the early stages the bones may appear normal, but eventual coarsening and widening of the cancellous trabeculations are the predominant features. Some of the bone trabeculae deossify; the remaining are thicker than normal and ill defined (Fig. 967). The long bone cortices become thin and the trabecular structures appear as a fine mesh. Fractures may occur. All bones are involved[4] but the skull may be spared.[1]

The differential diagnosis includes other forms of osteomalacia, multiple myeloma, and metastatic malignancy.

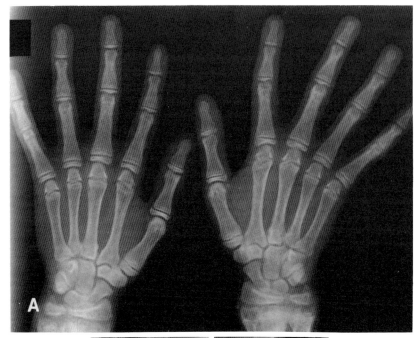
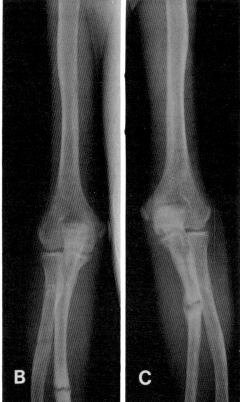

FIG. 966. RICKETS

Severe prolonged rickets with secondary hyperparathyroidism. The severity of this disease is unusual in the United States. The epiphyseal spaces are widened and there are multiple pseudofractures. (*A*) Hands and wrists: The trabecular pattern is disrupted and there is subperiosteal resorption of some of the middle phalanges. The metaphyseal areas of the wrist are irregularly ossified. (*B*) Fractures of the radius and ulna in the right upper extremity. (*C*) Fractures of radius and ulna in the left upper extremity. (*D*) Bilateral scapular fractures, radiolucencies in the metaphyses of both humeri and multiple rib fractures, particularly in the lower rib cage. (*E*) Multiple fractures of the pelvis with protrusio acetabuli. Both femoral necks and shafts are fractured. (*F*) Both lower extremities have fractures of the fibulas and the left tibia; the widened physes of the femur and tibia are evident. (Courtesy of Dr. M. Bajogkli, Teheran, Iran.)

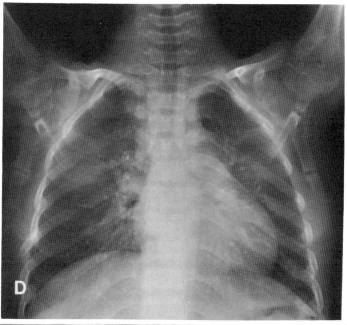
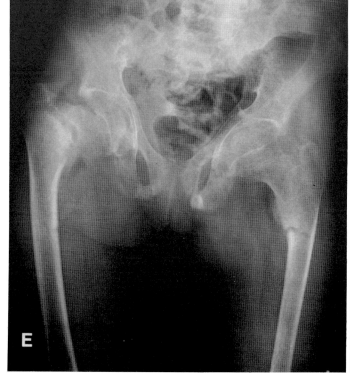
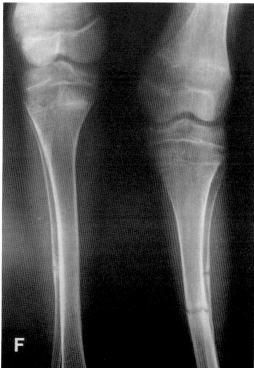

Fig. 966 (D–F)

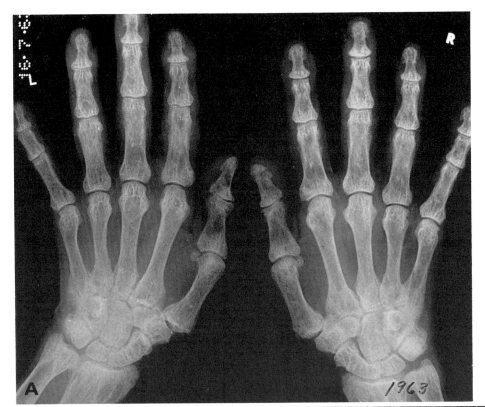
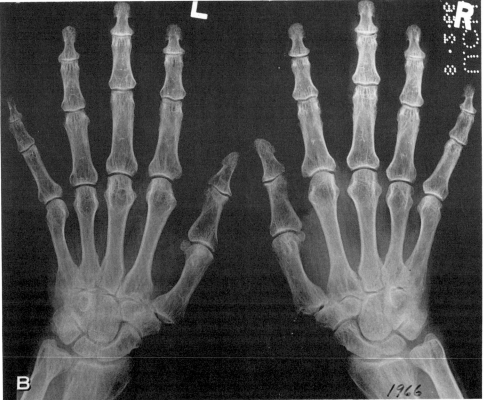

Fig. 967 (A and B)

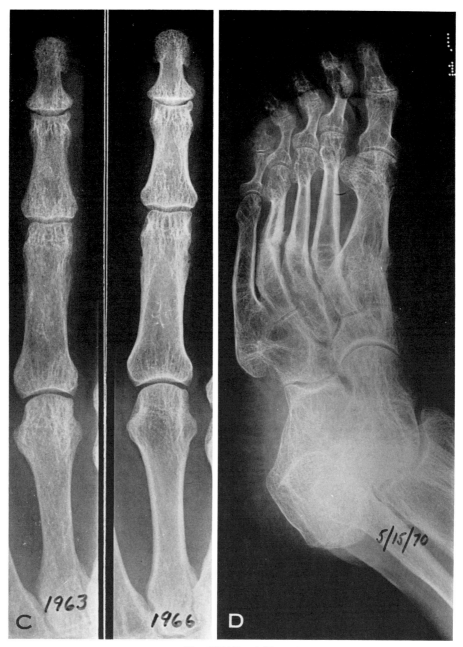

Fig. 967 (C and D)

Fig. 967. Fibrogenesis Imperfecta

This 53-year-old man had been incapacitated by generalized bone pain for 3 years. In 1963 he was started on large doses of vitamin D. He had exacerbation of bone pain within the week and then gradual improvement until he was able to walk. Films obtained in 1966 reveal marked improvement. The vitamin D was stopped in 1970, and he had recurrence of pain. He was restarted on high doses of vitamin D and improved. (A) Hands in 1963; thinning of all cortices and prominence of the trabeculations. (B) Hands in 1966; after high doses of vitamin D the cortices have regained their thickness and prominent trabeculations persist in the spongy areas of bone. (C) Comparison with finger in 1963 before treatment and in 1966 after high doses of vitamin D. In 1963 marked thinning of the cortices and prominences of trabecular structures. In 1966 the cortices have reformed and are smooth. The prominence of the trabecular pattern persists in the spongy bone. (D) Foot in 1970 after exacerbation of symptoms. The cortices remain thick. Multiple healing fatigue fractures of the metatarsals. Trabecular prominence in the spongy bone. (E) The knees in 1970 reveal prominence of the trabecula in the spongy areas of bone. (Courtesy of Dr. C. E. Dent: Fibrogenesis imperfecta osseum. *Journal of Bone and Joint Surgery, 48B:* 804, 1966, © British Orthopaedic Association, London.)

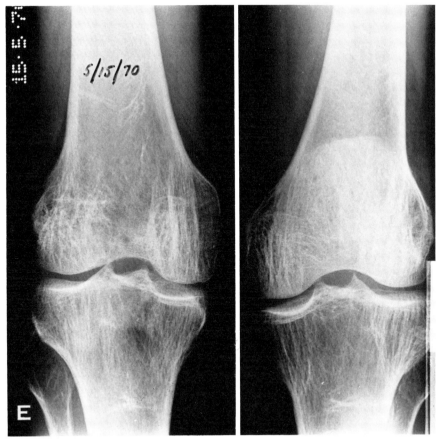

Fig. 967 (E)

HYPOPHOSPHATASIA

Hypophosphatasia[1] is a genetically determined metabolic disease characterized by three salient features: abnormal mineralization of bones, diminished alkaline phosphatase activity, and increased urinary excretion of phosphorylethanolamine.[2, 3]

Clinical Features. There are four clinical groups: Group I, newborns who develop manifestations in utero or in the first few days of life; Group II, infants who develop bone lesions within the first 6 months; Group III, infants and children who develop symptoms from 6 months to 13 years of age; and Group IV, adults.[4-11] As the age at onset increases, the symptoms decrease in severity.[4, 5]

GROUP I. These patients are the most severely affected and have the poorest prognosis. Many are stillborn; in those born alive, respiratory difficulty, cyanosis, and death may occur within the first hour. They may be asymptomatic for the first few days, but fail to gain weight and develop irritability and convulsions within 2 weeks. Inconstant cyanosis, anorexia, vomiting, constipation, continuous crying, and mild pyrexia are common. Blue sclerae are reported.[12] These infants appear normal at birth, but their skeletal changes are extensive, although less severe than in the stillborn infants. Practically all in this group will die within 6 months.

GROUP II. Clinical manifestations develop at 1-6 months of age, and gradually worsen until the 9th month, when the infant either dies or begins to improve. One half of the untreated cases die; the rest improve clinically in spite of persistent and progressive skeletal lesions. Clinical features are similar to those of Group I. In addition, cranial sutures are separated, the anterior fontanels bulge, and prominent scalp veins are present. If the infants live, symptoms gradually abate, but skeletal abnormalities delay weight-bearing, and moderate or severe dwarfism and genu valgum are frequent. Craniostenosis, in the second year, is common.[13] Clinical features of this somewhat older group are the same as in Group III. In most instances, mild manifestations may be traced to infancy or early childhood.

GROUP III. Orthopedic deformities of delayed weight-bearing, defective gait, genu valgum, and dwarfism are the main complaints. These children have an excellent prognosis, and most improve without treatment. Craniostenosis is rare.[13]

GROUP IV. Bone fragility and reduced serum alkaline phosphatase may be the sole features.[2]

Laboratory Findings. Characteristic laboratory findings include a reduction in serum alkaline phosphatase activity in the presence of phosphorylethanolamine in the urine. The serum alkaline phosphatase is consistently low,[1] and unrelated to the severity or fluctuation of the symptoms. Familial low phosphatase activity in a patient without clinical manifestations is reported.[2] Tissue alkaline phosphatase activity is consistently decreased in bones, kidneys, and intestinal tract.[1, 14]

Phosphorylethanolamine is usually present in the urine,[2, 13] but this finding is not limited to hypophosphatasia, appearing in celiac disease[15] and in normal adults with a high normal serum alkaline phosphatase. The mechanism is unknown, but it is suggested that phosphorylethanolamine is a naturally occurring substrate of alkaline phosphatase which, due to the deficiency, fails to break down completely.[2, 3, 16] In severe cases the total serum calcium is elevated. Serum inorganic phosphorus levels are normal, and renal insufficiency occurs only in severe cases.

Pathologic Features. The histologic features are marked in the growing regions of membranous and cartilaginous bones, and are indistinguishable from rickets.[17, 18] The resting and proliferating cartilage cells adjacent to the epiphysis are normal; however, there is disorganization of columnar orientation of chondrocytes adjacent to metaphyses. The zones of provisional calcification are disorganized and the columnar orientation is lost.[1, 12, 17] In the metaphysis, islands of uncalcified cartilage are developed by uncalcified bone, and there is excessive diaphyseal osteoid. Cortical bone is relatively normal, except for considerable irregular calcification and subperiosteal osteoid. Similar abnormalities in membranous bone consist of large amounts of uncalcified or poorly calcified osteoid.

Roentgenographic Features. Radiographic changes are due to the continued proliferation of osteoid and cartilaginous tissue that fails to calcify. The earlier the onset, the more striking the changes. The milder forms simulate rickets.

GROUP I. Most severe changes occur in this group. The calvaria are uncalcified, except for small centers of ossification (Fig. 969).[1] Long bones show lack of calcification at metaphyseal ends (Figs. 968 and 970). At the junction of the calcified and uncalcified bone, calcification is irregular, streaky, and spotty. Cupping of metaphyses is the rule. In severe cases only a small central calcification is present in diaphyses. Angulated shaft fractures and abundant callus formation may occur. Beading of the chondroid portions of ribs resembles the rachitic rosary.

GROUP II. Changes are less severe, and appear as small areas of uncalcified metaphyses (Figs. 971–974), as in rickets. There is a tendency to angulating fractures of shafts. The skull presents varying degrees of calcification.

GROUP III. Deformities such as clubfeet and genu valgum are the principal features. The uncalcified changes are not present at birth but may occur at 3 or 4 months[13] and progress through the same course as Group II. Metaphyseal rarefaction increases and the zone of provisional calcification becomes indistinct. The ossification centers of long bones and calvaria demineralize. Within a year, however, a level is reached, and after that abnormalities begin to improve. With calvarial recalcification, there may be premature closure of sutures. Healing in long bones, rapid at first, tapers off (Fig. 973). Metaphyses and ossification centers mineralize and shaft remodeling occurs. The healing may be complete.

GROUP IV. Symptoms may be traceable to infancy and early childhood, but were so mild as to go unrecognized. Bones are likely to be fragile. Many roentgen changes are residual, left from the incomplete healing of more severe disease.

The skeleton may show deossification and healed fractures. The chondral ends of ribs may be enlarged, suggesting rickets. The long bones may show lack of modeling with "Erlenmeyer flask" deformity of femur. A disturbance of primary dentition is usual.

The patients are smaller than normal in height and their pelvic measurements are small. Even though adult patients may present a normal skeletal appearance, the low serum alkaline phosphatase and presence of phosphorylethanolamine in the urine persists.

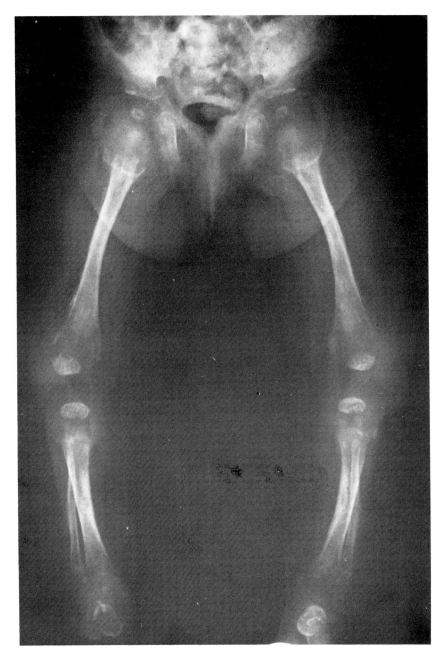

Fig. 968. Hypophosphatasia

Lack of calcification at the metaphyseal ends of the bones. At the junction of the calcified and uncalcified bone, calcification is irregular, streaky, and spotty. Cupping of the metaphyses is present. (Courtesy of Edward B. D. Neuhauser, The Children's Medical Center, Boston, Mass.)

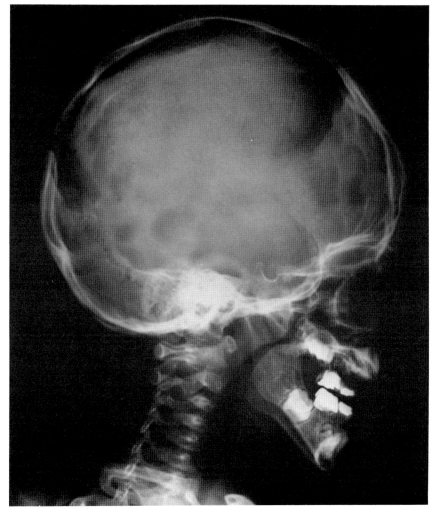

Fig. 969. Hypophosphatasia

This is a mild case; there is peculiar deossification of the superior and posterior portions of the parietal bone and the occipital bone.

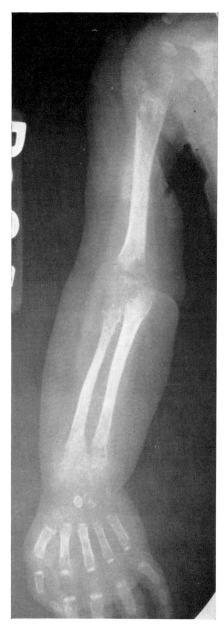

Fig. 970. Hypophosphatasia
Lack of calcification at the metaphyseal ends of the bones with irregular, streaky calcification at the junction with the shaft. (Same case as seen in Figure 968.) (Courtesy of Edward B. D. Neuhauser, The Children's Medical Center, Boston, Mass.)

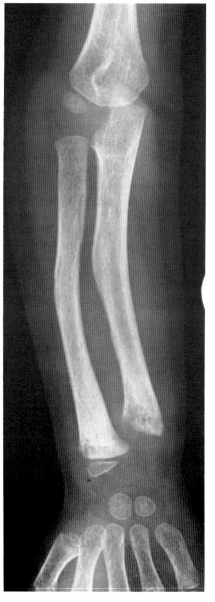

Fig. 971. Hypophosphatasia
The metaphyseal areas have deossified in this mild case of hypophosphatasemia. Compare this with the more severe case seen in Figure 970.

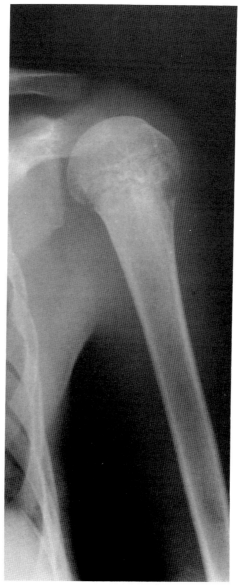

Fig. 972. Hypophosphatasia

Metaphyseal resorption of bone in the humerus. The ribs show marked deossification, but the clavicle appears normal.

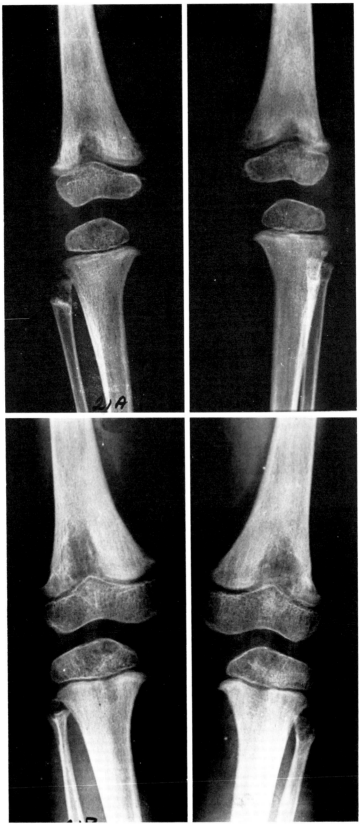

FIG. 973. HYPOPHOSPHATASIA

(*Top*) Central metaphyseal rarefaction. The epiphyses are deossified. Radiolucent line below the femoral and tibial epiphyses. Deossification of the proximal fibular ends. (*Bottom*) Two years later. The central metaphyseal deossification persists. The distal ends of the femurs are better ossified but have residual change. The epiphyses have areas of rarefaction but are better mineralized than 2 years earlier (above). (Courtesy of Dr. Howard Steinbach, Moffitt Hospital, San Francisco, Calif.)

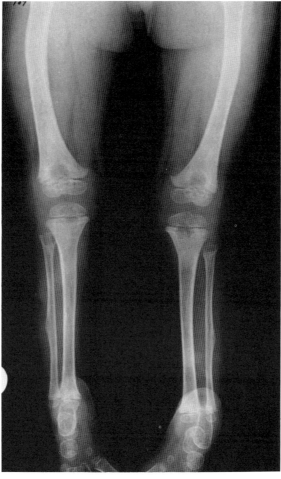

FIG. 974. HYPOPHOSPHATASIA

Note the similarity to the patient in Figure 973. There is notching of the metaphyseal areas of the femurs, tibias and fibulas; also there are fractures of the midportion of both fibulas.

HYPERPARATHYROIDISM

Primary hyperparathyroidism may be caused by a single parathyroid adenoma, parathyroid gland hypertrophy, or parathyroid carcinoma. It may occur in a newborn whose mother is hypoparathyroid. We have also observed patients with long standing pseudohypoparathyroidism develop hyperparathyroidism due to an adenoma (Fig. 975). Patients with Sipple syndrome, a familial disease of medullary thyroid carcinoma, pheochromocytoma, and parathyroid adenoma are reported.[1-6] It is distinct from multiple familial endocrine tumors.[7] Secondary hyperparathyroidism is a compensatory mechanism which may be due to rickets, osteomalacia, pregnancy, renal insufficiency, calcium deprivation, or maternal hypoparathyroidism. The most common cause of hyperparathyroidism is parathyroid adenoma. Albright and Reifenstein[8] believe that initial hypocalcemia stimulates the formation of parathyroid germination centers, which eventually lose the ability to respond to normal stimuli and produce parathyroid hormone in excess.

Parathyroid adenomas are sometimes associated with pituitary adenomas, pancreatic islet tumors,[9,10] acromegaly, hyperinsulinism, and the Zollinger-Ellison syndrome.

Clinical Features. Hyperparathyroidism is frequently encountered in the 3rd, 4th, and 5th decades, although it may occur in the newborn and in those over 70. It afflicts 3 females for every male.[11] The protean clinical manifestations of hyperparathyroidism are noteworthy. The

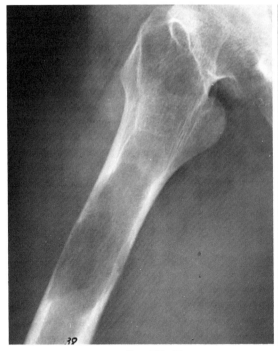
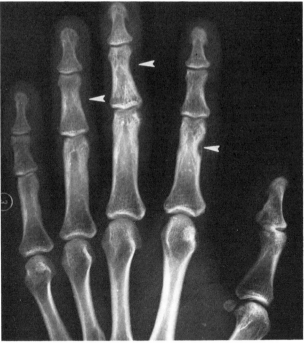

FIG. 975. HYPERPARATHYROIDISM AND PSEUDOHYPOPARATHYROIDISM

A 24-year-old man who complained of pain in the thigh. Serum calcium and phosphorus levels were normal. (*Left*) An osteolytic lesion is present in the upper end of the femur. (*Right*) Hand; shortening of the fourth and fifth metacarpals, a reflection of pseudohypoparathyroidism and subperiosteal erosions of the second proximal phalanx and the third and fourth middle phalanges (*arrows*). The latter changes are pathognomonic. The blood chemistries were normal because of the pseudohyperparathyroidism. (Courtesy of Dr. Ray Kilcoyne, Milwaukee, Wis.)

majority of patients will at one time or another develop urinary calculi, and many have either duodenal ulcers or gastric ulcers. "Brain tumors," psychiatric problems, swallowing difficulties (Fig. 976), and "arthritis" are frequent presenting symptoms. Other less generalized symptoms include change in bowel habits, weakness, lassitude, renal failure, fracture, soft tissue calcification, soft tissue mass, arteriosclerosis, and the failure of infants to thrive. Collectively, the symptoms of primary hyperparathyroidism may be grouped into those due to hypercalcemia, urinary tract disease, or bone disease.[8]

Hypercalcemia results in hypotonicity of both smooth and striated muscle. Hypotonicity of striated muscle leads to subjective weakness, but smooth muscle weakness leads to gastrointestinal complaints such as constipation, nausea and vomiting, and to dryness of the mucous membranes of the nose and throat, and difficulty in swallowing.[12] Thirty-three percent of the men and 5% of the women are reported to have active peptic ulcers.[13] Walsh and Howard[14] report calcium deposits in the deep conjunctival and superficial layers of the cornea; these sometimes regress on correction of hypercalcemia.

Urologic disturbances may include: (1) hypercalciuria and hyperphosphaturia, and polyuria and polydipsia, which may be mistaken for diabetes insipidus, and (2) kidney stones in the collecting systems (nephrocalculosis) and tubules (nephrocalcinosis).

Symptoms related to the *osseous system* are pronounced rheumatic bone pain and tenderness, which progressively increase, and usually disappear several weeks after treatment.

Pathophysiology. The absence of parathyroid hormone causes four sequential metabolic changes which are probably interrelated: (1) decrease in urinary phosphorous excretion, (2) rise in serum phosphorus level, with an almost simultaneous (3) dip in serum calcium level,[11] and (4) diminished urinary calcium excretion. An excess of parathyroid hormone causes changes in the opposite direction but in the same sequence; hyperphosphaturia, hypophosphatemia and hypercalcemia, and hypercalciuria.[8]

Three theories of the action of parathyroid hormone have been proposed. (1) The hormone acts directly on bone tissue to cause dissolution[7-14] with the electrolyte changes secondary to the bone disturbance. (2) The hormone acts directly on electrolyte equilibrium of body fluids and bone changes.[8] It is further hypothesized that the initial hormonal action is to decrease PO_4 tubular resorption by direct action, increasing the phosphate excretion.[15] With adequate calcium and phosphorus intake patients are predisposed to renal rather than to bone disease.[8] (3) Parathormone consists of two fractions, one acting on renal tubules and the other directly on bone.[16]

Laboratory Findings. Hypercalcemia coupled with hypophosphatemia indicates hyperparathyroidism, although occasionally multiple myeloma or metastatic malignancy cause these changes. Serum calcium and phosphorus levels may be normal in hyperparathyroid patients with low serum proteins. McLean and Hastings[17] published linear charts for evaluation of serum calcium in relation to total serum proteins. Severe renal damage may cause retention of phosphates which will mask the hypophosphatemia and reduce the hypercalcemia.

In patients with pseudohypoparathyroidism and hyperparathyroidism there is a low serum calcium and an elevated phosphorus which masks the hyperparathyroidism. The Sulkowitch urine test is an easy screening test used to detect urinary calcium. When high calcium excretion is shown, serum calcium and phosphorus should be evaluated. High alkaline phosphatase blood levels indicate skeletal involvement.

Roentgenographic Features. The skeletal changes of primary and secondary hyperparathyroidism are identical. Less than half of these patients manifest skeletal lesions.[8, 18, 19] The classic roentgenographic features are subperiosteal cortical bone erosions, generalized deossification, local destructive bone lesions ("brown tumor"), and calcification of the soft tissue. Although it was once necessary to demonstrate these bone changes radiographically to establish the diagnosis, refined biochemical techniques have made it possible to detect hyperparathyroidism in patients without skeletal changes.[12]

The incidence of bone lesions in hyperparathyroidism varies from 30 to 40%,[13, 19, 20] although a systemic disease like this should involve all bones. In advanced cases it does, but, in borderline cases, the skeleton may appear normal except for a single bone, perhaps with a cystlike lesion. Early changes should be sought where initial involvement typically appears. Less than 30% of the patients will show skeletal changes, although many will show deossification impossible to evaluate. Fine detail radiography may show skeletal changes in as high as 60% of all cases, as compared to less than 30% in standard radiography.[21, 22]

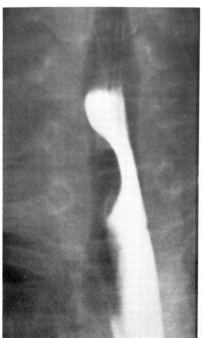
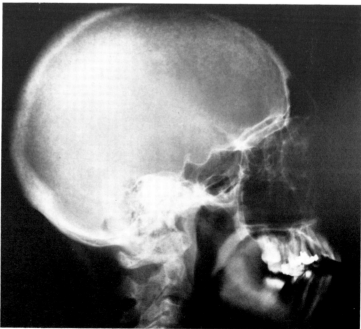

FIG. 976. HYPERPARATHYROIDISM
(*Left*) Swallowing function revealed a filling defect on the right side at the thoracic inlet. (*Right*) Skull roentgenogram reveals salt and pepper type of deossification. (Courtesy of Dr. Morris Ivker, Philadelphia, Pa.)

Erosive or cystic changes have a definite predilection for the small joints of the hands and wrists,[23] but may also appear in the large joints, particularly the knees. These changes and joint symptoms may be mistaken for a collagen disease or erosive arthritis, and the true nature of the condition be overlooked. Resnick[23] suggests that the changes accompanying these erosions aid in distinguishing the condition by (a) associated subperiosteal resorption along the radial aspects of the phalanges and metacarpals; (b) predilection for radiocarpal, inferior radioulnar, and metacarpophalangeal joints, with frequent involvement of distal joints including the interphalangeal joint of the thumb; (c) relative sparing of the proximal interphalangeal joints; (d) an ulnar side distribution to the erosive abnormalities; (e) irregular new bone formation or whiskering at

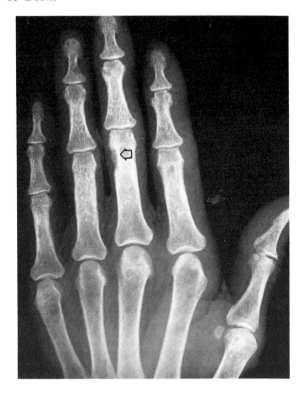

FIG. 977. HYPERPARATHYROIDISM
Subperiosteal resorption of the proximal phalanx of the third finger (*arrow*), a most characteristic sign of hyperparathyroidism. (Courtesy of George T. Wohl, Philadelphia General Hospital, Philadelphia, Pa.)

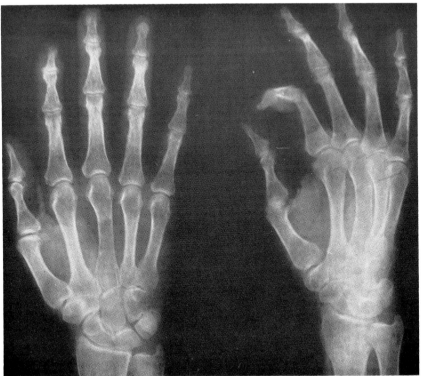

FIG. 978. HYPERPARATHYROIDISM
Subperiosteal resorption of bone is extensive and is shown by the narrowing of the midportion of the shaft of the middle phalanges. In the original roentgenogram, the lacelike character of the outer surface of the cortex could be seen. In this case, resorption of subperiosteal bone near the joints gives an appearance simulating an arthritic process. Cystlike lesion in the third metacarpal.

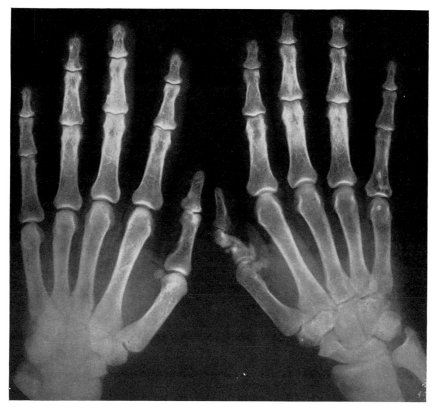

Fig. 979. Hyperparathyroidism

The changes in the hands are pathognomonic of the disease. Subperiosteal resorption of bone in the phalanges, resorption of the phalangeal tufts, cystlike lesion in the proximal phalanx of the right fifth finger, and the pathologic fracture of the proximal phalanx of the right thumb.

the joint margins; (f) absence of joint space narrowing at metacarpophalangeal and proximal interphalangeal joints; (g) associated cystic changes; and (h) presence of chondrocalcinosis and capsular or periarticular calcifications.

The early skeletal changes of hyperparathyroidism affect the hands, calvaria, and periodontal osseous tissue. In the hands there is subperiosteal bone resorption along the radial margin of middle phalanges; this is the most important and reliable radiographic sign of hyperparathyroidism[24] (Figs. 977–980). This does not mean that the entire cortex is deossified although cortical striations occur. Instead there is a peculiar lacelike decalcification of the outer border of cortex immediately beneath the periosteum (Fig. 981). The endosteal border of cortex remains intact until late in the course. Early, this change is manifested only in the middle phalanges or only in one phalanx, although ultimately all the phalanges and most long bones may be affected. Most patients with bone involvement show these changes in the hands. Other changes in the hands include destruction of the midportions of the distal phalanges with telescoping (Fig. 982),

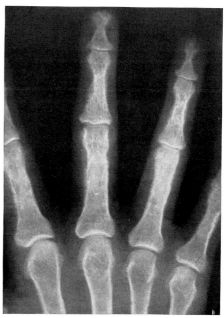

Fig. 980. Hyperparathyroidism

Extensive subperiosteal resorption of bone. The peculiar lacelike appearance of the margins of the bone is characteristic of hyperparathyroidism.

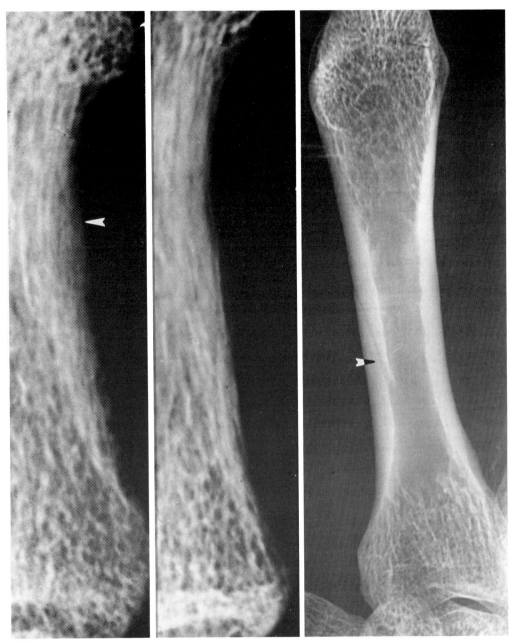

Fig. 981. Hyperparathyroidism

(*Left*) Phalangeal cortical striation as evidenced by vertical parallel radiolucent lines. Lacelike resorption of bone (*arrow*) which is characteristic of hyperparathyroidism. (*Center*) Hyperthyroidism. Phalangeal striation of the cortex but no lacelike resorption. (*Right*) Normal phalanx. No striations in the cortex. Oblique radiolucent line representing a nutrient canal (*arrow*). (Courtesy of H. E. Meema and D. L. Schatz: Simple radiologic demonstration of cortical bone loss in thyrotoxicosis. *Radiology, 97:* 9, 1970, © The Radiological Society of North America, Syracuse, N. Y.)

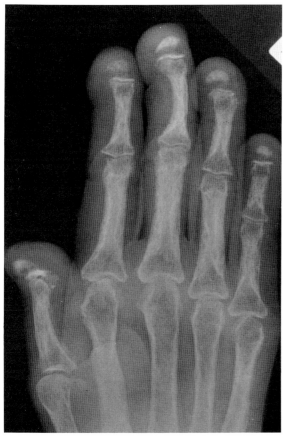

FIG. 982. HYPERPARATHYROIDISM
Destruction of the midportions of the distal phalanges with telescoping and soft tissue enlargement, and marked deossification.

brown tumors (Fig. 983), and erosions of the tufts (Fig. 984).

In long bones the endosteal surface of the cortex is resorbed as much as, or more than, the periosteal surface (Figs. 985 and 986). The subperiosteal resorption that occurs in the phalanges also occurs in the long bones, with a predilection for the concavities at the junctions of the metaphysis and epiphyseal areas. The most common site second to the phalangeal site is the medial border of the tibia, then its lateral border. It also occurs in the femur and humerus.

Slipped capital femoral epiphysis may occur.[25]

Absorption of the terminal tufts of the phalanges may be an early sign of bone disease and can occur before the subperiosteal resorption. If the tuft is extensively destroyed the finger becomes clubbed.

In the newborn, the changes are different from those of the adult. It may be secondary to maternal hypoparathyroidism, hyperplasia or adenoma. There is a radiolucent band below the epiphysis in most of the long bones, with fractures at these sites. There is marked deossification particularly the bone ends (Figs. 987 and 988). Occasionally, subperiosteal resorption of bone will occur at the concavity of the metaphysis in the long bones. It is more common in the long bones than the phalanges of the newborn.

Granular deossification, often associated with cystlike areas, may be found in the skull, but may be difficult to distinguish from normal bone (Figs. 989–992). This change does not establish the diagnosis since it is not characteristic. Typically, the outlines of both inner and outer tables become indistinct. Following removal of an adenoma, the diploic space becomes indistinct as fibrous tissue forms in the diploe.

Dental diagnosis of hyperparathyroidism is more accurate when all the periodontal structures are considered rather than just the lamina dura. The periodontal membrane space appears as a dark area surrounding the roots. The periodontal membrane is a specialized type of periosteum, and surrounding it is a dense white line of bone cortex around the tooth, the lamina dura. Subperiosteal resorption of cortex, therefore, appears as loss of lamina dura (Fig. 993). This change can occur in Paget disease and osteomalacia, and is not as specific as the changes in phalanges. Both loss of trabecular structure around the tooth and "cystic" lesions stand out dramatically because the normal density of teeth does not change throughout the disease. Normal anatomic landmarks such as the inferior dental canal, dental foramina, and sutures of the maxilla become indistinct or disappear (Fig. 994).

Resorption of the ends of the clavicles occurs, perhaps due to the stress of shoulder movement. The subchondral bone of the symphysis pubis may also be resorbed.[26] Indeed, many areas subject to stress may be resorbed, including the symphysis pubis, the sacroiliac joints, the calcaneus at the insertion of the plantar fascia, and at the midaxillary line of the ribs (usually at the upper border of the upper ribs). The above changes are all early manifestations that may occur without generalized deossification. Phalanges most frequently show the earliest change, but it may appear at any site.

In advanced cases there is rather obvious bone deossification (Figs. 995 and 996); trabeculae appear indistinct and normal trabecular architecture is deranged. The cortices of long bones become thin and subperiosteal resorption is easily demonstrated. The bones may have a ground

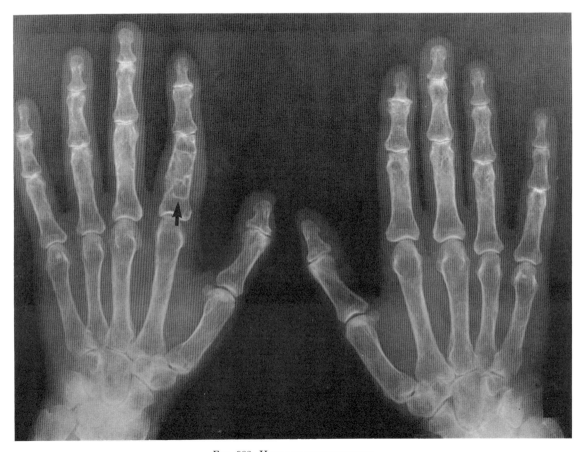

Fig. 983. Hyperparathyroidism

A brown tumor in the proximal phalanx of the index finger of the right hand (*arrow*). There is also marked demineralization and subperiosteal resorption of the radial aspect of the middle phalanges.

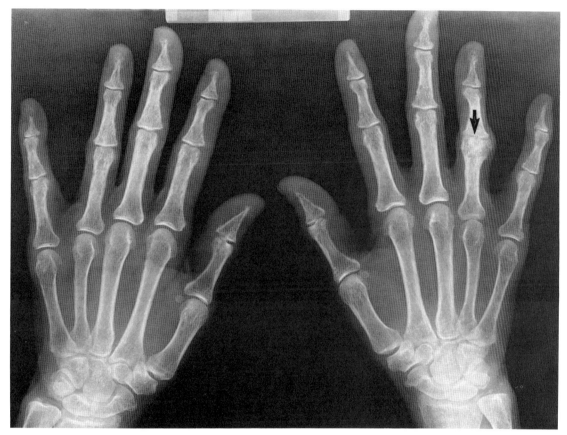

Fig. 984. Hyperparathyroidism

Erosive changes of the tufts of all the fingers causes the pointed appearance. Here in the left hand, there is deossification and arthrosis of the middle phalanx of the fourth finger (*arrow*).

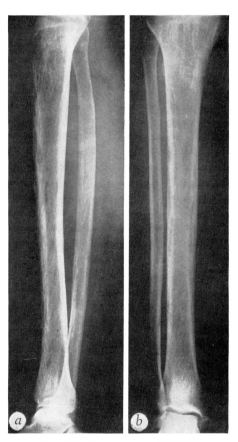

FIG. 985. HYPERPARATHYROIDISM

Cortices of the tibia reveal irregularity and resorption of bone. The change is more marked on the endosteal surface. The medial aspect of the upper third of the tibia shows subperiosteal resorption of bone.

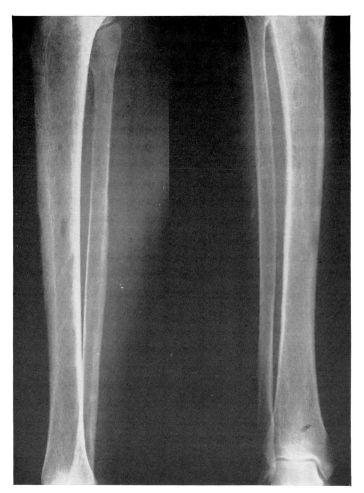

Fig. 986. Hyperparathyroidism
Appearance is rather characteristic, showing a combination of coarsened trabeculation and small cystlike lesions.

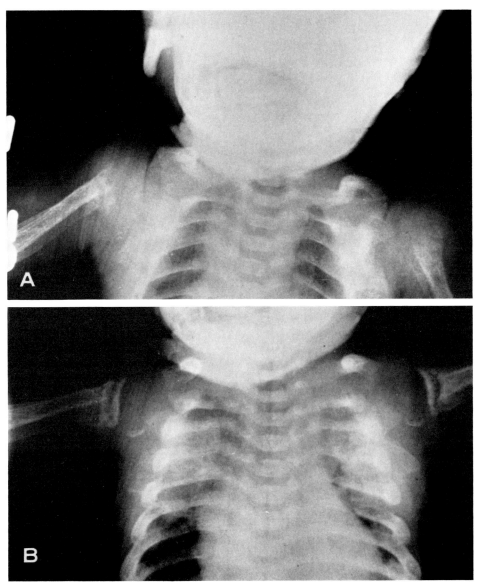

Fig. 987. Hyperparathyroidism

(A) Hyperparathyroidism in newborn that (B) healed spontaneously and probably due to maternal hypoparathyroidism. (C) Lower extremities. Resorption of the ends of the bones with fractures of the distal ends of the femurs. The bones are markedly deossified. (D) One month later without treatment there is recalcification of the bone ends and periosteal reaction and callus formation. (Courtesy of Dr. Kirkpatrick, St. Christopher's Hospital, Philadelphia, Pa.)

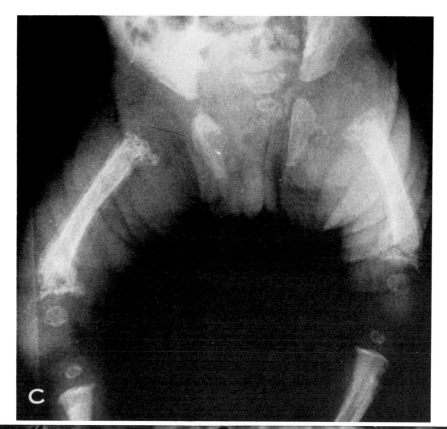
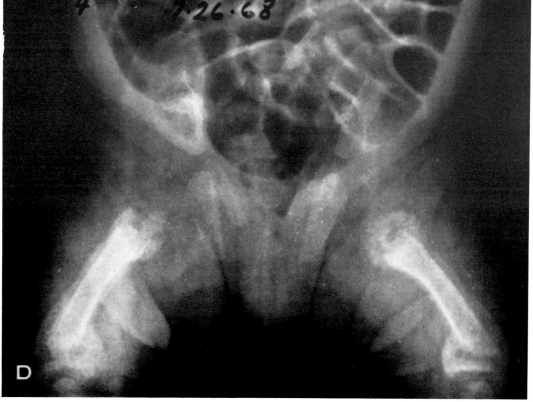

Fig. 987 (*C* and *D*)

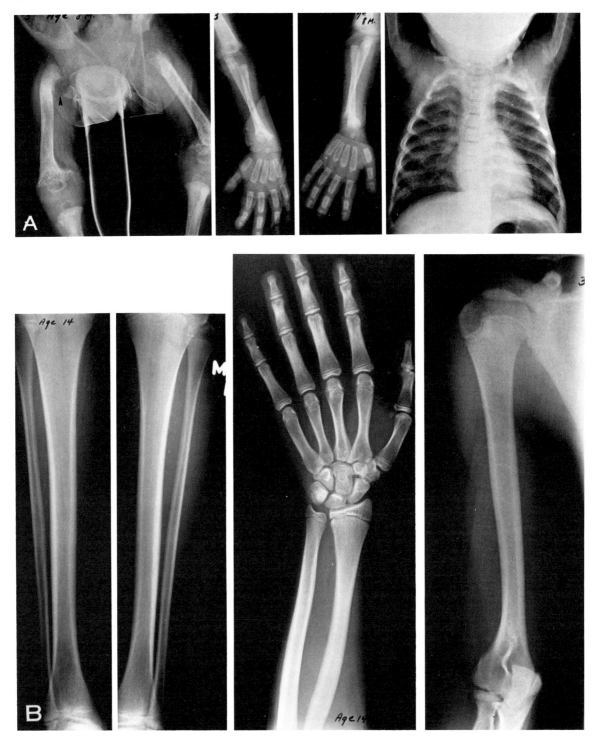

FIG. 988. HYPERPARATHYROIDISM OF NEWBORN DUE TO ADENOMA

(A) Resorption of the bone ends and fracture of the left femur. The bones are deossified and there is subperiosteal resorption at the upper end of the right femur (*arrow*). (B) The adenoma was removed at age 8 months, and at age 14 years the bones are normal. (Courtesy of Dr. John Kirkpatrick, St. Christopher's Hospital, Philadelphia, Pa.)

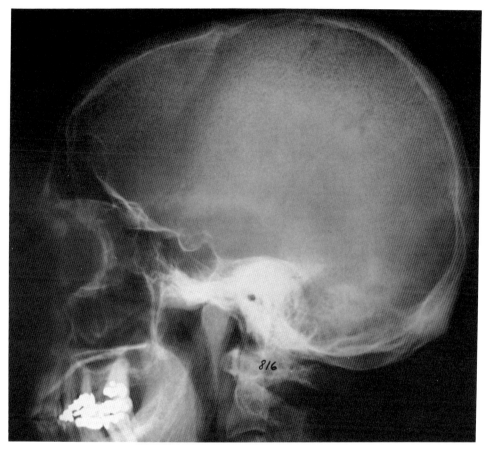

Fig. 989. Hyperparathyroidism
Granular deossification is most prominent in the outer third of the skull; it may be difficult to distinguish from normal bone.

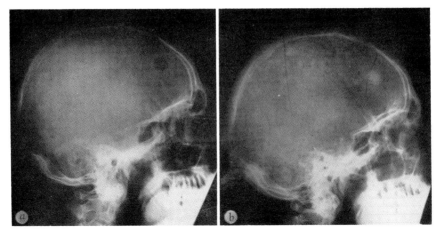

FIG. 990. HYPERPARATHYROIDISM

(a) Granular decalcification of the skull with cystlike lesion in the frontal bone. (b) Same case after removal of parathyroid adenoma. The bone is not normal in appearance but resembles somewhat bone as seen in Paget disease.

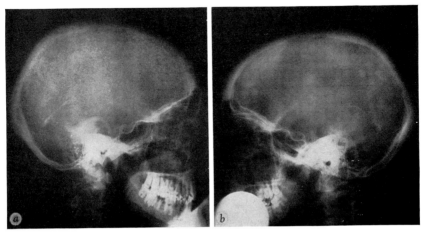

FIG. 991. HYPERPARATHYROIDISM

(a) Typical granular decalcification of the skull. (b) The same case 4 months after parathyroid adenoma was removed. Regions of increased density of bone which represent a healed stage of the disease.

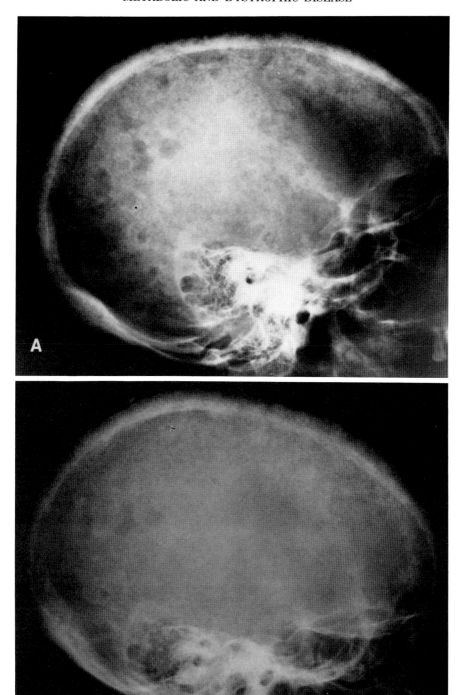

FIG. 992. HYPERPARATHYROIDISM

A granular deossification throughout the skull, with multiple radiolucent areas due to brown tumors. (*A*) Initial roentgenogram; (*B*) 1 month after removal of adenoma. The granular deossification has improved, especially in the parietal and frontal areas, although some brown tumors persist.

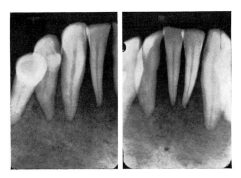

FIG. 993. HYPERPARATHYROIDISM

(*Top*) The lamina dura is gone. The trabecular structure is distorted and normal anatomical landmarks are indistinct. There has been no loss of calcium in the teeth and they therefore appear abnormally dense in contrast with the decalcified bones surrounding them. (*Bottom*) Same case 4 years after parathyroid adenoma was removed. The lamina dura is now present and the trabecular structure and ossification of the mandible are normal.

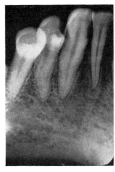 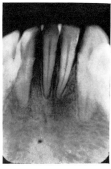

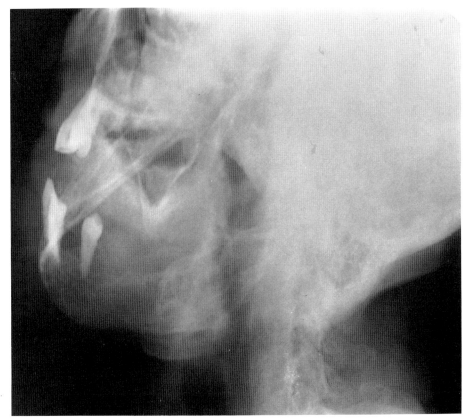

FIG. 994. HYPERPARATHYROIDISM

Normal anatomic landmarks such as the inferior dental canal, dental foramina, and cortices of mandible are indistinct due to deossification and brown tumor formation.

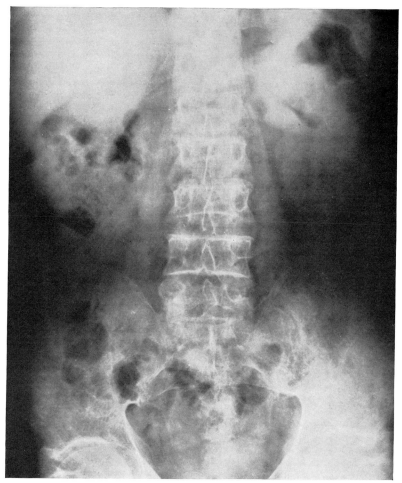

Fig. 995. Hyperparathyroidism

Extensive deossification and distortion of the trabecular pattern throughout the lumbar spine and pelvis. The appearance may be seen in osteomalacia.

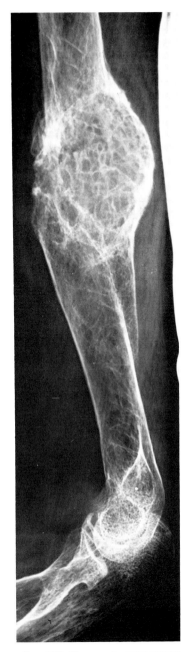

ing as large, eccentric, expanding destructive areas in the metaphyses of long bones (Figs. 998–1000, 1002, and 1004). Histologically and roentgenographically they may be difficult to distinguish from giant cell tumors and, at times, have characteristics of malignancy. Pathologic fractures and pseudofractures are infrequent.

Secondary bone changes of advanced disease are caused by marked bone deossification and softening (Figs. 1003, 1006, and 1007), which leads to basilar impression of the skull and wedging and flattening of the spine (Fig. 1007), which, in turn, causes kyphosis, scoliosis, and loss of stature, with biconcave deformities created by

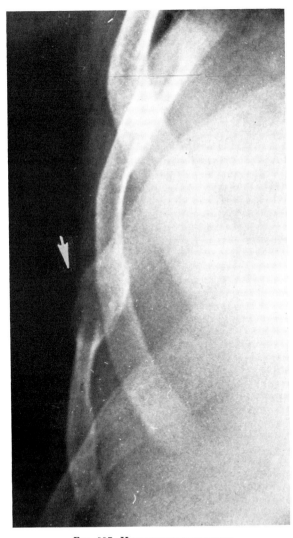

FIG. 996. HYPERPARATHYROIDISM
Decalcification, thinning of the cortex, and distorted trabecular pattern. A large brown tumor in the midshaft of the humerus. There is a healed pathologic fracture. (Courtesy of George T. Wohl, Philadelphia General Hospital, Philadelphia, Pa.)

glass appearance caused by indistinct trabeculae, or else the trabecular pattern may be coarse and prominent, as is often the case in the vertebrae and pelvis, as well as the long bones. Cystlike lesions (Figs. 997 and 1001) are common, appear-

FIG. 997. HYPERPARATHYROIDISM
Single brown tumor in the rib (*arrow*). The bones were otherwise normal. (Courtesy of George T. Wohl, Philadelphia General Hospital, Philadelphia, Pa.)

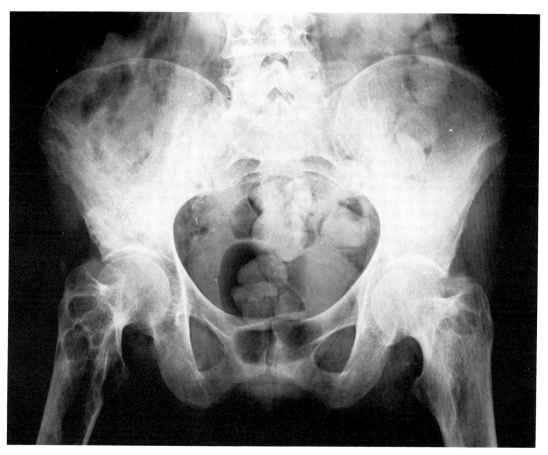

Fig. 998. Hyperparathyroidism
Brown tumors in both femurs and distorted trabecular pattern in the pelvis. (Courtesy of George T. Wohl, Philadelphia General Hospital, Philadelphia, Pa.)

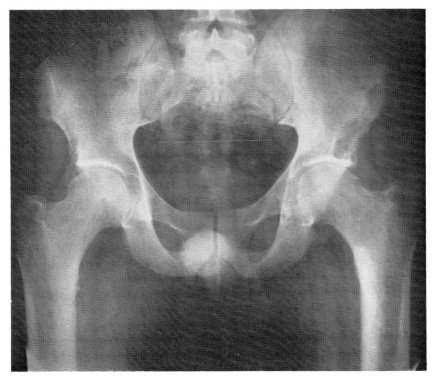

Fig. 999. Hyperparathyroidism

Brown tumor above the left acetabulum. The remainder of the pelvis and the upper ends of the femurs show no evidence of malacic disease.

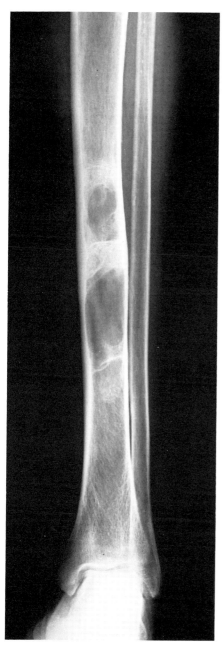

Fig. 1000. Hyperparathyroidism
Peculiar brown tumors in the shaft of the tibia. The remainder of the bone appears normal. (Courtesy of George T. Wohl, Philadelphia General Hospital, Philadelphia, Pa.)

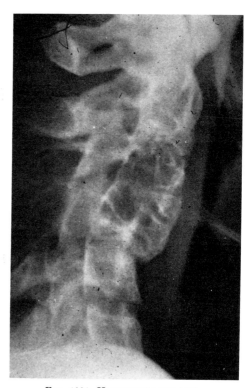

Fig. 1001. Hyperparathyroidism
Brown tumors in the third and fourth vertebrae with pathologic fracture dislocation.

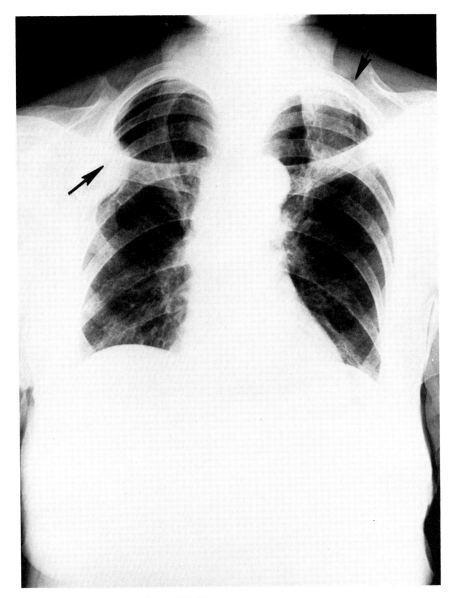

FIG. 1002. HYPERPARATHYROIDISM
Brown tumors in the first left rib and in the third rib on the right (*arrows*). Multiple small lytic lesions throughout the ribs.
(Courtesy of George T. Wohl, Philadelphia General Hospital, Philadelphia, Pa.)

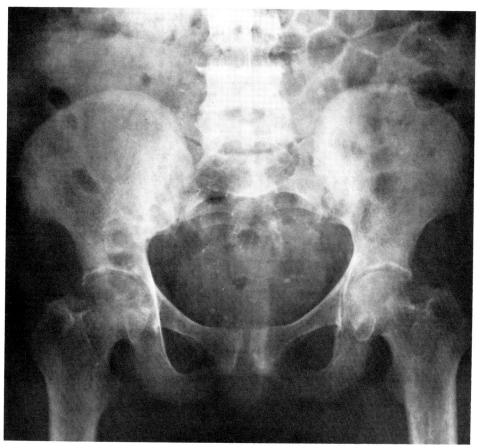

Fig. 1003. Hyperparathyroidism

Brown tumors throughout the pelvis and the femoral head. The trabecular pattern is ground glass and irregular and the cortices of the upper ends of the femurs show endosteal irregularity. (Courtesy of George T. Wohl, Philadelphia General Hospital, Philadelphia, Pa.)

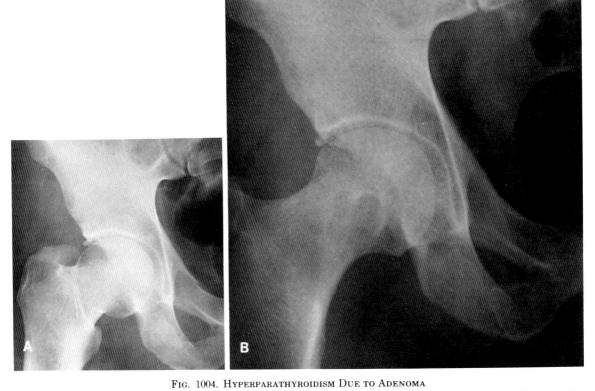

Fig. 1004. Hyperparathyroidism Due to Adenoma

(A) In 1969 the pelvis was normal; (B) after the development of hyperparathyroidism, a brown tumor in the neck of the femur. It has a sclerotic margin, is lobulated and might be mistaken for a cartilaginous tumor.

the unyielding nucleus pulposus (Fig. 1003). The pelvis may become distorted, and pathologic fractures from minimal trauma may occur, especially in the ribs and the femoral necks (Fig. 1005).

Increased bone density, either localized or diffuse, is reported. It is more commonly associated with secondary hyperparathyroidism caused by renal disease, but may be present in primary hyperparathyroidism without renal damage. The bone trabeculae are thickened and increased in density, and bones most active in hematopoiesis, such as the vertebral column, are most frequently involved.[27] In the centra, the thickening is conspicuous in zones adjacent to the superior and inferior end plates; the intervening bone is porous. This causes a horizontally striped vertebra, referred to as "rugger jersey spine."[28] The pelvis, ribs, and skull are also commonly involved; the tubular bones less frequently.

Spontaneous regression of bone lesions is reported,[29] but can be attributed to intermittent hyperparathyroidism. The bone lesions improve rapidly after the cause of hyperparathyroidism is removed, and marked changes are noted within 3 or 4 weeks after definitive treatment.

Recalcification begins at the periphery of the lesion. Where there is cortical resorption, bone formation can be detected at the most peripheral portion of cortex, so that an intervening radiolucent area composed of fibrous and osteoid tissue appears between new bone and cortex. Subsequently, all tissue is converted to dense bone. The localized destructive lesions also appear at the periphery and, eventually, the entire cystic area is replaced by bone.

New bone formation may extend along the tendons or ligaments and may result in fusion of symphysis pubis and sacroiliac joints.[20] Steinbach et al.[20] reported that bones do not revert completely to normal, but continue to increase in density at the site of the cyst.

Calcified articular or fibrocartilage in association with hyperparathyroidism[30, 31] occurs in 10–20% of patients.[31–35] The cartilage and meniscus are the most common sites and calcification of the triangular wrist cartilage is almost as frequent.[32–36] Calcification may occur without subperiosteal resorption of the phalanges.[36] It may be distinguished from calcification of degenerative joint disease by the absence of secondary degenerative change of the adjoining bone.[37] The roentgen appearance of calcification in primary chondrocalcinosis (pseudogout) and hypoparathyroidism is similar,[36] although it is reported that linear and punctate calcifications identify

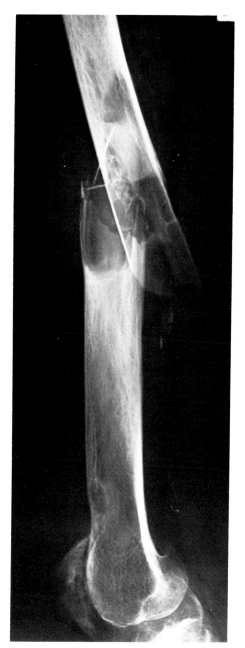

FIG. 1005. HYPERPARATHYROIDISM
Multiple brown tumors in the femur and a pathologic fracture in the midshaft. The cortices are thin with endosteal irregularity. Pathologic fractures are rare with hyperparathyroidism. (Courtesy of George T. Wohl, Philadelphia General Hospital, Philadelphia, Pa.)

pseudogout.[33, 38] Ectopic periarticular calcification may occur in primary hyperparathyroidism, but it is more common in the secondary form due to chronic renal failure. The metastatic cal-

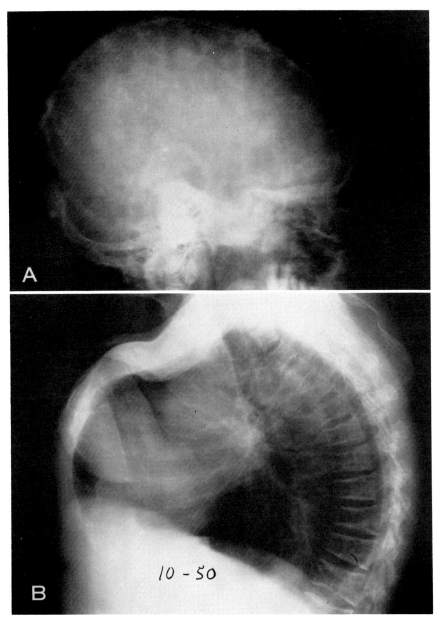

FIG. 1006. HYPERPARATHYROIDISM

A 35-year-old white woman was confined to a nursing home for many years totally incapacitated. Laboratory studies revealed increased serum calcium, decreased serum phosphate, and increased urinary calcium due to a parathyroid adenoma. (A) Skull; diffuse deossification; the lamina dura are absent. (B) Chest; marked deossification and platyspondyly, and a sternal deformity. (C) Hands; diffuse deossification and marked subperiosteal resorption of the middle phalanges. (D) Forearms; marked deossification, loss of the cortices, and small brown tumors. Healing fracture of the upper end of each ulna. (E) Pelvis; diffuse deossification and pelvic deformity. Fractures of both femoral necks and renal calcifications. (F) Both legs; marked deossification of both legs and fractures at the distal tibial ends. The cortex is thin and the bone diameters are wide. (Courtesy of Dr. John Anderson, Detroit, Mich.)

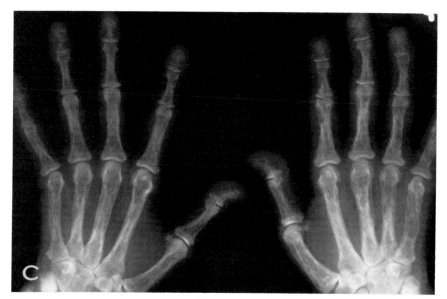

Fig. 1006C

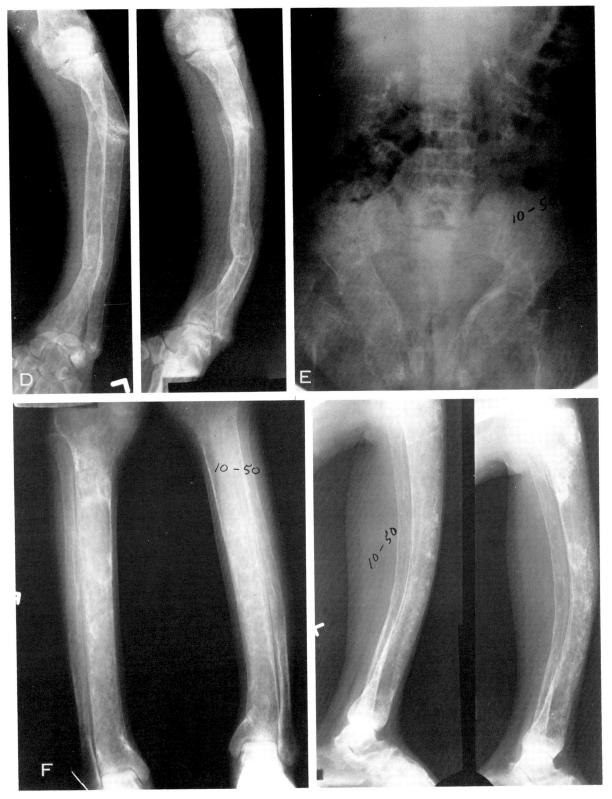

Fig. 1006 (D-F)

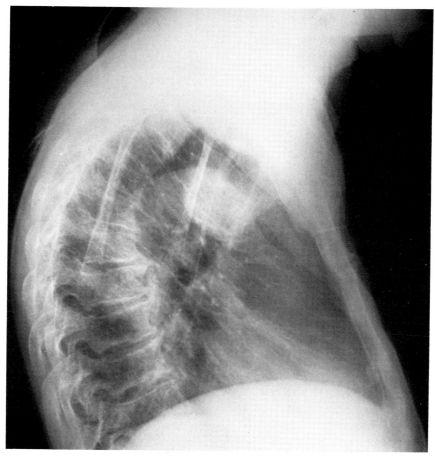

FIG. 1007. HYPERPARATHYROIDISM
The vertebrae are deossified and there is wedging of all centra. (Courtesy of George T. Wohl, Philadelphia General Hospital, Philadelphia, Pa.)

cification may reach considerable size, spreading into the surrounding soft tissues, especially around the shoulder joint.[39] Fluid levels within this periarticular calcification are reported.[40]

Calcium deposits in the kidneys are quite common. Albright estimates that 5% of patients with urinary calculi have hyperparathyroidism.[8] This incidence seems slightly high, but serves to emphasize the importance of considering hyperparathyroidism in the presence of urinary calcifications.

Chest roentgenograms may reveal a mediastinal mass when the adenomas are large.[41-43] Tracheal and esophageal displacement may occur (Fig. 975). When an adenoma is suspected, the slightest deviation of the trachea or esophagus may be of significance. Calcification of the capsule of the adenoma may be roentgenographically detected.[44-46] Parathyroid adenomas are usually found in the area of the parathyroid glands, but they may be anywhere between the parotid gland area and the mediastinum.[43, 47-51]

Selective inferior thyroid artery or internal mammary arteriography may localize the adenoma site, either by displacement of the arteries[52-55] or by staining of the adenoma.[54, 56-61] (Fig. 1008). Positive results are obtained in less than one-half of patients and do not distinguish a thyroid adenoma from a parathyroid tumor.

Parathyroid scintigraphy with ^{75}Se-selenomethionine may visualize the parathyroid adenoma in approximately one-half the patients.[62-76]

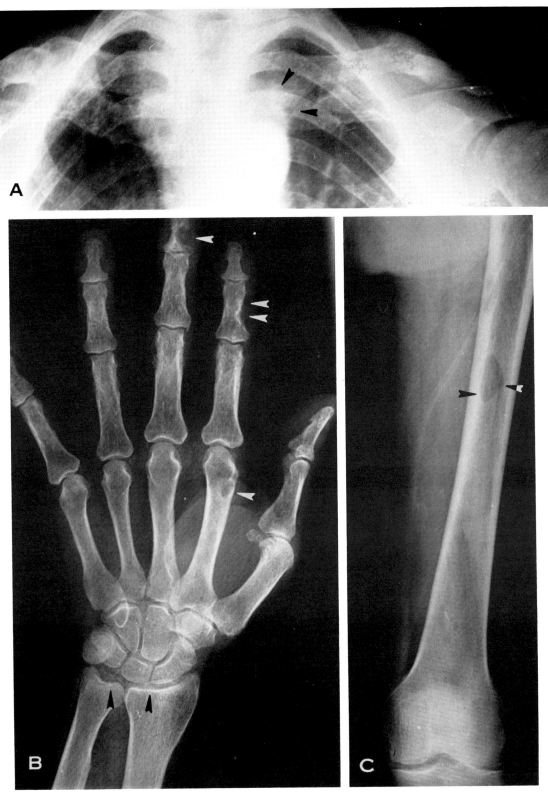

FIG. 1008. HYPERPARATHYROIDISM

A 70-year-old man who complained of bilateral "lumps" over the clavicles (due to fractures that were the first signs of hyperparathyroidism). (*A*) Clavicles; fractures of both clavicles, left mediastinal mass (*arrows*) immediately superior to the aorta and a prominent vascular shadow on the right. (*B*) Hand and wrist; diffuse deossification. Erosions of the distal third phalanx and in the second metacarpal (*horizontal arrow*). Subperiosteal resorption in the second middle phalanx (*double horizontal arrows*). Calcifications in the wrist joint (*vertical arrows*). (*C*) "Brown tumor" in the midfemur (*arrows*). (*D*) Calcification of the articular cartilage of the knee (*arrows*). (*E*) Selective arteriogram reveals vascularization of the adenoma. (*F*) Venous phase reveals tumor staining (*arrows*).

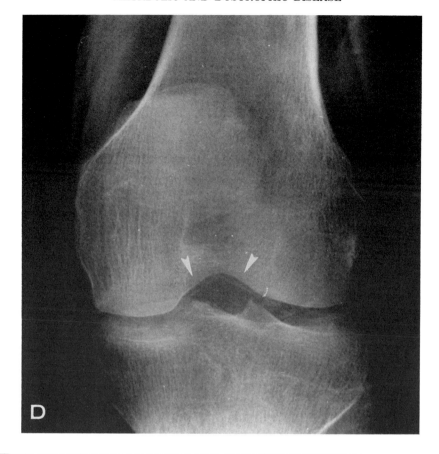
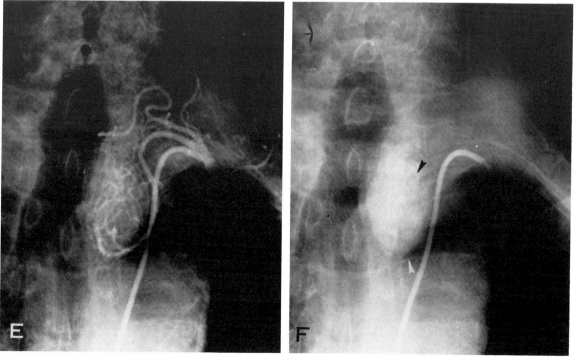

FIG. 1008 (D–F)

HYPOPARATHYROIDISM, PSEUDOHYPOPARATHYROIDISM, AND PSEUDO-PSEUDOHYPOPARATHYROIDISM

Hypoparathyroidism is an endocrine disorder; the two "pseudo" conditions are different manifestations of genetic and familial disturbances.

The chemical changes of hypocalcemia and hyperphosphatemia occur both in hypoparathyroidism and pseudohypoparathyroidism, but develop from different causes. In hypoparathyroidism they have an endocrine origin and are caused by the absence or insufficiency of parathyroid hormone; patients respond to the administration of parathyroid hormone. Pseudohypoparathyroid patients have normal or hyperplastic parathyroid glands capable of producing adequate parathyroid hormone and the chemical changes reflect, not insufficient production of parathyroid hormone, but the inability of the body to respond to it.

The genetic developmental abnormalities of pseudohypoparathyroidism and pseudo-pseudohypoparathyroidism are different expressions of the same disturbance. The cardinal features are skeletal developmental abnormalities and chemical changes of the blood due to genetic defects rather than to a primary endocrine disturbance. If both the typical chemical changes and the skeletal abnormalities are present, the disease is labeled pseudohypoparathyroidism; if only the skeletal changes are manifest, it is known as pseudo-pseudohypoparathyroidism. The presence of both syndromes in different members of a family is further evidence that these conditions are related, and represent genetic and familial disturbance.[1,2]

From a radiographic viewpoint there is no similarity between the bone changes seen in hypoparathyroidism and the two pseudohypoparathyroidisms. In the former, the radiographic changes are due exclusively to chemical changes in the blood, particularly hypocalcemia, whereas in the two pseudo types they are due to genetic defects. In very rare instances, however, pseudohypoparathyroid patients will also manifest some of the radiographic changes caused by hypocalcemia and, therefore, resemble patients with hypoparathyroidism. Thus, the radiographic changes seen in the two pseudo types of hypoparathyroidism represent a true chondrodystrophy.

Hypoparathyroidism

Idiopathic hypoparathyroidism was first reported in 1926 by Beumer and Falkenheim.[3] In 1939, Drake et al.[4] established the following criteria for its diagnosis: (1) low serum calcium; (2) high serum inorganic phosphorus level; (3) renal insufficiency, steatorrhea, chronic diarrhea, and alkalosis must not be present; and (4) rickets and osteomalacia must be excluded. Secondary hypoparathyroidism is due to a known cause and is more common. Aside from the biochemical changes, both types present the same clinical and roentgenographic features.

The most common cause of secondary hypoparathyroidism is the accidental removal of or damage to the parathyroid glands during thyroid surgery. If the glands are merely damaged they may regenerate in a few months. Idiopathic hypoparathyroidism itself is a rare condition[4] of unknown cause; Albright and Reifenstein[5] have found in at least 1 patient that the epithelial cells of the parathyroid glands were replaced by fat.

A relationship between hypoparathyroidism and moniliasis is suggested.[6,7] Although moniliasis often precedes hypoparathyroidism, no true relationship has been established. It is noteworthy that a tendency to idiopathic hypoparathyroidism appears to be associated with Addison's disease.[6,8]

Clinical Features. The classic patient with hypoparathyroidism is round faced, shorter than average, sometimes dwarflike, and may have cataracts. Often he complains of dry, coarse, and scaly skin, sparse hair, and atrophy of the fingernails and toenails.

Dental hypoplasia and aplasia develop, depending, probably, upon the patient's age at the time of onset of hypoparathyroidism. Thus, if a 10-year-old develops it, his teeth will be normal except for a blunting of the molar roots, which are the last parts of the teeth to form.

Hypocalcemia triggers the neuromuscular excitability of hypoparathyroidism. Tetany may then develop, with its numbness, cramps of the extremities, carpopedal spasm, laryngeal stridor, and generalized convulsions. The more severe symptoms may easily be mistaken for epilepsy. If untreated for years, mental retardation may result.

Laboratory Features. The characteristic laboratory findings include low serum phosphorus, and normal or low serum alkaline phosphatase. The urine shows little or no calcium.

Roentgenographic Features. Although the roentgenographic features of hypoparathyroidism may not be striking, they are sometimes highly significant. Particularly suggestive are intracranial calcifications of the basal ganglia[9] which, early, are discrete, coalescing later into homogeneous masses (Figs. 1009 and 1010). Occasionally these calcifications develop in the cerebellum.[10] Because they also occur in pseudohypoparathyroidism, they are not diagnostic. The deposition of calcium is related to the hypocalcemia.

Dental abnormalities include hypoplasia of enamel and dentine, blunting of dental roots, delay or failure of eruption, thickening of the lamina dura, and prominence of the dental membrane.[11]

The increased density of the skeleton described by some authorities often defies detection unless the changes are striking. Bronsky et al.[10] reported generalized increased bone density in 9% of 44 patients. Localized thickening of the skull, deformed hips with thickening and sclerosis of the femoral head and acetabulum, subcutaneous calcification, premature closure of the epiphyses, calcification of spinal ligaments, and even generalized deossification are reported in conjunction with idiopathic hypoparathyroidism. Ossification of muscle insertions, ectopic bone formation, ligamentous ossification, sacroiliac sclerosis, and vertebral hyperostosis may occur (Fig. 1011).

Noteworthy in hypoparathyroidism is the peculiar bandlike increase in density in the metaphysis of long bones (Fig. 1012), increased density of the iliac crest, and a double sclerotic line rimming the vertebral body as reported by Achenbach and Bohm[12] and by Taybi and Keele.[11] These changes suggest an abnormality in enchondral bone formation.

Pseudohypoparathyroidism

Pseudohypoparathyroidism, first described by Albright et al.[1] in 1942, is a congenital hereditary abnormality characterized by hypocalcemia and hyperphosphatemia. Although the symptoms and chemical findings are similar to those of hypoparathyroidism, it does not respond adequately to parathyroid hormone. There is a lack of phosphate diuresis, due to end organ resistance,[2] elaboration of a defective hormone,[3] or the presence of antienzymes in the tissues.[4] Studies indicate that there is a defect in the renal production of 1α,25-dihydroxyvitamin D that causes a decrease in the concentration of this hormone.[5, 6] There is impaired intestinal absorption of calcium, hypocalcemia, and secondary hyperparathyroidism which may be corrected by 1α,25-dihydroxyvitamin D_3.[5, 7] Long term use of this hormone in pseudohypoparathyroidism increases the serum 1α,25-dihydroxyvitamin D_3 which causes hypercalcemia. Cessation of the drug or decrease in the dose corrects the situation.[8] Osteitis fibrosis cystica is reported in some patients, probably secondary to excessive production of bone-resorbing hormone, possibly parathyroid hormone.[9, 10] We have observed several patients with pseudohypoparathyroidism and hyperparathyroidism. Lee et al.[11] suggest a possible excessive production of thyrocalcitonin. It is transmitted by a sex-linked dominant gene.

Most patients are short, obese, and mentally retarded, often with round faces and thick-set features. Corneal and lenticular opacity and brachydactylia may be encountered.[4, 6, 12-14]

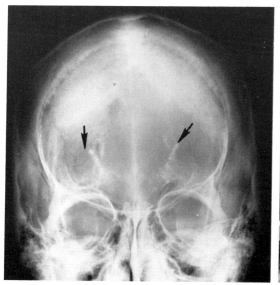
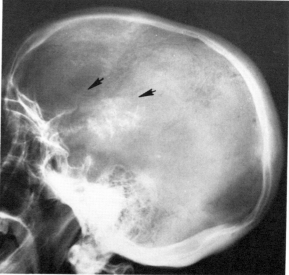

FIG. 1009. HYPOPARATHYROIDISM
Calcification in the basal ganglia (*arrows*).

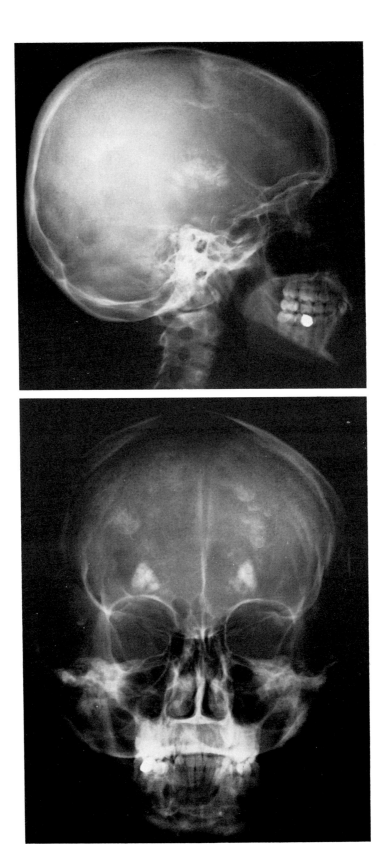

Fig. 1010. Hypoparathyroidism
Calcification of the basal ganglia. (*Top*) Lateral projection; (*bottom*) posteroanterior projection.

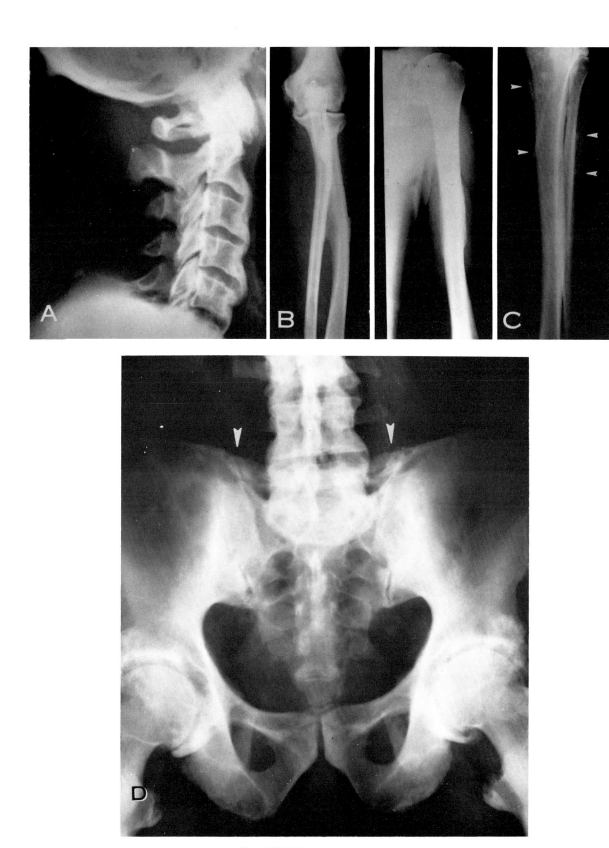

FIG. 1011. HYPOPARATHYROIDISM
(A) Cervical spine. Spur formation with relatively normal disk spaces. Hyperostosis may occur with hypoparathyroidism. (B) Bony prominences at ligamentous attachments are present. (C) Ectopic bone formation (*arrows*). (D) Sclerosis of both sacroiliacs and degenerative arthrosis of both hips. The lumboiliac ligaments are ossified (*arrows*).

Roentgenographic Features. The most striking change, seen in many but not all patients, is brachydactylia, usually of the metacarpal and metatarsal bones (Fig. 1013). The bones most likely to be short are those in which the epiphysis appears latest. The metacarpals develop epiphyses in the following sequence: II, III, IV, V, and I. Although all may be short, I, IV, and V are the most commonly affected. In addition, there are frequent disturbances of the epiphysis of other long bones, which may account for the apparent dwarfism. Early closure of epiphyses suggests an endocrine influence, but, since all epiphyses are not similarly affected, a genetic disturbance is indicated. Sometimes multiple exostoses will be found (Fig. 1014). The basal ganglion calcification which may also appear is probably related to the hypocalcemia.

Patients with pseudohypoparathyroidism tend to form bone or calcifications in the skin or subcutaneous tissues[15] and calcification of the basal ganglia and other parts of the brain.[16] The delayed, defective dentition also occurs in other

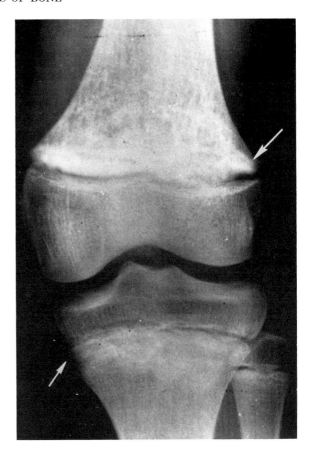

FIG. 1012. HYPOPARATHYROIDISM
Bandlike increase in density in the metaphyseal regions (*arrows*). (Courtesy of Hooshang Taybi and Doman Keele, Indiana University Medical Center, Indianapolis, Ind.)

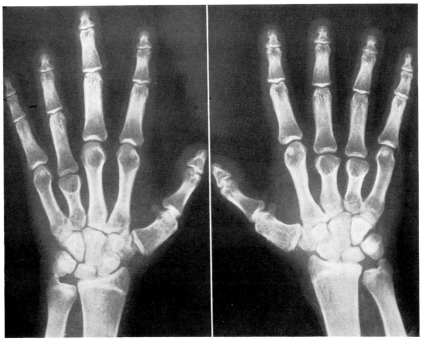

FIG. 1013. PSEUDOHYPOPARATHYROIDISM
Abnormally short metacarpals are found in this disease. The metatarsals also are usually affected.

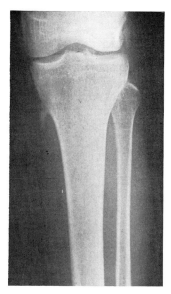

FIG. 1014. PSEUDOHYPOPARATHYROIDISM
Exostosis of the tibia. Multiple exostoses may be present in this condition.

skeletal deformity. Steinbach and Rudhe[16] reported a patient with significant blood chemistry at the age of 9 months, who developed skeletal lesions by the age of 49 months.

Skeletal deformities of the metacarpals and metatarsals occur with multiple familial exostoses, Turner syndrome, and other less common syndromes, but these conditions have other roentgen, clinical, and genetic findings that indicate the correct diagnosis.[18]

In some instances the skeletal density is diffusely decreased, which is usually ascribed to osteoporosis. In other cases the subperiosteal erosions typical of hyperparathyroidism (osteitis fibrosis cystica) are observed. We have observed 2 cases in which the classical signs of pseudohypoparathyroidism were present with low blood calcium and high blood phosphorus, and in which there were also subperiosteal erosions. These patients proved to have parathyroid adenoma as well as the signs of pseudohypoparathyroidism. The blood chemical changes did not reflect the hyperparathyroidism.

members of the family without the disease, and may represent an abnormal constitutional response to trauma. Pseudohypoparathyroidism is clearly related to pseudo-pseudohypoparathyroidism, the two affecting different individuals of the same family.[13]

Pseudohypoparathyroidism may be found in childhood or early infancy[17] without evidence of

Pseudo-pseudohypoparathyroidism

Pseudo-pseudohypoparathyroidism was the name selected by Albright and Reifenstein[1] to describe a syndrome in short-statured patients with rounded faces, whose roentgen features simulated exactly those of pseudohypoparathyroidism. In this group, there are no blood chemical changes.

HYPERVITAMINOSIS A

Excessive vitamin A produces generalized symptoms and skeletal changes.

Clinical Features. Anorexia and weight loss are the first manifestations, and persistent and generalized pruritus also occurs early. The hair becomes sparse and coarse, and the lips dry. Later, there is pain and swelling of the long bones. Yellow skin tinting occurs with high serum vitamin A levels. Hepatomegaly and splenomegaly are sometimes observed, and there may be an increase in serum lipids. The level of serum alkaline phosphatase is increased, and the serum protein level lowered.[1] A careful history will reveal excessive intake of vitamin A, and, with its decrease, the symptoms promptly disappear.[2, 3]

Roentgenographic Features. Periosteal proliferation, principally of the ulnae and metacarpals, is the main feature[4] (Fig. 1015), although other long bones may be affected.[5] The proliferation is solid or lamellated and greatest at the shaft center. Skeletal deossification and advanced bone maturation may occur.[6] Infantile cortical hyperostosis is differentiated by its thick lamellated periosteal reaction and mandibular involvement.

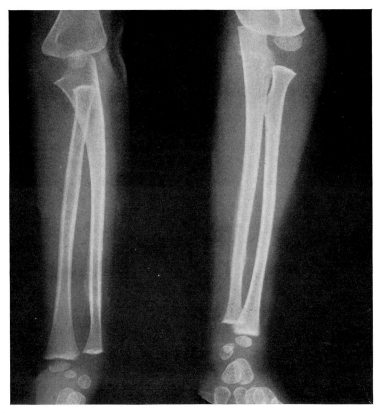

Fig. 1015. Hypervitaminosis A
Periosteal proliferation on the distal half of the ulna. (Courtesy of Clayton Hale and Tyree C. Wyatt.)

HYPERVITAMINOSIS D

Intoxication resulting from excessive ingestion of vitamin D is well recognized.[1] In infants and children it is due to overzealous or erroneous dosage. The action of vitamin D is unknown, but large doses seem to act as parathyroid hormone; results are excessive urinary phosphorus excretion and hypercalcemia, followed by hypercalciuria. Early, the serum phosphorus level may be low, but it elevates with renal damage.

Clinical Features. Symptoms are related to hypercalcemia, hypercalciuria, and renal insufficiency. Loss of appetite, drowsiness, and headache are early symptoms, accompanied by polyuria and polydipsia. Nausea, vomiting, abdominal cramps, and diarrhea are often present. As a result of renal damage, increased serum nonprotein nitrogen and creatinine occur.[2] A low specific gravity of urine and albuminuria eventually result. Microscopic examination of the urine reveals hyaline and granular casts.

Roentgenographic Features. The main feature is metastatic calcinosis of arterial walls, kidneys, and periarticular tissues.[3, 4] The periarticular tissues appear puttylike (Figs. 1016–1018). Extensive vascular calcification is noted in patients between 20 and 30 years of age. Nephrocalcinosis leads to renal insufficiency, but renal calcification is not often demonstrated.

Deossification is present but difficult to evaluate in elderly or partially immobilized patients. When vitamin D is discontinued, metastatic calcinosis resorbs completely. At times, partial resorption occurs and ossification replaces the puttylike deposits.

Initially, the long bones show widening of the provisional zones of calcification. Later, there is cortical thickening. Osteoporosis eventually produces deep zones of diminished density alternating with bands of increased density (Fig. 1019). The vertebral bodies are outlined by a dense band of bone, and adjacent radiolucent zones exaggerate the density (Figs. 1020 and 1021). The calvaria becomes dense and there is premature calcification of the falx cerebri (Fig. 1022), which is the most consistent sign of hypervitaminosis D.

Idiopathic Hypercalcemia. This disease is probably the result of hypervitaminosis D. The roentgen features are the same.

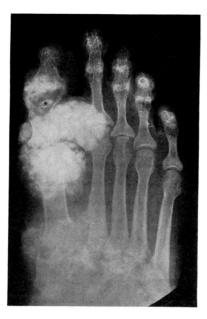

FIG. 1016. HYPERVITAMINOSIS D AND TOPHACEOUS GOUT
Presence of calcium in gouty tophi always suggests the possibility that hypervitaminosis D may be present.

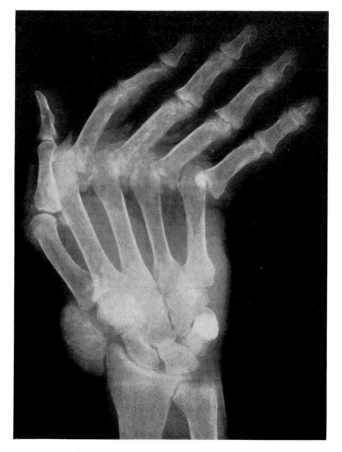

FIG. 1017. HYPERVITAMINOSIS D AND RHEUMATOID ARTHRITIS
Puttylike appearance of the periarticular deposits of calcium in hypervitaminosis D is quite distinctive.

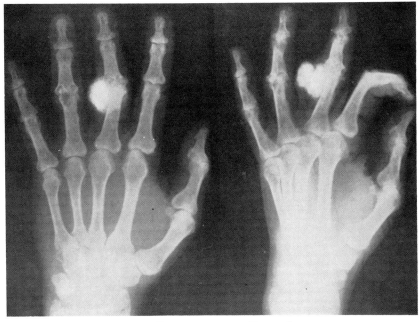

Fig. 1018. Hypervitaminosis D and Gout

Fig. 1019. Hypervitaminosis D
Widening of the provisional zone of calcification. Alternating bands of increased and decreased densities are present in the shafts near the epiphyses and in the epiphyses. (Courtesy of John A. Kirkpatrick, St. Christopher's Hospital, Philadelphia, Pa.)

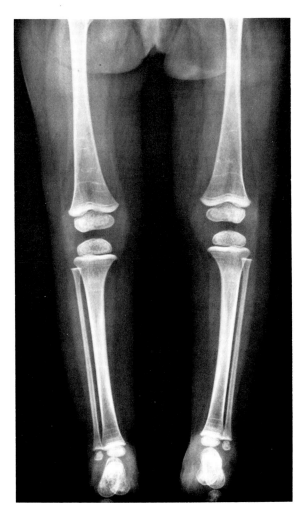

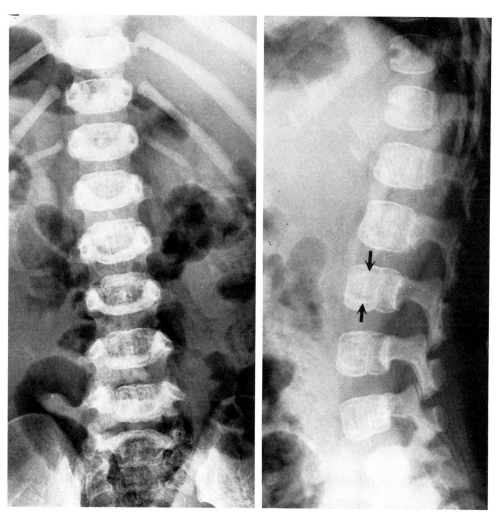

Fig. 1020. Hypervitaminosis D (Idiopathic Hypercalcemia)
The centra are dense at the periphery with a radiolucent area within (*arrows*). (*Left*) Anteroposterior projection; (*right*) lateral projection. (Courtesy of John A. Kirkpatrick, St. Christopher's Hospital, Philadelphia, Pa.)

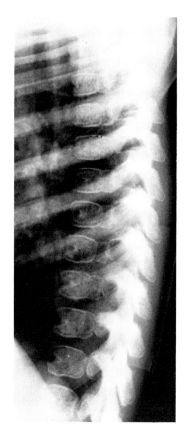

FIG. 1021. HYPERVITAMINOSIS D
Centra show a dense periphery wtih a radiolucent line below. (Same case as seen in Figures 1019 and 1020. (Courtesy of John A. Kirkpatrick, St. Christopher's Hospital, Philadelphia, Pa.)

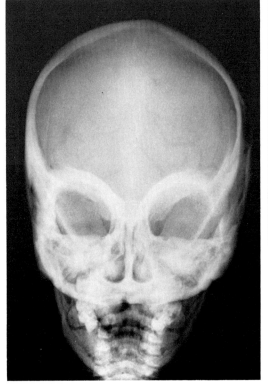
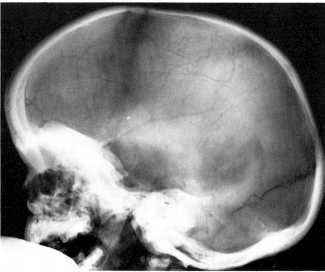

FIG. 1022. HYPERVITAMINOSIS D (IDIOPATHIC HYPERCALCEMIA)
Calvaria is dense but more important there is premature calcification of the falx cerebri. This is the most consistent sign of hypervitaminosis D. Posteroanterior projection and lateral projection of same patient. (Courtesy of John A. Kirkpatrick, St. Christopher's Hospital, Philadelphia, Pa.)

RENAL OSTEODYSTROPHY

Renal osteodystrophy is the skeletal response to long standing chronic renal disease. In the past, few patients with chronic renal failure survived a sufficient time for osseous abnormalities to become manifest. With improvement in dialyzing techniques and renal transplants, longevity is increased and renal osteodystrophy is a more frequent finding. Although dialysis prolongs life, the osseous abnormalities persist and progress, albeit at a slower rate. Since renal osteodystrophy will occur with increasing frequency, it is important to understand basic bone physiology and be cognizant of the roentgenographic features of the disease spectrum.

Renal osteodystrophy is a recognizable complex of skeletal changes and includes: (1) osteomalacia (adults) or rickets (children), (2) osteitis fibrosa (hyperparathyroidism), and (3) osteosclerosis.[1] Soft tissue calcifications are common. All of these features may not be present in the same individual. The skeletal changes may be due to several causes, classified by Dent[2] into glomerular and tubular forms. Glomerular renal disease is most often due to chronic glomerulonephritis.[3,4] Tubular forms include vitamin D-resistant rickets, the Fanconi syndrome, and renal tubular acidosis. "Glomerular" types are due to acquired renal disease, and tubular forms are due mainly to inborn metabolic errors. Therefore, we use the terms acquired (glomerular) and congenital (tubular) renal osteodystrophy.

Acquired Renal Osteodystrophy

Pathophysiology

Acquired renal osteodystrophy is caused by functional renal impairment. The basic mechanisms are controversial, and there are two main theories: (1) acquired insensitivity to all actions of vitamin D (ergosterol),[4-9] and (2) chronic acidosis and excessive secretion of phosphorus into the bowel.

Acquired Insensitivity to Vitamin D (Antivitamin D Factor)

Vitamin D has three known functions: to promote intestinal absorption of calcium and of phosphorus, and to exert a minor phosphaturic action. When these functions are lost, less circulating minerals are available to bone due to decreased intestinal absorption. Bone cannot utilize the circulating minerals present, because the direct effect on bone mineralization is lost. The result is osteomalacia. Eventually there is a decrease in serum calcium which stimulates the parathyroids to secrete large amounts of parathormone. Parathormone causes an increased rate of bone resorption (to release minerals in order to maintain serum calcium levels) and osteitis fibrosa results.

Two notable groups of studies support this theory of acquired insensitivity to vitamin D or an anti-vitamin D factor. Rachitic rat cartilage calcifies at a Ca-P product of 35, incubated with serum from control patients, but only at 55 with serum from chronic renal disease patients.[10] Also, large doses of vitamin D (20,000-50,000 units per day) will cause healing of renal osteodystrophy before the Ca-P product rises,[11] seemingly substantiating a direct mineralizing effect of vitamin D.[8,12,13]

Chronic Acidosis and Intestinal Phosphate Secretion

The older pathogenetic theory attributes the osseous changes to chronic acidosis.[14-18] Chronic acidosis occurs with renal failure and is associated with increased serum phosphorus levels. This results in the secretion of large amounts of phosphate into the bowel. Intestinal calcium is then bound to phosphate in the bowel lumen, forming insoluble calcium salts which cannot be absorbed. Hypocalcemia eventually ensues.

There is much evidence to refute the above explanation. Careful metabolic studies fail to show any increase in insoluble calcium-phosphate salts in the stool.[7] Moreover, aluminum hydroxide given orally to bind intestinal phosphate will not result in increased calcium absorption or blood levels,[19] and will not bring about bony healing. Chronic acidosis as the sole cause of renal osteodystrophy is unlikely, since alkalinization of the patients will not cause skeletal healing unless vitamin D is added.[7,12,16,18,19] Other chronic acidoses, e.g., diabetes mellitus and renal tubular acidosis, do not cause osteitis fibrosa.[20]

I. Osteomalacia and Rickets

Osteomalacia and/or rickets is the most common feature of renal osteodystrophy. The difference in adults (osteomalacia) and children (rickets) reflects only the stage of bone maturation present when the deficiency occurs.

Osteomalacia occurs in both diffuse and localized forms. The former is due to diminution in the number of trabeculae, and a compensatory thickening of primary stress trabeculae, thus accounting for the "increased trabecular pattern" often described. The focal form is known as "Milkman fracture," or "Looser nodes." These pseudofractures are really incomplete compression fractures with little or no callus response.[8, 21] Radiolucency persists due to the failure of calcification of osteoid callus. Local pain is the most common complaint, and the medial scapular border, ribs, and pubic and ischial rami are the most common sites. Steinbach and Noetzli[8] have observed a tendency toward bilateral symmetry of pseudofractures.

Rickets is caused by any mechanism which prevents calcium deposition in the zone of provisional calcification. Roentgenographically, the growth plate is widened due to continuing proliferation of cartilage which cannot calcify (Figs. 1023-1030). The average thickness of the rachitic growth plate is 3.2 mm in Steinbach and Noetzli's series, with the control thickness being 1.0 mm.[8] Metaphyseal cupping and fraying occur, as well as bowing of long bones and the "rachitic rosary"

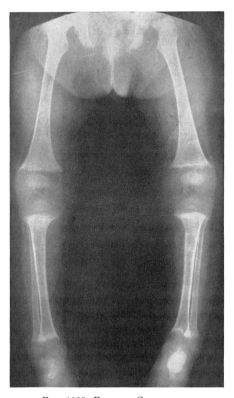

FIG. 1023. FANCONI SYNDROME
Changes in the knees and ankles cannot be distinguished roentgenologically from those seen in patients with infantile rickets.

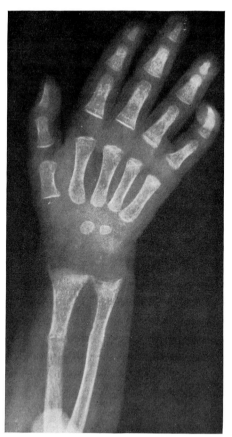

FIG. 1024. FANCONI SYNDROME
Abnormality of the bones resembles infantile rickets. Pathologic fracture in the radius and there were others elsewhere in the skeleton. Subperiosteal resorption of bone is not present.

of the ribs (Fig. 1031). The changes are similar to those of dietary rickets, although occasionally more severe due to their occurrence in older, heavier patients with a longer disease course.[13]

II. Osteitis Fibrosa

Osteitis fibrosa is caused by increased parathyroid activity. The stimulus for parathormone production is hypocalcemia, hyperphosphatemia, or both. The effect of parathormone is to increase osteoclasis. The bone defects produced by local resorption are replaced by fibrous tissue and giant cells. When this process becomes extensive, the calcium released may elevate the blood levels to normal, so that if a patient with chronic renal disease has a normal serum calcium, osteitis fibrosa is inevitably seen on the roentgenograms.[22] Osteitis fibrosa begins early in renal decompensation and may be observed in

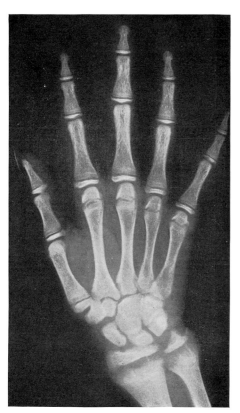 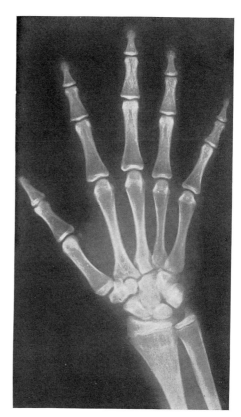

Fig. 1025. Renal Tubular Insufficiency with Acidosis

(*Left*) Roentgen changes are similar to those seen in rickets in older children, but result from chronic acidosis with excessive renal excretion of calcium rather than a deficient intake of calcium. There is no subperiosteal resorption of bone, which helps to distinguish this disease from secondary hyperparathyroidism. (*Right*) Rachitic changes have healed as a result of treatment which relieved the acidosis.

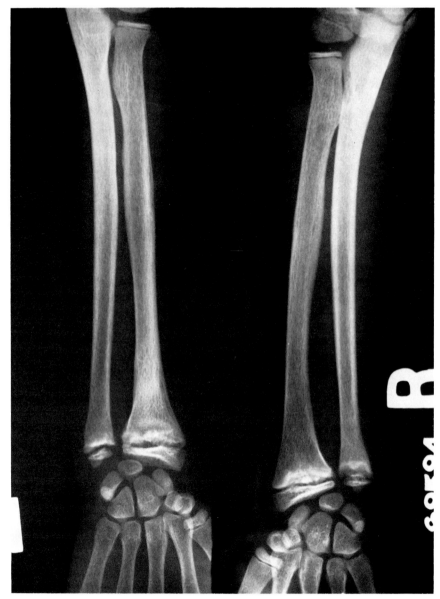

Fig. 1026. Renal Osteodystrophy
Widening of the epiphyseal line. Increased density in the metaphyseal end of the bone.

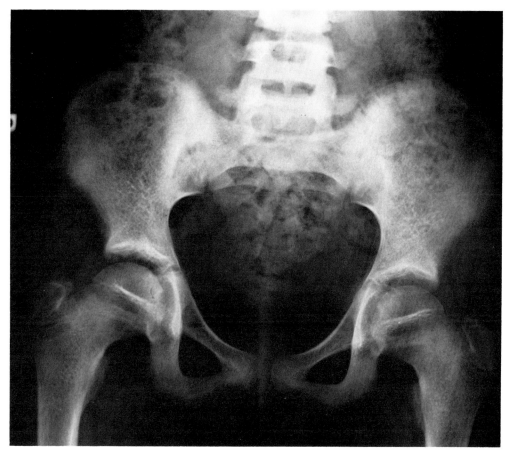

FIG. 1027. RENAL OSTEODYSTROPHY

Widened epiphyseal lines in the heads of the femur. Osteosclerosis of the bones adjacent to the sacroiliac and the lumbar spine. Atrophy of trabeculae in the remainder of the pelvis and femurs.

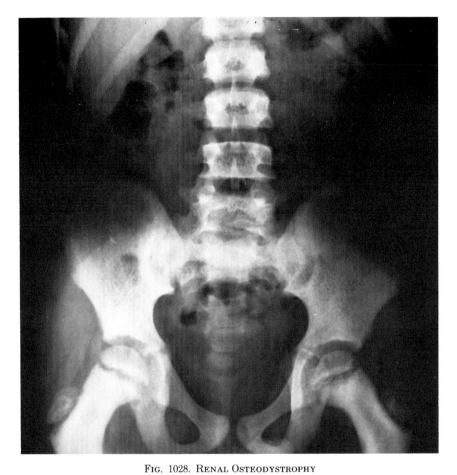

FIG. 1028. RENAL OSTEODYSTROPHY

Widening of the epiphyseal lines. Increased density in the bones adjacent to the sacroiliacs and in the femurs. Osteosclerosis is more common in renal osteodystrophy than in primary hyperparathyroidism.

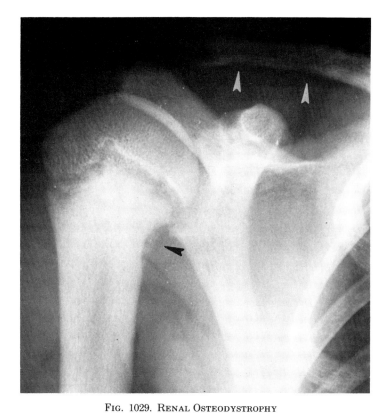

Fig. 1029. Renal Osteodystrophy
Widening of the growth plate (rickets), osteosclerosis of the metaphysis and subperiosteal resorption (*arrows*) (secondary hyperparathyroidism).

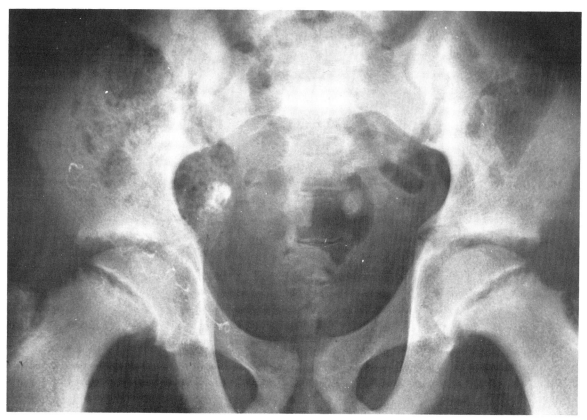

Fig. 1030. Renal Osteodystrophy
This patient had chronic pyelonephritis. Widening of the sacroiliac joints and the femoral growth plates, osteosclerosis adjacent to the sacroiliac joints, and bilateral slipping of the femoral capital epiphyses.

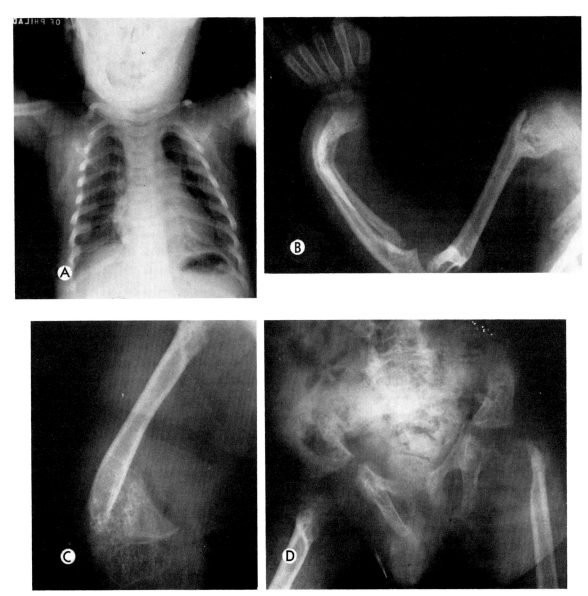

Fig. 1031. Renal Osteodystrophy

Renal failure was due to congenital absence of one renal artery and hypoplasia of the other. Note the similarity to primary hyperparathyroidism (Figs. 987 and 988). (A) Chest; pronounced rachitic rosary and marked deossification. (B) Humerus; fracture of the proximal humerus and bowing of the bones of the forearm and fractures at the distal ends due to osteomalacia. (C) Femur; marked deossification and fracture of the distal end. (D) Pelvis; marked deossification and deformity of the proximal femurs with abundant undermineralized osteoid at the proximal ends. (E) Absence of the left renal artery and hypoplasia of the right. (Courtesy of Dr. Patricia Borns, Children's Hospital, Philadelphia, Pa.)

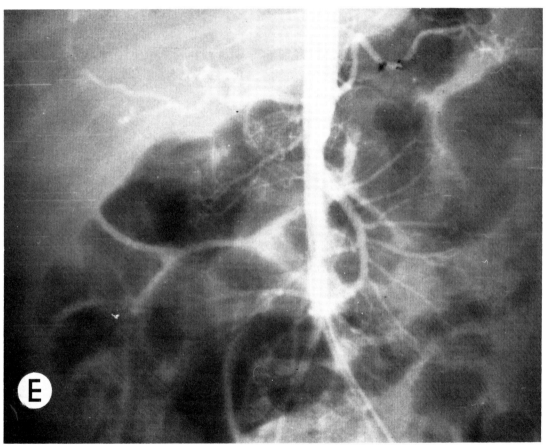

Fig. 1031E

patients dying of acute renal failure.[13] Osteitis fibrosa is far more common in acquired (glomerular) forms of renal disease than the tubular syndromes.[8]

Roentgenographic Features. The most constant and specific feature is subperiosteal bone resorption, most often seen at the radial surface of the middle phalanges (Fig. 1032). The resorp-

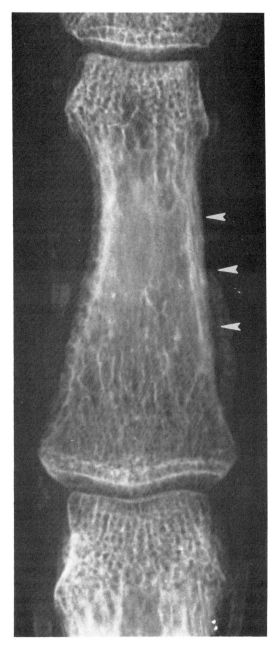

FIG. 1032. RENAL OSTEODYSTROPHY
Lacy subperiosteal bone resorption, vertical striation of the cortex (*arrows*), and arterial calcifications.

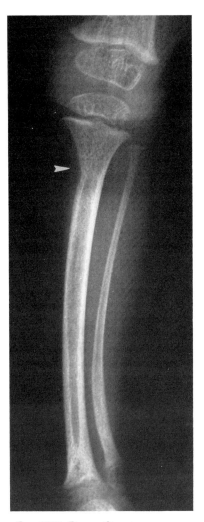

FIG. 1033. RENAL OSTEODYSTROPHY
Deossification, bowing and widening of the epiphyseal lines, subperiosteal resorption at the medial aspect of the tibia (*arrow*).

tion presents as a peculiar "lacelike" or "palisade" appearance diagnostic of the condition. Subperiosteal bone resorption also occurs in the clavicle (Figs. 1029 and 1033–1035), the medial aspect of the proximal tibia and humerus, and the distal ulna, in that order (Figs. 1036–1039).

In the skull, spotty deossification is a common feature of hyperparathyroidism, and causes a "wooly" or "salt and pepper" appearance (Figs. 1040 and 1041). Resorption of the lamina dura is a rather constant feature but it is not characteristic of the condition (Fig. 1042).

Metaphyseal fractures and slipped epiphyses may be prominent features of the disease (Figs. 1030 and 1043). "Brown tumors" are infrequent in secondary hyperparathyroidism. Destruction

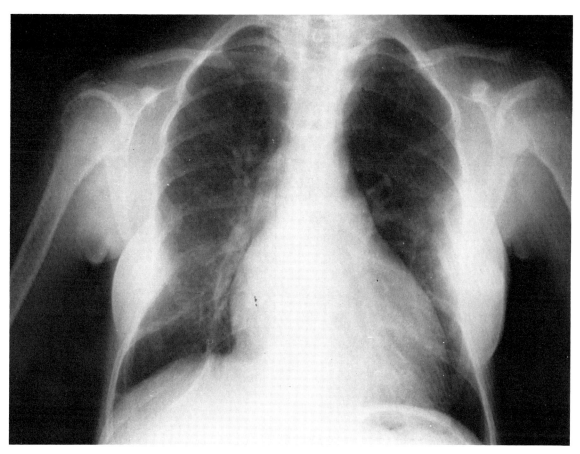

FIG. 1034. RENAL OSTEODYSTROPHY
Resorption of the distal ends of the clavicles, deossification, and the heart is enlarged.

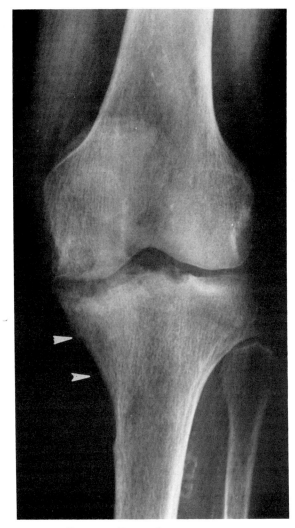

FIG. 1035. RENAL OSTEODYSTROPHY
Deossification and subperiosteal bone resorption (*arrows*) and irregularity of the medial joint surface due to compression fractures of the weakened bone.

of the phalangeal tufts may occur. The resultant telescoping of the soft tissues of the fingers causes a "pseudo-clubbing" appearance (Fig. 1044). Destruction of the interphalangeal joints occurs and may be confused with rheumatoid arthritis.[4, 20]

Soft tissue calcification[8, 23, 24] occurs with long standing hyperparathyroidism and may be found in soft tissue (Figs. 1045–1049), small and medium sized arteries (Fig. 1044), renal parenchyma, cartilage, and conjunctiva fornices.[18] Progressive small vessel calcification in the renal homotransplant patient usually indicates underlying secondary hyperparathyroidism.[25]

Ectopic periarticular calcification may reach a considerable size and spread into the surrounding soft tissues, especially around the shoulder joint.[26] Fluid levels within this periarticular calcification are reported.[27]

III. Osteosclerosis

Osteosclerosis (hyperostosis) most commonly occurs with chronic pyelonephritis.[28, 29] Many investigators believe that osteosclerosis results from an exaggerated osteoblastic response following bone resorption.[8, 11, 16, 28-30] There may be a separate fraction of parathormone, which stimulates osteoblastic activity,[19, 29, 31] especially early in renal disease with mild parathyroid hyperactivity and positive calcium balance.[3] In 1932, Selye[32] injected repeated small doses of parathyroid extract into rats and produced osteosclerosis, with or without osteitis fibrosa, depending on the dose. Osteosclerosis may occur early in the course of osteitis fibrosa when hormone levels are relatively low.[30, 32-34]

Thyrocalcitonin, elaborated by the parafollicular cells of the thyroid, acts to foster osteoblastic activity even in the face of hypercalcemia, and is thought by some to cause osteosclerosis perhaps via decreased excretion, increased production, or both.[35, 36] Osteosclerosis alone is reported with chronic renal disease[23] but is usually associated with osteomalacia and osteitis fibrosa.

Roentgenographic Features. Thickening and fusion of adjacent trabeculae produce an increased trabecular pattern on the roentgenograms, which progresses to a diffuse, chalky density, and obscuration of the normal bone architecture. In children, metaphyseal osteosclerosis is exaggerated because of adjacent deossification[8] (Figs. 1029, 1030, and 1043). In adults, the thoracolumbar spine is the most common site, occurring in 56 of the 63 cases in three combined series.[3, 28, 29] Dent and Hodson[21] termed this picture "rugger-jersey spine" due to the sclerosis at the superior and inferior vertebral borders (Figs. 1050 and 1051). Other common sites include the pelvis, ribs, femurs, and long bones. It may occur in the facial bones, simulating leontiasis ossea,[29] and in the base of the skull, especially in children (Fig. 1052).

Tubular Syndromes

Tubular syndromes cause renal osteodystrophy less commonly than acquired disease, and excretory impairment is not a prerequisite. The plethora of syndromes and eponyms often differ only in the abnormal excretion of specific metabolites. Skeletal changes are similar to acquired (glomerular) osteodystrophy, with a few exceptions. There are three major disease patterns: (1) vitamin D-resistant rickets, (2) Fanconi syndrome, and (3) renal tubular acidosis.

Vitamin D-resistant Rickets

These children present classic rachitic changes, bone pain, muscular weakness, occasional dwarfism, and persistent phosphaturia with no systemic acidosis, aminoaciduria, glycosuria, or proteinuria.[9] It is probably familial or genetic in origin, although spontaneous appearances are reported.[37] Hyperphosphaturia leads to hypophosphatemia, and hypocalcemia occurs secondary to intestinal malabsorption. The alkaline phosphatase is usually elevated.

Since large doses of vitamin D will cause healing without alleviating urinary phosphate loss, it is unlikely that the disease is solely due to resistance to vitamin D action.[4, 11] Fanconi, and earlier Johnson, suggested the term "phosphate diabetes" due to the persistent phosphaturia despite healing with ergosterol.

Partial to complete healing occurs with phosphate infusion and oral supplementation. However, the result is temporary unless calcium and vitamin D are added and maintained.[38, 39] A parallel would seem to exist between the pathophysiology of resistant rickets and acquired (glomerular) rickets since the primary defect in resistant rickets is probably a combination of a congenital renal phosphate leak combined with intestinal malabsorption of calcium.

The roentgenographic features resemble those of glomerular rickets, although dwarfism may be more marked.[39] Bowing of long bones, pseudofractures, and metaphyseal osteosclerosis all occur. Even with vitamin D and phosphate supplements, recurrences are common and deformities may result.[40]

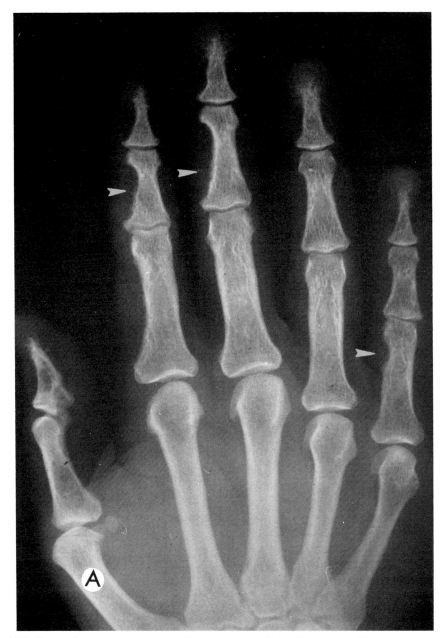

Fig. 1036. Renal Osteodystrophy with Characteristic Bone Changes of Hyperparathyroidism
(A) Subperiosteal resorption is most marked at the radial surface of the middle phalanges (*arrows*). Destruction of the tufts. (B) Shoulder; subperiosteal resorption of the clavicle and the humeral shaft (*arrows*). (C) Bone resorption of the pubic and ischial rami.

Fanconi Syndrome

The more severe childhood form of this disease has been termed Fanconi-De Toni-Debré or Fanconi-Lignac syndrome, whereas the milder adult form is commonly called the adult Fanconi syndrome. Although many variations exist due to differing tubular enzymatic defects, it is basically a triad of hyperphosphaturia, amino aciduria, and renal glycosuria with a normal blood glucose.[41] It is reported with hypokalemia and periodic weakness, nephrogenic diabetes insipidus,[41] and hyperuricuria.[7] Although mainly genetic, the Fanconi syndrome may complicate heavy metal intoxication and multiple mye-

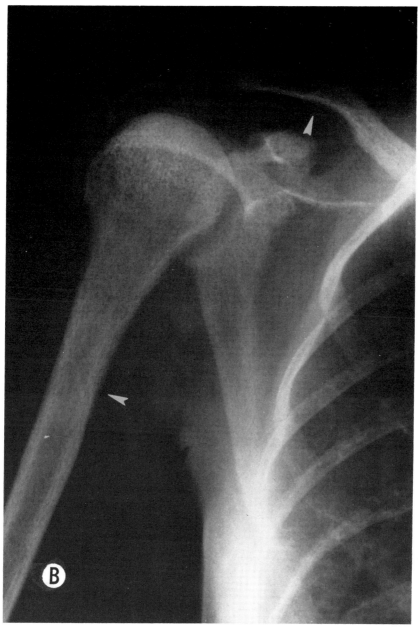

Fig. 1036B

loma.[41] Cystinosis is no longer considered the underlying cause.

When bone changes do occur, functional renal impairment is likely. Most cases heal with large doses of vitamin D,[32, 41] or alkalinization plus ergosterol, thus suggesting another link between acquired and tubular osteodystrophy. Rickets, osteomalacia, osteitis fibrosa,[7, 13] and osteosclerosis may be present,[8, 13] the latter two in more protracted cases. Bowing of long bones and epiphyseal collapse and malposition cause limb deformities. Peculiar spurs and ossicles at ligamentous insertions about the joints are described by Steinbach and Noetzli[8] in renal phosphaturia with or without glycosuria, and are presumably due to normal stresses placed on abnormally soft, undermineralized bone.

Renal Tubular Acidosis

The Lightwood syndrome or "salt-losing nephritis" is the self-limited form of the disturbance in infants and children and, while active, produces systemic acidosis and bone lesions but not nephrocalcinosis. The adult form of renal tubular acidosis (Butler-Albright syndrome), is the severe form. Systemic acidosis, bone lesion, and nephrocalcinosis are seen. The basic defect in both forms is failure of tubular epithelium to excrete the hydrogen ion, except through coupling with the ammonia radical. All the hydrogen that is filtered cannot be excreted through this pathway, and systemic hyperchloremic acidosis results. The serum phosphate rises as renal failure ensues. Therapy with vitamin D and alkalinization is only partially efficacious in causing bone healing.[2, 11, 13]

The roentgen features include rickets and/or osteomalacia, pseudofractures, and nephrocalcinosis. Osteitis fibrosa is rare,[13] and osteosclerosis is not reported.[42]

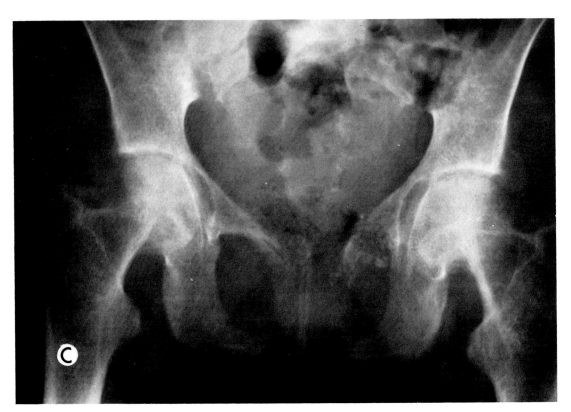

Fig. 1036C

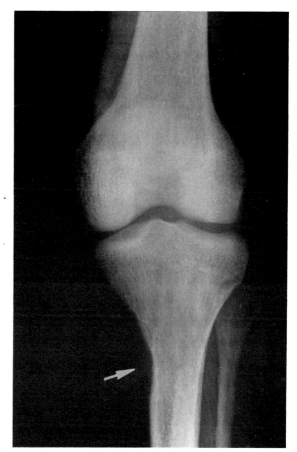

Fig. 1038. Renal Osteodystrophy with Secondary Hyperparathyroidism

There is subperiosteal resorption on the medial aspect of the tibial metaphysis (*arrow*).

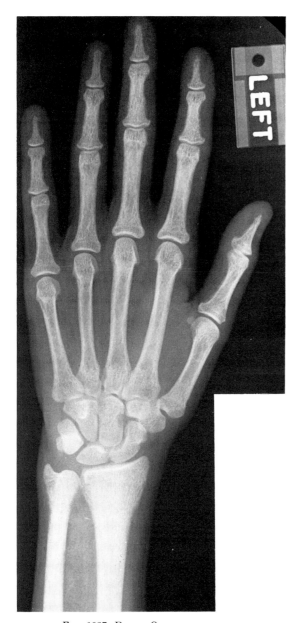

Fig. 1037. Renal Osteodystrophy

There is subperiosteal resorption of bone, particularly marked in the radial aspects of the second, third, fourth, and fifth middle phalanges. There is deossification of bone and also subperiosteal resorption on the medial aspect of the ulna.

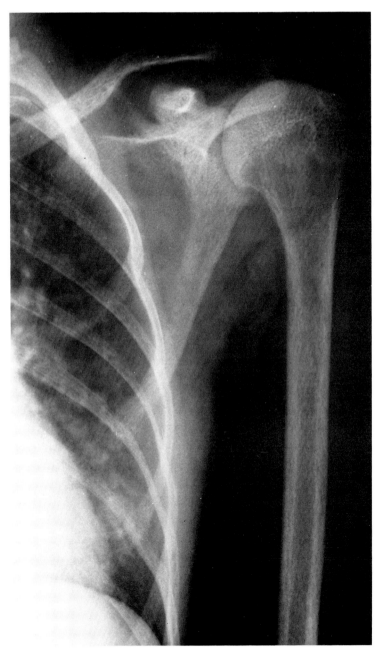

Fig. 1039. Renal Osteodystrophy with Secondary Hyperparathyroidism

Extreme deossification of the ribs and the humerus. Subperiosteal resorption of bone on the medial aspect of the humeral metaphysis.

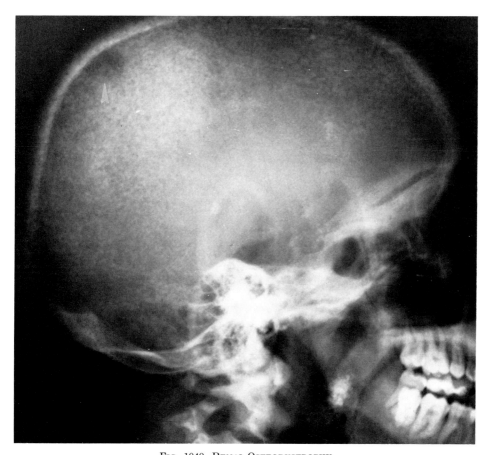

FIG. 1040. RENAL OSTEODYSTROPHY
Deossification producing a "salt and pepper" appearance. A small "brown tumor" is present in the posterior parietal bone (*arrow*).

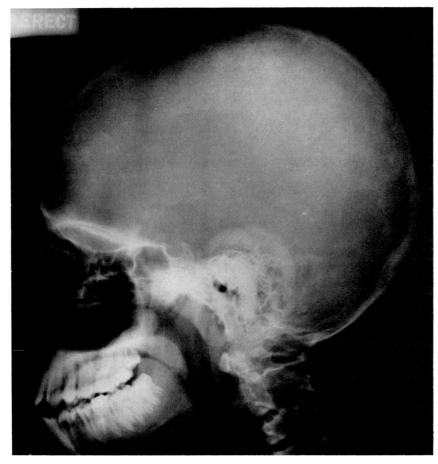

FIG. 1041. RENAL OSTEODYSTROPHY (SECONDARY HYPERPARATHYROIDISM)
Marked deossification of the calvaria.

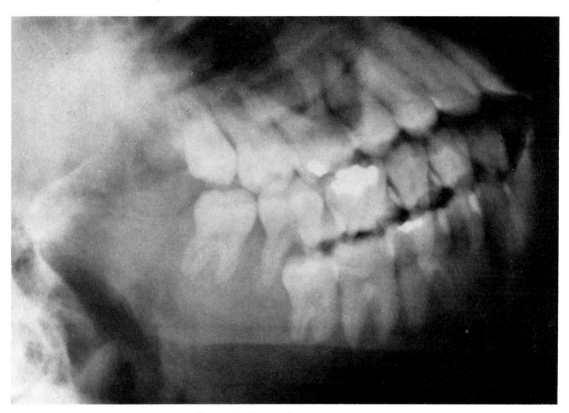

Fig. 1042. Renal Osteodystrophy
Resorption of the lamina dura.

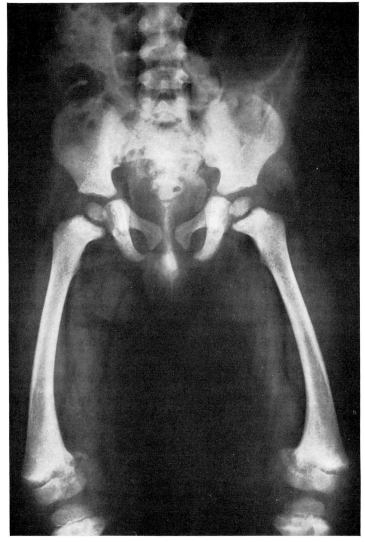

FIG. 1043. RENAL OSTEODYSTROPHY
Slipping of the distal epiphyses of the femurs. Slipping of the epiphyses is a frequent complication of this disease.

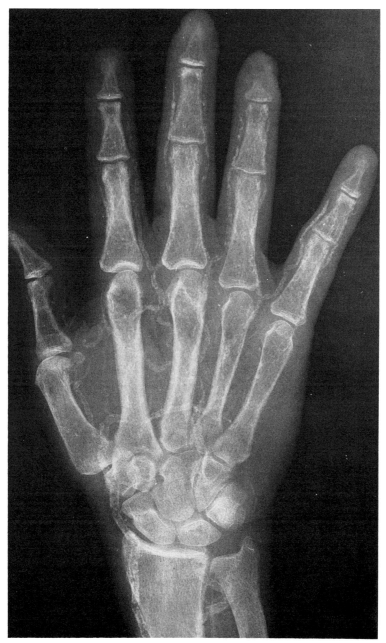

Fig. 1044. Renal Osteodystrophy

Destruction of the phalangeal tufts and pseudo-clubbing due to soft tissue telescoping. Arterial calcification is present. Subperiosteal bone resorption on the radial aspect of the middle phalanges.

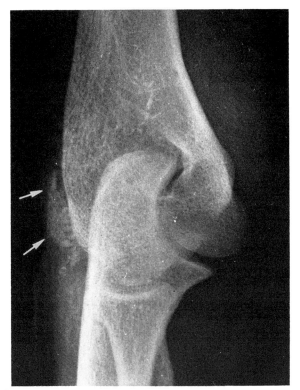

FIG. 1045. RENAL OSTEODYSTROPHY WITH SECONDARY HYPERPARATHYROIDISM

Considerable soft tissue calcification adjacent to the humerus (*arrows*). This patient has chronic glomerular nephritis.

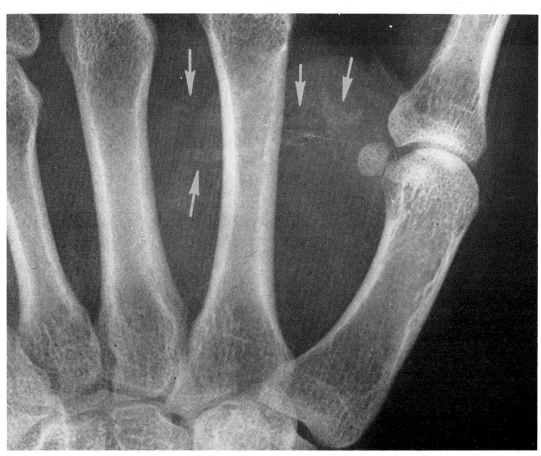

FIG. 1046. RENAL OSTEODYSTROPHY (SECONDARY HYPERPARATHYROIDISM)
Soft tissue calcification in the hand (*arrows*).

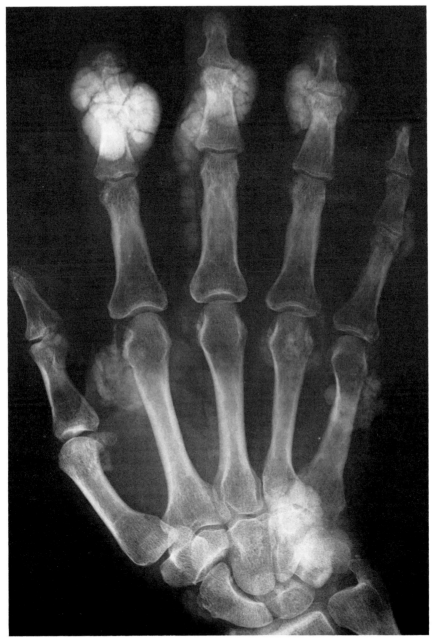

FIG. 1047. RENAL OSTEODYSTROPHY
Considerable soft tissue calcification. This is usually the result of administration of calcium with or without vitamin D.

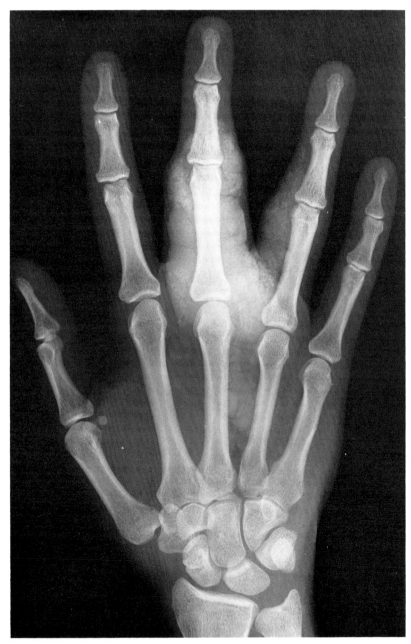

Fig. 1048. Renal Osteodystrophy
Soft tissue calcification.

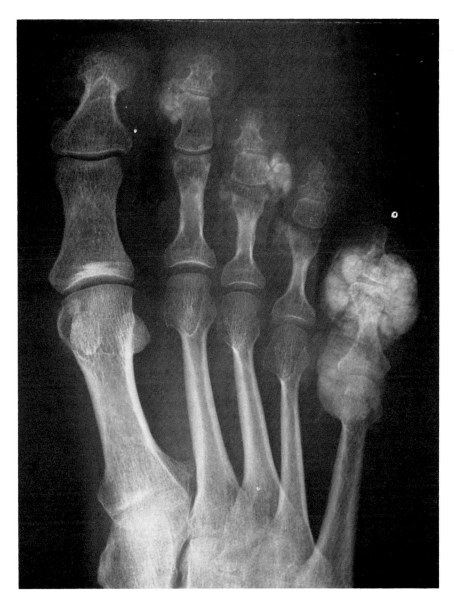

FIG. 1049. RENAL OSTEODYSTROPHY
Extensive soft tissue calcification.

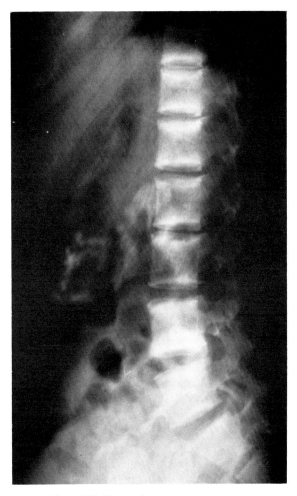

Fig. 1050. Renal Osteodystrophy with Osteosclerosis
This patient had chronic glomerular nephritis. Calcification in the kidneys and considerable sclerosis of the centra of the vertebrae.

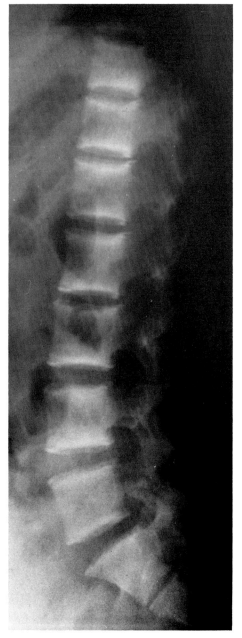

Fig. 1051. Renal Osteodystrophy
Osteosclerosis of the superoinferior margin of the vertebral centers produce a "rugger jersey" appearance.

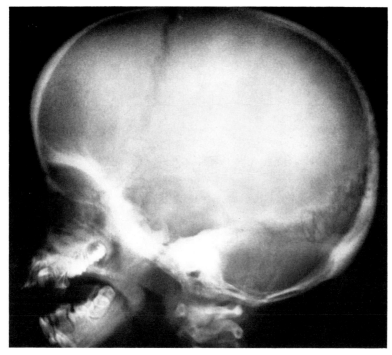

FIG. 1052. RENAL OSTEODYSTROPHY
Loss of definition of the skull tables pronounced in the frontal and occipital areas. Sclerosis of the base of the skull and mastoids.

HEREDITARY HYPERPHOSPHATASIA
(CHRONIC FAMILIAL HYPERPHOSPHATASEMIA)[1]

Synonyms: Juvenile Paget disease,[2] hyperostosis corticalis deformans juvenilis,[3] familial osteoectasia,[4] chronic progressive osteopathy with hyperphosphatasia,[5] osteochalasia desmalis familiaris,[6] chronic idiopathic hyperphosphatasia,[7] congenital hyperphosphatasia,[8] hereditary bone dysplasia with hyperphosphatasemia,[9] and chronic familial hyperphosphatasemia.[1]

Hereditary hyperphosphatasia is a rare disease characterized by sustained elevation of serum alkaline phosphatase and generalized bone thickening.

Clinical Features. Transmission is by an autosomal recessive mode.[4, 8, 10] It may appear toward the end of the 1st year of life; more often in the 2nd or 3rd year. Skeletal deformities of enlargement and bowing of the long bones and rapid calvarial enlargement are the usual initial features. Eventually, dwarfism develops and cranial nerve deficits appear, most frequently affecting the optic and auditory nerves.[8, 11] Respiratory infections are frequent.

Hypertension is reported in association with this condition. The one reported death was in a patient at age 18, due to cerebral vascular accident, secondary to intractable hypertension.[10] Pseudoxanthoma elasticum may occur with this condition.[12]

Alkaline phosphatase activity is consistently elevated, and in some patients acid phosphatase is elevated.[10] A high glycylproline urinary excretion is reported,[12-14] indicating an increased collagen turnover. The patients improve as they get older.

Clinical, radiographic, and histopathologic improvement is reported from prolonged treatment with human calcitonin.[9, 15-19] Bone scintigraphy is a valuable tool in diagnostic evaluation, revealing increased uptake of the nuclei and characteristic localization.[20]

Pathologic Features. Bone biopsies reveal a cortical thickening consisting of parallel trabeculae separated by loose, vascular, fibrous tissue replacing the haversian systems. Subperiosteal bone formation is prominent.[2, 4-6, 8, 12, 21] The pathogenesis is probably a rapid turnover of lamellar bone without compact cortical bone being laid down.

The condition is probably caused by hyperactive osteocytes producing the accelerated absorption. Immature woven bone is rapidly laid down, but simultaneous rapid destruction prevents normal maturation and mottling in membranous bone. Haversian bone is absent and there is evidence of accelerated metabolic activity. Thick osteoid seams cover the trabecula, which are devoid of a cartilaginous core and show intense osteocytic osteolysis.[9]

Roentgenographic Features. In children the hallmarks are bowing and thickening of the long bones, with decreased density of cortical bone. The cortices may be dense and thinner than normal, or markedly less dense and reveal loose trabecular rather than cortical bone (Figs. 1053–1062).

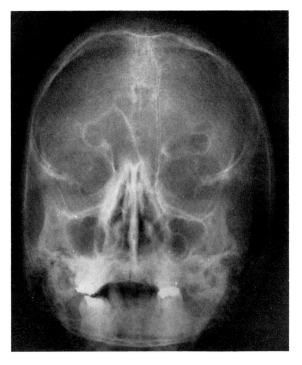

FIG. 1053. HYPERPHOSPHATASIA
The tables are widened and there is diffuse desossification. A metopic suture is present, and the paranasal sinuses are large.

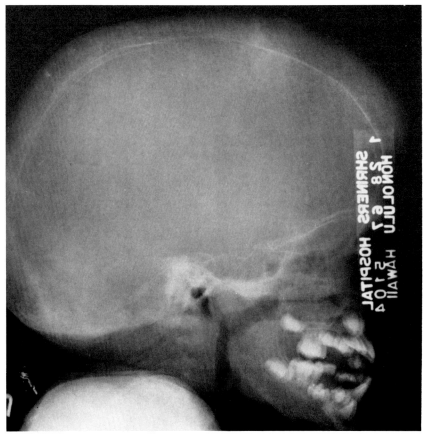

FIG. 1054. HYPERPHOSPHATASIA
The skull is thick. Generalized deossification.

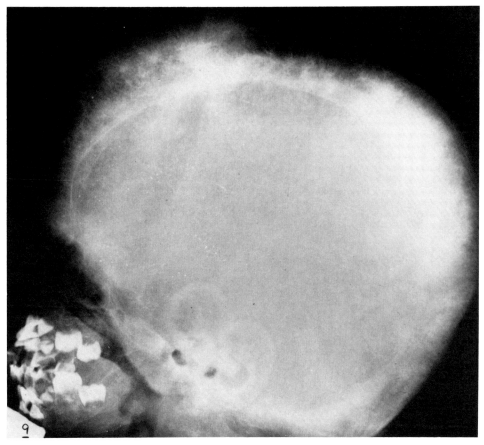

Fig. 1055. Hyperphosphatasia
The tables are thickened and there is irregular increased density. The similarity to Paget disease is remarkable.

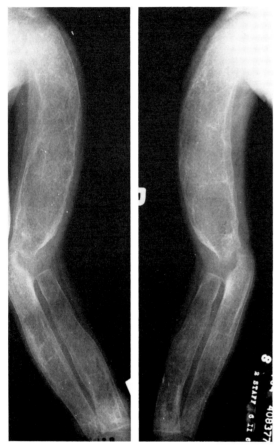

Fig. 1056. Hyperphosphatasia
Widening of the bones and marked thinning of the cortices and diffuse deossification.

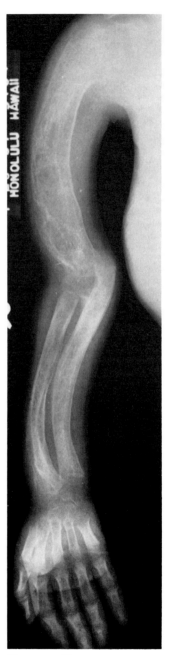

Fig. 1057. Hyperphosphatasia
Diffuse deossification, widening, and bowing. The cortices are thin.

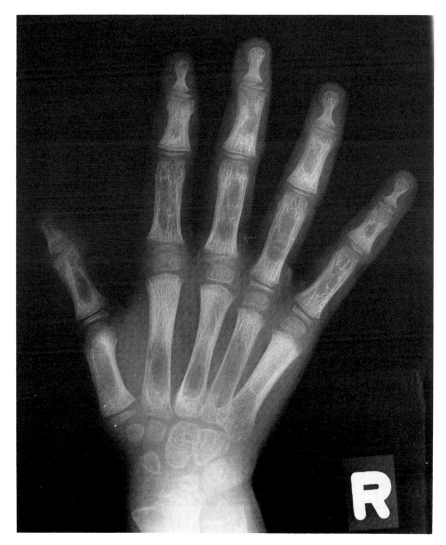

Fig. 1058. Hyperphosphatasia

The cortices are thinned and there is diffuse deossification. Some increase in density in the metacarpals and middle and distal phalanges. The bones are wider than normal.

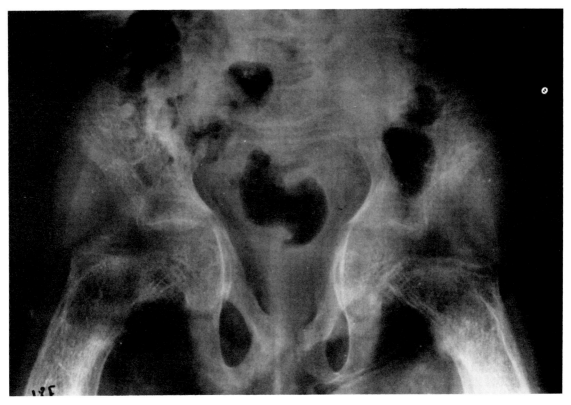

Fig. 1059. Hyperphosphatasia
Diffuse deossification and molding of the pelvis.

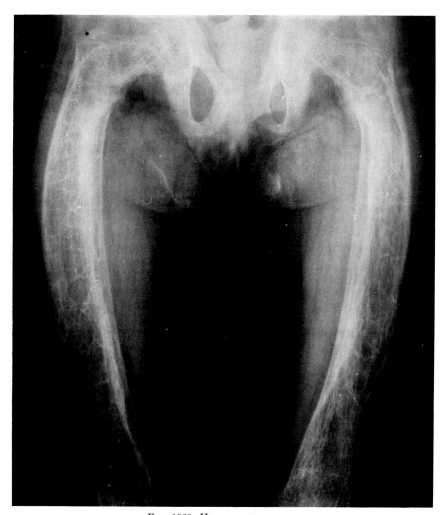

Fig. 1060. Hyperphosphatasia
The bones are deossified and bowed. The cortices are thinned and the bone diameter widened.

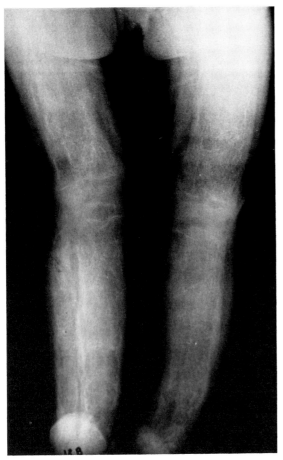

FIG. 1061. HYPERPHOSPHATASIA
Diffuse deossification and widening of the bones.

The skull is usually greatly thickened, with wide tables which may show complete loss of compact bone. Often there is a "cotton wool" appearance, with rounded areas of increased density interspaced with lessened density, and loss of definition of the skull tables. In early infancy the long bone changes may be dramatic without thickening of the skull.[15]

The resemblance to Paget disease is striking. The overproduction of lamellar bone, localized, may form a tumor easily mistaken for a malignant neoplasm. In the adult the bones are wider and irregularly ossified, and the cortices are slightly thin (Figs. 1063-1068).

The differential diagnosis includes: (1) Paget disease, which rarely, if ever, occurs before 20 years of age and is usually not generalized. (A biopsy taken in Paget disease will show characteristic mosaic pattern of bone formation.) (2) Van Buchem syndrome[22] (hyperostosis corticalis generalisata), which involves only the diaphysis and spares the epiphyseal ends. It occurs after 20 years of age, and long bone bowing is not reported. However, the serum alkaline phosphatase is persistently elevated. (3) Engelmann syndrome (progressive diaphyseal dysplasia) usually involves the lower limbs and is rarely generalized. The skull may be slightly thickened. Histologic studies reveal normal haversian systems. (4) Pyle disease (metaphyseal dysplasia) spares the midshafts of long bones. (5) Osteogenesis imperfecta may present a similar appearance, but the deformities are due to fractures. The presence of blue sclerae may aid in identifying it. (6) Polyostotic fibrous dysplasia is predominantly unilateral, rarely symmetrical and generalized. It differs histologically and is often associated with precocious puberty and abnormal skin pigmentation.

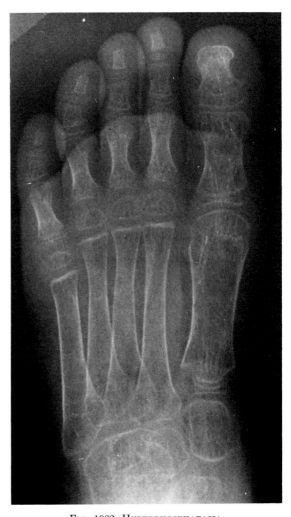

FIG. 1062. HYPERPHOSPHATASIA
The bones are diffusely deossified and slightly widened. The cortices are thinned.

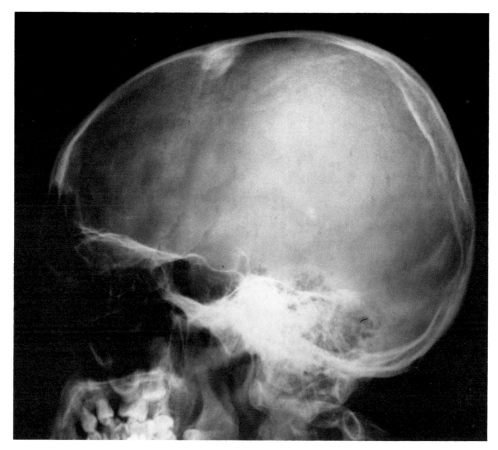

Fig. 1063. Hyperphosphatasia
Thickening of the tables of the skull in the posterior parietal and occipital areas. The changes are milder than in children.

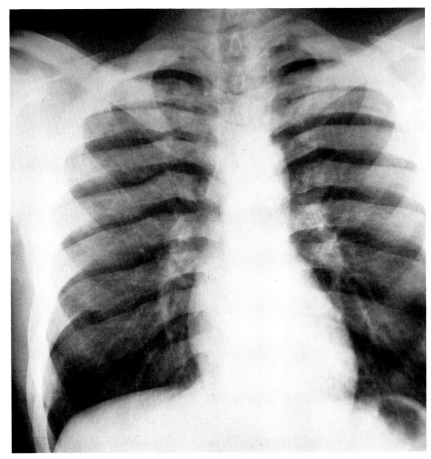

Fig. 1064. Hyperphosphatasia
The bones are wider and there is density increase. The cortices are thinner than normal.

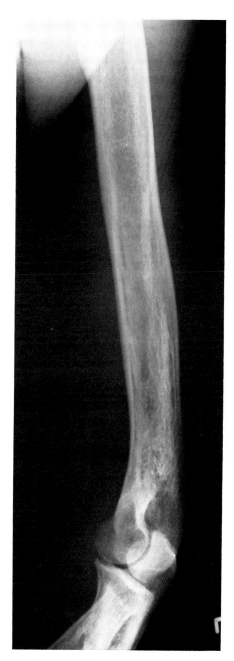

Fig. 1065. Hyperphosphatasia

The bones are wide and the cortices thin. Double cortices are noted. The disease is more severe in childhood and improves with age.

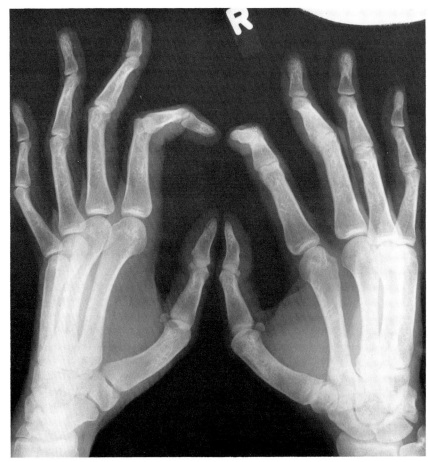

FIG. 1066. HYPERPHOSPHATASIA

Irregular ossification and widening of the bones. Loss of modeling of the middle phalanges. The cortices are thin.

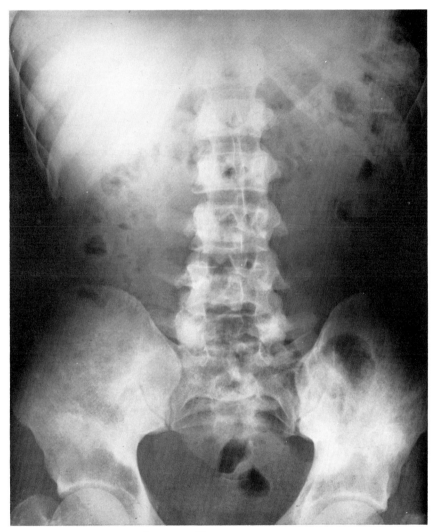

Fig. 1067. Hyperphosphatasia
Irregular increase in density throughout the spine and pelvis. (Courtesy of Dr. Harlan Spujt, Houston, Tex.).

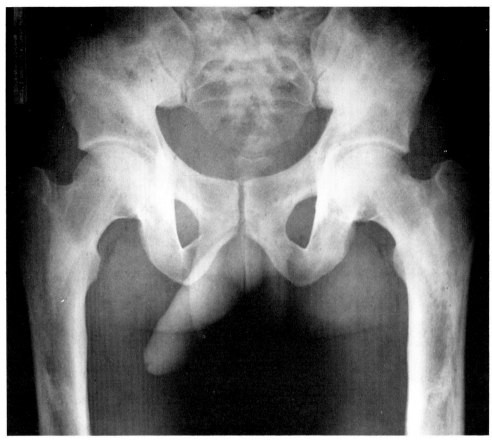

FIG. 1068. HYPERPHOSPHATASIA
Diffuse increase in bone density. The diameters of the bones are wide and the cortices thin.

PAGET DISEASE
(OSTEITIS DEFORMANS)

The advanced clinical form of osteitis deformans was described in 1877 by Sir James Paget.[1] Since then, little has been added to the clinical picture except that biochemical, histologic, and roentgenographic findings have made it possible to identify the early and the less classical examples of the disease.

Etiology. Many opinions exist concerning the cause of osteitis deformans; none have been substantiated and the etiology remains unknown. Paget thought the condition was due to chronic inflammation, an opinion consonant with the local heat often found in affected bones, particularly superficial ones such as the tibia. DaCosta et al.[2] referred to the work of Morpuse, Archangeli, and Fiocca, who claimed to have found a diplococcus in the affected bones. DaCosta[2] tried unsuccessfully to prepare a vaccine. Knaggs[3] believed that a bacterial toxin was responsible for the disease. Lannelongue[4] and many others suspected syphilis as the cause. This was refuted by Kay et al.,[5] Gutman and Kasabach,[6] and Dickson et al.,[7] who found evidence of syphilis in only a small percentage.

Lancereaux[8] suggested the central nervous system was to blame, a hypothesis probably based on the frequent occurrence of skeletal changes with tabes dorsalis and syringomyelia. De la Tourette et al., quoted by DaCosta,[2] found associated changes in the medulla and peripheral nerves, but concluded that they were the result of senility and played no part in the formation of the condition.

Originally, because Paget disease involved many bones, DaCosta[2] held osteitis deformans to be a metabolic disease. But Albright et al.[9] showed that, even though it can be polyostotic, it is not a generalized bone disease and, therefore, not of metabolic origin. It has also been contended that hyperparathyroidism is the cause of osteitis deformans, but, because the parathyroids are invariably normal and bone metabolic diseases are so different from osteitis deformans, there are few who still support this contention.

Because the disease early involves sites most susceptible to the wear and tear of aging, Schmorl[10] considered trauma the cause, a concept in keeping with the common involvement of the spine and sacrum. Schmorl's hypothesis is further supported by the higher incidence of osteitis deformans in the pelvis and lower extremities, compared to the shoulder girdles and upper extremities, but leaves unexplained its common occurrence in the calvaria.

Moehlig and associates[11, 12] reported a family history of diabetes mellitus in 5 of 12 patients with osteitis deformans; Dickson et al.,[7] however, found the association in only 18 of 367. Paget disease is more frequent in Great Britain and northern European countries than in the United States or Scandinavia. It is rare in Asian and African populations.[13] It is more common in the northern than in the southern United States.[13, 14]

Clinical Features. Osteitis deformans usually manifests itself during middle life; it is rarely encountered before the age of 30. The average age of onset has been reported variously as 39,[15] 46,[5] and 53.[7] Most reports reveal an incidence in males twice that in females, although Gutman and Kasabach[6] reported equal incidence in 116 patients.

At least 20% of patients with osteitis deformans are asymptomatic,[6, 7, 16] and the condition is usually accidentally discovered. With wider use of roentgenographic examinations, the discovery of asymptomatic Paget disease continues to rise.

Pain is the most common symptom, its severity depending on the site. Backache, rheumatic and stabbing, is the most frequent complaint. Pain in the thighs, knees, and legs, and headache are also common. Fatigue, too, is a common complaint,[7] especially when the changes are extensive. As basilar invagination of the skull takes place, due to the downward thrust of the heavy head upon the spine, numerous cranial nerve defects, including deafness, ensue.

Spinal cord and nerve root compression may result from the pressure of distorted soft bones or pathologic fractures.[17-22] It is most frequent in the upper thoracic spine[17] and in patients between the ages of 40 and 60. Back pain is the first symptom, followed by progressive numbness and paresthesias of the feet, difficulty in walking, and leg paresis. Spastic paraparesis develops, with bladder and bowel disturbances and upper thoracic sensory loss.[5, 7] Sarcomatous degeneration of Paget disease also occurs with significant frequency. High output cardiac failure may be associated with microscopic multiple arteriovenous malformations in bone involved with Paget disease (see Fig. 1088).

Laboratory Features. In Paget disease the serum calcium and plasma phosphorus are virtually normal.[7, 23] Kay[24] found the calcium slightly above normal, and was also the first to call attention to the striking elevation of the alkaline phosphatase, occasionally 20 times higher than normal. He pointed out that the extent of the disease and the elevation of alkaline phosphatase were roughly proportional. Dickson et al.,[7] correlating alkaline phosphatase with roentgenographic appearances, found it lower in the sclerotic type than in the combined osteolytic and sclerotic types. Early in the disease, or when monostotic, alkaline phosphatase levels were usually within normal limits. Chemical analysis of diseased bones[25] reveal a significant decrease in calcium and increase in fat content.

Pathologic Features. Schmorl,[10] studying monostotic osteitis deformans, found its histologic appearance to be exactly like that of the disseminated form. Irregular areas of lamellar bone interspaced by poorly stained cement lines in the form of an "irregular mosaic"[26] are characteristic. The mosaic appears to result from a cycle of bone absorption and new bone formation in which compact, spongy, and newly formed connective tissue bone all yield to this pathologic process.[27] The mosaic architecture is apparently only in the completely transformed bone and is not usually encountered in the rest of the host bone. At times, the resorption of bone is so rapid and extensive that new bone cannot form and the original bone is replaced by connective tissue.

Roentgenographic Features. Paget disease is primarily a destructive bone lesion, usually followed by repair, during which an excess of new bone is laid down, producing the characteristic roentgen appearance. Three stages of osteitis deformans are described classically: (1) active or destructive, (2) combined destructive and reparative, and (3) quiescent or sclerotic. The first or destructive stage is best seen in the skull. The second or combined destructive and reparative stage is usually best seen in the pelvis and long bones, and the third or quiescent stage in the pelvis and clavicle. The quiescent or sclerotic change usually remains inactive, but occasionally it may reactivate and revert to the destructive or combined destructive and reparative phases.[28]

The order of frequency of involvement by Paget disease is the pelvis, lumbar spine, sacrum, femurs, skull, and dorsal spine.[29, 30] The lower limb and girdle are more frequently involved than the upper limb, which suggests physical stress to bone as a factor,[3, 5, 15, 31-34] and the high incidence in the skull can be explained by the attachment of the muscles of mastication.

SKULL. The roentgenographic changes in the skull closely follow Knagg's description of all three stages.[3] The first stage is "vascular," with the skull enlarging, due to a surprising degree of vascularity in the connective tissue. The second stage is characterized by "advancing sclerosis," with the outer table of thickened bone smooth, yet perforated by innumerable vascular defects. The third stage of "complete diffuse sclerosis" finds all distinction between the diploe and calcarial bone tables lost and replaced by bone of similar texture. Knaggs[3] also described well defined islets of "ivory-like bone" which can be demonstrated radiographically.

The term "osteoporosis circumscripta" was first introduced by Schuller[35]; it was Sosman[36] who first related it to Paget disease. Osteoporosis circumscripta epitomizes the destructive, active, or monophasic first stage of Paget disease. Deossification usually begins in the frontal or occipital area, well demarcated and spreading slowly to encompass the major portion of the calvaria (Fig. 1069). The outer table is destroyed from within; the suture lines are no barrier (Fig. 1070). Physical stress has been mentioned as a factor in producing Paget disease, and involvement of the skull is perhaps explained by the attachments of the muscles of mastication. It is suggested that the greater frequency of osteoporosis circumscripta in the skull than elsewhere is due to relatively less stress than in the long bones.[37] Eventually the reparative process or second stage (biphasic) begins; its sclerotic changes affect first the inner table of the skull and the diploe. With healing and conversion of the de-

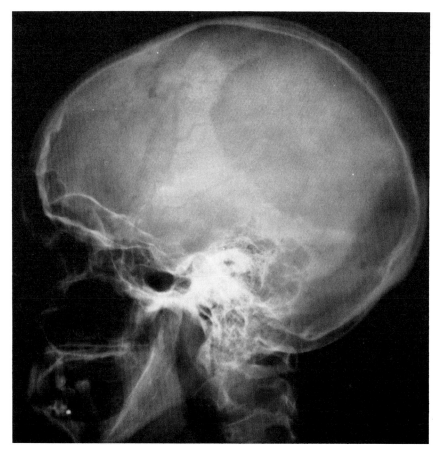

FIG. 1069. OSTEOPOROSIS CIRCUMSCRIPTA

The first or destructive stage of Paget disease. The well demarcated deossification spreads slowly to encompass the major portion of the calvaria.

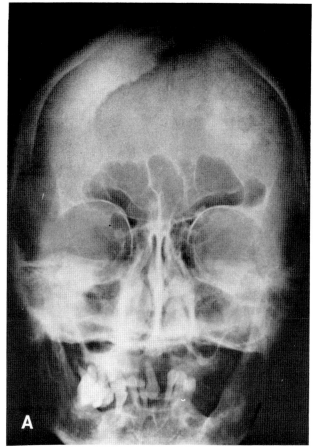
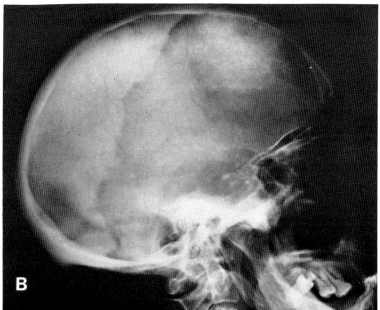

Fig. 1070. Osteoporosis Circumscripta

Osteoporosis of the frontal and parietal bones; it does not respect the suture line. (A) Posteroanterior projection; (B) lateral projection.

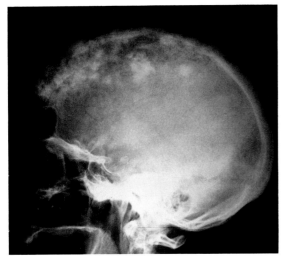 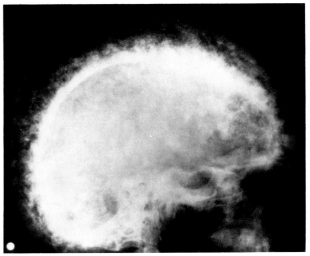

FIG. 1071. LATER STAGES OF PAGET DISEASE OF SKULL

(Left) Combined destructive and dense lesions of bone giving a cotton wool appearance. *(Right)* Advanced Paget disease of skull. Extensive involvement with thickening of the cortex and a wide diploe. Softening of the base of the skull with basilar invagination.

structive process, sclerosis of the inner surface of the outer table develops, gradually extending throughout the bones of the calvaria to produce a thickened sclerotic outer table. At this point the disease becomes quiescent, with large round areas of sclerosis appearing, surrounded by demineralization and causing the "cotton ball" appearance (Fig. 1071). These lesions may remain quiescent, or reactivate after months or years of apparent inactivity. Softening of the base of the skull leads to basilar invagination (Fig. 1071) and the complications of encroachment on the basilar foramina.

Stafne and Austin[38] have described the jaw changes in Paget disease. The maxilla is affected more frequently than the mandible. Evidence of Paget disease usually appears elsewhere in the skeleton, but occasionally the process affects only the jaw. The characteristic deossification contains areas of increased bone density with abnormal trabecular pattern. Early deossification is greater in the vicinity of the roots of the teeth. The bony wall of the affected tooth socket is obliterated, and the lamina dura is absent. The absorption of bone near the apices of the affected teeth resembles periapical granulomas, but the teeth are firm, not sensitive, and not devitalized. Later, hyperplasia of the cementum occurs around the affected teeth to a greater extent than in any other condition. As the disease progresses the picture becomes one of hypercementosis with increased density and coarse trabeculation distorting the hypertrophied jaw, displacing teeth with eventual malocclusion.

There is an asymmetric involvement of the disease in certain sites. In both sexes, the femurs and clavicles exhibit a right-sided bias.[39]

LONG BONES. Osteitis deformans of long bones almost invariably begins at one end and on rare occasions it is confined to the shaft[40–44] (Fig. 1072). In the pubic and ischial bones the abnormal area begins at the symphysis or acetabulum. Paget disease may begin away from the joint, although rarely, and usually in patients with extensive disease in other bones. It is unusual to demonstrate the early destructive or monophasic phase in a long bone, but it does occur and is usually manifested by a well defined V-shaped area of deossification rather characteristic of Paget disease (Figs. 1073–1076). In the smaller or flat bones a peculiar "bubbly" destruction occurs (Figs. 1077 and 1078). The advancing edge may be round with a sclerotic margin (Fig. 1076). The adjacent periosteum often reveals successive layering of new bone which eventually becomes sclerotic and fuses with the cortex to form the characteristic wide bone. Accordingly, the cortex, while proliferating along its external surface, may at the same time be destroyed along its endosteal surface. When this occurs the transverse diame-

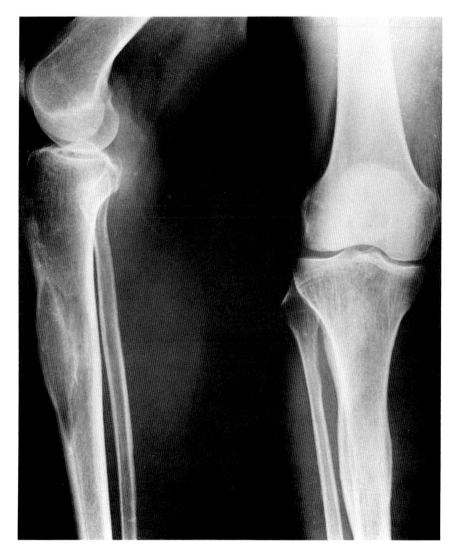

Fig. 1072. Paget Disease
The slight bone expansion, periosteal reaction and radiolucency are due to Paget disease. It is rare for it to originate away from the end of a bone, but when it does the tibia is the most frequent site.

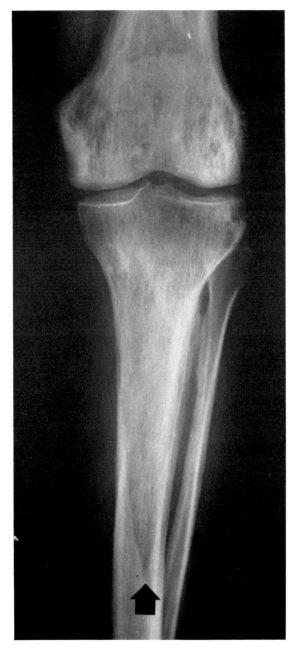

FIG. 1073. PAGET DISEASE
First or destructive stage. A well defined V-shaped area of deossification extends toward the midshaft of the tibia *(arrow)*. Second stage changes in the femur. Lesions extend to the joint surfaces.

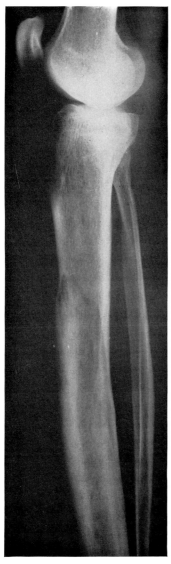

FIG. 1074. PAGET DISEASE
Destructive changes of Paget disease at the junction of middle and upper third of the tibia. V-shaped area of deossification at the advancing edge.

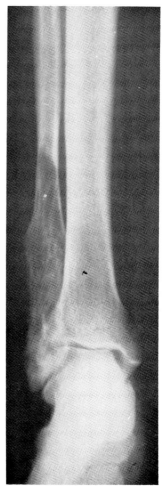

FIG. 1075. PAGET DISEASE
Destructive stage extending from distal end of fibula toward the midshaft. The well defined V-shaped area of deossification is characteristic.

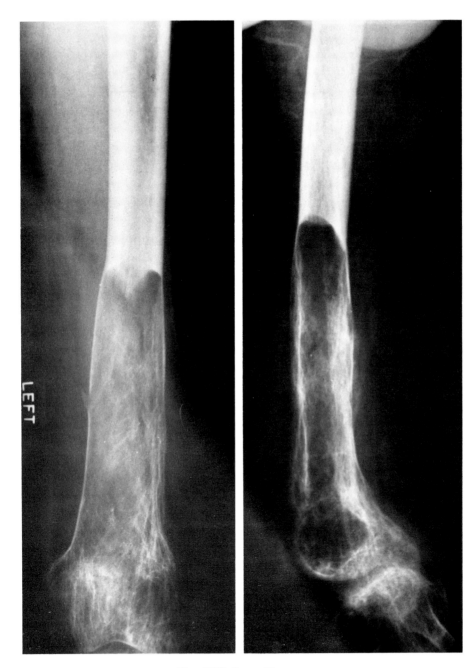

FIG. 1076. PAGET DISEASE

Destructive stage. Well demarcated deossification has extended from the knee to the junction of the middle and lower third of the femur. This patient had a paraplegia and the destructive change was exaggerated because of disuse. *(Left)* Anteroposterior projection; *(right)* lateral projection.

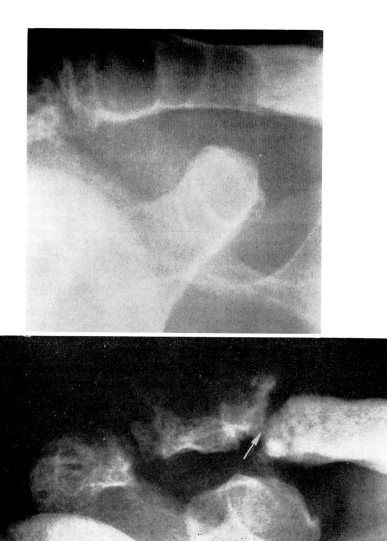

FIG. 1077. PAGET DISEASE

Destructive stage in the clavicle. Notice the bubbly appearance. This lesion was asymptomatic and was discovered on routine chest roentgenogram. *(Top)* Early lesion when first discovered. *(Bottom)* One year later. A pathologic fracture has occurred. Bone sclerosis and the lesion is still asymptomatic.

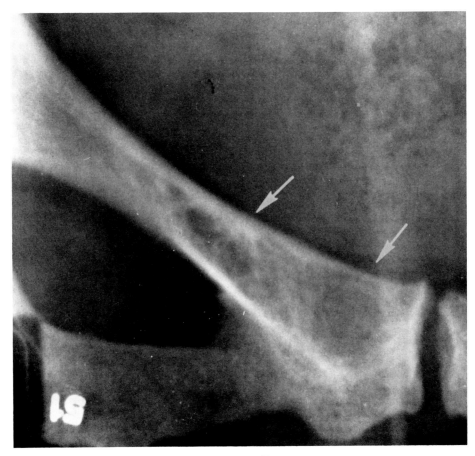

FIG. 1078. PAGET DISEASE

Destructive stage in horizontal ramus of the pubis *(arrows)*. Notice the tendency toward a bubbly configuration. The lesion extends to the joint surface and the first portion of the bone involved shows the second stage of combined sclerosis and destruction.

ter of the bone is increased without the cortex increasing in thickness (Figs. 1079 and 1080).

Deforming curvatures, lateral curvatures of the femur (Fig. 1081), anterior curvatures of the tibia (Fig. 1082), and invagination of the base of the skull in osteitis deformans can be explained

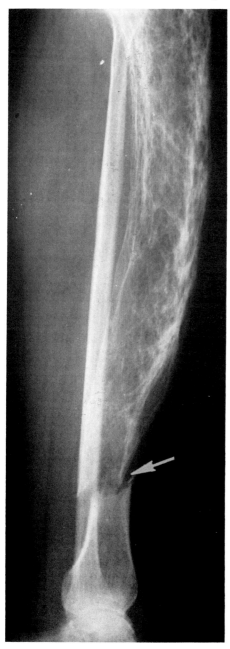

FIG. 1079. PAGET DISEASE
Second stage or combined stage. Marked widening of the bone but the cortex has not increased in thickness due to osteolysis on the internal border. Pathologic fracture at the distal end *(arrow)*.

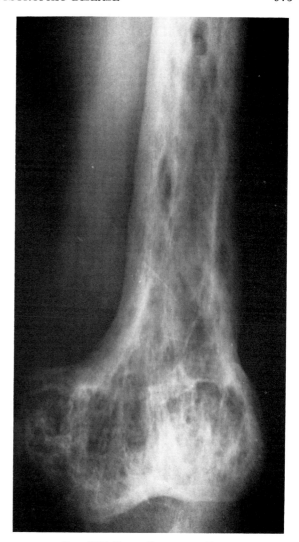

FIG. 1080. PAGET DISEASE OF FEMUR
Combined stage. Destructive areas with the thickening of the cortex and the increase in trabeculation in the condyle.

by the study of weight-bearing forces upon the abnormal bone. Weight-bearing produces tension stresses on the lateral aspect of the femur and the anterior aspect of the tibia, and compressing forces on the medial aspect of the femur and the posterior aspect of the tibia.[45] The bone affected by Paget disease responds by laying down periosteal bone on the surfaces under tension and by absorbing bone on the surfaces being compressed. The endosteal portions of each affected cortex usually show the reverse effect, attempting to maintain the width of the cortex and medullary canal in spite of the proliferation and destruction, an attempt which is not always successful.

Pathologic fractures, usually transverse and healed readily, are likely to occur in Paget disease, especially during the active phases.[45, 46] Fractures of the tibia or femur tend to be incomplete, especially when the bone is deformed. These are true fractures and should not be confused with the pseudofractures of osteomalacia, which are bilaterally symmetrical, and occur on the medial surfaces of the femoral shaft or the superior surfaces of the femoral neck. The cortical fractures of Paget disease often heal only at their periosteal and endosteal surfaces, leaving a permanent linear cortical defect (Fig. 1082). Femoral fractures are most frequent at the subtrochanteric region, but a few occur in the femoral neck. Union occurs after a normal length of time with immobilization.[47] Fractures do not lead to sarcoma, although sarcoma with fracture is frequent. Soft tissue masses may be associated with involvement of the long bones without tumor. This may be due to fracture with hematoma, an expanded cortex, or periosteal reaction without calcification.[40]

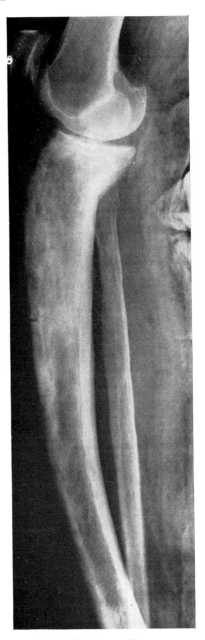

FIG. 1082. PAGET DISEASE
Combined stage. Anterior curvature of the tibia is characteristic of Paget disease. Small fracture on the anterior surface.

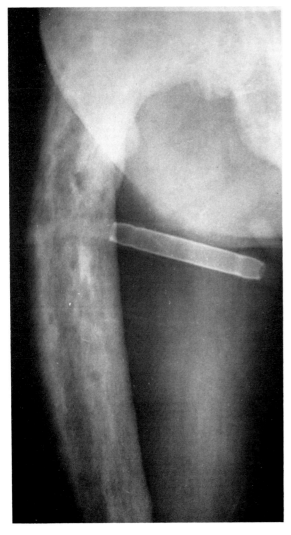

FIG. 1081. PAGET DISEASE
Combined stage with marked thickening of the cortex. Lateral bowing of the femur.

PELVIS. The changes in the pelvis are most often combined or biphasic (Figs. 1083–1088). The trabeculae are thickened and especially prominent in the sacrum. In the ilium they appear as multiple parallel curvilinear lines especially visible below the crest. Abnormal bone above the acetabulum occurs early in the disease. Rarefaction of bone is common in the central portions of the ilium and may contain irregular islets of increased bone density. The areas of radiolucency in the central portions of the ilium are prominent with Paget disease and serve to distinguish it from metastatic disease, especially from the prostate. Mechanical stress on the softened bones leads to acetabular protrusion. In approximately 45% of patients there is concentric narrowing of the hip joint to less than 3 mm[29, 48] (Figs. 1084 and 1086). Widening of the bone is especially noticeable when the pubis and ischial rami are diseased; osteitis deformans becomes truly "deforming."

The combined form of osteitis deformans usually can be differentiated from metastatic prostatic malignancy, especially when the process is "bone deforming" and trabeculations are present. This is particularly true when only half of the pelvis is affected. Metastatic processes extensive enough to involve one side of the pelvis rarely leave its other half intact, as in Paget disease. Confusion may exist, however, particularly in patients with prostatic cancer who also have Paget disease.

VERTEBRAE. The maximum frequency is, in order of involvement, the upper cervical spine, low dorsal, and midlumbar vertebrae.[39]

Vertebrae often show the combined or sclerotic phase of osteitis deformans (Figs. 1089 and 1090). Occasionally, the monophasic or destructive phase alone may occur (Figs. 1091 and 1092). When the biphasic or combined destructive and reparative process develops, the vertebral body shows increased trabeculation particularly prom-

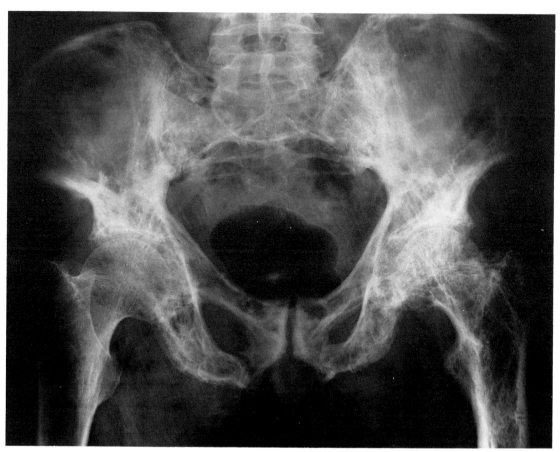

FIG. 1083. PAGET DISEASE OF PELVIS AND FEMUR
Combined stage with destructive lesions and repair with prominence of the trabecular pattern. The lateral and superior portions of the bodies of the iliac bones are not particularly involved.

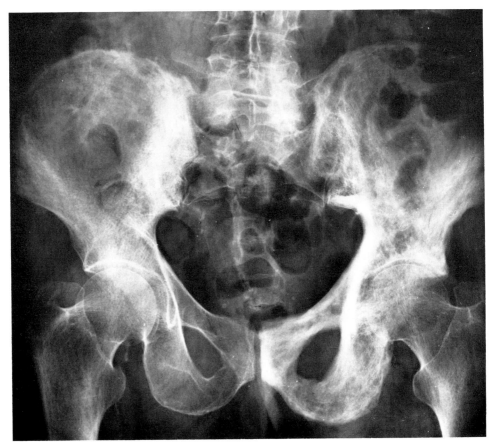

Fig. 1084. Paget Disease
Combined stage. The lesions are predominantly on one side of the pelvis.

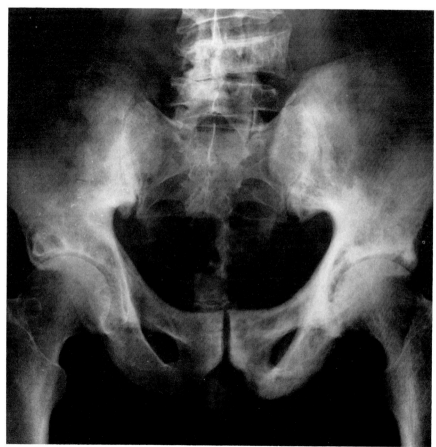

Fig. 1085. Paget Disease in Iliac Wings

Sclerotic stage is noted. In the rami of pubis and ischium there is the combined stage with destruction and bone repair.

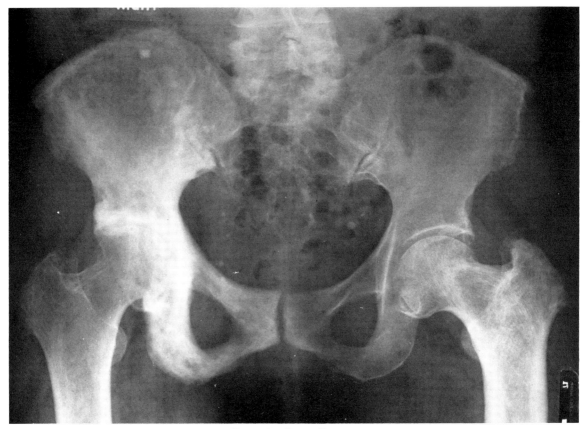

Fig. 1086. Paget Disease
Combined stage.

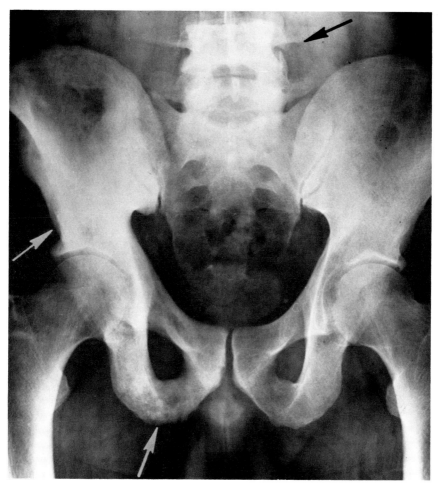

Fig. 1087. Paget Disease

Sclerotic or quiescent stage. This is an inactive stage of Paget disease. Eburnated sclerosis of the lumbar vertebrae and the right pelvis and rami of the ischium (*arrows*).

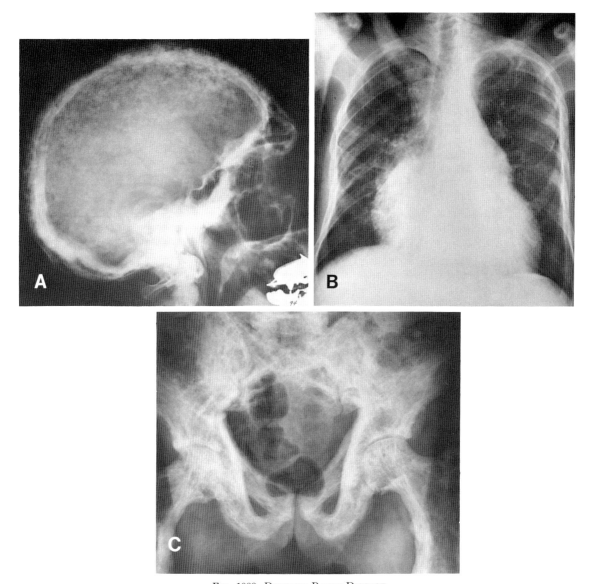

FIG. 1088. DIPHASIC PAGET DISEASE

(A) Skull; diphasic Paget disease with a cotton wool appearance. (B) Chest; changes in both clavicles and ribs. The patient was in failure with an enlarged heart due to microscopic multiple arteriovenous malformations in the bones. (C) Diphasic disease: pelvis. (D) Diphasic disease: Bones of the knee. (E) Diphasic disease; femur with anterior bowing. (F) Anteroposterior view of the tibia with marked thickening and diphasic disease; the fibula is spared.

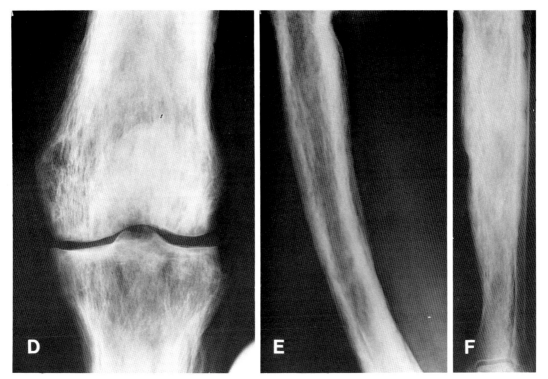

Fig. 1088 (D-F)

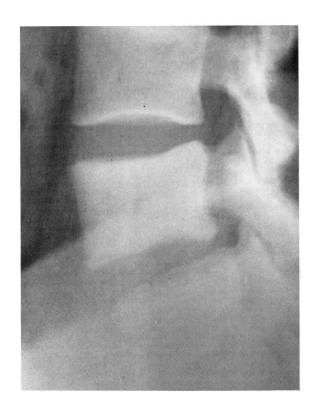

Fig. 1089. Paget Disease of Spine
Sclerotic or quiescent stage.

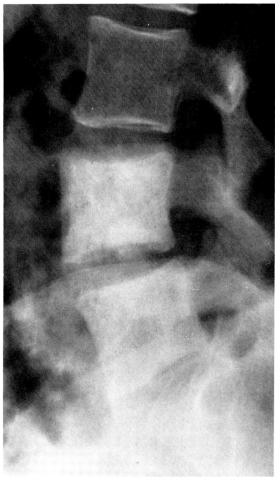

FIG. 1090. PAGET DISEASE

Squaring of the vertebral bodies with molding and increased anteroposterior dimensions compared to vertebral bodies above and below.

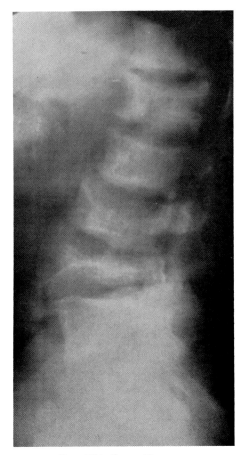

FIG. 1091. PAGET DISEASE

Predominantly the first or destructive stage. Very little attempt at bone repair and marked collapse of several of the lumbar vertebrae.

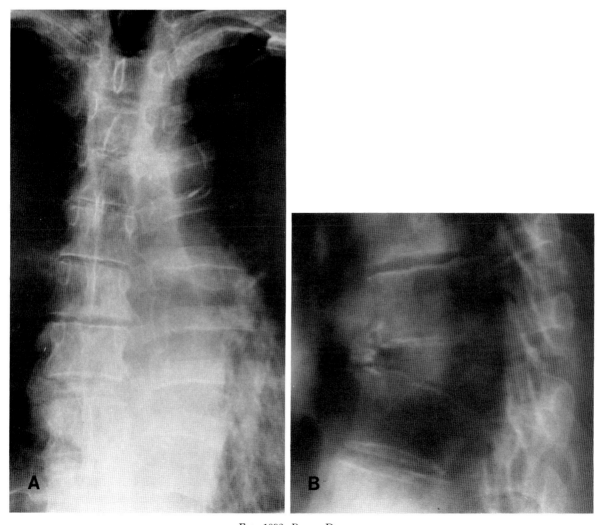

FIG. 1092. PAGET DISEASE

The first phase of Paget disease. There is almost complete collapse of a dorsal vertebra. There is no radiographic distinction from a malignancy, but biopsy proved it to be Paget disease. (A) Anteroposterior view; (B) lateral projection.

inent at the periphery of the bone, so that the borders of the centra of the vertebrae are exaggerated due to the multiple parallel trabecular lines (Fig. 1093). This distinguishes the condition from hemangioma of the vertebra, in which there is exaggerated vertical striation of the body, but not of the superior and inferior borders of the centrum. A "bone-within-a-bone" appearance may occur (Fig. 1094). Sometimes one vertebral body alone may be involved, to the exclusion of bone abnormalities elsewhere. It is usually sclerotic and reveals the effects of molding of bone. The body loses height, and its anteroposterior dimensions are greater than those of the adjacent vertebral bodies. Compression fractures of the vertebral bodies are not uncommon in Paget disease and may lead to spinal cord compression with the expanded or fractured vertebra protruding posteriorly into the spinal canal. The lamina may increase to 3 times normal size and compress the cord.[19, 49] The appendages may be involved (Fig. 1095). The vertebral body may be sclerotic

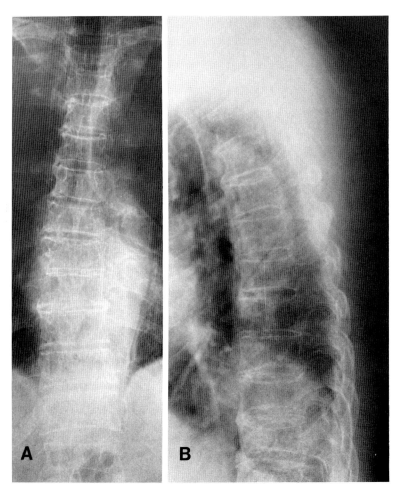

FIG. 1093. PAGET DISEASE

The spine, showing what appears to be osteoporosis; accentuation of the vertebral plates and the vertical striations in the bodies of the vertebrae; some loss of stature in multiple lumbar vertebrae and some dorsal vertebrae. The pelvis shows increased trabeculation and widening of the pubic rami; the left femur demonstrates considerable changes characteristic of Paget disease. Biopsy of the bodies of the vertebrae revealed Paget disease. The bone production on the lateral aspect of the femur was not malignant and probably the result of trauma. (A) Anteroposterior projection of dorsal spine. (B) Lateral projection of dorsal spine. (C) Anteroposterior projection of lumbar spine. (D) Lateral projection of lumbar spine. (E) Anteroposterior projection of pelvis.

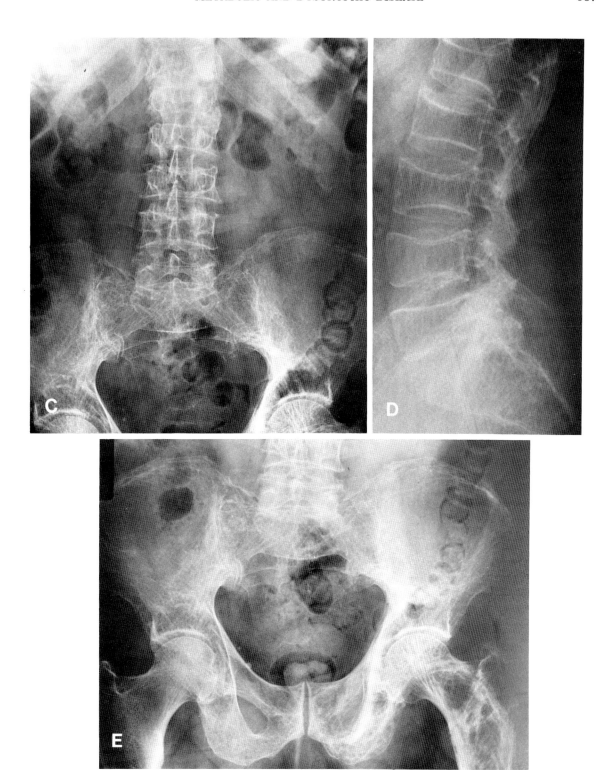

Fig. 1093 (C-E)

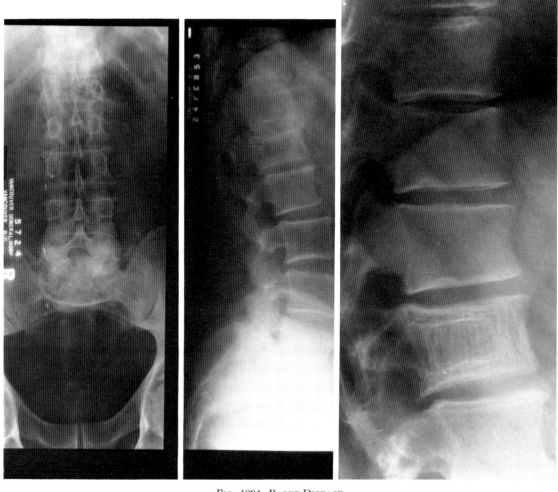

FIG. 1094. PAGET DISEASE
Changes in the first and fifth lumbar vertebrae in a patient with known Paget disease. (*Left*) Left anteroposterior view. (*Middle*) Lateral view. (*Right*) Coned-down view of the first lumbar vertebra reveals a "bone-within-a-bone" appearance; heavy trabeculations especially the horizontal trabecular pattern, and involvement of the posterior elements. (Courtesy of Dr. Lynette Thurber and Dr. Joachim Burhenne, University of British Columbia, Vancouver, Canada.)

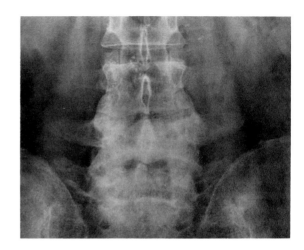

FIG. 1095. PAGET DISEASE
The fourth lumbar vertebra is dense and large. The transverse processes are dense and widened.

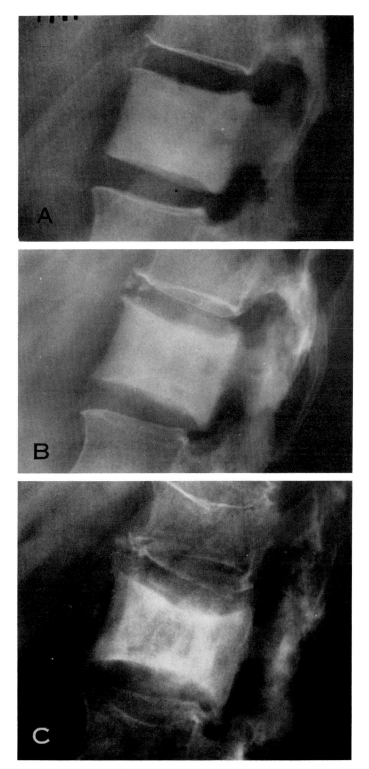

FIG. 1096. PAGET DISEASE

(A) In 1941 the vertebral body is sclerotic and normal in size. (B) In 1956 the body is sclerotic with loss of stature and the anteroposterior and vertical diameters are increased. (C) In 1967 there is deossification, further increase in anteroposterior diameter and stature loss. The progression partially reflects the generalized osteoporosis and active Paget disease, and indicates that sclerotic Paget disease may be active.

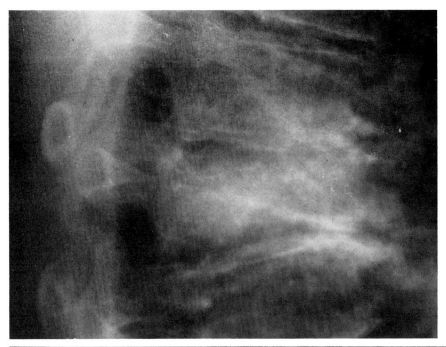

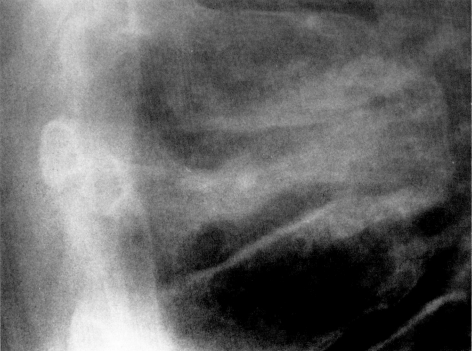

Fig. 1097. Paget Disease

(*Top*) In 1957, collapse, increase in density and increase in the anteroposterior diameter of the involved vertebral body.
(*Bottom*) In 1971 the collapse has progressed and there is density decrease; a part of the generalized process of osteoporosis.

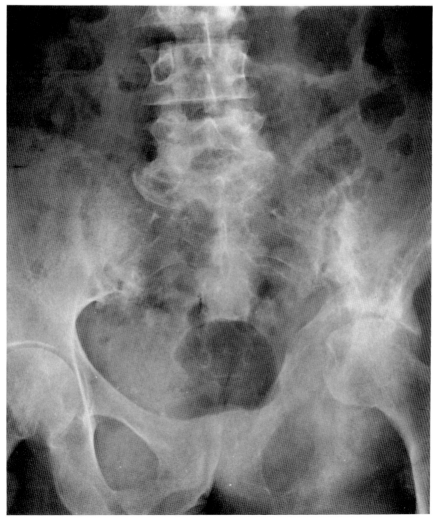

Fig. 1098. Paget Disease of the Left Ischial and Pubic Rami

Widening and softening of the bone with molding. This same appearance frequently occurs in the older patient, particularly in the spine as in Figure 1097.

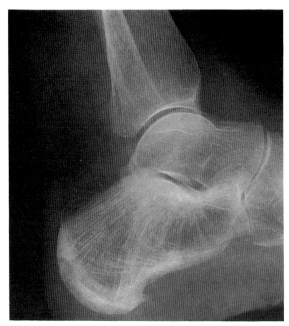

FIG. 1099. MONOSTOTIC PAGET DISEASE OF CALCANEUS

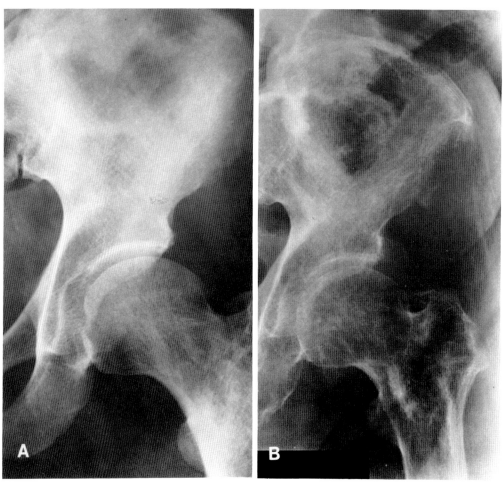

FIG. 1100. PAGET DISEASE WITH IMMOBILIZATION
(A) In 1972 there is extensive Paget disease of the left iliac bone, and head and neck of the femur. It is of the mixed type.
(B) Seven years later, after amputation of the right leg, marked demineralization has occurred in the pelvis, and head and neck of the femur. A well defined V-shaped area of deossification extends toward the shaft.

without enlargement. It may enlarge and eventually become osteoporotic and collapse—a part of generalized osteoporosis in the elderly (Figs. 1096–1098). A "bone-within-a-bone" appearance is sometimes present. This is not characteristic of Paget disease, since it may occur in other conditions where growth has ceased at an earlier age. Thoratrast administration or fluorosis in childhood are examples.[50–54]

Ossification may occur in the spinal ligaments, paravertebral soft tissue, anterior rib ends and disk spaces.[55, 56] Calcified paravertebral soft tissue masses may result.

Extramedullary hematopoiesis may occur and cause a soft tissue mass in the paravertebral area. The fractures may be due to extrusion of hematopoietic marrow and result in tumorous extramedullary hematopoiesis secondary to fracture, rather than to anemia.[57–61]

RIBS. The ribs are commonly affected in Paget disease and usually the changes are biphasic. Only one or two ribs may be involved, perhaps showing only increased bone density resembling osteoblastic metastatic involvement. Again, the widening of the bone (deforming osteitis) is an important distinguishing point; the latter usually is not seen in osteoblastic metastasis. When the ribs are affected by the combined or biphasic Paget disease, the diagnosis is usually clear.

OTHER BONES. The clavicle may be affected by either the monophasic or the biphasic characteristics. At times, monostotic disease appears in a small bone of the hand or foot, which is usually larger and more dense than the surrounding small bones (Fig. 1099).

Stress and strain are important factors in the reparative process of Paget disease, and patients should never be immobilized unless unavoidable. Immobilization leads to inactivity of the osteoblasts, so that the destructive process continues uninhibited and the affected bones become extremely deossified (Figs. 1076 and 1100). In addition, there will be an appreciable flow of calcium from the bone to the serum which may lead to hypercalcemia and hypercalciuria. Renal damage with nephrocalcinosis or renal calculi may ensue. For the same reason patients with Paget disease should not be given large quantities of calcium and vitamin D.

Reportedly, sarcomatous degeneration develops in 5–14% of patients with Paget disease (Fig. 1101); however, this incidence is high, and in our experience it occurs in less than 1% of patients with polyostotic disease. It may occur with monostotic Paget disease. The sarcoma is usually osteosarcoma, but giant cell tumor of bone is reported.[62–66] There is little doubt, however, that osteitis deformans, especially the disseminated form, predisposes to sarcoma in older age groups.[2, 15, 67–69]

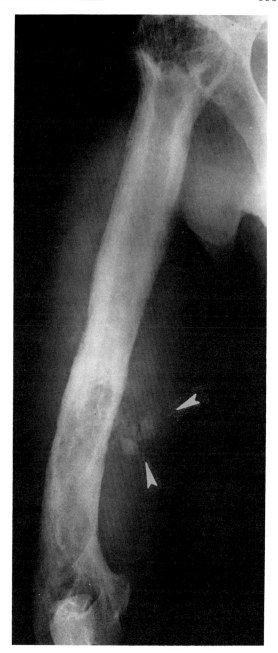

FIG. 1101. PAGET DISEASE WITH SARCOMATOUS CHANGE
New bone formation in the soft tissue indicating osteosarcoma (arrows).

FIBROUS DYSPLASIA

Synonyms: Polyostotic fibrous dysplasia, monostotic fibrous dysplasia, Albright syndrome, osteitis fibrosa disseminata.

Fibrous dysplasia is characterized by fibrous replacement of portions of the medullary cavities of a bone or bones. It was first recognized as distinct from hyperparathyroidism in 1937 when Albright et al.[1] described it as osteitis fibrosa disseminata, characterized by (1) bone lesions, which tend to be unilateral; (2) "cafe-au-lait" spots, which tend to be on the same side as the bone lesions; and (3) endocrine dysfunction, associated in females with precocious puberty. This triad became known as "Albright syndrome." Lichtenstein[2] later called it polyostotic fibrous dysplasia and extended it to include similar skeletal lesions unaccompanied by cafe-au-lait spots or precocious puberty. Although Lichtenstein and Jaffe[3] suggested a relationship between monostotic osseous lesions and multiple osseous lesions, it was Schlumberger[4] in 1946 who described fibrous dysplasia of single bones (monostotic fibrous dysplasia). He believed that no relationship existed between monostotic and polyostotic fibrous dysplasia, but most authorities now believe that they are indeed related.[5] Because they are now considered a single entity, the term fibrous dysplasia has become its preferred name. The eponym Albright syndrome distinguishes the disease with extraskeletal manifestations from the purely osseous affectation.

The cause of fibrous dysplasia is unknown. The most widely accepted concept is that a developmental abnormality of the bone-forming mesenchyme results in replacement of the spongiosa and filling of the medullary cavity by fibrous tissue in which trabeculae of poorly calcified primitive new bone are developed by osseous metaplasia.[2] Other concepts, including trauma, neuroendocrine disturbance,[6] and xanthogranuloma,[7] have not been widely accepted. No proof exists that this process is either neoplastic or inflammatory.

Clinical Features. Both sexes are equally affected by this disease. It begins in young individuals, often in infancy, but, because it is frequently asymptomatic, often remains unrecognized until adulthood. There is no familial tendency.

In 148 patients reported by Pritchard,[8] ages ranged from 4 months to 70 years. Of these, in 7 the disease could be traced to infancy, in 131 to childhood, and in 10 to adolescence. In 13 patients between 20 and 57 years of age, the onset of symptoms appeared before the 20th year in 2, after the 20th year in 9, and 2 patients remained without symptoms. In 5 patients between 40 and 60 years, the duration of symptoms was from 1 to 30 years.

The initial symptoms vary from vaginal bleeding in an infant to limping in an older patient; over half the cases began with limp, pain in the leg, or fracture.[9] In females it is recognized earlier because of the vaginal bleeding.

The most common physical deformity is a difference in leg lengths often caused by the shepherd's crook deformity of the neck of the femur. Forty percent of patients with polyostotic fibrous dysplasia suffer fractures at some time, but some escape this tendency despite extensive involvement.

While Albright syndrome is usually chronic and seldom fatal, some patients may die after a few months or even years. Harris et al.[9] attributed 3 deaths to the disease in a study of 37 cases; 2 patients died before the age of 16, and the third at 41 after a progressive and extensive 34-year course of the disease. The condition tends to become quiescent at puberty but has been known to remain active and progressive long beyond it. It may become reactivated, especially during pregnancy.[10-12]

Schlumberger[4] studied 67 patients with monostotic fibrous dysplasia and found that a large majority were asymptomatic. Local swelling was common, especially in superficial bones such as the tibia or clavicle. Sometimes arthritic pain occurred, often accompanied by local tenderness. Often pathologic fractures were the first sign of the disease. None of Schlumberger's patients showed cafe-au-lait spots or abnormalities of the serum calcium, phosphorus, or phosphatase.

The prognosis for monostotic lesions is uniformly good. There is no recorded case of this form becoming polyostotic nor does endocrine disturbance appear later.

The cafe-au-lait spots merit particular mention. Of 90 patients with polyostotic fibrous dysplasia reviewed by Lichtenstein and Jaffe,[3] 32 had this pigmentation (35%); it did not appear at all in the monostotic form of the disease.[4] The pigmentation, although usually on the same side as the bone lesions, is often on the opposite side, and does not tend to occur directly over the affected bone.[13] The contour of the cafe-au-lait spots of fibrous dysplasia is irregular ("coast of

Maine" outline) which distinguishes them from the smooth ("coast of California") outline of the pigmentations of neurofibromatosis.[1]

Sexual precocity is found in about 20% of females with the polyostotic disease; the menarche has been reported to begin as early as the second day of life.[14] Advanced skeletal and somatic maturation often accompany this sexual precocity, and during childhood and early adolescence these patients are large for their age, but, because of early epiphyseal closure, they mature as shorter-than-average adults. There is no consistent abnormality of gonadal function in males. Sexual precocity in males, when it does occur, is mild.[15-17]

Albright and Reifenstein[18] believed that the onset of puberty is triggered by the release of gonadotrophic hormone from the anterior pituitary which results from stimuli traveling through the hypothalamic-pituitary-nervous-humoral pathway. The mechanism for the sexual precocity of fibrous dysplasia probably arises from a disturbance in the region of the hypothalamus. Thyroid dysfunction may occur in association with Albright syndrome; its mechanism is unknown. Hyperthyroidism may develop in either sex before or after the bone lesions.[3, 19, 20]

Congenital anomalies have been reported in conjunction with Albright syndrome, including coarctation of the aorta, rudimentary left kidney,[21] and congenital arteriovenous aneurysm,[22] but their incidence is no greater than in the general population. No evidence suggests that patients with fibrous dysplasia are mentally deficient.[5]

The skeletal lesions associated with Albright syndrome may be multiple and extensive, affecting predominantly one side of the body, but the syndrome may also be accompanied by only a minimum of skeletal lesions.[1] Both polyostotic and monostotic fibrous dysplasia (without endocrine or dermatologic lesions) may appear first through fractures or skeletal deformities. Deformity in the skull may infrequently distort normal foramina, causing neurologic abnormalities.

Laboratory studies are usually within normal values, except for an occasional elevation of serum phosphatase, unrelated to the extent of skeletal disease.[9] The literature mentions bone sarcoma in association with fibrous dysplasia[23-26]; Harris et al.[9] reported 2 patients with sarcoma which occurred 20 and 22 years following irradiation.

Pathologic Features. On gross examination, bones affected by fibrous dysplasia show that the medullary cavity has been replaced by white or gray-white tissue of rubbery consistency, usually gritty with spicules of new bone.

Microscopically, the basic tissue is fibrous, with varying amounts of osseous tissue scattered throughout in disorganized fashion. Large areas may contain no osteoid. Occasional nests of cartilage and of multinucleated giant cells may be found, suggesting Ollier's disease or giant cell tumor. Also, occasionally, foam cells may permeate these lesions. The cement lines between the newly formed trabeculae are erratic, suggesting Paget disease despite the absence of a true generalized mosaic pattern. The histologic picture of primary fibrous replacement may simulate other conditions, including osteitis fibrosa cystica, agnogenic myeloid metaplasia, and metastatic malignancy.

Fibro-osseous defects of the craniofacial region have long been confused with Paget disease, meningioma, metastatic disease, blood dyscrasias, metabolic disease, and osteodystrophies. These lesions having been eliminated, there remains the large group of fibro-osseous lesions which may range from the purely osseous to the purely fibrous. Depending on the ratio of these components, they include osteoma, fibrous osteoma (osteofibroma, fibroosteoma), ossifying fibroma (central fibroma), osteogenic fibroma (odontogenic fibroma), and fibrous dysplasia. Some advise considering all of these lesions as fibrous dysplasia, differing only in the ratio of their fibrous and osseous elements. The possible exception is the osteoma, which should be considered separately.

Roentgenographic Features. The skeletal lesions of fibrous dysplasia are not usually present at birth. They usually appear several years before puberty and may progress through the entire life of the patient. Even in the presence of Albright syndrome, the bone age may not be advanced in infancy, but usually after age 5 it is (Fig. 1102). All the bones may be involved at an early age, including the hands, and densities within the paranasal sinuses often distinguish this condition (Fig. 1103).

Although the skeletal lesions of fibrous dysplasia present a variety of roentgen appearances, their features are usually sufficiently distinctive to assure the diagnosis, especially in the absence of symptoms (Figs. 1104–1111). In the long bones, the basic change is replacement of the medullary cavity, which produces lesions varying from completely radiolucent to homogeneous "ground

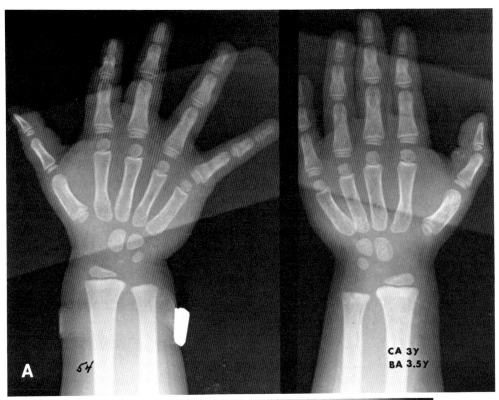

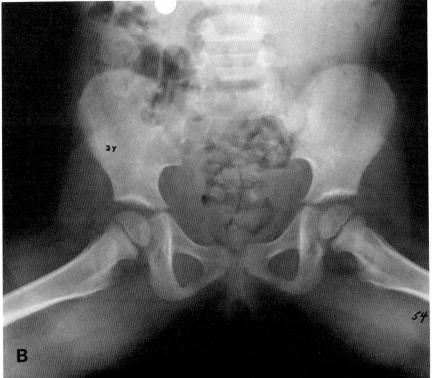

Fig. 1102. Albright Syndrome (Fibrous Dysplasia)

This child had precocious puberty; roentgenograms revealed a bone age almost within normal limits for the chronological age at 3 years. There were no significant bony lesions except for a slight osteolytic area in the right humerus near the midshaft. ((A) Hands, (B) pelvis, (C) humeri, and (D) lower extremities at age 3 years.) At the age of 9 years the bone age was 13, and there were significant lesions in the left femur and left tibia. ((E) Hands, (F) left femur, and (G) tibias.) (Courtesy of Dr. Marie Capitanio, St. Christopher's Hospital, Philadelphia, Pa.)

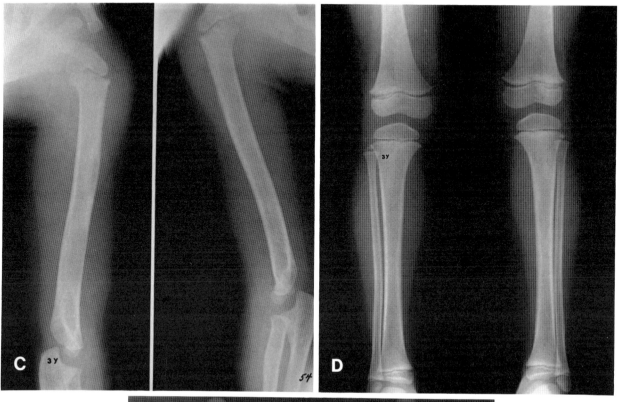

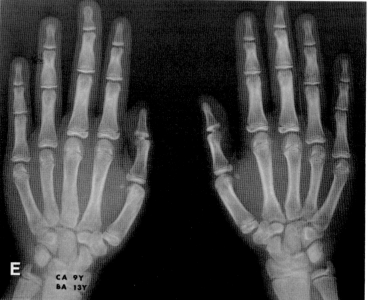

Fig. 1102 (*C–E*)

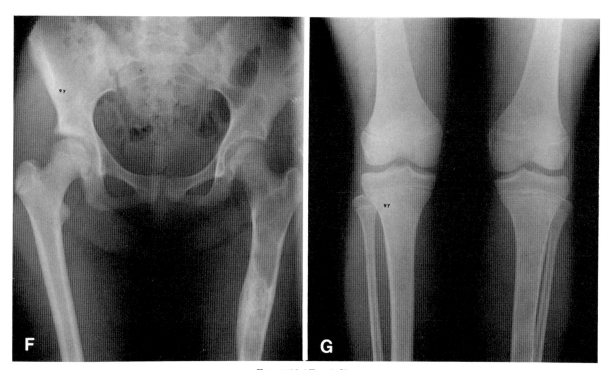

Fig. 1102 (*F* and *G*)

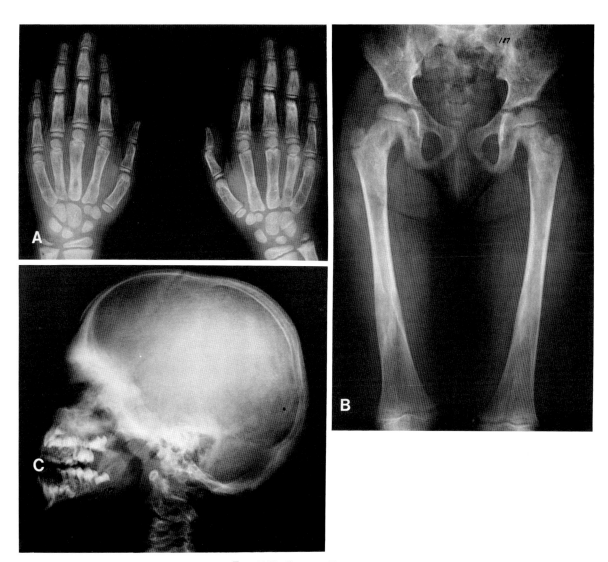

Fig. 1103. Fibrous Dysplasia

(A) Extensive bony involvement. The hands show trabecular abnormalities throughout the metacarpals and phalanges; the bone age was slightly increased. (B) Both femurs show cortical irregularity especially on the right. (C) The skull shows densities throughout all paranasal sinuses. (Courtesy of Dr. Fargin Eftekhari, Teheran, Iran.)

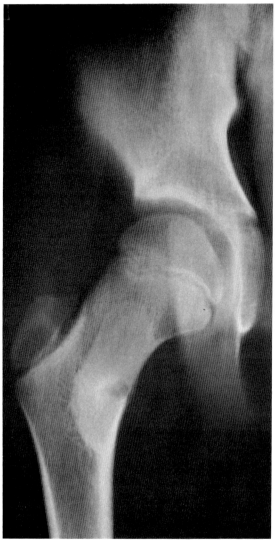

FIG. 1104. FIBROUS DYSPLASIA
Anteroposterior view of the right proximal end of the femur shows a well defined area of increased density with a radiolucency within it. This might be considered an abscess or an osteoid osteoma but the absence of pain suggests the correct diagnosis of fibrous dysplasia.

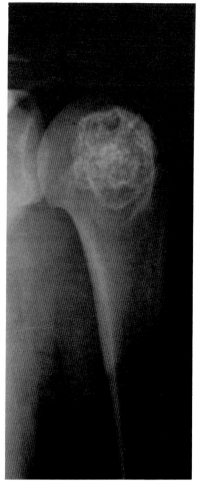

FIG. 1105. FIBROUS DYSPLASIA
Increased density in the upper end of the humerus suggests either a chondroma or an infarction; because of the border surrounding the lesion it seemed unlikely to be an enchondroma. Biopsy proved fibrous dysplasia.

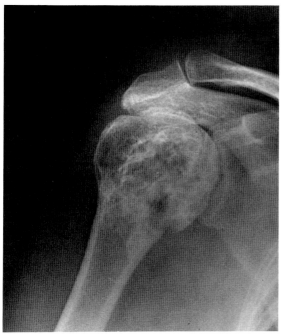

FIG. 1106. FIBROUS DYSPLASIA
The appearance of this somewhat aggressive lesion suggested chondroblastoma. There is a ground glass appearance in part of the lesion, but a break in the cortex suggested an aggressive lesion and biopsy proved this to be fibrous dysplasia.

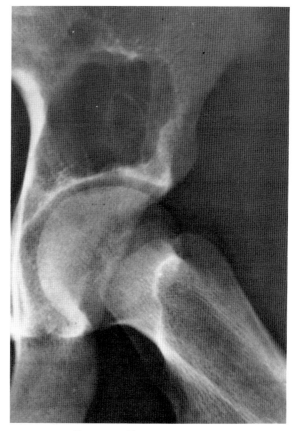

FIG. 1107. FIBROUS DYSPLASIA
The osteolytic area has a well defined border but not a rind. It appeared to be a benign lesion and eosinophilic granuloma was suggested. Biopsy showed it to be fibrous dysplasia. (Courtesy of Dr. Gerald D. Dodd, M.D. Anderson Hospital and Tumor Institute, Houston, Texas.)

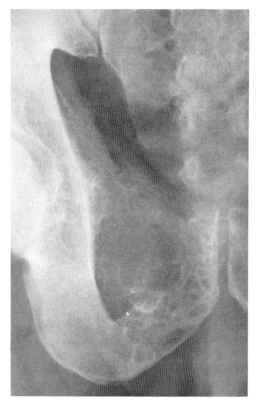

FIG. 1108. FIBROUS DYSPLASIA

This bubbly lesion has caused almost complete destruction of the horizontal ramus of the symphysis with extension into the ischial ramus and the acetabulum. There is no associated soft tissue mass in the pelvis; however, the obturator foramen appears filled with homogeneous density. This was an asymptomatic lesion which should have suggested the diagnosis. Biopsy proved it to be fibrous dysplasia. Note its similarity to Figure 1128.

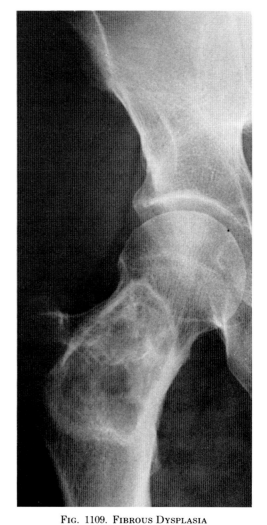

FIG. 1109. FIBROUS DYSPLASIA

The thick increase in density surrounding the lesion (rind) suggests the diagnosis in an asymptomatic patient. The rind is caused by a longstanding lesion and could well be an enchondroma. Biopsy proved this to be fibrous dysplasia.

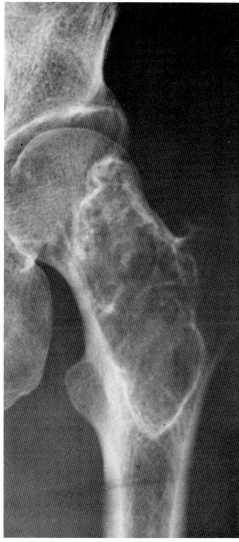

FIG. 1110. FIBROUS DYSPLASIA
The most common site of fibrous dysplasia is the neck of the femur and the thick increase in density (rind) suggests the diagnosis.

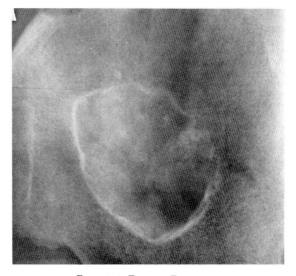

FIG. 1111. FIBROUS DYSPLASIA
The thick rind suggests the diagnosis which was proven by biopsy.

glass" increased density, depending on the amount of fibrous or osseous tissue deposited in the medulla (Figs. 1113–1116). When densities are increased, normal cortical definition is lost and the entire transverse diameter of the bone appears to be of one texture (Fig. 1117). This can be seen in the ribs and also at times in the long bones (Fig. 1118). In radiolucent lesions the cortex is thinned along its medullary surface (Figs. 1119–1121), but at times, especially after healed fractures, the cortex is considerably thickened. It may also be simultaneously thickened and eroded (Figs. 1122–1125). The deossified lesions always reveal well defined sclerotic margins; the inner surface is dense; the outer surface fades gradually into the normal bone. Variations in bone density do not depend on the age of the lesion or of the patient, but depend on the amount of dysplastic material in the bone and its calcium content. Sequestration may form in the absence of osteomyelitis.[27]

Expansion of bone occurs in the ribs and skull, and is also commonly seen in long bones, always more frequent in the presence of radiolucent

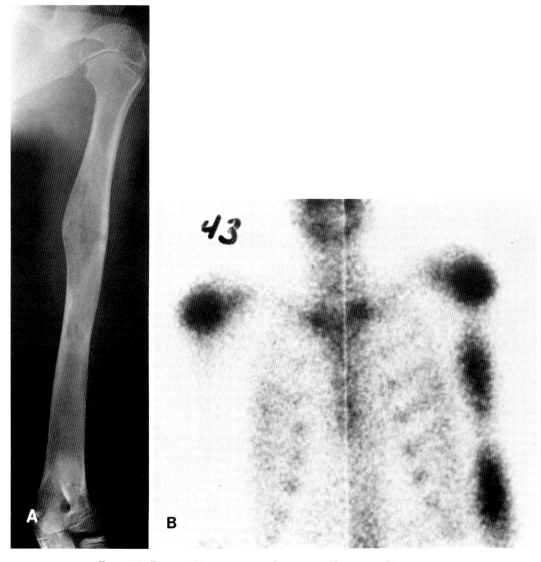

FIG. 1112. FIBROUS DYSPLASIA WITH INCREASED UPTAKE OF TECHNETIUM

(A) Roentgenogram of humerus reveals a homogeneous increase in density throughout the upper portion of the humerus. (B) Technetium scan shows increased uptake, indicating activity of the lesion. (Courtesy of Dr. Murali Sundaram, St. Louis University Hospital, St. Louis, Missouri.)

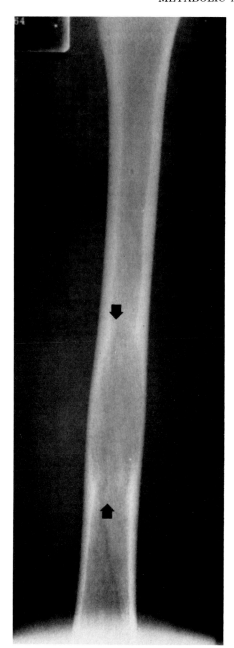

FIG. 1113. FIBROUS DYSPLASIA
Radiolucent lesion lies between the confines of the *arrows*. There is no osseous tissue within the lesion. The cortex is narrowed and there is only slight ballooning of the involved section of bone.

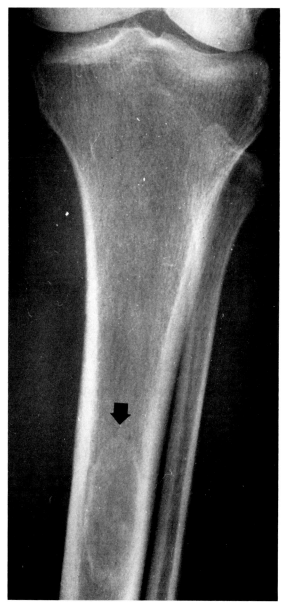

FIG. 1114. FIBROUS DYSPLASIA
Lesion within the medullary cavity has partially ossified producing a clearly defined abnormality (*below arrow*). The other tibia was extensively involved.

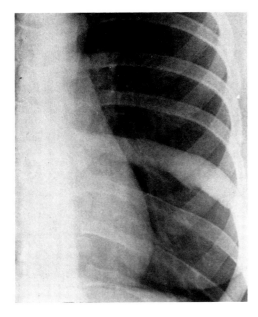

Fig. 1115. Fibrous Dysplasia
The fifth rib is sclerotic and enlarged.

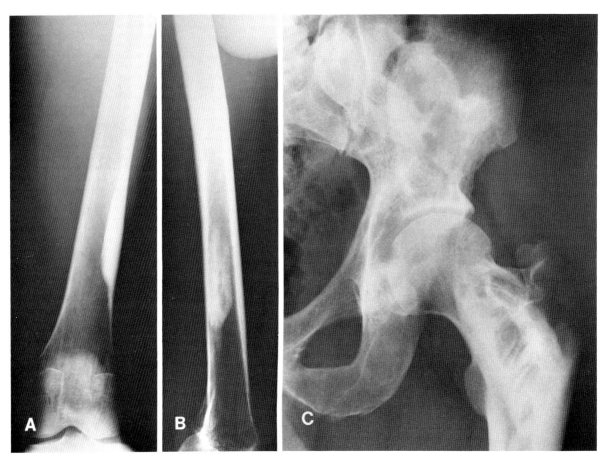

Fig. 1116. Fibrous Dysplasia
Many of the lesions in this patient are dense, due to deposition of calcium. (A) Anteroposterior projection of left femur. (B) Later projection of left femur. (C) Anteroposterior projection of left hip and pelvis. (Courtesy of Dr. Beth Edeiken, M. D. Anderson Hospital and Tumor Institute, Houston, Texas.)

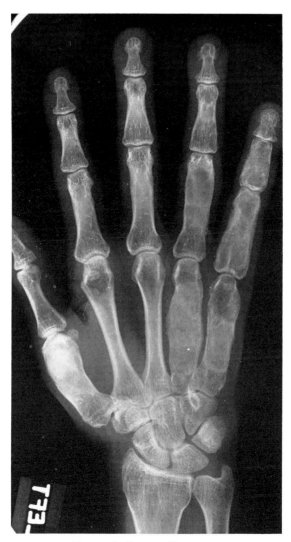

FIG. 1117. FIBROUS DYSPLASIA
Changes in the fourth and fifth metacarpals are characteristic of the ground glass increase in density of fibrous dysplasia. The cortical definition is lost and the entire transverse diameter of the bone appears to be of one texture.

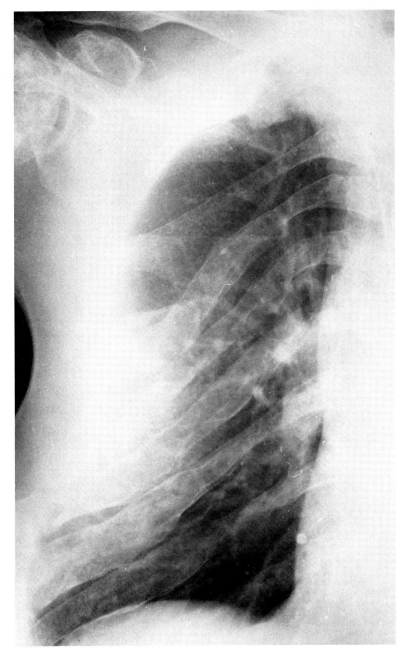

Fig. 1118. Fibrous Dysplasia
Normal cortical outlines of most of the ribs are lost due to replacement by fibro-osseous tissue.

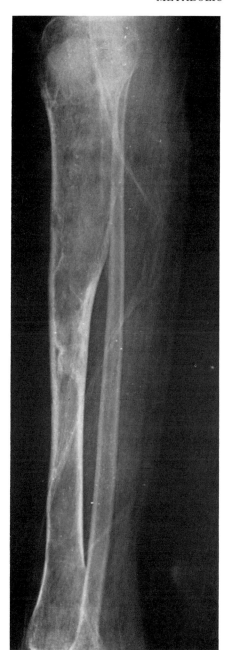

FIG. 1119. FIBROUS DYSPLASIA
Extensive fibrous displacement of the medullary canal and the upper portion of the tibia causing thinning of the cortex from the medullary side.

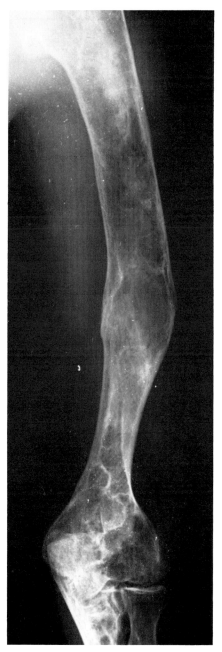

FIG. 1120. FIBROUS DYSPLASIA
Extensive fibrous replacement in the medullary cavity has caused narrowing of the cortical bone from the medullary side. Several radiopaque areas within the diseased bone representing ossification. Deformity at the midshaft due to a previous fracture.

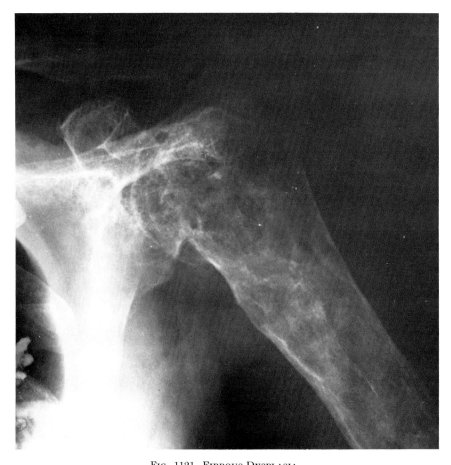

Fig. 1121. Fibrous Dysplasia

Extensive replacement of medullary cavity by fibrous tissue with small areas of punctate ossification. The cortex is markedly thin.

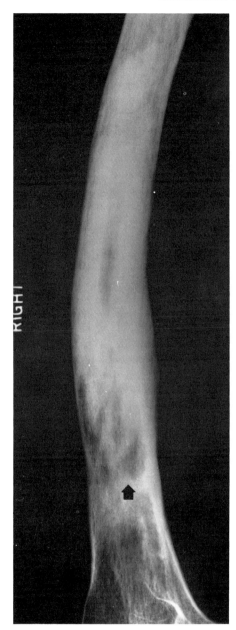

FIG. 1122. FIBROUS DYSPLASIA
Cortex is very thick and at the inferior aspect of the lesion (*arrow*) there is an area of erosion. The thickened cortex may be the result of an old fracture or of the disease process.

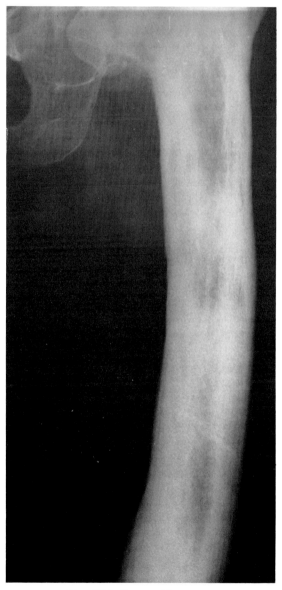

FIG. 1123. FIBROUS DYSPLASIA
Markedly thickened cortex with encroachment on the medullary cavity.

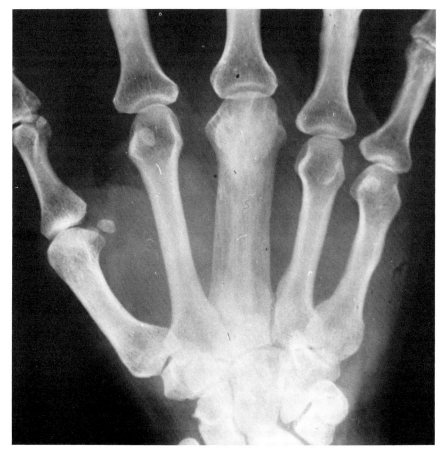

Fig. 1124. Fibrous Dysplasia
Markedly thickened cortex. The original bone may be seen within the thickened metacarpal.

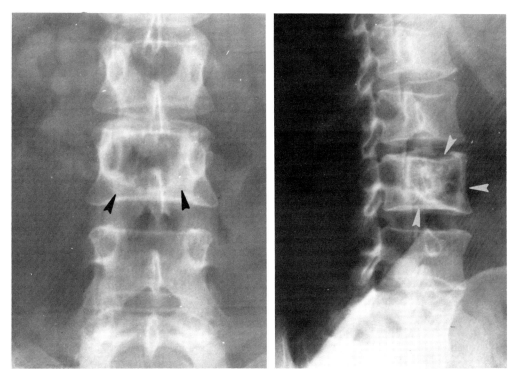

Fig. 1125. Fibrous Dysplasia

Osteolytic lesion with a dense sclerotic margin in the vertebral body (*arrows*). (*Left*) Anteroposterior projection; (*right*) oblique projection. (Courtesy of Dr. Robert M. Peck, Atlantic City, N.J.)

than dense lesions (Figs. 1126–1129). The metaphysis is the primary site of involvement, and a portion of the diaphysis may also expand to include the midportion of the shaft; it rarely extends the entire length of the bone.

A well developed lesion in an infant, especially in the tibia, which histologically represents fibrous dysplasia, is probably osteofibrous dysplasia. It is monostotic and often has a pseudoarthrosis. It has a different connotation because these lesions will usually heal by the age of 10 (see Fig. 1140).

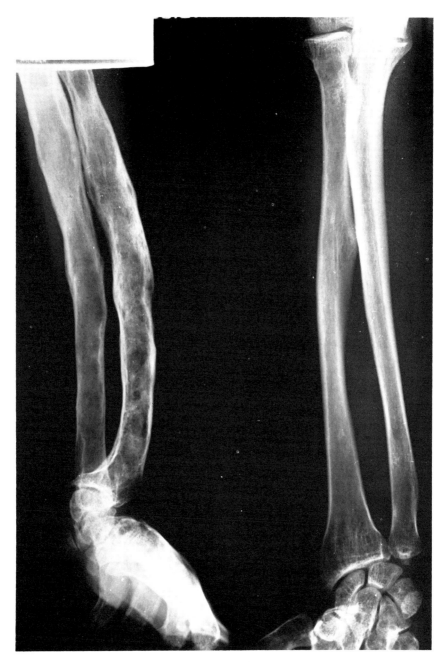

FIG. 1126. FIBROUS DYSPLASIA

Extensive involvement of the right radius reveals irregular ballooning of the bone, commonly associated with radiolucent lesions. Stippling indicates areas of ossification. The right ulna also shows narrowing of the cortex due to fibrous replacement. In the left forearm, only small areas of increased density are noted, indicating the presence of ossified lesions.

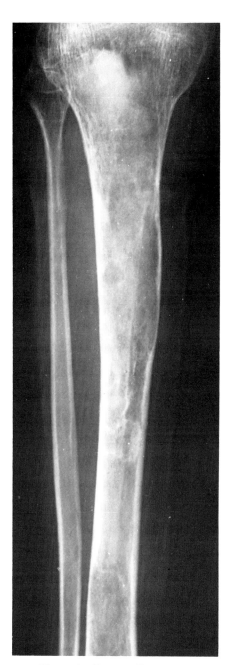

FIG. 1127. FIBROUS DYSPLASIA
Fibrous replacement of the upper end of the tibia has caused thinning of the cortex and minimal ballooning. In the proximal end of the lesion, a dense area of ossification is noted.

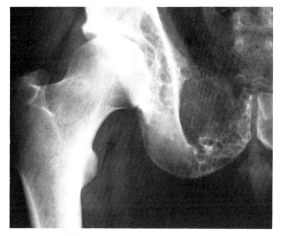

FIG. 1128. FIBROUS DYSPLASIA
Expansile lesion in the right iliac and pubic bones. Sclerotic margins surround most of the osteolytic areas; however, one large expanding lesion could be mistaken for tumor. They are usually asymptomatic.

When multiple bones of the lower half of the body are affected, the femur is almost invariably involved and shows the greatest abnormality (Fig. 1130). The usual picture is involvement of the neck and metaphysis of one femur, with the iliac bone and the tibia on the same side involved in varying degrees (Fig. 1131). Bowing deformities in the areas of localized cortical thickening

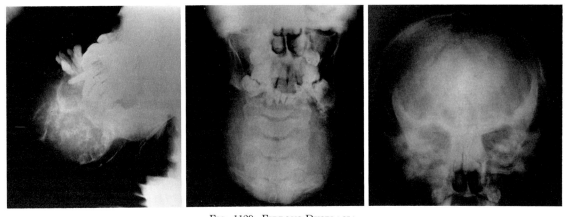

FIG. 1129. FIBROUS DYSPLASIA

(*Left*) Lateral projection; huge expanding mandibular lesion with ossification and displacement of the teeth (*Center*) Anteroposterior projection; the mass extends inferiorly. (*Right*) Posteroanterior projection; sclerotic ethmoid and frontal sinuses and thickening of the greater wings of the sphenoid and orbits.

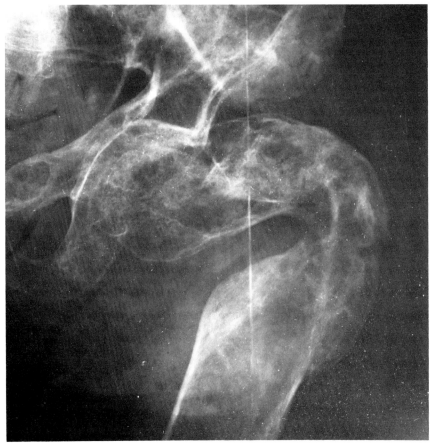

FIG. 1130. FIBROUS DYSPLASIA
Marked molding of the softened femur. The upper end is almost totally replaced by fibrous tissue.

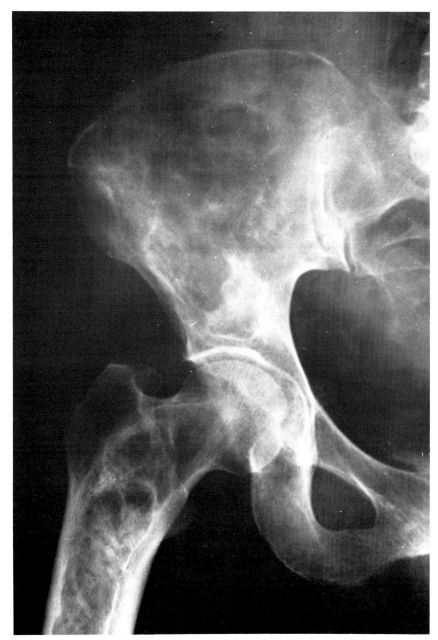

FIG. 1131. FIBROUS DYSPLASIA

Lesions are predominantly on one side of the pelvis and the femur. Well defined sclerotic margin surrounding most of the lesions. Some stippled calcification is noted within the lesion. The lesion does not extend to the proximal end of the femoral head (which would aid in differentiating from Paget disease.)

are usually the result of healed fractures (Figs. 1132–1134). Pseudoarthrosis may occur. However, pseudoarthrosis of the tibia occurring in the infant, with or without a significant bone lesion, should be considered osteofibrous dysplasia (see Fig. 1140).[28–30]

The changes of fibrous dysplasia progress throughout the patient's life and have increase in activity with technetium scanning (Fig. 1112). Sometimes, small spotty lesions may be noted on the contralateral side.

The bone not affected by fibrous replacement, as well as the remainder of the skeleton, maintains a normal appearance; this aids differentiat-

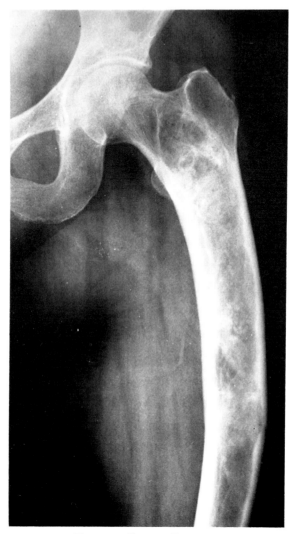

FIG. 1132. FIBROUS DYSPLASIA
Extensive involvement of the femur with mixed radiolucent and dense lesions. Bowing of the femur partially due to an old fracture at the midportion.

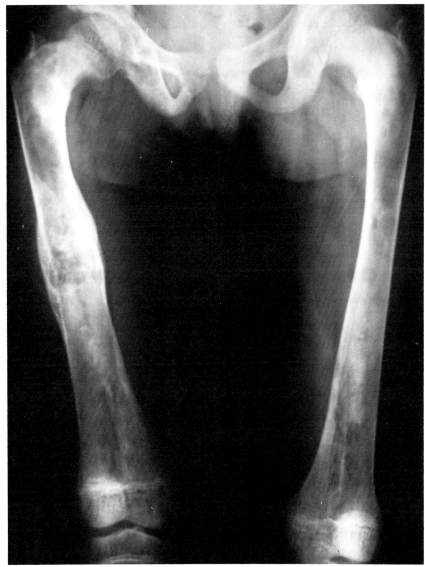

Fig. 1133. Fibrous Dysplasia
Lesions in both femurs. Shortening of the right femur. An old fracture in the midportion of the right femur.

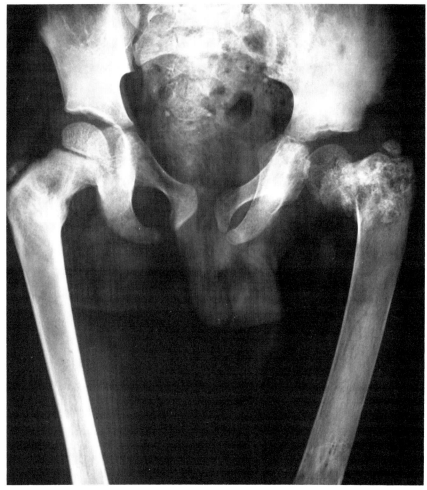

FIG. 1134. FIBROUS DYSPLASIA
Fracture of the left femoral neck which is healed. The femoral shaft is replaced by a ground glass radiopacity with thinning of the cortex. This is characteristic of the condition. Some changes are present in the left iliac wing.

ing the disease from general metabolic disturbances.

Skull lesions may often be confused with those of Paget disease (Fig. 1135). Tangential projections reveal expansion of the outer table of the calvaria with the inner table but minimally affected (Figs. 1136 and 1137). The base of the skull is affected and presents a sclerotic "ground glass" appearance. Facial lesions commonly extend into the paranasal sinuses, which may become filled with an expanding mass of considerable density (Fig. 1138). Fibrous dysplasia and meningioma en plaque may be confused, since both may produce hyperostosis of the frontal and sphenoid bones, exophthalmos, and optic nerve involvement.[31–33] However, cerebral arteriography reveals the absence of meningoarterial supply in fibrous dysplasia, distinguishing it from meningioma.[18, 33]

The lesions of monostotic and polyostotic fibrous dysplasia differ only in the number of bones they involve and should be considered as part of the same disease process unless further study proves otherwise. Rarely will malignant degeneration occur (Fig. 1139). Sarcomatous change occurs more often in the polyostotic than in the monostotic type,[34] and with or without irradiation.[26] Sarcoma, usually osteosarcoma or fibrosarcoma, may develop in young adults. Chondrosarcoma and giant cell tumor may occur.[26]

Hyperparathyroidism, so long confused with fibrous dysplasia, is now easily identified by its chemical changes, as well as by the roentgen features of generalized deossification of the skeleton and subperiosteal resorption of bones.

Neurofibromatosis will rarely produce osseous manifestations, except for congenital malformations and deformities, with or without contiguous neurofibromas. The rare intraosseous neurofibroma does not resemble fibrous dysplasia. Neurofibromas are not features of fibrous dysplasia

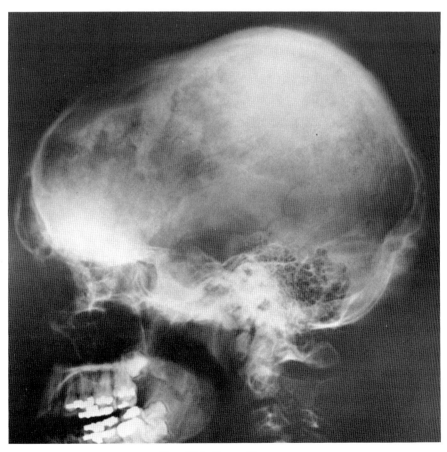

FIG. 1135. PAGET DISEASE

The increased density of the frontal bone and the parietal bones, with rounded radiolucencies between the two densities, suggest the diagnosis of Paget disease. The density in the ethmoidal and sphenoidal sinuses should suggest the proper diagnosis of fibrous dysplasia. (Courtesy of Dr. Beth Edeiken, M.D. Anderson Hospital and Tumor Institute, Houston, Texas.)

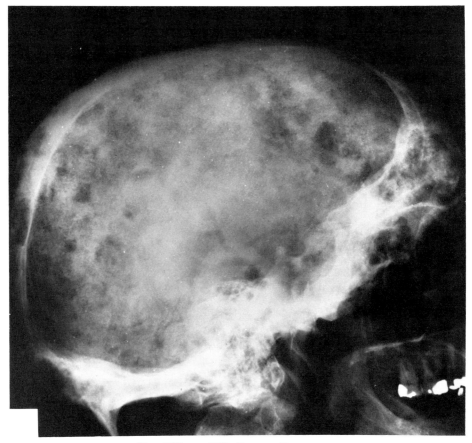

Fig. 1136 and 1137. Fibrous Dysplasia

Multiple osteolytic areas throughout the skull may be confused with metastatic disease. Usually the inner table is not eroded to the same extent as the outer table.

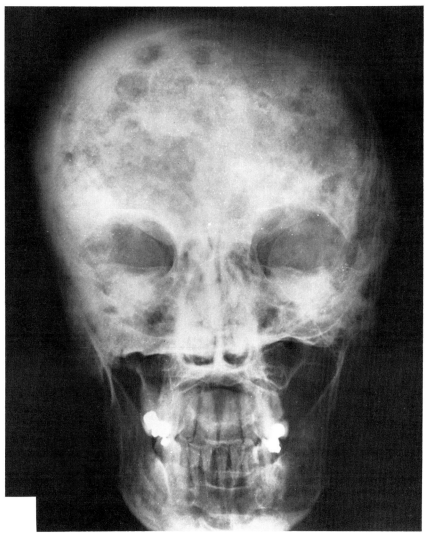

Fig. 1137

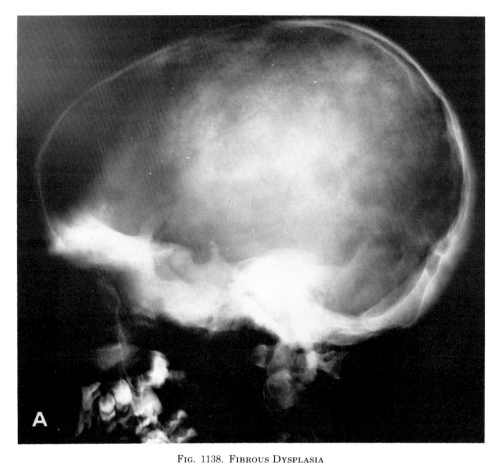

Fig. 1138. Fibrous Dysplasia

(*A* and *B*) Increased bone density in the base. The sinuses are opaque; especially the ethmoids, maxillaries and sphenoids. The frontals are not developed.

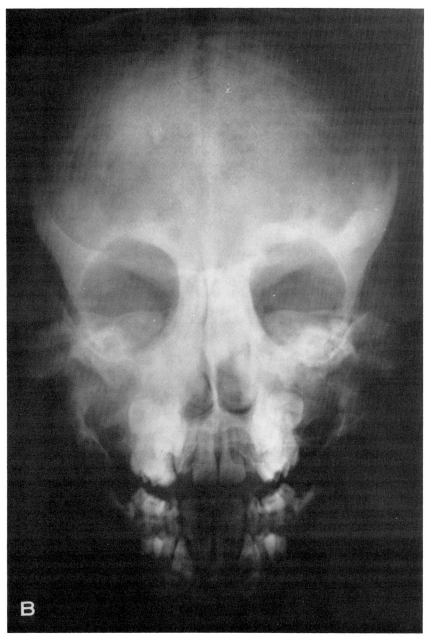

Fig. 1138B

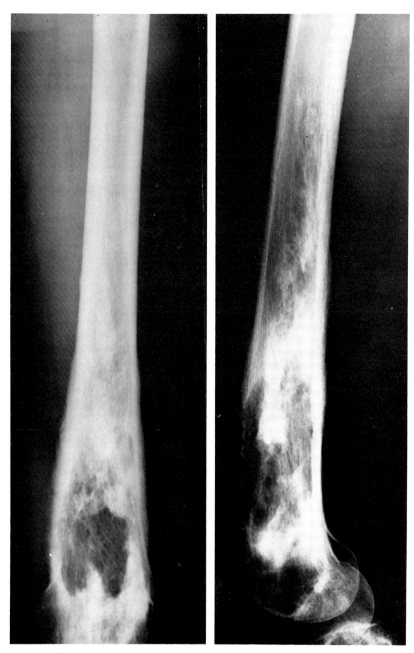

Fig. 1139. Fibrous Dysplasia with Malignant Degeneration
This patient had long standing fibrous dysplasia and the malignant tumor which developed was a mesenchymoma. (*Left*) Anteroposterior projection; (*right*) lateral projection. (Courtesy of Irwin M. Freundlich, Hahnemann Medical College and Hospital, Philadelphia, Pa.)

and their osseous lesions are usually caused by slowly growing adjacent tumors. "Cystic" bony lesions resulting from intraosseous neurofibromas are rare. The cafe-au-lait spots differ in the two diseases, and neurofibromatosis tends to be familial.[35, 36]

Paget disease may be confused histologically if insufficient tissue for biopsy is obtained, but with adequate material there should be no difficulty in distinguishing between the two. Radiographically, the distinction between cranial fibrous dysplasia and Paget disease may be quite difficult, especially in the case of monostotic disease in an adult.

Osteofibrous dysplasia[28-30] occurs almost exclusively in the tibia in infants and children below the age of 5. The histologic pattern is frequently mistaken for fibrous dysplasia, but the age and the monostotic involvement in a lesion beginning in the cortex of bone identifies it.

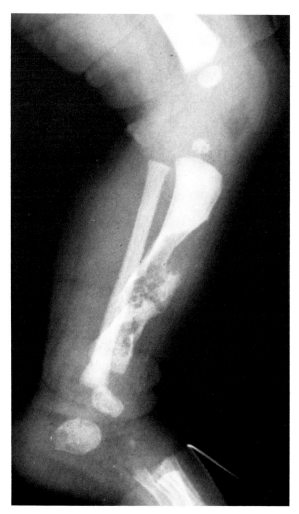

FIG. 1140. OSTEOFIBROUS DYSPLASIA WITH PSEUDOARTHROSIS

This lesion was originally believed to be a fibrous dysplasia, but it was monostotic and occurred in a newborn; review of the slides showed it to be osteofibrous dysplasia.

OSTEOFIBROUS DYSPLASIA

Osteofibrous dysplasia is a distinct clinicopathologic entity previously mistaken for fibrous dysplasia. It occurs predominantly in the tibia and is usually complicated by anterior bowing; pathologic fractures and pseudoarthroses are less frequent.

Campanacci[1-3] originally described this condition with 22 of his cases and found 17 similar cases in the literature under the following diagnoses: congenital fibrous dysplasia, congenital fibrous defect of the tibia, and ossifying fibroma.

Clinical Features. Osteofibrous dysplasia occurs in newborns or infants up to the age of 5 years. There is a slight male predominance.

The first sign is enlargement of the tibia and/or anterior bowing. A pathologic fracture, present in 25% of patients, will cause pain. The lesion rarely progresses after the age of 5 years. Because repeated recurrences take place after excision and resection in one-half the cases, particularly in the younger age groups, surgery should be avoided before the age of 5, and preferably before 10, except when fracture is imminent or actual fracture or pseudoarthrosis has occurred. However, without resection the fractures will usually heal with immobilization. The lesions will heal spontaneously in one-third of the cases, and recur during infancy in one-half. Correction of the curvature should be delayed until the age of 10–12 years.

Pathologic Features. The characteristic features are fibrous tissue surrounding bone trabeculae lined by osteoblasts, and "zonal" architecture. The fibrous tissue is less cellular than in fibrous dysplasia, and may have a whirled storiform pattern, similar to histiocytic fibroma. The fibroblasts are well differentiated.

Giant cells are often present, possibly related to microhemorrhages and bone resorption.

The zonal architecture progresses from the center of the lesion, which contains immature trabeculae that are sparse, small, and woven, and increase in size to large lamellar-type bone trabeculae that finally merge with the bone cortex.

Radiographic Features. Campanacci states that it is usually confined to the tibia, but that the ipsilateral fibula is affected in 20% of patients.[1, 18, 19] There may be bilateral involvement of both tibias and fibulas. One patient developed lesions in one tibia, the heads of two metatarsals, and the terminal phalanx of the first toe on the same side. There are no histologically proven cases in other bones.

The mid-diaphysis of the tibia is the site in one-half of the patients. The lesion may extend to both ends of the bone, involving the metaphyses and at times the epiphysis. The remainder of the lesions in the tibia are equally divided between the proximal and distal thirds of the diaphysis with occasional multiple separate foci throughout. Anterior curvature of the tibia is usually present (Fig. 1141).

The lesion begins in the cortex, usually at the anterior aspect (Fig. 1142), and in the tibia does not involve the entire bone circumference, although it may in the fibula and smaller bones. The cortex is usually thin or invisible and the periosteal aspect is "expanded." The extension into the medullary cavity is defined by a sclerotic margin resembling that of non-osteogenic fibroma or chondromyxoid fibroma. There may be a single small lesion, or multiple coalesced bubbly lesions, or multiple separate osteolytic lesions. They may have a ground glass appearance.

Pathologic fractures, usually incomplete, occur in 25%, and will heal with immobilization. Pseu-

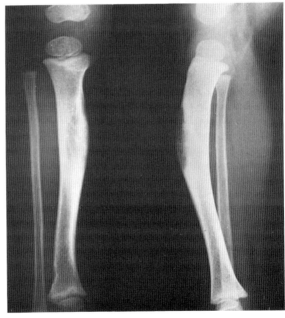

FIG. 1141. OSTEOFIBROUS DYSPLASIA
Swelling of the tibia at 18 months noted by mother. The radiograph was obtained at age 26 months, showing intracortical osteolysis with swelling and anterior bowing. (Courtesy of Dr. M. Campanacci, Bologna, Italy. Previously published in the *Italian Journal of Orthopaedics and Traumatology*, 2: 221, 1976.[1])

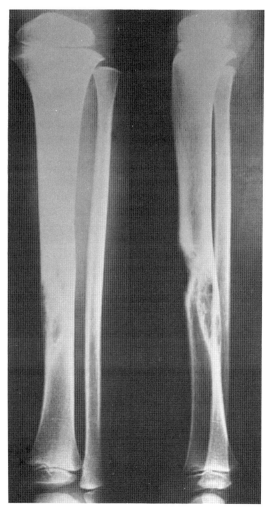

FIG. 1142. OSTEOFIBROUS DYSPLASIA
Swelling in the tibia had been noted before the age of 2 months; there is a double area of intracortical osteolysis. (Courtesy of Dr. M. Campanacci, Bologna, Italy. Previously published in the *Italian Journal of Orthopaedics and Traumatology*, 2: 221, 1976.[1])

doarthrosis is infrequent, affecting the tibia and/or the fibula,[1, 17] and will heal with fixation and/or bone grafting (Fig. 1140).

The lesion does not usually progress but may spread rapidly in infants and then become static. In older children and adults the appearance closely simulates Paget disease with widening of the bone and anterior curvature (Fig. 1143). Small lesions may regress spontaneously.

Osteofibrous dysplasia is almost invariably mistaken for fibrous dysplasia, but its distinctive features serve to identify it (Table 25).[1]

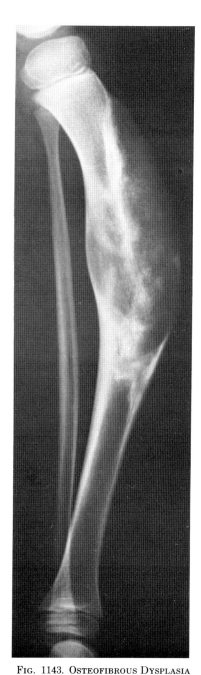

FIG. 1143. OSTEOFIBROUS DYSPLASIA
A bubbly expanding lesion with bowing of the tibia and involvement of the posterior and lateral cortex. This child had had swelling of the tibia from 15 months of age, and had previous surgery with recurrence of the osteolysis. (Courtesy of Dr. M. Campanacci, Bologna, Italy. Previously published in the *Italian Journal of Orthopaedics and Traumatology*, 2: 221, 1976.[1])

TABLE 25. DIFFERENCES BETWEEN OSTEOFIBROUS DYSPLASIA OF THE LONG BONES AND FIBROUS DYSPLASIA OF BONE

Feature	Osteofibrous Dysplasia of Long Bones	Fibrous Dysplasia of Bone
Site	Almost exclusively tibia and fibula	Almost any part of the skeleton
Distribution	Tibia and sometimes fibula	Often poliostotic
Age	0–10 years	Generally over 10 years in the monostotic types
Radiographic features	Intracortical swelling	Development generally endomedullary
Associated with pseudarthrosis of the tibia	Rare	Absent
Spontaneous regression	Yes	No
Progression in infancy	Little	Marked
Tendency to recur before the age of 10 years	Moderate	High
Histological features	1. More mature connective tissue	1. Connective tissue more cellular
	2. Bone trabeculae bordered by osteoblasts	2. Bone trabeculae without osteoblasts
	3. Zonal architecture	3. No zonal architecture
	4. Merging with the cortex	4. No merging with the cortex

AMYLOIDOSIS

Amyloidosis is characterized by an accumulation and infiltration of a protein polysaccharide complex in body tissues. The primary form is distinguished from the secondary by specific staining characteristics, an absence of chronic disease, and a predilection for specific sites, such as mesenchymal tissues (Table 26).[1, 2]

The most frequently involved organs are heart (90%), gastrointestinal tract (70%), tongue (40%), spleen (40%), liver (35%), kidneys (35%), lungs (30%), and the skin and subcutaneous tissue (25%). The adrenals, lymph nodes, muscles, and peripheral nerves may be infiltrated. When the bones and periarticular structures are infiltrated, they offer a distinctive pattern. Primary amyloidosis occurs in 10–15% of patients with multiple myeloma, and a similarity between the abnormal protein patterns suggest a relationship.[3, 4] Other disorders associated with amyloidosis are listed in Table 27.[2]

The amyloid infiltrates and replaces normal tissues, causing organ enlargement and tissue atrophy. The symptoms vary according to the organ involved, but cardiac failure and macroglossia are frequent.[5, 6]

Roentgenographic Features. Bone lesions appear with amyloid deposited either in and around the joints or within the marrow spaces.[7] Amyloid infiltrating the periarticular and articular structures is usually in the synovium, joint capsule, and adjacent tendons and ligaments. A prominent soft tissue swelling results and there may be secondary invasion and erosion of articular bone, causing small radiolucencies usually without sclerotic margins. A massive invasion of the joint by amyloid tends to cause subluxation.[7-11] Osteolytic lesions and subluxations frequently occur in the proximal humerus and femoral neck.[8, 9, 11] Multiple osteolytic lesions are usually 1–2 cm in diameter. They may have a sclerotic border or may coalesce to form large osteolytic areas (Fig. 1144). Pathologic fractures may occur (Fig. 1144). Destruction of the olecranon and coronoid process is described.[8] Solitary osteolytic lesions may occur in the ribs and other bones. Vertebral collapse may result from spinal lesions.

A coarse trabecular pattern of the hand and bones may mimic sarcoid. Osteoporosis[12] may develop when there is diffuse bone marrow infiltration.

With multiple myeloma, the tumorous bone lesions may contain amyloid,[13] but myeloma cells predominate within the mass.[12, 14-16] When multiple myeloma is associated with amyloidosis, it is generally only the myelomatous bone lesions that contain amyloid.[17] The amyloid may not cause osseous lesions, but may cause multiple joint swellings and bone erosion simulating rheumatoid arthritis[7, 18] (Fig. 1145).

TABLE 26. A CLINICAL CLASSIFICATION OF AMYLOIDOSIS

1. *Primary amyloidosis* (with no apparent predisposing cause)
   a. Sporadic with atypical distribution
   b. Sporadic with typical distribution
   c. Tumor forming amyloid
2. *Heredofamilial amyloidoses*
   a. Familial amyloid polyneuropathy (various types)
   b. Amyloidosis with familial Mediterranean fever
   c. Familial amyloid cardiopathy
   d. Familial amyloid nephropathy
   e. Familial cutaneous amyloid
   f. Familial medullary thyroid carcinoma with amyloid
3. *Secondary amyloidosis* (with associated inflammatory, neoplastic, and other diseases)
4. *Amyloidosis with lymphoproliferative disorders*
   a. Plasma cell (multiple) myeloma
   b. Waldenstrom macroglobulinemia
   c. Heavy chain disease
5. *Amyloidosis of aging*
   a. Senile cardiac amyloid
   b. Other senile amyloidoses

TABLE 27. DISORDERS ASSOCIATED WITH AMYLOIDOSIS

1. *Chronic inflammatory disease*
   a. Tuberculosis
   b. Leprosy
   c. Syphilis
   d. Schistosomiasis
   e. Osteomyelitis
   f. Bronchiectasis
   g. Pyelonephritis
   h. Reiter disease (rarely)
   i. Whipple disease (rarely)
   j. Regional enteritis
   k. Ulcerative colitis
   l. Chronic cholecystitis
2. *Dermatoses*
   a. Hidradenitis suppurativa
   b. Stasis ulcer
   c. Psoriatic arthritis
   d. Dystrophic epidermolysis bullosa
3. *Connective tissue disorders*
   a. Rheumatoid arthritis
   b. Dermatomyositis
   c. Scleroderma
   d. Disseminated lupus erythematosus
4. *Paraplegia*
5. *Diabetes mellitus*
6. *Plasma cell dyscrasias and neoplasias*
   a. Plasma cell (multiple) myeloma
   b. Waldenstrom macroglobulinemia
   c. Heavy chain (Franklin) disease
7. *Other neoplasms*
   a. Hodgkin disease
   b. Renal cell carcinoma
   c. Medullary thyroid carcinoma
   d. Calcifying odontogenic tumor of Pindborg
   e. Others

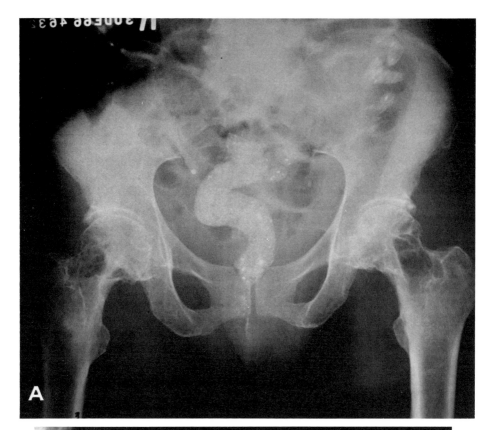
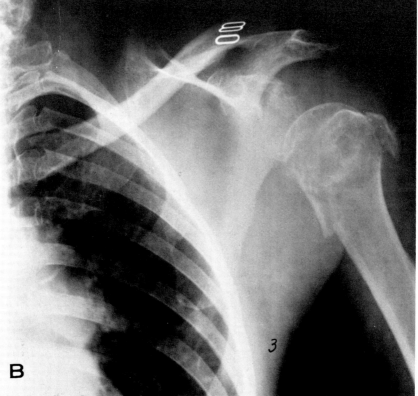

Fig. 1144. Primary Amyloidosis

(A) Pelvis; multiple osteolytic lesions with sclerotic margins of the femoral necks. (B) Humerus; pathologic fracture-dislocation of the humerus. Osteolytic lesions with sclerotic margins are seen within the head of the humerus. (Courtesy of Dr. Raymond Kilcoyne, Milwaukee, Wis.)
Fig. 1144B

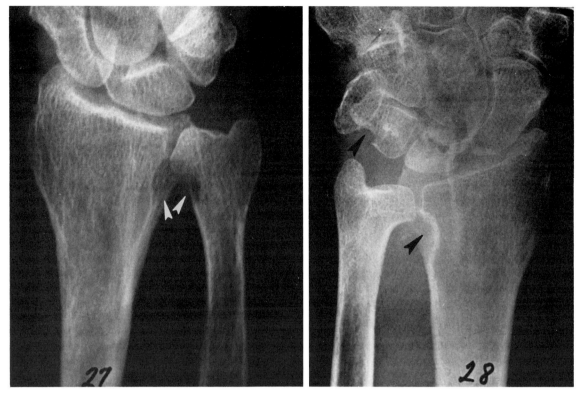

Fig. 1145. Primary Amyloidosis
Considerable synovial thickening and erosion of the triquetrum and the radius due to synovial deposits and secondary bone erosion. (Courtesy of Dr. Bernard Ostrum, Einstein Medical Center, Philadelphia, Pa.)

OXALOSIS
(CALCIUM OXALOSIS)

This rare disorder is caused by an inborn error of metabolism and is characterized by deposits of oxalite crystals in the kidneys with progressive renal failure. Widespread deposits of oxalite crystals appear throughout the body, including the bones. Excessive amounts of oxalic acid are formed and combine with calcium to form opaque calcium oxalite.[1, 2]

Oxalosis may occur secondary to either chronic renal failure[3-7] or excessive oxoid intake. Regular hemodialysis treatments appear to decrease the incidence of secondary oxalosis.[8, 9]

Roentgenographic Features. There are cystic rarefactions with sclerotic margins in multiple tubular bones, the bones of the hand and feet, and the long bones. The lesions occur in the metaphysis (Fig. 1146), but may extend throughout the diaphysis. Cystic changes may not be present.[10] Erosions appear on the concave sides of the metaphyses near the epiphyses, suggesting the changes which occur in hyperparathyroidism. There is a "bone-within-a-bone" appearance of the spine. Extensive kidney calcification may also be present.

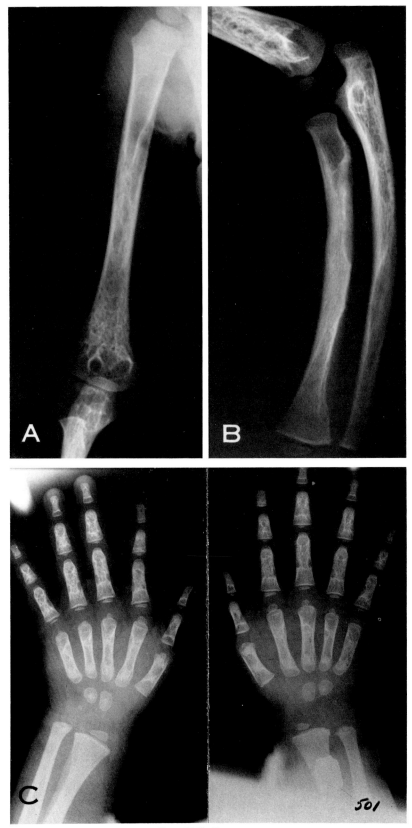

Fig. 1146. Oxalosis

Multiple metaphyseal rarefied areas with sclerotic margins. Most extend into the diaphysis. Subperiosteal erosions are present at the upper ends of the femora and tibiae, and the distal end of the radius. The spine has a bone-within-a-bone appearance. (A) Humerus, (B) elbow and forearm, (C) hands, (D) spine and pelvis, (E) lateral spine, (F) femurs, (G) legs, and (H) feet.

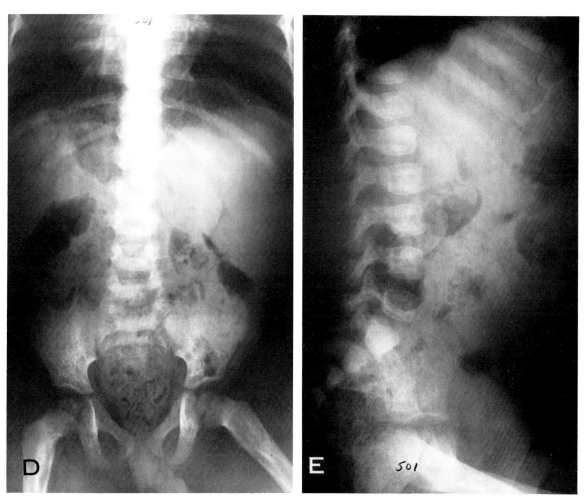

Fig. 1146 (*D* and *E*)

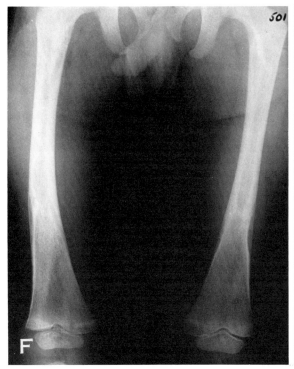
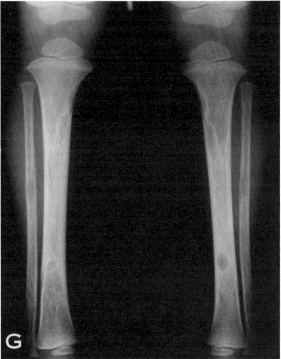
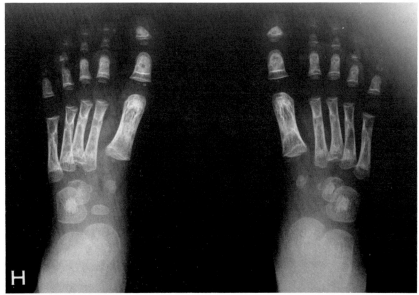

Fig. 1146 (*F–H*)

REFERENCES

Introduction and Osteoporosis

1. Shapiro, R.: Metabolic bone disease—a basic review. Clin Radiol, *13:* 238, 1962.
2. Snure, H., and Maner, G. D.: Roentgen-ray evidence of metastatic malignancy in bone. Radiology, *28:* 172, 1937.
3. Wagoner, G. W., Hunt, A. D., Jr., and Pendergrass, E. P.: A study of the relative importance of the cortex and spongiosa in the production of the roentgenogram of the normal vertebral body. AJR, *53:* 40, 1945.
4. Reifenstein, E. C., Jr.: Definitions, terminology and classification of metabolic bone disorders. Clin Orthop, *9:* 30, 1957.
5. Jowsey, J., Kelly, P. J., Reggs, B. L., Bianco, C. J., Scholy, D. A., and Gerdon-Cohen, J.: Quantitative microradiographic studies of normal and osteoporotic bone. J Bone

Joint Surg, *47A:* 785, 1965.
6. Meema, H. E., and Meema, S.: Cortical bone mineral density versus cortical thickness in the diagnosis of osteoporosis; roentgenologic-densitometric study. J Am Geriatr Soc, *17:* 1969.
7. Garn, S. M., Rohmann, C. G., and Wagner, B.: Bone loss as a general phenomenon in man. Fed Proc, *26:* 1729, 1967.
8. Doyle, F. H.: Radiologic assessment of endocrine effects on bone. Radiol Clin North Am, *5:* 289, 1967.
9. Cameron, J. R., and Sorenson, J.: Measurement of bone mineral in vivo; an improved method. Science, *142:* 230, 1963.
10. Albright, F., and Reifenstein, E. C.: *The Parathyroid Glands and Metabolic Bone Disease.* Williams & Wilkins, Baltimore, 1948.
11. Reifenstein, E. C., and Albright, F.: The metabolic effects of steroid hormones in osteoporosis. J Clin Invest, *26:* 24, 1947.
12. Saville, P. D.: Observations on 80 women with osteoporotic spine fractures. In *Osteoporosis,* edited by U. S. Barzel. Grune & Stratton, New York, 1970.
13. Deitrick, J. E., Whedon, G. D., Shorr, E., and Barr, D. P.: Effects of bed rest and immobilization upon various physiological and chemical functions of normal men. Conference on metabolic aspects of convalescence including bone and wound healing. Transactions of the ninth meeting, February 2-3, 1945, pp. 62-81, distributed by Josiah Macy, Jr., Foundation, New York.
14. Williams, R. H., and Morgan, H. J.: Thyrotoxic osteoporosis. Int Clin. *2:* 48, 1940.
15. Griffith, G. C., Nichols, G., Jr., Asher, J. D., and Flanagan, B.: Heparin osteoporosis. JAMA, *193:* 85, 1965.
16. Goldhaber, P.: Heparin enhancement of factors stimulating bone resorption in tissue culture (abstract). Science, *147:* 407, 1965.
17. Larner, J.: Inborn errors of metabolism. Ann Rev Biochem, *31:* 569, 1962.
18. Poppel, M. H., Gruber, W. F., Silber, R., Holder, A. K., and Christmas, R. O.: Roentgen manifestations of urticaria pigmentosa (mastocytosis). AJR, *82:* 239, 1959.
19. Steinbach, H. L.: The roentgen appearance of osteoporosis. Radiol Clin North Am, *2:* 191, 1964.

Regional Migratory Osteoporosis

1. Duncan, H., Frame, B., Frost, H. M., and Arnstein. A. R.: Migratory osteolysis of the lower extremities. Ann Intern Med, *66:* 1165, 1967.
2. Gupta, R. C., Popovtzer, M. M., Huffer, W. E., and Smyth, C. J.: Regional migratory osteoporosis. Arthritis Rheum, *16:* 363, 1973.
3. Lequesne, M.: Transient osteoporosis of the hip; a nontraumatic variety of Sudeck's atrophy. Ann Rheum Dis, *27:* 463, 1968.
4. Swezey, R. L.: Transient osteoporosis of the hip, foot and knee. Arthritis Rheum, *13:* 858, 1970.
5. DeMarchi, E., Santacroce, A., and Solarino, G. B.: Su di una peculiare artropatia rarefacente dell'anca. Arch Putti Chir Organi Mov, *21:* 62, 1966.
6. Renier, J. C.: Les algodystrophies du membre inferieur et leur traitement. Rev Patricien, *8:* 3835, 1958.
7. Lejeune, E., Bouvier, M., Maitrepierre, J., et al.: Le pied decalcifie douloureux ou algodystrophie reflexe du pied. Rheumatologie, *18:* 377, 1966.
8. Duncan, H., Frame, B., Frost, H., et al.: Regional migratory osteoporosis. South Med J, *62:* 41, 1969.
9. Steiner, R. M., and McKeever, C.: Regional migratory osteoporosis. J Assoc Can Radiol, *24:* 70, 1973.
10. Lequesne, M. G.: Personal communication, 1979.

Transitory Demineralization of Femoral Head

1. Rosen, R. A.: Transitory demineralization of the femoral head. Radiology, *94:* 509, 1970.

Jaundice Osteoporosis

1. Atkinson, M., Norden, B. E., and Sherlock, S.: Malabsorption and bone disease in prolonged obstructive jaundice. Q J Med, *25:* 299, 1956.

Sudeck Atrophy

1. Sudeck, P.: Ueber die akute (trophoneurotische) Knochenatrophie nach Entzundungen und Traumen der Extremitaten. Dtsch Med Wochenschr, *28:* 336, 1902.
2. Herrmann, L. G., Reineke, H. G., and Caldwell, J. A.: Post-traumatic painful osteoporosis; clinical and roentgenological entity. AJR, *47:* 353, 1942.
3. DeLorimier, A. A., Minear, W. L., and Boyd, H. B.: Reflex hyperemic deossifications regional to joints of the extremities. Radiology, *46:* 227, 1946.

Idiopathic Juvenile Osteoporosis

1. Houang, M. T. W., Brenton, D. P., Renton, P., and Shaw, D. G.: Idiopathic juvenile osteoporosis. Skeletal Radiol, *3:* 17, 1978.
2. Berlung, G., and Lindquist, B.: Osteopenia in adolescence. Clin Orthop, *17:* 259, 1960.
3. Dent, C. E., and Friedman, M.: Idiopathic juvenile osteoporosis. Q J Med, *34:* 177, 1965.
4. Schippers, J. C.: Spontaneous generalized osteoporosis in a girl ten years old. Maandschr Kindergeneeskd, *8:* 108, 1938.
5. Jowsey, J., and Johnson, K. A.: Juvenile osteoporosis; bone findings in seven patients. J Pediatr, *81:* 511, 1972.
6. Exton-Smith, A. N., Millard, P. H., Payne, T. R., and Wheeler, E.: Method for measuring quantity of bone. Lancet, *2:* 1153, 1969.

Infantile Scurvy (Barlow Disease)

1. Grewar, D.: Infantile scurvy. Clin Pediatr (Phila), *4:* 82, 1965.
2. Robertson, W., and Van, B.: The biochemical role of ascorbic acid in connective tissue. Ann NY Acad Sci, *92:* 159, 1961.
3. May, C. D., et al.: Experimental megaloblastic anemia and scurvy in the monkey. J Nutr, *49:* 121, 1953.
4. Moore, C. V.: Importance of nutritional factors in pathogenesis of iron-deficiency anemia. Am J Clin Nutr, *3:* 3, 1955.
5. Mazur, A.: Role of ascorbic acid in the incorporation of plasma iron to ferritin. Ann NY Acad Sci, *92:* 223, 1961.
6. Mazur, A.: Scurvy and blood coagulation. Nutr Rev, *18:* 242, 1960.
7. Lee, R. E.: Ascorbic acid and the peripheral vascular system. Ann NY Acad Sci, *92:* 295, 1961.
8. LoPresti, J. M., et al.: Grand rounds; scurvy. Clin Proc Child Hosp, *20:* 119, 1964.
9. Burus, R. R.: The unusual occurrence of scurvy in an 8-week-old infant. AJR, *89:* 923, 1963.
10. Shorhe, H. B.: Infantile scurvy. Clin Orthop, *1:* 49, 1963.
11. Follis, R. H., Parks, E. A., and Jacobson, D.: The prevalence of scurvy at autopsy during the first 2 years of age. Am J Dis Child, *87:* 569, 1950.
12. McIntosh, R.: Infantile scurvy. In *Brennemann's Practice of Pediatrics,* Vol. 1, Ch. 35. W. T. Prior Co., Hagerstown, Md., 1973.
13. McCann, P.: The incidence and value of radiological signs in scurvy. Br J Radiol, *35:* 683, 1962.

14. Brailsford, J. F.: *The Radiology of Bones and Joints*, Ed. 5, p. 647. Williams & Wilkins, Baltimore, 1953.
15. MacLean, A. D.: Spinal changes in a case of infantile scurvy. Br J Radiol, *41:* 385, 1968.

Osteomalacia

1. Frost, H. M.: *Bone Remodelling Dynamics.* Charles C Thomas, Springfield, Ill., 1963.
2. Albright, F., Burnett, C. H., Parson, W., Reifenstein, E. C., Jr., and Roos, A.: Osteomalacia and late rickets; the various etiologies met in the United States, with emphasis on that resulting from a specific form of renal acidosis, the therapeutic indications for each etiological subgroup, and the relationship between osteomalacia and Milkman's syndrome. Medicine, *25:* 399, 1946.
3. Hannon, R. R., Liu, S. H., Chu, H. L., Wang, S. H., Chen, J. C., and Chou, S. J.: Calcium and phosphorus metabolism in osteomalacia; I. The effect of vitamin D, and its apparent duration. Chin Med J, *48:* 623, 1934.
4. Albright, F., Butler, A. M., and Bloomberg, E.: Rickets resistant to vitamin D therapy. Am J Dis Child., *54:* 529, 1937.
5. Harris, W. H., and Heaney, R. P.: Skeletal renewal and metabolic bone disease. N Engl J Med, *280:* 193, 1969.
6. Arnstein, A. R., Frame, B., and Frost, H. M.: Recent progress in osteomalacia and rickets. Ann Int Med, *67:* 1296, 1967.
7. Looser, E.: Uber pathologische Formen von Infraktionen und Callusbildungen bei Rachitis und Osteodmalakie und anderen Knockenerkrankungen. Zentralbl Chir, *47:* 1470, 1920.
8. Milkman, L. A.: Pseudofractures (hunger osteopathy, late rickets, osteomalacia); report of case. AJR, *24:* 29, 1930.
9. Milkman, L. A.: Multiple spontaneous idiopathic symmetrical fractures. AJR, *32:* 622, 1934.
10. Edeiken, L., and Schneeberg, N. G.: Multiple spontaneous idiopathic symmetrical fractures (Milkman's syndrome). JAMA, *122:* 865, 1943.
11. Fulkerson, J. P., and Ozonoff, M. B.: Multiple symmetrical fractures of bone of unresolved etiology. AJR, *129:* 313, 1977.
12. Fleisch, H.: Role of nucleation and inhibition of calcification. Clin Orthop, *32:* 170, 1964.

Rickets or Vitamin D Deficiency

1. Mellanby, E.: *Experimental Rickets.* H. M. Stationary Office (Privy Council Medical Research Council, Special Reports Series No. 61), London, 1921.
2. McCollum, E. V., Simmonds, N., Becker, J. E., and Shipley, P. G.: Studies on experimental rickets; XXI. An experimental demonstration of the existence of a vitamin which promotes calcium deposition. J Biol Chem, *53:* 293, 1922.
3. Albright, F., and Reifenstein, E. C.: *The Parathyroid Glands and Metabolic Bone Disease.* Williams & Wilkins, Baltimore, 1948.
4. Eek, S., Gabrielsen, L. H., and Halvorsen, S.: Prematurity and rickets. Pediatrics, *20:* 63, 1957.
5. Caffey, J.: *Pediatric X-ray Diagnosis.* Year Book Medical Publishers, Chicago, 1961.
6. Schmorl, G.: Die pathologische Anatomie der rachitischen Knochenerkrankung, etc. Ergeb Inn Med Kinderheilkd, *4:* 403, 1909.
7. Howland, J., and Kramer, B.: Calcium and phosphorus in the serum in relation to rickets. Am J Dis Child, *22:* 105, 1921.
8. Hess, A. F., and Unger, L. J.: Infantile rickets; significance of clinical, radiographic and chemical examinations in its diagnosis and incidence. Am J Dis Child, *24:* 327, 1922.
9. Park, E. A.: Observations on the pathology of rickets with particular reference to the changes at the cartilage-shaft junctions of growing bones. Harvey Lect, *34:* 157, 1938-1939.
10. Pommer, G.: *Untersuchungen uber Osteomalacie and Rachitis.* F. C. W. Vogel, Leipzig, 1885.
11. Birtwell, W. M., Magsamen, B. F., Fenn, P. A., Torg, J. S., Tourtellotte, C. D., and Martin, J. H.: An unusual hereditary osteomalacic disease—pseudo-vitamin-D deficiency. J Bone Joint Surg., *52A:* 1222, 1970.

Steatorrhea

1. Albright, F., Butler, A. M., and Bloomberg, E.: Rickets resistant to vitamin D therapy. Am J Dis Child., *54:* 529, 1937.
2. Bauer, W., Marble, A., and Claflin, D.: Studies on the mode of action of irradiated ergosterol; I. Its effects on the calcium, phosphorus and nitrogen metabolism of normal individuals. J Clin Invest, *11:* 1, 1932.

Fibrogenesis Imperfecta

1. Baker, S. L., Dent, C. E., Friedman, M., and Watson, L.: Fibrogenesis imperfecta ossium. J Bone Joint Surg, *48B:* 804, 1966.
2. Baker, S. L.: Fibrogenesis imperfecta ossium. J Bone Joint Surg, *38B:* 378, 1956.
3. Baker, S. L., and Turnbull, H. M.: Two cases of a hitherto undescribed disease characterized by a gross defect in the collagen of the bone matrix. J Pathol Bacteriol, *62:* 132, 1950.
4. Golding, F. C.: Fibrogenesis imperfecta. J Bone Joint Surg, *50B:* 619, 1968.

Hypophosphatasia

1. Rathbun, J. C.: Hypophosphatasia, a new developmental anomaly. J Dis Child, *75:* 822, 1948.
2. Fraser, D., Vendt, E. R., and Christie, F. H.: Metabolic abnormalities in hypophosphatasia. Lancet, *1:* 286, 1955.
3. McCance, R. A.: The excretion of phosphoethanolamine and hypophosphatasia. Lancet, *1:* 131, 1955.
4. Weingarten, R. J.: Tropical eosinophilia. Lancet, *1:* 103, 1943.
5. Beaver, P. C., and Danarj, T. J.: Ascariasis associated with Löffler's syndrome. Am J Trop Med, *7:* 100, 1958.
6. Liebow, A. A.: Hypersensitivity Disease of the Lung. Read at the Thomas Dent Mutter Annual Lecture, Philadelphia, February 2, 1966.
7. Spencer, H.: Allergic granuloma of lung. In *Pathology of the Lung*, pp. 576-578. Macmillan, New York, 1962.
8. Klinghoffer, J. F.: Löffler's syndrome following the use of a vaginal cream. Ann Intern Med, *40:* 343, 1954.
9. Reeder, W. H., and Goodrich, E. B.: Pulmonary infiltration with eosinophilia (PIE syndrome). Ann Intern Med, *36:* 1217, 1952.
10. Buckley, J. J.: Microfilariasis causing tropical eosinophilia. Trans R Soc Trop Med Hyg, *52:* 335, 1958.
11. Birtwell, W. M., Riggs, B. L., Peterson, L. F. A., and Jones, J. D.: Hypophosphatasia in an adult. Arch Intern Med, *120:* 1967.
12. Fraser, D.: Hypophosphatasia. Am J Med, *22:* 730, 1957.
13. Currarino, G., Neuhauser, E. B. D., Reyersbach, G. C., and Sobel, E. H.: Hypophosphatasia. AJR, *78:* 392, 1957.
14. Anspach, W. E., and Clifton, W. M.: Hyperparathyroidism in children. Am J Dis Child, *58:* 540, 1939.
15. Fisher, O. D., and Neill, D. W.: Excretion of ethanolamine

phosphoric acid in coeliac disease. Lancet, 1: 334, 1955.
16. Dent, C. E.: Discussion on surgical aspects of disordered calcium metabolism. Proc R Soc Med, 49: 715, 1956.
17. McCance, R. A., Fairweather, D. V. I., Barrett, A. M., and Morrison, A. B.: Genetic, clinical, biochemical and pathological features of hypophosphatasia. Q J Med, 25: 523, 1956.
18. Park, E. A.: Discussion, quoted by E. H. Sobel, L. C. Clark, and M. Robinow, Am J Dis Child, 83: 411, 1952.

Hyperparathyroidism

1. Sipple, J. H.: The association of pheochromocytoma with carcinoma of the thyroid gland. Am J Med, 31: 163, 1961.
2. Williams, E. D.: A review of 17 cases of carcinoma of the thyroid and pheochromocytoma. J Clin Pathol, 18: 288, 1965.
3. Ljungberg, O., Cederquist, E., and von Studnitz, W.: Medullary thyroid carcinoma with pheochromocytoma; a familial chromaffinomatosis. Br Med J, 1: 279, 1967.
4. Sarosi, G., and Doe, R. P.: Familial occurrence of parathyroid adenomas, pheochromocytoma, and medullary carcinoma of the thyroid with amyloid stroma (Sipple's syndrome). Ann Intern Med, 68: 1305, 1968.
5. Schmike, R. N., et al.: Syndrome of bilateral pheochromocytoma, medullary thyroid carcinoma and multiple neuromas. N Engl J Med, 279: 1, 1968.
6. Paloyan, E., Scann, A., Straus, F. H., Pickleman, J. R., and Paloyan, D.: Familial pheochromocytoma, medullary thyroid carcinoma, and parathyroid adenomas. JAMA, 214: 1443, 1970.
7. Schimke, R. N., and Hartmann, W. H.: Familial amyloid-producing medullary thyroid carcinoma and pheochromocytoma. Ann Intern Med, 63: 1027, 1965.
8. Albright, F., and Reifenstein, E. C.: *The Parathyroid Glands and Metabolic Bone Disease.* Williams & Wilkins, Baltimore, 1948.
9. Lloyd, P. C.: Case of hypophyseal tumor with associated tumorlike enlargement of the parathyroids and islands of Langerhans. Bull Johns Hopkins Hosp, 45: 1, 1929.
10. Shelburne, S. A., and McLaughlin, C. W.: Coincidental adenomas of islet-cells, parathyroid gland and pituitary gland. J Clin Endocrinol, 5: 232, 1945.
11. Castleman, B., and Mallory, T. B.: The pathology of the parathyroid gland in hyperparathyroidism; study of 25 cases. Am J Pathol, 11: 1, 1935.
12. Albright, F., Aub, J. C., and Bauer, W.: Hyperparathyroidism; common and polymorphic condition as illustrated by 17 proven cases from one clinic. JAMA, 102: 1276, 1934.
13. Hellstrom, J.: Experience from 105 cases of hyperparathyroidism. Acta Chir Scand, 113: 501, 1957.
14. Walsh, F. B., and Howard, J. E.: Conjunctival and corneal lesions in hypercalcemia. J Clin Endocrinol, 7: 644, 1947.
15. Tepperman, H. M., L'Heaureux, M. V., and Wilhelmi, A. E.: The estimation of parathyroid hormone activity by its effect on serum inorganic phosphorus in the rat. J Biol Chem, 168: 151, 1947.
16. Munson, P. L.: Studies on role of parathyroids in calcium and phosphorus metabolism. Ann NY Acad Sci, 60: 776, 1955.
17. McLean, F. C., and Hastings, A. B.: Clinical estimation and significance of calcium-ion concentration in the blood. Am J Med Sci, 189: 601, 1935.
18. Luck, V. J.: *Bone and Joint Diseases,* pp. 311–327. Charles C Thomas, Springfield, Ill., 1950.
19. Pugh, D. G.: Roentgenologic diagnosis of diseases of bone. In R. Golden, *Diagnostic Roentgenology, Vol. II.* Williams & Wilkins, Baltimore, 1952.
20. Steinbach, H. L., Gordon, G. S., Eisenberg, E., Crane, J. T., Silverman, S., and Goldman, L.: Primary hyperparathyroidism; a correlation of roentgen, clinical, and pathologic features. AJR, 86: 329, 1961.
21. Weiss, A.: Incidence of subperiosteal resorption in hyperparathyroidism studied by fine detail bone radiography. Clin Radiol, 25: 273, 1974.
22. Meema, H. E., and Schatz, D. L.: Simple radiologic demonstration of cortical bone loss in thyrotoxicosis. Radiology, 97: 9, 1970.
23. Resnick, D. L.: Erosive arthritis of the hand and wrist in hyperparathyroidism. Radiology, 110: 263, 1974.
24. Pugh, D. G.: Subperiosteal resorption of bone; roentgenologic manifestations of primary hyperparathyroidism and renal osteodystrophy. AJR, 66: 577, 1951.
25. Chiroff, R. T., Sears, K. A., and Slaughter, W. H.: Slipped capital femoral epiphyses and parathyroid adenoma. Case report. J Bone Joint Surg, 56A: 1063, 1974.
26. Teng, C. T., and Nathan, M. H.: Primary hyperparathyroidism. AJR, 83: 716, 1960.
27. Crawford, T., Dent, C. E., Lucas, P., Martin, N. H., and Narsim, J. R.: Osteosclerosis associated with chronic renal failure. Lancet, 2: 981, 1954.
28. Dent, C. E., and Hodson, C. J.: Radiologic changes associated with certain metabolic bone diseases. Br J Radiol, 27: 65, 1954.
29. Whitby, L. G.: Intermittent hyperparathyroidism. Lancet, 1: 883, 1958.
30. Avioli, L. V., McDonald, J. E., and Singer, R. A.: Excretion of pyrophosphate in disorders of bone metabolism. J Clin Endocrinol Metab, 25: 912, 1965.
31. Bywaters, E. G. L., Dixon, A. St.J., and Scott, J. T.: Joint lesions of hyperparathyroidism. Ann Rheum Dis, 22: 171, 1963.
32. Dodds, W. J., and Steinbach, H. I.: Primary hyperparathyroidism and articular cartilage calcification. AJR, 104: 1968.
33. Parlee, D. E., Freundlich, I. M., and McCarty, D. J.: Comparative study of roentgenographic techniques for detection of calcium pyrophosphate dihydrate deposits (pseudogout) in human cartilage. AJR, 99: 688, 1967.
34. Vix, V. A.: Articular and fibrocartilage calcification in hyperparathyroidism; associated hyperuricemia. Radiology, 83: 468, 1964.
35. Bywaters, E. G. L., Discussion of S. W. Stanbury, A. J. Popert, and J. Ball: Stimulation of rheumatic disorders by metabolic bone disease. Ann Rheum Dis, 18: 63, 1959.
36. Dodds, W. J., and Steinbach, H. I.: Triangular cartilage calcification in the wrists; its incidence in elderly persons. AJR, 105: 850, 1969.
37. Urist, M. R.: Recent advances in physiology of calcification. J Bone Joint Surg, 46A: 889, 1964.
38. McCarty, D. J., Hogan, J. M., Gatter, R. A., and Grossman, M.: Studies on pathological calcifications in human cartilage; I. Prevalence and types of crystal deposits in menisci of 215 cadavers. J Bone Joint Surg, 48A: 309, 1966.
39. Katz, A. I., Hampers, C. L., and Merrill, J. P.: Secondary hyperparathyroidism and renal osteodystrophy in chronic renal failure. Medicine, 48: 333, 1969.
40. Smith, F. W., and Junor, B. J. R.: Periarticular calcification with fluid levels in secondary hyperparathyroidism. Br J Radiol, 51: 741, 1978.
41. Hanson, D. J.: Unusual radiographic manifestations of parathyroid adenoma; report of a case. N Engl J Med, 267: 1080, 1962.
42. Hardy, J. D., Snavely, J. R., and Langford, H. G.: Low

intrathoracic parathyroid adenoma; large functioning tumor representing fifth parathyroid, opposite eighth dorsal vertebra with independent arterial supply and opacified at operation with arteriogram. Ann Surg, 159: 310, 1964.
43. Maurer, W. J., Johnson, J. R., and Mendenhall, J. T.: Mediastinal parathyroid adenoma; case report. J Thorac Cardiovasc Surg, 49: 657, 1965.
44. Polga, J. P., and Balikian, J. P.: Partially calcified functioning parathyroid adenoma; case demonstrated roentgenographically. Radiology, 99: 55, 1971.
45. Castleman, B.: Tumors of the parathyroid glands. In *Atlas of Tumor Pathology, Section IV, Fasc. 15.* Washington Registry of Pathology, Armed Forces Institute of Pathology, Washington, D. C., 1952.
46. Hanson, D. J.: Unusual radiographic manifestations of parathyroid adenoma; report of a case. N Engl J Med. 267: 1080, 1962.
47. Cope, O.: Story of hyperparathyroidism at Massachusetts General Hospital. N Engl J Med, 274: 1174, 1966.
48. Dodds, W. J., Newton, T. H., and Enloe, L. T.: Parathyroid adenoma of anterior mediastinum demonstrated by preoperative selective arteriography; report of a case. Radiology, 91: 923, 1968.
49. Black, B. M.: Problems in the treatment of hyperthyroidism. Surg Clin North Am, 41: 1061, 1961.
50. Pachter, M. R., and Lattes, R.: Uncommon mediastinal tumors; report of two parathyroid adenomas, one nonfunctional parathyroid carcinoma and one "bronchial-type-adenoma." Dis Chest, 43: 519, 1963.
51. Pyrah, L. N., Hodghinson, A., and Anderson, C. K.: Primary hyperparathyroidism. Br J Surg., 53: 245, 1966.
52. Canivet, J., et al.: Adenome parathyroidien inferieur droit localise avant l'interfention par arteriographie. Presse Med, 75: 2329, 1967.
53. Lang, E. K.: Arteriographic demonstration of parathyroid adenoma. J Indiana State Med Assoc, 60: 1656, 1967.
54. Newton, T. H., and Eisenberg, E.: Angiography of parathyroid adenomas. Radiology, 86: 843, 1966.
55. Seldinger, S. I.: Localization of parathyroid adenomata by arteriography. Acta Radiol, 42: 353, 1954.
56. Borm, D., and Wener, H.: Angiographische Lokalisation von Epithelkorperchen-adenomen. Zentralbl Chir, 89: 1537, 1964.
57. Hardy, J. D., Snavely, J. R., and Langford, H. G.: Low intrathoracic parathyroid adenoma; large functioning tumor representing fifth parathyroid, opposite eighth dorsal vertebra with independent arterial supply and opacification at operation with arteriogram. Ann Surg, 159: 310, 1964.
58. Lohr, B., and Borm, D.: Diagnostik und Lokalisation von Epithelkorperchentumoren. Zentralbl Chir, 91: 316, 1966.
59. Steiner, R. E., Fraser, E., and Aird, I.: Operative parathyroid arteriography for location of parathyroid tumor. Br Med J, 2: 400, 1956.
60. Wanke, R.: Epithelkorperchen chirurgie bei primaren hyperparathyroidismus. Chirurgie, 33: 53, 1962.
61. Doppman, J. L., Hammond, W. G., Melson, G. L., Evens, R. G., and Ketcham, A. S.: Staining of parathyroid adenomas by selective arteriography. Radiology, 92: 527, 1969.
62. Colella, A. C., and Pigorini, F.: Experience with parathyroid scintigraphy. AJR, 109: 1970.
63. Sisson, J. C., and Beierwaltes, W. H.: Radiocyanocobalamine (^{57}Co-B$_{12}$) concentration in the parathyroid glands. J Nucl Med, 3: 160, 1962.
64. Workman, J. B., and Connor, T. B.: Preoperative localization of parathyroid adenomata. J Nucl Med, 5: 372, 1964.
65. Potchen, E. J.: Isotopic labeling of rat parathyroid as demonstrated by autoradiography. J Nucl Med, 4: 480, 1963.
66. Haynie, T. P., Otte, W. K., and Wright, J. C.: Visualization of hyperfunctioning parathyroid adenoma using ^{75}Se-selenomethionine and photoscanner. J Nucl Med, 5: 710, 1964.
67. Potchen, E. J., and Sodee, D. B.: Selective isotopic labeling of human parathyroid; preliminary case report. J Clin Endocrinol, 24: 1125, 1964.
68. Buhring, H., and Prevot, J. H.: Lokalisation von Nebenschiddrusen-adenomen mit 75-Selen-L-Methionin. Nucl Med Suppl, 6: 397, 1967.
69. Colella, A. C., and Pigorini, F.: La rappresentazione scintigrafica di tessuto paratiroideo; su un caso di morbo di recklinghausen con adenoma a sede antipica. Nunt Radiol, 31: 1349, 1965.
70. Colella, A. C., and Pigorini. F.: Sulla rappresentazione scintigrafica di adenomi paratiroidei. *Proceedings of the Tenth Congress of Italian Society of Nuclear Biology and Medicine,* Pisa, p. 19, 1967.
71. Conte, N. Ziliotto, D., and Scandellari, C.: Localizazione scintigrafica delle neoplasie paratiroidee con seleniometionina-^{75}Se. Acta Isot 5: 337, 1965.
72. Garrow, J. S., and Smith, R.: Detection of parathyroid tumors by selenomethionine scanning. Br J Radiol, 41: 307, 1968.
73. Gottschalk, A., Ranninger, K., Paloyan, E., Paloyan, D., and Harper, P. V.: Bilateral intra-arterial injection in thyrocervical trunk; technique to facilitate localization of parathyroid adenomas with selenium-75-methionine. J Nucl Med, 7: 374, 1966.
74. Haubold, V., Zonntag, A., Pabst, H. W., Frey, K. W., and Karl, H. J.: Zum Problem der szintigraphischen Darstellung von Epithelkorperchenadenomen mit Hilfe von 75-Se-Selenomethionin. In *Radioistope in der Lokalisations Diagnostik,* pp. 389–395. F. K. Schattauer, Stuttgart, 1967.
75. McGeown, M. G., Bell, T. K., Soyannwo, M. A. O., Fenton, S. S. A., and Oreopoulos, D.: Parathyroid scanning in human with selenomethionine-^{75}Se. Br J Radiol, 41: 300, 1968.
76. Sack, H., Petry, R., and Duwell, H. J.: Darstellung eines Nebenschilddrusenadenoma mit 75-Selen-Methionin under der Szintillationskamera. Dtsch Med Wochenschr, 90: 25353, 1965.

Hypoparathyroidism

1. Kinard, R. E., Walton, J. E., and Buckwalter, J. A.: Pseudohypoparathyroidism; report on a family with four affected sisters. Arch Intern Med, 139: 204, 1979.
2. Mann, J. B., Alterman, S., and Hills, A. G.: Albright's hereditary osteodystrophy comprising pseudohypoparathyroidism and pseudopseudohypoparathyroidism. Ann Intern Med, 56: 315, 1962.
3. Beumer, H., and Falkenheim, C.: Idiopathische Tetanie, hamokrinin und epithelk Orperchen-hormon. Munch Med Wochenschr, 73: 818, 1926.
4. Drake, T. G., Albright, F., Bauer, W., and Castleman, B.: Chronic idiopathic hypoparathyroidism; report of 6 cases with autopsy findings in 1. Ann Intern Med, 12: 1751, 1939.
5. Albright, F., and Reifenstein, E. C.: *The Parathyroid Glands and Metabolic Bone Disease.* Williams & Wilkins, Baltimore, 1948.

6. Talbot, N. B., Butler, A. M., and MacLachlan, E. A.: The effect of testosterone and allied compounds on the mineral, nitrogen, and carbohydrate metabolism of a girl with Addison's disease. J Clin Invest, 22: 583, 1943.
7. Sutphin, A., Albright, F., and McCune, D. J.: Five cases (3 siblings) of idiopathic hypoparathyroidism associated with moniliasis. J Clin Endocrinol, 3: 625, 1943.
8. Leonard, M. F.: Chronic idiopathic hypoparathyroidism with superimposed Addison's disease in a child. J Clin Endocrinol, 6: 493, 1946.
9. Eaton, L. M., and Haines, S. F.: Symmetrical cerebral calcification associated with parathyroid insufficiency; preliminary report. Proc Staff Meet Mayo Clin, 14: 48, 1939.
10. Bronsky, D., Kushner, D. S., Dubin, A., and Snapper, I.: Idiopathic hypoparathyroidism and pseudohypoparathyroidism; case reports and review of literature. Medicine, 37: 317, 1958.
11. Taybi, H., and Keele, D.: Hypoparathyroidism; review of the literature and report of 2 cases in sisters, one with steatorrhea and intestinal pseudo-obstruction. AJR, 88: 432, 1962.
12. Achenbach, W., and Bohm, A.: Skelettveranderungen bei parathyreogenen Tetanien. Fortschr Geb Rontgenstr Nuklearmed, 79: 95, 1953.

Pseudohypoparathyroidism

1. Albright, F., Burnett, C. H., Smith, P. H., and Parson, W.: Pseudo-hypoparathyroidism, an example of "Sebright-Bantam syndrome"; report of 3 cases. Endocrinology, 30: 922, 1942.
2. Albright, F., and Reifenstein, E. C., Jr.: *The Parathyroid Glands and Metabolic Bone Disease.* Williams & Wilkins, Baltimore, 1948.
3. Costello, J. M.: Hypo-hyperparathyroidism. Arch Dis Child, 38: 397, 1963.
4. Mann, J. B., Alterman, S., and Hills, A. G.: Albright's hereditary osteodystrophy comprising pseudohypoparathyroidism and pseudo-pseudohypoparathyroidism; with a report of 2 cases representing the complete syndrome occurring in successive generations. Ann Intern Med, 56: 315, 1962.
5. Bell, N. H., Khairi, M. R. A., Johnston, C. C. et al.: Effects of 1,25-dihydroxyvitamin D_3 on calcium metabolism and quantitative bone histology in pseudohypoparathyroidism. In *Endocrinology of Calcium Metabolism: Proceedings of the Sixth Parathyroid Conference, Vancouver, Canada*, pp. 33–38. Excerpta Medica, Amsterdam, 1977.
6. Drezner, M. K., Neelon, F. A., Haussler, M., et al.: 1,25-Dihydroxycholecalciferol deficiency; the probable cause of hypocalcemia and metabolic bone disease in pseudohypoparathyroidism. J Clin Endocrinol Metab, 42: 621, 1976.
7. Sinha, T. K., Deluca, H. F., and Bell, N. H.: Evidence for a defect in the formation of $1\alpha,25$-dihydroxyvitamin D in pseudohypoparathyroidism. Metabolism, 26: 731, 1977.
8. Bell, N. H., and Stern, P. H.: Hypercalcemia and increases in serum hormone value during prolonged administration of $1\alpha,25$-dihydroxyvitamin D. N Engl J Med, 298: 1241, 1978.
9. Kolb, F. O., and Steinbach, H. L.: Syndrome of pseudohypohyperparathyroidism. Acta Endocrinol Suppl, 51: 475, 1960.
10. Kolb, F. O., and Steinbach, H. L.: Pseudohypoparathyroidism with secondary hyperparathyroidism and osteitis fibrosa. J Clin Endocrinol, 22: 59, 1962.
11. Lee, J. B., Tashjian, A. H., Streeto, J. M., and Frantz, A. G.: Familial pseudohypoparathyroidism; role of parathyroid hormone and thyrocalcitonin. N Engl J Med, 279: 1179, 1968.
12. Bergstrand, C. G., Ekengren, K., Filipsson, R., and Haggert, A.: Pseudohypoparathyroidism; familial incidence and comparison with idiopathic hypoparathyroidism. Acta Endocrinol, 29: 201, 1958.
13. Bronsky, D., Kushner, D. S., Dubin, A., and Snapper, I.: Idiopathic hypoparathyroidism and pseudohypoparathyroidism; case reports and review of the literature. Medicine, 37: 317, 1958.
14. Papaioannou, A. C., and Marsas, B. E.: Albright's hereditary osteodystrophy (without hypocalcemia) (brachymetacarpal dwarfism without tetany, or pseudo-pseudohypoparathyroidism); report of a case and review of the literature. Pediatrics, 31: 599, 1963.
15. Cohen, M. L., and Donnell, G. N.: Pseudohypoparathyroidism with hypoparathyroidism; case report and review of the literature. J Pediatr, 56: 369, 1960.
16. Steinbach, Howard L., Rudhe, U. L. F., Jonsson, M., and Young, D. A.: Evolution of skeletal lesions in pseudohypoparathyroidism. Radiology, 85: 670, 1965.
17. Taitz, L. S.: Pseudohypoparathyroidism in infancy. Arch Dis Child, 35: 506, 1960.
18. Steinbach, H. L., and Young, D. A.: The roentgen appearance of pseudohypoparathyroidism (PH) and pseudo-pseudohypoparathyroidism (PPH); differentiation from other syndromes associated with short metacarpals, metatarsals, and phalanges. Radiology, 97: 49, 1966.

Pseudo-pseudohypoparathyroidism

1. Albright, F., and Reifenstein, E. C.: *The Parathyroid Glands and Metabolic Bone Disease.* Williams & Wilkins, Baltimore, 1948.

Hypervitaminosis A

1. Josephs, H. W.: Hypervitaminosis A and carotenemia. Am J Dis Child, 67: 33, 1944.
2. Toomey, J. A., and Morisette, R. A.: Hypervitaminosis A. Am J Dis Child, 73: 473, 1947.
3. Mellanby, E.: Vitamin A and bone growth; the reversibility of vitamin A-deficiency changes. J Physiol, 105: 382, 1947.
4. Caffey, J.: *Pediatric X-ray Diagnosis.* Year Book Medical Publishers, Chicago, 1961.
5. Rothman, P. E., and Leon, E. E.: Hypervitaminosis A; report of 2 cases in infants. Radiology, 51: 368, 1948.
6. Wolbach, S. B.: Vitamin A deficiency and excess in relation to skeletal growth. J Bone Joint Surg (N.S.), 29A: 171, 1947.

Hypervitaminosis D

1. Paul, W. D.: Toxic manifestations of large doses of vitamin D as used in the treatment of arthritis. J Iowa Med Soc, 36: 141, 1946.
2. Danowski, T. S., Winkler, A. W., and Peters, J. P.: Tissue calcification and renal failure produced by massive dose vitamin D therapy of arthritis. Ann Intern Med, 23: 22, 1945.
3. Bauer, J. M., and Greyberg, R. H.: Vitamin D intoxication with metastatic calcification. JAMA, 130: 1208, 1946.
4. Covey, G. W., and Whitlock, H. M.: Intoxication resulting from the administration of massive doses of vitamin D; with report of 5 cases. Ann Intern Med, 25: 508, 1946.

Renal Osteodystrophy

1. Baker, S. L.: General softening of bone due to metabolic causes. I. Histopathology of porotic and malacic conditions of bone. Br J Radiol, 27: 604, 1954.
2. Dent, C. E.: Rickets and osteomalacia from renal tubule defects. J Bone Joint Surg, 34B: 266, 1952.
3. Kaye, M., Pritchard, J. E., Halpenny, G. W., and Light, W.: Bone disease in chronic renal failure with particular reference to osteosclerosis. Medicine, 39: 157, 1960.
4. Stanbury, S. W.: Azotaemic renal osteodystrophy. Br Med Bull, 13: 57, 1957.
5. Davies, D. R., Dent, C. E., and Willcox, A.: Hyperparathyroidism and steatorrhoea. Br Med J, 2: 1113, 1956.
6. Editorial: Renal osteodystrophy. N Engl J Med, 268: 617, 1963.
7. Stanbury, S. W., and Lumb, G. A.: Metabolic studies of renal osteodystrophy; I. Calcium phosphorus and nitrogen metabolism in rickets, osteomalacia, and hyperparathyroidism complicating chronic uremia and in osteomalacia of adult Fanconi syndrome. Medicine, 41: 1, 1962.
8. Steinbach, H. L., and Noetzli, M.: Roentgen appearance of skeleton in osteomalacia and rickets. AJR, 91: 955, 1964.
9. Tapia, J., Stearns, G., and Ponseti, I. V.: Vitamin-D resistant rickets; long term clinical study of 11 patients. J Bone Joint Surg, 46A: 935, 1964.
10. Yendt, E. R., Connor, T. B., and Howard, J. E.: In vitro calcification of rachitic rat cartilage in normal and pathological human sera with some observations on pathogenesis of renal rickets. Bull Johns Hopkins Hosp, 96: 1, 1955.
11. Fanconi, G.: Physiology and pathology of calcium and phosphate metabolism. Adv Pediatr, 12: 307, 1962.
12. Fletcher, R. F., Jones, J. H., and Morgan, D. B.: Bone disease in chronic renal failure. Q J Med, 32: 321, 1963.
13. Stanbury, S. W.: In *Renal Disease*, p. 508. D. A. Black (ed.). F. A. Davis, Philadelphia, 1962.
14. Bloomer, H. A., Canary, J. J., Kyle, L. H., and Auld, R. M.: Fanconi syndrome with renal hyperchloremic acidosis; sequential development of multiple tubular dysfunctions in child. Am J Med, 33: 141, 1962.
15. Davis, J. G.: Osseous radiographic findings of chronic renal insufficiency. Radiology, 60: 406, 1953.
16. Karani, S.: Secondary hyperparathyroidism; primary renal failure. Proc R Soc Med, 48: 527, 1955.
17. Mitchell, A. G.: Nephrosclerosis (chronic interstitial nephritis) in childhood; with special reference to renal rickets. Am J Dis Child, 40: 101, 1930.
18. Joiner, C. L., and Thorne, M. G.: On occurrence of skeletal disorders in cases of longstanding renal failure. Guys Hosp Rep, 102: 1, 1953.
19. Webster, G. D., Jr.: Azotemic renal osteodystrophy. Med Clin North Am, 47: 985, 1963.
20. Pugh, D. G.: Subperiosteal resorption of bone; roentgenologic manifestations of primary hyperparathyroidism and renal osteodystrophy. AJR, 66: 577, 1951.
21. Dent, C. E., and Hodson, C. J.: General softening of bone due to metabolic causes; II. Radiologic changes associated with certain metabolic bone diseases. Br J Radiol, 27: 605, 1954.
22. Anderson, W. W., Mann, J. B., Kenyon, N., Farrell, J. J., and Hills, A. G.: Subtotal parathyroidectomy in azotemic renal osteodystrophy. N Engl J Med, 268: 575, 1963.
23. Baird, I. M., and Lees, F.: Renal osteodystrophy in adults. Arch Intern Med, 98: 16, 1956.
24. Herbert, F. K., Miller, H. G., and Richardson, G. O.: Chronic renal disease, secondary parathyroid hyperplasia, decalcification of bone and metastatic calcification. J Pathol Bacteriol, 53: 161, 1941.
25. Peterson, R.: Small vessel calcification and its relationship to secondary hyperparathyroidism in the renal homotransplant patient. Radiology, 126: 627, 1978.
26. Katz, A. I., Hampers, C. L., and Merrill, J. P.: Secondary hyperparathyroidism and renal osteodystrophy in chronic renal failure. Medicine, 48: 333, 1969.
27. Smith, F. W., and Junor, B. J. R.: Periarticular calcification with fluid levels in secondary hyperparathyroidism. Br J Radiol, 51: 741, 1978.
28. Wolf, H. L., and Denko, J. V.: Osteosclerosis in chronic renal disease. Am J Med Sci, 235: 33, 1958.
29. Zimmerman, H. B.: Osteosclerosis in chronic renal disease; report of 4 cases associated with secondary hyperparathyroidism. AJR, 88: 1152, 1962.
30. Dreskin, E. A., and T. A. Fox: Adult renal osteitis fibrosa with metastatic calcification and hyperplasia of one parathyroid gland; report of a case. Arch Intern Med, 86: 533, 1950.
31. Bartelheimer, H., and Schmitt-Rohde, J. M.: Quoted by G. Fanconi in Adv Pediatr, 12: 307, 1962.
32. Selye, H.: On stimulation of new bone formation with parathyroid extract and irradiated ergosterol. Endocrinology, 16: 547, 1932.
33. Beveridge, B., Vaughan, B. F., and Walters, M. N. I.: Primary hyperparathyroidism and secondary renal failure with osteosclerosis. J Fac Radiol, 10: 197, 1959.
34. Wills, M. R., Richardson, R. E., and Paul, R. G.: Osteosclerotic bone changes in primary hyperparathyroidism with renal failure. Br Med J, 1: 252, 1961.
35. Copp, D. H., Cameron, E. C., Cheney, B. A., Davidson, A. G. F., and Henze, K. G.: Evidence for calcitonin; new hormone from parathyroid that lowers blood calcium. Endocrinology, 70: 638, 1962.
36. Doyle, F. H.: Some quantitative radiological observations in primary and secondary hyperparathyroidism. Br J Radiol, 39: 161, 1966.
37. Dent, C. E., and Harris, H.: Hereditary forms of rickets and osteomalacia. J Bone Joint Surg, 38B: 204, 1956.
38. Fraser, D., Geiger, D. W., Munn, J. D., Slater, P. E., Jahn, R., and Liu, E.: Calcification studies in clinical vitamin D deficiency and hypophosphatemic vitamin D refractory rickets; induction of calcium deposition in rachitic cartilage without administration of vitamin D. Am J Dis Child, 96: 460, 1958.
39. Pierce, D. S., Wallace, W. M., and Herndon, C. H.: Long-term treatment of vitamin-D resistant rickets. J Bone Joint Surg, 46A: 978, 1964.
40. Stamp, W. G., Whitesides, T. E., Field, M. H., and Scheer, G. E.: Treatment of vitamin-D resistant rickets; long term evaluation of its effectiveness. J Bone Joint Surg, 46A: 965, 1964.
41. Harrison, H. E.: Fanconi syndrome. J Chronic Dis, 7: 346, 1958.
42. Weller, M. P., Edeiken, J., and Hodes, P. J.: Renal osteodystrophy. AJR, 104: 1968.

Hereditary Hyperphosphatasia

1. Bakwin, H., and Eiger, M. S.: Fragile bones with macrocranium. J Pediatr, 49: 558, 1956.
2. Choremis, C., Yannakos, D., Papadatos, C., and Baroutsou, E.: Osteitis deformans (Paget's disease) in an 11-year-old boy. Helv Paediatr Acta, 13: 185, 1968.
3. Swoboda, J. W.: Hyperostosis corticalis deformans juvenilis. Helv Paediatr Acta, 13: 292, 1958.
4. Stemmermann, G. N.: An histologic and histochemical, study of familial osteoectasia. Am J Pathol, 48: 641, 1966.
5. Marshall, W. C.: A chronic progressive osteopathy with

hyperphosphatasia. Proc R Soc Med, 55: 238, 1962.
6. Fanconi, G., Moreira, G., Uehlinger, E., and Giedion, A.: Osteochalasia desmalis familiaris. Helv Paediatr Acta, 19: 279, 1964.
7. Caffey, J.: *Pediatric X-ray Diagnosis*, Ed. 4, p. 1042. Year Book Medical Publishers, Chicago, 1961.
8. Eyring, E. J., and Eisenberg, E.: Congenital hyperphosphatasia. J. Bone Joint Surg, 50A: 1099, 1968.
9. Whalen, J. P., Horwith, M., Krook, L., et al.: Calcitonin treatment in hereditary bone dysplasia with hyperphosphatasemia; a radiographic and histologic study of the bone. AJR, 129: 29, 1977.
10. Thompson, R. C., Jr., Gauli, G. E., Horowitz, S. J., and Schenk, R. K.: Hereditary hyperphosphatasia; studies of 3 siblings. Am J Med, 47: 209, 1969.
11. Van Buchem, F. S. P., Hadders, H. N., Hansen, J. F., and Woldring, M. G.: Hyperostosis corticalis generalisata. Proc K Ned Akad Wet (Biol Med), 65: 205, 1962.
12. Mitsudo, S. M.: Chronic idiopathic hyperphosphatasia associated with pseudoxanthoma elasticum. J Bone Joint Surg, 53A: 1971.
13. Seakins, J.: Peptiduria in an unusual bone disorder; isolation of two peptides. Arch Dis Child, 38: 215, 1963.
14. Alderman, M. H., and Frimpter, G.: Inherited bone disease with glycyl-proline peptiduria. Clin Res Proc, 16: 295, 1968.
15. Doyle, F. H., Woodhouse, N. J. Y., Glen, A. C. A., et al.: Healing of the bones in juvenile Paget's disease treated by human calcitonin. Br J Radiol, 47: 9, 1974.
16. Caffey, J.: Therapeutic value of thyrocalcitonin. AJR, 129: 175, 1977.
17. Woodhouse, N. J. Y., Fisher, M. T., Sigurdsson, G., et al.: Paget's disease in a 5-year-old; acute response to human calcitonin. Br Med J, 4: 267, 1972.
18. Horwith, M., Suh, S. M., Torun, B., et al.: Synthetic human calcitonin in the treatment of hereditary bone dysplasia (hyperphosphatasemia). In *Human Calcitonin and Paget's Disease*, pp. 207–215, edited by I. Macintyre. Hans Huber, Stuttgart, 1976.
19. Dunn, V., Condon, V. R., and Rallison, M. L.: Familial hyperphosphatasemia; diagnosis in early infancy and response to human thyrocalcitonin therapy. AJR, 132: 541, 1979.
20. Iancu, T. C., Almagor, G., Friedman, E., Hardoff, R., and Front, D.: Chronic familial hyperphosphatasemia. Radiology, 129: 669, 1978.
21. Caffey, J.: Chronic idiopathic hyperphosphatasia. In *Pediatric X-ray Diagnosis*, Ed. 4, pp. 983–987. Year Book Medical Publishers, Chicago, 1967.
22. Van Buchem, F. S. P., Hadders, H. N., Hansen, J. F., and Woldring, J. W.: Hyperostosis corticalis generalisata. Am J Med, 33: 387, 1963.

Paget Disease

1. Paget, J.: On a form of chronic inflammation of bones (osteitis deformans). Med Chir Trans, 60: 37, 1877.
2. DaCosta, J. C., Funk, E. H., Bergeim, O., and Hawk, P. B.: Osteitis deformans; report of 5 cases, with complete metabolism studies in two instances and a review of the literature. Publ Jefferson Med Coll Hosp, 6: 1, 1915.
3. Knaggs, R. L.: On osteitis deformans (Paget's disease) and its relation to osteitis fibrosa and osteomalacia. Br J Surg, 13: 206, 1925.
4. Lannelongue: Note sur la syphilis osseuse héréditaire chez les nouveaux-nés (maladie de Parrot), ches les enfants et les adolescents, chez les adultes et les vieillards (maladie de Paget). Bull Acad Med, 49: 299, 1903.
5. Kay, H. D., Simpson, S. L., and Riddoch, G.: Osteitis deformans. Arch Intern Med, 53: 208, 1934.
6. Gutman, A. B., and Kasabach, H.: Paget's disease (osteitis deformans); analysis of 116 cases. Am J Med Sci, 191: 361, 1936.
7. Dickson, D. D., Camp, J. D., and Ghormley, R. K.: Osteitis deformans; Paget's disease of the bone. Radiology, 44: 449, 1945.
8. Lancereaux, E.: *Traite d'anatomie pathologique, Vol. III*. V. A. Delahaye, Paris, 1883.
9. Albright, F., Aub, J. C., and Bauer, W.: Hyperparathyroidism; a common and polymorphic condition as illustrated by 17 proved cases from one clinic. JAMA, 102: 1276, 1934.
10. Schmorl, G.: Ueber ostitis deformans Paget. Virchows Arch Pathol Anat, 283: 694, 1932.
11. Moehlig, R. C., and Adler, S.: Carbohydrate metabolism disturbance in osteoporosis and Paget's disease. Surg Gynecol Obstet, 64: 747, 1937.
12. Moehlig, R. C., and Murphy, J. M.: Paget's disease (osteitis deformans). Endocrinology, 19: 515, 1935.
13. Rosenbaum, H. D.: Geographic variation in the prevalence of Paget's disease of bone. Radiology, 92: 959, 1969.
14. Lackey, R. S.: The geographical incidence of Paget's disease of bone. South Med J, 53: 602, 1960.
15. Packard, F. A., Steele, J. D., and Kirkbride, T. S., Jr.: Osteitis deformans. Am J Med Sci, 122: 552, 1901.
16. Stein, I., Stein, R., and Beller, M.: *Living Bone in Health and Disease*. J. B. Lippincott, Philadelphia, 1955.
17. Latimer, F. R., Webster, M. D., and Gurdjian, J. E.: Osteitis deformans with spinal cord compression; report of 3 cases. J Neurosurg, 10: 583, 1953.
18. Schwarz, G. A., and Reback, S.: Compression of the spinal cord in osteitis deformans (Paget's disease) of the vertebrae. AJR, 42: 345, 1939.
19. Turner, J. W. A.: The spinal complications of Paget's disease. Brain, 63: 321, 1940.
20. Wyllie, W. G.: The occurrence in osteitis deformans of lesions of the central nervous system, with a report of 4 cases. Brain, 46: 336, 1923.
21. Siegelman, S. S., Levine, S. A., and Walpin, L.: Paget's disease with spinal cord compression. Clin Radiol, 19: 421, 1968.
22. Hartman, J. T., and Dohn, D. F.: Paget's disease of the spine with cord or nerve-root compression. J Bone Joint Surg, 48A: 1079, 1966.
23. Hunter, D.: Studies in calcium and phosphorus metabolism in generalized diseases of bone. Proc R Soc Med, 28: 1619, 1935.
24. Kay, H. D.: Plasma phosphatase in osteitis deformans and in other diseases of bone. Br J Exp Pathol, 10: 253, 1929.
25. Locke, E. A.: Osteitis deformans with sarcoma of the humerus. Med Clin North Am, 1: 947, 1918.
26. Freund, E.: Zur frage der osteitis deformans Paget. Virchows Arch Pathol Anat, 274: 1, 1929.
27. Jaffe, H. L.: Paget's disease of bone. Arch Pathol, 15: 83, 1933.
28. Burgener, F. A., and Perry, P. E.: Pitfalls in the radiographic diagnosis of Paget's disease of the pelvis. Skeletal Radiol, 2: 231, 1978.
29. Guyer, P. B., and Dewbury, K. C.: The hip joint in Paget's disease (Paget's "coxopathy"). Br J Radiol, 51: 574, 1978.
30. Barker, J. D. P., Clough, P. W. L., Guyer, P. B., and Gardner, M. J.: Paget's disease of bone in 14 British towns. Br Med J, 1: 1181, 1977.
31. Albright, F., and Reifenstein, E. C.: *The Parathyroid Glands and Metabolic Bone Disease*. Balliere, Tindall & Cox, London, 1948.
32. Collins, D. H.: Paget's disease of bone; incidence and

subclinical forms. Lancet, 2: 51, 1956.
33. Edeiken, J., DePalma, A. F., and Hodes, P. J.: Paget's disease; osteitis deformans. Clin Orthop, 46: 141, 1966.
34. McKusick, V. A.: *Hereditable Disorders of Connective Tissue*, Ed. 2, C. V. Mosby, St. Louis, 1966.
35. Schüller, A.: Dysostosis hypophysaria. Br J Radiol, 31: 156, 1926.
36. Sosman, M. C.: Radiology as an aid in the diagnosis of skull and intracranial lesions. Radiology, 9: 396, 1927.
37. Jacobs, P.: Osteolytic Paget's disease. Clin Radiol, 25: 137, 1974.
38. Stafne, E. C., and Austin, L. T.: A study of dental roentgenograms in cases of Paget's disease (osteitis deformans), osteitis fibrosa cystica and osteoma. J Am Dent Assoc, 25: 1202, 1938.
39. Guyer, P. B., and Clough, P. W. L.: Paget's disease of bone; some observations on the relation of the skeletal distribution to pathogenesis. Clin Radiol, 29: 421, 1978.
40. Bowerman, J. W., Altman, J., Hughes, J. L., and Zadek, R. E.: Pseudo-malignant lesions in Paget's disease of bone. AJR, 124: 57, 1975.
41. Brailsford, J. F.: Paget's disease of bone; its frequency, diagnosis and complications. Br J Radiol, 2: 507, 1938.
42. Kasabach, H. H., and Dyke, C. G.: Osteoporosis circumscripta of skull as form of osteitis deformans. AJR, 28: 192, 1932.
43. Meyer-Borstel, H.: Die zirkumskripte Osteoporose des Schadels als Fruhsymptom der Pagetschen Knochenerkrankung. Fortschr Geb Roentgenstr, 42: 589, 1930.
44. Seaman, W. B.: Roentgen appearance of early Paget's disease. AJR, 66: 587, 1961.
45. Evans, F. G.: *Stress and Strain in Bones*. Charles C Thomas, Springfield, 1957.
46. Grainger, R. G., and Laws, J. W.: Paget's disease—active or quiescent? Br J Radiol, 30: 120, 1957.
47. Barry, H. C.: Fractures of the femur in Paget's disease of bone in Australia. J Bone Joint Surg, 49A: 1359, 1967.
48. Goldman, A. B., Bullough, P., Kammerman, S., and Ambos, M.: Osteitis deformans of the hip joint. AJR, 128: 601, 1977.
49. Serre, H., Labange, A., Simon L., and Danau, M.: Spinal cord compression of Paget's disease; apropos of 3 cases. Rev Rhum Mal Osteoartic, 29: 307, 1962.
50. Teplick, J. G., Head, G. L., Kricun, M. E., and Haskin, M. E.: Ghost infantile vertebrae and hemipelves within adult skeleton from Thorotrast administration in childhood. Radiology, 129: 657, 1978.
51. Janower, M. L., Miettinen, O. S., and Flynn, M. J.: Effects of long-term thorotrast exposure. Radiology, 103: 13, 1972.
52. Symposium on distribution, retention and late effects of thorium dioxide. Ann NY Acad Sci, 145: 523, 1967.
53. Looney, W. B.: An investigation of the late clinical findings following thorotrast (thorium dioxide) administration. AJR, 83: 163, 1960.
54. Looney, W. B.: Late clinical changes following the internal deposition of radioactive materials. Ann Intern Med, 42: 378, 1955.
55. DeSeze, S., Guiot, G., Hubault, A., and Dujarrier, L.: A case of medullary complication in Paget's disease surgically treated with favorable result (presentation of a patient). Rev Rhum Mal Osteoartic, 28: 672, 1961.
56. Taveras, J. M., and Wood, E. H.: *Diagnostic Neurology*. Williams & Wilkins, Baltimore, 1964.
57. Kadir, S., Kalisher, L., and Schiller, A. L.: Extramedullary hematopoiesis in Paget's disease of bone. AJR, 129: 493, 1977.
58. Barry, H. C.: *Paget's Disease of Bone*. Livingstone, Edinburgh, 1969.
59. Blaisdell, J. L.: Extramedullary hematopoiesis in a retroperitoneal tumor. Arch Pathol, 16: 643, 1933.
60. Lyall, A.: Massive extramedullary bone marrow formation in case of pernicious anemia. J Pathol, 41: 469, 1935.
61. Dodge, O. G., and Evans, D. M. D.: Hemopoiesis in a presacral fatty tumor (myelolipoma). J Pathol, 72: 313, 1956.
62. Pearlman, A. W., and Friedman, M.: Radiation therapy of benign giant cell tumor arising in Paget's disease of bone. AJR, 102: 1968.
63. Hutter, R. V. P., Foote, F. W., Jr., Frazell, E. L., and Francis, K. C.: Giant cell tumors complicating Paget's disease of bone. Cancer, 16: 1044, 1963.
64. Russell, D. S.: Malignant osteoclastoma and association of malignant osteoclastoma with Paget's osteitis deformans. J Bone Joint Surg, 31B: 281, 1949.
65. Schajowicz, F., and Slullitel, I.: Giant cell tumor associated with Paget's disease of bone; a case report. J Bone Joint Surg, 48A: 1340, 1966.
66. Sklar, G., and Meyer, I.: Giant cell tumor of maxilla in area of osteitis deformans (Paget's disease of bone). Oral Surg, 11: 835, 1958.
67. Gruner, O. C., Scrimger, F. A., and Foster, L. S.: A clinical and histologic study of a case of Paget's disease of the bones with multiple sarcoma formation. Arch Intern Med, 9: 641, 1912.
68. Speiser, F.: Sarkomatose entartung bei der ostitis deformans. Arch Klin Chir, 149: 274, 1928.
69. Coley, B. L., and Sharp, G. S.: Paget's disease; a predisposing factor to osteogenic sarcoma. Arch Surg, 23: 918, 1931.

Fibrous Dysplasia

1. Albright, F., Butler, A. M., Hampton, A. O., and Smith, P.: Syndrome characterized by osteitis fibrosa disseminata, areas of pigmentation and endocrine dysfunction, with precocious puberty in females; report of 5 cases. N Engl J Med, 216: 727, 1937.
2. Lichtenstein, L.: Polyostotic fibrous dysplasia. Arch Surg, 36: 874, 1938.
3. Lichtenstein, L., and Jaffe, H. L.: Fibrous dysplasia of bone. Arch Pathol, 33: 777, 1942.
4. Schlumberger, H. G.: Fibrous dysplasia of single bones (monostotic fibrous dysplasia). Milit Surgeon, 99: 504, 1946.
5. Daves, M. L., and Yardley, J. H.: Fibrous dysplasia of bone. Am J Med Sci, 234: 590, 1957.
6. Thannhauser, S. J.: Neurofibromatosis (von Recklinghausen) and osteitis fibrosa cystica localisata et disseminata (von Recklinghausen). Medicine, 23: 105, 1944.
7. Snapper, I.: On lipoid granulomatosis of bones without symptoms of Schüller-Christian's disease. Chin Med J, 56: 303, 1939.
8. Pritchard, J. E.: Fibrous dysplasia of the bones. Am J Med Sci, 222: 313, 1951.
9. Harris, W. H., Dudley, H. R., Jr., and Barry, R. J.: The natural history of fibrous dysplasia. J Bone Joint Surg, 44A: 207, 1962.
10. Hunter, D., and Turnbull, H. M.: Hyperparathyroidism; generalized osteitis fibrosa, with observations upon bones, parathyroid tumors, and normal parathyroid. Br J Surg, 19: 203, 1931.
11. Dockerty, M. B., Ghormley, R. K., Kennedy, R. L. J., and Pugh, D. G.: Albright's syndrome (polyostotic fibrous dysplasia with cutaneous pigmentation in both sexes and gonadal dysfunctions in females). Arch Intern Med, 75: 357, 1945.
12. Bonduelle, M., and Claisse, R.: Dysplasie fibreuse des os

et syndrome d'Albright; leur place nosologique. Sem Hop Paris, 24: 514, 1948.
13. Sante, L. R., Bauer, W., and O'Brien, R. M.: Polyostotic dysplasia (Albright's syndrome) and its comparison with dyschondroplasia (Ollier's disease); a correlation of the radiological and pathological findings. Radiology, 51: 676, 1948.
14. Hacket, L. J., Jr., and Christopherson, W. M.: Polyostotic fibrous dysplasia. J Pediatr, 35: 767, 1949.
15. Falconer, M. A., Cope, C. L., and Robb-Smith, A. H. T.: Fibrous dysplasia of bone with endocrine disorders and cutaneous pigmentation (Albright's disease). Q J Med, 11: 121, 1942.
16. Warrick, C. K.: Polyostotic fibrous dysplasia—Albright's syndrome; review of the literature and report of 4 male cases, 2 of which were associated with precocious puberty. J Bone Joint Surg, 31B: 175, 1949.
17. Delannoy, E., and Ingelrans, P.: Dysplasie fibreuse polyostotique avec modifications crâniennes a type d'hyperostose. Mem Acad Chir, 74: 535, 1948.
18. Albright, F., and Reifenstein, E. C., Jr.: *The Parathyroid Glands and Metabolic Bone Disease.* Williams & Wilkins, Baltimore, 1948.
19. McCune, D. J., and Bruch, H.: Osteodystrophia fibrosa; report of a case in which the condition was combined with precocious puberty, pathologic pigmentation of the skin and hyperthyroidism. Am J Dis Child, 54: 806, 1937.
20. Summerfeldt, P., and Brown, A.: Osteodystrophia fibrosa. Am J Dis Child, 57: 90, 1939.
21. Coleman, H. M.: Polyostotic fibrous dysplasia. Can Med Assoc J, 56: 318, 1947.
22. Stauffer, H. M., Arbuckle, R. K., and Aegerter, E.: Polyostotic fibrous dysplasia with cutaneous pigmentation and congenital arteriovenous aneurysms. J Bone Joint Surg, 23: 323, 1941.
23. Coley, B. L., and Stewart, F. W.: Bone sarcoma in polyostotic fibrous dysplasia. Ann Surg, 121: 872, 1945.
24. Sutro, C. J.: Osteogenic sarcoma of the tibia in a limb affected with fibrous dysplasia. Bull Hosp Joint Dis, 12: 217, 1951.
25. Jaffe, H. L.: *Tumors and Tumorous Conditions of the Bones and Joints.* Lea & Febiger, Philadelphia, 1958.
26. Schwartz, D. T., and Alpert, M.: The malignant transformation of fibrous dysplasia. Am J Med Sci, 247: 1, 1964.
27. Pratt, A. D., Felson, B., Wiot, J. F., and Paige, M.: Sequestrum formation in fibrous dysplasia. AJR, 106: 162, 1969.
28. Badgley, C. E., and O'Connor, S. J.: Congenital kyphoscoliotic tibia. J Bone Joint Surg, 34A: 349, 1952.
29. Buttner, A., and Eysholdt, K. G.: Die angeborenen Verbiegungen und Pseudarthrosis des Unterschenkels. Ergeb Chir Orthop, 36: 165, 1950.
30. Campanacci, M., Giunti, A., Leonessa, C., Pagani, P. A., and Trentani, C.: Pathological fractures in osteopathies and bony dysplasias. Ital J Orthop Traumatol Suppl 1, 1975.
31. Lin, J. P., Goodkin, R., Chase, N. E., and Kricheff, I. I.: The angiographic features of fibrous dysplasia of the skull. Radiology, 92: 1275, 1969.
32. Fries, J. W.: The roentgen features of fibrous dysplasia of the skull. AJR, 77: 71, 1957.
33. Sassin, J. F., and Rosenberg, R. N.: Neurological complications of fibrous dysplasia of the skull. Arch. Neurol, 18: 363, 1968.
34. Gross, C. W., and Montgomery, W. W.: Fibrous dysplasia and malignant degeneration. Arch Otolaryngol, 85: 1967.
35. Albright, F.: Polyostotic fibrous dysplasia; a defense of the entity. J Clin Endocrinol, 7: 307, 1947.
36. Jaffe, H. L.: Fibrous dysplasia of bone: a disease entity and specifically not an expression of neurofibromatosis. J Mt Sinai Hosp, 12: 364, 1945.

Osteofibrous Dysplasia

1. Campanacci, M.: Osteofibrous dysplasia of long bones; a new clinical entity. Ital J Orthop Traumatol, 2: 221, 1976.
2. Badgley, C. F., and O'Connor, S. J.: Congenital kyphoscoliotic tibia. J Bone Joint Surg, 34A: 349, 1952.
3. Buttner, A., and Eysholdt, K. G.: Die angeborenen verbiegunen und pseudoarthrosen des unterschenkels. Ergeb Chir Orthop, 36: 165, 1950.
4. Compere, S. L.: Localized osteitis fibrosa in the newborn and congenital pseudoarthrosis. J Bone Joint Surg, 18: 513, 1936.
5. Eyre-Brook, A. L., Baily, R. A. J., and Price, C. H. G.: Infantile pseudoarthrosis of the tibia. Three cases treated successfully by delayed autogenous by-pass graft, with some comments on the causative lesion. J Bone Joint Surg, 51B: 604, 1969.
6. Fevre, M.: Les pseudoarthroses de jambe du nouveau-né secondaires aux dystrophies kystiques congenitales. Rev Chir Orthop, 40: 305, 1954.
7. Franghenheim, P.: Angeborene ostitis fibrosa als ursache einer intrauterinen unterschenkel fraktur. Arch Klin Chir, 117: 22, 1921.
8. Frigo, M., and Novello, A.: Sulla pseudartrosi congenita della tibia (studio clinico-radiografico su 4 casi). Chir Ital, 19: 72, 1967.
9. Gruca, A.: Operationsmethodik bei kongenitalen unterschenkel pseudoarthrosen (vorlaufige Mitteilung). Beitr Orthop Traumatol, 15: 138, 1968.
10. Guilleminet, M.: Pseudarthroses congenitale du tibia. Rev Chir Orthop, 39: 670, 1953.
11. Guilleminet, M.: Presentation de documents sur les pseudarthroses congenitales de jambe chez l'enfant. Lyon Chir, 50: 723, 1955.
12. Guilleminet, M., and Ricard, R.: *Pseudarthrose congenitale du tibia et son traitement.* Masson Ed., Paris, 1958.
13. Kempson, R. L.: Ossifying fibroma of the long bones. A light and electron microscopic study. Arch Pathol, 82: 218, 1966.
14. Lloyd-Roberts, G. C., and Shaw, N. E.: The prevention of pseudarthrosis in congenital kyphosis of the tibia. J Bone Joint Surg, 54B: 100, 1969.
15. McFarland, B.: Birth fracture of the tibia. Br J Surg, 27: 706, 1939-40.
16. Semian, D. W., Willis, J. B., and Bove, K. E.: Congenital fibrous defect of the tibia mimicking fibrous dysplasia. J Bone Joint Surg, 57A: 854, 1975.
17. Stewart, M. J., Gilmer, W. S., and Edmonson, A. S.: Fibrous dysplasia of bone. J Bone Joint Surg, 44B: 302, 1962.
18. Campanacci, M., Giunti, A., Leonessa, C., Pagani, P. A., and Trentani, C.: Pathological fractures in osteopathies and bony dysplasias. Ital J Orthop Traumatol Suppl 1, 1975.
19. Campanacci, M., and Leonessa, C.: Displasia fibrosa dello scheletro. Chir Organi Mov, 59: 195, 1970.

Amyloidosis

1. Dahlin, D. C.: Classification and general aspects of amyloidosis. Med Clin North Am, 34: 1107, 1950.
2. Pear, B. L.: Radiographic studies of amyloidosis. CRC Crit Rev Radiol Sci, pp. 425–449, August 1972.
3. Bayrd, E. D., and Bennett, W. A.: Amyloidosis complicating myeloma. Med Clin North Am, 34: 1151, 1950.
4. Patrassi, G.: Primary amyloidosis. Postgrad Med J, 41:

247, 1965.
5. Cassidy, J. T.: Cardiac amyloidosis; 2 cases with digitalis sensitivity. Ann Intern Med, 55: 989, 1961.
6. Symmers, W.: Primary amyloidosis; a review. J Clin Pathol, 9: 187, 1956.
7. Weinfeld, A., Stern, M. H., and Marx, L. H.: Amyloid lesions of bone. AJR, 108: 1970.
8. Koletsky, S., and Stecher, R. M.: Primary systemic amyloidosis; involvement of cardiac valves, joints and bones, with pathologic fracture of femur. Arch Pathol, 27: 267, 1939.
9. Gardner, H.: Bone lesions in primary systemic amyloidosis; report of a case. Br J Radiol, 34: 778, 1961.
10. Golden, R.: Amyloidosis of small intestine. AJR, 72: 401, 1954.
11. Grossman, R. E., and Hensley, G. T.: Bone lesions in primary amyloidosis. AJR, 101: 1967.
12. Lindsay, S., and Knorp, W. F.: Primary systemic amyloidosis. Arch Pathol, 39: 315, 1945.
13. Bauer, W. H., and Kuzma, J. F.: Solitary tumors of atypical amyloid (paramyloid). Am J Clin Pathol, 19: 1097, 1949.
14. Case Records of Massachusetts General Hospital. N Engl J Med, 260: 288, 1959.
15. Lowell, D. M.: Amyloid-producing plasmacytoma of pelvis. Arch Surg, 94: 899, 1967.
16. Rosenblum, A. H., and Kirshbaum, J. D.: Multiple myelomas with tumor-like amyloidosis; clinical and pathologic study. JAMA, 106: 988, 1936.
17. Jaffe, H. L.: *Tumors and Tumorous Conditions of Bones and Joints.* Lea & Febiger, Philadelphia, 1958.
18. Goldberg, A., Brodsky, I., and McCarty, D.: Multiple myeloma with paramyloidosis presenting as rheumatoid disease. Am J Med, 37: 653, 1964.

Oxalosis

1. Dunn, H. G.: Oxalosis; report of a case with review of the literature. Am J Dis Child, 90: 58, 1955.
2. Caffey, J.: *Pediatric X-ray Diagnosis*, Ed. 5, p. 1059. Year Book Medical Publishers, Chicago, 1967.
3. Bednar, R., Jirasek, A., Stejskal, J., and Chytil, M.: Die sekundare uramische Oxalose. Zentralbl Allg Pathol, 102: 289, 1961.
4. Bennett, B., and Rosenblum, C.: Identification of calcium crystals in the myocardium in patients with uremia. Lab Invest, 10: 947, 1961.
5. Bennington, J. L., Haber, S. L., Smith, J. V., and Warner, N. E.: Crystals of calcium oxalate in the human kidney; studies by means of electron microprobe and x-ray diffraction. Am J Clin Pathol, 41: 8, 1964.
6. Macaluso, M. P., and Berg, N. O.: Calcium oxalate crystals in kidneys in acute tubular nephrosis and other renal disease with functional failure. Acta Pathol Microbiol Scand, 46: 197, 1959.
7. Milgram, J. W., and Salyer, W. R.: Secondary oxalosis of bone in chronic renal failure; a histopathological study of three cases. J Bone Joint Surg, 56A: 387, 1974.
8. Salyer, W. R., and Keren, D.: Oxalosis as a complication of chronic renal failure. Kidney Int, 4: 61–66, 1973.
9. Zarembski, P. M., Hodgkinson, A., and Parsons, F. M.: Elevation of the concentration of plasma oxalic acid in renal failure. Nature, 212: 511, 1966.
10. Carsen, G. M., and Radkowski, M. A.: Calcium oxalosis; a case report. Radiology, 113: 165, 1974.

9

The Anemias

HEREDITARY ANEMIAS

The hereditary anemias are a group of diseases resulting from molecular abnormalities of hemoglobin, abnormalities in the shape of red blood cells, or both. The abnormal hemoglobins (S, C, D, and G) are transmitted as Mendelian dominants.[1] They are molecularly distinct from normal hemoglobin (A) and may be identified by electrophoresis[2]; examples are sickle cell anemia (S-S hemoglobin) and sickle cell-SC disease. The abnormality of the shape of red blood corpuscles is responsible for the anemia of spherocytosis. Thalassemia (Cooley or Mediterranean anemia) and sickle cell-thalassemia are examples of anemias due to combinations of both abnormal hemoglobin and abnormalities in the shape of red blood corpuscles.

Sickle Cell Anemia and Variants

The presence of sickle cell erythrocytes is a hereditary trait transmitted as a Mendelian dominant. In the genetic analysis of families with the sickle cell trait, Neel[1] found that both parents must transmit the sickling character for an offspring to show the clinical picture of sickle cell anemia (homozygous due to S-S hemoglobin). If one parent has normal A hemoglobin and the other S hemoglobin, the offspring will show the sickling trait (S-A hemoglobin) but not anemia. Hemoglobin S in combination with abnormal hemoglobin C results in sickle cell S-C disease, a less severe form. Occasionally the combination of hemoglobin S-thalassemia is seen, but combinations with hemoglobin D and G are rare.[3]

Clinical Features. The incidence of sickling factor in American Negroes is 7.0–7.5%; the incidence of hemoglobin C in Negroes is 2%.[4] Of those with the sickle cell trait only 1 in 40 will manifest sickle cell anemia,[5, 6] which is 3 times as common as hemoglobin S-C disease.

The disease manifests itself in infants as severe hemolytic anemia, resulting in chronic fatigue and jaundice.[4] Abdominal crises are common, resembling an acute surgical abdomen. Bone crises also are common, and the pain may be quite severe, probably representing infarction. Often, rheumatic-like joint pains are troublesome and frequent. Chronic leg ulcers, priapism, and hematuria may be manifest. Splenomegaly is found before the age of 10 but usually not thereafter. Cardiomegaly and congestive heart failure are common, and their rarity in variants of hereditary anemia makes them excellent differential criteria. Many patients die before age 30; most are dead by 40. Manifestations after 40, especially those of bone, are usually due to variants.

The clinical course of the variants of sickle cell anemia is milder, with infrequent or no abdominal or bone crises. The anemia is much milder in the S-C form. Often its first clinical sign is evidence of aseptic necrosis of the hip or hematuria due to multiple infarctions of the kidneys. Because bone infarctions are more common in the S-C form of the disease, bone and joint pain are present more frequently.

Sickle cell trait (S-A) rarely causes clinical manifestations; however, splenic infarction during aerial flight[7] and recurrent gross hematuria have been reported.[8] Combinations of sickle cell-thalassemia may cause a severe anemia with all the manifestations of both thalassemia and sickle cell anemia.

Pathogenesis. The sickle cell is an abnormally shaped erythrocyte which is rapidly destroyed by the reticuloendothelial system. When there is a high percentage of the cells in blood, anemia and jaundice result. Anemia stimulates erythropoietic hyperplasia, which causes crowding of the bone marrow elements.

The mechanism of visceral and osseous infarction is unknown but related to the molecular abnormality, which, under lowered oxygen tension, results in sickling shape of red blood corpuscles.[9] Sickling leads to increased blood viscosity, stasis, and, eventually, thrombosis and infarction.

Roentgenographic Features. With the accumulation of cases studied by electrophoresis, it has become apparent that sickle cell anemia and its variants produce roentgenographically similar bone changes.[10] These can be conveniently divided into four groups[11]: (1) deossification due to marrow hyperplasia, (2) thrombosis and infarction, (3) secondary osteomyelitis, and (4) growth defects. Sickle cell anemia is most likely to produce marrow hyperplasia and deossification, whereas the variants are more often associated with bone infarction.

Deossification due to marrow hyperplasia is most commonly seen with sickle cell anemia but may be found with the variants.[9, 12, 13] Marrow hyperplasia crowds and thins trabeculae, with resultant osteoporosis. Bone softening of the centra occurs in 70% of patients with S-S disease; the centrum becomes biconcave, the result of disks bulging into softened bone (Figs. 1147 and 1148).[14] These changes persist throughout life, unlike the regressive changes in long bones.

Erythropoietic hyperplasia in long bones causes widening of the medullary spaces with thinning of cortices and coarsening of the trabecular pattern (Fig. 1149). Diaphyseal trabeculae become sparse while the remaining trabeculae are thickened. Metaphyseal bone remains relatively dense. The transverse diameter of the diaphysis does not increase, as often happens in thalassemia. In children, the changes in the long bones and the small tubular bones of hands and feet are usually pronounced (Fig. 1150). Osteoporosis found in the long bones tends to disappear in adolescence, probably an effect of the conversion of red marrow to yellow marrow in the extremities. The deossification is not as marked as in thalassemia. Early fusion of the proximal humeral epiphyses may occur.

The flat bones, especially the ribs, will also present a thickened trabecular pattern with cortical narrowing (Fig. 1151). There may be rib notching, "bone within a bone," and a "spatula" appearance. The skull may show obvious radiographic features, such as an indistinct trabecular

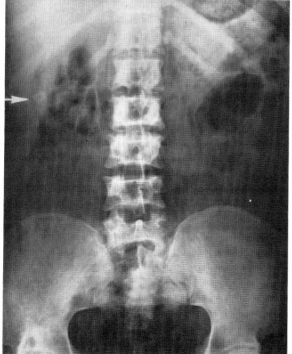
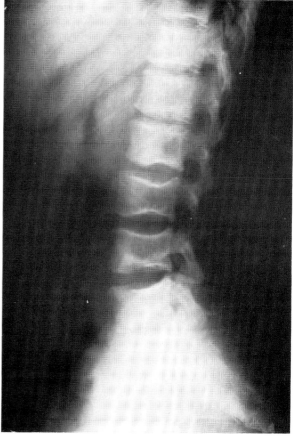

FIG. 1147. SICKLE CELL ANEMIA

(*Left*) Anteroposterior projection reveals thickened trabeculations throughout the vertebrae and pelvis. Changes in the right hip due to old aseptic necrosis. A gallstone is present (*arrow*). (Courtesy of Paul K. Berg, Doctors Hospital, Staten Island, N. Y.) (*Right*) Biconcavity of the vertebral centra and increased prominence of trabeculations.

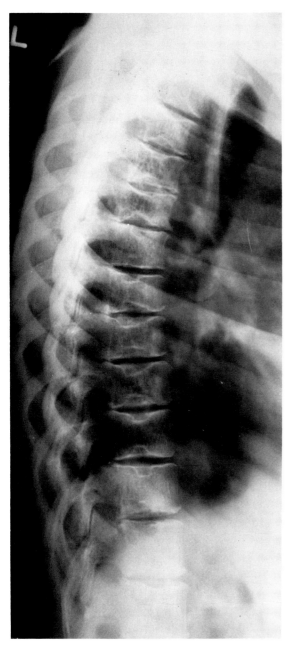

Fig. 1148. SICKLE CELL ANEMIA
Softening of the centra and multiple protrusions of the nucleus pulposa. The trabeculations are sparse, and the remaining vertical trabeculations are increased in density. (Courtesy of George T. Wohl, Philadelphia General Hospital, Philadelphia, Pa.)

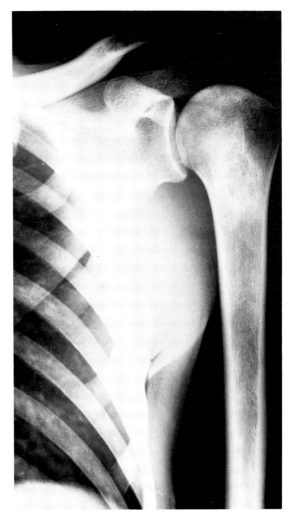

Fig. 1149. SICKLE CELL ANEMIA
Erythropoietic hyperplasia in long bones causes widening of the medullary spaces with thinning of the cortices. Thickening of the remaining trabeculae. (Courtesy of George T. Wohl, Philadelphia General Hospital, Philadelphia, Pa.)

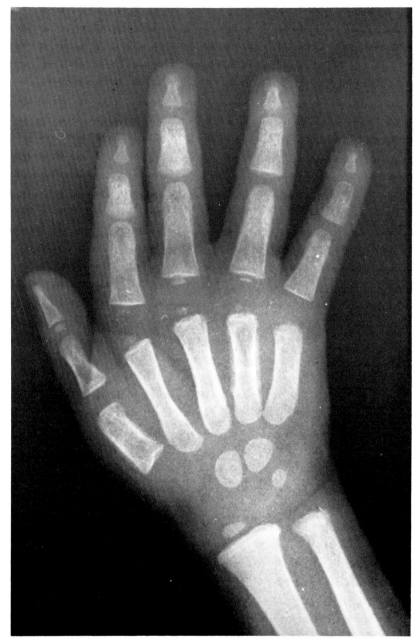

Fig. 1150. Sickle Cell Anemia
Widening of the medullary spaces and bone infarctions of the first and fourth metacarpals and first proximal phalanx. The infarctions are indistinguishable from lesions caused by infection. (Courtesy of George T. Wohl, Philadelphia General Hospital, Philadelphia, Pa.)

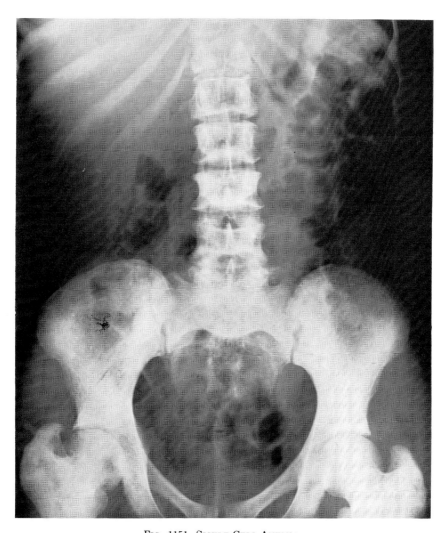

FIG. 1151. SICKLE CELL ANEMIA
Coarse trabeculations throughout the spine, hips, and pelvis. (Courtesy of George T. Wohl, Philadelphia General Hospital, Philadelphia, Pa.)

pattern with a "ground glass" appearance and the outer table appearing thin and partially absorbed.[15] Radial trabeculation causing a "hair-on-end" appearance is occasionally seen, but not as commonly as in thalassemia (Fig. 1152). Older patients show symmetrical parietal thickening due to lamellations of new bone.[16]

Skull radiographs of 194 patients from four months to 55 years old revealed porous decrease in bone density in 25%, widening of the diploe associated with a relative decrease in width of the outer table in 22%, and vertical hair-on-end striations in 5%. The youngest patient with vertical striations was 5 years old, and the oldest was 39. Serial examinations in 62 patients revealed no decrease of the skull width or disappearance of the striations with age.[17]

In the spine, a steplike end-plate depression is frequently seen.[18] Although this has been considered pathognomonic of sickle cell anemia, it may also appear in thalassemia major,[19] Gaucher disease,[20, 21] congenital hereditary spherocytosis,[22] and unexplained osteopenia.[22]

Infarctions of Bone. In infants and children, bone infarcts most frequently occur in the diaphyseal portions of small tubular bones (Figs. 1153 and 1154); in adults, in the metaphyseal and subchondral portions of long bones (Figs. 1155–1158). The entire long bone may show evidence of the infarction, or only small areas of bone may be involved. Vertebral infarctions may occur in both children and adults. The radiographic patterns of bone infarction include articular disintegration, juxta-articular sclerosis, dystrophic medullary calcification and periosteal reaction (bone within a bone), osteolysis, and in the vertebrae a central "step down" sign.[23] Bone infarction is less common in children; few infarctions occurred in those under age 25.[23]

Infarction in children is often preceded by a painful soft-tissue swelling of finger or toe, then, within 10 days, varying degrees of periostitis.

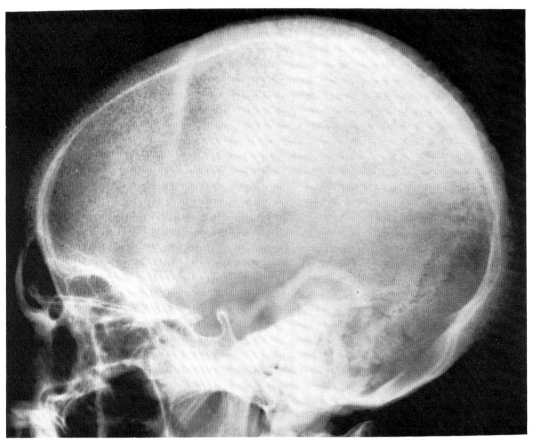

FIG. 1152. SICKLE CELL ANEMIA

Thickening of the tables in the posterior parietal area with a tendency toward radical trabeculation. This hair-on-end appearance is not as common or as marked as in cases of thalassemia. (Courtesy of George T. Wohl, Philadelphia General Hospital, Philadelphia, Pa.)

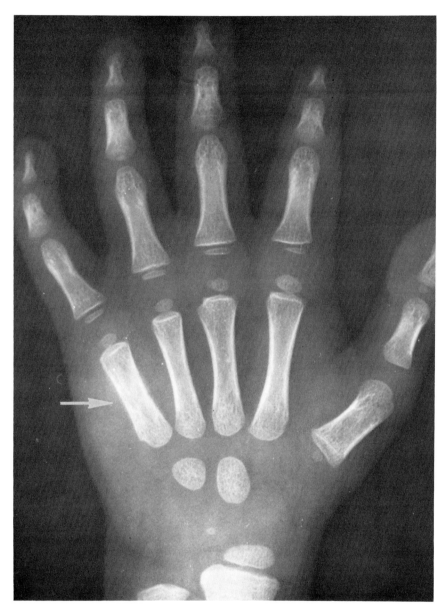

FIG. 1153. SICKLE CELL ANEMIA
Bone infarct in the fifth metacarpal (*arrow*) is indistinguishable from a lesion caused by infection.

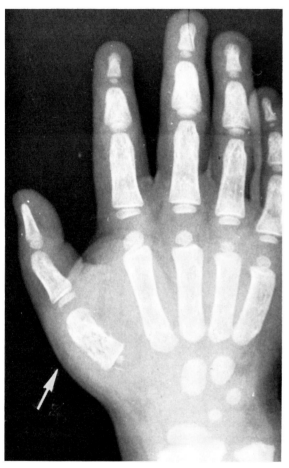

Fig. 1154. Sickle Cell Anemia

Bone infarct in the first metacarpal (*arrow*) is indistinguishable from a lesion caused by infection. Prominent trabeculae are noted throughout the remainder of the bones in the hand.

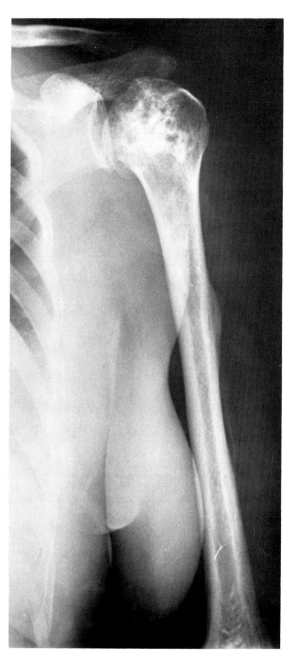

Fig. 1155. Sickle Cell Anemia

Bone infarction in the subchondral portion of the humeral head.

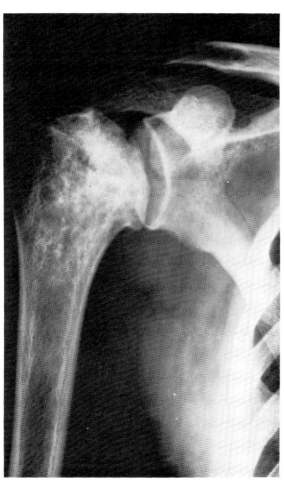

Fig. 1156. Sickle Cell Anemia

Infarction of the subchondral portion of the humeral head with deformity and irregularity of the articular surface. In the shaft of the humerus there is irregularity of the endosteal portion of the cortex, the residual of an old infarction. (Courtesy of Paul K. Berg, Doctors Hospital, Staten Island, N.Y.)

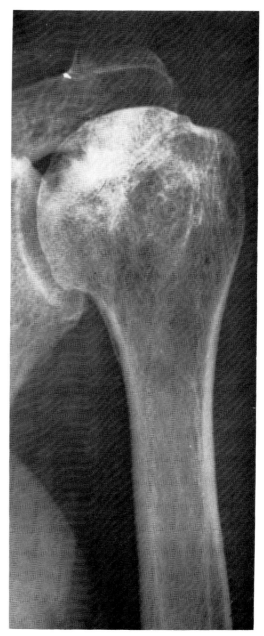

Fig. 1157. Sickle Cell Anemia

Increased bone density in the subchondral portion of the humeral head due to bone infarction. (Courtesy of George T. Wohl, Philadelphia General Hospital, Philadelphia, Pa.)

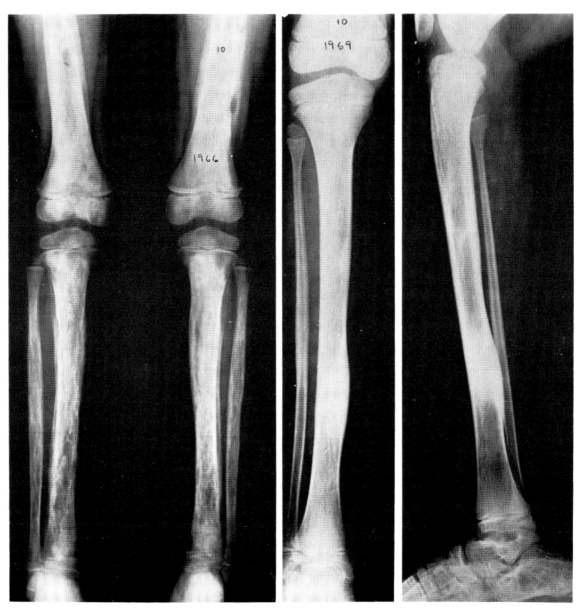

FIG. 1158. SICKLE CELL ANEMIA WITH BONE INFARCTION

(*Left*) A 6-year-old child with massive infarctions of the lower extremity. Periosteal reaction and sclerosis. The entire shafts are enveloped in an involucrum. (*Right*) Three years later, remodeling and repair have occurred. Residual sclerosis and deformity at the junction of the middle and lower thirds of the tibia. (Courtesy of Dr. Carlos Eduardo Vallim Telles, Sao Paulo, Brazil.)

Single thin periosteal laminations appear and progress to become thick areas of new bone formation. At the same time, small foci of bone destruction develop within the shaft. The changes may not progress unless massive infarction has occurred, when the entire shaft may be enveloped in a prominent involucrum, producing a rectangular shaft outline.[5] Then massive osteolysis followed by sclerosis may be expected. The periosteum lays down new bone, which gradually merges with the original bone. By the process of remodeling and repair, bone may be restored to normal; sometimes there is permanent residual bone thickening. These features may also occur in the large tubular bones. A striking similarity to the roentgenographic features of osteomyelitis causes considerable difficulty.[10, 24, 25] Osteomyelitis may be superimposed on infarcts.

Infarctions in medullary portions of bone usually occur in the metaphysis and appear, initially, as ill defined radiolucent areas (Figs. 1159 and 1160). Infrequent symptoms account for failure to discover early infarctions. Eventually the necrotic area becomes surrounded by a shell of reactive sclerosis, and later, the entire area calcifies and presents as the characteristic irregular dense zone within bone.

Infarctions of the vertebral bodies also occur, and occasionally, massive infarction may result in partial or complete collapse of the centra,[6] which regain their stature if weight-bearing is avoided.[11] A central vertebral end-plate depression frequently occurs.[5] This has been termed the "fish vertebral" sign, but it may occur without hemoglobinopathy.[22] It also occurs in thalassemia and Gaucher disease.[21, 26]

Ischemic necrosis of bone ends is common in sickle cell disease, often appearing in the humeral or femoral capital epiphysis just at the time of fusion to shaft. The process is frequently focal or multifocal instead of affecting the entire epiphysis.[2] Its appearance in subchondral bone var-

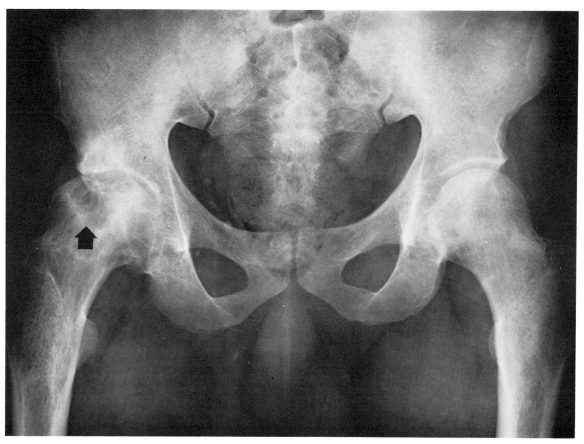

FIG. 1159. SICKLE CELL ANEMIA

Radiolucent area in the head of the right femur (*arrow*) represents an acute infarction. Increased bone density adjacent to this area represents an old infarct. Collapse of the femoral head and changes in the right acetabulum. The remainder of the bones reveal increased trabeculation. (Courtesy of Paul K. Berg, Doctors Hospital, Staten Island, N. Y.)

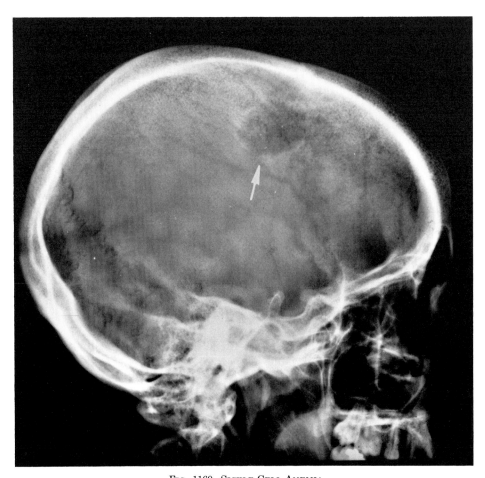

Fig. 1160. Sickle Cell Anemia

Osteolytic area in the posterior portion of the frontal bone indicates a bone infarction (*arrow*). (Courtesy of Herbert M. Stauffer, Temple University Hospital, Philadelphia, Pa.)

ies, from a small lytic area with a sclerotic margin, to multiple destroyed areas surrounded by considerably increased density (Fig. 1161). Collapse of the femoral head is not unusual. A small fragment of subchondral bone may sequestrate and simulate an osteochrondritis. These changes in the femoral head may be distinguished from Perthes disease by their occurrence at an earlier age, usually just as the epiphysis is fusing; by the more local and limited extent of their lesions; and by their failure to involve the metaphyseal area, always involved in Perthes disease.

In children with sickle cell anemia, the distal femur is a common site for acute long-bone diaphyseal infarcts.[27] The resulting changes may be coned metaphyses or various differences with osteolytic lesions in the center of the femur and peripheral overgrowth of the metaphyses (Fig. 1162). Bohrer[27] has classified these changes in Table 28.

Pathologic fractures are often a complication in patients with sickle cell anemia, perhaps because of cortices due to marrow hyperplasia. Underlying factors could also be bone infarcts and/or osteomyelitis.[28]

Moseley[10] has called attention to a diffuse increase in subchondral bone density, especially noticeable in the head of the femur or humerus, which he attributes to infarction. It may escape notice because of overlapping of acetabulum or acromion. Moseley[10] reports a peculiar thickening of the endosteal border of cortex which encroaches on the marrow cavity, and may even be separated from the original cortex to produce a "bone within a bone" appearance (Fig. 1163).[15] Infarction in other bones, when it does occur, is not severe and is characterized by localized osteoporotic patches that later become sclerosed.[11] Besides avascular necrosis, dislocation of the contralateral hip may occur, due to infarction with effusion and stretching of the capsule.[11]

Salmonella Infections. Salmonella infection strikes with unusual frequency with sickle cell anemias,[29–32] although not as frequently as

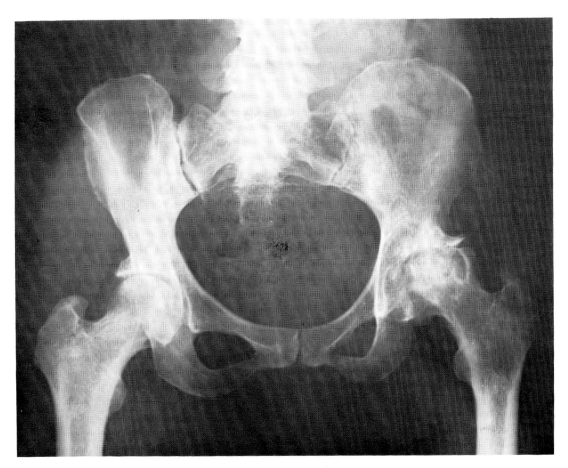

FIG. 1161. SICKLE CELL ANEMIA
Left femoral head shows multiple subchondral infarctions with increase in bone density and deformity. Secondary changes in the acetabulum. The bones of the pelvis and right femur reveal increase in trabeculations.

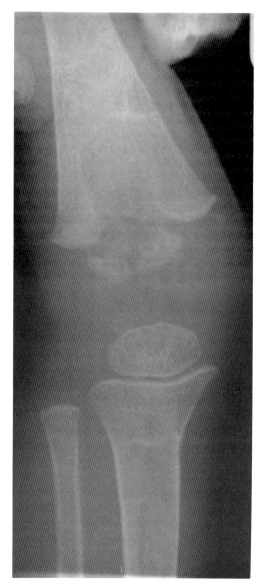

Fig. 1162. Sickle Cell Anemia with Coned Epiphysis and Metaphysis

The defect in the center of the femur is due to an infarct, with overgrowth of the metaphyses to compensate for the central defect due to undergrowth. (Courtesy of Dr. Stanley Bohrer, Ibadan, Nigeria. Previously published by S. P. S. Bohrer: *Clinical Radiology, 25:* 221, 1974.[27])

TABLE 28. GROWTH DISTURBANCES OF THE DISTAL ILIUM

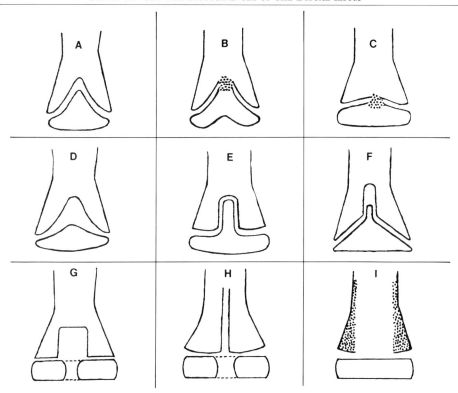

Schematic drawings showing some of the types of EM growth disturbances and "channel" defects. Variations of these are included, such as oblique rather than central metaphyseal defects, sclerosis or lack of it at the margin of the defects, EM fusion (*stippled central EM areas*) in those shown not fused and vice versa, different widths and lengths of the metaphyseal defects, etc. (A) Classic "cup" defect, or metaphysis with overgrowth of the EOC with a normal (may be narrowed or fused) growth plate. (B) Same as A but also shows notching of the articular side of the EOC and central EM fusion. (C) Central EM fusion. (D) "Cup" defect of metaphysis, overgrowth of the EOC, and a widened growth plate. (E) Variation of A with a "peg-in-hole" appearance. (F) Variation of D. A "cup" defect of the metaphysis changes at its apex to a "channel" with parallel sides. (G) Rectangular central metaphyseal defect with or without split EOC. The sides and end are sclerotic. (H) Marginated narrow metaphyseal "channel" with or without split EOC. (I) Unmarginated wide (could be narrow) metaphyseal "channel." (Modified from S. P. S. Bohrer: Clinical Radiology, 25: 221, 1974.[27])

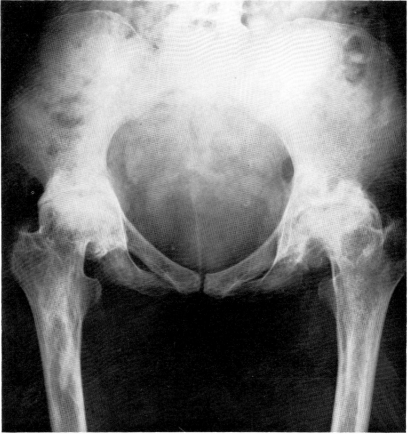

FIG. 1163. SICKLE CELL ANEMIA

Bone-within-a-bone appearance of the upper shaft of the left femur. This is due to an old infarction. Changes in the upper shaft of the right femur also indicate an old infarction. Both hips reveal multiple subchondral infarctions with deformity. (Courtesy of George T. Wohl, Philadelphia General Hospital, Philadelphia, Pa.)

staphylococcic infections. It is assumed that the avascular zones are appropriate foci for the settling of organisms during bacteremia. In infants and younger children, the small bones of the hands and feet may show symmetrical osteomyelitis (Figs. 1164 and 1165), but in older patients the large tubular bones are most likely to be affected. Roentgenographic changes are usually those of periostitis, with single or double layering of periosteal new bone reaction, and may prog-

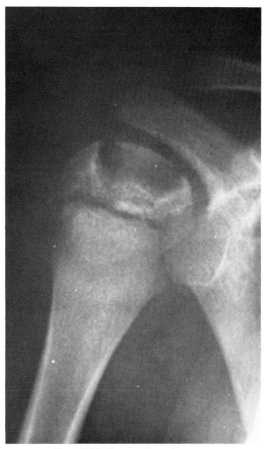

FIG. 1165. SICKLE CELL ANEMIA
Destructive lesion in the epiphysis of the humerus due to salmonella infection.

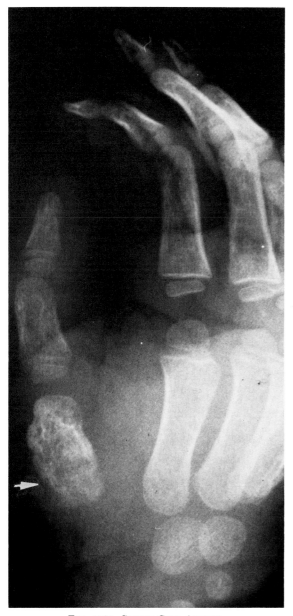

FIG. 1164. SICKLE CELL ANEMIA
Destruction of the first metacarpal due to salmonella infection (*arrow*).

ress to bone destruction with massive involucrum formation. The roentgenographic appearance is indistinguishable from bone infarction, and since it occurs concomitantly, it must be differentiated by its clinical features. The course is chronic, with little local reaction and only minor constitutional upset.

Growth Effects. Bone infarction that affects the epiphyseal cartilaginous plates may lead to growth disturbances of bone, with shortening and epiphyseal deformity, particularly in the hand. These changes usually appear several months after infarction, most commonly in a metacarpal or phalanx. Barton and Cockshott[12] believe that the central epiphyseal portions recover more slowly than the peripheral portions, owing to variations in local blood supply. The change often involves several phalanges or metacarpals.

Chronic anemia per se may cause retardation of bone growth. Vertebral cupping and diminution in the height of centra may cause shortening of stature and kypholordosis.

Thalassemia

Synonyms: Cooley anemia, Mediterranean anemia, leptocytosis.

Thalassemia is a hereditary disorder of hemoglobin synthesis which produces a severe anemia. Thalassemia major, known as Mediterranean disease or Cooley anemia, is the homozygous form of the condition, i.e., inherited from both parents. Thalassemia minor occurs when the trait is from only one parent. Twenty inherited abnormal variants of hemoglobin have been described, and it is possible to find any of these abnormal genes in combination with the thalassemia gene. Eventually, many variations of thalassemia may be identified. Some that have already been described include thalassemia-Lepore hemoglobin, thalassemia-hemoglobin H, thalassemia-hemoglobin S, thalassemia-hemoglobin C, thalassemia-hemoglobin E, and sickle cell-thalassemia. Many of these have been only incom-

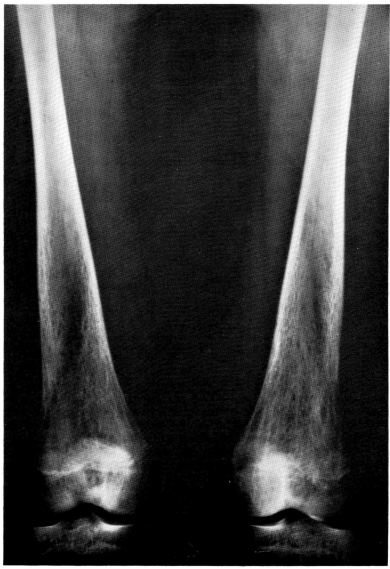

FIG. 1166. THALASSEMIA

Erythroid hyperplasia of the bone marrow has produced widening of the medullary spaces and thinning of the cortices, especially in the metaphyseal regions. Many trabeculae have been resorbed, and the remaining trabeculae are coarsened. Modeling of the distal ends of the femora has not occurred. (Same case as Figure 1169.) (Courtesy of Herbert M. Stauffer, Temple University Hospital, Philadelphia, Pa.)

pletely studied, and their roentgenographic patterns are not clearly defined.

The hemoglobin molecule seems to contain 19 different amino acids; the thalassemia mutant genes lead to amino acid substitutions which affect the rate of hemoglobin synthesis but do not alter the normal electrophoretic pattern.[1]

Clinical Features. Thalassemia major is a severe anemia of infants and children, few of whom survive beyond adolescence. Mongoloid facies, present in all cases, give a striking facial resemblance among the patients. Splenomegaly and hepatomegaly are the rule, often sufficient to produce protuberant abdomen. Cardiac enlargement becomes evident late in childhood and slowly increases. Retarded growth is noticeable at 8–10 years of age[2]; and the final stature of those with thalassemia major is short. Secondary sexual characteristics are retarded, and normal menstruation is rare.

Important hematologic features are hypochromic microcytic anemia, nucleated red blood cells, reticulocytosis, leukocytosis, target cells, decrease in red cell survival, elevated serum bilirubin, and marked increase of fecal urobilinogen.

Clinical features of thalassemia minor vary considerably, from a barely perceptible erythrocytic anomaly to severe anemia resembling thalassemia major.

Roentgenographic Features. Roentgenographic features of thalassemia derive from erythroid hyperplasia of bone marrow, more marked than in other hereditary anemias. Medullary spaces are widened, resulting in the thinning of cortices by pressure atrophy (Figs. 1166–1171). Many trabeculae are resorbed, and the

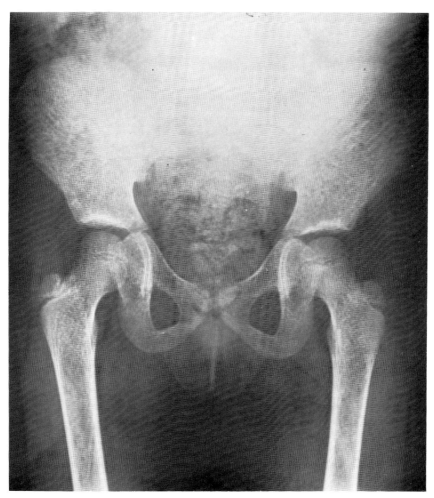

FIG. 1167. THALASSEMIA
Atrophy of trabeculae. The remaining trabeculae are coarsened due to narrow hyperplasia. (Same case as Figures 1171, 1172, and 1175.)

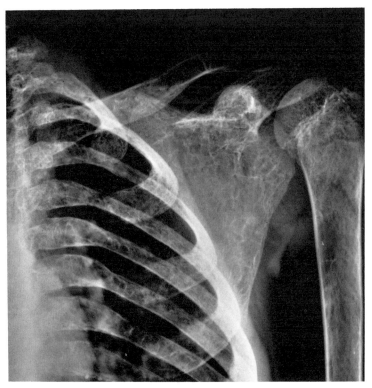

FIG. 1168. THALASSEMIA
Marked erythroid hyperplasia has produced trabecular atrophy and thinning of the cortices. The remaining trabeculae are thickened. (Same case as Figure 1770.) (Courtesy of Herbert M. Stauffer, Temple University Hospital, Philadelphia, Pa.)

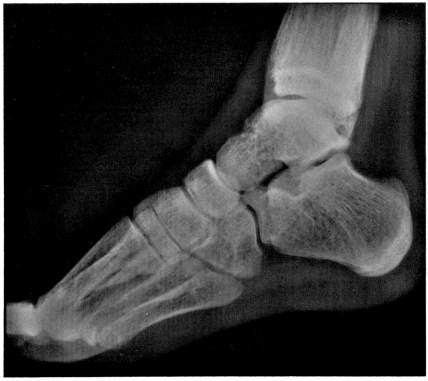

FIG. 1169. THALASSEMIA
Coarsening of remaining trabeculae. (Same case as Figure 1166.) (Courtesy of Herbert M. Stauffer, Temple University Hospital, Philadelphia, Pa.)

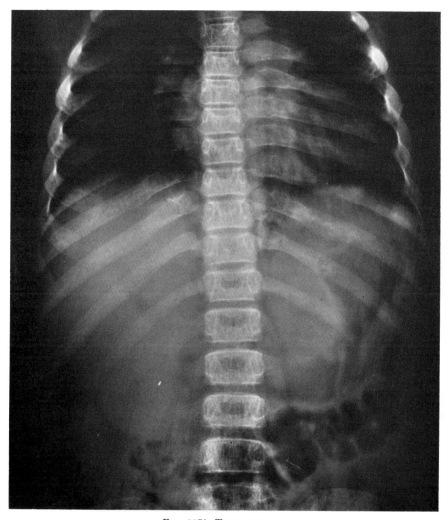

FIG. 1170. THALASSEMIA

Atrophy and coarsening of the trabeculae due to thalassemia. (Same case as Figure 1168.) (Courtesy of Herbert M. Stauffer, Temple University Hospital, Philadelphia, Pa.)

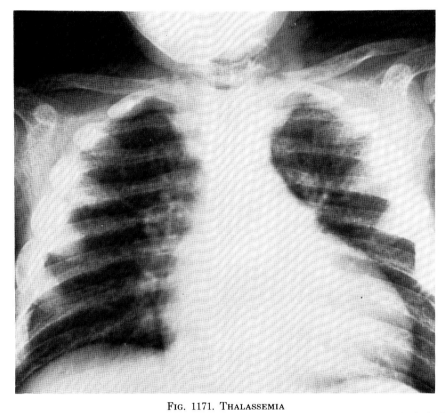

FIG. 1171. THALASSEMIA

Atrophy of most of the trabeculae with coarsening of the remaining trabeculae. The ribs are particularly dense. The heart is enlarged. (Same case as Figures 1167, 1172, and 1175.)

remaining trabeculae coarsened. Modeling of long bones does not occur; instead, there is bulging of the normally concave outlines of long bones. The child's entire skeleton is involved, since red marrow is found in all areas in this age group. Although skeletal manifestations are occasionally seen before 6 months of age, they do not usually become noticeable until the end of the first year. The earliest changes are found in the small bones of the hands and feet (Figs. 1172 and 1173).

Skull changes consist of the widening of diploic spaces and displacement and thinning of the outer table. At times, the diploic trabeculae orient themselves perpendicular to the tables, producing a radial pattern called "hair-standing-on-end" (Fig. 1174). The frontal bones show the earliest and most striking changes.[3] There is no involvement of the occipital squamosa inferior to the internal occipital protuberance. Iron deficiency anemias produce changes in the skull strikingly similar to those of thalassemia but do not affect the peripheral skeleton.

Caffey[3] has described the classic changes of the frontal, temporal, and facial bones. Overgrowth of marrow in these bones consistently impedes pneumatization of paranasal and mastoid sinuses (Fig. 1175). Only the ethmoidal sinuses are spared, due to their lack of red marrow. These changes are not usually seen in any other anemia.

Marrow overgrowth in the maxillary bone may cause lateral displacement of the orbits and leads to ventral displacement of the central incisors, producing the "rodent facies" which, with the sinus changes, is pathognomonic of thalassemia.

Patients who survive show regression of the peripheral skeleton changes as the red marrow becomes yellow marrow.[4] The dramatic changes in the tubular bones regress with age; adults bones appear perfectly normal, with no sign of previous change.[4] In the spine, skull, and pelvis the red marrow remains active throughout life, and changes here are accentuated. Osteoporosis, with cupping of vertebrae, begins in younger adults but is not seen in children. The vertebral

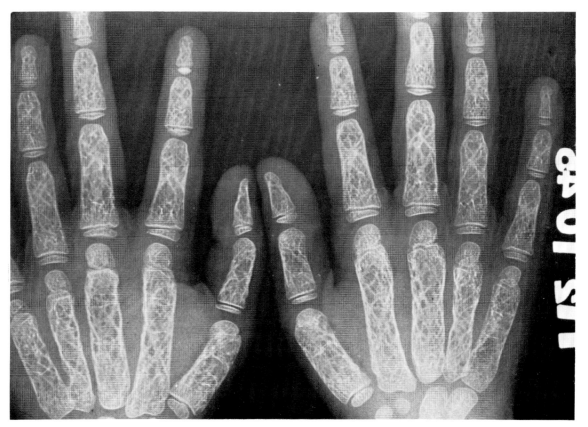

FIG. 1172. THALASSEMIA
Changes in the hands are marked in children. These changes regress at puberty. (Same case as Figures 1167, 1171, and 1175.)

body and end-plate deformities are attributed to ischemia of the midportion of the growing end bone and are seen most commonly in the lower thoracic spine. A rectangular depression in the center of the end plate, with no changes in the anterior and posterior margins, is almost specific for the infarctions of sickle cell anemia.[5, 6]

Anterior vertebral vascular notches on the anterior aspect of the lower thoracic spine are normal, but they are more prominent, persist later, and are readily seen in the lateral chests of children with sickle cell anemia.[6] A double-contour cortical tract of the long bones often remains and should suggest the diagnosis. It probably represents an old infarct, with the previous cortex unabsorbed and the new cortex forming over it, and its characteristic appearance can be seen in almost any disease with long-bone infarct, but of course, with sickle cell this is the most common cause.[7]

Bone changes are rare in thalassemia minor and the other variants and are never as marked in thalassemia major. Since patients with thalassemia minor live longer and the anemia is less

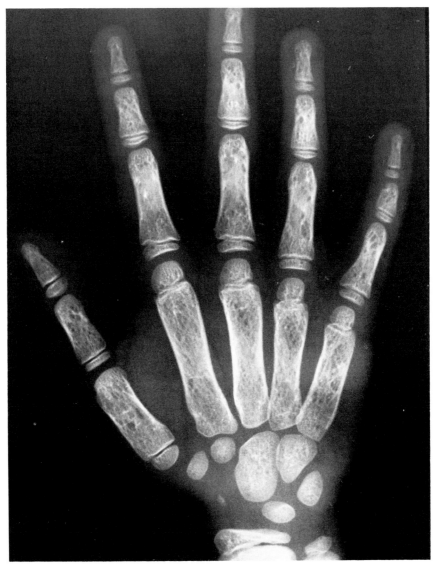

FIG. 1173. THALASSEMIA
The changes in the tubular bones are beginning to regress, although cortical thinning and multiple radiolucent areas with thickened trabeculations persist. They will completely regress at puberty. (Courtesy of Children's Hospital, Philadelphia, Pa.)

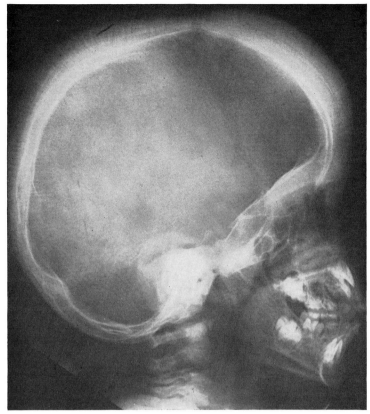

FIG. 1174. THALASSEMIA
Radiating spicules of bone running at right angles to the tables of the skull. The diploic space is unusually wide.

severe, they are usually studied as young adults. Thus, it is not uncommon for bone changes to be found in the central skeleton only and to consist of vertebral osteoporosis and diploic widening.

Hereditary Spherocytosis (Congenital Hemolytic Anemia)

Hereditary spherocytosis is a chronic hemolytic disease characterized by spherocytes in the peripheral blood and transmitted by either parent as an autosomal dominant trait. The genetic defect is responsible for the abnormal shape of the red blood cells, which increases their susceptibility to hemolysis by the spleen. Splenectomy corrects the anemia even though spherocytemia persists.

Clinical Features. Clinical features vary considerably in time of onset and intensity. Anemia usually begins in late childhood and early infancy or not until late adulthood. It is rarely severe. Jaundice usually occurs in late childhood and progresses, but it may be present in the newborn, simulating fetal erythroblastosis.[1,2]

Roentgenographic Features. Bone changes are rare because the anemia is usually mild. When they occur, they result from marrow hyperplasia and most commonly affect the skull, which shows evidence of widening of the diploe, with displacement and thinning of the outer table. The formation of perpendicular diploic trabeculae may give the "hair-on-end" appearance. Long-bone deossification is rare, since the anemia appears toward adolescence, when the long-bone red marrow changes to yellow marrow. In severe forms, however, when the onset is in infancy, there may be long-bone changes of hyperplastic bone marrow. Improvement in skeletal alterations follows splenectomy.

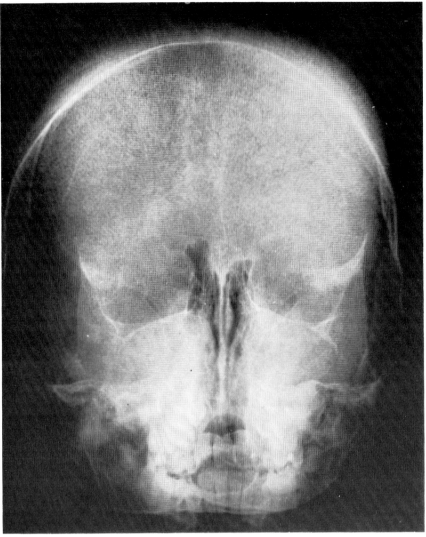

Fig. 1175. Thalassemia
Overgrowth of the marrow in the facial bones and calvaria has impeded the pneumatization of the paranasal sinuses and mastoids. Only the ethmoidal sinuses are well developed. (Same case as Figures 1167, 1171, and 1172.)

THE ACQUIRED ANEMIAS

Iron Deficiency Anemia

The most important causes of iron deficiency anemia in infants and children are (1) inadequate iron stores at birth, (2) inadequate intake due to deficient diet, (3) impaired gastrointestinal absorption of iron, and (4) excessive demands from blood loss. It is reported also in patients with congenital cyanotic heart disease[1] and polycythemia vera.[2]

In the normal newborn infant there is a gradual decline of hemoglobin and red blood cells for 2–3 months, referred to as the physiologic anemia of the newborn[3] and resulting from the destruction of red blood cells, inactive erythropoiesis is stimulated, and the hemoglobin regenerates and utilizes iron stored in the body. The faster the infant grows, the faster the iron supply is depleted. With anemia, the erythropoietic response is exaggerated and manifested by marrow hyperplasia.

Roentgenographic Features. Only in the skull do the roentgenographic features[4] differ from those of other anemias. There is widening of the diploic space and thinning of the tables (Fig. 1176). Occasionally, perpendicular radiating diploic trabeculae produce a "hair-on-end" appearance. Caffey[5] has noted an absence of hyperplasia in the occipital squamosa inferior to the internal occipital protuberance in cases of hemolytic anemia. This is true also in iron deficiency anemia and, presumably, is due to the absence of red marrow at this site.

Osteoporosis is the usual finding in the long bones and is most prominent in the hands. In most patients the cortex is thinned, and atrophy of the spongiosa is present in one-half.[6] Rickets are found in some patients.

Iron deficiency anemia may be distinguished from thalassemia by its absence of facial bone involvement, and the less frequent involvement of the long bones.

Agnogenic Myeloid Metaplasia (Myelosclerosis)

Synonyms: Myeloid sclerosis, megakaryocytic myelosis, myeloid metaplasia, pseudoleukemia and myeloproliferative syndrome, leukanemia, and osteosclerotic anemia.

Clinical Features. It usually appears in patients over 50, but sometimes in those between 30 and 50.[1] Myelofibrosis is rare in children.[2-4] It occurs idiopathically at all ages, and it has been reported in association with metastatic carcinoma,[5] chemical poisoning,[6] and chronic infection, such as tuberculosis.[7] In children, it has been associated with acute myelogenous leukemia,[2] McCune-Albright syndrome,[8] and histiocytosis X.[9] There is no sex predominance. Initial complaints are the insidious onset of dyspnea, weakness, fatigue, and loss of weight; marked splenomegaly produces a feeling of fullness in the abdomen. There are, occasionally, initial hemorrhagic symptoms.

Hematologic Features. Hematologic features are normochromic normocytic anemia, with immature forms of red blood cells and leukocytes. Platelet count is normal or low. Leukocyte count varies from leukopenia to leukocytosis. Polycythemia may precede myelosclerosis.

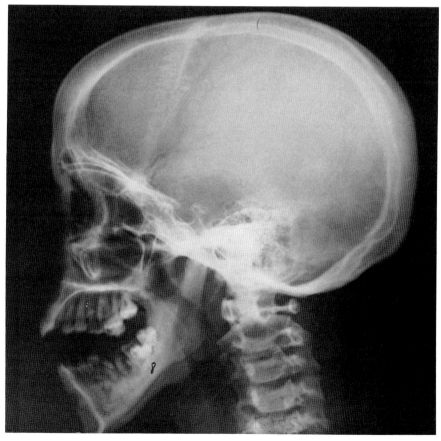

FIG. 1176. IRON DEFICIENCY ANEMIA
Widening of the diploic space and thinning of the tables. The occipital area is relatively unharmed.

Out of 61 cases, Wasserman found that 59% had antecedent polycythemia. Bone marrow aspiration often fails due to thick sclerotic bone, which is often the first diagnostic clue. Surgical removal of a portion of bone marrow will reveal marrow fibrosis or sclerosis.

Pathologic Features. The histologic appearance of bone marrow varies from fibrous to bony replacement types. Windholz and Foster[11] suggest that the fibrous tissue is more mature than the fibrosis of leukemia. Marrow cellular depopulation of unknown cause accompanies the fibrosis. Extramedullary hematopoiesis occurs in the liver, spleen, skin, adrenals, lung, choroid plexus, and lymph nodes.[12]

Roentgenographic Features. There is roentgenographic evidence of osteosclerosis in 40% of patients.[1, 13-16] The thoracic cage and pelvis are most frequently affected; the femora, humeral shafts, and lumbar spine are usually involved, and often the skull and peripheral bones are affected.[16-19]

The characteristic roentgenographic feature is a widespread, diffuse, variable increase in bone density (Figs. 1177-1179). A uniform increase in density "ground glass," is frequent. Mottled density also occurs (Figs. 1180-1182). The proximal humerus and the femur display this change particularly well, and coalescing islands of increased density are readily apparent. The bone ends may be sclerotic, with mottled densities toward the shaft (Figs. 1183-1185). Interspersed round radiolucent areas throughout are due to fibrotic foci. Fine trabecular markings may be completely obliterated, and the ribs may show striking sclerosis, appearing as "jail bars" crossing the thorax (Fig. 1186). With advancing sclerosis, the trabecular pattern is lost, and the endosteal cortical margins become indistinct, blending with the spongiosa, especially in the end plates of the spine described as "sandwich" or "rugger jersey." The cortices may be thickened endosteally and encroach on the medullary cavities, eventually obliterating them, especially in the vertebrae and long bones. This takes place endosteally, and the external contour is not usually altered; however, irregular thin or thick periosteal new bone formation may occur, most often in the medial

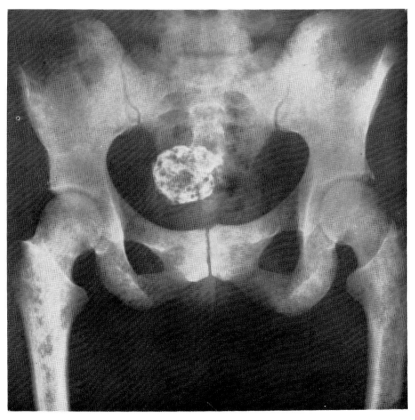

FIG. 1177. AGNOGENIC MYELOID METAPLASIA
Sclerosis of the pelvis is uniform and can be distinguished from the changes in Paget disease or osteoblastic metastasis. Large calcified fibromyoma of the uterus.

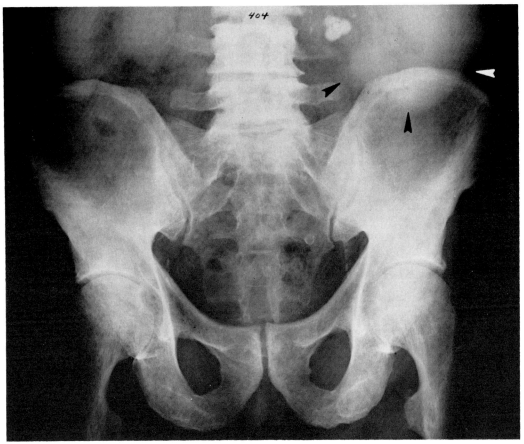

Fig. 1178. Agnogenic Myeloid Metaplasia
Homogeneous increased density of the iliac bones, femoral heads, and lower lumbar vertebrae. Mottled increased density in the descending rami of the ischium. The lower end of the markedly enlarged spleen overlies the left iliac wing (*arrows*).

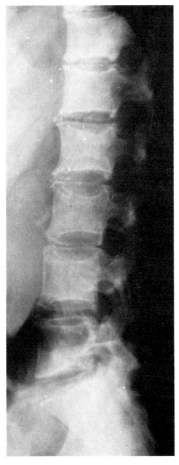

Fig. 1179. Agnogenic Myeloid Metaplasia
Diffuse "ground glass" density of the lumbar vertebrae.

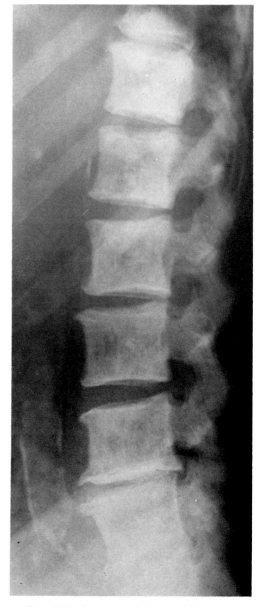

Fig. 1180. Agnogenic Myeloid Metaplasia
Diffuse mottled increased density throughout the lumbar vertebrae.

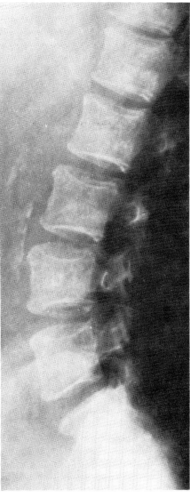

Fig. 1181. Agnogenic Myeloid Metaplasia
Spotty increased density throughout the lumbar spine. (Calcification of the abdominal aorta. (Same case as Figure 1182.)

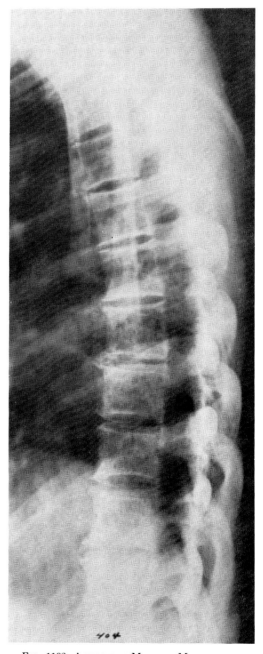

Fig. 1182. Agnogenic Myeloid Metaplasia
Spotty increased density throughout the centra of the thoracic vertebrae. (Same case as Figure 1181.)

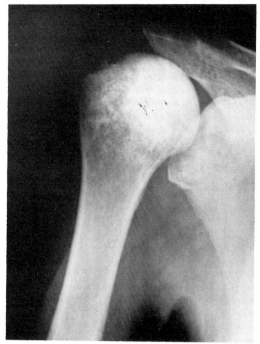

FIG. 1183. AGNOGENIC MYELOID METAPLASIA
Homogeneous increased density in the proximal end of the humerus and mottled densities extending toward the shaft in the lateral aspect of the humerus.

margins of the distal femoral shafts, the lateral margins of the distal femoral shafts, and the lateral margins of the upper tibial shafts. Similar changes are less common in other long bones.[18] Periosteal proliferation occasionally occurs in the pelvis, but this is a normal finding in some older individuals (Fig. 1187).

Skull changes are of three types: (1) generalized increase in bone density, with obliteration of the diploic space and decreased vascular markings, (2) scattered, small, rounded radiolucent lesions throughout, and (3) increased sclerotic density and radiolucent lesions combined.[18] There is no correlation between the duration of the disease, the patient's age, splenomegaly, and extent of osteosclerosis. The diagnosis may be suspected from osteosclerosis in a middle-aged adult with splenomegaly. Chronic leukemia, lymphoma, and mastocytosis are distinguished by their histologic and clinical features.

If splenomegaly is not apparent, or if the spleen has been removed, the differential diagnosis must include osteoblastic metastasis, fluorine poisoning, osteopetrosis, and chronic renal disease, but these conditions are easily distinguished by clinical and laboratory means.

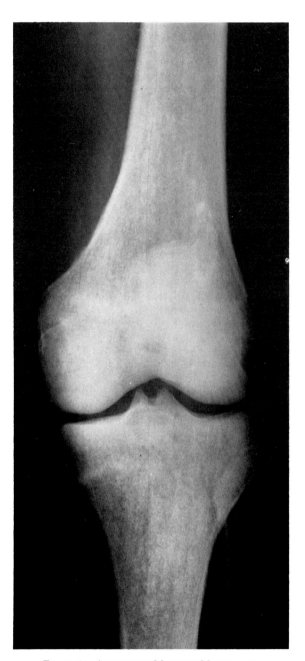

FIG. 1184. AGNOGENIC MYELOID METAPLASIA
The distal femur and the proximal tibia show diffuse homogeneous increased density. Mottled sclerosis toward the shaft of the tibia.

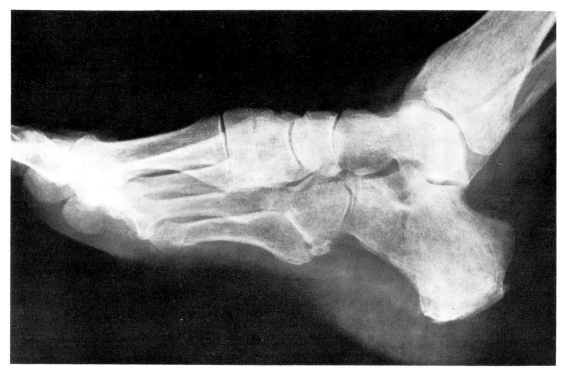

FIG. 1185. AGNOGENIC MYELOID METAPLASIA
Mottled sclerosis through the bones of the ankle, particularly noticeable in the calcaneus and the distal end of the tibia.

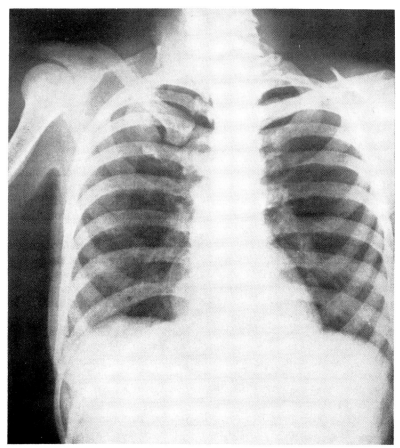

FIG. 1186. AGNOGENIC MYELOID METAPLASIA
Diffuse uniform sclerosis of the bones of the thorax.

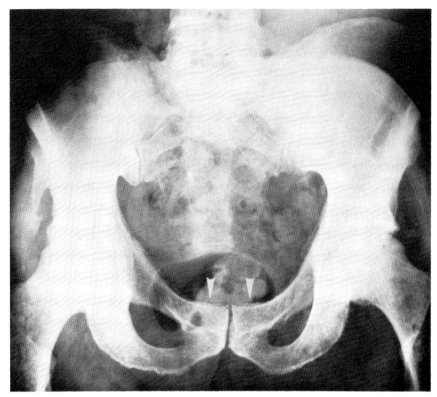

FIG. 1187. AGNOGENIC MYELOID METAPLASIA WITH PERIOSTEAL REACTION
Diffuse sclerosis of the iliac bones, lower lumbar vertebrae, and the femoral heads. The pubic rami show periosteal reaction (*arrows*). The ischial rami also show sclerosis. The lower border of the markedly enlarged spleen overlies the left iliac crest.

EXTRAMEDULLARY HEMATOPOIESIS

Extramedullary hematopoiesis results from prolonged erythrocyte deficiency due to the destruction of red blood cells, or the inability of normal blood forming organs to produce them. To compensate, blood formation occurs in the areas of fetal erythropoiesis, the liver, spleen, adrenal, heart, thymus, lung, lymph nodes, renal pelvis, gastrointestinal lymphatics, and dura mater.[1,2] The lesions are very vascular, and biopsy may lead to copious bleeding. Extramedullary hematopoiesis is often due to a congenital hemolytic anemia, most frequently sickle cell anemia. It is also reported in patients with severe anemia of unknown cause, myelofibrosis,[3] polycythemia,[4] erythroblastosis fetalis,[5] leukemia,[6] Hodgkin disease,[7] carcinomatosis,[8] hyperparathyroidism,[9] and rickets.[10] It sometimes appears, however, in patients without anemia or known disease.[11,12]

Roentgenographic Features. Roentgenographically, extramedullary hematopoiesis has been demonstrated in the thorax and spinal canal.[5,13,14]

Thoracic extramedullary hematopoiesis is usually located in the paraspinal area at the level between T8 and T12 and may extend from T2.[15] The paravertebral mass may be bilateral. The lateral borders are well defined, with rounded margins often lobulated. The medial border cannot be distinguished from the mediastinum (Fig. 1188).

Intraosseous hematopoiesis with erosion is reported. In the lateral projection, the mass overlies the spinal shadows and is often difficult to define. If the mass extends anterior to the spine, a well defined lobulated or rounded anterior border can be distinguished. The absence of pain, bone erosion, or calcification aids in differentiating it from other mediastinal masses.

Spinal extramedullary hematopoiesis may compress the spinal cord and be demonstrated myelographically.[16,17] Renal hematopoiesis may produce a peripelvic mass demonstrated by urography. Extramedullary hematopoietic tissue is reported over the brain,[18] in the falx cerebri,[1] and in the renal pelvis.[19-24]

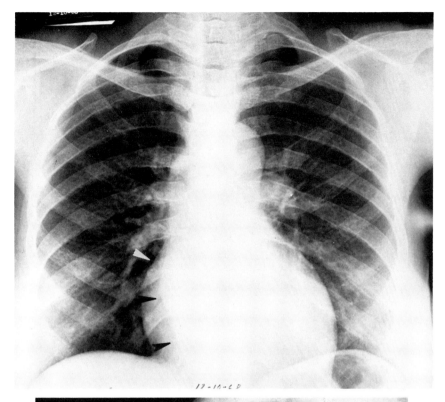
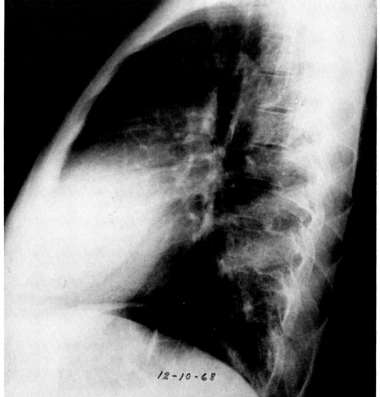

FIG. 1188. EXTRAMEDULLARY HEMATOPOIESIS IN PATIENT WITH SICKLE CELL ANEMIA
(*Top*) Posteroanterior projection. A mass is located in the right paraspinal area with well defined lateral borders (*arrows*). The medial border cannot be distinguished from the mediastinum. (*Bottom*) Lateral projection. The mass overlies the spine and cannot be seen.

REFERENCES

Hereditary Anemias

1. Neel, J. V.: The inheritance of sickle cell anemia. Science, *110:* 64, 1949.
2. Pauling, L., Itano, H. A., Singer, S. J., and Wells, I. C.: Sickle cell anemia, a molecular disease. Science, *110:* 543, 1949.
3. LeWald, L. T.: Roentgen evidence of osseous manifestations in sickle cell (drepanocytic) anemia and in Mediterranean (erythroblastic) anemia. Radiology, *18:* 792, 1932.
4. Hewett, B. V., and Nice, C. M., Jr.: Radiographic manifestations of sickle cell anemia. Radiol Clin North Am, *2:* 249, 1964.
5. Carroll, D. S., and Evans, J. W.: Roentgen findings in sickle cell anemia. Radiology, *53:* 834, 1949.
6. Legant, O., and Ball, R. P.: Sickle cell anemia in adults; roentgenographic findings. Radiology, *51:* 665, 1948.
7. Smith, E. W., and Conley, C. L.: Sicklemia and infarction of the spleen during aerial flight. Bull Johns Hopkins Hosp, *96:* 35, 1955.
8. Smith, E. W., and Conley, C. L.: Clinical features of the genetic variants of sickle cell disease. Bull Johns Hopkins Hosp, *94:* 289, 1954.
9. Reynolds, J.: Roentgenographic and clinical appraisal of sickle cell hemoglobin-C disease. AJR, *88:* 512, 1962.
10. Moseley, J. E. *Bone Changes in Hematologic Disorders (Roentgen Aspects).* Grune & Stratton, New York, 1963.
11. Golding, J. S., MacIver, J. E., and Went, L. N.: Bone changes in sickle cell anemia and its genetic variants. J Bone Joint Surg, *41B:* 711, 1959.
12. Barton, C. J., and Cockshott, W. P.: Bone changes in hemoglobin S-C disease. AJR, *88:* 523, 1962.
13. Moseley, J. E.: Patterns of bone change in sickle cell states. J Mt Sinai Hosp, *26:* 424, 1959.
14. Henkin, W. A.: Collapse of the vertebral bodies in sickle cell anemia. AJR *62:* 395, 1949.
15. Cole, W. R.: Sickle cell anemia—radiological features. In *British Surgical Practice, Surgical Progress 1955,* p. 126. Butterworth, London, 1955.
16. Hamburg, A. E.: Skeletal changes in sickle cell anemia; report of an unusual case. J Bone Joint Surg, *32A:* 893, 1950.
17. Sebes, J. I., and Diggs, L. W.: Radiographic changes of the skull in sickle cell anemia. AJR, *132:* 373, 1979.
18. Reynolds, J.: A reevaluation of the "fish vertebra" sign in sickle cell hemoglobinopathy. AJR, *97:* 693, 1966.
19. Cassady, J. R., Berdon, W. E., and Baker, D. H.: The "typical" spine changes of sickle cell anemia in a patient with thalassemia major (Cooley's anemia). Radiology, *89:* 1065, 1967.
20. Junghagen, S.: Rontgenologische skelettveranderungen bei morbus Gaucher. Acta Radiol, *5:* 506, 1926.
21. Hansen, G. C., and Gold, R. H.: Central depression of multiple vertebral end plates; a "pathognomonic" sign of sickle hemoglobinopathy in Gaucher's disease. AJR, *129:* 343, 1977.
22. Rohlfing, B. M.: Vertebral end-plate depression; report of two patients without hemoglobinopathy. AJR, *128:* 599, 1977.
23. Ennis, J. T.: The radiologic pattern of bone infarction in the haemoglobinopathies. Br J Radiol, *51:* 71, 1978.
24. Rowe, C. W., and Haggars, M. E.: Bone infarcts in sickle cell anemia. Radiology, *68:* 661, 1957.
25. Buchman, J.: Sickle cell disease simulating osteomyelitis. Bull Hosp Joint Dis, *10:* 239, 1949.
26. Moseley, J. E.: Skeletal changes in the anemias. Semin Roentgenol, *9:* 169, 1974.
27. Bohrer, S. P. S.: Growth disturbances of the distal femur following sickle cell bone infarcts and/or osteomyelitis. Clin Radiol, *25:* 221, 1974.
28. Bohrer, S. P.: Fracture complicating bone infarcts and/or osteomyelitis in sickle cell disease. Clin Radiol, *22:* 83, 1971.
29. Ellenbogen, N. E., Raim, J., and Grossman, L.: *Salmonella sp.* (type Montevideo) osteomyelitis. Am J Dis Child, *90:* 275, 1955.
30. Vandepitte, J., Colaert, J., Lambotte-Legrand, J., Lambotte-Legrand, C., and Perin, F.: Les osteites a salmonella chez les sicklanemiques. Ann Soc Belg Med Trop, *33:* 511, 1953.
31. Carrington, H. T., Ferguson, A. D., and Scott, R. B.: Studies in sickle-cell anemia; XI. Bone involvement simulating aseptic necrosis. Am J Dis Child, *95:* 157, 1958.
32. Burch, J. E.: Paratyphoid osteomyelitis. South Med J, *42:* 135, 1949.

Thalassemia

1. Ingram, V. M., and Stretton, A. O. W.: Genetic basis of the thalassemia disease. Nature, *184:* 1903, 1959.
2. Smith, C. H.: *Blood Disorders of Infancy and Childhood.* C. V. Mosby, St. Louis, 1960.
3. Caffey, J.: Cooley's anemia; a review of the roentgenographic findings in the skeleton. AJR, *78:* 381, 1957.
4. Caffey, J.: Cooley's erythroblastic anemia; some skeletal findings in adolescents and young adults. AJR, *65:* 547, 1951.
5. Reynolds, J.: Re-evaluation of fish "vertebra" sign in sickle cell hemoglobinopathy. AJR, *97:* 693, 1966.
6. Riggs, W., Jr., and Rockett, J. F.: Roentgen chest findings in childhood sickle cell anemia; a new vertebral body finding. Radiology, *104:* 838, 1968.
7. Bohrer, S. P.: Acute long bone diaphyseal infarcts in sickle cell disease. Br J Radiol, *43:* 685, 1970.

Hereditary Spherocytosis (Congenital Hemolytic Jaundice)

1. Moseley, J. E.: *Bone Changes in Hematologic Disorders (Roentgen Aspects).* Grune & Stratton, New York, 1963.
2. Snelling, C. C., and Brown, A.: Case of hemolytic jaundice with bone changes. J Pediatr, *8:* 330, 1936.

Iron Deficiency Anemia

1. Ascenzi, A., and Marinozzi, V.: Sur le crane en bosse au cours des polyglobulies secondaries a l'hypoxemie chronique. Acta Haematol (Basel), *19:* 253, 1958.
2. Dykstra, O. H., and Halbertsma, T.: Polycythemia vera in childhood; report of case with changes in skull. Am J Dis Child, *60:* 907, 1940.
3. Smith, C. H.: *Blood Diseases of Infancy and Childhood.* C. V. Mosby, St. Louis, 1960.
4. Lie-Injo Luan Eng.: Chronic iron deficiency anemia with bone changes resembling Cooley's anemia. Acta Haematol (Basel), *19:* 263, 1958.
5. Caffey, J.: Cooley's anemia; a review of the roentgenographic findings in the skeleton. AJR, *78:* 381, 1957.
6. Agarwal, K. N., Dhar, N., Shah, M. M., and Bhardwaj, P.: Roentgenologic changes in iron deficiency anemia. AJR *110:* 1970.

Agnogenic Myeloid Metaplasia (Myelosclerosis)

1. Leigh, T. H., Corley, C. C., Jr., Huguley, C. M., Jr., and Rogers, J. V., Jr.: Myelofibrosis; the general and radio-

logic findings in 25 proved cases. AJR, 82: 183, 1959.
2. Say, B., and Berke, I.: Idiopathic myelofibrosis in an infant. J Pediatr, 64: 580, 1964.
3. Boxer, L. A., Camitta, B. M., Berenberg, W., and Fanning, J. P.: Myelofibrosis; myeloid metaplasia in childhood. Pediatrics, 55: 861, 1975.
4. Tebbi, K., Zarkowsky, H. S., Siegel, B. A., McAllister, W. H.: Childhood myelofibrosis and osteosclerosis without myeloid metaplasia. J Pediatr, 84: 860, 1974.
5. Kiely, J. M., and Silverstein, M. N.: Metastatic carcinomas simulating agnogenic myeloid metaplasia and myelofibrosis. Cancer, 24: 1041, 1969.
6. Feldman, F.: Myelosclerosis in agnogenic myeloid metaplasia. Semin Roentgenol, 9: 195, 1974.
7. Samuelson, S. M., Killander, A., Werner, I., and Stenkvist, B.: Myelofibrosis associated with tuberculosis lymphadenitis. Acta Med Scand, 179: 326, 1966.
8. Shantharaj, S., Gilman, S., Maurer, H. S., and Rosenthal, I. M.: Hyperthyroidism in an infant with McCune-Albright syndrome; report of a case with myeloid metaplasia. J Pediatr, 80: 275, 1972.
9. Pinckney, L., and Parker, B. R.: Myelosclerosis and myelofibrosis in treated histiocytosis-X. AJR, 129: 521, 1977.
10. Wasserman, L. R.: Polycythemia vera—its course and treatment; relation to myeloid metaplasia and leukemia. Bull NY Acad Med, 39: 343, 1954.
11. Windholz, F., and Foster, S. E.: Bone sclerosis in leukemia and in non-leukemic myelosis. AJR, 61: 61, 1949.
12. Wyatt, J. P., and Sommers, S. C.: Chronic marrow failure, myelosclerosis and extramedullary hematopoiesis. Blood, 5: 329, 1950.
13. Sussman, M. L.: Myelosclerosis with leukoerythroblastic anemia. AJR, 57: 313, 1947.
14. Mulchahy, F.: Bone changes in myelosclerosis. Proc R Soc Med, 50: 100, 1957.
15. Bouroncle, B. A., and Doan, C. A.: Myelofibrosis; clinical and hematologic and pathologic study of 110 patients. Am J Med Sci, 243: 697, 1962.
16. Pettigrew, J. D., and Ward, H. P.: Correlation of radiologic, histologic, and clinical findings in agnogenic myeloid metaplasia. Radiology, 93: 541, 1969.
17. Jacobson, H. G., Fateh, H., Shapiro, J. H., Spaet, T. H., and Poppel, M. H.: Agnogenic myeloid metaplasia. Radiology, 72: 716, 1959.
18. Moseley, J. E.: *Bone Changes in Hematologic Disorders (Roentgen Aspects)*. Grune & Stratton, New York, 1963.
19. Killman, S. A.: Myelofibrosis. Clin Orthop, 52: 95, 1967.

Extramedullary Hematopoiesis

1. Brannan, D.: Extramedullary hematopoiesis in anemias. Bull Johns Hopkins Hosp, 41: 104, 1927.
2. Close, A. S., Taira, Y., and Cleveland, D. A.: Spinal cord compression due to extramedullary hematopoiesis. Ann Intern Med, 48: 421, 1958.
3. Donhauser, J. L.: Human spleen as haematoplastic organ exemplified in case of splenomegaly with sclerosis of bone marrow. J Exp Med, 10: 559, 1908.
4. Wintrobe, M. M.: *Clinical Hematology*, Ed. 4. Lea & Febiger, Philadelphia, 1956.
5. Covey, G. W.: Erythroblastosis; report of case presenting erythroblastic tumor in thoracic cavity. Am J Pathol, 11: 551, 1935.
6. Schiller, W.: Local myelopoiesis in myeloid leukemia. Am J Pathol, 19: 809, 1943.
7. Wintrobe, M. M.: Relation of disease of liver to anemia; type of anemia, response to treatment, and relation of type of anemia to histopathologic changes in liver, spleen and bone marrow. Arch Intern Med, 57: 289, 1936.
8. Jordan, H. E.: Extramedullary blood production. Physiol Rev, 22: 375, 1942.
9. Young, J. K., and Cooperman, M. B.: Von Recklinghausen's disease or osteitis fibrosa, with report of case presenting multiple cysts and giant cell tumors. Ann Surg, 75: 171, 1922.
10. Jordan, H. E.: Extramedullary erythrocytopoiesis in man. Arch Pathol, 18: 1, 1934.
11. Foster, J. B. T.: Primary thoracic myelolipoma; case report. Arch Pathol, 65: 295, 1958.
12. Saleeby, E. R.: Heterotopia of bone marrow without apparent cause. Am J Pathol, 1: 69, 1925.
13. Ask-Upmark, E.: Tumor simulating intrathoracic heterotopia of bone marrow. Acta Radiol, 26: 425, 1945.
14. Katz, I., and Dziadiw, R.: Localized mediastinal lymph node hyperplasia; a report of case with roentgen findings simulating posterior mediastinal neurofibroma. AJR, 84: 206, 1960.
15. Marinozzi, V.: Aspetti insoliti dell'iperplasia midollare nelle anemie emolitiche. Haematologica, 43: 737, 1958.
16. Ross, P., and Logan, W.: Roentgen findings in extramedullary hematopoiesis. AJR, 106: 604, 1969.
17. Sorsdahl, O. S., Taylor, P. E., and Noyes, W. D.: Extramedullary hematopoiesis; mediastinal masses, and spinal cord compression. JAMA, 189: 343, 1964.
18. Hu, C. H., and Cash, J. R.: Erosion of inner table of skull by hyperplasia of bone marrow in kala-azar, with extramedullary formation of blood on surface of dura. Far East Asia Trop Med Trans Seventh Congr, 3: 80, 1930.
19. Fawcett, J., and Boycott, H. E.: Bone marrow in hilus of kidney. J Pathol Bacteriol, 14: 404, 1909–1910.
20. Herzenberg, J.: Zur Frage der Heterotropie des Knochenmarkes. Arch Pathol Anat, 239: 145, 1922.
21. Matxunaga, R.: Uber myeloide Zellherde im Nierenhilusbindegewebe bei Leukamie. Zentralbl Allg Pathol, 29: 337, 1918.
22. McKenzie, I., Browning, C. H., and Dunn, J. S.: Occurrence of bone marrow in hilus of kidney in children. J Pathol Bacteriol, 14: 139, 1909–1910.
23. Schultze, W. H.: Uber tumorformige Bildung Myeloiden Gewebes im Bindegewebe des Nierenhilus. Verh Dtsch Ges Pathol, 15: 45, 1912.
24. Tanaka, T.: Uber Knochenmarkgewebsentwicklung in Nierenhilusbindegewebe bei Anaemia Splenica (Anaemia Pseudoleucaemia infantum); Beitrage zur Kenntnis dieser Krankheit. Beitr Pathol Anat, 53: 338, 1912.

10

Congenital Hip Dislocation

Until recently, dislocation was generally believed to occur several months after birth.[1,2] It is now realized, however, that congenital hip dislocations are all present at birth. Abnormal hip position in utero leads to stretching of the soft-tissue elements and subsequent dislocation. If untreated, the degree of dislocation increases.[3–5] If recognized early and treated properly, they may be completely corrected. The acetabulum will develop normally only if the femoral head is held in position, subjecting it to normal muscular stress. The femoral head should be maintained in the acetabulum to form a normal cup.

A laxity or tear of the capsule and ligaments is the prime structural factor in the instability.[6–9] Other factors may play a role, and there may be a mesodermal abnormality, for there is a tendency toward generalized ligamentous laxity.[10] Andren and Borglin[11] believe that hip dislocation is often associated with instability of the symphysis pubis, which they ascribe to an inborn error of estrogen metabolism. A hereditary tendency occurs in approximately 20%.[9,12,13] Wynne-Davies[14,15] believes there are two gene systems reacting: One relates to acetabular dysplasia, and the other to capsular laxity. They may be separate or combined, but there is a preponderance of the laxity type in the newborn and of the acetabular type in the later-diagnosed patients.

Hip dysplasia occurs with almost equal frequency in both sexes, but frank dislocation is 6 times more frequent in girls.[12] Hip dislocation is 10 times more frequent in breech-delivered children[12]; 23% of children with congenital hip disease were breech deliveries,[10] and of these, the girl to boy ratio is only 2 to 1.[12] Hip dislocations occur in families in the higher income groups.[15] There is a preponderance of first-born children,[16,17] and there is a higher incidence of second-born children than of later-born children.[12] Children born in the winter months have a higher incidence.[15]

Anteversion has been considered a causative factor. It is probably the result of dislocation, since it spontaneously improves after reduction.[9] The Ortolani sign[18] of the jerk, click, or snap is the earliest positive and pathognomonic evidence of hip dysplasia with subluxation and may be useful in the neonatal period when roentgen studies are difficult to interpret and the femoral capital epiphysis is unossified. The "click" is produced when the femoral head fovea suddenly meets and rides over the acetabular ridge. The abduction limitation immediately disappears, indicating reduction. If reduction is performed on all newborns, the late sequelae may be almost completely abolished, for the sign is present in a high percentage of patients. If the dislocation is overlooked in the neonatal period, a progression of hip changes ensue. In the first few days, the Ortolani sign may be the only evidence of hip dislocation. At this stage the hip joint is lax and reversibly dislocatable with an adduction contracture. Many new-born hips are unstable; most stabilize, but all should be treated. On the first day of life, 1 infant in 60 has instability of one or both hips. Without treatment, 68% stabilize within 1 week, and 88% by third month, leaving 12% with instability. This is 1.5 per 1,000 live births, approximately the incidence of dislocation.[4] Most dislocations occur within the first 2 weeks. Some occur later when weight-bearing begins,[9] and no acetabular dysplasia will be evident.

Roentgenographic Features. Dysplastic hips are not always dislocated, but they are on their way. In the first 2 or 3 months, the femoral capital epiphysis is unossified, and roentgen interpretation, with or without dislocation, is difficult.[19] Adequate neonatal clinical examination will uncover most congenital hip dysplasias; routine mass neonatal x-ray studies without clinical examination lead only to confusion and unnecessary exposure.[19]

The most useful position is that of the hips extended and of each leg abducted to 45° in full internal rotation.[20] If the hip is not dislocated, a line through the long axis of the femur points to the lateral part of the acetabulum and crosses the spine at the level of the lumbosacral joint (Fig. 1189). With dislocation, the line extends lateral to the acetabulum and crosses the spine at a higher level.

Shenton's line is an imaginary continuous curve formed by the inferior surface of the femoral neck and the superior pubic ramus (Fig.

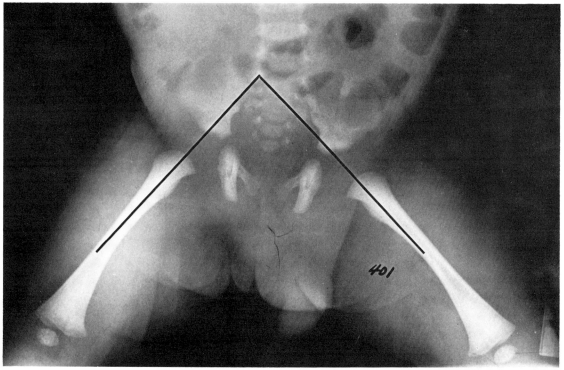

FIG. 1189. NORMAL HIPS

The hips are extended, and each extremity is adducted 45° and in full internal rotation. A line through the long axis of the femur overlies the lateral portion of the acetabulum and crosses the spine at the level of the lumbosacral joint. With dislocation, the line extends lateral to the acetabulum and crosses the spine at a higher level.

1190); disturbance of this line is evidence of hip dysplasia.

Garavaglia[21] describes changes in the upper portion of the acetabular roof and soft-tissue alterations. A small triangular area of sclerosis forms in the upper end of the acetabular roof,[22] and a small fossa or groove appears in the upper end of the acetabulum.[23] An arcuate radiolucent shadow normally extends from the superior acetabular roof to the great trochanter, probably representing the articular capsule surrounding the femoral head and providing information on its shape and position before it ossifies. Normally it forms a continuous arc with the acetabular roof and acetabular portions of the pubis, but with dislocation the arc is interrupted and loses its sharp circular appearance. It may become enlarged and irregular in contour. A triangular radiolucent area may form beneath the superior acetabular roof.[22]

A steep acetabular inclination occurs in the first month.[5] It is measured by the acetabular index, the angle formed between the roof or iliac portion of the acetabulum and a horizontal line passing through the Y or triradiate cartilage. In a child the normal acetabular index is 20°; 30° or more indicates dysplasia (Fig. 1190).

A relative increase in distance from the upper end of the femur to the acetabular floor is also an indication. The ossification center of the femoral capital epiphysis is delayed, and hypoplastic when it does develop. Normal femoral head ossification does not occur until 3-6 months or later. It is not of diagnostic value during infancy but is helpful after the femoral head has begun to form (Fig. 1191).

Bilateral dislocation is frequent; Jacobs[5] reports 50% are bilateral, and Ortolani[24] believes it is always bilateral. The dislocations are rarely if ever symmetrical, and one side is usually more marked.

In older children, flattening of the acetabulum becomes evident, and measurements are unnecessary. After weight-bearing, subluxation and dislocation frequently occur. The femoral head is displaced cephalad and posteriorly, lying on the dorsal aspect of the ilium where it rests in a shallow depression or false acetabulum (Fig. 1192). When the dislocation is long standing, the femoral head becomes conical, and the femoral

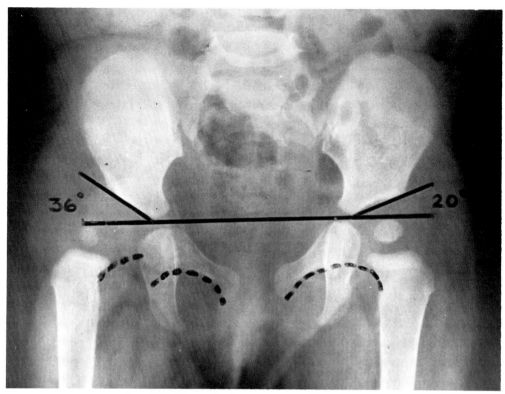

Fig. 1190. Right Hip Dysplasia with Dislocation

The right hip is dislocated, and the femoral capital epiphysis is smaller than the left. The acetabular roof on the abnormal side is steeply inclined and measures 36° compared to the normal 20° on the left side. The *solid lines* measure the acetabular index, which is the angle formed between the roof or iliac portion of the acetabulum and the horizontal line passing through the Y or triradiate cartilages. The *dotted lines* represent Shenton's line, which is an imaginary continuous curve formed by the inferior surface of the femoral neck and the superior pubic ramus. On the left side, this line is continuous, whereas on the right side, it is broken due to the dislocation.

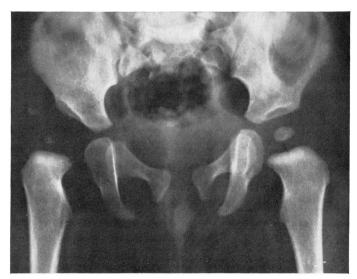

Fig. 1191. Congenital Hip Dislocation

Congenital hip dislocation of the right hip. The acetabulum is shallow, and the femur is displaced upward. The epiphysis of the femoral head is hypoplastic. The angle of inclination of the acetabulum is acute.

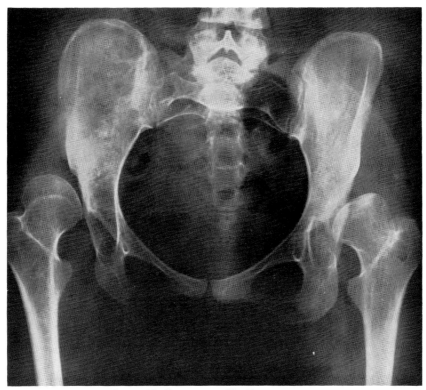

FIG. 1192. CONGENITAL HIP DISLOCATION

Bilateral congenital dislocation of the hips. The femoral heads lie in shallow false acetabulae. The iliae are hypoplastic as the result of abnormal weight-bearing.

neck is shortened and displaced horizontally. The neck of the femur is twisted forward and foreshortened in the anteroposterior radiographic projection.

With minor subluxation, secondary osteoarthritic changes occur. Murray[25] believes the etiology of primary hip osteoarthritis may be minor anatomical abnormalities responsible for articular incongruity. These abnormalities may be due to minimal or subclinical subluxation of the hip and of a shallow acetabulum. Osteochondrosis of the capital femoral epiphysis is reported, probably related to manipulative procedures causing compression in reduction.[23]

REFERENCES

1. Caffey, J., Ames. R., Silverman, W. A., Ryder, C. T., and Hough, G.: Contradiction of the congenital dysplasia—predislocation hypothesis of congenital dislocation of the hip through a study of the normal variation in acetabular angles at successive periods in infancy. In *Pediatrics*. Charles C Thomas, Springfield, Ill., 1956.
2. Wittenborg, M. H.: Malposition and dislocation of the hip in infancy and childhood. Radiol Clin North Am, *2:* 235, 1964.
3. von Rosen, S.: Diagnosis and treatment of congenital dislocation of the hip joint in the newborn. J. Bone Joint Surg, *44B:* 284, 1962.
4. Barlow, T. G.: Early diagnosis and treatment of congenital dislocation of the hip. J Bone Joint Surg, *44B:* 292, 1962.
5. Jacobs, P.: Detection of early congenital dislocation of the hip. Proc R Soc Med, *59:* 1225 (Section of Radiology, pp. 35–39), 1966.
6. Andren, L.: Instability of the pubic symphysis and congenital dislocation of the hip in newborns. Acta Radiol, *54:* 123, 1960.
7. Carter, C., and Wilkinson, J.: Persistent joint laxity and congenital dislocation of the hip. J Bone Joint Surg, *46B:* 40, 1964.
8. Howorth, B.: The etiology of congenital dislocation of the hip. Clin Orthop, *29:* 164, 1963.
9. Salter, R. B.: Etiology, pathogenesis and possible prevention of congenital dislocation of the hip. Can Med Assoc J, *98:* 933, 1968.
10. Wilkinson, J. A.: Prime factors in the etiology of congenital dislocation of the hip. J Bone Joint Surg, *45B:* 268, 1963.

11. Andren, L., and Borglin, N. E.: A disorder of oestrogen metabolism as a causal factor of congenital dislocation of the hip. Acta Orthop Scand, *30:* 169, 1960.
12. Andren, L.: Aetiology and diagnosis of congenital dislocation of the hip in newborns. Radiologe, *1:* 89, 1961.
13. Cox, D. W.: Unpublished data.
14. Wynne-Davies, R.: Acetabular dysplasia and familial joint laxity; two etiological factors in congenital dislocation of the hip. J Bone Joint Surg, *52B:* 704, 1970.
15. Wynne-Davies, R.: A family study of neonatal and late-diagnosis congenital dislocation of the hip. J Med Genet, *7:* 315, 1970.
16. Wessel, A. B.: Laaghalte slegter i finmarken. Tidsskr Nor Laegeforen, *8:* 337, 1918.
17. Record, R. G., and Edwards, J. H.: Environmental influences related to the aetiology of congenital dislocation of the hip. Br J Prev Soc Med, *12:* 8, 1958.
18. Ortolani, M.: *La Lussazione Congenita Dell 'anca: Nuovi Criteri Diagnostici E Profilattico-Correttivi.* Cappelli, Bologna, 1948.
19. Hart, V. L.: Congenital dysplasia of the hip joint and sequelae. In *The Newborn and Early Post-natal Life.* Charles C Thomas, Springfield, Ill., 1952.
20. Andren, L., and von Rosen, S.: Diagnosis of dislocation of hip in newborns and primary results of immediate treatment. Acta Radiol, *49:* 89, 1958.
21. Garavaglia, C.: Early diagnosis of congenital dysplasia of the hip; new roentgenologic signs. AJR, *110:* 587, 1970.
22. Garavaglia, C.: Diagnosi precoce della displasia congenita dell 'anca. Conferenza di aggiornamento XXIV Congresso della S.I.R.M.N., Palermo, April, 1970.
23. Conrad, M. B.: Congenital dislocation of the hip. Instructional Course Lectures, *18:* 207, 1961.
24. Ortolani, M.: Proceedings and reports of councils and associations. J Bone Joint Surg, *43B:* 194, 1961.
25. Murray, R. O.: The aetiology of primary osteoarthritis of the hip. Br J Radiol, *38:* 810, 1965.

11
Spondylolisthesis

Spondylolisthesis is forward displacement of a vertebral body, resulting from a defect in the neural arch. The pars interarticularis, that part of the neural arch lying between the superior and inferior articulating processes, is the site of the defect. Usually the neural arch is open bilaterally, although the defect may be unilateral. The neural arch fault in spina bifida is in the lamina, but this will not cause spondylolisthesis. A defect in the pars interarticularis without forward slipping is known as spondylolysis[1] or pre-spondylolisthesis, which is not accurate, since slipping may never occur. Spondylolisthesis occurs most commonly at the lumbosacral junction or the fourth lumbar interspace, but it may occur at any lumbar interspace and occasionally in the cervical region (Fig. 1193). Congenital spondylolisthesis of the sixth cervical vertebra has been described.[2, 3] In these instances the posterior element of the sixth cervical vertebra is small and bifid. The previously published cases have all been in male patients; neurologic abnormalities were absent. It is still possible that it is not a congenital spondylolisthesis but is a developmental abnormality. It is rare in the thoracic region. The defect of the neural arch has not been found in infants, and trauma with fracture is probably the mechanism; a primary congenital weakness may be a factor. Trauma is clearly involved as far as symptoms are concerned, since at the level of spondylolisthesis the spinal column is unstable and, therefore, more prone to injury.

In many instances, spondylolisthesis is asymptomatic. The symptoms are not related to the degree of spondylolisthesis; some patients with low-grade spondylolisthesis have great disability, while others with Grade IV deformity have little difficulty. The most frequent symptom is low backache, which may be progressive. Eventually, neurologic symptoms may develop, including pain referred to one or both legs. Low back pain is probably due to muscle spasm secondary to instability. Neurologic complications may result from degeneration of the intervertebral disk, with protrusion and impingement on nerves.

Roentgenographic Features. The roentgenologic diagnosis is based on the recognition of the defect in the neural arch and anterior displacement of the vertebral body.[4] The anteroposterior view is not reliable for diagnosis, although a bilateral neural arch defect may be visible (Fig. 1194). Sometimes the body of the fifth lumbar vertebra is superimposed on the first sacral segment, but this appearance can also

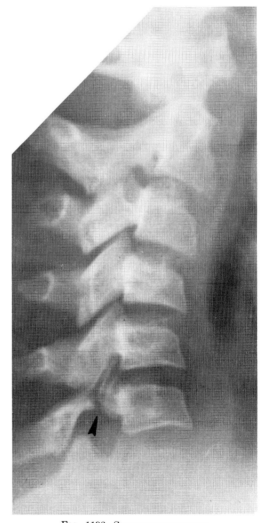

FIG. 1193. SPONDYLOLISTHESIS
Defect in the neural arch of C6 (*arrow*). No slipping is evident in this view. Cervical spondylolisthesis is rare. (Courtesy of Dr. Robert M. Peck, Tripler General Hospital, Hawaii.)

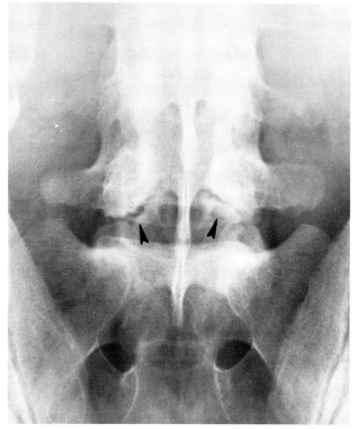

Fig. 1194. Spondylolisthesis

Bilateral neural arch defect (*arrows*). The anteroposterior view with angulation is not usually reliable for diagnosis.

result from sacral lordosis. When the degree of spondylolisthesis is great, the body of the fifth lumbar vertebral body may be so superimposed on the sacrum that only the upper four lumbar vertebrae are demonstrable. Tilting of the transverse and spinous processes of the upper lumbar vertebrae is also present, and the ribs appear to be abnormally close to the pelvis.

The lateral view (with flexion) is most helpful[5] and often shows separation of the neural arch (Figs. 1195–1197). With forward displacement of the vertebral body, the degree of spondylolisthesis can be graded according to the Meyerding method (Fig. 1198). The body of the fifth lumbar vertebra may be displaced forward and downward so as to lie anteriorly to the first sacral segment. Lateral views are also helpful in demonstrating the degree of disk degeneration, which causes hypertrophic changes, and frequently the posterior aspect of the affected vertebral body will appear small and flattened.

Oblique views sometimes demonstrate separation of the neural arch (Figs. 1199–1201). The shadow of the pedicle and lamina are likened to a Scotty dog's head and neck; the defect occurs in the neck. These views may aid in evaluating spondylolysis and Grade I spondylolisthesis, but they are usually not necessary in more severe grades. Normally, each lumbar apophyseal joint is posterior to the one below and, in the oblique projection, gives a "stepladder" appearance. With spondylolisthesis, this relationship is lost at the level of the defect (Fig. 1202).

The term "reverse spondylolisthesis" is the cause of considerable controversy. It does not result from separation of the neural arch. A false appearance of reverse spondylolisthesis is seen with scoliosis or with rotation in the lateral views. With a degenerated disk, the downward displacement of the vertebral body allows it to swing slightly posteriorly, so that the posteroinferior angle lies posterior to the vertebral body beneath. Such cases are described as reverse spondylolisthesis, but the symptoms depend on disk degeneration and not on the luxation of the vertebra.

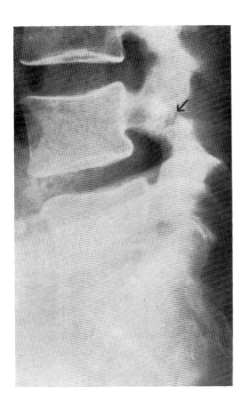

Fig. 1195. Grade I Spondylolisthesis at Fourth Lumbar Interspace

Defect in the neural arch of the fourth lumbar vertebra (*arrow*).

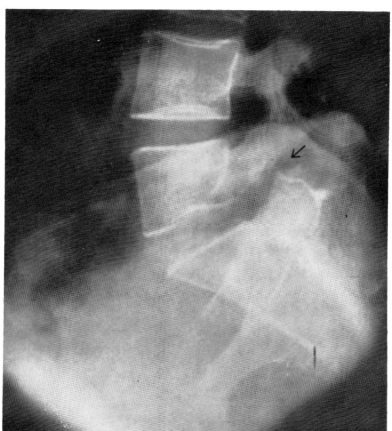

Fig. 1196. Grade I Spondylolisthesis at Lumbosacral Interspace

Defect in the neural arch of the fifth lumbar vertebra (*arrow*).

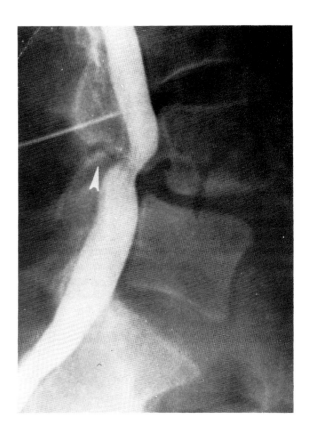

Fig. 1197. Spondylolisthesis
Defect in the neural arch (*arrow*). Slipping of the fourth lumbar vertebra on the fifth. Extradural defect on the opaque column, indicating a herniated disk.

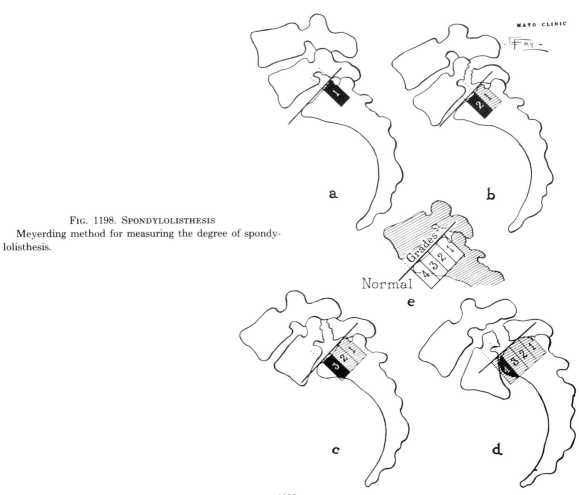

Fig. 1198. Spondylolisthesis
Meyerding method for measuring the degree of spondylolisthesis.

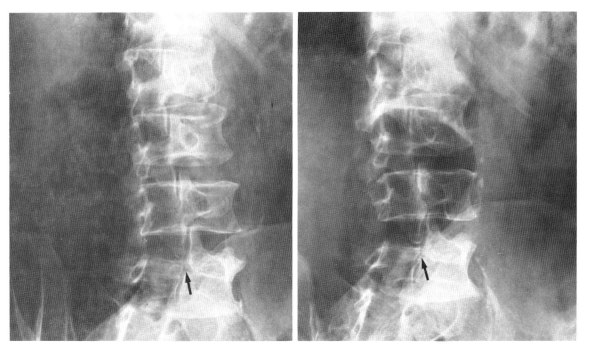

FIG. 1199 AND 1200. SPONDYLOLISTHESIS
Bilateral defect in the pars interarticularis of the neural arch of the fifth lumbar vertebra (*arrows*).

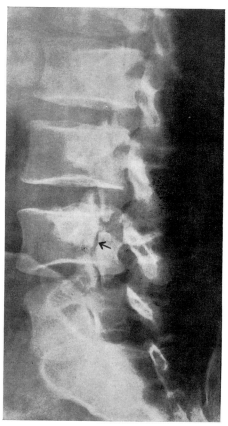

FIG. 1201. SPONDYLOLISTHESIS
Defect in the pars interarticularis of the neural arch of the fourth lumbar vertebra (*arrow*).

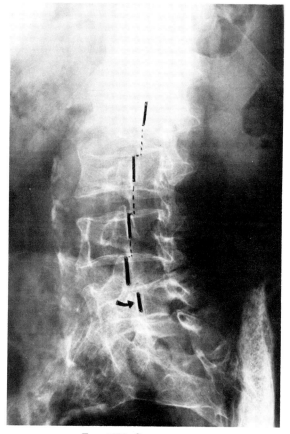

FIG. 1202. SPONDYLOLISTHESIS
The *solid lines* represent the apophyseal joints, which descend in stepladder fashion anteriorly. Just below the spondylolisthesis (*arrow*), the normal relationship is lost, and the plane of the apophyseal joint is posterior to the one above.

REFERENCES

1. Bailey, W.: Observations on the etiology and frequency of spondylolisthesis and its precursors. Radiology, *48:* 107, 1947.
2. Bellamy, R., Lieber, A., and Smith, S. D.: Congenital spondylolisthesis of the sixth cervical vertebra. J Bone Joint Surg, *56:* 405, 1974.
3. Niemeyer, T., and Penning, L.: Functional roentgenographic examination in a case of cervical spondylolisthesis. J Bone Joint Surg, *45:* 1671, 1963.
4. Meyerding, H. W.: Diagnosis and roentgenologic evidence in spondylolisthesis. Radiology, *20:* 108, 1933.
5. Pugh, D. G.: *Roentgenologic Diagnosis of Diseases of Bones.* Williams & Wilkins, Baltimore, 1952.

12

Calcinosis, Calcifications, and Myositis Ossificans

Calcification of the soft tissues accompanies many diseases, trauma, and inflammations. The formation of calcium salts under any circumstances is better understood by the following equation[1]:

$$\text{Calcium ions} + \text{phosphate ions} \xrightarrow[\text{Alkaline phosphatase}]{\text{Alkaline medium}} \text{calcium phosphate}$$

Calcium salt deposition is favored by an alkaline environment and is catalyzed by the enzyme, alkaline phosphatase. The formation of calcium phosphate is a function of the production of the concentrations of calcium and phosphate ions. When the concentration of circulated calcium in milligrams per 100 milliliters is multiplied by the concentration of circulating phosphate in milligrams per 100 milliliters, and this exceeds 75, the serum becomes oversaturated, and metastatic or metabolic calcification occurs.[2] Hypervitaminosis D and the milk-alkali syndrome cause calcification because of the excess of circulating calcium and phosphate. The excess is due to increased absorption from the gastrointestinal tract, associated with impaired excretion from the kidneys. Metastatic calcification occurs with renal failure in secondary hyperparathyroidism because of the marked elevation of serum phosphate, even though the serum calcium is normal.[1]

Dystrophic calcification is due to excessive local deposition of calcium salts and devitalized tissues in patients without systemic disorders of calcium or phosphorus metabolism.[1, 3]

Chondroitin sulfate of connective tissue normally inhibits calcification by binding the calcium and limiting its diffusion.[4, 5] In connective tissue diseases in which chondroitin sulfate is damaged, calcium may be released into the skin and subcutaneous tissue.

Calcification in injured tissue may occur because it acts as a substrate for a phosphate-releasing enzyme.[6] Intracellular alkaline phosphatase does not contribute to calcification, but when it is liberated by the dying cell, the enzyme promotes calcification.[7]

TUMORAL CALCINOSIS

Synonyms: Lipocalcinogranulomatosis,[1] calcifying endothelioma,[2] and calcareous granulomas.[3]

The condition consists of large painless juxta-articular calcified masses, most frequently in the soft tissues of the shoulders, hips, and elbows.

Clinical Features. Manifestations begin in children as soft-tissue masses occurring most frequently in the hips (near the trochanters and buttocks) and the elbows but also occurring in the toes, wrists, ankles, shoulders, ribs, and ischial spines and only rarely in the knees. These masses grow progressively, sometimes resulting in infected skin ulcerations, and occur concurrently or at intervals. They tend to recur after incomplete excision.

There is a familial tendency,[4-10] and most cases have been reported in Negroes.[11-13] The etiology is unknown, but a metabolic defect and trauma are suspected.[9, 14] There is no known associated renal, metabolic, or collagen disease. Retinal streaks are reported.[10]

Pathologic Features. The tumors are surrounded by a connective tissue capsule with nodular projections and contain a white milky fluid with white particles. Histologically, the capsule is composed of fibroblasts and compressed collagen fibers. Multilocular spaces are lined with epithelioid and multinucleated giant cells containing fine and coarse calcium granules.[15, 16]

All laboratory studies are normal except for occasional elevation of alkaline phosphatase[16] and serum phosphate.[16, 17]

Roentgenographic Features. Densely loculated homogeneous masses of calcification occur in the periarticular soft tissues. They grow progressively and recur after incomplete excision (Figs. 1203 and 1204). Irregular in shape, they vary 1–20 cm in size.[18] Radiolucent septa separate aggregates of calcification, giving a multiloculated appearance (Fig. 1205). Large periarticular calcifications may have fluid levels.[19] The bones are normal. Tumors in the buttocks tend to be the largest, those in the fingers and toes

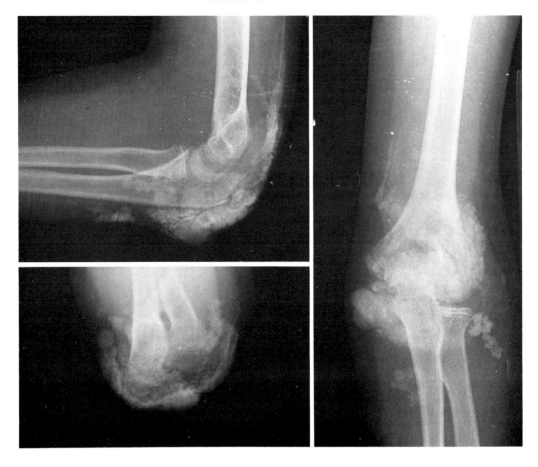

FIG. 1203. TUMORAL CALCINOSIS
Extensive calcification in the tissues adjacent to the joint. The cause of the calcification is unknown.

FIG. 1204. TUMORAL CALCINOSIS
(A) 1974: small calcifications over the forearm. (B) Fourteen months later the calcifications have increased considerably. (Courtesy of Dr. Burt Schaffer, Underwood Memorial Hospital, Woodbury, N.J.)

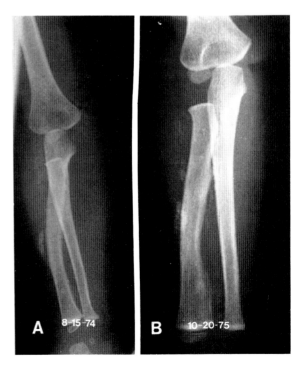

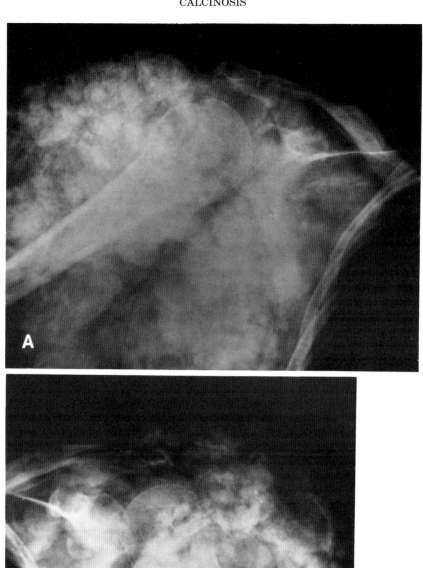

FIG. 1205. TUMORAL CALCINOSIS
Extensive calcium deposits over most of the large joints. The calcifications are separated by septa, giving a globular effect. (*A*) Right shoulder. (*B*) Left shoulder. (*C*) Left elbow. (*D*) Pelvis. (*E*) Radionuclide scan shows dense uptake in the areas of calcification. (Courtesy of Dr. Burt Schaffer, Underwood Memorial Hospital, Woodbury, N.J.)

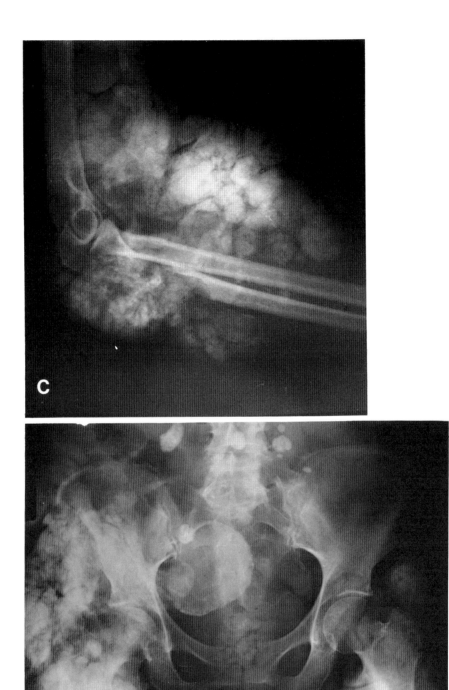

Fig. 1205 (*C* and *D*)

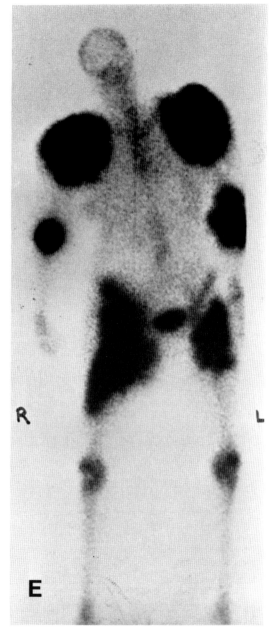

Fig. 1205E

the smallest. In the elbows and other extremities they tend to occur on the extensor surfaces and are related to tendon sheaths. Calcinosis universalis occurs in the skin, subcutaneous tissue, and muscles but not in tendon sheaths. The most common sites are the lateral and superior aspects of the shoulders, the posterior aspects of the elbows, and the lateral hips and buttocks. Less common are the lateral aspects of the feet, retroscapular and acromioclavicular areas, and sacral and ischial regions. Occasionally, deposits in the neck and distal femurs are found.

Radionuclide scans will show dense increase in uptake over all areas of involvement (Fig. 1205).

SOFT-TISSUE CALCIFICATION AND OSSIFICATION

Soft-tissue calcification and ossification can be produced by either systemic disease or local phenomena secondary to trauma or local destruction. Systemic diseases causing calcification include metabolic disease, collagen disease, vascular disease, infestations, infections, arthritides, and dysplasias[1] (Table 29). Most of these conditions are discussed in other chapters.

Calcification invariably appears dense on the roentgenograms, but a thin layer of calcium may seem dense when perpendicular to the x-rays and may disappear when viewed en face. The best example is the pleural calcification which may be plaquelike in one projection and not discernible in another. Calcification generally has a homogeneous appearance and, when newly formed, seems fluffy or cloudy compared to the very compact density of a mature collection.

Soft-tissue ossification usually has a distinct trabecular pattern and may even develop a cortex, a medullary cavity, and spongiosa. The mechanisms of calcification remain poorly understood.[2-8]

TABLE 29. SOFT-TISSUE CALCIFICATIONS IN EXTREMITIES IN SYSTEMIC DISEASE

A. Metabolic diseases
   1. Hypercalcemia
      a. Primary hyperparathyroidism
      b. Hypervitaminosis D
      c. Idiopathic hypercalcemia
      d. Milk-alkali syndrome
      e. Widespread bone destruction
   2. Nonhypercalcemia
      a. Secondary hyperparathyroidism
      b. Chronic renal disease
      c. Hypoparathyroidism
      d. Pseudohypoparathyroidism
      e. Pseudo-pseudohypoparathyroidism
      f. Gout
      g. Pseudogout (chondrocalcinosis)
      h. Ochronosis (alkaptonuria)
      i. Diabetes mellitus
      j. Myositis ossificans progressiva
B. Collagen disease
C. Vascular diseases
   1. Atherosclerosis
   2. Medial sclerosis (Mönckeberg)
   3. Venous calcifications
D. Infestations
   1. Cysticercosis
   2. Dracunculosis (guinea worm)
   3. Loiasis
   4. Bancroft filariasis
   5. Hydatid disease
   6. Leprosy
E. Miscellaneous
   1. Ehler-Danlos syndrome
   2. Pseudoxanthoma elasticum
   3. Werner syndrome
   4. Neuropathic calcification
   5. Calcinosis
      a. Calcinosis circumscripta
      b. Calcinosis universalis
      c. Tumoral calcinosis

INTERSTITIAL CALCINOSIS

Calcinosis Universalis

Calcinosis universalis is a disease of unknown origin, usually found in children and young adults.[1] It often is progressive and may culminate in altered body function and death. Calcium is deposited in the skin and subcutaneous tissue as plaquelike areas of increased density. Tendons and muscles are sometimes affected, suggesting myositis ossificans. Observation will show that no true bone forms.

Calcinosis Circumscripta

Deposits of calcium in subcutaneous tissues occur in a variety of diseases, most frequently in or around joints; the cause may never be determined.[2] Systemic diseases should be suspected when circumscribed calcinosis is present. In acrosclerosis, granular deposits accumulate around the joints of the fingers and toes and the tips of the distal phalanges (Fig. 1206). Absorption of the ends of the distal phalanges occurs with scleroderma,[3] probably a result of the combination of vascular change and pressure on bone (Figs. 1207–1209). With dermatomyositis, extensive subcutaneous deposits of calcium sometimes occur (Fig. 1210). Varicosities may cause extensive calcification, particularly in the calf (Fig. 1211). Primary hyperparathyroidism is infrequently associated with periarticular calcinosis. In renal osteodystrophy with secondary hyperparathyroidism, calcific deposits occur more frequently, with extensive vascular calcification appearing even in young patients. Occasionally, hypoparathyroidism causes interstitial calcinosis, especially around the joints. Symmetrical intracranial calcification of hypoparathyroidism and pseudohypoparathyroidism is discussed elsewhere. Periarticular calcinosis is frequently caused by vitamin D intoxication, either with rheumatoid arthritis, as a periarticular puttylike deposit, or with calcium deposited in tophi. Ordinarily, calcium is not laid down in gouty tophi, and its appearance is suggestive of vitamin D intoxication. Although there is no such condition as "calcium gout," the term is used to describe periarticular calcinosis.[4]

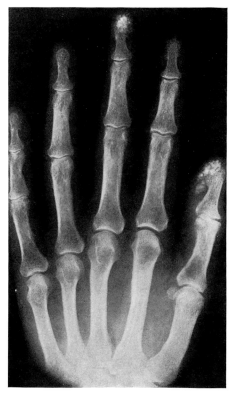

FIG. 1206. ACROSCLEROSIS WITH CALCINOSIS

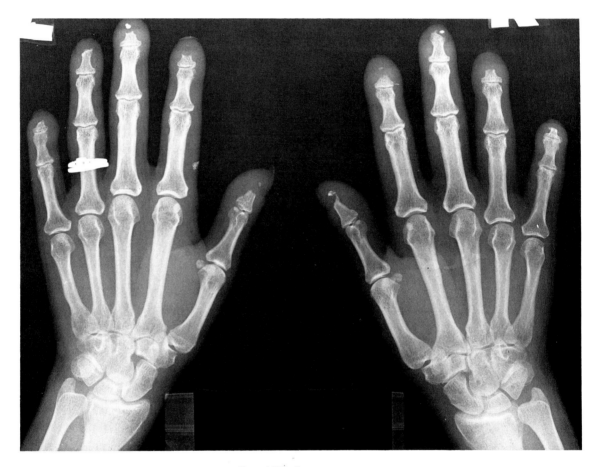

Fig. 1207. Scleroderma

Calcinosis circumscripta in the soft tissues near the tufts of the phalanges. The distal phalanges show absorption of bone. The changes are probably the result of the combined effect of vascular change and pressure on the bone. (Courtesy of Paul K. Berg, Doctors Hospital, Staten Island, N.Y.)

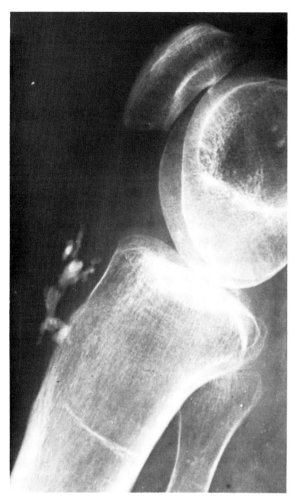

Fig. 1208. SCLERODERMA
Calcinosis circumscripta anterior to the tibia.

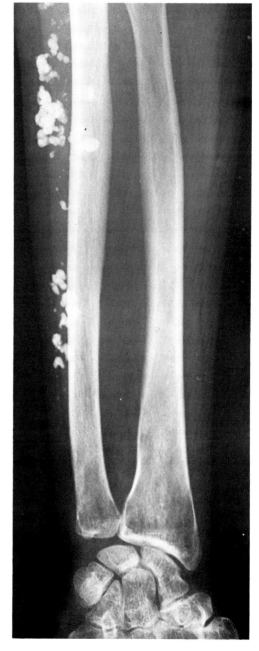

Fig. 1209. SCLERODERMA
Extensive calcification in the soft tissues.

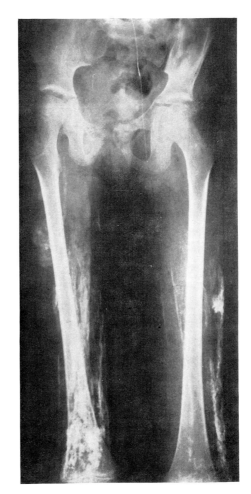

Fig. 1210. Dermatomyositis with Calcinosis

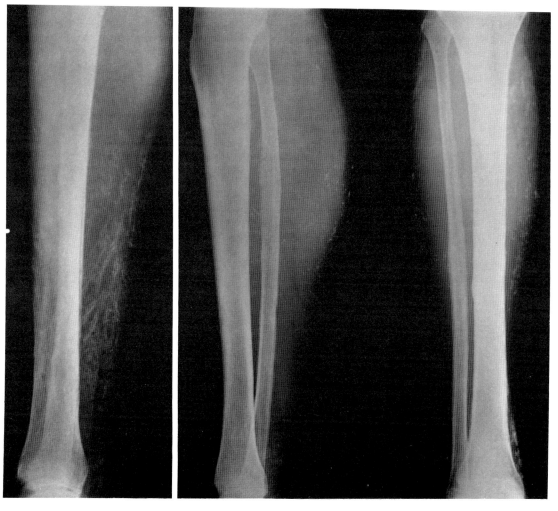

Fig. 1211. Varicosities with Calcification
Extensive calcification in the soft tissues of the leg and irregular periosteal reaction of the tibia and fibula. (*Left*) Lateral view of the right leg. (*Center*) Lateral view of the left leg. (*Right*) Anteroposterior view of the left leg.

CALCIFICATION (BURSITIS) OF TENDONS OF SHOULDER

Calcification of tendons is probably the most common cause of shoulder pain and disability. This is not likely to be secondary to an infectious process. It rises from a degenerative change in the tendon, the result of attrition, and is exacerbated by certain types of work or play. There is evidence that without abnormal stress, repeated sudden motions of adduction or rotation cause calcification of tendons.

The tendons of the subscapularis, supraspinatus, infraspinatus, and teres minor muscles converge and fuse with the joint capsule about 1 inch from its distal margin. Thus, a common tendon-capsule cuff forms, attached to the superior margin of the tuberosities and anatomic humeral neck. The subdeltoid bursa lies on the tendon cuff near the great tuberosity and is covered by the deltoid muscle, acromion, and acromioclavicular ligament.[1]

Calcific deposits seldom occur in the subdeltoid bursa but are found in cavities in the tendons.[1] These cavities may be smooth or irregular, single or multiple, and sometimes communicate with each other but are usually separate and distinct. The consistency of calcareous material varies considerably. In the early lesions the contents are milklike, but when chronic, they become pasty or puttylike[2] and eventually gritty and sandlike. The affected tendon is soft, stringy, and degenerated, and the calcific material may infiltrate the tendon. Calcium is usually in the supraspinatus tendon and less frequently in the tendons of the infraspinatus, teres minor, or subscapularis muscles. Calcareous deposits are rare in the subdeltoid bursa, but with bursitis the bursa is usually edematous, hyperemic, and thickened. If calcific material ruptures from the subdeltoid bursa, pain is often spontaneously relieved.[3]

Roentgenographic Features. The calcification is seen as a well-circumscribed shadow between the humeral head of the greater tuberosity and the acromion (Fig. 1212). Sometimes, there is only a thin, curvilinear shadow, but usually the accumulated calcium is of considerable thickness. The deposits vary greatly in shape and size and may be smooth or irregular. When calcification is extensive, multiple foci are seen. Accumulations of calcium with smooth outlines and homogeneous density are likely to be liquid or pasty in consistency, whereas those which are broken up, rough, and show uneven density are usually gritty and sandlike. Their location depends on the tendon affected; calcification of the supraspinatus tendon is found just above the juncture of the head and humeral tuberosity (Fig. 1212); in the infraspinatus and teres minor tendons, it is lower in position and is superimposed on the tuberosity of the humerus except when the arm is in internal rotation. In the subscapularis tendon, it is superimposed on the joint when the arm is in internal rotation and on the head of the humerus when the arm is in external rotation. When it is in the subdeltoid bursa, it extends down over the tuberosity and is likely to be faint and homogeneous.

In order to demonstrate calcification of the tendons of the shoulder adequately, a sufficient number of roentgenographic projections must be obtained, including anteroposterior roentgenograms with the arms (1) in the neutral position, (2) with 45° internal rotation, (3) with 45° external rotation, and (4) with 90° external rotation.

Calcification (bursitis) can occur at any site of prominent bone. Usually, bursae form, then inflammation may occur and may result in calcification. Bursitis occurs most commonly in the shoulder; the second most frequent site is the great trochanter. An unusual site is the prepatellar area (Fig. 1213).

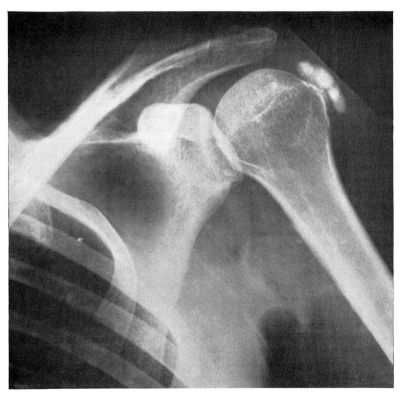

FIG. 1212. CALCIFICATION OF SHOULDER
Calcification of the supraspinatus tendon.

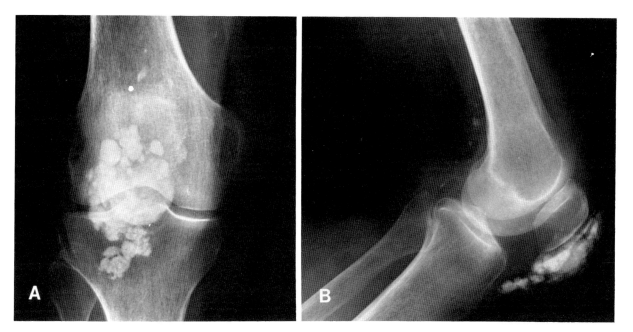

Fig. 1213. Bursitis of the Prepatellar Bursa Tendons and Synovia of Tendons
(A) Anteroposterior projection reveals flocculent calcifications overlying the patella. (B) Lateral projection shows them to be anterior to the patella in a bursa and the synovia of tendinous attachments.

RETROPHARYNGEAL TENDINITIS

Synonyms: Calcific tendinitis of the neck, tendinitis of the longus colli muscle, bursitis of the retropharyngeal tendon.[1-5]

Retropharyngeal tendinitis is characterized by a small calcareous deposit just inferior to the anterior arch of the first cervical vertebra. It is in the region of the longus colli muscle which inserts into the tubercle of the atlas and the second, third, and fourth cervical vertebrae. When this tendon becomes inflamed, soft-tissue prominence of the retropharyngeal structures may be present.

Retropharyngeal tendinitis is accompanied by severe neck pain and rapidly developing rigidity. The calcification is usually present at the onset of symptoms (Fig. 1214). Dysphagia may accompany it. The condition is self-limited; pain improves and completely disappears in 1–3 weeks, and the calcareous deposit also disappears in 3 weeks to 3 months.

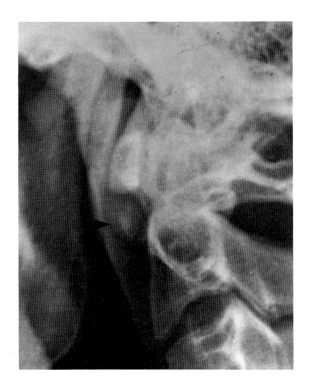

Fig. 1214. Retropharyngeal Tendinitis
It was a calcification below the anterior arch of the first cervical vertebra (arrow) with soft-tissue swelling accompanied by severe neck pain and rapidly developing rigidity and dysphagia. It is a self-limited condition; the symptoms abate in 1–3 weeks, and the calcareous deposits disappear by 3 months.

MYOSITIS OSSIFICANS

Heterotopic formation of bone in soft tissue, especially in the muscles, results from a variety of conditions; myositis ossificans should be considered a symptom, and its cause sought. A simple classification of myositis ossificans is: (1) myositis ossificans progressiva; (2) localized myositis ossificans secondary to direct trauma; (3) myositis ossificans associated with neurologic disorders, especially paraplegia; and (4) localized myositis ossificans of unknown origin.[1]

Localized Myositis Ossificans Secondary to Direct Trauma (Post-Traumatic Ossification)

Myositis ossificans is a poor designation because there is no inflammation, often no bone, and in some instances, no muscle included in the pathologic process.[2]

In 60% of the patients, localized myositis ossificans arises from trauma, but whether trauma is the sole factor is debatable.[3-7] It is possible that where there was no history of trauma, it indeed existed, but it was so slight as to pass unrecognized.[8]

Certain occupations and sports are likely to lead to myositis ossifications. Heterotopic bone arising in the adductor longus muscle, for instance, is known as "rider's bone." In the brachialis muscle it is "fencer's bone," and in the soleus it is "dancer's bone."

Pellegrini-Stieda disease, regarded as calcification of the tibial collateral ligament, is actually a localized myositis ossificans,[9] often following trauma—a sprained ankle or a direct blow over the internal femoral condyle. The lesion consists of a deposit of heterotopic bone medial to the adductor tubercle of the femur but unconnected to it except by fibrous tissue; it is not a chip fracture. The deposit of heterotopic bone lies beneath a fibrous tissue membrane overlying the medial femoral condyle, usually between this membrane and the periosteum of the adductor tubercle, but it may extend upward to lie between the membrane and the tendon of the adductor of magnus muscle. In the roentgenograms it is seen as an elliptical or curvilinear shadow of ossification lying adjacent to the medial femoral condyle (Fig. 1215).

Pathologic Features. After injury, there is a definable histologic evolution.[6] In the first month, it develops a highly cellular pseudosarcomatous appearance and is easily mistaken for a malignant neoplasm.[10] From the second week to the second month, a centrifugal pattern of maturation emerges. The central portion remains cellular, and the periphery is composed of immature osteoid that organizes primitive fiber bone and eventually matures into lamellar bone.[2, 6, 10] The result is "somal phenomenon."[10] The somal pattern distinguishes it from a malignant tumor.[2]

Roentgenographic Features. The roentgenographic evolution parallels the histologic evolution.[6] A soft-tissue mass develops shortly after injury. In the third to fourth week, a faint calcification begins within it. If it is juxtacortical, a periosteal reaction also becomes visible. At 6-8 weeks a sharply circumscribed cortical periphery forms around the central lacy pattern of new bone formation (see Fig. 1221). The soft-tissue central core may encyst, the cavity may enlarge and hollow the peripheral cortex, and the end state may appear as an eggshell calcified cyst[6] (Figs. 1216-1218). Maturity is reached in 5-6

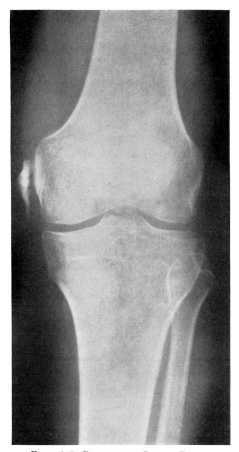

FIG. 1215. PELLEGRINI-STIEDA DISEASE
A form of localized myositis ossificans secondary to direct trauma.

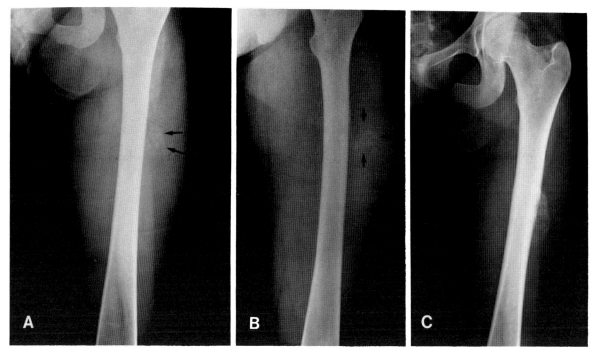

FIG. 1216. POST-TRAUMATIC MYOSITIS OSSIFICANS

(A) Three weeks after injury to the thigh: faint calcification in the soft tissues (*arrows*). (B) Three weeks after (A) the calcification is larger (*arrows*), with periosteal reaction proximal to the calcification. (C) Six months after injury the calcification appears attached to the bone, and the periosteal reaction has resorbed.

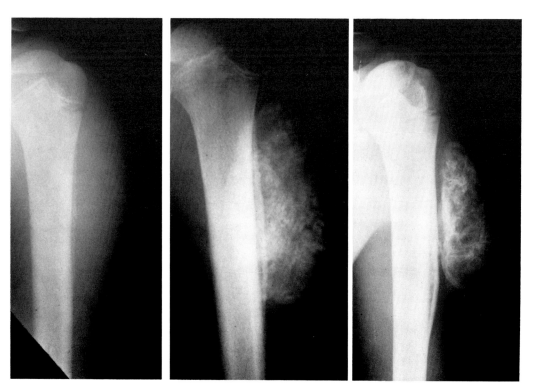

FIG. 1217. MYOSITIS OSSIFICANS

(*Left*) Three and one-half weeks after injury: poorly defined flocculated densities in the soft tissues, and periosteal new bone reaction. (*Center*) Seven and one-half weeks after injury the mass is heavily ossified and better circumscribed. The periosteal reaction is more apparent. (*Right*) Six and one-half months after injury the focus of myositis ossificans is well organized. A peripheral cortex surrounds a relatively less dense center, and a radiolucent cleft separates the lesion from adjacent bone. (Courtesy of A. Norman and H. D. Dorfman, *Radiology*, 96: 301, 1970,[6] © The Radiological Society of North America, Syracuse, N.Y.)

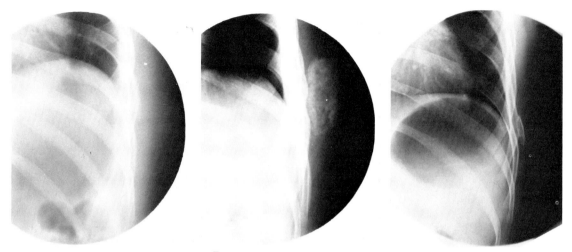

FIG. 1218. MYOSITIS OSSIFICANS

(*Left*) Three and one-half weeks after injury: faint densities in the soft tissues of the lateral chest wall, periosteal response in the adjacent seventh, eighth, and ninth ribs, and pleural reaction. (*Center*) Five weeks after injury the mass is heavily ossified and well circumscribed, and the central portion is less dense. A distinct cleft of radiolucency separates it from the bone. (*Right*) Twenty-nine months later the focus of myositis ossificans is much smaller, the ribs are remodeled and the pleural reaction is resorbed. (Courtesy of A. Norman and H. D. Dorfman, *Radiology*, 96: 301, 1970,[6] © The Radiological Society of North America, Syracuse, N.Y.)

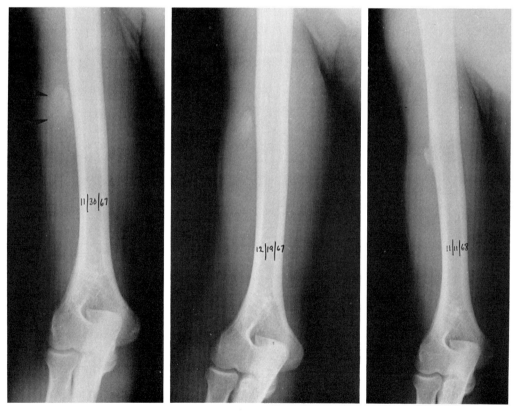

FIG. 1219. MYOSITIS OSSIFICANS

(*Left*) Six weeks after injury: well circumscribed soft-tissue calcification (*arrows*). (*Center*) Three weeks later the calcification is smaller. (*Right*) One year later the calcification is much smaller, and it appears attached to bone. Radiolucency separates the density of the exostosis from the host bone. (Courtesy of Dr. Barnett Finkelstein, M. D. Anderson Hospital, Houston, Tex.)

months, and the calcified mass begins to regress and becomes smaller. The juxtacortical lesions are separated from the periosteal reaction and the bone by a continuous radiolucent zone. In a year the mass usually disappears completely, but there may be residual exostosis attached to bone without the radiolucent line (see Figs. 1219, 1222, and 1223). Extensive soft-tissue injury may result in diffuse punctate and linear calcifications which may suggest dermatomyositis, vascular calcifications, or cartilaginous tumor (see Fig. 1224).

Myositis ossificans can be distinguished from periosteal sarcoma in the early stages by the radiolucent zone that completely separates it from the bone, whereas the latter is attached by a sessile base and a discontinuous radiolucent zone. Osteochondroma and the host bone have a common cortex and medullary cavity and are easily distinguished.

Myositis Ossificans Associated with Neurologic Disorders

Myositis ossificans occurs in various types of lesions affecting the spinal cord, but the incidence is greatest in paraplegia. Between 33% and 49% of paraplegics show myositis ossificans in the paralyzed part.[11] Osseous deposits occur in muscles, tendons, and ligaments. Although they are often attached to the periosteum, they do not arise from it. The heterotopic bone is seen in largest quantity around the large joints, especially the hip.

Roentgenographic Features. The heterotopic bone associated with paraplegia may be compact or spongy in composition. When the bone is formed along fascial planes, it is seen as small irregular fragments or in the shape of spicules. Large bone masses formed in muscle have a mature cancellous structure. Bony growths are greatest around the hip and sometimes arise from the margins of the acetabulum or greater trochanter of the femur (Fig. 1220). Often they are formed at the site of muscular attachment; in the knee the heterotopic bone is most frequently near the tibial collateral ligament. Bone is commonly present along the fascial planes of the thigh, especially the medial aspects, and the buttocks. When muscle is affected, the shadow of the heterotopic bone runs parallel or

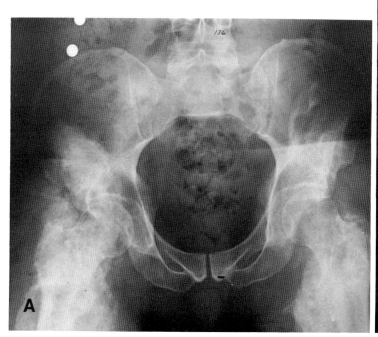
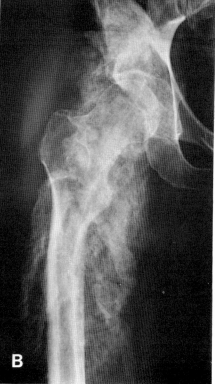

FIG. 1220. MYOSITIS OSSIFICANS IN A PARAPLEGIC
Large bone masses have formed in the muscles and connective tissues around the hip and thighs. (*A*) Anteroposterior view of the pelvis. (*B*) Anteroposterior view of the hip and upper femur.

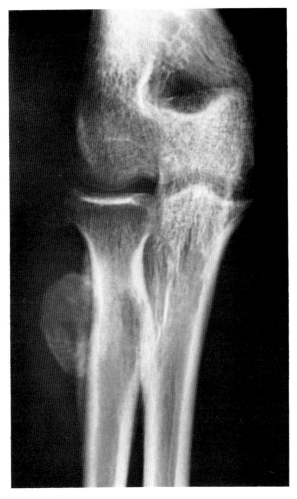

FIG. 1221. MYOSITIS OSSIFICANS 2 MONTHS AFTER INJURY

The bone density attached to the radius is associated with thick periosteal reaction separated slightly from the cortex of the radius.

oblique to the adjacent tubular bone. Sometimes, formation of bone around a joint may be so massive as to immobilize it.

In paraplegia, decubitus ulcers almost always develop, occurring over the sacrum, femoral trochanters, and ischial tuberosities. Paraplegics, because of soft-tissue infections and hyperemia, not only develop myositis ossificans but are subject to absorption of the greater trochanters and ischial tuberosities. As the greater trochanter is absorbed, its shape is altered, and the tip is prolonged upward, which limits the motion in the hip joint, but actual involvement by a neuropathic or infectious process is never seen. Osteoporosis of the lower extremities and pelvis is a constant finding in paraplegia. This is another manifestation of osteoporosis caused by immobilization, with resultant elevation of the serum calcium level and excessive excretion of calcium in the urine. As a consequence, urinary calculi are a common complication of paraplegia. The abnormal calcium and phosphorus balance may be an initiating factor in heterotopic bone formation.

Localized Myositis Ossificans of Unknown Origin

Sometimes, myositis ossificans is found in a localized form affecting one or several large muscle bundles. By the time it is recognized, progression may have halted permanently. The cause of this type is unknown but may be due to trauma (Fig. 1225). It differs from myositis ossificans progressiva because of its localized nature and the normality of the tubular bones of the hands and feet. It may represent a healed phase of dermatomyositis.

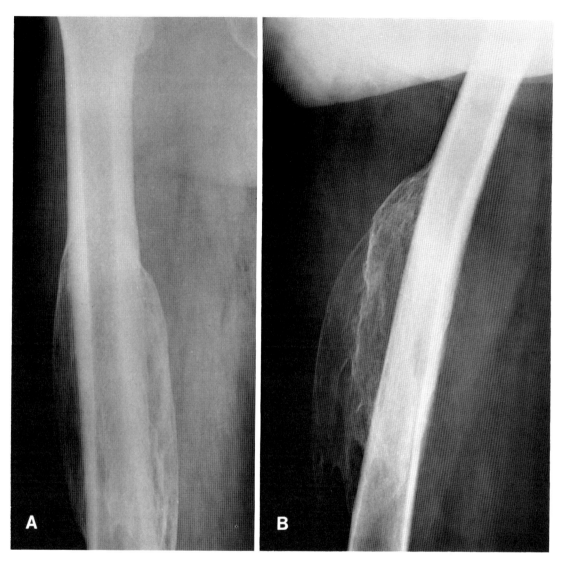

Fig. 1222. Myositis Ossificans
A very old injury. The bone may never be resorbed. (A) Anteroposterior view. (B) Lateral view.

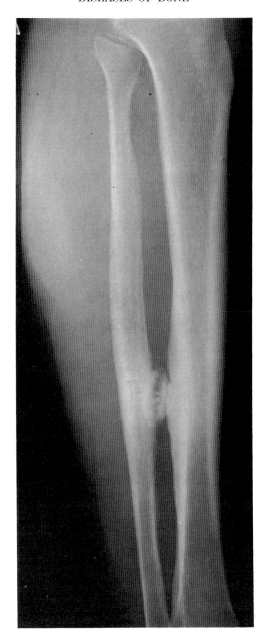

Fig. 1223. Myositis Ossificans of Interosseous Structures

A forearm injury many years old: residual change of bone formation in the interosseous space attached to both the radius and ulna, and thickening of the bone due to old periosteal reaction.

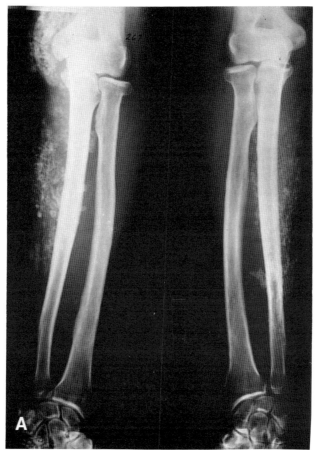

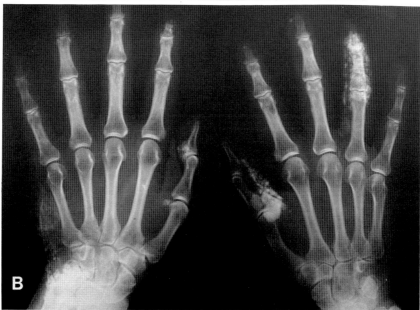

Fig. 1224. Myositis Ossificans (Post-Traumatic Calcification)

This patient fell from a height onto both elbows and hands. He did not seek medical help for many months. (*A*) Anteroposterior view of both elbows shows extensive calcifications in the soft tissues of the forearm and the right wrist and elbow and healed fracture of the right radial head. (*B*) Calcifications in both hands in the regions of the wrists and some of the fingers due to the previous trauma. (Courtesy of Dr. Jerome Wiot, Cincinnati General Hospital, Cincinnati, Ohio.)

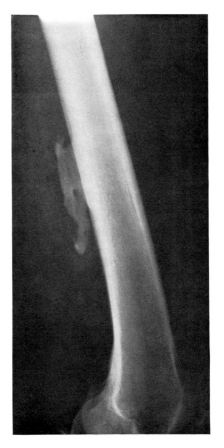

FIG. 1225. MYOSITIS OSSIFICANS

Localized myositis ossificans secondary to direct trauma. The heterotopic bone probably developed in an old hematoma. It is distinguished from an exostosis by the irregular outline and by the fact that it does not represent an outward projection of the adjacent bone but merely seems to be attached to the periosteum.

CALCIFICATION OF THE ACHILLES TENDON

Calcification of the Achilles tendon is a relatively common finding. It is probably due to trauma with hemorrhage, and it has a very characteristic appearance, rectangular, and usually beginning 1 cm above the insertion into the calcaneus. It is usually 3–4 cm in length, with spotty calcification proximal, but it may be as large as 10 cm (Fig. 1226).

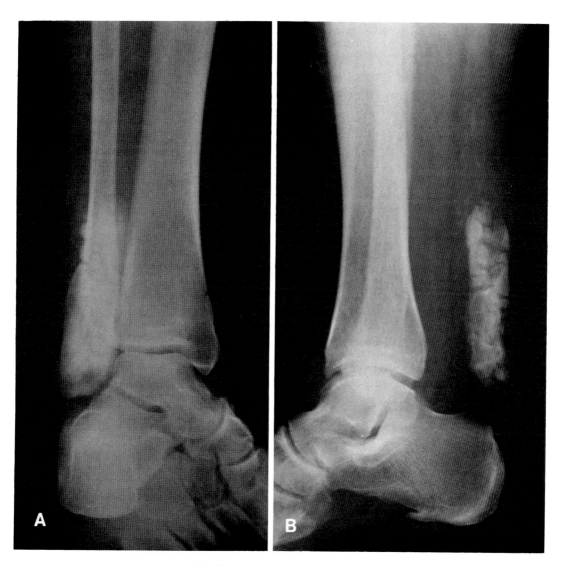

FIG. 1226. CALCIFICATION OF ACHILLES TENDON

Large rectangular calcification in the Achilles tendon and spotty calcification proximal to the main portion. (*A*) Oblique view. (*B*) Lateral View. (Courtesy of Dr. Theodore Keats, University of Virginia, Charlottesville, Va.)

REFERENCES

1. Siegelman, S. S., and Jacobson, H. G.: The foot in acquired systemic disease. Semin Roentgenol, 5: 436, 1970.
2. Stanbury, S. W., and Lumb, G. A.: Parathyroid function in chronic renal failure: a statistical survey of the plasma biochemistry in azotaemic renal osteodystrophy. Q J Med, 35: 1, 1966.
3. Hilbish, T. F., and Bartter, F. C.: Roentgen findings in abnormal deposition of calcium in tissues. AJR, 87: 1128, 1962.
4. Bowness, J. M.: Present concepts of the role of ground substance in calcification. Clin Orthop, 59: 233, 1968.
5. Urist, M. J., Speer, D. P., Ibsen, K. J., and Strates, B. S.: Calcium binding by chondroitin sulfate. Calcif Tissue Res, 2: 253, 1968.
6. Wheeler, C. E., Curtis, A. C., Cawley, E. P., Grekin, R. H., and Zheutlin, B.: Soft tissue calcification with special reference to its occurrence in the "collagen diseases." Ann Intern Med, 36: 1050, 1952.
7. Henrichsen, E.: Alkaline phosphatase and the calcification in tissue cultures. Exp Cell Res, 11: 403, 1956.

Tumoral Calcinosis

1. Teutschlaender, O.: Uber progressive Lipogranulomatose der Muskulatur. Klin Wochenschr, 14: 451, 1935.
2. Duret, M. H.: Tumeurs multiples et singulaires des bourses seruses. Bull Mem Soc Anat Paris, 74: 725, 1899.
3. Inclan, A.: Tumoral calcinosis. JAMA, 121: 490, 1943.
4. Hacuhanefioglu, U.: Tumoral calcinosis; a clinical and pathological study of eleven unreported cases in Turkey. J Bone Joint Surg, 60A: 1131, 1978.
5. Baldursson, H., Evans, E. B., Dodge, W. F., and Jackson, W. T.: Tumoral calcinosis with hyperphosphatemia; a report of a family with incidence in four siblings. J Bone Joint Surg, 51A: 913, 1969.
6. Ghormley, R. K., and McCrary, W. E.: Multiple calcified bursae and calcified cysts in soft tissues. Trans Western Surg Assoc, 51: 292, 1942.
7. Lambotte, C., Israel, E., and Durenne, J. M.: La calcinose tumorale; affection hereditaire d'incidence variable. Ann Soc Belg Med Trop, 55: 47, 1975.
8. Thomson, J. E. M., and Tanner, F. H.: Tumoral calcinosis. J Bone Joint Surg, 31A: 132, 1949.
9. Harkess, J. W., and Peters, H. J.: Tumoral calcinosis; a report of 6 cases. J Bone Joint Surg, 49A: 721, 1967.
10. Barton, D. I., and Reeves, R. J.: Tumoral calcinosis; report of 3 cases and review of the literature. AJR, 86: 351, 1961.
11. Agnew, C. H.: Tumoral calcinosis. J Kans Med Soc, 62: 100, 1961.
12. von Albertini, A.: Spezielle Pathologie der Sehen, Sehnenscheiden und Schleimbeutel. In *Handbuch der Speziellen Pathologischen Anatomie und Histologie*, F. Henke and O. Lubarsch (Eds.). Julius Springer, Berlin, 1929.
13. Andreas, E.: Die Lipocalcinogranulomatose—eine neue Lipoidose. Med Klin, 44: 913, 1949.
14. Selye, H.: *Calciphylaxis*, p. 331. University of Chicago Press, Chicago, 1962.
15. Albright, F., and Reifenstein, E. C., Jr.: *The Parathyroid Glands and Metabolic Bone Disease*. Williams & Wilkins, Baltimore, 1948.
16. Poppel, M. H., and Zeitel, B. E.: Roentgen manifestations of milk drinker's syndrome. Radiology, 67: 195, 1956.
17. Christensen, W. R., Liebman, C., and Sosman, M. C.: Skeletal and periarticular manifestations of hypervitaminosis D. AJR, 65: 27, 1951.
18. Jaffe, H. J.: Primary and secondary (renal) hyperparathyroidism. Surg Clin North Am, 22: 621, 1942.
19. Hug, I., and Guncaga, J.: Tumoral calcinosis with sedimentation sign. Br J Radiol, 47: 734, 1974.

Soft-Tissue Calcification and Ossification

1. Gayler, B. W., and Brogdon, B. G.: Soft tissue calcification in the extremities in systemic disease. Am J Med Sci, 249: 590, 1965.
2. Albright, F., and Reifenstein, E. C.: *The Parathyroid Glands and Metabolic Bone Disease*. Williams & Wilkins, Baltimore, 1948.
3. Dixon, T. F., and Perkins, H. R.: The chemistry of calcification. In *The Biochemistry and Physiology of Bone*, Ch. 10, pp. 287–307, G. H. Bourne (Ed.). Academic Press, New York, 1956.
4. Hass, G. M.: Pathological calcification. In *The Biochemistry and Physiology of Bone*, Ch. 24, pp. 767–809, G. H. Bourne (Ed.). Academic Press, New York, 1947.
5. Howard, J. E.: Normal and abnormal function of the parathyroids. Trans Stud Coll Physicians Phila, 30: 55, 1962.
6. Howard, J. E., and Thomas, W. C.: Clinical disorders of calcium homeostasis. Medicine, 42: 25, 1963.
7. Klotz, O.: Studies upon calcareous degeneration; I. The process of pathological calcification. J Exp Med, 7: 633, 1905.
8. Selye, H.: *Calciphylaxis*. University of Chicago Press, Chicago, 1962.

Interstitial Calcinosis

1. Steinitz, H.: Calcinosis circumscripta ("Kolkgicht") und Calcinosis Universalis. Ergeb Inn Med Kinderheilkd, 39: 216, 1931.
2. Swanson, W. W., Forster, W. G., and Iob, L. V.: Calcinosis circumscripta. Am J Dis Child, 45: 590, 1933.
3. Moran, F. T.: Calcinosis; a brief review of the literature and report of 2 cases. South Med J, 40: 840, 1947.
4. Pugh, D. G.: *Roentgenologic Diagnosis of Diseases of Bones*. Williams & Wilkins, Baltimore, 1954.

Calcification (Bursitis) of Tendons of Shoulder

1. Codman, E. A.: *The Shoulder*. E. A. Codman, Boston, 1934.
2. Howorth, M. B.: Calcification of the tendon cuff of the shoulder. Surg Gynecol Obstet, 80: 337, 1945.
3. Wilson, C. L.: Lesions of the supraspinatus tendon; degeneration, rupture and calcification. Arch Surg, 46: 307, 1943.

Retropharyngeal Tendinitis

1. Hartley, J.: Acute cervical pain associated with retropharyngeal calcium deposit. J Bone Joint Surg, 46A: 1753, 1964.
2. Sutro, C. J.: Calcification of the anterior atlanto-axial ligament as cause for painful swallowing and for painful neck. Bull Hosp Joint Dis, 28: 1, 1967.
3. Bernstein, S. A.: Acute cervical pain associated with soft tissue calcium deposition anterior to the interspace of the first and second cervical vertebrae. J Bone Joint Surg, 57A: 426, 1975.
4. Fahlgren, H., et al.: Retropharyngeal tendinitis. Acta Neurol Scand [Suppl 31], 43: 188, 1967.
5. Newmark, H., Forrester, D. M., Brown, J. C., Robinson,

A., Olken, S. M., and Bledsoe, R.: Calcific tendinitis of the neck. Radiology, *128:* 355, 1978.

Myositis Ossificans

1. Pugh, D. G.: *Roentgenologic Diagnosis of Diseases of Bones.* Williams & Wilkins, Baltimore, 1954.
2. Ackerman, L. V.: Extra-osseous localized non-neoplastic bone and cartilage and pathological confusion with malignant neoplasms. J Bone Joint Surg, *40A:* 279, 1958.
3. Aegerter, E. E., and Kirkpatrick, J. A., Jr.: *Orthopedic Disease: Physiology, Pathology, Radiology,* ed. 3. Saunders, Philadelphia, 1968.
4. Geschickter, C. F., and Maseritz, I. H.: Myositis ossificans. J Bone Joint Surg, *20:* 661, 1938.
5. Goldman, A. B.: Myositis ossificans circumscripta; a benign lesion with a malignant differential diagnosis. AJR, *126:* 32, 1976.
6. Norman, A., and Dorfman, H. D.: Juxtacortical circumscribed myositis ossificans; evolution and radiographic features. Radiology, *96:* 301, 1970.
7. Paterson, D. C.: Myositis ossificans circumscripta; report of four cases without history of injury. J Bone Joint Surg, *52B:* 296, 1970.
8. Gilmer, W. S., and Anderson, L. D.: Reactions of soft somatic tissue which may progress to bone formation; circumscribed (traumatic) myositis ossificans. South Med J, *52:* 1432, 1959.
9. Nachlas, I. W., and Olpp, J. L.: Para-articular calcification (Pellegrini-Stieda) in affections of the knee. Surg Gynecol Obstet, *81:* 206, 1945.
10. Johnson, L. C.: Histogenesis of myositis ossificans (abstract). Am J Pathol, *24:* 681, 1948.
11. Heilbrun, N., and Kuhn, W. F., Jr.: Erosive bone lesions and soft tissue ossifications associated with spinal cord injuries (paraplegia). Radiology, *48:* 579, 1947.

13

Osseous Manifestations of Metal Poisoning

LEAD POISONING

In the United States, lead poisoning in children has declined sharply since its dangers have been publicized, and lead is no longer used in paint for children's furniture and toys. Lead nipple shields, lead foil and food wrappings, lead cooking utensils, and lead toys have been largely eliminated. Lead-containing ointments, lotions, and dusting powders are still possible sources of lead poisoning in infants. Poisoning may occur not only by ingestion but by the inhalation of volatile lead as well.[1] Several lead poisoning epidemics have resulted from the use of storage battery casings for fuel.

In adults with lead poisoning, lead is deposited in the bones, but it does not appear in roentgenographic studies because bone growth has ceased. Before endochondral bone growth has ceased, changes are often pathognomonic of the condition.

Clinical Features. Early symptoms of lead poisoning are loss of appetite, vomiting, constipation, and abdominal cramps. Peripheral neuritis occurs frequently in adults but only occasionally in children. The most dangerous manifestation of lead poisoning in children is meningoencephalitis. Anemia almost always occurs, but basophilic stippling of red blood cells is usually not present, nor is it pathognomonic of plumbism when encountered. A lead line at the gum margins is a frequent finding in adults but is seldom found in children. The determination of the quantity of lead excreted daily in the urine is the best and most exact method of confirming the diagnosis, but this analysis is difficult and, due to various technical factors, should be attempted only by those completely familiar with analytical chemical technique. When no such laboratory aid is available, the roentgenologist can sometimes offer helpful evidence to support the clinical diagnosis of plumbism. A negative roentgenologic examination, however, does not rule out lead poisoning in children.

Roentgenographic Features. Because lead and calcium are utilized interchangeably by bone, lead tends to be deposited in bone in rather high concentrations. In growing bones, lead is concentrated most heavily in the metaphyses. Therefore, the greatest roentgenologic change occurs in the distal end of the femur, both ends of the tibias, and the distal ends of the radii, since these are the most rapidly growing portions of the skeleton. The increased density of bone, so characteristic of lead poisoning, results from condensation of the trabeculae of the metaphysis and deposition of lead there.[2]

The roentgenologic manifestation of lead poisoning in children consists of bands of increased density at or near the metaphyses of the shafts of tubular bones (Figs. 1227 and 1228).[2] In flat and long bones, the margins, which represent their metaphyses, are sometimes also increased in density. The density and width of the lead line depends on the concentration of lead ingested and inhaled, as well as on the duration of intake. In infants with severe lead poisoning, the band of increased density may be quite wide, and the pathologic change will prevent normal remodeling, leaving the distal ends of the diaphyses somewhat clubbed. Periosteal thickening and reaction never are seen in plumbism. After ingestion or inhalation of lead ceases, normal bone is laid down on the epiphyseal side of the lead line. As recovery takes place, the lead line becomes broader and less dense and eventually disappears, although if the condition has been prolonged, there may be lack of bone modeling (Fig. 1229). Lead lines or dense growth lines may persist throughout the patient's life (Fig. 1230). Any condition which causes a prolonged growth disturbance may produce similar growth lines.

In normal infants of less than 3 years of age, increased density of metaphyseal ends of tubular bones is not unusual and is sometimes conspicuous enough to be mistaken for the band of increased density caused by plumbism. Healed rickets also presents an increased density of metaphyses.

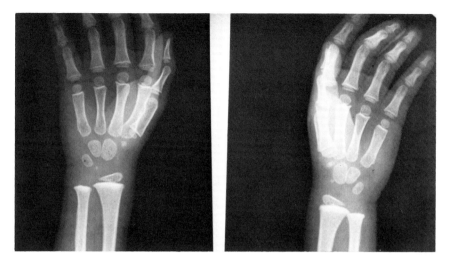

Fig. 1227. Lead Poisoning
Bands of increased density in the distal ends of the radii and ulnas. The metacarpals also show lead lines.

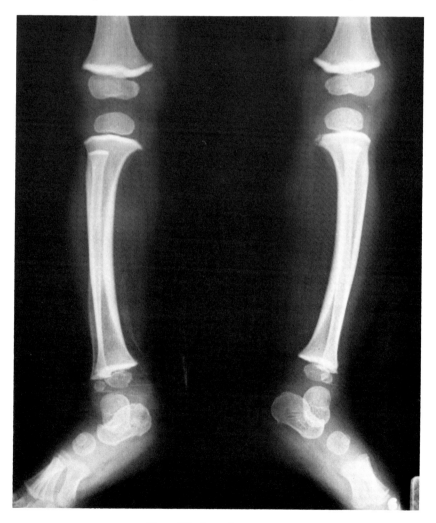

Fig. 1228. Lead Poisoning
Bands of increased density in the metaphyses of the femurs, tibias, and fibulas.

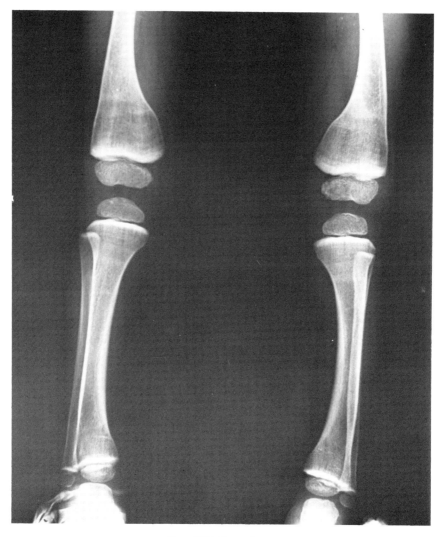

Fig. 1229. Lead Poisoning
Bands of increased density in the metaphyseal ends of the long bones, and a lack of modeling at the distal ends of the femurs because of the long standing condition. (Courtesy of George T. Wohl, Philadelphia General Hospital, Philadelphia, Pa.)

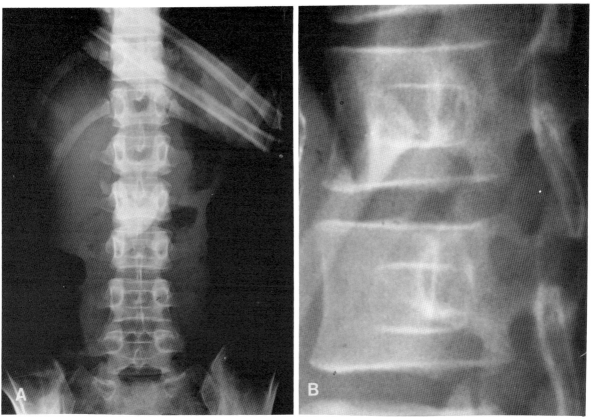
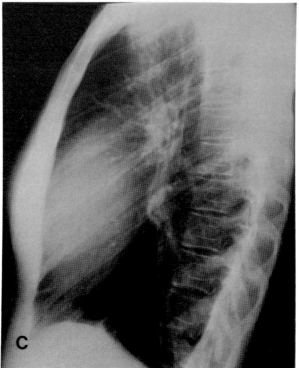

Fig. 1230. Lead Poisoning
(A) Anteroposterior projection of lumbar spine reveals the bone-within-a-bone appearance due to lead poisoning in infancy.
(B) Lateral projection of the lumbar spine. (C) Lateral projection of the dorsal spine.

PHOSPHORUS POISONING

The ingestion of metallic phosphorus (yellow phosphorus) causes bands of increased density in the metaphyses of long bones, roentgenologically indistinguishable from those produced by lead and bismuth.[3] Phosphorus poisoning is seldom seen in children now, but in the past these changes were found occasionally in rachitic or tuberculous children treated with phosphorized cod liver oil. Because of intermittency of medication and variation in growth rate, phosphorus lines often are seen as multiple transverse lines of increased density at intervals at the metaphysis and in the adjacent diaphysis. In flat bones, such as the ilium, phosphorus lines were seen as single or multiple curvilinear bands of increased density near the periphery of the bone (Fig. 1231). After phosphorus ingestion ceases, the bone formation is normal (Fig. 1232). The bands of increased density gradually fade and, after a few years, disappear.

Phosphorus poisoning in adults is almost never seen in the United States since red phosphorus has been substituted for yellow phosphorus in the manufacture of matches. The first manifes-

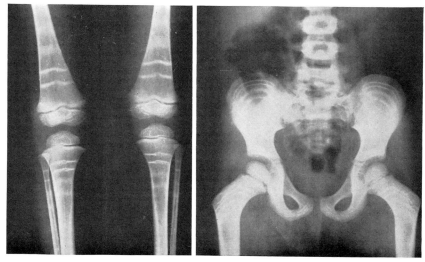

FIG. 1231. PHOSPHORUS POISONING
Phosphorus lines resulting from the ingestion of phosphorized cod liver oil.

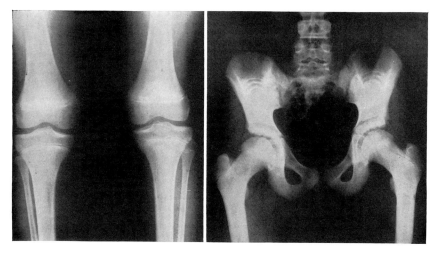

FIG. 1232. PHOSPHORUS POISONING
Same case as shown in Figure 1231. The phosphorus lines have disappeared except for those which remained in the ilia. Eventually they disappeared completely.

tations of this type of phosphorus poisoning are gingivitis and periostitis, and severe cases sometimes cause resistant osteomyelitis. The entire mandible may be destroyed, and there may be extensive sequestration and involucrum formation.

BISMUTH POISONING

The osseous manifestations of bismuth poisoning are usually seen in the long bones of syphilitic babies treated with bismuth, or in babies born to syphilitic women treated by injections of bismuth during pregnancy.[4] Histologically, the bismuth zone is said to be indistinguishable from the changes produced by lead, but there is some difference between the changes of lead and bismuth and those of phosphorus. Roentgenologically, however, the zones of increased density caused by lead, bismuth, and phosphorus cannot be differentiated.[5]

COMPLICATIONS OF RADIATION

Radium Poisoning

The term "radium poisoning" should be restricted to the harmful effects of ingestion, inhalation, or injection of radium or some other radioactive substances. Radium poisoning does not include the harmful effects of external irradiation from radium, radioactive substances, and roentgen rays. The principal victims of radium poisoning were workers who painted luminous dials on watches and clocks and ingested the radioactive luminous compound by licking their brushes to point them. After this was recognized, in 1925, the incidence of radium poisoning was greatly decreased.[6] Martland[7,8] has summarized these cases. During World War I there was an enormous demand for luminous dials, and since no satisfactory mechanical method of applying the radium paint had been developed, hundreds of workers were used to hand-paint them, and radium poisoning became an important industrial hazard. Scientific workers who handle radium or other radioactive substances are also in danger of inhaling or ingesting them, a current and definite hazard.

Radium Osteitis and Radium-induced Neoplasia

Radium is deposited in the skeleton, like calcium. It is an alpha emitter with a half-life of 1620 years and, hence, emits continuous radiation. Bone that contains a large concentration of radium for 20 years or more after its ingestion or injection consists mostly of dead osseous tissue. Physiologic bone formation and resorption become erratic, producing large resorption cavities filled with gelatinous material or osteoid-like matrix. Radiation may cause (1) immediate or delayed cell death, (2) cellular injury with recovery, (3) arrest of cellular division, and (4) abnormal repair with neoplasia.[9]

Radiographic changes may occur in the skull (Fig. 1233) and are usually present in the long bones (Fig. 1234). They consist of (1) *punched-out lesions* resembling multiple myeloma which may increase in size,[10] (2) *bone sclerosis* occurring particularly in the metaphyses and due to bone ischemia, (3) *ischemic necrosis* usually in the head of the femur and humerus (Fig. 1235, *A* and *B*), (4) *pagetoid changes* of coarse trabeculations (Figs. 1236 and 1237), *pathologic fracture*, and (6) *neoplasm* (Fig. 1238). The average latent period is 23 years, and osteosarcoma and fibrosarcoma are the usual tumors.

Radiation Necrosis

The threshold of radiation changes in bone is believed to be 3000 rads; cell death occurs with 5000 rads.[11,12]

Radiation Necrosis of the Mandible. Radiation necrosis of the mandible occurs in one-third of the patients treated for intraoral cancer, but the risk must be accepted to achieve a high cure rate.[13] The loss of salivary gland function and changes secondary to radiation therapy predispose to infection, and poor dental hygiene adds to the risk. Dental extractions are necessary prior to radiation therapy. The roentgen features of mandibular necrosis are (1) *ill defined destructive lesion* without sequestrum or soft-tissue mass (Fig. 1239); (2) *destruction of the cortex*; (3) *possible pathologic fracture*; (4) *frequent superimposed infection*, rarely with sequestrum formation; (5) *the lesion progressing slowly or healing* with conservative therapy; and (6) *necrosis confined* to the irradiated area.

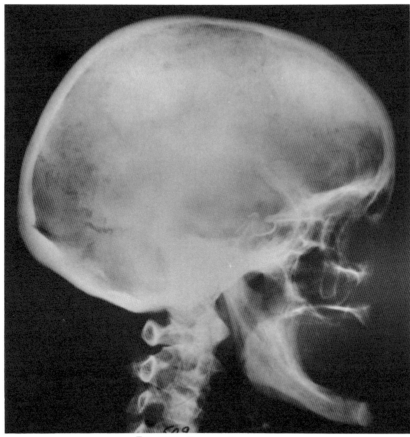

FIG. 1233. RADIUM POISONING
Radium dial worker. Multiple lytic defects, particularly in the posterior parietal area, resembling multiple myeloma. (Previously published in M. K. Dalinka, J. Edeiken, and J. B. Finkelstein: *Seminars in Roentgenology, 9:* 29, 1974.[17])

FIG. 1234. RADIUM POISONING
Multiple lytic defects in the tibial and fibular shafts bilaterally in a former radium dial worker. (Courtesy of Dr. Charles Zimmerman, New York, N.Y. Previously published in M. K. Dalinka, J. Edeiken, and J. B. Finkelstein: *Seminars in Roentgenology, 9:* 29, 1974.[17])

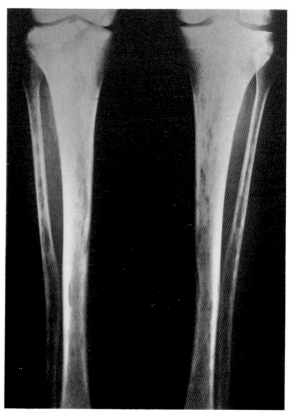

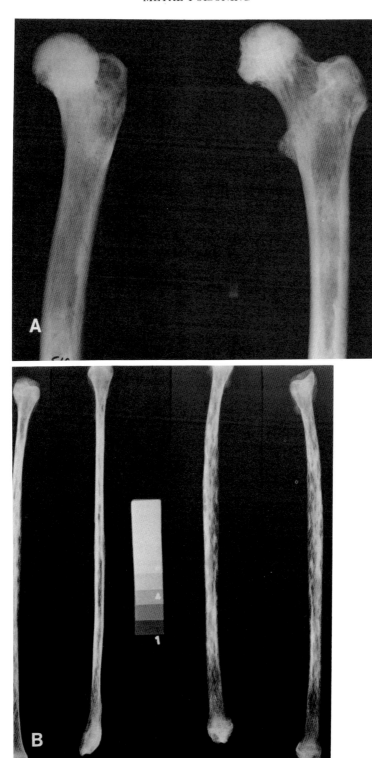

Fig. 1235. Radium Poisoning

(A) Autopsy specimen of radium dial worker. Two views of left femur reveal coarse trabeculation, ischemic necrosis of the femoral head, and increased density in the metaphyseal region. (B) Autopsy specimens of the fibulas in a radium dial worker, with coarse trabeculations and areas of increased density and osteolytic lesions.

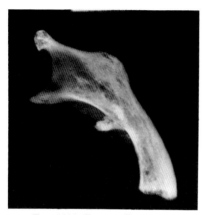

Fig. 1236. Radium Poisoning
Autopsy specimen of radium dial worker. A portion of the mandible shows an increase in density and osteolytic areas with coarse trabeculation.

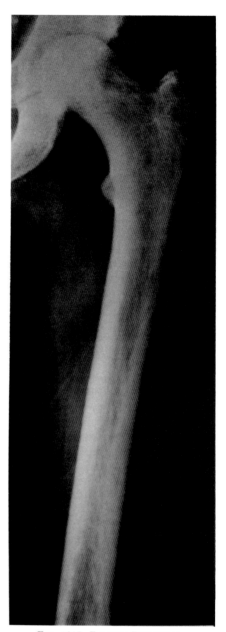

Fig. 1237. Radium Poisoning
Radium dial worker. Sclerosis of left femoral head with cortical thickening and trabecular prominence of the shaft resembling Paget disease, with similar changes on the right. (Courtesy of Dr. C. B. Henle, Horace Martland Hospital, Newark, N.J. Previously published in M. K. Dalinka, J. Edeiken, and J. B. Finkelstein: Seminars in Roentgenology, 9: 29, 1974.[17])

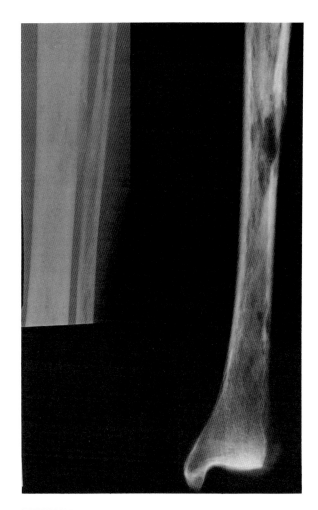

Fig. 1238. Radium-induced Osteosarcoma
(*Left*) Destructive lesion in tibial shaft with periosteal new bone formation. (*Right*) Pathologic specimen of the same lesion with the destructive neoplasm more easily seen. In addition, multiple lucencies in the tibial shaft represent resorption cavities caused by the radium. (Courtesy of Dr. C. B. Henle, Horace Martland Hospital, Newark, N.J. Previously published in M. K. Dalinka, J. Edeiken, and J. B. Finkelstein: *Seminars in Roentgenology, 9:* 29, 1974.[17])

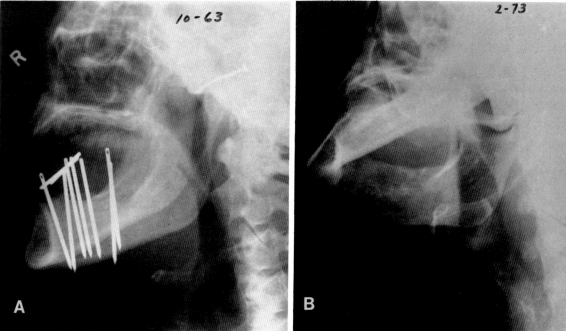

Fig. 1239. Radiation Necrosis of Mandible
(*A*) Radium implant for carcinoma of the left side of the mouth. For 6 days in 1963, 6500 rads were given. The patient was treated again in 1971 for carcinoma of the midportion of the tongue with an implant and external irradiation. (*B*) Poorly defined destructive mandibular lesion with evidence of sclerosis in 1973. Two biopsies showed no evidence of recurrent tumor. (Previously published in M. K. Dalinka, J. Edeiken, and J. B. Finkelstein: *Seminars in Roentgenology, 9:* 29, 1974.[17])

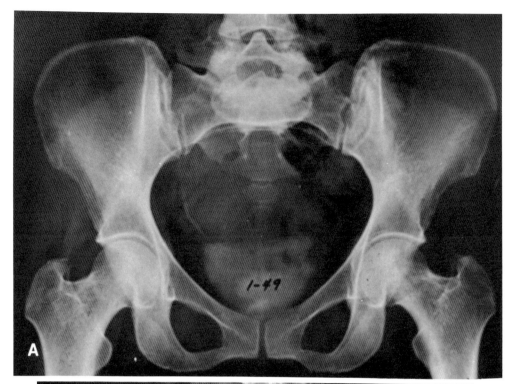
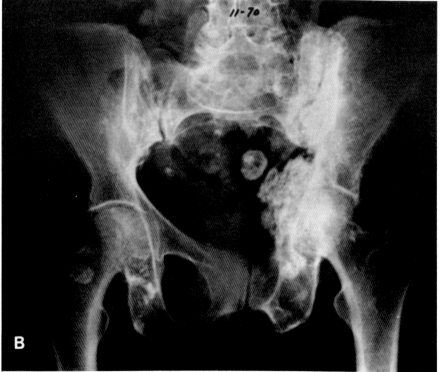

FIG. 1240. RADIATION NECROSIS
Treatment for carcinoma of the cervix in 1949 with orthoirradiation and radium. (A) Before treatment the pelvis is normal. (B) Twenty-one years after therapy: sclerosis of the sacroiliacs and calcification in the left pelvis. Fractures were present in the left pubic rami. (Previously published in C. A. Gooding and A. Margulis (Eds.): *Diagnostic Radiology, 1979*, Masson Publishing, New York, 1979.[41])

Necrosis may be distinguished from recurrence by the absence of a mass, the well defined lesion, and the failure to progress after a short period of time.

Radiation Necrosis in Femoral Neck. The signs of radiation necrosis are fracture as early as 5 months after therapy, sclerotic bone, and ischemic necrosis of the head of the femur. The joint spaces of the hips and shoulders are frequently narrowed, with an appearance similar to that of osteoarthritis (see Figs. 1257 and 1258).

Radiation Necrosis of the Pelvis. The roentgen features of radiation necrosis of the pelvis are (1) irregularities of the sacroiliac joint which may appear with sclerosis, widening of the joint space, and calcification (Figs. 1240–1243); (2) fractures of the ischium and pubis (Fig. 1240); and (3) destruction of bone in the region of radon or gold seed implants (see Fig. 1248). Fracture of the femoral neck occurs after radiation therapy[11, 14–17] (see Figs. 1245–1247) and has been reported as early as 5 months following therapy, frequently with pain preceding its radiographic evidence.[14, 16] Prior to fracture, sclerotic changes in the femoral neck and ischemic necrosis of the head may be present and, indeed, may indicate impending fracture (Fig. 1244).

Radiation Necrosis of the Ribs, Clavicle and Humerus. Fractures of the ribs, clavicle, scapula, and humerus are frequent after orthotherapy for breast carcinoma (Figs. 1249 and 1250). They rarely heal and may be painless. Resorption of the fracture fragments may occur, particularly in the clavicle and ribs. Multiple lesions all confined within the therapy portals help to differentiate it from metastatic disease (Fig. 1251).

Radiation Necrosis of Other Bones. Radiation dermatitis with underlying osseous changes is reported in physicians and dentists who handle radium or work closely with x-rays. Bragg et al.[11] described faint periosteal new bone formation of phalangeal lesions.

Radiation-induced Tumors

The following criteria are necessary to establish the diagnosis of radiation-induced tumor[18]:

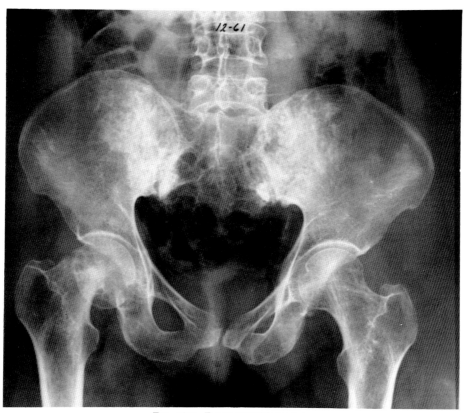

FIG. 1241. RADIATION NECROSIS

Fracture of the left femoral neck. The patient was treated for carcinoma of the cervix 4 years earlier, with 2500 rads of external irradiation plus two applications of radium. Note the calcification and other changes about the sacroiliac joint.

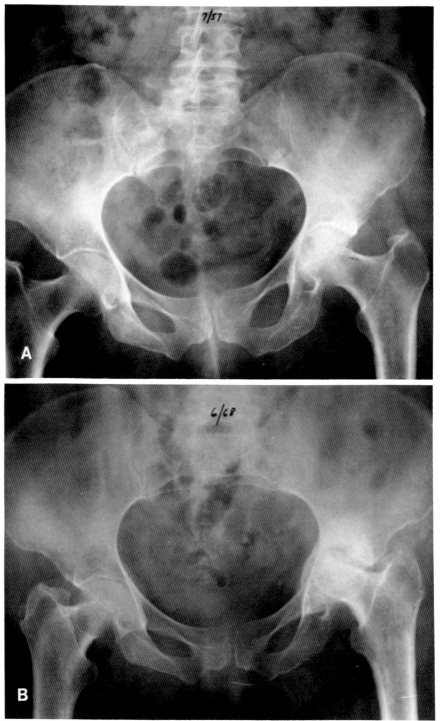

FIG. 1242. RADIATION NECROSIS OF THE LEFT HIP

The patient was treated postoperatively for carcinoma of the endometrium in 1955 and was retreated for recurrence in the left pelvis in 1956. (A) July 1957: sclerosis of the left femoral head with ischemic necrosis. (B) June 1968: fracture of the acetabulum with cyst formation and flattening of the left femoral neck. (Courtesy of Dr. Richard Rosen, New York, N.Y. Previously published in M. K. Dalinka, J. Edeiken, and J. B. Finkelstein: Seminars in Roentgenology, 9: 29, 1974.[17])

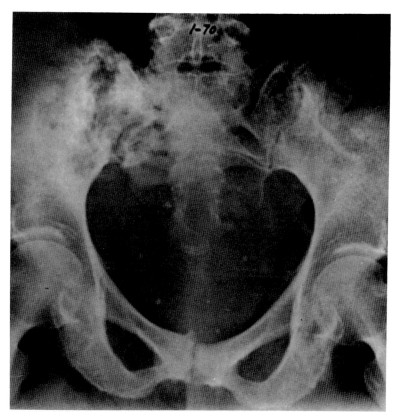

Fig. 1243. Extensive Radiation Necrosis of the Right Sacroiliac Joint

Widening, irregularity, and sclerosis of the sacroiliac joint. Ten years earlier the patient had been given 7000 rads externally with the Betatron and 3290 mg of local radiation therapy per hour for carcinoma of the cervix. The patient died of a ruptured necrotic bladder. She also had radiation necrosis to the bowel. (Previously published in M. K. Dalinka, J. Edeiken, and J. B. Finkelstein: *Seminars in Roentgenology*, 9: 29, 1974.[17])

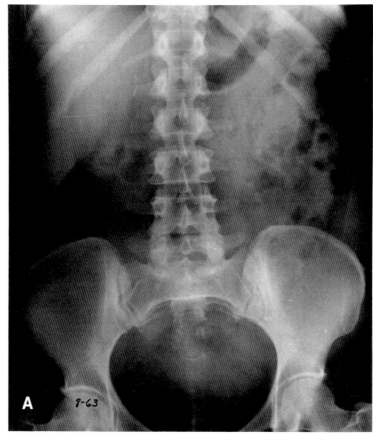
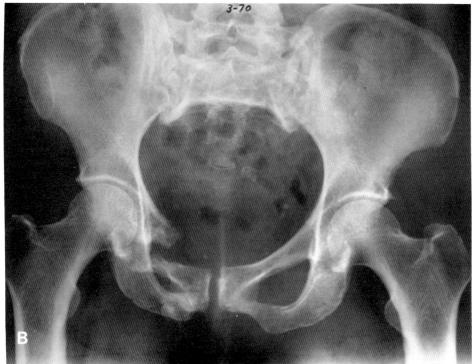

Fig. 1244. Radiation Necrosis

Radiation necrosis. (A) Normal pelvis, August 1963: the patient was treated with 6100 rads to the total pelvis on the Betatron and 4680 mg of radium per hour for Stage IIIA carcinoma of the cervix. (B) March 1970: irregularity and widening of the sacroiliac joints bilaterally. Fractures through the superior aspect of the right ischium and inferior aspect of the pubis. The margins of the adjacent bone are irregular with increased density. (Previously published in M. K. Dalinka, J. Edeiken, and J. B. Finkelstein: *Seminars in Roentgenology, 9:* 29, 1974.[17])

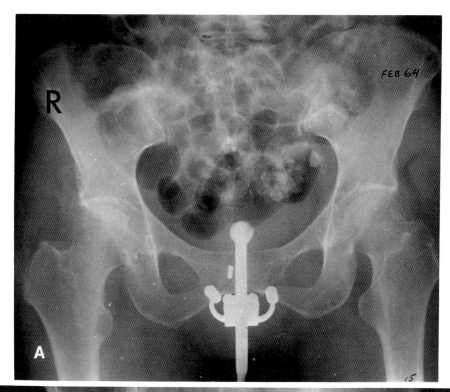
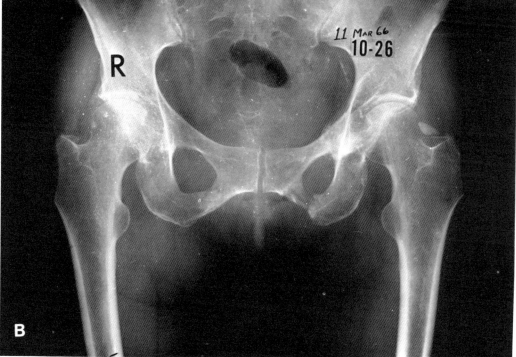

FIG. 1245. RADIATION NECROSIS
The patient was treated for carcinoma of the cervix. (A) Notice that the lower capsules lie against the ischial rami. (B) Two years later there are fractures and resorption on the left side and increased density of the ischial ramus on the right side at the site of the lower capsules of the applicator.

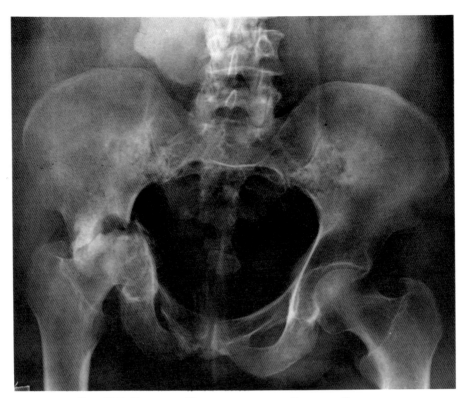

FIG. 1246. RADIATION NECROSIS FOLLOWING EXTERNAL IRRADIATION

Fracture of the acetabulum, resorption of the head of the femur, and some increased density. Note the absence of soft-tissue mass. Changes have occurred in both sacroiliacs and at the symphysis pubis.

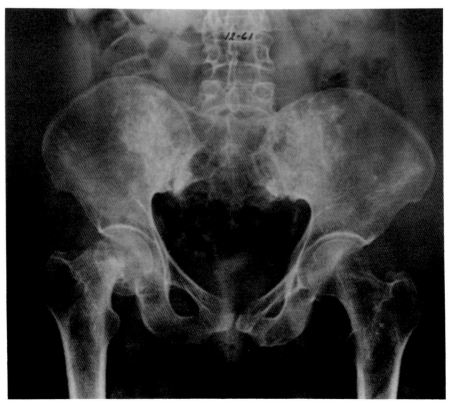

FIG. 1247. RADIATION NECROSIS

Fracture of left femoral neck. The patient was treated for carcinoma of the cervix 4 years earlier with 2500 rads of external irradiation plus two applications of radium. Note the calcification and other changes about the sacroiliac joints.

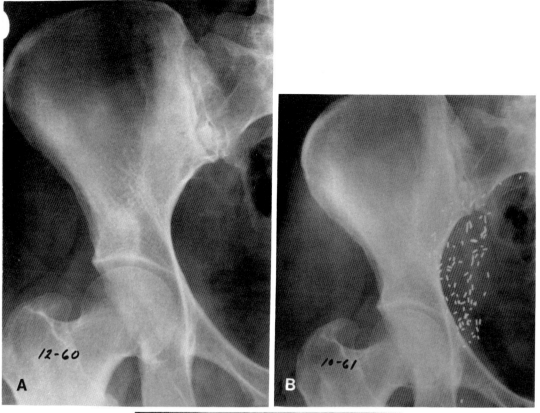
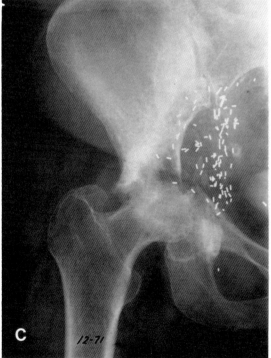

FIG. 1248. RADIATION NECROSIS
Treatment of the pelvis for carcinoma of the cervix. (A) Before treatment with 4200 mg of radium and 3000 rads with 250 kv to the pelvis. (B) Ten months later, after gold seeds had been implanted for recurrence and 200 rads of 250 kv therapy had been administered. (C) Fourteen months later: increased densities of the ilium with fractures in the acetabulum, with resorption of the head and ischemic necrosis.

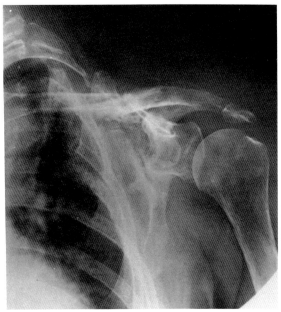

FIG. 1249. RADIATION NECROSIS

Fractures of the scapula, clavicle, and ribs in a patient treated with orthovoltage radiotherapy for carcinoma of the breast. Notice the almost complete resorption of the upper ribs. (Courtesy of Dr. Stanley Siegelman, New York, N.Y. Previously published in M. K. Dalinka, J. Edeiken, and J. B. Finkelstein: *Seminars in Roentgenology, 9:* 29, 1974.[17])

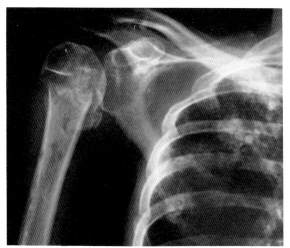

FIG. 1250. RADIATION NECROSIS

Examination of the shoulder 10 years following irradiation for carcinoma of the breast had revealed pathologic fractures of the right humerus and multiple ribs. The lucencies in the proximal humeral shaft are also secondary to radiation necrosis. (Previously published in M. K. Dalinka, J. Edeiken, and J. B. Finkelstein: *Seminars in Roentgenology, 9:* 29, 1974.[17])

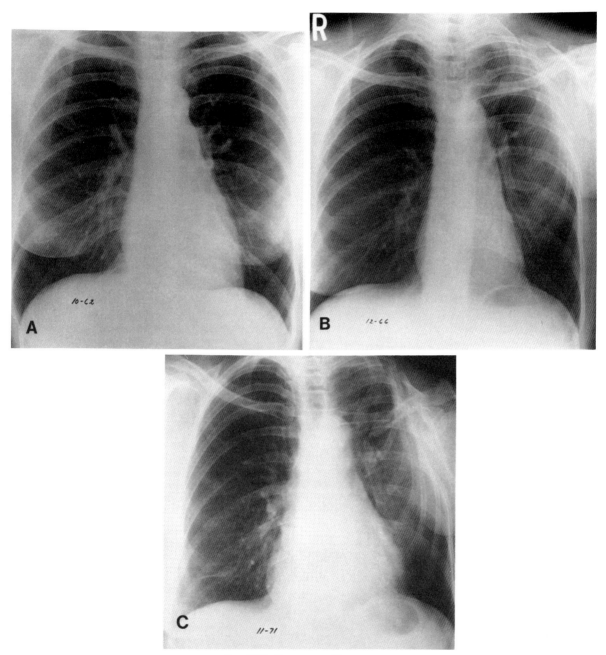

FIG. 1251. RADIATION NECROSIS

Radiation for carcinoma of the breast. (*A*) Before therapy. (*B*) Four years after therapy: increased density of the lung and fractures of the upper ribs in the midaxillary line. There was no pain or mass. (*C*) Nine years after therapy: collapse of the upper rib cage, resorption of the ends of the ribs, and a fracture of the clavicle. These lesions were painless, and there was no evidence of soft-tissue tumor mass. (Previously published in C. A. Gooding and A. Margulis (Eds.): *Diagnostic Radiology, 1979*, Masson Publishing, New York, 1979.[41])

(1) roentgen or microscopic evidence of the benign primary condition, (2) a sarcoma arising within the field of radiation therapy, (3) a latent period of at least 4 years, and (4) histologic proof of sarcoma.

In half of the cases, areas of radiation osteitis are present in the bone adjacent to a radiation-induced tumor.[19] Pain and mass are consistent symptoms[11, 20] (Figs. 1252-1254).

The sarcoma is usually an undifferentiated fibrosarcoma; osteosarcoma and chondrosarcoma are also reported. Osteosarcoma has been reported following Thoratrast injection,[21] and after irradiation for pituitary tumor.[22]

Radiation Changes in Children. The irradiation changes that occur in children are related to the age of the child and to the portion of the bone treated. If half of the spine is treated, a

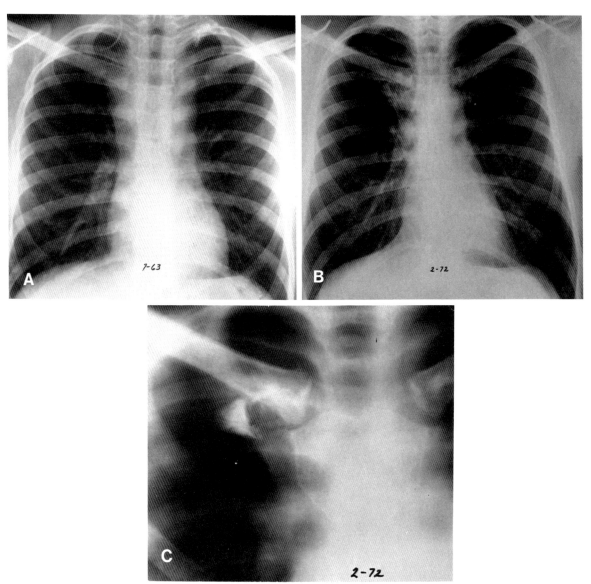

FIG. 1252. RADIATION-INDUCED OSTEOSARCOMA
(A) Films obtained in 1963 reveal a mediastinal mass bilaterally. A total of 4000 rads was delivered to the mediastinum for Hodgkin disease. (B) Nine years later, the calcific density of the anterior portion of the first rib proved to be osteosarcoma. (C) Body section films reveal the tumor new bone formation in the osteosarcoma. (Previously published in C. A. Gooding and A. Margulis (Eds.): *Diagnostic Radiology, 1979.* Masson Publishing, New York, 1979.[41])

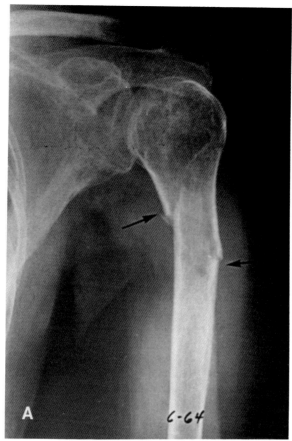
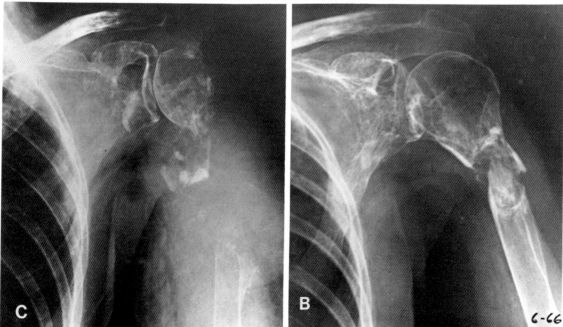

FIG. 1253. RADIATION-INDUCED SARCOMA

The patient was treated with orthovoltage radiotherapy for carcinoma of the breast in 1958 and presented with pathologic fracture in the humerus in 1964. (A) 1964: pathologic fracture in the upper end of the humerus (arrows). Note the radiation changes in the scapula and head of the humerus. (B) 1966: pathologic fractures have failed to heal after 2 years of conservative therapy. (C) 1972: undifferentiated sarcoma involving bones and soft tissues. Note the sclerosis secondary to radiation necrosis of the scapula. (Previously published in M. K. Dalinka, J. Edeiken, and J. B. Finkelstein: Seminars in Roentgenology, 9: 29, 1974.[17])

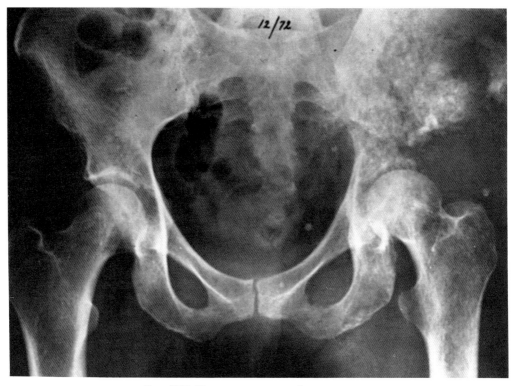

FIG. 1254. RADIATION-INDUCED OSTEOSARCOMA
Osteosarcoma of the left iliac crest 17 years following therapy for cervical carcinoma. (Previously published in M. K. Dalinka, J. Edeiken, and J. B. Finkelstein: *Seminars in Roentgenology*, 9: 29, 1974.[17])

scoliosis will result; and the younger the child treated, the more severe the curvature (Fig. 1255).

Treatment to epiphyseal areas may cause the death of part or all of the epiphyseal cartilage, resulting in deformity and shortening (Fig. 1256).

Treatment to the abdomen with inclusion of the ilium will cause hypoplasia (Fig. 1255).

Pathologic fractures, areas of increased density, and radiolucency are frequent and are due to the same causes as those described in the adult.

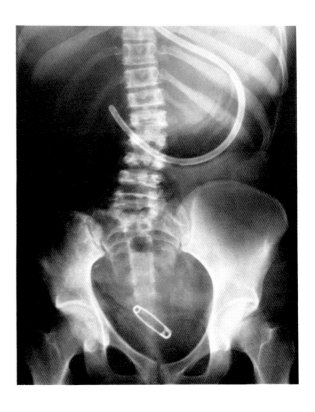

FIG. 1255. RADIATION COMPLICATION
This patient was treated for right-sided Wilms tumor in childhood. Note the severe decrease in the height of the vertebral bodies and scoliosis convex to the left. The ribs and iliac crest on the treated side are hypoplastic. The long tube was present because of intestinal obstruction which was later shown to be secondary to radiation enteritis. (Courtesy of Dr. Murray K. Dalinka, Philadelphia, Pa.)

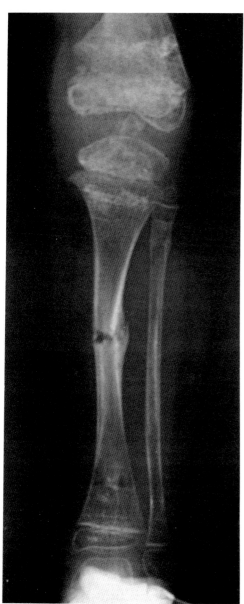

FIG. 1256. RADIATION NECROSIS
Patient irradiated approximately 2 years prior to this roentgenogram for leukemia knee pain. The portal included most of the tibia and the knee. The roentgenogram shows a fracture through the midshaft of the tibia with periosteal reaction; periosteal reaction also at the midshaft of the fibula; metaphyseal irregularity of the distal femur, with deformity secondary to irradiation therapy; and diffuse osteopenia. (Courtesy of Dr. Murray K. Dalinka, Philadelphia, Pa.)

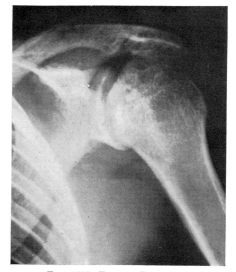

FIG. 1257. RADIUM POISONING
Irregularity of the articular surface of the humerus secondary to devitalization of the articular cartilage and aseptic necrosis of portions of the humeral head.

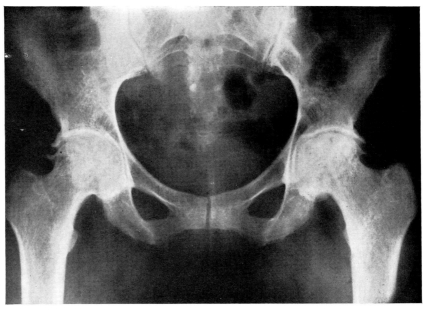

FIG. 1258. RADIUM POISONING
Appearance is that of osteoarthritis of the hip. The diagnosis of radium poisoning could not be made roentgenologically.

FLUORINE POISONING

Although fluorides are general protoplasmic poisons, the incidence of fluorine intoxication is low. The distribution of fluorine is so widespread in nature that a small intake is practically unavoidable and not deleterious. For practical purposes it is only from the following sources that fluorine can be ingested in quantities sufficient to cause intoxication: (1) drinking water containing 1 part per million or more of fluorine; (2) fluorine compounds used as insecticidal sprays for fruits and vegetables (cryolite and barium fluorosilicate); (3) the mining of phosphate rock and conversion to superphosphate for fertilizer; (4) fluorides used in the smelting of metals, such as steel and aluminum, and in the production of glass, enamel, and brick; and (5) excessive fluoride intake for treatment of myeloma, Paget disease, etc.

The principal pathologic effects produced by chronic fluorine poisoning are hypoplasia of the teeth ("mottled enamel") and bone sclerosis.[23]

It was originally assumed that fluorine was stored directly in the bones and teeth,[24] forming an abnormal inorganic matrix with subsequent dense irregular calcium deposits. Others[25, 26] have suggested, however, a derangement of the osteoid mucopolysaccharides. Plasma alkaline phosphatase is increased, indicating osteoblastic activity,[27, 28] but there is also increased bone resorption,[29] a situation that resembles Paget disease.

Roentgenographic Features. The degree of bone change does not correlate well with the bone fluoride content,[30] and roentgenograms may be normal when the fluoride bone content is 6 times above normal.[24]

Endemic fluorine poisoning rarely produces bone changes. Osteoporosis may be an early sign in young patients.[31–33] Fluorine is more readily incorporated into active bone growth areas,[25] and it is possible that the fluorotic tendency is increased at puberty.[34] Only 15% of individuals drinking, over a period of years, water with a fluoride content of more than 8 parts per million develop x-ray signs of fluorosis.[24]

Three roentgenographic stages of fluorine poisoning are described.[35]

STAGE I. Trabecular roughening and blurring appear in the vertebrae and pelvis.

STAGE II. The thick trabeculae merge to cause a diffuse, structureless appearance. The bone contours become uneven, especially the ribs, pelvis, and spine. The appendicular skeleton is less affected, but the medullary cavities may be narrowed by endosteal cortical thickening. Ligamentous calcification begins most frequently in the paraspinous, sacrospinous, and sacrotuberous ligaments.

STAGE III. The axial bones lose most morphologic features and appear as white marble. Cortical and trabecular definition are lost and appear "wooly." The long bone cortices are dense and thick due to amorphous subperiosteal new bone formation. The medullary cavities are reduced by endosteal cortical thickening.[36] Calcification occurs in the periarticular ligaments, musculocutaneous attachments, and interosseous membranes (Fig. 1259).

Most changes occur in the axial skeleton until

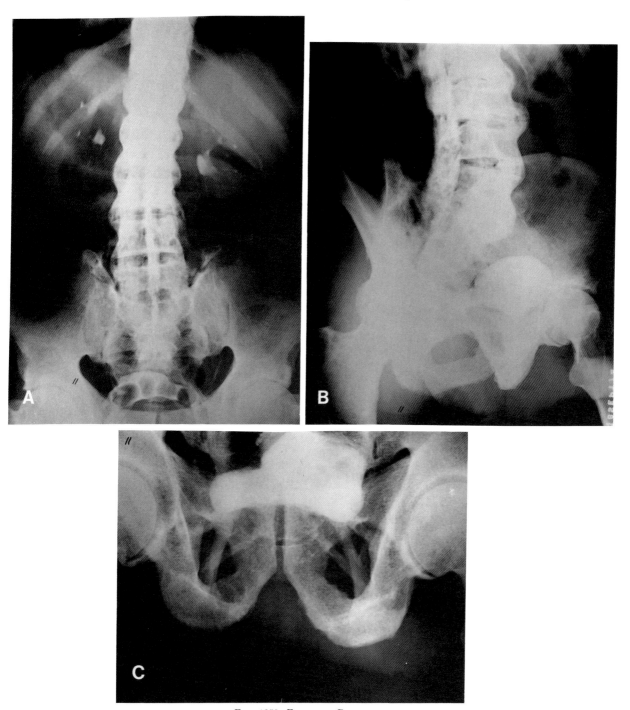

FIG. 1259. FLUORINE POISONING

(A) Anteroposterior lumbar spine: marked osteophyte formation and increased density. Calcification of the ligaments of the sacroiliac and pelvis was found. (B) Oblique view of lumbar spine: marked calcification of the posterior ligament and posterior elements of the spine and the osteophytosis. (C) View of the pelvis: calcification of the ligaments and fluffy reactive bone change at the ischial spines.

late in the disease. Vertebral osteophyte formation is frequent[37] (Fig. 1259). The ribs may have needlelike calcifications extending into the intercostal muscles. The skull may be thickened.[38, 39] Calcaneal spurring and plantar cortical thickening develop in the late stages.[32] The bones of the hands and feet infrequently show cortical thickening.[34] Acute or subacute fluorosis (periostitis deformans) may occur in operating-room personnel who are addicted to the fluoride-containing anesthetic Penthrane (methoxyflurane) and in individuals habituated to wine containing a fluorine preservative.[40] There is increased density throughout the spine and marked periostitis in the long bones, especially in the hands. Considerable bone pain is present, and it is of a relapsing nature, depending on the amounts and times taken. The history is not volunteered and should be suspected in hospital personnel who have access to anesthetics.

The skeletal changes of fluorine poisoning are usually more diffuse and uniform than in osteoblastic metastasis. Paget disease, with its typical increased bone trabeculation, seldom resembles fluorine poisoning roentgenographically or clinically. Myelosclerosis causes a diffuse increased density of most of the skeleton which, roentgenologically, is often indistinguishable from fluorine poisoning, and anemia is found in both conditions. Myeloid metaplasia, however, is always present in myelosclerosis and is never present in fluorine poisoning; the blood should be examined to distinguish between the two.

The fluorotic mechanism is uncertain and may result from reactive bone response rather than increased fluoride concentration.[3] The increased bone density may be quantitative rather than qualitative, due to increased bone thickness with no increase in mineralization, as evidenced by less dense and less calcified exostosis of fluorosis.[25]

REFERENCES

1. Cooper, G., Jr.: An epidemic of inhalation lead poisoning with characteristic skeletal changes in the children involved. AJR, 58: 129, 1947.
2. Caffey, J.: Clinical and experimental lead poisoning; some roentgenologic and anatomic changes in growing bones. Radiology, 17: 967, 1931.
3. Pugh, D. G.: *Roentgenologic Diagnosis of Diseases of Bones.* Williams & Wilkins, Baltimore, 1962.
4. Whitridge, J., Jr.: Changes in long bones of newborn infants following the administration of bismuth during pregnancy. Am J Syph Gonor Venerol Dis, 24: 223, 1940.
5. Caffey, J.: Changes in the growing skeleton after the administration of bismuth. Am J Dis Child, 53: 56, 1937.
6. Leake, J. P.: Radium poisoning. JAMA, 98: 1077, 1932.
7. Martland, H. S.: Occupational poisoning in manufacture of luminous watch dials. JAMA, 92: 466, 1929.
8. Martland, H. S.: Occupational poisoning in manufacture of luminous watch dials. JAMA, 92: 552, 1929.
9. Vaughan, J.: The effects of skeletal irradiation. Clin Orthop, 56: 283, 1968.
10. Hasterlik, R. J., and Finkel, A. J.: Diseases of bones and joints associated with intoxication by radioactive substances, principally radium. Med Clin North Am, 49: 285, 1965.
11. Bragg, D. G., Shidnia, H., Chu, F. C. H., et al.: The clinical and radiographic aspects of radiation osteitis. Radiology, 97: 103, 1970.
12. Cade, S.: Radiation induced cancer in man. Br J Radiol, 30: 393, 1957.
13. Grant, B. P., and Fletcher, G. H.: Analysis of complications following megavoltage therapy for squamous cell carcinomas of the tonsillar area. AJR, 96: 28, 1966.
14. Gratzek, F. R., Holmstrom, E. G., and Rigler, L. G.: Postirradiation bone changes. AJR, 53: 62, 1945.
15. McCrorie, W. D. C.: Fractures of the femoral neck following pelvic irradiation; a review of 10 cases. Br J Radiol, 23: 587, 1950.
16. Slaughter, D. P.: Radiation osteitis and fractures following irradiation with report of five cases of fractured clavicle. AJR, 48: 201, 1942.
17. Dalinka, M. K., Edeiken, J., and Finkelstein, J. B.: Complications of radiation therapy; adult bone. Semin Roentgenol, 9: 29, 1974.
18. Cahan, W. G., Woodard, H. Q., Higinbotham, N. L., et al.: Sarcoma arising in irradiated bone. Cancer, 1: 3, 1948.
19. Arlen, M., Higinbotham, N. L., Huvos, A. G., et al.: Radiation induced sarcoma of bone. Cancer, 28: 1087, 1971.
20. Case records of the Massachusetts General Hospital; case 39 1966. N Engl J Med, 275: 496, 1966.
21. Altner, P. C., Simmons, D. J., Lucas, J. F., et al.: Osteogenic sarcoma in a patient injected with Thoratrast. J Bone Joint Surg, 54A: 670, 1972.
22. Sparagana, M., Eells, R. W., Stephani, S., et al.: Osteogenic sarcoma of the skull; a rare sequela of pituitary irradiation. Cancer, 29: 1376, 1972.
23. Fishbein, M.: Chronic fluorine intoxication. JAMA, 123: 150, 1943.
24. Leone, N. C., Stevenson, C. A., Hilbish, T. F., and Sosman, M. C.: Roentgenologic study of human population exposed to high fluoride domestic water; 10 year study. AJR, 74: 874, 1955.
25. Weatherall, J. A., and Weidmann, S. M.: The uptake and distribution of fluorine in bones. J Pathol, 78: 243, 1959.
26. Weatherall, J. A., and Weidmann, S. M.: The effect of fluoride on bone. Proc Nutr Soc, 22: 105, 1963.
27. Bernstein, D. S., Guri, D., Cohen, P., Collins, J. J., and Tamyakopoulos, S.: The use of sodium fluoride in metabolic bone disease (abstract). J Clin Invest, 42: 916, 1963.
28. Rich, C., and Ensinck, J.: Effect of sodium fluoride on calcium metabolism of human beings. Nature, 191: 184, 1961.
29. Weidmann, S. M., Weatherall, J. A., and Jackson, D.: The effect of fluoride on bone. Proc Nutr Soc, 22: 105,

1963.
30. Morris, J. W.: Skeletal fluorosis among Indians of the American southwest. AJR, *94:* 608, 1965.
31. Anderson, W. A. D.: In *Pathology*. Mosby, St. Louis, 1948.
32. Kumar, S. P., and Harper, R. A. Kemp: Fluorosis in Aden. Br J Radiol, *36:* 497, 1963.
33. Middlemiss, H.: In *Tropical Radiology*, p. 68. Heinemann, London, 1961.
34. Cook, P. L., and Carbone, P. P.: Myeloma of bone treated with sodium fluoride. Clin Radiol, *19:* 379, 1968.
35. Rohoim, K.: *Fluoride Intoxication. A Clinical Hygienic Study*. Lewis, London, 1937.
36. Azar, H. A., Nukko, C. K., Bayyuk, S. I., and Bayyuk, W. B.: Skeletal sclerosis due to chronic fluoride intoxication; cases from an endemic area of fluorosis in the region of the Persian Gulf. Ann Intern Med, *55:* 193, 1961.
37. Singh, A., Jolly, S. S., Bansal, S. C., and Methur, M.: Endemic fluorosis; epidemiological, clinical and biochemical study of chronic fluorine intoxication in Panjab (India). Medicine, *42:* 229, 1963.
38. Calenoff, L.: Osteosclerosis from intentional ingestion of hydrofluoric acid. Radiology, *87:* 1112, 1962.
39. Linsman, J. F., and McMurray, C. A.: Fluoride osteosclerosis from drinking water. Radiology, *40:* 474, 1943.
40. Soriano, M., and Manchon, F.: Radiological aspects of a new type of bone fluorosis; periostitis deformans. Radiology, *87:* 1089, 1966.
41. Edeiken, J.: Radiation changes in bone. In *Diagnostic Radiology, 1979*, C. A. Gooding, and A. Margulis (Eds.). Masson Publishing, New York, 1979.

14

Skeletal Maturation

METHODS OF DETERMINING MATURATION

Physical development and growth during childhood may be measured according to weight, height, and skeletal maturation. Skeletal maturation is the best indicator of the individual constitution, since weight and height vary according to familial characteristics. Weight also varies with nutritional state and height is affected temporarily by fluctuations in health. Skeletal maturation is little affected by nutrition and health unless the variations are extreme.

Generally, growth, as measured by weight and height, should be distinguished from maturation since a child may grow and still be immature, as in endocrine disturbances. Because it can be studied with relative ease, the skeleton is referred to most frequently in maturation studies; however, not only the skeleton, but also general body form and behavior give evidence of retarded or precocious maturation. Todd clearly differentiated skeletal maturation from growth when he stated "maturation progress is evident in every part of the skeleton, but it is in the transformation of fibrous tissue and cartilage in the bone that the most easily identifiable criteria present themselves." Thus, maturation is defined as a process of transformation and it is distinguished from growth by the creation of new cells and tissues. It is necessary to measure this transformation from cartilage to bone until the fully matured osseous skeleton of the adult is attained. This transformation may be divided into three distinct phases: (1) in utero ossification of the long bone diaphyses; (2) ossification of the epiphyses and long bone centers, which begins shortly before birth and is almost completed at puberty; and (3) the ossification and bony fusion of the epiphyseal growth plates to the shaft.

J. W. Pryor[1,2] first studied the transformation of cartilage to bone and made three fundamental observations: (1) "the bones of the female ossify in advance of the male," (2) "ossification is bilaterally symmetrical," and (3) "variation in the ossification of bone is a hereditable trait." Thereafter, many methods for assessment of skeletal maturation are described, but none are completely satisfactory.

Early attempts were made to assess skeletal maturation by recording the initial appearance of the carpal ossification centers and the distal epiphyses of the radius and ulna.[3,4] Later, the amount of bone in the carpals was measured directly.[5-7] Then Sontag, Snell, and Anderson[8] made charts which considered only the initial appearance of ossification centers, probably the most variable of the entire maturation process.

Charts compiled by Camp and Cilley[9] (Figs. 1260 and 1261) and Hodges[10] (Table 30) also used the appearance of ossification centers throughout the skeleton and also the time of fusion of some of these centers. Although these charts were useful for obtaining gross estimates of skeletal maturation, they ignored the sequential intervening changes in the maturing ossification center.

In 1937, Todd published the *Atlas of Skeletal Maturation of the Hand*, later revised by Greulich and Pyle.[11] Todd distinguished between growth and maturity. He recognized the importance of the serial changes of maturing ossification centers and considered them maturity "determinators." He considered the successive changes in outline and contour which occurred regularly and in predictable and irreversible order, thereby precluding a search for newly formed ossification centers. The observer can merely study the radiographic centers.

Todd's series of radiographs are the left hands of healthy, upper middle class, white children in Cleveland, Ohio. Most of the children were radiographed near their birthdays and at 6-month intervals for 5 years. The most suitable films, for each age group, were then chosen as a standard for the *Atlas*. Separate films for male and female were included and these were assigned a skeletal age determined by the age of the children upon whom the standard was based (Figs. 1262 and 1263). Time, then, is the independent variable on which the skeletal changes of maturation are considered dependent. One year in a normal child's life was now equivalent to 1 year of development in skeletal age.

In addition, separate atlases based upon the population of children have been developed for the foot[12] and knee.[13] These do not separate radiographic standards for male and female, but use two separate ages for each radiographic standard. Study has shown that although the median trends of maturity for a population were close to the hand and knee atlas standard, there was a difference of up to 1.2 years between the skeletal maturity levels of the hands and knee.[14] This may reflect regional differences, or the fact that different children were used for the maturity standard.

These atlases, despite the above disadvantages, are perhaps the greatest advance in the field of assessing skeletal maturity; however, other problems remain. Todd believed that illness and deprivation could disrupt the order of appearance of ossification centers and that one center could therefore have a different skeletal age than the others. The greater the disparity among these ossification centers, the greater would be the insults the child had suffered. On the other hand, the closer the centers were to each other in skeletal age, the healthier the child. In pursuing this concept, Todd may have overlooked Pryor's finding of genetic determination of skeletal maturing patterns. This has led to a difficulty in interpreting. Since the radiograph in question will rarely match the exact standard radiograph in the atlas, it must be aligned between standards. Because of differences in genetic patterns of maturation, some of the hand and wrist bones of healthy children will match one standard while others will match another. Thus, some weighted, biased judgments will have to be made.

Pyle proposed a solution by assigning a skeletal age to each ossification center and then averaging the individual bones to determine the skeletal age. If a center appears unusually late, for a variety of reasons, its eventual ossification will lower the mean age. It is estimated that 95% confidence level in the single assessment of a hand radiograph, using the Gruelich-Pyle *Atlas*, may be in the range of 8-12 months.[15] Each observer brings his own bias to these readings and thereby sets his own level of reading. Increased training narrows the gap between observers.[15] Assessment of skeletal maturation on the same child by a pair of trained observers, with the result being the mean of their reading, further diminishes intraobserver differences.[16] *Despite the disadvantages, the Greulich-Pyle* Atlas *remains the best method of assessing the present developmental status of an individual child.* It seems to distinguish the quick from the slow developer. However, the concept of skeletal age developed in the *Atlas* has given rise to some difficulties: (1) Since maturation is distinct from growth it should have a separate scale of measurements. (2) Maturation does not proceed at the same rate in all healthy children, therefore, the skeletal year cannot have the same mean for any two healthy children, nor can it have the same mean for two bones in the same child. (3) As females mature more rapidly than males, the skeletal year in females will contain more maturity changes than males. (4) Skeletal maturation in any given child does not proceed at a constant rate throughout the time of development; e.g., the prepubertal and pubertal spurts. Therefore, with a single sex, the rate of maturation in a skeletal year will depend upon the chronologic age.

In 1954, Acheson[17] attempted to devise a scoring system that would rate maturation on a scale independent of chronologic age. In this, the Oxford method, the roentgenograms are not characterized by age and sex. For each ossification center, the radiographs are lined from the last to the most mature. A series of maturity indicators are then chosen arbitrarily out of the developmental continuum so that each standard represents a stage in the invariable, irreversible and universal process. The standard is clearly definable in ossification centers and ought to be recognizable at site by a trained observer. An ossified center is scored as zero, and the successive stages from onset to maturity are ranked on an ordinate scale from 1 to 8 and then summed (Fig. 1264).

The Oxford method never deals with the problem of different skeletal ages other than the ossification centers of the hands and wrists. Two hands may thus have the same total score; while one has well developed epiphyseal centers in the phalanges and a relatively immature carpus, the other may have the opposite conditions. It is not known if these hands are equally mature. It is certain that this type of anachronism tends to diminish as puberty approaches. The method has been described as mathematically illogical, because the same maturation score can be reached in different ways.[18]

The second widely used modern method is based on the total score of maturity indicators in

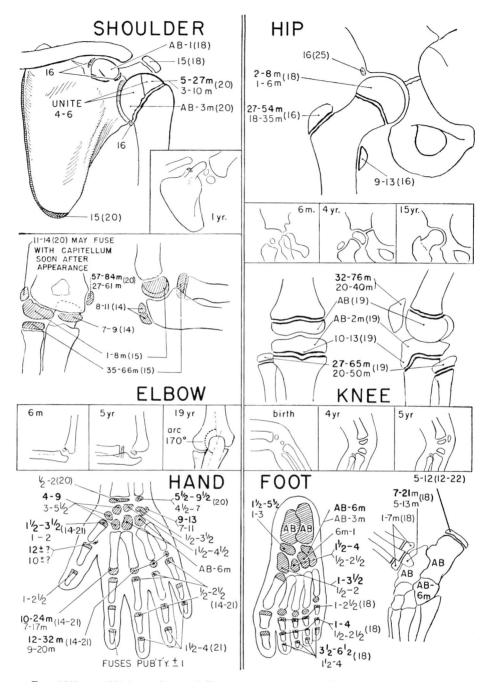

Figs. 1260 and 1261 (*opposite page*). Time of Appearance of Centers of Ossification

The chart published by Camp and Cilley (1931)[9] has been revised by Dr. Bertram R. Girdany, Radiological Service of the Babies Hospital, New York. The figures indicate the range of time of appearance of centers of ossification from the 10th to the 90th percentile, obtained from the studies on bone growth available in 1950. The centers of ossification appearing after the 6th year have not been as carefully studied as those appearing earlier. Figures followed by (*m*) mean months, otherwise all numbers indicate years. Where two sets of numbers are given for one center of ossification, the upper heavier figures refer to males and the lower lighter figures refer to females. A single set of figures applies to both sexes. (*AB*) indicates that the ossification center is visible at birth. Figures in parentheses give *approximate* time of fusion.

individual ossification centers. In 1959, Tanner and Whitehouse published criteria for the rating of individual bones of the hand and wrist throughout the range of immaturity. All the bones were rated from 1 to 8, except for the radius, rated from 1 to 9. A zero was assigned to a nonossified center. In 1962, Tanner, Whitehouse, and Healy devised a system to estimate total maturity by weighing the stages of development in each bone, thereby making allowance for the relative importance of the individual stages.[19] Half of the total score was derived from

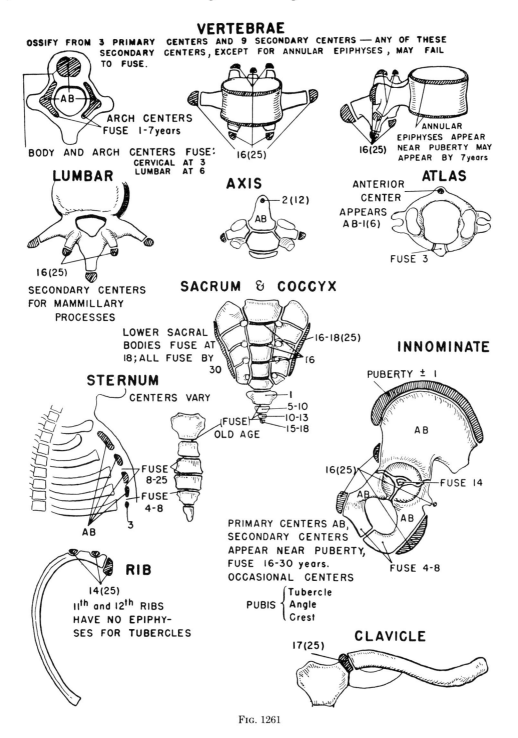

Fig. 1261

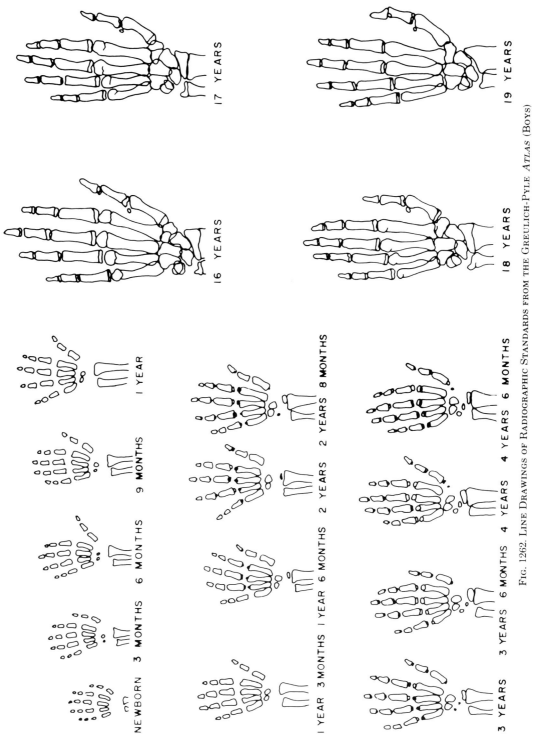

Fig. 1262. Line Drawings of Radiographic Standards from the Greulich-Pyle Atlas (Boys)

SKELETAL MATURATION

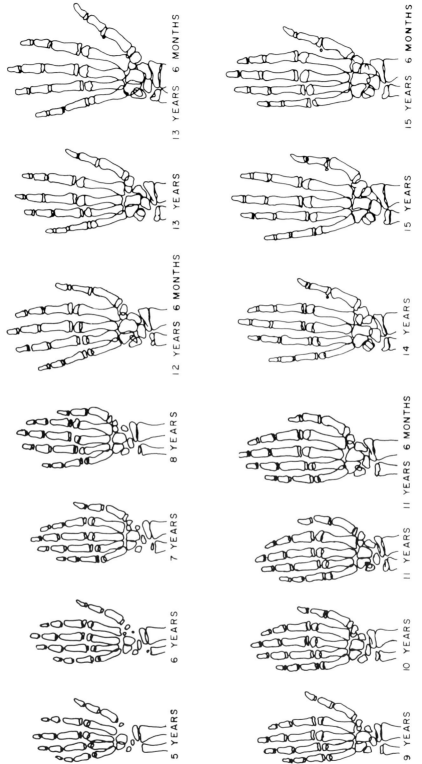

FIG. 1262 (continued)

the round bones of the carpus and half from the long and short bones. Only 20 of the 28 possible bones of the hand and wrist were involved in this weighted assessment (Fig. 1264). The total score of a fully matured adult hand was 1000 points.

An objection to the Tanner-Whitehouse method is that too much weight is given to the carpal bones, poor indicators of maturity. Later investigators claim that the method is more accurate if the carpus is ignored.[20] Another objection is that, for some indicators, a ratio between the width of the epiphysis and diaphysis is used. This can be seriously distorted by pathologic conditions such as rickets. The method is still being evolved and the reweighting of the ossification center stages may avoid some of these difficulties.

A comparison of the Greulich-Pyle *Atlas* and the Tanner-Whitehouse method reveals that the mean skeletal age values for the former are 1 year lower than the latter. The 95% confidence limit, or random error, of a single reading was smaller when the Tanner-Whitehouse method was used, probably because of weighted totals of judgments. The between-observer variation, or systemic error, was smaller within the Greulich-Pyle *Atlas*, probably because only one judgment is required for the *Atlas*, compared to 20 separate judgments in the Tanner-Whitehouse method.[21, 22]

If the carpus is ignored because of its difficulty of assessment and dubious value in overall rating, the Tanner-Whitehouse method appears better because of its low random error. For practical purposes, the Greulich-Pyle *Atlas* is simpler and quicker.

The most recent advances in the field of skeletal maturation come from Garn et al.[23] Extensive computerized analysis indicates the ossification centers with the greatest value in maturity

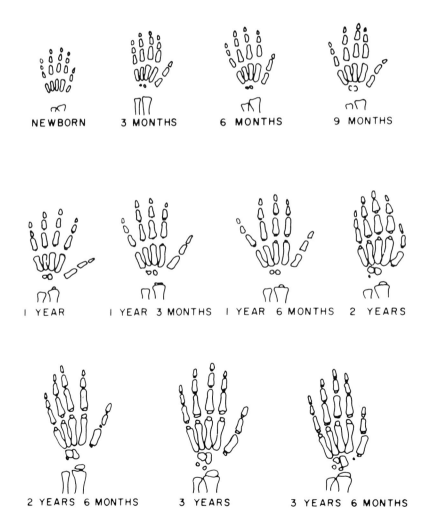

FIG. 1263. LINE DRAWINGS OF RADIOGRAPHIC STANDARDS FROM THE GREULICH-PYLE ATLAS (GIRLS)

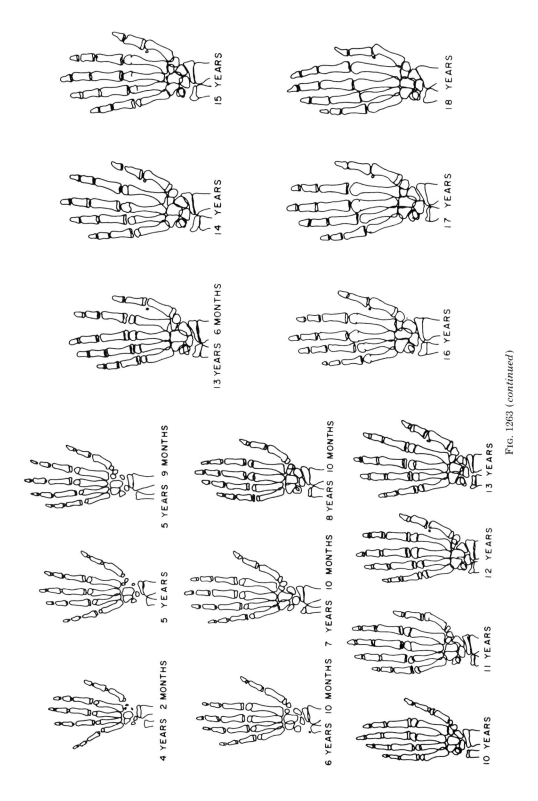

FIG. 1263 (continued)

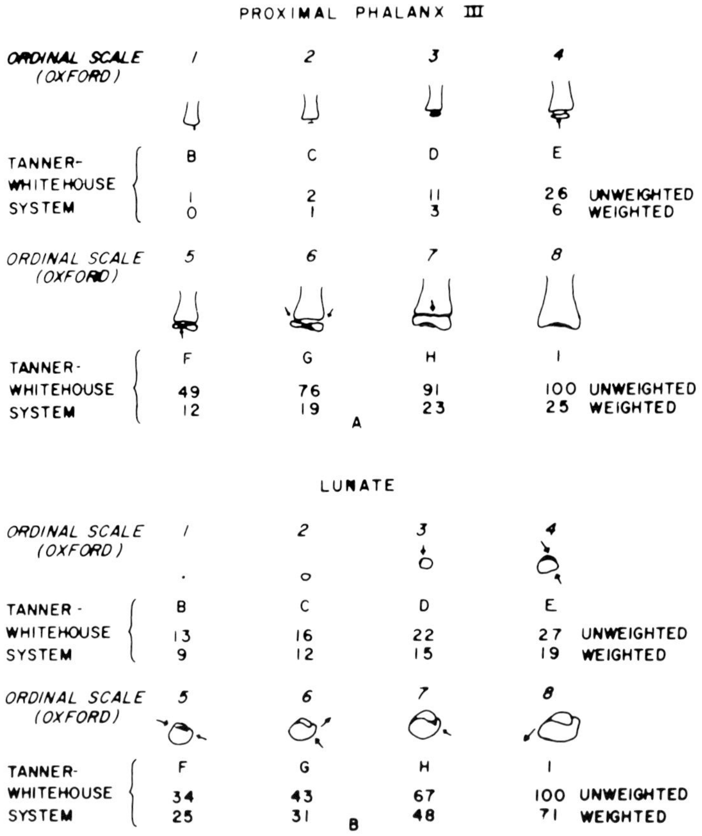

PROXIMAL PHALANX III

ORDINAL SCALE (OXFORD)	1	2	3	4	
TANNER-WHITEHOUSE SYSTEM	B 1 0	C 2 1	D 11 3	E 26 6	UNWEIGHTED WEIGHTED
ORDINAL SCALE (OXFORD)	5	6	7	8	
TANNER-WHITEHOUSE SYSTEM	F 49 12	G 76 19	H 91 23	I 100 25	UNWEIGHTED WEIGHTED

A

LUNATE

ORDINAL SCALE (OXFORD)	1	2	3	4	
TANNER-WHITEHOUSE SYSTEM	B 13 9	C 16 12	D 22 15	E 27 19	UNWEIGHTED WEIGHTED
ORDINAL SCALE (OXFORD)	5	6	7	8	
TANNER-WHITEHOUSE SYSTEM	F 34 25	G 43 31	H 67 48	I 100 71	UNWEIGHTED WEIGHTED

B

FIG. 1264. TANNER-WHITEHOUSE STAGES AND MATURITY INDICATORS FOR THIRD PROXIMAL PHALANX AND LUNATE, TOGETHER WITH WEIGHTED AND UNWEIGHTED SCORES FOR EACH STAGE

Also shown (*in italics*) is the ordinal scale, the summing of which is the basis of the Oxford system. The textural descriptions of the Tanner-Whitehouse stages are as follows:

Third Proximal Phalanx: "Stage B. The center is just visible as a single deposit of calcium, or more rarely as multiple deposits. The border is frequently ill-defined.

"Stage C. The center is distinct in appearance and disc-shaped, with a smooth continuous border. (The maximum diameter is less than half the width of the adjacent phalangeal shaft.)

"Stage D. The epiphysis is half or more the width of the end of the adjacent shaft.

"Stage E. The proximal border of the epiphysis is concave and distinctly thickened. (This is the forerunner of the development of the metacarpal articular surface, which usually takes place only in the next stage. Sometimes in Stage E, however, some differentiation into palmar and dorsal surfaces, as described in Stage F, can be seen.) The epiphysis is not yet as wide as the shaft.

"Stage F. The epiphysis is as wide as its shaft and follows closely the shape of the shaft, although it does not yet cap it at the edges. (Further development of the metacarpal articular surface has taken place since the last stage and, at least on the second and third metacarpals at this stage, although not always on the fourth and fifth, a distinct differentiation of palmar and dorsal edges can be seen. The palmar surface is visible as the proximal border of the epiphysis. The dorsal edge is represented by the thickened white line which runs in an arc concentric with the end of the metacarpal head from one proximal corner of the epiphysis to the other. In some positions of the hand, however, the palmar edge may coincide with the dorsal, and the dorsal thickened concave white line is all that can be seen.)

"Stage G. The epiphysis caps its shaft.

"Stage H. Fusion of epiphysis and shaft has now begun. A line is still visible across the shaft, composed partly of black areas where the epiphyseal cartilage remains and partly of dense white areas where fusion is proceeding.

"Stage I. Fusion of epiphysis and shaft is completed. Over the majority of its length the line of fusion has entirely disappeared, but some thickened remnant of it may still be visible."

Lunate: "Stage B. The center is just visible as a single deposit of calcium or more rarely as multiple deposits. The border is frequently ill-defined.

"Stage C. The centre is distinct in appearance and oval in shape, with a smooth continuous border. (The maximum diameter is less than half the width of the end of the ulnar shaft.)

"Stage D. The maximum diameter is half or more the width of the end of the ulnar shaft.

assessment. The age of appearance of each ossification center is correlated with every other center, showing which centers yield the most information. These are then listed accordingly, offering a basis for skeletal appraisal and a weighted system. The 20 highest rankings of 73 centers were chosen for each sex (Fig. 1265). The 20 highest bones were then correlated among themselves and their coefficients rose. None of the round bones of the hand or foot were included

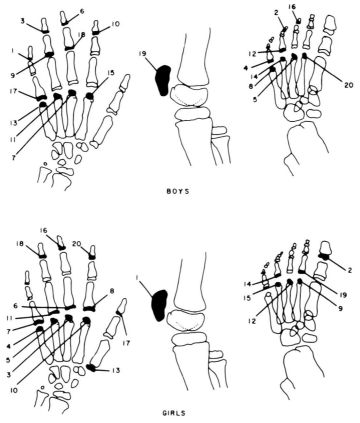

FIG. 1265. METHOD OF GARN ET AL.[23]

The 20 centers of maximum predictive value in boys and in girls. The postnatal ossification centers that have the highest statistical "communality" and hence the greatest predictive value in skeletal assessment are located in the hand, the foot, and the knee. Thus, three radiographs can actually provide more diagnostically useful information than the larger number often made. The numbers indicate the order of the greatest correlation value.

"The distal border of the bone is now thickened (but palmar and dorsal surfaces there are not yet visible).

"Stage E. The palmar and dorsal surfaces of the distal part of the bone are not clearly defined with one or the other, or both, projecting distal to the thickened white line which marks their area of confluence. The dorsal surface may project towards the navicular but no proper saddle, as in the next stage, is yet formed.

"The navicular and triquetral borders are not flat and slightly thickened.

"Stage G. The dorsal surface of the capitate saddle has further enlarged since the last stage and now covers more than half the distance from the palmar edge of the saddle to the navicular.

"Stage H. The dorsal surface of the capitate saddle now extends laterally to touch or overlap the edge of the navicular. (Either palmar or dorsal surface, or both, depending on individual shape and positioning, touch or overlap the capitate.) The navicular border is now concave.

"Stage I. Since the last stage, growth has been relatively greatest in the direction of the ulnar styloid process, so that there is now a definite projection in this area culminating in a relatively sharp point.

"The gap between lunate and navicular is almost or entirely obliterated, and that between lunate and triquetral reduced to its adult appearance."

Note: With the weighted system, matured centers do not total 100; the total score of 1000 for a fully matured hand and wrist is attained by summing 20 separate centers.

TABLE 31. AGE-AT-APPEARANCE (YEARS-MONTHS) PERCENTILES FOR SELECTED OSSIFICATION CENTERS[a]

Centers[b]	Boys			Girls		
	5th	50th	95th	5th	50th	95th
1. Humerus, head	—	0- 0	0- 4	—	0- 0	0- 4
2. Tibia, proximal	—	0- 0	0- 1	—	0- 0	0- 0
3. Coracoid process of scapula	—	0- 0	0- 4	—	0- 0	0- 5
4. Cuboid	—	0- 1	0- 4	—	0- 1	0- 2
5. Capitate +	—	0- 3	0- 7	—	0- 2	0- 7
6. Hamate +	0- 0	0- 4	0-10	—	0- 2	0- 7
7. Capitellum of humerus +	0- 1	0- 4	1- 1	0- 1	0- 3	0- 9
8. Femur, head +	0- 1	0- 4	0- 8	0- 0	0- 4	0- 7
9. Cuneiform 3 +	0- 1	0- 6	1- 7	—	0- 3	1- 3
10. Humerus, greater tuberosity +	0- 3	0-10	2- 4	0- 2	0- 6	1- 2
11. Toe phalanx 5M	—	1- 0	3-10	—	0- 9	2- 1
12. Radius, distal	0- 6	1- 1	2- 4	0- 5	0-10	1- 8
13. Toe phalanx 1D	0- 9	1- 3	2- 1	0- 5	0- 9	1- 8
14. Toe phalanx 4M	0- 5	1- 3	2-11	0- 5	0-11	3- 0
15. Finger phalanx 3P	0- 9	1- 4	2- 2	0- 5	0-10	1- 7
16. Toe phalanx 3M	0- 5	1- 5	4- 3	0- 3	1- 0	2- 6
17. Finger phalanx 2P	0- 9	1- 5	2- 2	0- 5	0-10	1- 8
18. Finger phalanx 4	0-10	1- 6	2- 5	0- 5	0-11	1- 8
19. Finger phalanx 1D	0- 9	1- 6	2- 8	0- 5	1- 0	1- 9
20. Toe phalanx 3P	0-11	1- 7	2- 6	0- 6	1- 1	1-11
21. Metacarpal 2	0-11	1- 7	2-10	0- 8	1- 1	1- 8
22. Toe phalanx 4P	0-11	1- 8	2- 8	0- 7	1- 3	2- 1
23. Toe phalanx 2P	1- 0	1- 9	2- 8	0- 8	1- 2	2- 1
24. Metacarpal 3	0-11	1- 9	3- 0	0- 8	1- 2	1-11
25. Finger phalanx 5P	1- 0	1-10	2-10	0- 8	1- 2	2- 1
26. Finger phalanx 3M	1- 0	2- 0	3- 4	0- 8	1- 3	2- 4
27. Metacarpal 4	1- 1	2- 0	3- 7	0- 9	1- 3	2- 2
28. Toe phalanx 2M	0-11	2- 0	4- 1	0- 6	1- 2	2- 3
29. Finger phalanx 4M	1- 0	2- 1	3- 3	0- 8	1- 3	2- 5
30. Metacarpal 5	1- 3	2- 2	3-10	0-10	1- 4	2- 4
31. Cuneiform 1 +	0-11	2- 2	3- 9	1- 6	1- 5	2-10
32. Metatarsal 1	1- 5	2- 2	3- 1	0	1- 7	2- 3
33. Finger phalanx 2M	1- 4	2- 2	3- 4	0- 8	1- 4	2- 6
34. Toe phalanx 1P	1- 5	2- 4	3- 4	0-11	1- 7	2- 6
35. Finger phalanx 3D	1- 4	2- 5	3- 9	0- 9	1- 6	2- 8
36. Triquetrum	0- 6	2- 5	5- 6	0- 3	1- 8	3- 9
37. Finger phalanx 4D	1- 4	2- 5	3- 9	0- 9	1- 6	2-10
38. Toe phalanx 5P	1- 6	2- 5	3- 8	1- 0	1- 9	2- 8
39. Metacarpal 1	1- 5	2- 7	4- 4	0-11	1- 7	2- 8
40. Cuneiform 2	1- 2	2- 8	4- 3	0-10	1-10	3- 0
41. Metatarsal 2	1-11	2-10	4- 4	1- 3	2- 2	3- 5
42. Femur, greater trochanter	1-11	3- 0	4- 4	1- 0	1-10	3- 0
43. Finger phalanx 1P	1-10	3- 0	4- 7	0-11	1- 9	2-10
44. Navicular of foot +	1- 1	3- 0	5- 5	0- 9	1-11	3- 7
45. Finger phalanx 2D	1-10	3- 2	5- 0	1- 1	2- 6	3- 3
46. Finger phalanx 4D	2- 1	3- 3	5- 0	1- 0	2- 0	3- 5
47. Finger phalanx 5M	1-11	3- 5	5-10	0-11	2- 0	3- 6
48. Fibula, proximal +	1-10	3- 6	5- 3	1- 4	2- 7	3-11
49. Metatarsal 3	2- 4	3- 6	5- 0	1- 5	2- 6	3- 8
50. Toe phalanx 5D	2- 4	3- 1	6- 4	1- 2	2- 4	4- 1
51. Patella +	2- 7	4- 0	6- 0	1- 6	2- 6	4- 0
52. Metatarsal 4	2-11	4- 0	5- 9	1- 9	2-10	4- 1
53. Lunate	1- 6	4- 1	6- 9	1- 1	2- 7	5- 8
54. Toe phalanx 3D	3- 0	4- 4	6- 2	1- 4	2- 9	4- 1
55. Metatarsal 5	3- 1	4- 4	6- 4	2- 1	3- 3	4-11
56. Toe phalanx 4D	2-11	4- 5	6- 5	1- 4	2- 7	4- 1
57. Toe phalanx 2D	3- 3	4- 8	6- 9	1- 6	2-11	4- 6
58. Radius, head +	3- 0	5- 3	8- 0	2- 3	3-10	6- 3
59. Navicular of wrist	3- 7	5- 8	7-10	2- 4	4- 1	6- 0

TABLE 31—Continued

Centers[b]	Boys			Girls		
	5th	50th	95th	5th	50th	95th
60. Greater multiangular	3- 6	5-10	9- 0	1-11	4- 1	6- 4
61. Lesser multiangular	3- 1	6- 3	8- 6	2- 5	4- 2	6- 0
62. Medial epicondyle of humerus +	4- 3	6- 3	8- 5	2- 1	3- 5	5- 1
63. Ulna, distal	5- 3	7- 1	9- 1	3- 3	5- 4	7- 8
64. Calcaneal apophysis +	5- 2	7- 7	9- 7	3- 6	5- 4	7- 4
65. Olecranon of ulna +	7- 9	9- 8	11-11	5- 7	8- 0	9-11
66. Lateral epicondyle of humerus +	9- 3	11- 3	13- 8	7- 2	9- 3	11- 3
67. Tibial tubercle +	9-11	11-10	13- 5	7-11	10- 3	11-10
68. Adductor sesamoid of thumb	11- 0	12- 9	14- 7	8- 8	10- 9	12- 8
69. Os acetabulum	11-11	13- 6	15- 4	9- 7	11- 6	13- 5
70. Acromion +	12- 2	13- 9	15- 6	10- 4	11-11	15- 9
71. Iliac crest +	12- 0	14- 0	15-11	10-10	12- 9	14- 4
72. Coracoid apophysis +	12- 9	14- 4	16- 4	10- 4	12- 3	16- 4
73. Ischial tuberosity +	13- 7	15- 3	17- 1	11- 9	13-11	0

[a] Garn's table modified by C. B. Graham: *Radiologic Clinics of North America*, 10: 185, 1972.[26]

[b] P = proximal, M = middle, D = distal. + = particularly important events.

in the top 20. Also, the correlations were slightly better for girls than for boys. This information has yet to be converted into a simple method for practical use.

The objections to it are that it relies solely on the first appearance of ossification and on epiphyseal closure, and that it involves radiographs of other parts of the body beside the hand, thereby increasing radiation hazards.

The radiologic evaluation of skeletal maturation includes the standards, the technique, the interpretation, and clinical indication for bone age studies.[24]

The most acceptable standards are: *The Radiographic Atlas of Skeletal Development of the Hand and Wrist* by Greulich and Pyle,[11, 25] and *Radiographic Standard of Reference for the Growing Hand and Wrist* by Pyle, Waterhouse and Greulich,[25] (Table 31), Graham's assessment of bone maturation, and (Table 32) Graham's bone age sampling method chart.[26]

The hand is the only bone age determination necessary except when the patient is under the age of 2 years, or between ages of 6 and 10,[24] or if it does not coincide with the chronological age. In these cases additional areas to be radiographed are determined by consulting Graham's bone age sampling method (Table 32), and these radiographs compared to Graham's modification of the original Garn data (Table 31) for reliable assessment of skeletal age.[24] In children less than 2 years of age, the entire right upper extremity, including the hand, and the lower extremity and an anteroposterior view of the foot are required. Ultrasonography is of use in the evaluation of fetal age in utero. The newborn presents a special problem in evaluating physiologic maturity.

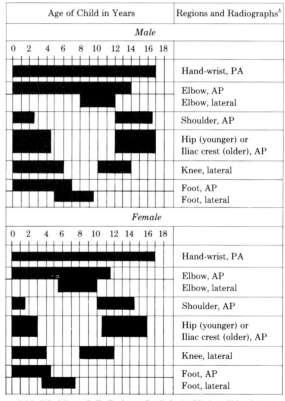

TABLE 32. BONE AGE SAMPLING METHOD[a]

[a] Modified from C. B. Graham: *Radiologic Clinics of North America*, 10: 185, 1972.[26]

[b] PA, posteroanterior; AP anteroposterior.

Tooth calcification is more dependable than bone age for evaluating fetal maturity, and for detail the article by Kuhns and Finnestrom[27] should be consulted. Tooth calcification during gestation occurs in the following sequence: cusp of the first deciduous molar at 33 weeks; the second deciduous molar and cuspid at 36 weeks. The normal sequence of intrauterine ossification is calcaneus, talus, distal femoral epiphysis and proximal tibial epiphysis; humeral head and cuboid follow and appear at approximately the same time.

The bone age is an index of physiologic maturity and the interpretation should be skewed toward the normal. One deviant ossification center should not be heavily weighed.[24]

The presence of the ossification centers is more reliable than their size and shape, especially in the 6-10 age group. The assessment of phalangeal ossification centers is more reliable than carpal centers.

Advanced bone age in a boy and retarded bone age in a girl are more likely to indicate disease than the reverse.[24]

The indications for bone age examination are many; and the probable findings and indications are summarized by Born and Moshang (Table 33).

TABLE 33. INDICATIONS FOR BONE AGE ESTIMATION AND PROBABLE FINDINGS[a]

Short Stature and/or Delayed Pubescence	Tall Stature and/or Accelerated Pubescence
A. *Genetic short stature*—In this group the skeletal maturation matches the chronologic age. These are the small children of small parents	A. *Genetic tall stature*—In these children skeletal maturity and chronologic age match. These are the tall children of tall parents
B. *Constitutional delay*—These children are usually males. Their mildly retarded bone age is compatible with their height age. They have a potential for adequate stature	B. *Constitutional advancement*—These are the children with a premature growth spurt. The bone age and chronologic age are proportional. One must remember that the tall child is frequently a short adult.
C. *Disease states:* 1. Hypothyroidism—This produces the most severe bone age retardation. The presence of epiphyseal dysgenesis is an important clue to its presence 2. Growth hormone deficiency—Congenital or acquired. This produces a retardation of bone age slightly less severe than the accompanying retardation in height 3. Adrenal steroid excess both exogenous and endogenous produces retardation of bone age accompanied by osteoporosis 4. Turner syndrome presents characteristically as a short female with mild retardation of bone age, mild osteoporosis and subtle shortening of the third and fourth metacarpals 5. Chronic systemic disease such as Crohn disease can produce growth arrest which precedes the onset of gastrointestinal symptoms. Other chronic illnesses seem to have less effect on skeletal maturation than on somatic growth 6. Generalized affectation of the skeleton such as achondroplasia, the mucopolysaccharidoses and a myriad of unusual entities can present as growth failure, but produce characteristic skeletal disorganization identifiable on the hand films	C. *Disease states:* 1. Adrenogenital syndrome is the most common syndrome producing severe precocity 2. Idiopathic sexual precocity occurs chiefly in females. It is sometimes part of the rare McCune-Albright syndrome of fibrous dysplasia and sexual precocity 3. Functioning endocrine tumors are rare but advanced maturation is their usual presentation 4. Obesity is usually accompanied by mild advancement of bone age 5. Cerebral gigantism is a syndrome of advanced somatic growth, mental retardation and advanced bone age proportional to height 6. Marfan syndrome has the characteristic scoliosis and arachnodactyly

Summary

Bone age determinations are tedious, not very challenging procedures, that yield invaluable clinical information to the clinician. Most of the children examined will be normal but the interpretation should be performed with a finicky attention to detail because inappropriate treatment may be undertaken if the bone age assessment is unreliable.

Height prediction based on evidence of skeletal maturation is fraught with error.

An advanced bone age in a boy or a retarded bone age in a girl is more likely to indicate disease than is the converse.

[a] Orthopaedic conditions requiring an estimation of growth potential for management. For example, scoliosis and leg length inequality. These are usually normal children, near the 50th percentile and a hand film is an adequate examination. (Modified from P. F. Born and T. Moshang: The Radiologic Evaluation of Skeletal Maturation. No. 37. Radiology Science Update, 1977.[24])

Conclusions

There are no absolute biologic indicators of maturity. One bone, or one joint, does not perfectly reflect the development of others. The osseous system does not perfectly measure the developmental status of other tissue systems. No strict rules are available on the normal limits of skeletal maturity, although a skeletal age differing from chronologic age by more than 1 year is generally regarded as potentially outside normal limits. Serial roentgenographic examinations to ascertain the rate of maturity are more rewarding than the examination of a single film.

The hand and wrist are studied by the Greulich-Pyle *Atlas* in assessing the skeletal age, and, although there is considerable variation in individuals, the consideration of this part of the skeleton is the most reliable and the simplest method. The hands may differ in bone age in an individual without apparent cause (Fig. 1266). It is noteworthy that the distal epiphysis of the femur is almost always present at birth, and its absence is presumptive evidence of prematurity.

Although the diagnosis of disturbed skeletal maturity should not be made too quickly, neither should it be ignored, since real disturbance requires corrective measures before the age of 10 years. Clark[28] stated that the following variations from the normal should be checked by thorough roentgenographic examinations to be sure that the rate of progress is satisfactory:

Age of Patient	Maximal Variation in Skeletal Maturation Considered Normal
Birth to 1 year	3 months
1–3 years	6 months
3–6 years	9 months
6–12 years	1 year
12–20 years	2 years

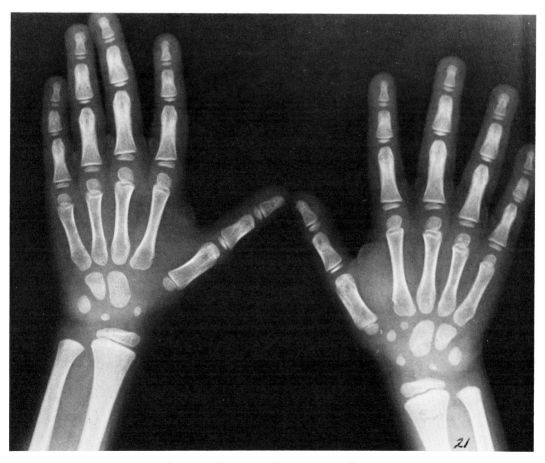

FIG. 1266. BONE AGE DIFFERENCE IN HANDS
The carpal ossification centers show different bone age of no known cause.

FACTORS AFFECTING RATE AND PATTERN OF SKELETAL MATURATION

Roentgenologic studies of skeletal maturation consist of determining the rate and time of appearance and the contours of ossification centers and the epiphysis (Table 30).[29] The age when the union of the epiphysis occurs is also important. Normal epiphyseal development depends on adequate endocrine function, particularly on the pituitary, thyroid, and gonads. If there is excessive glandular secretion, growth is accelerated. If one or more of the glands is deficient, growth stimulants are lacking, and the ossification will be retarded (Fig. 1267). If the organic matrix of the ossification center lacks stimulus for development, the enchondral bone formation is affected and the appearance of the ossification center is delayed or prevented. The results are similar when there is inadequate protein available to lay down organic matrix and bone is not laid down in the ossification center. This may occur in renal disease, especially nephrosis. In other cases, deficiency of mineral salts, as in malnutrition, rickets, and celiac disease, prevents or delays ossification. The appearance is the same whether the basic difficulty is endocrinologic (lack of growth response), abnormal protein metabolism (deficiency of organic matrix), or disturbance in mineral metabolism (deficiency of ossification). The roentgenologist may not be able to determine the cause of skeletal immaturity, but only indicate the degree.

Sex clearly influences the rate and pattern of skeletal maturation; females mature at a faster rate in all ossification centers. The hip and pelvis show prominent pattern differences between the sexes. In the females the femur matures more rapidly than the innominate bones. The onset of ossification is more variable in the male. This disparity is probably due to endocrine differences. The epiphyses of the male eunuch remain unfused, and growth continues into the 3rd and 4th decades. In females with ovarian dysgenesis, the epiphyses may remain unfused until the 3rd or 4th decade, but growth ceases and dwarfism results. In both eunuchism and ovarian dysgenesis, epiphyseal closure may be therapeutically induced by appropriate steroids.

Retardation of skeletal maturation for extrinsic reasons is more likely to affect the male, the female being more resilient. But, just as the male is more influenced by adverse environment, so is he more responsive to improved circumstances.

There are hereditary similarities in the pattern and appearance of ossification centers and of epiphyseal fusion.[9, 11, 30]

Certain congenital abnormalities are reflected in skeletal maturation. Patients with ovarian dysgenesis have brachymetacarpalism, a useful sign in detecting the condition. With mongolism, deformed epiphyses may mask skeletal maturity. The patient with cretinism shows a retarded rate of skeletal maturation which thyroxin raises to normal, but an overdose may cause an advanced rate.

The thyroid hormone initiates and maintains osteogenesis of skeletal cartilage until the age of puberty, when influence of sex hormones completes the process and full skeletal maturity is achieved. Skeletal maturation can be adversely affected by poor health, and a chronic or severe acute illness can slow the rate and later the pattern of maturation. Rickets, Still disease, scurvy, osteomyelitis, osteosarcoma, poliomyelitis, celiac disease and other conditions of malabsorption, hyperthyroidism, adrenal tumors, and other endocrine diseases all have disordering effects, resulting in a retarded skeleton due to

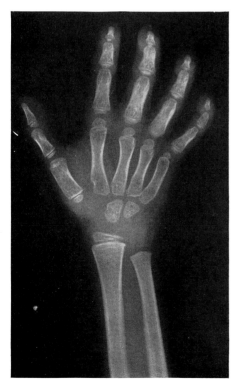

FIG. 1267. HAND 8-YEAR-OLD CHILD: MENTAL AND SKELETAL RETARDATION
The cause could not be determined but was apparently genetic and not endocrine.

lack of nutrients and/or a change in metabolic state. Inflammatory processes close to growth cartilage may accelerate local maturation and cause premature epiphyseal closure.

Therefore, although certain aspects of skeletal maturation are genetically controlled, these can be altered and modified by environmental factors, such as disease, malnutrition, and endocrine abnormality. The endocrinologic abnormalities which influence not only skeletal maturation but also growth are outlined in greater detail in the following section.

ENDOCRINE DISTURBANCE AFFECTING SKELETAL GROWTH AND MATURATION

Hypothyroidism

Hypothyroidism in childhood may reflect either a congenital thyroid deficiency, called cretinism, or one developing after birth, called juvenile myxedema. The effect on the skeleton is the same in either case, although the changes are usually greater in cretinism. Cretinism can usually be diagnosed clinically without difficulty but, at times, must be distinguished from mongolism and other genetic disturbances, and the roentgenologic examination is important in making this distinction.[31-33] Myxedema in childhood is more difficult to diagnose than cretinism, and it is here that the roentgenologic examination is of greatest value. The determination of the basal metabolic rate (BMR) is difficult in children and the results unreliable, therefore, roentgenologic examinations to determine skeletal growth and maturation have tended to supplant these studies.

Both cretinism and myxedema cause delayed appearance of ossification centers and, when they do appear, act to retard their growth rate (Fig. 1268). There is delayed closure of the epiphyses, probably the result of the hypothyroid effect on the gonads, since this closure depends on normal gonadal function (Fig. 1269). Also, since the endocrine gland functions are closely interrelated, it is probable that some of the hypothyroid changes result from low pituitary activity secondary to the low metabolic rate.[28]

The ossification centers, when they finally develop, are often irregular and somewhat deformed. The deformity of the epiphyseal centers is not so great, however, as is seen in such protoplasmic disturbances as mongolism and osteochondrodystrophy (Figs. 1270 and 1271).

In hypothyroidism there are often bands of increased density at the metaphyseal ends of the tubular bones which may be quite marked in

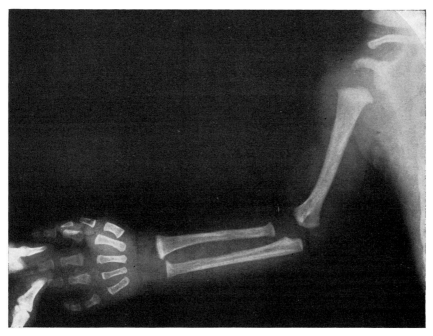

FIG. 1268. RETARDED SKELETAL MATURATION IN CRETIN 8 MONTHS OF AGE
Normally, two carpal ossification centers should be visible at this age.

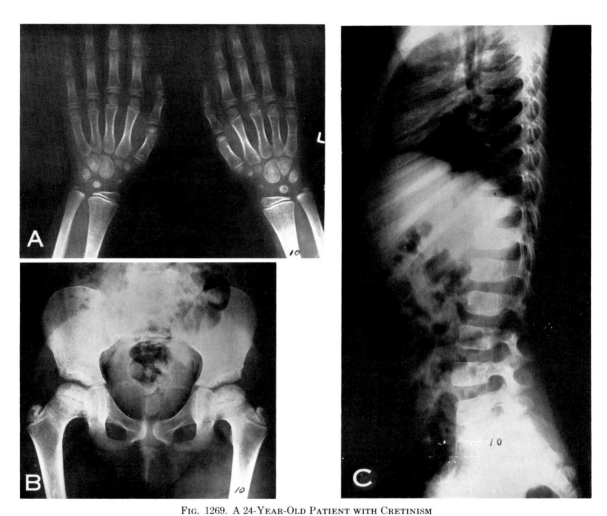

Fig. 1269. A 24-Year-Old Patient with Cretinism

(A) Hands and wrists. Marked retardation of bone age and open epiphyses. The hands are small. (B) Hips and pelvis. Increased density and slight flattening of the femoral capital epiphyses. The epiphyses are open. The acetabula are slightly irregular. (C) The vertebrae are small, flattened, with notching anteriorly.

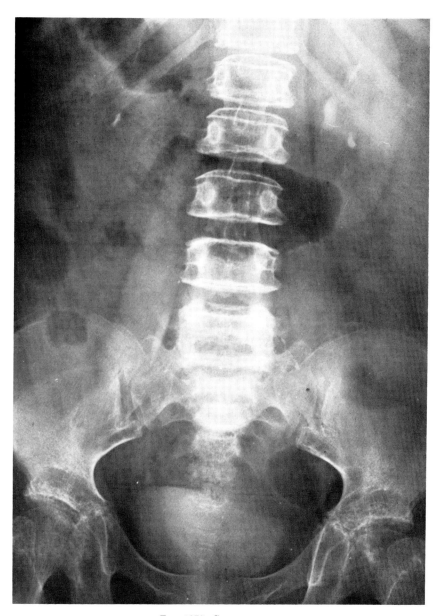

Fig. 1270. Cretin

This patient is 19 years of age with minor changes in both hips due to epiphyseal irregularity. The changes are not as marked as seen in mongolism. All bones are smaller than normal. Opaque material is present in the urinary tract.

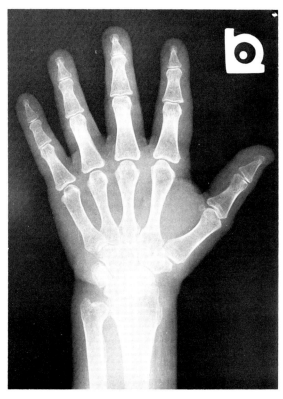

FIG. 1271. CRETIN AGED 19 YEARS
Bones are dwarfed but all epiphyseal lines are closed.

untreated cases. The epiphyses of the tubular bones are sometimes formed from multiple ossification centers. This is most conspicuous at the femoral capital epiphysis, where its spotty appearance has been described as the cretinoid epiphysis, but this is not necessarily diagnostic of hypothyroidism, since it also occurs in some cases of osteochondrodystrophy.

The base of the skull does not develop normally in cretinism and is abnormally short so that the base of the nose is often depressed, as in mongolism. In severe hypothyroidism the bones of the cranium are often thicker than normal. The fontanels remain open for an abnormally long time, and wormian bones sometimes develop along the sutures.

In marked cases the vertebral bodies are often reduced in height and appear flattened. It has been reported that some cases of so-called vertebral epiphysitis are the result of hypothyroidism, although most cases are not caused by deficiency of thyroid hormone.

An abnormally short middle phalanx of the fifth finger is common in mongolism and is seen in some cases of osteochondrodystrophy, so it should suggest a genetic rather than an endocrine abnormality.

If hypothyroidism is suspected in childhood, the diagnosis can be confirmed easily by administration of desiccated thyroid, since the response is spectacular, with the ossification centers appearing promptly and developing at a normal rate. If this response does not follow, the diagnosis should be seriously questioned, and some other condition suspected, such as mongolism or osteochondrodystrophy. In some cases of mental deficiency there is associated skeletal retardation without evidence of endocrine insufficiency. Such cases should be recognized as genetic disturbances, and endocrine therapy abandoned.

Hyperthyroidism

Hyperthyroidism seldom occurs in children and when it does its course is usually rapid, with no apparent bone abnormality. In rare instances when hyperthyroidism in children has a chronic course, the appearance and growth of ossification centers are accelerated. In older children or adults, the greatest change seen in the bone is generalized decalcification. This is most conspicuous at the distal ends of the femurs, where the thin cortical bone and rarefaction of the spongiosa somewhat resemble that seen in Gaucher disease. The majority of patients with thyrotoxicosis show cortical bone changes on roentgenograms obtained by magnification techniques or industrial film viewed through a magnifier.[34] The metacarpal cortical bone has a longitudinal striated pattern (Fig. 1272). The second metacarpal is the most sensitive to bone loss. If marked, this change may be appreciated with routine techniques and routine films, but it is better identified with other techniques.[34] Phalangeal cortical striations may also occur in normal subjects.

Cortical striations may appear in patients with hyperparathyroidism with an accompanying lacelike subperiosteal resorption that distinguishes it from hyperthyroidism (Fig. 1273).[34]

Thyroid Acropachy

Thyroid acropachy is a form of osteoarthropathy present in approximately 1% of patients with Graves disease.[35, 36] The incidence is probably higher than reported since it may be asymptomatic. It consists of finger and toe clubbing, phalangeal and long bone periosteal reaction, and soft tissue swelling almost always associated with current or previous hyperthyroidism, exo-

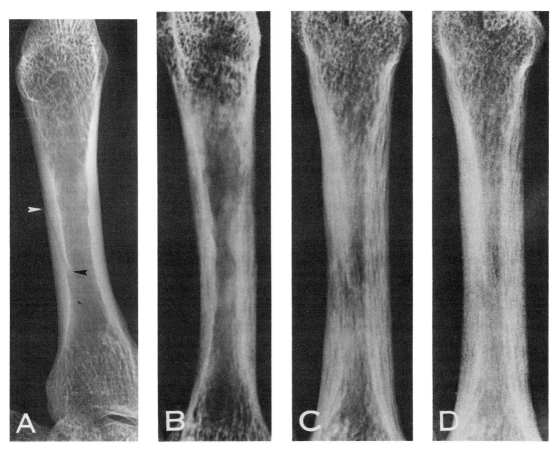

Fig. 1272. Thyrotoxicosis

(A) A magnified view of the normal metacarpal (×3) taken with a 0.3 focal spot tube. There are no vertical striations in the cortex. A nutrient canal (*arrows*) extends from the periosteal surface to the endosteal surface of the cortex. This oblique radiolucency should not be confused with the parallel vertical striations of the next figure. (B) Magnified radiographs (×3) of the second metacarpal bones in thyrotoxic patients. Striations are moderate (industrial film). (C and D) A thyrotoxic patient with marked striations. (C = industrial film; D = regular RP film) Note the granular appearance in the regular film. (Courtesy of H. E. Meema and D. L. Schatz: *Radiology, 97:* 9, 1970,[34] © The Radiological Society of North America, Syracuse, N.Y.)

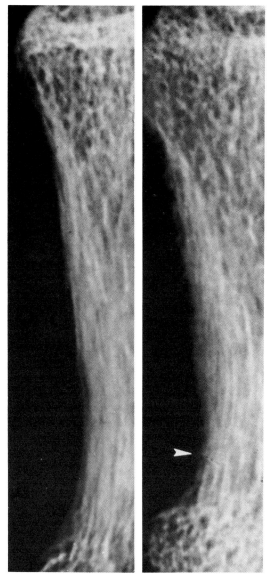

FIG. 1273. CORTICAL STRIATION
(*Left*) Hyperthyroidism. Vertical striation of the cortical margins and the periosteal surface is smooth. (*Right*) Hyperparathyroidism. Cortical striations and irregular lacelike resorption (*arrow*). (Courtesy of H. E. Meema and D. L. Schatz: *Radiology*, 97: 9, 1970,[34] © The Radiological Society of North America, Syracuse, N.Y.)

phthalmus, and pretibial myxedema.[37] Sex incidence is equal. It is described in patients treated with thyroidectomy and radioactive iodine but not in those treated with antithyroid drugs.[38] Although it may appear without soft tissue edema or clubbing, it is not reported in the absence of exophthalmus.

Roentgenographic Features. A bubbly or lacy periosteal proliferation is the early sign. Usually this occurs 18–24 months after treatment for Graves disease. The phalangeal diaphyses are the most frequent site (Fig. 1274), but involvement of the distal long bones is common. The appearance is indistinguishable from other neurovascular forms of periosteal reaction, e.g., varicose veins or pulmonary osteoarthropathy. In the later stages, a thick, smooth periosteal cloaking occurs. Soft tissue swelling overlaying the periosteal reaction may be evident but the joints are normal.

Hypopituitarism

The pituitary gland not only acts directly to affect bone growth, but also controls the gonads and the thyroid gland, so that its disturbance causes abnormal function of these other endocrines (Figs. 1275–1278). Typical hypopituitary dwarfism is known as Lorain-Levi infantilism and has a classic appearance.[32] Those affected are slender and well proportioned and, except for being abnormally small, appear quite normal. Their mentality is unaffected, but they remain sexually immature. Probably as a result of secondary hypogonadism, the epiphyses are late in uniting, so that bone age is much retarded; in some cases, they never unite at all. The sutures of the skull remain open, and the teeth erupt late.

Pituitary dwarfism should be distinguished from so-called primordial dwarfism, a congenital growth disturbance unrelated to pituitary insufficiency. In primordial dwarfism the appearance and fusion of the ossification centers are normal, which is important in distinguishing it from pituitary dwarfism. The bones appear perfectly normal roentgenologically, except that they are unusually small. These patients are dwarfs at birth, and never attain normal stature, but they are sexually normal and capable of normal reproduction, frequently transmitting dwarfism to their children, which would not occur if hypopituitarism were the cause.

Hyperpituitarism (Acromegaly)

Hyperpituitarism results from eosinophilic adenoma or hyperplasia of the eosinophilic cells of the anterior lobe of the pituitary gland. Its development during endochondral bone growth causes giantism. Developing after endochondral bone growth has stopped, it produces acromegaly. There is little difference between the two, except that, with the former, the bones can greatly increase in length as well as in transverse diameter. The trunk is proportionately shorter

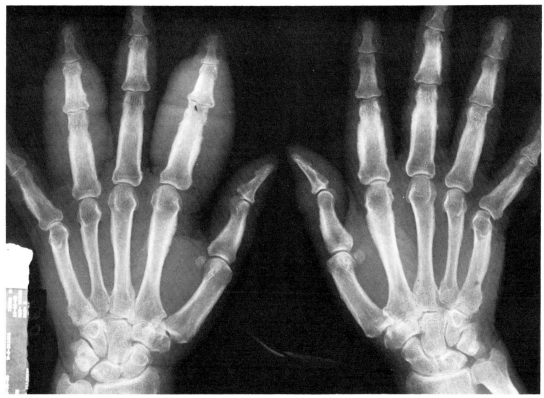

FIG. 1274. GRAVES DISEASE

This patient had Graves disease and thyroidectomy 11 years previously. The thyroid acropachy had persisted for over 10 years. Soft tissue swelling in the second and fourth fingers of the left hand. Periosteal reaction on most of the proximal phalanges, the second metacarpals and the fifth right metacarpal. Periosteal reaction of the radii. (Courtesy of Dr. T. W. Staple, Mallinkrodt Institute, St. Louis.)

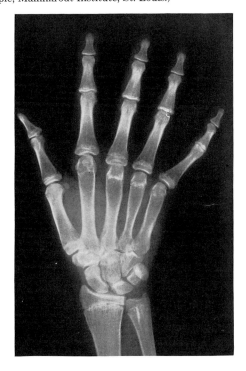

FIG. 1275. PITUITARY INSUFFICIENCY WITH SECONDARY HYPOGONADISM

This is a 39-year-old man. Many of the epiphyses are ununited.

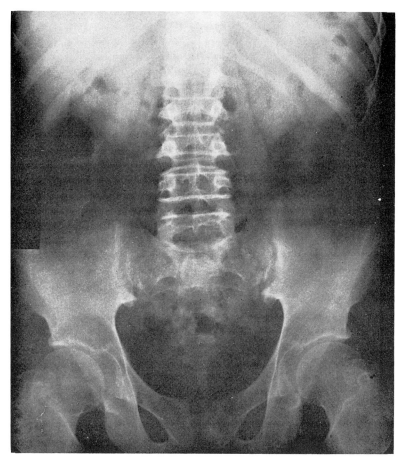

Fig. 1276. Pituitary Insufficiency with Secondary Hypogonadism
This is a 39-year-old man. The sacroiliacs are indistinct and resemble those of an adolescent.

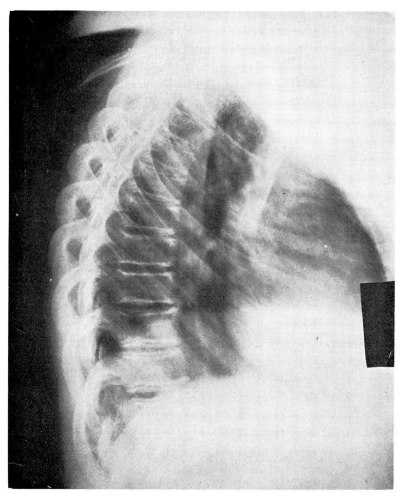

Fig. 1277. Pituitary Insufficiency with Secondary Hypogonadism
This is a 39-year-old man. Secondary ossification centers of vertebral bodies are ununited.

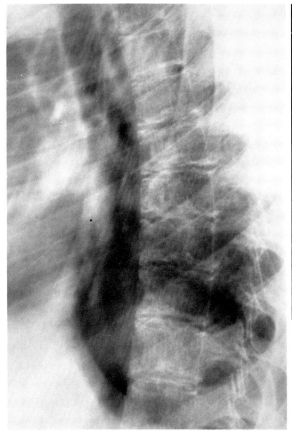
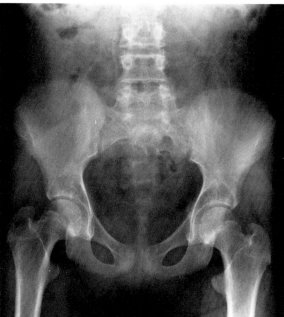

Fig. 1278. Pituitary Insufficiency
Patient is age 24. Bodies of the vertebrae are osteoporotic with irregularity of the epiphyseal centers. (*Left*) Lateral projection. (*Right*) Anteroposterior projection.

than the extremities, and the span is greater than the height.

With the latter condition, acromegaly, occurring after the endochondral bone growth has ceased, the bones thicken but cannot lengthen to any great extent. There is, however, some increase in bone length which results from the activity of articular cartilage cells proliferating and assuming a function that only the epiphyseal cartilages ordinarily have. The articular cartilages are stimulated to cause endochondral growth that slightly increases the length of the bones, but this is noticeable only in the hands. Because the phalanges and metacarpal bones have seven articular cartilages for each digit, the slight increase in the length of each bone leads to perceptible lengthening of the fingers.

The tubular bones become larger in diameter and this also is especially noticeable in the hands. The ungual tufts become unusually prominent (Fig. 1279). All of the bony ridges of the bone shafts in the hands are prominent in hyperpituitarism, and coarse trabeculation of all bones is quite apparent.

The measurement of the sesamoid index was believed to be different in normal and acromegalic groups.[39] However, the wide range and significant overlap between the two groups[40, 41] and the male sex bias[40, 42] make it useless as an aid in diagnosis.

The mandibular angle is increased, and the mandible is lengthened, causing prognathism. The teeth are separated and maloccluded. The vault of the skull thickens, and the paranasal sinuses are greatly increased in size.

The increased growth of bone at the costochondral junctures lengthens the ribs, producing an unusually great anteroposterior measurement of the thorax. Rib thickening may occur (Fig. 1280). The vertebral bodies increase in size through new peripheral bone formation, particularly noticeable in the lateral roentgenographic projection, where the normal vertebral body can sometimes be seen lying inside the shell of newly laid down bone (Fig. 1281).

Thickening of the heel pad (above 20 mm) is a significant feature of individuals with long standing acromegaly.[43, 44] Heel pad thickening

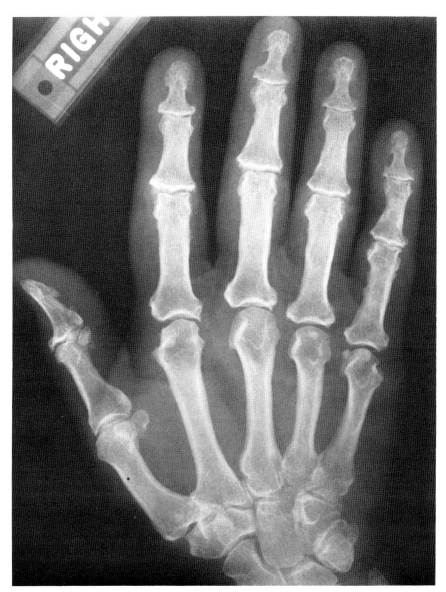

FIG. 1279. ACROMEGALY

Phalangeal tufts are conspicuous and phalanges are abnormally thick. The metaphyses of the phalanges show hyperostosis and osteophyte formation. The joints are widened, especially the metacarpophalangeals.

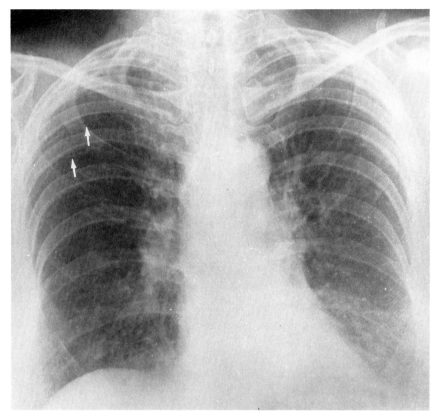

Fig. 1280. Acromegaly
The ribs are thickened particularly the lower margins (*arrows*).

may also be caused by obesity,[45] injury, infection, myxedema, peripheral edema and prolonged Dilantin therapy.[46] Dilantin may also cause calvarial thickening which can be confused with acromegaly.[47, 48]

Enlargement of the sella turcica may not be seen in giantism, since there is often only hyperplasia of the eosinophilic cells of the anterior lobe of the pituitary. In acromegaly, eosinophilic adenoma of the pituitary gland is usually present and the sella turcica is enlarged and eroded (Fig. 1282). When enlargement occurs, the ballooning of the sella turcica is similar to that caused by chromophobe adenoma, but the other skull changes of acromegaly make the diagnosis obvious in most cases. It is important to keep in mind that hyperpituitarism may occur without roentgenologic evidence of an intrasellar tumor.

Eunuchoidism

Eunuchoidism is seen in both sexes and results from the surgical removal of the gonads in childhood, or from some disorder which causes atrophy or destruction of the gonads before puberty. The action of the gonads has much to do with the union of epiphyses, which is delayed when there is a deficiency or absence of gonadal secretion. For that reason endochondral bone growth continues over an abnormally long period in cases of eunuchoidism, producing unusually long and slender tubular bones in the extremities. At times there is generalized osteoporosis, thought to be from deficient osteoblastic activity secondary to deficient gonadal secretion.

Ovarian Agenesis

Ovarian agenesis is probably a generalized genetic disturbance in protoplasm, since not only are the ovaries undeveloped (Fig. 1283) but other abnormalities are commonly associated with it,[11] such as webbed neck, coarctation of the aorta, hypertension, osteoporosis, ocular defects, congenital deafness, and cubitus valgus. The epiphyseal development is usually slightly retarded, and the fusion of the epiphyses delayed,[31] although in some cases these are entirely normal. There-

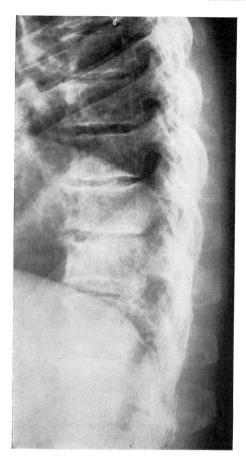

Fig. 1281. Acromegaly
The vertebral bodies are large due to peripheral new bone formation.

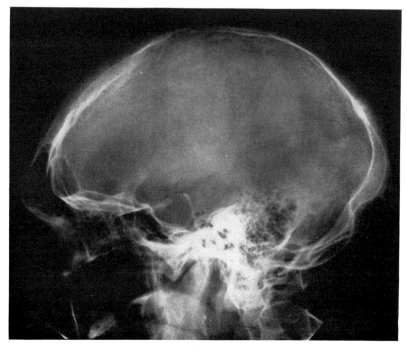

Fig. 1282. Acromegaly
The sella and paranasal sinuses are large.

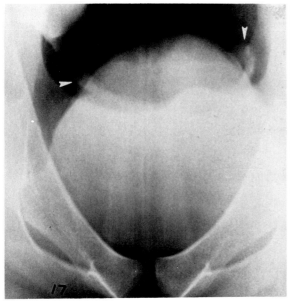

FIG. 1283. OVARIAN AGENESIS

(*Left*) An air study reveals atrophic or absent ovaries (*arrows*). (*Right*) Hand. Delayed epiphyseal fusion and a positive metacarpal sign indicated by the short fourth and fifth metacarpals. Phalangeal preponderance and a positive carpal sign.

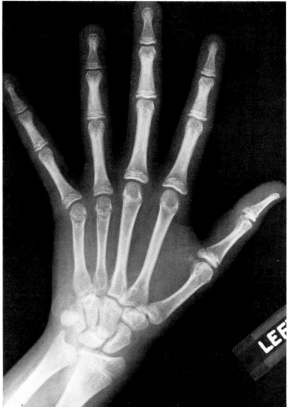

fore, it is probable that the dwarfism, present in most cases, indicates genetic abnormality and not just a lack of ovarian secretion to stimulate normal growth. Precocious senility has been found in patients with ovarian agenesis; additional evidence of generalized dysplasia of tissue. Vertebral epiphysitis is reported as a frequent complication.

Hypergonadism

Sexual precocity is accompanied by early appearance of the ossification centers and rapid growth of bone, but the early closure of the epiphyses stops growth, so that the final result is often dwarfing, rather than increase in height.[32]

Often the cause of true precocious puberty is unknown (Fig. 1284). It may be part of the genetic disturbance known as Albright syndrome, in which case it is accompanied by fibrous dysplasia of bone and regions of pigmentation of the

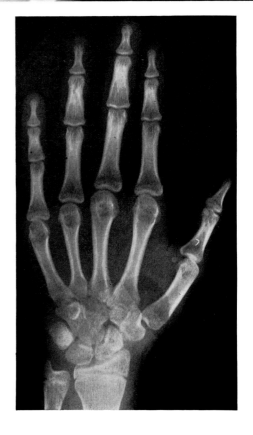

FIG. 1284. PRECOCIOUS PUBERTY WITH ACCELERATED SKELETAL MATURATION

Chronologic age is 8 years; the bone age is approximately 14 years.

skin called cafe-au-lait spots. This syndrome occurs only in females. Another condition that causes sexual precocity in females and deserves special consideration is the granulosa cell tumor of the ovary. Five percent of patients with this condition are under 15 years of age and experience precocious pseudopuberty, precocious menstruation, accelerated skeletal development, and advanced bone age. If this tumor can be removed entirely, the prognosis is excellent and the child returns to a normal state.

In males, pineal tumors cause precocious sexual and skeletal development, and the prognosis in these cases is grave indeed. Why neither females nor adults of either sex manifest similar sexual and somatic stimulation is not known. Fortunately, most cases of precocious puberty do not involve pinealomas.

Hyperfunction of the adrenal cortex in childhood, caused by either adrenal cortical hyperplasia or tumor, may cause sexual and somatic precocity. The epiphyses appear early and growth is rapid, but, since the epiphyses unite prematurely, the final result is small stature. Hyperfunction of the adrenal cortex in females causes virilism, with masculine physical and mental traits. In males, male characteristics are intensified and the genitalia resemble those of an adult.[33]

REFERENCES

1. Pryor, J. W.: The hereditary nature of variation in the ossification of bones. Anat Rec, *1*: 84, 1907.
2. Pryor, J. W.: Differences in the time of development of centers of ossification in the male and female skeleton. Anat Rec, *25*: 257, 1923.
3. Rotch, T. M.: A study of the development of the bones in childhood by the roentgen method, with the view of establishing a developmental index for the grading of and the protection of early life. Trans Assoc Am Physicians, *24*: 603, 1909.
4. Bardeen, C. R.: The relation of ossification to physiological development. J Radiol, *2*: 1, 1921.
5. Lowell, F., and Woodrow, H.: Some data on anatomical age and its relation to intelligence. Redagog Semin, *29*: 1, 1922.
6. Carter, T. M.: Techniques and devices in radiographic study of the wrist bones of children. J Educ Psychol, *17*: 27, 1926.
7. Flory, C. D.: Osseous development in the hand as an index of skeletal development. Monogr Soc Res Child Dev, *1*: 1936.
8. Sontag, L. W., Snell, D., and Anderson, M.: Rate of appearance of ossification centers from birth to age 5 years. Am J Dis Child, *58*: 949, 1939.
9. Camp, J. D., and Cilley, E. I. L.: Diagramatic chart showing time of appearance of the various centers of ossification and period of union. AJR, *26*: 905, 1931.
10. Hodges, P. C.: Development of the human skeleton. AJR, *30*: 809, 1933.
11. Greulich, W. W., and Pyle, S. I.: *Radiographic Atlas of Skeletal Development of the Hand and Wrist*. Stanford University Press, Palo Alto, Calif., 1950 and 1959.
12. Hoerr, N. L., Pyle, S. I., and Francis, C. C.: *Radiographic Atlas of Skeletal Development of the Foot and Ankle*. Charles C Thomas, Springfield, Ill., 1962.
13. Pyle, S. I., and Hoerr, N. L.: *Radiographic Atlas of Skeletal Development of the Knee*. Charles C Thomas, Springfield, Ill., 1955.
14. Roche, A. F., and French, N. Y.: Differences in skeletal maturity levels between the knee and hand. AJR, *109*: 307, 1970.
15. Acheson, R. M., et al.: Studies in the reliability of assessing skeletal maturity from x-rays. I. The Greulich-Pyle Atlas. Hum Biol. *35*: 317, 1963.
16. Roche, A. F., et al.: Some factors influencing the replicability of assessments of skeletal maturity (Greulich-Pyle). AJR, *109*: 299, 1970.
17. Acheson, R. M.: The Oxford method of assessing skeletal maturity. Clin Orthop, *10*: 19, 1954.
18. Masse, G., and Hunt, E. E., Jr.: Skeletal maturation of the hand and wrist in West African children. Hum Biol, *35*: 3, 1963.
19. Tanner, J. J., Whitehouse, R. H., and Healy, M., Jr.: A new system for estimating skeletal maturity from the hand and wrist, with standards derived from a study of 2,000 healthy British children; II. The scoring system. International Children's Center, Paris, 1962.
20. Acheson, R. M., Vicinus, J. H., and Fowler, G. B.: Studies in the reliability of assessing skeletal maturity from x-rays: II. The bone-specific approach. Hum Biol, *36*: 211, 1964.
21. Acheson, R. M., Vicinus, J. H., and Fowler, G. B.: Studies in the reliability of assessing skeletal maturity from x-rays; III. Greulich-Pyle Atlas and Tanner-Whitehouse method contrasted. Hum Biol, *38*: 204, 1966.
22. Fry, E. I.: Assessing skeletal maturity; comparison of the atlas and individual bone techniques. Nature, *220*: 496, 1968.
23. Garn, S. M., Rohmann, C. G., and Silverman, F. N.: Radiographic standards for postnatal ossification and tooth calcification. Med Radiogr Photogr, *43*: 45, 1967.
24. Born, P. F., and Moshang, T.: The Radiologic Evaluation of Skeletal Maturation. No. 37. Radiology Science Update. Biomedics, Inc., Princeton, N.J. 1977.
25. Pyle, S. I., Waterhouse, A. M., and Greulich, W. W.: *Radiographic Standard of Reference for the Growing Hand and Wrist*. Case Western Reserve, Cleveland, Ohio (distributed by Year Book, Chicago) 1971.
26. Graham, C. B.: Assessment of bone maturation—methods and pitfalls. Radiol Clin North Am, *10*: 185, 1972.
27. Kuhns, L. R., and Finnestrom, O.: New standards of ossification of the new born. Radiology, *119*: 655, 1976.
28. Clark, D. M.: The practical value of roentgenography of the epiphyses in the diagnosis of pre-adult endocrine disorders. AJR, *35*: 752, 1936.
29. Todd, T. W.: *Atlas of Skeletal Maturation. Part I: The Hand*. C. V. Mosby, St. Louis, 1937.
30. Leonard, D. W.: Revised "ossification index" for the detection of endocrine disorders in childhood. AJR, *56*: 716, 1946.
31. Johnston, J. A.: Physical growth and development. In *The Child in Health and Disease; a Textbook for Students and Practitioners of Medicine*, C. G. Grulee and R. C. Eley (eds.). Williams & Wilkins, Baltimore, 1952.

32. Greenblatt, R. B., and Nieburgs, H. W.: Some endocrinologic aspects of retarded growth and dwarfism. Med Clin North Am, *31:* 712, 1947.
33. Lisser, H., Curtis, L. E., Escamilla, R. F., and Goldberg, M. B.: The syndrome of congenitally aplastic ovaries with sexual infantilism, high urinary gonadotropins; short stature and other congenital abnormalities; tabular presentation of 25 previously unpublished cases. J Clin Endocrinol, *7:* 665, 1947.
34. Meema, H. E. and Schatz, D. L.: Simple radiologic demonstration of cortical bone loss in thyrotoxicosis. Radiology, *97:* 9, 1970.
35. Gimlette, T. M. D.: Localized myxedema and thyroid acropachy. In *The Thyroid Gland*, Ed. 2, R. Pitt-Rivers and W. R. Trotter (eds). Butterworths, Washington, 1964.
36. Thomas, H. M., Jr.: Secondary subperiosteal new bone formation. Arch Intern Med, *51:* 511, 1933.
37. Gimlette, T. M. D.: Thyroid acropachy. Lancet, *1:* 22, 1960.
38. Kinsella, R. A., Jr., and Back, D. K.: Thyroid acropachy. Med Clin North Am, *52:* 393, 1968.
39. Kleinberg, D. L., Young, I. S., and Kupperman, H. S.: The sesamoid index. An aid in the diagnosis of acromegaly. Ann Intern Med, *64:* 1075, 1966.
40. Ling, S. R., and Lee, K. F.: Relative value of some radiographic measurements of the hand in the diagnosis of acromegaly. Invest Radiol, *6:* 426, 1971.
41. Duncan, T. R.: Validity of the sesamoid index in the diagnosis of acromegaly. Radiology, *115:* 617, 1975.
42. Anton, H. C.: Hand measurements in acromegaly. Clin Radiol *23:* 445, 1972.
43. Steinbach, H. L., and Russell, W.: Measurement of heel-pad as aid to diagnosis of acromegaly. Radiology, *82:* 418, 1964.
44. Kho, K. M., Wright, A. D., and Doyle, F. H.: Heel-pad thickness in acromegaly. Br J Radiol, *43:* 119, 1970.
45. Jackson, D. M.: Heel-pad thickness in obese persons. Radiology, *90:* 129, 1968.
46. Kattan, K. R.: Thickening of the heel-pad associated with long-term Dilantin therapy. AJR, *124:* 29, 1975.
47. Kattan, K. R.: Calvarial thickening after Dilantin medication. AJR, *110:* 102, 1970.
48. Lefebvre, E. B., Haining, R. G., and Labbe, R. F.: Coarse facies, calvarial thickening and hypophosphatasia associated with long-term anticonvulsant therapy. New Engl J Med, *286:* 1301, 1972.

15

Dysplasias

APPROACH TO PATTERNS OF MALFORMATION

The diagnosis of patients with multiple defects and anomalies requires the knowledge of a multitude of recognizable patterns or syndromes. Since the diagnosis is based on the total pattern, it is necessary to consider many syndromes in the differential diagnosis. No one can remember all the changes of a single syndrome, and not all patients will manifest the full core pattern of disease. The etiology and basic abnormalities are unknown in many of the dysplasias and in other diseases simulating dysplasias, and the diagnosis of most will depend on the clinical findings until the basic abnormalities are discovered. In some diseases, such as mucopolysaccharidosis, the lipidoses, and the chromosomal abnormalities, the diagnosis can be made by laboratory tests; however, since the clinical and roentgenologic features of these diseases closely simulate dysplasias, they must be considered as a whole. Congenital malformations may be classified as follows: (1) ectodermal dysplasias, (2) dysplasia with altered skeletal morphogenesis, (3) connective tissue disorders, (4) mucopolysaccharidoses, (5) chromosomal abnormality syndromes, (6) hamartomas, (7) abiotrophies, and (8) lipidoses.

The ectodermal dysplasias involve disease predominantly derived from embryonic ectoderm; e.g., skin, nails, teeth, and lens. Nonectodermal involvement may also occur.

Dysplasias with altered skeletal morphogenesis include the spectrum of chondro-osseous dysplasias (hypoplasia to hyperplasia) and multiple-system disorders in which the skeletal system is used for the classification. The mucopolysaccharidoses are not included in this category.

Chromosomal abnormalities are due to morphologic chromosomal changes and the core patterns are useful to determine the syndromes meriting type/pattern studies.

Connective tissue disorders indicate an involvement of fibrous tissue which may reflect skeletal change.

Mucopolysaccharidoses are characterized by excess tissue storage and/or urinary excretion of mucopolysaccharides. Other manifestations are altered bone configuration, joint limitation, corneal opacity, hepatosplenomegaly, mental deterioration, and cardiovascular changes. These appear after birth.

Hamartomas are a group of syndromes with abnormal mixture of normal tissue, including fibromas, adenomas, and the malignant potential of Peutz-Jegher and Gardner syndromes. Skeletal dysmorphogenesis occurs in the basal cell nevus syndrome and the Goltz syndrome.

The abiotrophies occur in individuals who are usually normal at birth, with subsequent deterioration of normally developed tissues. Progeria is the prototype of this category.

Lipidoses are a group of diseases causing accumulation of lipid in the reticuloendothelial and/or nervous system, and they may reflect skeletal changes.

Smith[1] listed the anomalies and composed a core pattern and hereditary modes where applicable. This core pattern of disease aids in the distinction and may be used when a peculiar skeletal abnormality is demonstrated and the cause is unrecognized. He also listed these diseases alphabetically, to aid in checking the charts for the core pattern of disease. This method is excellent for screening the dysplasias and related diseases.

INTRODUCTION TO HEREDITARY DYSPLASIAS

Hereditary bone dysplasias have long caused confusion due to their great variety of changes, etiologic heterogeneity, complexity, and variability. Further confusion is caused by limited knowledge of the mechanisms of bone growth and development, the classification and nomenclature, specific biochemical and morphologic diagnostic aids, and basic morphologic and biologic chromosomal abnormalities. Until recently the dysplasias were evaluated by the clinical and roentgenographic manifestations, which explains the various classifications of the same condition according to its severity and the age of the patient.

Recent advances have improved the diagnoses and understanding of bone dysplasias:

1. The clinical and radiologic changes of many syndromes have been precisely described.
2. Family pedigrees and hereditary transmission modes have been carefully evaluated.
3. Biochemical advances have established a relationship between certain bone abnormalities and enzymatic deficiency and metabolic disorders involving phosphorus, calcium, mucopolysaccharides, lipids, and amino acids.
4. Fibroblast cell culture has improved the identification of specific mucopolysaccharides, a diagnosis which may now be made in utero.
5. Recent histologic studies of growing non-weight-bearing cartilage have opened new diagnostic avenues.

Correct identification of constitutional (intrinsic) disease of bone is important, even though most of the diseases are rare. Exact identification permits prediction of progression and complication, and optimal treatment. Most of the conditions are heritable, and genetic counseling is indicated. Several of the diseases previously classified as dysplasias may now be identified by chemical or laboratory means, rather than by clinical and roentgenographic evaluation.

A nomenclature was formulated in Paris in November 1969 and approved by the European Society of Pediatric Radiology and the Society of Pediatric Radiology.[2] McKusick and Scott[3] modified it to conform to American usage. An attempt has been made to use one synonym and try to eliminate eponyms. This is not feasible in all cases, but as far as practical this classification should be used.

Because of the rapid progress in the delineation and classification of these disorders, the International Nomenclature was revised in May 1977. This new list includes clearly identified forms of a single disease or defect. "Other forms" means that other forms of the disorder exist which, as yet, have not been defined well enough to include.[4]

1977 International Nomenclature of Constitutional Diseases of Bone

Osteochondrodysplasias

I. Abnormalities of cartilage and/or bone growth and development
II. Defects of growth of tubular bones and/or spine
  A. *Identifiable at birth*
   1. Achondrogenesis Type I, Parenti-Fraccaro[5]
   2. Achondrogenesis Type II, Langer-Saldino
   3. Thanatophoric dysplasia[6]
   4. Thanatophoric dysplasia with cloverleaf skull
   5. Short rib-polydactyly syndrome Type I, Saldino-Noonan (perhaps several forms)
   6. Short rib-polydactyly syndrome Type II, Majewski
   7. Chondrodysplasia punctata[7]
      a. Rhizomelic form
      b. Dominant form
      c. Other forms, excluding symptomatic stippling in other disorders (e.g., Zellweger syndrome, warfarin embryopathy)
   8. Campomelic dysplasia
   9. Other dysplasias with congenital bowing of long bones (several forms)
   10. Achondroplasia[8]
   11. Diastrophic dysplasia[9]
   12. Metatrophic dysplasia (several forms)[10]
   13. Chondroectodermal dysplasia, Ellis-van Creveld[11]
   14. Asphyxiating thoracic dysplasia, Jeune[12]
   15. Spondyloepiphyseal dysplasia congenital[13]
      a. Type Spranger-Wiedemann
      b. Other forms (see B, 11-12)
   16. Kniest dysplasia
   17. Mesomelic dysplasia
      a. Type Nievergelt[14]
      b. Type Langer (probable homozygous dyschondrosteosis)[15]
      c. Type Robinow
      d. Type Rheinhardt
      e. Other forms
   18. Acromesomelic dysplasia
   19. Cleidocranial dysplasia[16]
   20. Larsen syndrome
   21. Otopalatodigital syndrome
  B. *Identifiable in later life*
   1. Hypochondroplasia[17]
   2. Dyschondrosteosis[18]
   3. Metaphyseal chondrodysplasia type Jansen[19]
   4. Metaphyseal chondrodysplasia type Schmid[20]
   5. Metaphyseal chondrodysplasia type McKusick[21]
   6. Metaphyseal chondrodysplasia with exocrine pancreatic insufficiency and cyclic neutropenia[22]

7. Spondylometaphyseal dysplasia
   a. Type Kozlowski[24]
   b. Other forms
8. Multiple epiphyseal dysplasia[25, 26]
   a. Type Fairbanks
   b. Other forms
9. Arthroophthalmopathy, Stickler[27]
10. Pseudoachondroplasia[28]
11. Spondyloepiphyseal dysplasia tarda[29]
12. Spondyloepiphyseal dysplasia, other forms (see A, 15–16)
13. Dyggve-Melchior-Clausen dysplasia
14. Spondyloepimetaphyseal dysplasia (several forms)
15. Myotonic chondrodysplasia, Catel-Schwartz-Jampel
16. Parastremmatic dysplasia
17. Trichorhinophalangeal dysplasia
18. Acrodysplasia with retinitis pigmentosa and nephropathy Saldino-Mainzer[30,33]

III. Disorganized development of cartilage and fibrous components of skeleton
   1. Dysplasia epiphyseal hemimelica[34]
   2. Multiple cartilaginous exostoses[35]
   3. Acrodysplasia with exostoses, Giedion-Langer
   4. Enchondromatosis, Ollier[34]
   5. Enchondromatosis with hemangioma, Maffucci
   6. Metachondromatosis[36]
   7. Fibrous dysplasia, Jaffe-Lichtenstein[37]
   8. Fibrous dysplasia with skin pigmentation and precocious puberty, McClune-Albright[38]
   9. Cherubism[39] (familial fibrous dysplasia of the jaws)
   10. Neurofibromatosis[40]

IV. Abnormalities of density of cortical diaphyseal structure and/or metaphyseal modeling
   1. Osteogenesis imperfecta congenita[41] (several forms)
   2. Osteogenesis imperfecta tarda[42] (several forms)
   3. Juvenile idiopathic osteoporosis[43]
   4. Osteoporosis with pseudoglioma
   5. Osteopetrosis with precocious manifestations[44]
   6. Osteopetrosis with delayed manifestations (several forms)[45]
   7. Pycnodysostosis[46]
   8. Osteopoikilosis[47]
   9. Osteopathia striata
   10. Melorheostosis[48]
   11. Diaphyseal dysplasia, Camurati-Engelmann[49]
   12. Craniodiaphyseal dysplasia[50]
   13. Endosteal hyperostosis
      a. Autosomal dominant, Worth
      b. Autosomal recessive, Van Buchem[51]
   14. Tubular stenosis, Kenny-Caffey[51]
   15. Pachydermoperiostosis[53]
   16. Osteodysplasty, Melnick-Needles[54]
   17. Frontometaphyseal dysplasia[55]
   18. Craniometaphyseal dysplasia (several forms)[56, 57]
   19. Metaphyseal dysplasia, Pyle[58]
   20. Sclerosteosis
   21. Dysosteosclerosis[60]
   22. Osteoectasia with hyperphosphatasia

Dysostoses

V. Malformation of individual bones singly or in combination
VI. Dysostoses with cranial and facial involvement
   1. Craniosynostosis (several forms)[61, 62]
   2. Craniofacial dysostosis, Crouzon[63]
   3. Acrocephalosyndactyly, Apert (and others)[64]
   4. Acrocephalopolysyndactyly,[65] Carpenter (and others)
   5. Mandibulofacial dysostosis[66]
      a. Type Treacher Collins, Franceschetti
      b. Other forms[67]
   6. Oculomandibulofacial syndrome,[68] Hallermann-Streiff-Francois
   7. Nevoid basal cell carcinoma syndrome[69]
VII. Dysostoses with predominant axial involvement
   1. Vertebral segmentation defects, including Klippel-Feil[70]
   2. Cervicooculoacoustic syndrome, Wildervanck[71]
   3. Sprengel anomaly[72]
   4. Spondylocostal dysostosis[73, 74]
      a. Dominant form
      b. Recessive forms
   5. Oculovertebral syndrome, Weyers[75]
   6. Osteoonychodysostosis[76]
   7. Cerebrocostomandibular syndrome
VIII. Dysostoses with predominant involvement of extremities[7]
   1. Acheiria
   2. Apodia
   3. Ectrodactyly syndrome
   4. Aglossia-adactyly syndrome[78]

5. Congenital bowing of long bones (several forms) (see also osteochondrodysplasias)
   6. Familial radioulnar synostosis[79]
   7. Brachydactyly (several forms)[80]
   8. Symphalangism[81]
   9. Polydactyly (several forms)
   10. Syndactyly (several forms)
   11. Polysyndactyly (several forms)
   12. Camptodactyly
   13. Poland syndrome
   14. Rubinstein-Taybi syndrome
   15. Pancytopenia-dysmelia syndrome, Fanconi
   16. Thrombocytopenia-radial-aplasia syndrome
   17. Orodigitofacial syndrome
       a. Type Papillon-Leage
       b. Type Mohr
   18. Cardiomelic syndrome, Holt-Oram[82] (and others)
   19. Femoral facial syndrome
   20. Multiple synostoses (includes some forms of symphalangism)
   21. Scapuloiliac dysostosis, Kosenow-Sinios
   22. Hand-foot-genital syndrome
   23. Focal dermal hypoplasia, Goltz

Idiopathic Osteolyses

1. Phalangeal (several forms)[83]
2. Tarsocarpal[84]
   a. Including Francois form (and others)
   b. With nephropathy
3. Multicentric[85]
   a. Hajdu-Cheney form
   b. Winchester form
   c. Other forms

Chromosomal Aberrations

Specific entities not listed

Primary Metabolic Abnormalities

I. Calcium and/or phosphorus
   1. Hypophosphatemic rickets[86,87]
   2. Pseudodeficiency rickets, Prader, Royer[88]
   3. Late rickets, McCance[89]
   4. Idiopathic hypercalcuria
   5. Hypophosphatasia (several forms)[90,91]
   6. Pseudohypoparathyroidism[92] (normo- and hypocalcemic forms, include acrodysostosis)
II. Complex carbohydrates
   1. Mucopolysaccharidosis, type 1[93] (α-L-iduronidase deficiency)
      a. Hurler form
      b. Scheie form
      c. Other forms
   2. Mucopolysaccharidosis, Type II,[93] Hunter (sulfoiduronate sulfatase deficiency)
   3. Mucopolysaccharidosis, Type III,[93] Sanfilippo
      a. Type A (heparin sulfamidase deficiency)
      b. Type B (N-acetyl-α-glucosaminidase deficiency)
   4. Mucopolysaccharidosis, Type IV,[93] Morquio (N-acetylgalactosamine-6-sulfate-sulfatase deficiency)
   5. Mucopolysaccharidosis, Type VI,[93] Maroteaux-Lamy (aryl sulfatase B deficiency)
   6. Mucopolysaccharidosis, Type VII, (β-glucuronidase deficiency)
   7. Aspartylglucosaminuria (aspartylglucosaminidase deficiency)
   8. Mannosidosis (α-mannosidase deficiency)
   9. Fucosidosis[94] (α-fucosidase deficiency)
   10. GM_1-gangliosidosis (β-galactosidase deficiency)
   11. Multiple sulfatase deficiency,[95,96] Austin, Thieffry
   12. Neuraminidase deficiency (formerly mucolipidosis I)
   13. Mucolipodosis II
   14. Mucolipidosis III
III. Lipids
   1. Niemann-Pick disease
   2. Gaucher disease[10]
IV. Nucleic acids
   1. Adenosine-deaminase defining and others
V. Amino acids
   1. Homocystinuria and others
VI. Metals
   1. Menkes kinky hair syndrome and others

CONSTITUTIONAL (INTRINSIC) DISEASES OF BONES
(Modification by McKusick and Scott[3])

Constitutional Diseases of Bones with Unknown Pathogenesis

Osteochondrodysplasias (Abnormalities of Cartilage or Bone Growth, or Both, and Development)

I. Defects of growth of tubular bones or spine, or both
   A. Manifested at birth
      1. Achondrogenesis[5]
      2. Thanatophoric dwarfism[6]
      3. Achondroplasia[8]
      4. Chondrodysplasia punctata (formerly stippled epiphysis, or chondrodystrophia calcificans congenita), several forms[7]
      5. Metatrophic dwarfism[10]
      6. Diastrophic dwarfism[9]
      7. Chondroectodermal dysplasia (Ellis-van Creveld)[11]
      8. Asphyxiating thoracic dysplasia (Jeune)[12]
      9. Spondyloepiphyseal dysplasia congenita[13]
      10. Mesomelic dwarfism
         a. Nievergelt type[14]
         b. Langer type[15]
      11. Cleidocranial dysplasia (formerly cleidocranial dysostosis)[16]
   B. Manifested in later life
      1. Hypochondroplasia[17]
      2. Dyschondroosteosis[18]
      3. Metaphyseal chondrodysplasia,* Jansen type[19]
      4. Metaphyseal chondrodysplasia, Schmid type[20]
      5. Metaphyseal chondrodysplasia, McKusick type (formerly cartilage-hair hypoplasia)[21]
      6. Metaphyseal chondrodysplasia with malabsorption and neutropenia[22]
      7. Metaphyseal chondrodysplasia with thymolymphopenia[23]
      8. Spondylometaphyseal dysplasia (Kozlowski)[24]
      9. Multiple epiphyseal dysplasia (several forms)[25,26]
      10. Hereditary arthroophthalmopathy.[27]
      11. Pseudoachondroplastic dysplasia (formerly pseudoachondroplastic type of spondyloepiphyseal dysplasia)[28]
      12. Spondyloepiphyseal dysplasia tarda[29]
      13. Acrodysplasia
         a. Trichorhinophalangeal syndrome (Giedion)[30]
         b. Epiphyseal (Thiemann)[31,32]
         c. Epiphysometaphyseal (Brailsford)[33]

II. Disorganized development of cartilage and fibrous components of the skeleton
   1. Dysplasia epiphysealis hemimelica[34]
   2. Multiple cartilaginous exostoses[35]
   3. Enchondromatosis (Ollier)[33]
   4. Enchondromatosis with hemangioma (Maffucci)[36]
   5. Fibrous dysplasia (Jaffe-Lichtenstein)[37]
   5. Fibrous dysplasia with skin pigmentation and precocious puberty (McCune-Albright)[38]
   7. Cherubism[39]
   8. Multiple fibromatosis[40]

III. Abnormalities of density, or cortical diaphyseal structure or of metaphyseal modeling, or both
   1. Osteogenesis imperfecta congenita (Vrolik; Porak-Durante)[41]
   2. Osteogenesis imperfecta tarda (Lobstein)[42]
   3. Juvenile idiopathic osteoporosis[43]
   4. Osteopetrosis with precocious manifestations[44]
   5. Osteopetrosis with delayed manifestations[45]
   6. Pycnodysostosis[46]
   7. Osteopoikilosis[47]
   8. Melorheostosis[48]
   9. Diaphyseal dysplasia (Camurati-Engelman)[49]
   10. Craniodiaphyseal dysplasia[50]
   11. Endosteal hyperostosis (van Buchem and other forms)[51]
   12. Tubular stenosis (Kenny-Caffey)[52]
   13. Osteodysplasty* (Melnick-Needles)[54]

---

* Hyphens were used in the original: chondro-ectodermal, spondylo-epiphyseal, chondro-dysplasia, rhino-trichophalangeal, etc. (Incidentally, we have used "trichorhinophalangeal," rather than "rhinotrichophalangeal.") These have been removed, leaving only the hyphens in arthro-ophthalmopathy, cervico-oculo-acoustic, osteo-onychodysostosis, acro-osteolysis.

* In English, or at least in American usage, acrocephalosyndactyly, brachydactyly, polydactyly, etc., seem to enjoy preference over the forms ending in -ia. In the revision of this nomenclature we have been admittedly inconsistent, in retaining amelia, phocomelia, etc. The decision to do so has

14. Pachydermoperiostosis[59]
15. Osteoectasia with hyperphosphatasia[59]
16. Metaphyseal dysplasia (Pyle)[58]
17. Craniometaphyseal dysplasia (several forms)[56,57]
18. Frontometaphyseal dysplasia[55]
19. Oculo-dental-osseous dysplasia (formerly oculo-dento-digital syndrome)[97]

Dysostoses (Malformation of Individual Bone, Singly or in Combination)

I. Dysostoses with cranial and facial involvement
   1. Craniosynostosis, several forms[61,62]
   2. Craniofacial dysostosis (Crouzon)[63]
   3. Acrocephalosyndactyly (Apert)[64]
   4. Acrocephalopolysyndactyly (Carpenter)[65]
   5. Mandibulofacial dysostosis (Treacher, Collins, Franceschetti and others)[66]
   6. Mandibular hypoplasia (includes Pierre Robin syndrome)[67]
   7. Oculomandibulofacial syndrome (Hallerman-Streiff-Francois)[68]
   8. Nevoid basal cell carcinoma syndrome[69]
II. Dysostoses with predominant axial involvement
   1. Vertebral segmentation defects (including Klippel-Feil)[70]
   2. Cervico-oculo-acoustico syndrome (Wildervanck)[71]
   3. Sprengel deformity[72]
   4. Spondylocostal dysostosis (several forms)[73,74]
   5. Oculovertebral syndrome (Weyers)[75]
   6. Osteo-onychodysostosis (formerly nail-patella syndrome)[76]
III. Dysostoses with predominant involvement of extremities[77]
   1. Amelia
   2. Hemimelia (several types)
   3. Acheiria
   4. Apodia
   5. Adactyly and oligodactyly
   6. Phocomelia
   7. Aglossia-adactylia syndrome[78]
   8. Congenital bowing of long bones (several types)
   9. Familial radioulnar synostosis[79]
   10. Brachydactyly (several types)[80]
   11. Symphalangism[81]
   12. Polydactyly (several types)
   13. Syndactyly (several types)
   14. Polysyndactyly (several types)
   15. Camptodactyly
   16. Clinodactyly
   17. Biedl-Bardet syndrome[98,99]
   18. Popliteal pterygium syndrome[100]
   19. Pectoral aplasia-dysdactyly syndrome (Poland)[101]
   20. Rubinstein-Taylor syndrome[102]
   21. Pancytopenia-dysmelia syndrome (Fanconi)[103]
   22. Thrombocytopenia-radial-aplasia syndrome[104]
   23. Orofaciodigital (OFD) syndrome (Papillon-Leage)[105,106]
   24. Cardiomelic syndrome (Holt-Oram and others)[82]

Idiopathic Osteolyses

1. Acro-osteolysis
   a. Phalangeal type[83]
   b. Tarso-carpal form with or without nephropathy[84]
2. Multicentric osteolysis[85]

Primary Disturbances of Growth

1. Primordial dwarfism (without associated malformation)[107,108]
2. Cornelia de Lange syndrome[109]
3. Bird-headed dwarfism (Virchow, Seckel)[110]
4. Leprechaunism[111]
5. Russel-Silver syndrome[112]
6. Progeria[113]
7. Cockayne syndrome[114]
8. Bloom syndrome[115]
9. Geroderma osteodysplastica[116]
10. Spherophakia-brachymorphia syndrome (Weill-Marchesani)[117]
11. Marfan syndrome[93]

Constitutional Diseases of Bones with Known Pathogenesis

I. Chromosomal aberrations
II. Primary metabolic abnormalities
   A. Calcium phosphorous metabolism
      1. Hypophosphatemic familial rickets[86,87]
      2. Pseudo-deficiency rickets (Royer, Prader)[88]
      3. Late rickets (McCance)[89]
      4. Idiopathic hypercalciuria
      5. Hypophosphatasia (several forms)[90,91]
      6. Idiopathic hypercalcemia[118]
      7. Pseudohypoparathyroidism (normo- and hypocalcemic forms)[92]

---

been based entirely on usage. Similarly, usage dating back to Harvey Cushing, who coined the term, dictates the use of symphalangism rather than symphalangy.

B. Mucopolysaccharidosis[93]
   1. Mucopolysaccharidosis I (Hurler)
   2. Mucopolysaccharidosis II (Hunter)
   3. Mucopolysaccharidosis III (Sanfilippo)
   4. Mucopolysaccharidosis IV (Morquis)
   5. Mucopolysaccharidosis V (Ullrich-Scheie)
   6. Mucopolysaccharidosis VI (Maroteaux-Lamy)
C. Mucolipidosis[119] and lipidosis
   1. Mucolipidosis I (Spranger-Wiedemann)[120]
   2. Mucolipidosis II (Leroy-Opitz)[121]
   3. Mucolipidosis III (pseudo-Hurler polydystrophy)[94]
   4. Fucosidosis[94]
   5. Mannosidosis[123]
   6. Generalized G_{M1} gangliosidosis (several forms)
   7. Sulfatidosis with mucopolysacchariduria (Austin, Thieffry)[95,96]
   8. Cerebrosidosis including Gaucher disease
D. Other metabolic extraosseous disorders
III. Bone abnormalities secondary to disturbances of extraskeletal systems
   1. Endocrine
   2. Hematologic
   3. Neurologic
   4. Renal
   5. Gastrointestinal
   6. Cardiopulmonary

It is hoped this classification will standardize the terms used in medical publications. The subdivisions are used to avoid a haphazard listing without logical order. The subdivisions, or groups, are open to criticism. The dysplasias and dysostoses may overlap, since certain diseases may be associated with both types of abnormalities, such as pycnodysostosis with a mandibular malformation (dysplasia) and generalized increased skeletal density (dysostosis).

The use of eponyms is limited to those with universal usage, and some have been introduced in the hope of clarifying previous inaccuracies and the confusion of certain words. The nomenclature will probably be revised, but it is hoped that it will form the basis of a standard terminology in future publications.

With new techniques for evaluating bone anatomy and physiology, it is now possible to separate specific growth functions, assigning them to anatomic sites in bone: epiphysis, physis, metaphysis, diaphysis. Disturbances in growth (constriction, tubulation, modeling, longitudinal and transverse growth, and other deformities) can now be related to specific areas of the bone that caused the abnormality.

Rubin[34] has proposed a dynamic classification of bone diseases based on the growth mechanism, rather than on physical or clinical characteristics. Although this classification does not explain all dysplasias of bone, it does answer many questions. It seems flexible enough to accommodate new diseases and new ideas, provides an initial logical approach to the dysplasias, and defines the morbid changes in physiologic terms. It is suggested that this classification be used in the initial approach to a bone dysplasia, followed by Smith's classification.

Rubin's statement, "there is logic to the order of bone growth" is the basis of this classification of bone dysplasias. Its use requires an understanding of normal development.

I. Epiphyseal dysplasias
   A. Epiphyseal hypoplasias
      1. Failure of articular cartilage: spondyloepiphyseal dysplasia, congenita and tarda
      2. Failure of ossification of center: multiple epiphyseal dysplasia, congenita and tarda
   B. Epiphyseal hyperplasia
      1. Excess of articular cartilage: dysplasia epiphysealis hemimelica
II. Physeal dysplasias
   A. Cartilage hypoplasias
      1. Failure of proliferating cartilage: achondroplasia, congenita and tarda
      2. Failure of hypertrophic cartilage: metaphyseal dysostosis, congenita and tarda
   B. Cartilage hyperplasias
      1. Excess of proliferating cartilage: hyperchondroplasia
      2. Excess of hypertrophic cartilage: enchondromatosis
III. Metaphyseal dysplasias
   A. Metaphyseal hypoplasias
      1. Failure to form primary spongiosa: hypophosphatasia, congenita and tarda
      2. Failure to absorb primary spongiosa: osteopetrosis, congenita and tarda.
      3. Failure to absorb secondary spongiosa: craniometaphyseal dysplasia, congenita and tarda
   B. Metaphyseal hyperplasia
      1. Excessive spongiosa: multiple exostoses
IV. Diaphyseal dysplasias
   A. Diaphyseal hypoplasias

1. Failure of periosteal bone formation: osteogenesis imperfecta, congenita and tarda
2. Failure of endosteal bone formation: idiopathic osteoporosis, congenita and tarda
B. Diaphyseal hyperplasias
1. Excessive periosteal bone formation: progressive diaphyseal dysplasia
2. Excessive endosteal bone formation: hyperphosphatasemia

Common dysplasias presenting recognizable patterns are easily diagnosed, but in the rare and less familiar ones core patterns must be studied as well as clinical and hereditary characteristics. Here, Smith's classification helps point to the reasonable conclusion.

The patient with multiple defects often presents a diagnostic challenge. Genetic counsel depends on it. The multiple defects may be due to a single, primary, localized defect resulting in secondary malformations, as in meningomyelocele with clubfoot and hydrocephalus, the primary defects being due to a localized early neural tube closure. In morphogenesis, major or minor, we are concerned more with the patient with multiple primary defects. The radiologist must decide whether the pattern of malformation is recognizable, and if his previous experience leads to an evaluation, prognosis, management, and genetic counsel. The diagnosis is based on the total pattern and often requires consideration or elimination of many syndromes. Because of the multitude of possible dysplasias, one physician cannot expect to be familiar with all the patterns, and so Smith[1] has designed tables to help in the initial differential diagnosis. He says:

".. the formulation of the tables were based on the following precept: Individual defects are, with rare exception, nonspecific in terms of being pathognomonic for a particular syndrome entity; however, the proper selection of two or more anomalies may provide a diagnostic core pattern for a particular syndrome. For each condition a core pattern of two or more clinically detectable abnormalities has been selected which is highly suggestive or even diagnostic of that syndrome as contrasted to other recognized conditions. For some conditions, such as the mucopolysaccharidoses, the core pattern includes normal features which are of value in the differential diagnosis. The anomalies selected are those which are common phenomena in that disorder and serve to distinguish it from the other multiple defect syndromes. For example, web neck was not included under XO (the Turner syndrome) because it is not one of the more frequent features of that syndrome. Besides the core pattern of defects, the presence of small stature or mental deficiency and the mode of etiology are set forth, and these may be of further diagnostic value. Finally, a reference source is indicated for obtaining more complete information about each disorder. The general references are listed alphabetically and the specific references are listed numerically in accordance with the number of the syndrome in the tables."

The somewhat arbitrary order of presentation is based on similarities in system involvement, type of abnormality, and etiology in the case of chromosomal aberrations. Within each group the order is predominately set by clinical similarities. The group order of presentation proceeds as follows: ectodermal dysplasia, connective tissue disorders, mucopolysaccharidoses, altered skeletal morphogenesis, chromosomal abnormalities, miscellaneous disorders, hamartomas, and abiotrophies. The subtle graduation from one group to the next completes a full circle, there being features of ectodermal dysplasia in the group classified as abiotrophies.

Several cautionary comments are indicated. First, not all patients with a given disorder will have the full core pattern of anomalies, there being rather wide individual variation in manifestations, especially in the autosomal dominant disorders and those due to a chromosomal abnormality. The core patterns simply act as leads toward a particular diagnosis which is dependent on an appraisal of the total pattern of abnormality.

Secondly, many patients with multiple defects have conditions which have not been recognized in the past or are not included in these tables. The author has limited the conditions to those which appear to be a concise entity of *multiple primary* defects, omitting such questionable disease entities as Klippel-Feil syndrome, the Grieg syndrome, and the lissencephaly syndrome.

Some of the disorders such as Pierre Robin syndrome, Poland syndrome, and the triad syndrome may be the consequence of a single localized primary defect in early morphogenesis. However, present knowledge of the developmental pathology of these patterns of malformation does not allow for a clear interpretation as to whether the total pattern of defect in these syndromes was the consequence of a single early localized defect or multiple primary defects.

The clinician should be extremely cautious in making a diagnosis. Vague or partial similarity is generally insufficient for a diagnosis and an incorrect diagnosis may give rise to serious errors in management, prognosis, and genetic counsel.

Genetic counsel may be inferred from the

mode of etiology stated for each disorder. In utilizing these data it is important to realize that many of the multiple defect patients with a single altered gene have a fresh mutation and their parents have virtually no risk of having another affected child. For example, about 85% of patients with achondroplasia, an autosomal dominant disorder, are born of normal parents whose risk for recurrence is negligible. Therefore genetic counsel should only be rendered after a careful assessment of the family.

When the etiology is unknown, such as for the Cornelia de Lange syndrome, genetic counsel is dependent on the actual recurrence risk found in affected families. Though the recurrence risk appears to be of low magnitude for most of these syndromes of unknown etiology, the presently available data are not adequate for definitive counseling.

Besides those children seen with established syndromes, there are many who have multiple malformations which do not comprise a recognized syndrome. A small percentage of such patients may be found to have a chromosomal abnormality, most commonly a structural aberration with extra chromosomal material. In such cases genetic counsel should be withheld until it has been determined whether either parent is a balanced translocation carrier. In the absence of chromosomal abnormality or familial data suggesting a particular mode of inheritance, it is generally impossible to state any accurate risk of recurrence. It is particularly presumptuous to inform the parents that "this is a rare condition and therefore, quite unlikely to recur in your future children." Under these circumstances the author's present approach to genetic counseling is to tell the parents that "the lowest recurrence risk is zero and the highest risk with each pregnancy could be 25%. The latter figure is predicated on the possibility of autosomal recessive or X-linked recessive inheritance."

ANOMALIES OF HAND, FOOT, AND ARM

Certain neonatal anomalies occur in the hand, foot, and arm of the newborn and alert the radiologist to certain dysplasias. They may occur as isolated anomalies. The more common anomalies and diseases in which they occur are listed below:

I. Clinodactyly of Fifth Finger:
  1. Carpenter syndrome
  2. Cornelia de Lange syndrome
  3. Cri-du-chat syndrome
  4. Down's syndrome
  5. Laurence-Moon-Biedl syndrome
  6. Moar syndrome
  7. Myositis ossificans congenita
  8. Myotonic dystrophy
  9. Nail-patella syndrome
  10. Oral-facial-digital syndrome
  11. Prader-Willi syndrome
  12. Rubinstein-Taybi syndrome
II. Thumb—Hypoplasia or Aplasia:
  1. Aminopterin-induced syndrome
  2. Cornelia de Lange syndrome
  3. Fanconi syndrome
  4. Holt-Oram syndrome
  5. Myositis ossificans congenita
  6. Radial aplasia-thrombocytopenia
  7. Trisomy 18
III. Radial Hypoplasia or Aplasia:
  1. Cornelia de Lange syndrome
  2. Fanconi syndrome
  3. Holt-Oram syndrome
  4. Radial aplasia-thrombocytopenia
  5. Seckel syndrome
  6. Trisomy 13
  7. Trisomy 18
IV. Metacarpal Hypoplasia—Third, Fourth, and/or Fifth:
  1. Basal cell nevus syndrome
  2. Biemond's syndrome
  3. Brachydactyly Type E
  4. Chondrodystrophic calcificans congenita
  5. XO syndrome
V. Polydactyly:
  1. Bloom syndrome
  2. Carpenter syndrome
  3. Chondrodystrophy calcificans congenita
  4. Ellis-van Creveld syndrome
  5. Goltz syndrome
  6. Noack syndrome
  7. Oral-facial-digital syndrome
  8. Robinson type ectodermal dysplasia
  9. Rubinstein-Taybi syndrome
  10. Smith-Lemli-Opitz syndrome
  11. Trisomy 13 syndrome
VI. Syndactyly—Cutaneous and/or Osseous:
  1. Acrocephalosyndactyly (Apert, Vogt, Type III)
  2. Waardenburg type (Pfeiffer type)
  3. Aminopterin-induced syndrome
  4. Bloom syndrome
  5. Chondrodystrophy calcificans congenita
  6. Cornelia de Lange syndrome

7. Down syndrome
8. Fanconi syndrome
9. Fraser syndrome
10. Goltz syndrome
11. Hallermann-Streiff syndrome
12. Laurence-Moon-Biedl syndrome
13. Neurofibromatosis
14. Oculodentodigital syndrome
15. Oral-facial-digital syndrome
16. Otopalatodigital syndrome
17. Popliteal web syndrome
18. Prader-Willi syndrome
19. Radial aplasia-thrombocytopenia
20. Robinson type ectodermal dysplasia
21. Silver syndrome
22. Smith-Lemli-Opitz syndrome
23. Trisomy 13
24. Trisomy 18

VII. Broad Thumb and/or Toe:
1. Acrocephalosyndactyly (Apert, Carpenter, etc.)
2. Leri plenostosis
3. Rubinstein-Taybi syndrome

Clinodactyly, Camptodactyly, and Crooked Fingers[1]

Finger deformities occur as isolated anomalies (Fig. 1285), or are inherited along with a host of congenital anomalies associated with widespread skeletal disease, connective tissue and cardiovascular disease, and chromosomal abnormalities. A hand deformity may be evidence of underlying disease and merits skeletal survey and clinical evaluation.

Clinodactyly refers to a finger curvature in either the medial or lateral plane.[2-4] Some authors confine it to the fifth finger,[5] but it should be used in a more general sense.[4] The deformity is easily evaluated clinically, since it is unassociated with short phalanges and fixed flexion must be demonstrated. A shortened middle phalanx is sometimes associated.

Camptodactyly[6-11] indicates a permanent finger flexion at the proximal interphalangeal joint. It is due to a fascial abnormality[7, 10] or muscle contracture.[9] More easily evaluated clinically, it

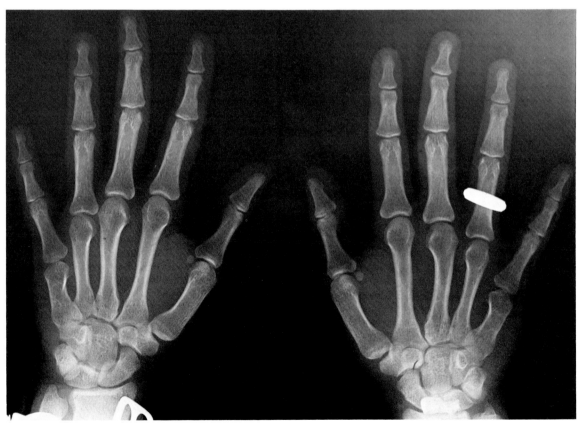

FIG. 1285. BILATERAL SHORTENING OF FIFTH METACARPALS IN OTHERWISE HEALTHY WOMAN
Isolated finger deformities may occur without associated skeletal disease or it may be evidence of underlying disease. Skeletal survey and clinical evaluation are merited.

is usually unassociated with short phalanges; fixed flexion must be demonstrated. If isolated, it may be sporadic, acquired, or traumatic, or it may be transmitted by an autosomal dominant mode.[12] A shortened middle phalanx may be an isolated anomaly.[13]

Factitious curvature of the fingers may be due to poor positioning and may be suspected when adjacent phalanges are also curved, and there are periarticular soft tissue wrinkles that do not curve with true clinodactyly.

Clinodactyly may be associated with different forms of familial brachydactyly.[12, 14-16] Epiphyseal trauma may result in unilateral curvature of a finger. Contractures of arthrogryposis and Dupuytren contracture usually cause camptodactyly.

Congenital diseases or syndromes that consistently produce clinodactyly and camptodactyly include: mongolism,[17-19] Silver syndrome,[20, 21] Cornelia de Lange syndrome,[22-25] Klinefelter syndrome,[26] and several craniofacial syndromes.[27-32] Other conditions and associated findings are shown in Table 34.[1]

The specificity of crooked fingers varies in the various syndromes. Clinodactyly and camptodactyly may occur alone or in combination. Kirner deformity occasionally occurs in Silver syndrome and Cornelia de Lange syndrome.[20, 23] Clinodactyly and associated skeletal anomalies may be specific, or suggest the correct syndrome.[1] Clinodactyly and childhood dwarfism usually indicate Silver syndrome. Clinodactyly and abnormal thumbs suggest Cornelia de Lange syndrome. Oculodentodigital syndromes may produce clinodactyly (fifth) and a thickened mandible. Clinodactyly is rarely present in XO Turner syndrome,[4, 33, 34] but appears in male Turner syndrome (Noonan).[34-36] Many other chromosomal disorders are associated with clinodactyly.

Anomalous fingers and other anomalies occur

TABLE 34. CROOKED FINGERS ASSOCIATED WITH OTHER ABNORMALITIES

Abnormality	Incidence of Crooked Fingers	Type of Crooked Fingers	Commonly Associated Radiologic Findings in Hand
A. Hand-Foot Abnormality			
1. Symphalyngism	Occasional	Clinodactyly 5	Fusion between phalanges; carpal fusions
2. Familial brachydactyly	Frequent	Clinodactyly 5	Anomalies of phalangeal epiphyses; varying size of bones
3. Dupuytren contracture		Flexion deformity	None
4. Other hand anomalies	Frequent	Variable	Variable
B. Chromosomal Disorders			
1. Trisomy 21 (mongolism)	43-68%	Clinodactyly 5	Short 5th middle phalanx; retardation of skeletal maturation, particularly ulnar side
2. XXXXY Klinefelter	84%	Clinodactyly 5	Elongated ulna; pseudoepiphysis; short metacarpals; corner defect in capitate
3. Trisomy 18	Occasional	Clinodactyly and others	Hand held clenched; "V" between 3rd and 4th fingers
4. Trisomy 13-15	Frequent	Camptodactyly; flexion deformity	Flexion contractures of hands; polydactyly.
5. Other chromosomal anomalies	Occasional	Variable	Depends on type anomaly.
C. Craniofacial Syndromes			
1. Otopalatodigital	Frequent	Clinodactyly 5	Large epiphyses 5th proximal phalanges; short broad distal phalanges; teardrop lesser multangular; capitate with long transverse axis
2. Orodigitofacial	14/22 showed hand anomalies	Clinodactyly 5; camptodactyly	Brachydactyly; syndactyly; extra digits
3. Oculodentodigital	Frequent	Clinodactyly 5; camptodactyly	Hypoplastic phalanges; syndactyly
4. Treacher Collins syndrome	Occasional	Clinodactyly 5	
5. Ankyloglossia superior	Unknown	Clinodactyly; camptodactyly	Syndactyly; hypoplasia of digits; lobster-claw hand.

TABLE 34—Continued

Abnormality	Incidence of Crooked Fingers	Type of Crooked Fingers	Commonly Associated Radiologic Findings in Hand
D. Bone Dysplasias and Dysostoses			
1. Multiple exostoses	Occasional	Variable	Due to local effect on epiphyses
2. Mucopolysaccharidoses	Occasional	Flexion deformity	Coarse thick bones; pointed metacarpals; whole hand held in flexion
3. Osteo-onychodysplasia	Occasional	Camptodactyly 5; clinodactyly 5	
E. Trauma			
Epiphyseal injury, frostbite	Occasional	Variable	
F. Miscellaneous Syndromes			
1. Silver syndrome	76%	Clinodactyly 5; occasional	Occasional syndactyly; asymmetry
2. Cornelia de Lange syndrome	88%	Clinodactyly 5; occasional Kirner	Proximally placed thumb
3. Congenital heart disease, including Holt-Oram	Frequent	Clinodactyly 5; camptodactyly	Variable, abnormal fingerlike thumbs and carpal anomalies in Holt-Oram syndrome
4. Turner phenotype male or female	7/8	Clinodactyly 5; Kirner	Short 5th middle phalanges; short 4th metacarpal
5. Prader-Willi syndrome	4/14	Clinodactyly 5	Variable; retarded maturation
6. Seckel bird-headed dwarf	Frequent	Clinodactyly 5	Absent thumb, sometimes
7. Fanconi aplastic anemia	Occasional	Clinodactyly 5	Hypoplastic absent thumb; slender phalanges; radial hypoplasia
8. Focal dermal hypoplasia (Goltz)	Occasional	Clinodactyly 5; camptodactyly	Syndactyly; polydactyly; absence of fingers; hypoplastic cleft hand.
9. Absence of pectoral muscle	Unknown	Clinodactyly 5; camptodactyly	Syndactyly; microdactyly; brachydactyly; supernumerary digits
10. Léri pleonosteosis	Unknown	Flexion curvature	Short broad metacarpals; broad thumb; flexion contractures
11. Arthrogryposis	Frequent	Flexion deformity	Carpal fusion; characteristic position
12. Arthritides; in particular, rheumatoid arthritis	Frequent	Variable	Variable
13. Marfan syndrome	Unknown	Clinodactyly; camptodactyly	Long fingers
14. Progeria	Unknown	Flexion	Short, pointed distal phalanges; arthritis
15. Popliteal pterygium syndrome	Unknown	Clinodactyly 5	Syndactyly
16. Myositis ossificans progressiva	Unknown	Clinodactyly 5	Other digital anomalies

in such undefined syndromes as camptodactyly and radial ulnar synostosis,[12] foot anomalies and camptodactyly,[11] and clinodactyly and absent thymus.[4]

Hereditary Bowing of Terminal Phalanges of Fifth Fingers (Kirner Deformity)[1]

Synonym: dystelphalangy.[2]

Kirner deformity is a hereditary,[3] painless, symmetrical bowing of the terminal phalanges of the fifth fingers. The age of onset is unknown; the youngest patient reported was 5 years old,[4] and it is not reported in the newborn. Blank and Girdany[5] reported a 14-year-old and a 16-year-old with Kirner deformity whose fingers were normal at 1½ years and at 5½ years, respectively.

Roentgenographic Features. There is bowing of the fifth terminal phalanges, with a normal epiphysis (Fig. 1286). The tuft and shaft of the terminal phalanges may be globular. The metaphysis of the terminal fifth phalanx remains bowed after epiphyseal closure. Bone maturation may be delayed before the deformity appears. Silver syndrome[6] and Cornelia de Lange syndrome[7] occasionally show Kirner deformity.

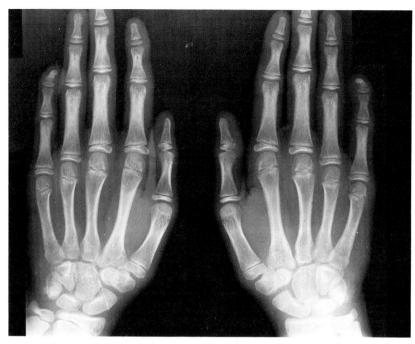
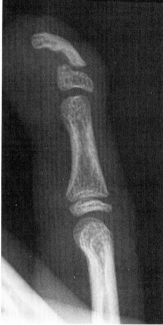

FIG. 1286. KIRNER DEFORMITY (HEREDITARY BOWING OF TERMINAL PHALANGES OF FIFTH FINGERS) (*Left*) Bowing of the fifth terminal phalanges and normal epiphyses. (*Right*) Magnified view. Bowing of the distal phalangeal metaphysis. The epiphysis is not involved. (Courtesy of Dr. Robert M. Peck, Tripler General Hospital, Hawaii.)

ACHONDROPLASIA

Synonyms not now in use: chondrodystrophia fetalis, micromelia, chondrodystrophic dwarfism, chondrodystrophia.

Achondrodysplasia is a congenital abnormality of unknown etiology. Parrot[1] first used the name "achondroplasia" in 1878, and Kaufman[2] in 1892 suggested the term "chondrodystrophia fetalis."

Clinical Features. Achondroplasia is hereditary and transmitted by an autosomal dominant mode; Phemister[3] has traced the condition through six generations of one family. Fresh mutations are frequent.[4] All long bones are symmetrically shortened at birth. Elbow deformities may be asymmetrical and prevent limitation of supination and/or full extension.[5] The length of the vertebral column is relatively normal, so that dwarfism is due primarily to shortened extremities. Achondroplastics may die at birth because of difficult delivery by an achondroplastic mother, a small contracted foramen magnum, or a constricted thorax. These cases may be confused with those of thanatophoric dwarfism.

Brachycephaly is the rule, and the head is often enlarged due to nonprogressive hydrocephalus.[6] The frontal bones are prominent and the bridge of the nose recedes due to hypoplasia of the base of the skull. These features distinguish achondroplasia from many other types of dwarfism.

The buttocks are prominent, giving the false impression of lumbar lordosis. The hands are trident because of the separation of the second and third digits.[7] The abdomen is protuberant. The musculature is very well developed. The characteristic rolling gait is caused by a backward tilt of the pelvis and posterior displacement of the hip joints. Sexual and mental development are normal.

Pathologic Features. Biopsies of non-weight-bearing chondroosseus junctions (ribs and iliac crest) are histologically normal.[8] There is well organized enchondral ossification; although there may be a slight decrease in the height of the cartilage cell columns. In past reports, marked changes in the columnar orientation or cartilaginous cells are described,[9] but it seems likely that these abnormalities represent conditions other than achondroplasia, such as metatrophic dwarfism, metaphyseal dysostosis, and thanatophoric dwarfism.

Periosteal ossification proceeds normally and periosteal growth at the ends of the long bones outstrips epiphyseal growth. Occasionally a fibrous band grows from the periosteum to interpose the ossifying cartilage and the remainder of the epiphysis, which precludes further longitudinal growth of bone.

Roentgenographic Features. Symmetrical shortening of all long bones is the main characteristic. (Asymmetrical shortening of long bones creates an illusion of widened shafts.) Proximal segments of extremities may be more severely involved than the distal portions: the humerus and femur may be disproportionately shorter than the radius and tibia. The ends of the long bones are splayed because of an increase in the appositional growth. This is in response to retarded or absent enchondral bone growth. The radial heads may be enlarged, irregular and subluxed and interfere with elbow function.[5] Appositional growth at the bone end is often deficient toward the metaphyseal center and forms a V-shaped surface in which the epiphysis is incorporated (Fig. 1287).[10]

The digits of the short, thick hands tend to be the same length (Fig. 1288). Separation of ring and middle finger results in a "trident hand."[7] The fibula is less affected than the tibia, which results in a higher position of the fibular head, often incorporated into the knee joint (Fig. 1289). The ulna is shorter than the radius and the proximal end is thick, the distal slender.[9] The clavicle is not usually affected. The ribs are short and the dimensions of the thorax correspond-

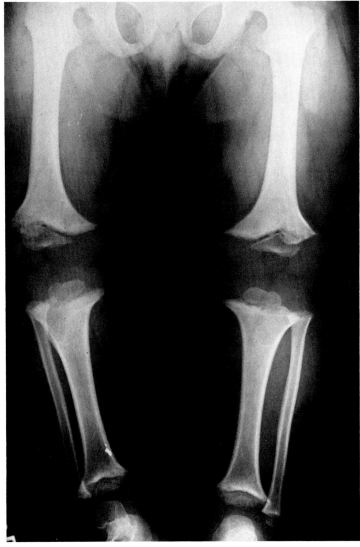

FIG. 1287. ACHONDROPLASIA

Long bones are shortened. Appositional growth at the ends of the femur is deficient toward the center and forms a V-shaped surface into which the epiphysis is incorporated. (Courtesy of John W. Hope, Children's Hospital, Philadelphia, Pa.)

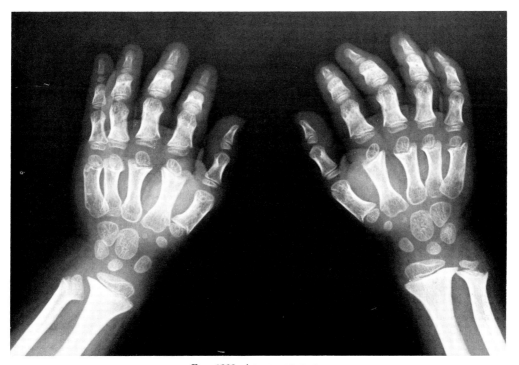

FIG. 1288. ACHONDROPLASIA
Hands are short and thick and the digits tend to be the same length. (Courtesy of John W. Hope, Children's Hospital, Philadelphia, Pa.)

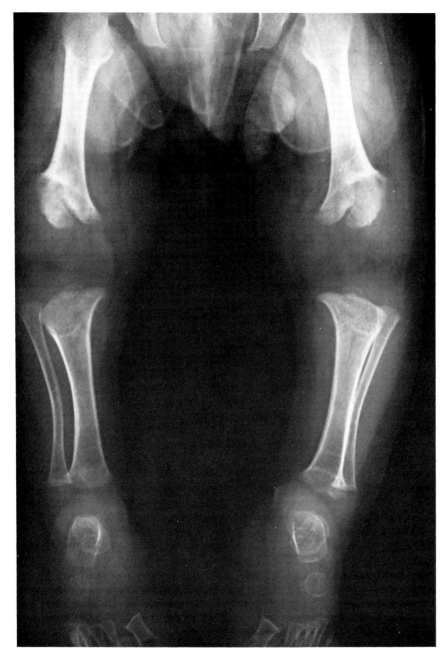

Fig. 1289. Achondroplasia

Fibula is less affected than the tibia which results in a relatively higher position of the fibular head. (Courtesy of John W. Hope, Children's Hospital, Philadelphia, Pa.)

ingly small. "Squaring of the inferior margin of the scapula is characteristic of achondroplasia."[11] The scapular glenoid is small.

Although the pelvis shows characteristic changes emphasized by Caffey,[12] these are similar in chondroectodermal dysplasia, asphyxiating thoracic dystrophy, and thanatophoric dwarfism.[13-17] The entire pelvis is smaller than normal, with the iliac bones the most reduced in size. The ilium is most markedly shortened in the caudal segment. The great sciatic notches are small, and the acetabular angles are obliterated. There is excessive cartilage at the Y-cartilage (Fig. 1290). Although the characteristic appear-

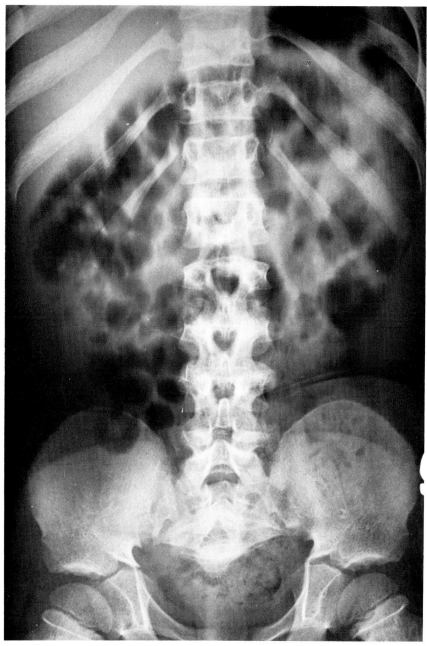

FIG. 1290. ACHONDROPLASIA

Interpediculate spaces and spinal canal progressively narrow from the first to the fifth segments. Serious neurologic deficits may result later in life. Note the hypoplastic caudal portion of the iliac wings with the narrow sacrosciatic border. The normal appearance of the hips helps to differentiate achondroplasia from Morquio's disease. (Courtesy of John W. Hope, Children's Hospital, Philadelphia, Pa.)

ance of the pelvis may be confused in the infant with Ellis-van Creveld syndrome, there are minor differences, as well as differences in the rest of the skeleton, which distinguish the two.

At birth the vertebral column is not shortened, although slight platyspondyly may occur. Marked platyspondyly indicates thanatophoric dwarfism and not achondroplasia. At maturity the vertebral centra may remain slightly flattened but otherwise normal (Fig. 1291). A narrowing of approximately 50% in the ventrodorsal measurements of the spinal canal occurs in the region of the fifth lumbar vertebra, and may lead to serious neurologic defects later in life (Figs. 1292 and 1293). Fairbanks describes angular kyphosis at the upper lumbar level, particularly L2 and L3. In this region, the anterior surface of the vertebra appears to be worn away in the superior-inferior aspects, which results in a rounded beaking (Fig. 1294).

The skull reveals brachycephaly, frontal bossing, and a recession of the bridge of the nose characteristic of achondroplasia. The foramen magnum is small. Aegerter and Kirkpatrick[18] believe that communicating hydrocephalus in achondroplasia is due to deficient brain space in the posterior fossa, caused by the early closure of intersphenoid and sphenoid occipital synchondroses. The brain seems to grow upward, often bulging in the forehead, an appearance not seen in many other types of dwarfism. Because the mandible grows normally, its relatively larger size gives the appearance of prognathism. The sinuses are often large.[9]

Of historical interest, three types of achondroplasia were recognized by Kaufmann.[2] (1) Hypoplastic achondroplasia is the most common type. The ends of the bones are slightly flared but nearly normal in width. The epiphyseal line is smooth and the epiphyses are smooth in outline. (2) Hyperplastic achondroplasia is much less common and is characterized by marked irregularity of epiphyseal line and fragmented and irregular epiphyses. Joints are large and bulbous, and there is often a large space between the epiphysis and the metaphyseal end of the bone. (It is likely that this type of achondroplasia is actually either metatrophic dwarfism or pseudoachondroplasia (spondyloepiphyseal dysplasia congenita), confused in the past with achondroplasia.) (3) The malacic form, which is extremely rare, is characterized by bone deformity due to softening of abnormal cartilage. This type, usually seen in the newborn, is often lethal. It probably represents thanatophoric dwarfism and not achondroplasia.

Achondroplasia has often been confused with other cases of dwarfism, most commonly mucopolysaccharidoses, trisomy, and other forms of chondrodystrophic dwarfism. Each dwarf should have a karyotype evaluation and urinalysis for presence of mucopolysacchariduria. If these are normal, there still remains a large group of chondrodystrophic dwarfs to be differentiated by clinical and roentgenographic means (Table 35).

Although recently the changes of achondroplasia have been more clearly defined,[19-21] the distinctive manifestations are present only in the growing bones, and after epiphyseal fusion differentiation from other chondrodystrophies may be impossible by roentgenographic evaluation alone.

Differential Diagnosis. The mucopolysaccharidoses have been recently defined and the

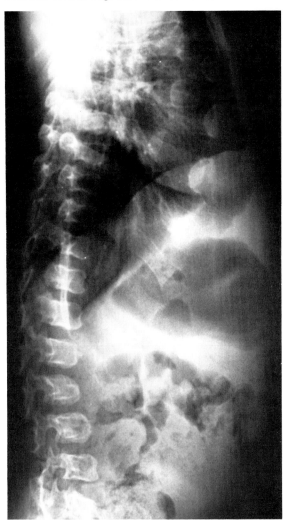

FIG. 1291. ACHONDROPLASIA
Vertebral bodies have lost the flattening, prominent in infancy, and appear almost normal. (Courtesy of John W. Hope, Children's Hospital, Philadelphia, Pa.)

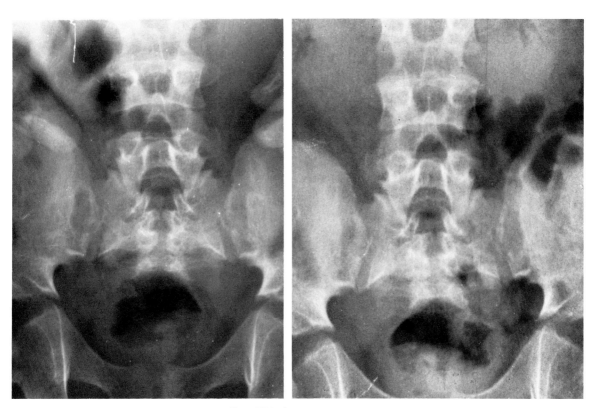

FIG. 1292. ACHONDROPLASIA

(*Left*) Shortening of the caudal end of the pelvis with a small sacrosciatic notch. (*Right*) Progressive decrease in the interpediculate spaces which may lead to serious neurologic defects later in life. (Courtesy of Dr. Robert M. Peck, Atlantic City, N.J.)

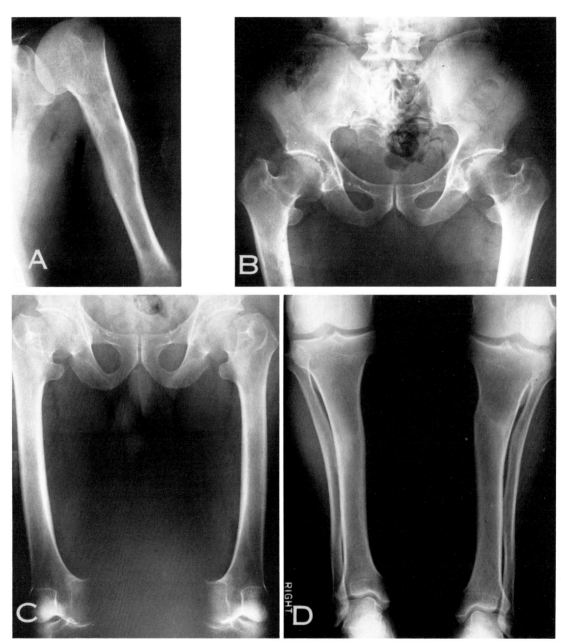

Fig. 1293. Achondroplasia

This adult began having serious neurologic difficulties due to the narrowing of the spinal canal in the lumbosacral region. All the long bones were symmetrically shortened and the fifth interpediculate measurement was approximately 50% of normal. (A) Humeri, (B) pelvis, (C) femurs, (D) tibias and fibulas.

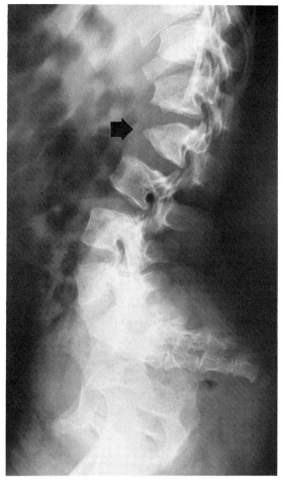

FIG. 1294. ACHONDROPLASIA
Angular kyphosis of the upper lumbar vertebrae. The anterior surface of the second lumbar vertebra appears to be worn away in the superior and inferior aspects, resulting in central beaking (*arrow*). (Courtesy of John W. Hope, Children's Hospital, Philadelphia, Pa.)

presence of mucopolysacchariduria separates this form of dwarfism from achondrodystrophic dwarfs.

Since achondroplasia serves as a model for the other chondrodystrophic types of dwarfism, the differential diagnosis is an important consideration. Achondroplastic dwarfs have long been confused with cases of chondrodystrophic dwarfism.

The bone changes more closely resemble spondyloepiphyseal dysplasia (pseudoachondroplasia). The pointed proximal metacarpals and metatarsal ends and widespread epiphyseal irregularity distinguishes Morquio disease from achondroplasia. The pelvis shows a slightly hypoplastic caudal segment, but the iliac rami are pointed and the acetabular flattening and irregularity are quite distinctive. The skull is usually normal. The sella may show a J-shaped abnormality. The interpediculate spaces are normal.

Clubfeet and scoliosis differentiate *diastrophic dwarfism*[22] from achondroplasia. In addition, there may be external ear abnormality and deformities of the thumb (hitchhiker's thumb) and great toe.[23] Other metacarpal bones may be irregular in length and distorted. The forearms and legs are more affected than the arms and thighs. Stippled epiphyses may be present at birth. Autosomal mode of transmission is postulated.[21]

Cartilage-hair hypoplasia,[24] a severe form of dwarfism, is distinguished from achondroplasia by metaphyseal irregularities, hypoplasia of cartilage and hair, normal skull and spine, and polydactyly. The sacrosciatic notch is normal or only mildly reduced. Inheritance is autosomal recessive. Histopathology of the ribs (costochondral junction) reveals fewer cartilage cells than normal and disorderly arrangement of the cartilage cell columns.[8, 21] The rib histology of achondroplasia is normal.

Pseudoachondroplastic form of spondyloepiphyseal dysplasia may be the most difficult differential diagnosis.[25-28] The dwarfism is very severe, with short massive limbs, but there are no craniofacial changes and the pelvis is not as severely involved. There may be slight shortening of the caudal segment of the pelvis (Fig. 1295). The interpediculate spaces are of normal width.

Hypochondroplasia, once considered a mild form of achondroplasia, is probably a separate entity[21]; however, one family has been studied with both achondroplastic dwarfs and hypochondroplasia.[29] Rubin[30] refers to it as achondroplasia tarda. It differs from achondroplasia in that there is less disproportion between the shortened limbs

TABLE 35. CHONDRODYSTROPHIC DWARFISMS

A. *Nonlethal:*
   Achondroplasia
   Diastrophic dwarfism
   Cartilage-hair syndrome
   Spondyloepiphyseal dysplasia
       (extreme of pseudoachondroplastic type)
   Hypochondroplasia
   Metatrophic dwarfism
   Ellis-van Creveld syndrome
   Asphyxiating thoracic dystrophy
   Stippled epiphyses
   Metaphyseal dysostosis
B. *Often lethal:*
   Thanatophoric dwarfism
   Achondrogenesis

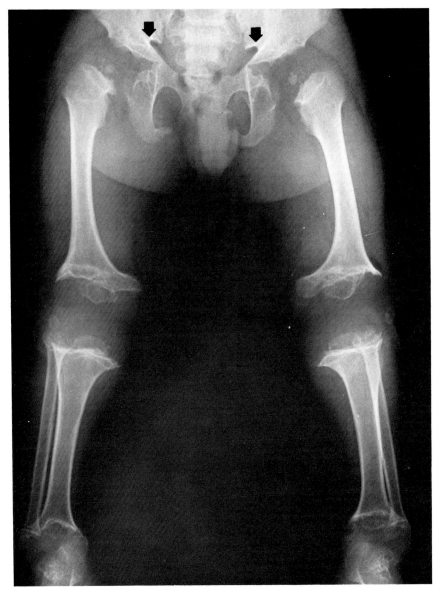

Fig. 1295. Pseudoachondroplastic Dwarfism

All long bones are shortened, the metaphyses widened and irregular, and the caudal end of the pelvis (*arrows*) hypoplastic. There is similarity to achondroplasia.

and trunk. There is no frontal bossing, although the cranium may be large. Narrowing of the interpediculate spaces occurs, but the pelvis is either normal or only mildly affected.

Metatrophic dwarfism[31] resembles achondroplasia in the infant, who is short limbed but with normal thoracic length (Fig. 1296). Later, there is transformation to a relatively short trunk and relatively long limbs, giving an apelike appearance. The spinal shortening is due to a progressive kyphoscoliosis which may be present at birth (Fig. 1297). During childhood, the epiphyses and joints become enlarged, resembling those seen in Morquio's syndrome rather than in achondroplasia (Fig. 1296).[21] There may be a tail-like appendage at the distal end of the gluteal cleft due to a skin fold overlying the coccyx. Flat vertebral centra persist throughout adulthood. Although the shortening of the tubular bones resembles achondroplasia, the marked flaring at the ends of the bones without notching for the epiphysis and constriction of the diaphyses should distinguish this condition from achondroplasia.

Kniest syndrome is a distinct chondrodystrophy with disproportionate dwarfism, distinctive from metatrophic dwarfism and the spondyloepiphyseal dysplasias.[32-36] It has sometimes been referred to as a metatrophic dwarfism, Type II and pseudometatrophic dwarfism.[37]

Clinical Features. There is disproportionate dwarfism with a shortened kyphotic trunk, deformed extremities with limited joint motion and a characteristically round, flattened facies. The deformities are present at birth and are most noticeable by the enlarged joints, especially the knees. The fingers are long and knobby, with limited flexion at the metacarpophalangeal and proximal interphalangeal joints resulting in the inability to form a fist. The chest is short, wide and barrel-shaped with prominent dorsal scoliosis.

Pathologic Features. Resting cartilage contains large chondrocytes in a loosely woven matrix. The numerous empty spaces cause a Swiss cheese appearance, "Swiss cheese cartilage syndrome" has been proposed for this disorder.[35, 38, 39] The Swiss cheese appearance of the resting cartilage is distinctive for this disorder.[39, 40] The growth plate contains extremely hypercellular cartilage with large chondrocytes and little intervening matrix, which causes broad, short irregular spicules of calcified cartilage and bone.

Roentgenographic Features. The major

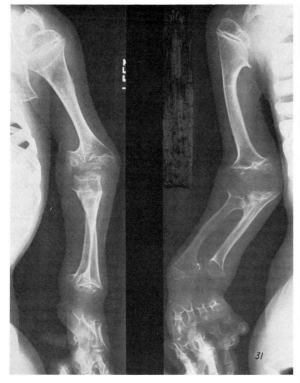 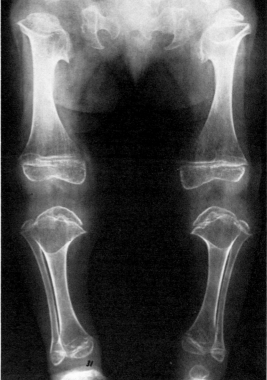

FIG. 1296. METATROPHIC DWARFISM

(*Left*) Forearms; all bones are short and there is widening and irregularity of the metaphyses. (*Right*) Legs; all bones are short. Widening of the metaphyseal areas. (Courtesy of Dr. John Dorst, Johns Hopkins Hospital, Baltimore, Md.)

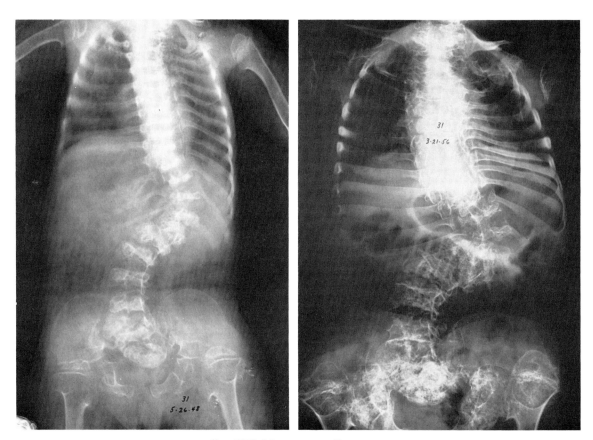

FIG. 1297. METATROPHIC DWARFISM

(*Left*) Age 3; the kyphoscoliosis is marked. This may be present at birth or develop later. The left hip is dislocated. The hip epiphyses and metaphyses are widened. The caudal end of the pelvis is deficient. (*Right*) Age 10; the kyphoscoliosis has progressed. Dislocation of both hips. Widening of the metaphyses and epiphyses. (Courtesy of Dr. John Dorst, Johns Hopkins Hospital, Baltimore, Md.)

roentgenographic features consist of[32]: (a) flat, markedly elongated irregular vertebral bodies with superior-inferior defects in their midportion during infancy and early childhood (Fig. 1298); (b) dumbbell-shaped long bones which later have epiphyseal ossification delay and irregularity, and expanded metaphysis with cloudlike effects on both sides of the epiphyseal plate; and (c) flattened and squared-off epiphyses of the hands with joint space narrowing.

Ellis-van Creveld syndrome is chondroectodermal dysplasia, with abnormalities of the nails and teeth. Polydactyly is the most distinctive sign. The head is normal. The dwarfism is acromelic rather than rhizomelic as in achondroplasia.[21] Adults cannot make a closed fist due to the exaggerated shortening of the distal phalanges in relation to the others.[41] Congenital heart disease may be present. The inheritance is autosomal recessive.[41] The skull and spine are normal. The pelvic changes in infancy may resemble achondroplasia, but the trident acetabular roof and the premature appearance of the femoral capital epiphysis indicate the Ellis-van Creveld syndrome.[15] The pelvic abnormalities improve in later childhood and have usually disappeared in older children and adults.

Asphyxiating thoracic dysplasia[13, 14] is easily confused with achondroplasia in the neonatal period. The outstanding feature, rib hypoplasia and shortening, however, leads to a distinctive stenotic thorax. The pelvic abnormalities resemble Ellis-van Creveld syndrome[42] in the newborn and may be confused with achondroplasia. Accelerated development of the femoral capital epiphyses, in contrast to their delayed appearance in achondroplasia, and occasional polydactyly[16, 42, 43] distinguish it. With survival, the long bone shortening and pelvic changes improve.[21]

With *dysplasia epiphysialis punctata*[44] the stippled epiphyses, articular cartilages and ligaments, due to calcifications, distinguish the condition from achondroplasia in the infant. A saddle-nose deformity and skull base involvement are similar to achondroplasia, but the presence of congenital cataracts and respiratory cartilage calcifications do not occur with achondroplasia. Humeri and femora are more severely affected than the bones of the forearm and legs, but the tarsal and carpal bones are deformed and in the older child may resemble multiple epiphyseal dysplasia.[45] The small tubular bones in the hands and feet are usually normal, unlike the stubby digits of achondroplasia. Asymmetrical extremity shortening is common. It is probably transmitted by autosomal recessive modes, although Silverman[21] has observed it in siblings with normal parents and in the offspring of one sib with a normal male.

Metaphyseal dysostosis[46-48] is a short limbed dwarfism with relatively long upper extremities and shortened and bowed lower extremities causing an apelike appearance. The normal cranium and pelvis differentiate it from achondroplasia. The rachiticlike metaphyseal defects are distinctive.

Thanatophoric dwarfism[49] is a lethal micromelia long confused with achondroplasia. The infants are often stillborn or die soon after birth. There is extreme disproportion between the length of the trunk and the extremities. The markedly shortened long bones are bowed and curved. The vertebral centra are extremely flat, a most striking difference from achondroplasia. The pelvic findings, however, and the narrowing of the interpediculate spaces closely resemble achondroplasia.

Achondrogenesis is easily diagnosed by the almost total absence of long bone ossification, although the extremely short limbs and large skull may simulate achondroplasia. The patients are stillborn or die soon after birth.

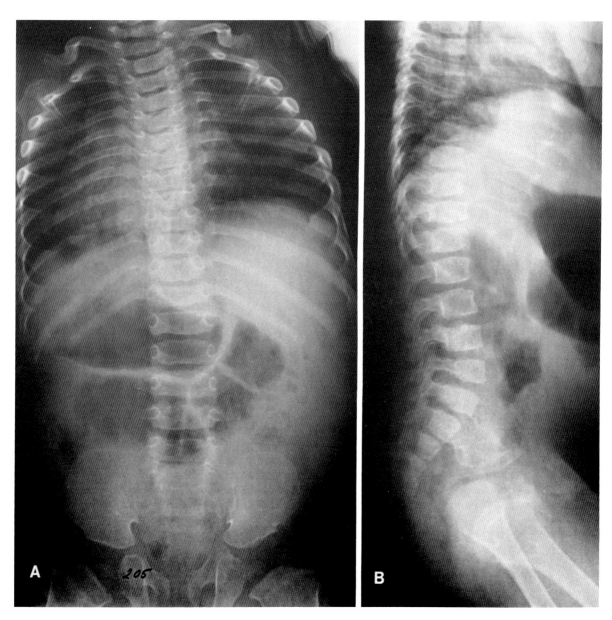

Fig. 1298. KNIEST SYNDROME

The vertebral bodies are flat, with superoinferior defects on some of the lumbar vertebrae in the midportions. Dumbbell-shaped long bones show epiphyseal ossification delay and irregularity and expanded metaphyses with cloudlike effects at the epiphyseal plates. (A) Anteroposterior lumbar spine, (B) lateral lumbar spine, (C) lower extremities.

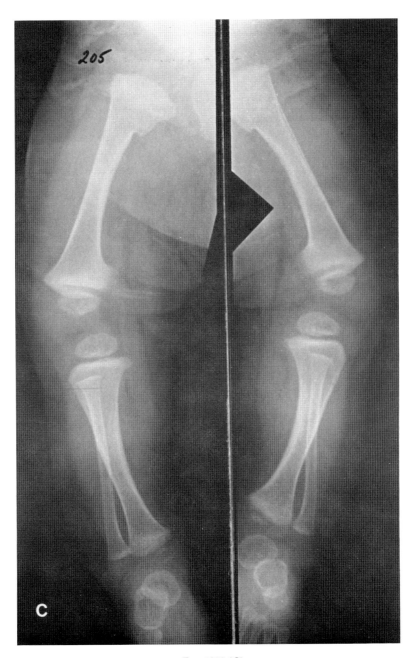

Fig. 1298 (C)

THANATOPHORIC DWARFISM

Thanatophoric dwarfism is a severe micromelic type that usually results in death soon after birth.[1] Although it may be an extreme expression of achondroplasia, certain genetic, clinical, and radiographic differences justify a separate clinical classification.

Clinical Features. Thanatophoric dwarfism probably results from a dominant gene mutation, since there are no documented cases with a family history or parental consanguinity.[1-3] One reported case had a normal fraternal twin.[3]

Clinically, the disease is so severe that it may be recognized in utero.[2, 4-7] Maternal history of polyhydramnios and previous spontaneous abortions is reported.[8] Lethal aspects probably result from the inability of infants to inflate their lungs, due to the extreme narrowness of the chest. Longest survival is 25 days.[2]

At birth, there is a striking disproportion between the micromelic extremities and the essentially normal length of the trunk. The head is relatively large, the thorax small. The fontanels are large and the forehead is prominent. Infants are hypotonic and primitive reflexes are absent. The foramen magnum is small and impingement on the spinal cord is reported.[2, 8]

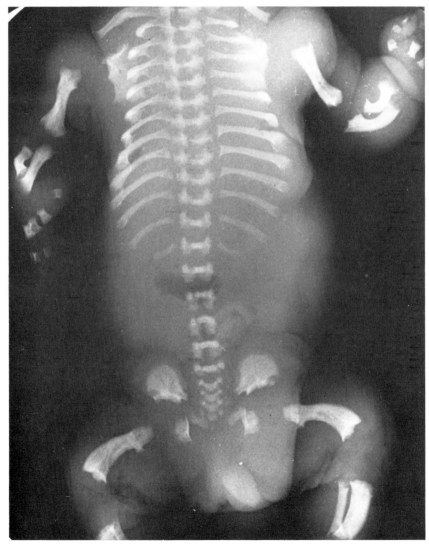

FIG. 1299. THANATOPHORIC DWARFISM
All long bones are symmetrically shortened and bowed. The caudal portions of the pelvic bones are deficient. The thoracic cage is deformed.

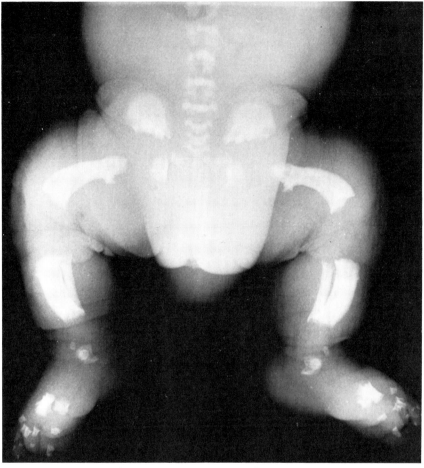

FIG. 1300. THANATOPHORIC DWARFISM
Symmetrical shortening and bowing of all long bones. The caudal portion of the pelvis is deficient.

Roentgenographic Features. There is severe micromelia with bowing of the extremities (Figs. 1299–1301). The skull has a short base and prominent frontal bone (Fig. 1303), with depression of the root of the nose. A cloverleaf skull deformity ("Kleeblattschadel") occurs in some patients with this condition.[9-14] Its appearance is that of a trilobe skull deformity due to huge coronal and lambdoidal sutures and downward displacement of the middle cranial fossa, producing a cloverleaf appearance on the frontal projection. The chest is narrow and the ribs are short (Fig. 1299). There is a reduction of the interpediculate spaces of the last few lumbar vertebrae. The iliac wings are small and square, the sacrosciatic notch narrow, and the acetabular roof horizontal. All of these changes are similar to those seen in achondroplasia.

The features distinguishing thanatophoric dwarfism from achondroplasia are severe universal vertebral plana (Fig. 1302) and excessive intervertebral space height, preserving a normal trunk length. The short tubular bones are bowed. The metaphyseal areas are irregular and frequently cupped. Although these latter features usually serve to differentiate the condition from achondroplasia, there are cases of achondroplasia with flat vertebrae and a somewhat narrowed chest. However, the degree of change is not as marked as in thanatophoric dwarfism, and, even with a narrowed chest, the achondroplastic infants survive.

The roentgenographic features serve to distinguish this condition from asphyxiating thoracic dystrophy, achondrogenesis, hypophosphatasia, and Ellis-van Creveld syndrome.

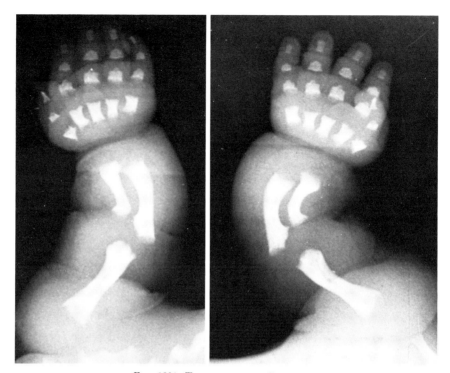

Fig. 1301. Thanatophoric Dwarfism
There is symmetrical shortening and bowing of all long bones.

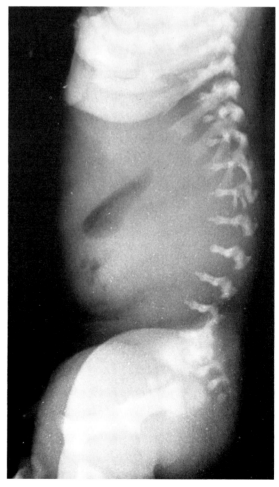

Fig. 1302. Thanatophoric Dwarfism
In infants there is platyspondyly and poorly formed vertebrae. Wide spaces are present between the vertebrae. Achondroplastic vertebrae are much better formed.

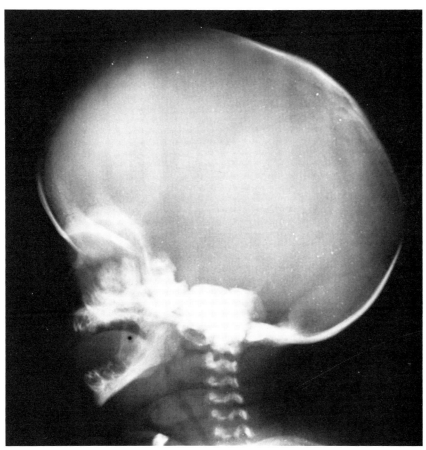

FIG. 1303. THANATOPHORIC DWARFISM
The skull is enlarged and brachycephalic.

FAMILIAL ASPHYXIATING THORACIC DYSTROPHY

Synonyms: thoracic asphyxiating dystrophy, infantile thoracic dystrophy, Jeune disease and thoracic-pelvic-phalangeal dystrophy (TPPD).

In 1954 and 1955, Jeune et al.[1,2] first reported familial asphyxiating thoracic dystrophy. It is predominately a thoracic deformity resulting in a functionally immobile chest. It may easily be confused with Ellis-van Creveld syndrome, a spondyloepiphyseal dysplasia, or achondroplasia. (See differential diagnosis of achondroplasia on page 1196). It is probably transmitted by an autosomal recessive mode.

Clinical Features. Clinically, the thoracic dystrophy appears isolated except for the hands, which are short and flat, and the first three fingers are almost equal in length. There may be polydactyly. The face, spine, and long bones appear normal. The chest is elongated and cylindrical, narrowed in the upper and middle areas but flared at the bases by the abdominal organs. There is reduced thoracic mobility during respiration. Pulmonary infections are common and have caused death in a number of patients, usually between the ages of 6 and 18 months.[1,2] The thorax approaches normal size with increasing age.[3]

Progressive renal failure and hypertension may lead to uremia and death in some patients, usually before the age of 5 years. Renal biopsies reveal interstitial fibrosis, tubular atrophy, round cell infiltration, and periglomerular fibrosis.[3]

Roentgenographic Features. The combination of thoracic and pelvic abnormalities strongly suggest the diagnosis in infancy. The clavicles are horizontal and lie well above the first rib at the level of the sixth cervical vertebra (Fig. 1304). The ribs are short and horizontally oriented with bulbous expansion at the costo-

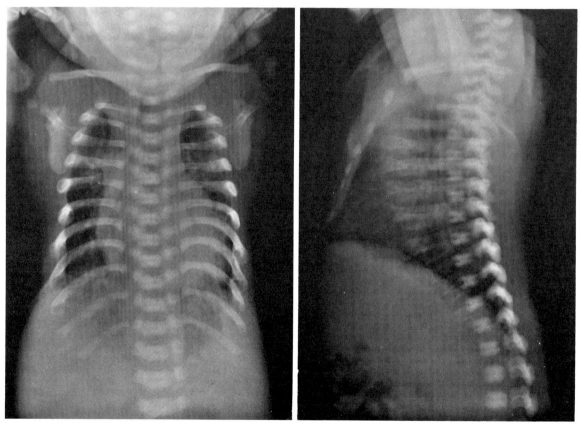

FIG. 1304. FAMILIAL ASPHYXIATING THORACIC DYSTROPHY
Clavicles are horizontal and lie well above the first rib at the level of the sixth cervical vertebra. The ribs are of normal length, but ossification ends at the midaxillary line. The upper ribs are horizontal and the lower ribs are progressively oblique, The thoracic cavity is narrow and cylindrical. (*Left*) Posteroanterior projection. (*Right*) Lateral projection. (Courtesy of John A. Kirkpatrick, St. Christopher's Hospital for Children, Philadelphia, Pa.)

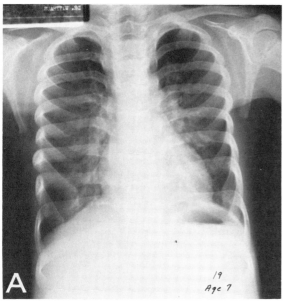 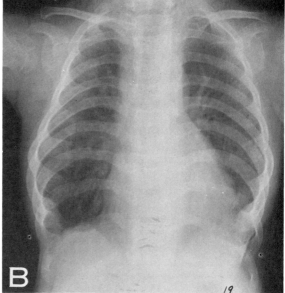

FIG. 1305. FAMILIAL ASPHYXIATING THORACIC DYSPLASIA

(A) Age 7. The ribs and clavicles are horizontal. (B) Age 9. The bell-shaped flaring of the lower ribs has recurred. The ribs are horizontal and there is bulbous expansion of the costochondral ends. The clavicles are horizontal and higher than normal. (C) Age 2. Flaring of the superior portion of the iliac bones and shortening of the cephalocaudal diameter of the pelvis. These changes resemble achondroplasia. There may be premature ossification of the capital femoral epiphyses and, as the infant grows, the configuration of the pelvis may approach normal. (D) Hands at age 2. Considerable shortening, especially of the distal phalanges. The epiphyses fit into cone-shaped deformities of the metaphyses and apparent fusion centrally between the epiphyses and metaphyses of the middle phalanges. Similar changes are present in the epiphyses of the distal phalanges. The middle phalanges are short and broad. The metacarpals are short. (E) At age 9. The distal phalanges remain short and the middle phalanges broad. Less deformity of the epiphyses of the proximal phalanges. The metacarpals remain short. (F) The foot at age 2. Cone-shaped epiphyses are more frequent in the feet than hands of individuals without other detectable abnormalities. In both the hands and feet complete obliteration of the growth plate occurs earlier with cone-shaped epiphyses than with normally shaped epiphyses. The changes present in the foot are similar to the hand changes (D). (Courtesy of L. O. Langer, Jr., M.D.: Thoracic-pelvic-phalangeal dystrophy; asphyxiating thoracic dystrophy of the newborn, infantile thoracic dystrophy. *Radiology*, 91: 447, 1968, © The Radiological Society of North America, Syracuse, N.Y.)

chondral junction (Fig. 1305). The lower ribs flare out around the upper abdominal viscera, giving a bell-shaped configuration in the frontal projection. The narrow cylindrical chest and the normal sized heart leave very little room for the lungs. In the older child, although the chest maintains its narrowed configuration, respiratory distress is not clinically evident, in spite of reduced pulmonary function.

The pelvic deformities in the infant consist of flaring of the superior portion of the iliac bone and shortening in the cephalocaudad diameter, resembling the changes seen in achondroplasia (Fig. 1305). The ischial and pubic bones are short. Usually a downward hook configuration extends from the ilium at the great sciatic notch and there may be a second downward projection at the lateral margin of the acetabulum. The interpediculate spaces are normal. There is premature ossification of the capital femoral epiphysis. As the infant grows, the configuration of the pelvis may approach normal, and the adult pelvis is normal.

Limb dwarfism is variable. In the hands, there may be considerable shortening of the long bones, especially the distal phalanges (Fig. 1305), and occasionally polydactyly. The soft tissues of the extremities are increased, possibly due to hydrops accompanying general circulatory failure, but perhaps from their failure to decrease in formation in proportion to the skeletal shortening.

Langer reports that at 2 years of age cone-shaped epiphyses occur in the phalanges. This is due to fusion between the epiphyseal centers and the central portions of the adjacent metaphyses, most prominent in the distal and middle phalanges. The distal phalanges tend to be short, the middle phalanges broad and short. Cone-shaped epiphyses may involve the proximal phalanges,

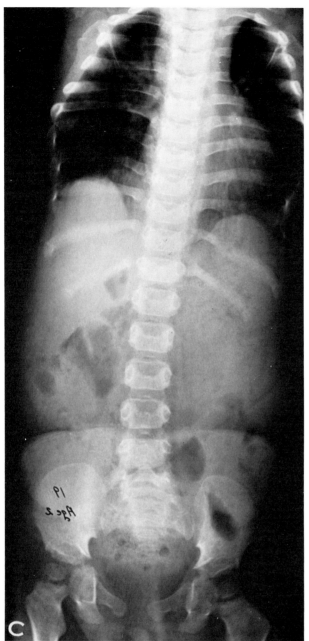
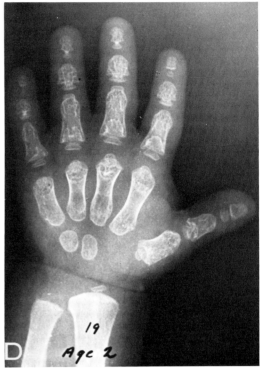

Fig. 1305 (C and D)

and there may be some invagination of the epiphyseal centers from the metacarpals into the adjacent metaphysis (Fig. 1305). Cone-shaped epiphyses have also been reported in chondroectodermal dysplasia and achondroplasia.

Intravenous urograms may show enlarged kidneys and linear opaque streaking of the papillae due to collections of opaque media and dilated tubules.[3] These renal changes, not described in other dysplasias, may serve to differentiate this condition from other osteochondrodystrophies.

The differential diagnosis in a dwarfed individual with a narrow chest should include hypophosphatasia, conditions affecting the external muscles of respiration, and the short limbed dysplasias that manifest in the newborn: achondroplasia, chondrodystrophia calcificans congenita, diastrophic dwarfism, spondyloepiphyseal dysplasia congenita, metatrophic dwarfism, cartilage-hair hypoplasia, and the Ellis-van Creveld syndrome.

In infants with polydactyly, the combination of changes in the thorax and pelvis is diagnostic of familial asphyxiating dystrophy. With polydactyly, it is not possible to distinguish this dystrophy radiographically from Ellis-van Creveld syndrome, but the visceral and ectodermal manifestations of Ellis-van Creveld syndrome are distinctive.[4] Carter suggests that patients with polydactyly may be a separate entity, because of their lower infant mortality.

In later infancy and childhood, the characteristic configuration of the knee in chondroectodermal dysplasia serves to differentiate it[3]: the proximal tibial metaphysis slants cephalad on both sides to a central apex, ossification of the proximal tibial epiphysis begins adjacent to the medial slope, and a knock-knee deformity develops. This has not been described in familial asphyxiating thoracic dystrophy.

The presence of cone-shaped epiphyses in the hand is reported in trichorhinophalangeal dysplasia[4]; however, there is fine, sparse hair which grows slowly, a prominent broad nose with a bulbous end, and other congenital anomalies not present with familial asphyxiating thoracic dystrophy. The cone-shaped epiphyses are rather scattered and usually involve only the middle phalanges, rarely the proximal and distal. Although deformity of the femoral capital epiphyses is frequent in this condition, the pelvic configuration is different from that of familial asphyxiating dystrophy.

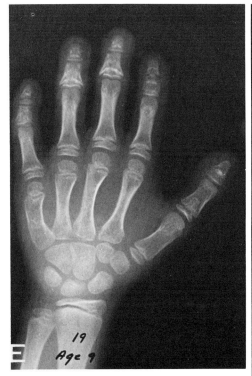
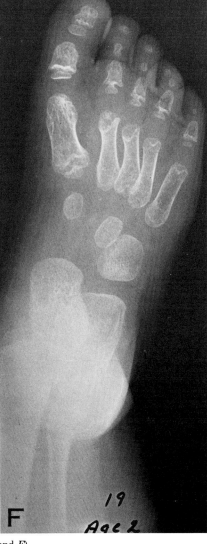

Fig. 1305 (*E* and *F*)

DIASTROPHIC DWARFISM

Synonym: epiphyseal dysostosis.[1]

Diastrophic dwarfism is a rare hereditary disorder of mesodermal structures. In 1960, Lamy and Maroteaux[2] recognized diastrophic dwarfism as a distinct entity. They reported 3 cases and found 11 others in the literature under various titles (atypical achondroplasia, Morquio syndrome, and other osseous dysplasias). The term diastrophic means tortuous and twisted, and describes the severe clubfoot deformity and scoliosis characteristic of the syndrome. Other authors[3-11] confirm the surprisingly constant clinical and roentgenographic features that establish the validity of a separate classification.

Clinical Features. The clinical features are: (1) dwarfism, (2) scoliosis, (3) clubfeet, (4) deformed hands, (5) both lax and rigid joints with contractures, (6) dislocation of multiple joints, (7) deformed external ears, and (8) cleft palate.

The clinical features, except scoliosis, are present in the newborn.[5, 10] Dwarfism is marked, and is often exaggerated by kyphosis and knee and hip contractures. The hands present remarkably constant findings of *hypermobile and abducted thumbs,* extension contractures of the interphalangeal joints, short fingers, and ulnar deviation of the wrists. Talipes equinovarus deformity is present in most patients, and is peculiarly resistant to corrective surgery.

In early infancy, blisterlike masses develop on the external ears which spontaneously resolve, leaving residual auricular cartilage deformity. Approximately half of the patients manifest cleft palate.

Most exhibit limitation of mobility of the shoulders, elbows, and hips, but a significant number show hypermobility of the knees and other joints.[12] Eventually, there is a variable degree of joint contracture, causing stiff joints and dislocations, especially of the hips. Wilson[13] separated the patients into stiff and lax joint types, but most other characteristics do not vary. Limitation of joint motion and stiffness result from osseous deformity, but abnormality of tendon and joint capsule also contributes to the progressive deformity.[11]

Scoliosis, not present at birth, appears within the first few months and progresses rapidly. Mental retardation has not been noted. In the first few months of life, mortality is high,[2, 10, 14] but thereafter the prognosis for life is good.

Roentgenographic Features. The roentgenologic features of diastrophic dwarfism are distinctive in the newborn. *The first metacarpal is oval and usually hypoplastic,* a most distinctive feature (Figs. 1306–1309). Other metacarpals are broad, with the distal ends wider than the proximal. The proximal and middle phalanges are broader than the other small bones. The first metatarsal may be oval-shaped and the broadest metatarsal. The thumb is usually abducted (Fig. 1306) (hitchhiker's thumb). There is accelerated bone maturation of the carpal and tarsal bones, and delay of secondary ossification centers of the metatarsals, metacarpals, and phalanges (Figs. 1306 and 1309).[13] The majority show talipes equinovarus (Figs. 1307 and 1309), sometimes associated with varus of the great toe.

The large tubular bones are short with broad metaphyses. The humeri and femora are shorter than the distal long bones. The distal femoral epiphyses are flattened and smaller than the proximal tibial epiphyses, which are medially placed on broad tibial metaphyses (Figs. 1307 and 1309). Lateral patella dislocation is common. Enlargement and flattening of the humeral head and glenoid enlargement are present in most patients.

The bony pelvis is short and broad, with a definite basal segment of the innominate bone. The lumbar interpediculate space measurements either increase or remain the same from L1 to L5, and do not narrow as in achondroplasia. Often a posterior tilt of the sacrum is present.[11] Scoliosis, not present at birth, is progressive and severe thereafter (Fig. 1306). Cervical kyphosis may be present in the newborn.[5, 10] Cervical spina bifida occulta is present in most patients.[15]

Diastrophic dwarfism is most often mistaken for achondroplasia and Morquio syndrome. The oval first metacarpal of diastrophic dwarfism is the most distinguishing feature. The presence of a basal segment of the innominate bone and the lack of narrowing of the lumbar interpediculate spaces should serve to distinguish diastrophic dwarfism from achondroplasia. The absence of mucopolysaccharulduria distinguishes diastrophic dwarfism from Morquio syndrome. The multiple stiff joints of arthrogryposis multiplex superficially resemble changes of diastrophic dwarfism, but are easily identified roentgenographically by decreased muscle mass, not present in diastrophic dwarfism.

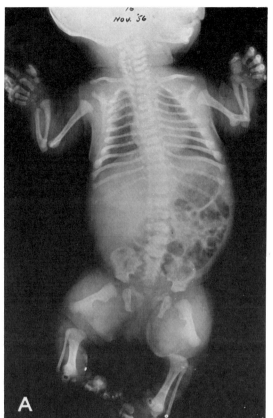
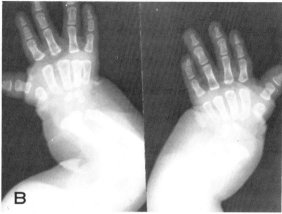
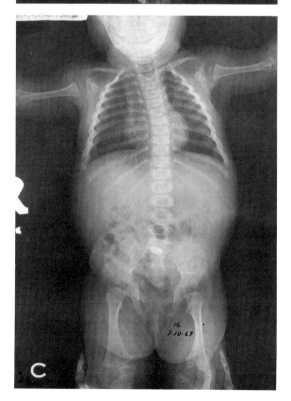

Fig. 1306. Diastrophic Dwarfism

(A) Newborn. Evidence of short limb dwarfism and marked deformity of the feet. Accelerated bone maturation is present in the carpals. (B). Hands at age 8 months. Accelerated carpal maturation. The first metacarpal is oval and hypoplastic. (C) At 8 months. Scoliosis is beginning to develop. The femoral and humeral capital epiphyses are flat. (Courtesy of Dr. Robert M. Peck, Tripler General Hospital, Hawaii.)

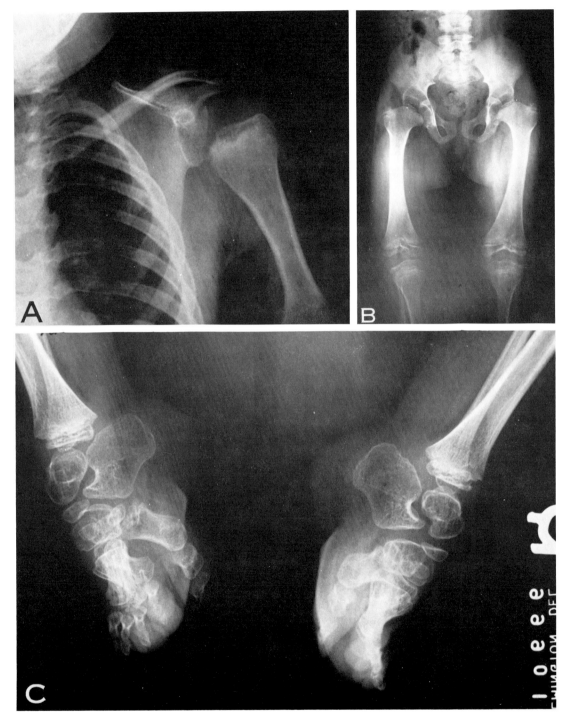

FIG. 1307. DIASTROPHIC DWARFISM
(A) Flattening of the left humeral epiphysis and dislocation of the shoulder. (B) Flattening of the femoral capital epiphyses, the distal femoral epiphyses, and proximal tibial epiphyses. (C) Talipes equinovarus deformity.

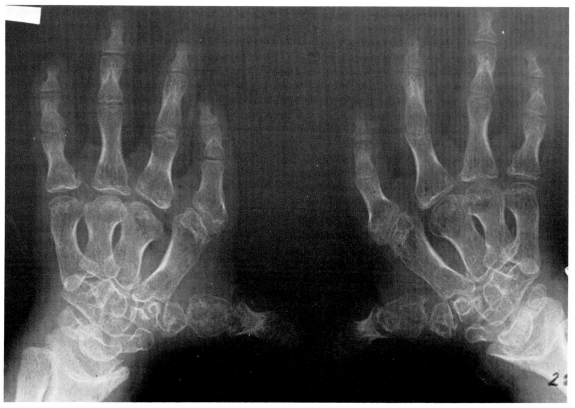

Fig. 1308. Diastrophic Dwarfism

All bones are short. The first metacarpal is short and oval, a characteristic sign of diastrophic dwarfism. Both thumbs are abducted (hitchhiker's thumbs). The metaphysis of the radius is cone-shaped. (Courtesy of Dr. Howard Steinbach, Moffitt Hospital, San Francisco, Calif.)

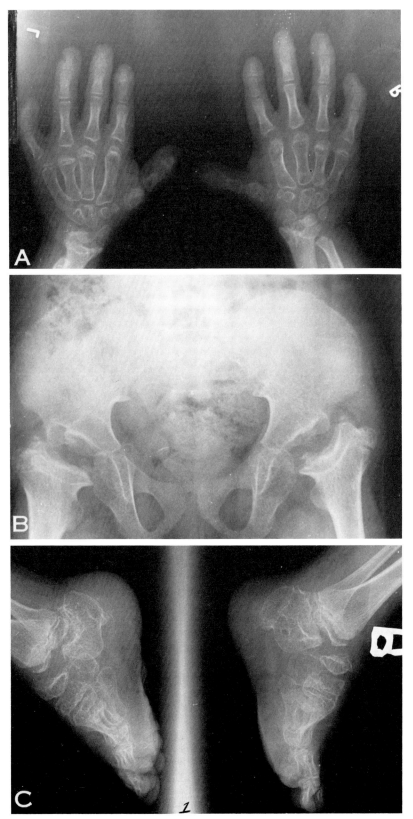

Fig. 1309. Diastrophic Dwarfism

(A) Both hands reveal hypoplastic oval metacarpals which are characteristic. Clinodactyly of both fifth fingers. All long bones are short. Cone-shaped radial metaphyses. Hypoplasia of the distal end of the left ulna. (B) The femoral capital epiphyses are flattened and irregular and there is subluxation of the left. (C) There is a talipes equinovarus deformity and flattened epiphyses. (Courtesy of Dr. Wm. Greendyke, Wilford Hall Hospital, Lackland Air Force Base, Tex.)

HYPOCHONDROPLASIA[1, 2]

This is a hereditary dysplasia transmitted by an autosomal dominant mode.[3-6] Fathers are above average age in sporadic cases.[6-8] It closely resembles achondroplasia but is a distinct type of dwarfism. The changes may not be marked and the abnormality overlooked.[3, 9]

Clinical Features. There is a short stature with relatively long trunk and disproportionate short limbs. The head is normal which distinguishes it from achondroplasia. There is usually a lumbar lordosis and sacral tilt.

Some believe that dwarfing is always mild,[10, 11] but patients are usually considerably shorter than the norm, and often as small as achondroplastics.[6] Mental retardation is frequent.[3, 6] Neurologic complications are unreported.

Roentgenographic Features. The skull is normal. The limbs are short compared to the relatively long trunk (Fig. 1310). The bones of each limb are proportionately short, unlike those in a chondroplasia, which is marked by a disproportionate shortening of the humeri compared to the forearm bones. Occasionally the styloid ulnar process and fibula are slightly longer.

There may be metaphyseal flaring. A mild V-shaped metaphyseal indentation occurs in children, especially at the distal femur. The hand bones are short but normally formed.

The pelvis is small and a lordotic sacrum is frequent (Fig. 1310). The caudal portion may be mildly shortened and the sacrosciatic notch reduced in size. The lumbar interpediculate space measurements decrease from the first to the fifth, but the narrowing is less marked than in achondroplastics, and neurologic disturbances are unreported.[6]

In early infancy and childhood hypochondroplasia must be distinguished from the following dysplasias (Fig. 1310): achondroplasia, diastrophic dwarfism, cartilage-hair hypoplasia, stippled epiphyses, spondyloepiphyseal dysplasia, metatrophic dwarfism, thanatophoric dwarfism, chondroectodermal dysplasia (Ellis-van Creveld), osteogenesis imperfecta, mesomelic dwarfism, achondrogenesis, and lymphopenic agammaglobulinemia with short limbed dwarfism. In older patients it must be distinguished from achondroplasia (Figs. 1311 and 1312),[12] spondyloepiphyseal dysplasia of the pseudoachondroplastic type, metaphyseal dysostosis, multiple epiphyseal dysplasia, hypophosphatemia, pseudohypoparathyroidism, pseudo-pseudohypoparathyroidism, and cartilage-hair hypoplasia. Roentgen features alone will usually establish the diagnosis.

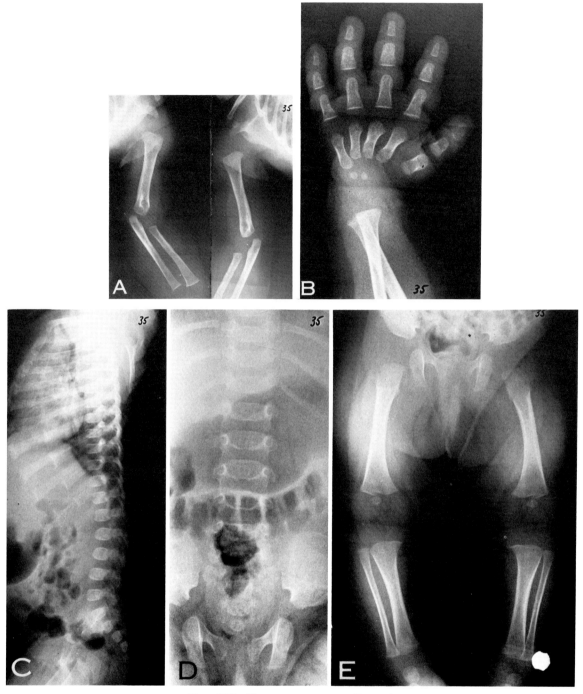

FIG. 1310. HYPOCHONDROPLASIA (INFANT)

The limbs are short compared to the relatively long trunk. The long bones are proportionately short, unlike those seen in achondroplasia in which a disproportionate shortening of the humeri occurs compared to the forearm bones. The iliac crest is flared and there is a short caudal segment which will develop normally as the child grows. The interpediculate space measurements may decrease from the first to the fifth lumbar area but narrowing is less marked than in achondroplasia. (A) Forearms, (B) hand, (C) lateral projection of the dorsolumbar spine, (D) lumbar spine and pelvis, and (E) lower extremities. (Courtesy of Dr. John Dorst, Johns Hopkins Hospital, Baltimore, Md.)

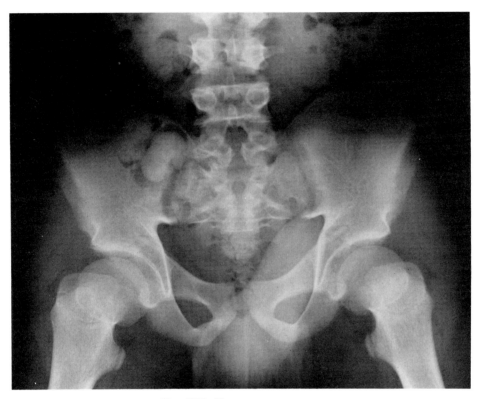

FIG. 1311. HYPOCHONDROPLASIA
Narrowing of the fifth interpediculate space. The pelvis is smaller than normal but otherwise normally formed. All long bones are short but of normal configuration. (Courtesy of Dr. John Dorst, Johns Hopkins Hospital, Baltimore, Md.)

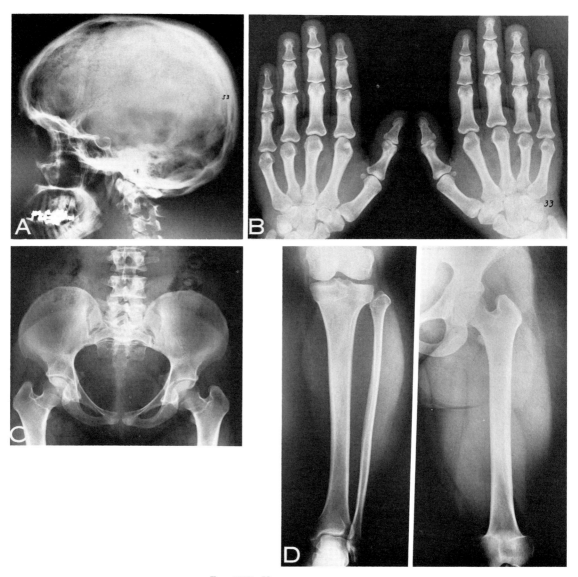

FIG. 1312. HYPOCHONDROPLASIA
All long bones are symmetrically short. The pelvis has fully developed. Minor narrowing of the fifth lumbar interpediculate space. (A) Skull. There is no frontal bossing that occurs with achondroplasia. (B) Hands. Shortening of the long bones but normal configuration. (C) Pelvis. Normal configuration with normal development of the caudal end. Slight narrowing of the fifth interpediculate space. (D) Slight symmetrical shortening of all long bones but normal configuration. (Courtesy of Dr. John Dorst, Johns Hopkins Hospital, Baltimore, Md.)

MESOMELIC DWARFISM (HYPOPLASTIC ULNA, FIBULA, MANDIBLE TYPE)

Mesomelic dwarfism is a distinct heritable bone dysplasia, characterized by shortening of all the long bones associated with hypoplasia of the ulna, fibula, and mandible.

Clinical Features. Transmission is probably by an autosomal recessive mode.[1,2] Except for dwarfism and hypoplasia of the jaw, no serious physical or mental abnormalities are reported.

Roentgenographic Features. Although all long bones are shortened, the most distinctive characteristic is hypoplasia of the fibula. The ulna is usually short and thick with a configuration suggesting, in many instances, absence of the distal end and resulting in a marked deviation of the hand. The fibula may be only half the length of the tibia, although it varies considerably in length (Figs. 1313 and 1314).

Hypoplasia of the mandible with short condyles is common. The humerus and femur are short but with normal modeling. Later, there may be prominence of the deltoid attachment in the humerus. As a child grows older, premature fusion of the medial aspect of the proximal humeral epiphysis may cause a varus deformity. The femur may develop a small exostosis on the medial aspect of the diaphysis. There is early closure of the distal femoral epiphysis.

The radius and tibia are proportionally shorter than the humerus and femur, and the latter may show a mild curvature. The proximal radial epiphysis may be absent, and premature fusion of the distal radial epiphysis may produce angulation in later childhood. The tibia may also develop angulation of the articular surface, bowing of the shaft, and mild medial angulation of the proximal epiphysis. There may be spinal anomalies, usually hypoplasia of a body (Fig. 1314).

The differential diagnosis is discussed by Langer,[2] and includes:

1. *Dyschondrosteosis*, a mesomelic form of dwarfism, is usually milder in degree than mesomelic dwarfism of the hypoplastic ulna, fibula, and mandible type. Madelung deformity of the wrist is the most striking feature of dyschondros-

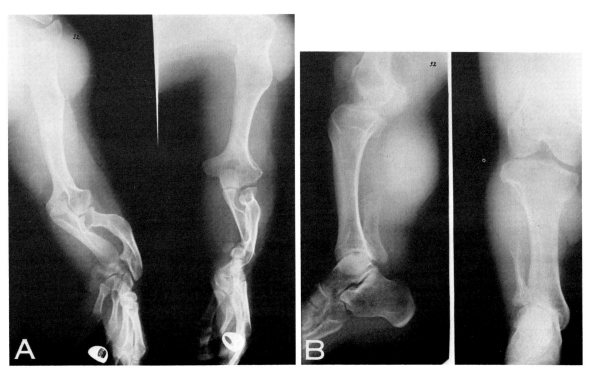

FIG. 1313. MESOMELIC DWARFISM

(A) Upper extremity. The humerus is short and there is prominence of the deltoid tubercle. The ulna is proportionately short and hypoplastic at its distal end. The radius is short and curvature accommodates the shorter ulna. Distortion of the proximal row of carpals. (B) Lower extremities. Shortening of all the bones. The tibia is proportionately shorter than the femur. The proximal end of the fibula is hypoplastic and the lateral malleolus is absent. (Courtesy of L. O. Langer, M.D.: *Radiology*, 89: 654, 1967,[2] © The Radiological Society of North America, Syracuse, N.Y.)

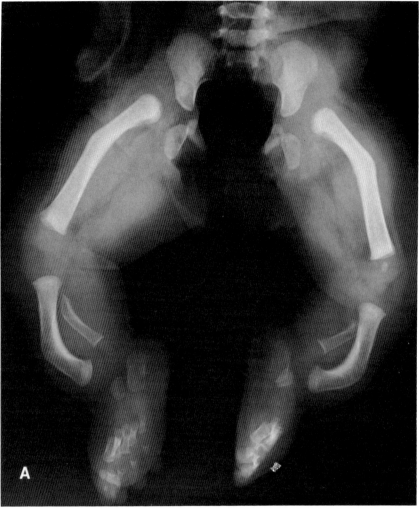

FIG. 1314. MESOMELIC DWARFISM

(A) Marked shortening of the fibulas and bowing of the tibias and femurs. (B) Only minor shortening of the ulna and the radius. (C) The centra of the seventh cervical vertebra is hypoplastic (arrow).

teosis. There is no mandibular hypoplasia and the fibula is proportioned to the tibia. It is an autosomal dominant condition.[3]

2. In a form of mesomelic dwarfism with a dominant mode of genetic transmission and characterized by a striking hypoplasia or absence of the fibula, the proximal segments of the fibula, the proximal segments of the extremities are normal in length and ligamentous relaxation, especially in the lower extremities, is a striking feature.[4]

3. *Nievergelt syndrome* is an autosomal dominant condition with marked shortening of the middle segments in addition to rhomboid-shaped tibia and fibula, and elbow and hind-foot deformity.[5,6] Radial ulnar synostosis is present in the majority of cases.

4. *The Ellis-van Creveld syndrome* (chondroectodermal dysplasia), exhibits a relative shortening of both the middle and distal segments. Polydactyly of the hands, characteristic knee, shoulder, and pelvic changes, and dysplasia of the teeth and nails will be noted here. Congenital heart disease is common. There is an autosomal recessive mode of genetic transmission.[7]

5. Although there is little to confuse *achondroplasia* with mesomelic dwarfism, there is a tendency to use achondroplasia as a wastebasket for all short-limb bone dysplasias. In achondroplasia, there is shortening of the proximal segment

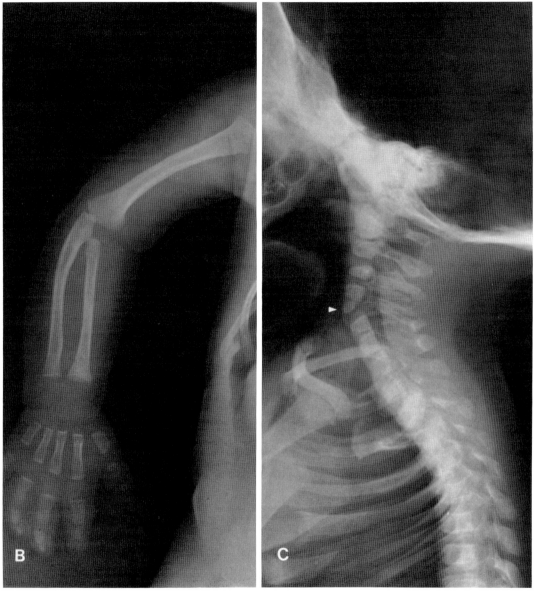

FIG. 1314 (*B* and *C*)

of the upper extremities, and characteristic changes in the lumbar spine, pelvis, and skull distinguish it from other short limb conditions.[8, 9]

Perhaps, in the future, other mesomelic forms of dwarfism will be found to be similar to the mesomelic dwarfism of the hypoplastic ulna, fibula and mandible type, but at present this dwarfism must be classified separately.

ACHONDROGENESIS

There are two unrelated dysplasias termed achondrogenesis: Type I (lethal type or Patenti-Fraccaro type)[1-3] and Type II (Brazilian or Grebe type)[1, 4-6]

Roentgenographic Features. TYPE I. The child is usually stillborn or dies soon after birth. Deficient vertebral ossification is most marked in the lumbar area, with absence of ossification

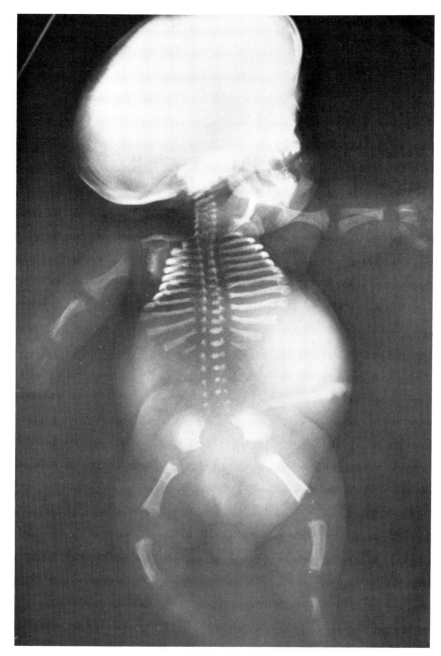

Fig. 1315. Achondrogenesis, Type I (Parenti-Fraccaro or Lethal Type)

A stillborn infant with deficient ossification of the lumbar vertebrae and absent ossification of the sacral, pubic, and ischial bones. The proximal and distal metaphyses of the humeri are flared. (Courtesy of L. O. Langer, M.D.: Mesomelic dwarfism of the hypoplastic ulnar, fibular and mandibular type. *Radiology, 89:* 654, 1967, © The Radiologic Society of North America, Syracuse, N.Y.)

of the sacrum, pubic, and ischial bones (Fig. 1315). The long bones are shortened, and metaphyseal broadening occurs in both ends of the humerus.[2, 7] Membranous bones are involved, and the skull is affected, which would serve to distinguish it from Type II.

TYPE II. This type is not lethal and usually occurs in children from consanguinous parents. The shortness occurs in the distal extremities. The humeri and femora are only slightly shortened, and the bones of the legs and forearms are markedly shortened and bowed. The distal or middle segments of the ulna and fibula may be absent. The carpals, tarsals, metacarpals, metatarsals, and proximal phalanges are not ossified at birth and ossify late. Polydactyly of the hands and feet is common. Membranous bones are not involved, nor is the skull, serving to distinguish it from Type I.

ELLIS-VAN CREVELD SYNDROME (CHONDROECTODERMAL DYSPLASIA)

McIntosh,[1] in 1933, described a patient with chondrodysplasia and ectodermal dysplasia. Ellis and van Creveld,[2] in 1940, added 2 patients who also manifested congenital heart disease and nail changes, and named the syndrome chondroectodermal dysplasia. They suggested that the disease was transmitted by an autosomal recessive mode, which was subsequently confirmed by Metrakos and Fraser.[3]

Clinical Features. Chondrodysplasia and polydactyly of the hands are the most constant findings. Polydactyly of the feet occurs in 20% of the patients.[4-6] The majority show ectodermal dysplasia affecting both teeth and nails. The nails may be absent, or small, dystrophic, brittle, and spoon-shaped. The teeth are irregular and pointed. Premature eruption of the teeth may occur, and this is one of the few conditions in which teeth may even be present at birth.[7] A distinctive neonatal finding is obliteration of the maxillary mucobuccal space by thick frenula between the alveolar mucosa and the buccal mucosa of the upper lip.[2, 4-6] As the infant matures, the maxilla increases in height and the heavy frenula become thin and gradually recede. The mucobuccal space reforms by 3 years of age.[2]

Over one-half of the patients exhibit congenital heart disease, usually an interatrial or interventricular septal defect.[8-11] Bilocular and trilocular hearts,[5] including total anomalous venous return, are reported.[12-14] Other less common anomalies include: malformation of the genitalia, undescended testes, cleft palate, cleft lip, strabismus, coloboma of the iris, and hepatosplenomegaly.[5]

The life expectancy depends largely on the severity of the congenital heart disease, but death may also result from respiratory failure due to absence of the bronchial cartilage or severe funnel chest deformity.[15]

Roentgenographic Features. The most constant feature is a shortening of all long bones, typically more severe in the distal segments of the extremities (Figs. 1316 and 1317). There may be enlargement of the distal ends of the humeri, femurs, and radii, and the proximal end of the tibia. There may be dislocation of the radial head due to disproportionate shortening of the radius and ulna.

The tibia is usually markedly shortened (50% of normal) and the proximal end is widened and pointed (Fig. 1317).[11] Exostoses tend to form on the tibia.[2] The tibial epiphyses are often hypoplastic and medially displaced. The phalanges may be shortened and the terminal phalangeal epiphyses may be absent. Frequently, there is fusion of two or more carpal and tarsal bones, particularly the hamate and capitate.[6] The spine and calvaria are normal.

Polydactyly is usually manifested by a sixth digit on the ulnar[5] or the radial[4] side of the hand. The supernumerary digits may be partially or completely fused (Fig. 1317). The pelvis shows shortening of the caudal segment which may eventually return to normal. The acetabulum often has an indentation in the roof, forming a "trident" shape considered almost pathognomonic (Fig. 1317). Later, this also may develop normally. Exostoses tend to form on the acetabula.[2] Wormian bones of the skull may occur (Fig. 1316). The most commonly confused condition is achondroplasia, but this is marked by a symmetrical shortening of all bones most pronounced in the proximal, rather than the distal, segments. Absence of polydactyly and abnormal skull formation point to achondroplasia. The short caudal segment of the pelvis may suggest it, but a trident acetabulum and the eventual normalization of the pelvis with maturity indicate chondroectodermal dysplasia.

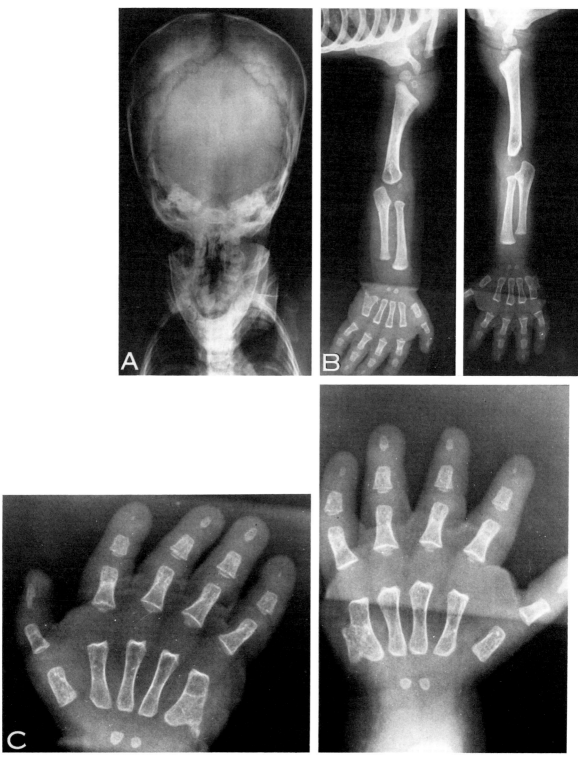

Fig. 1316. Ellis-van Creveld Syndrome (Chondroectodermal Dysplasia)

(*A*) Skull. Wormian bones are present. (*B*) Upper extremity. Shortening of the long bones and proportional shortening of the bones of the forearm. The distal phalanges are hypoplastic. A fused supernumerary fifth metacarpal. (*C*) Hands. Shortening of the bones and greater shortening of the distal phalanges. A supernumerary fused fifth metacarpal bilaterally. (Courtesy of Dr. Patricia Borns, Children's Hospital, Philadelphia, Pa.)

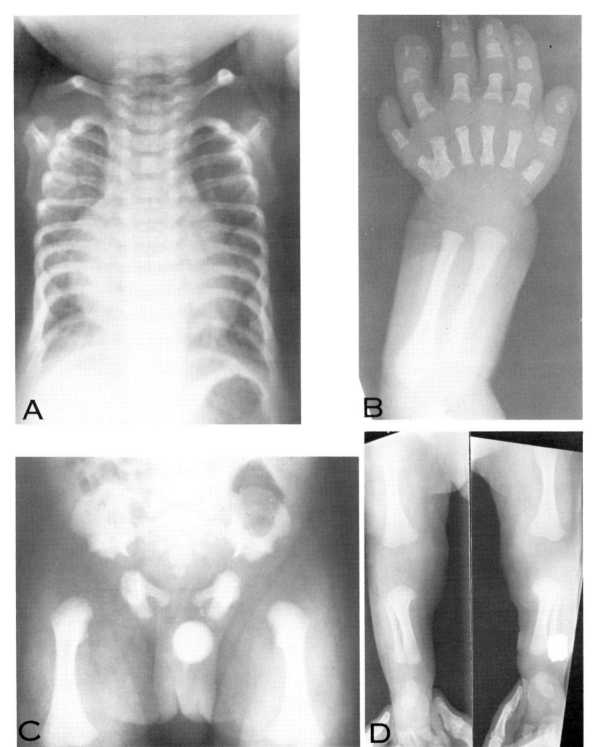

FIG. 1317. ELLIS-VAN CREVELD SYNDROME (CHONDROECTODERMAL DYSPLASIA)
(A) Chest reveals horizontal ribs, elevated clavicles and an enlarged heart. The thoracic cage deformity simulates familial thoracic dystrophy. Over one-half of the patients exhibit congenital heart disease. (B) The hand and forearm reveal shortening of the long bones. Polydactyly and syndactyly of the metacarpals. Marked hypoplasia of the terminal phalanges. (C) Pelvis. There is shortening of the caudal segment which will later develop normally. A trident acetabular shape with medial and lateral prominences or spurs. (D) The lower extremities show a disproportionate shortening of the tibia and fibula compared to the short femurs. (Courtesy of Dr. Robert M. Peck, Atlantic City, N.J.)

CHONDRODYSTROPHIA CALCIFICANS CONGENITA[1]
(STIPPLED EPIPHYSES)

Synonyms: stippled epiphyses, dysplasia epiphysialis punctata,[2] chondrodystrophia fetalis calcificans, epiphyseal dysplasia puncticularis,[3] chondrodystrophia fetalis hypoplastica.[4]

This rare condition is characterized by multiple punctate opacities present at birth or shortly thereafter. It was first described by Conradi[4] in 1914, who considered it to be chondrodystrophia fetalis.

Clinical Features. It is transmitted by an autosomal recessive mode.[5-7] There is a female preponderance.[8] The condition begins in utero, and may be apparent in the newborn; most cases are discovered in the 1st year of life, and rarely after 2 years of age. The infants are weak and subject to frequent infections and, although death rate from infection within the first 2 years is high, the prognosis is not universally poor.[9]

One or more limbs are often dwarfed but rarely all limbs symmetrically. Severest shortening occurs in the humeri and femora.[8-10] The condition may regress completely without deforming sequela, and no dwarfism results.

Chondrodystrophic calcificans congenita may be divided into a benign, nonrhizomelic group and a potentially lethal, recessively inherited rhizomelic group (Table 36).[11]

The nonrhizomelic group has a good prognosis with little residual deformity, but some are lethal, especially when there is excessive stippling.[12] This group may represent several distinct genetic disorders from a mild type with no residual deformity[13] to a type lethal in males and transmitted as an X-linked dominant mode.[14] The rhizomelic type is almost invariably lethal, usually through respiratory failure[15] and occasionally through tracheal stenosis[16] or spinal cord compression.[17] This group may live as long as 7 years.[12]

Flexion contractures occur in the hips, knees, and elbows. Other associated anomalies include clubfoot, congenital dislocation of the hips, saddle nose, microcephaly, oxycephaly, mental deficiency, cleft palate, cardiac defect, and cataracts.

It may occur in babies whose mothers have received antivitamin K drugs.[18]

Pathologic Features. Harris[19] described a mucoid degeneration in the epiphyseal center near the articular surface; epiphyseal hypervascularity is also noted. The mucoid degeneration apparently leads to fragmentation of the epiphysis, and the fragments become foci for calcareous cartilaginous deposits and punctate ossification.[20]

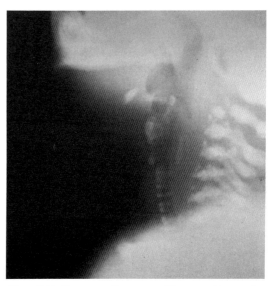

FIG. 1318. STIPPLED EPIPHYSES (CONRADI DISEASE)
Laryngeal cartilage calcification, with irregular calcification of the vertebral epiphyses.

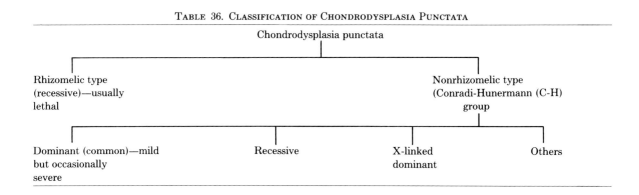

TABLE 36. CLASSIFICATION OF CHONDRODYSPLASIA PUNCTATA

Chondrodysplasia punctata

- Rhizomelic type (recessive)—usually lethal
- Nonrhizomelic type (Conradi-Hunermann (C-H) group)
  - Dominant (common)—mild but occasionally severe
  - Recessive
  - X-linked dominant
  - Others

Roentgenographic Features. The condition is characterized by multiple small punctate calcifications of varying size, occurring in the respiratory cartilages (Fig. 1318) and epiphyses before the normal time for appearance of ossification centers. In severe cases, the calcifications extend beyond the joints and into the soft tissues and ligaments. The respiratory cartilage calcification may also extend into the soft tissues (Fig. 1319). The densities are discrete in the early changes and may either disappear by age 3[1, 21, 22] or gradually increase in size and coalesce to form a normal-appearing single ossification center. The knee, hip, shoulder, and wrist are involved

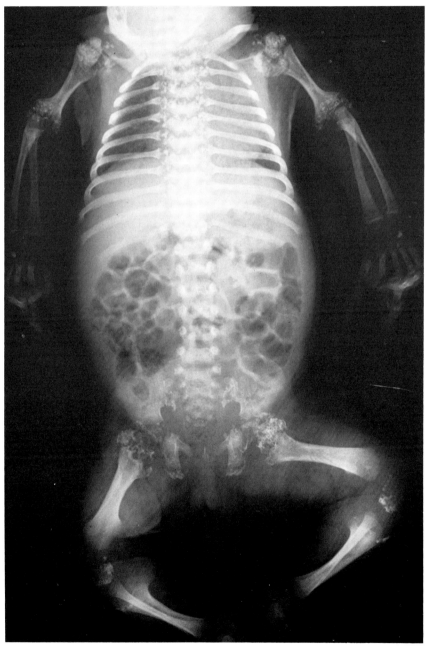

FIG. 1319. CHONDRODYSTROPHIA CALCIFICANS CONGENITA (RHIZOMELIC TYPE)
This condition is characterized by multiple small varying sized punctate densities occurring in the epiphyses of the long bones. There is stippling at the costovertebral junctions. There is rhizomelic shortening of the upper limbs and the lower limbs are not as severely affected. (Courtesy of John W. Hope, Children's Hospital, Philadelphia, Pa.)

most frequently. Other epiphyses are affected less often, but any combination of epiphyses may show stippling. Stippling also has been reported in the ends of the ribs, the thyroid cartilage, the hyoid,[8] and the cartilage at the base of the skull.[21] The ossification centers may appear earlier than expected, even though no stippling occurs.

The nonrhizomelic type has mild to severe epiphyseal and para-epiphyseal stippling with occasional asymmetric shortening of a long bone. Vertebral and paravertebral calcifications are usually present and may be severe. Vertebral body anomalies may occur and scoliosis can develop. Laryngeal and tracheal cartilage calcifications are frequently present (Fig. 1320).

The rhizomelic type has symmetrical shortening of the upper limbs; the lower limbs may be unaffected (Fig. 1319). The metaphyses are splayed and surrounded by stippling of variable degrees. Vertebral and paravertebral stippling is usually absent or mild; coronal cleft vertebral bodies are always present. Laryngeal and tracheal cartilage calcifications are rarely seen (Table 37).[11]

Although the epiphyses may gradually develop a normal appearance, one or more limbs may be permanently shortened (Fig. 1320). Also other bones, without evident stippling, may later develop chondrodystrophic changes.[7]

The characteristic radiographic appearance in infants of small punctate calcifications of varying size in one or more epiphyses and adjacent soft tissues establishes the diagnosis.[23] In the stippling stage, the distinctive appearance is characteristic and should be confused with no other condition. In the later stages, shortened limb

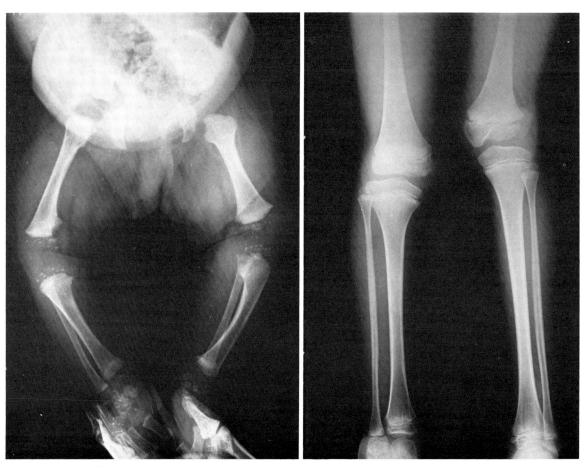

FIG. 1320. CHONDRODYSTROPHIA CALCIFICANS CONGENITA (NONRHIZOMELIC TYPE)
(*Left*) Stippled epiphyses characterize the condition. There is asymmetric shortening of the left femur. (*Right*) Small child, 6 years later. The epiphyses have lost the stippling and most appear normal. There is shortening of the left femur and the right tibia and fibula. This should differentiate the condition from achondroplasia in which dwarfing is symmetrical and the rhizomelic type which usually does not live after the first year. (Courtesy of Dr. John W. Hope, Children's Hospital, Philadelphia, Pa.)

TABLE 37. DIFFERENTIATING FEATURES BETWEEN RHIZOMELIC AND USUAL CONRADI-HUNERMANN (C-H) TYPES OF CHONDRODYSPLASIA[a]

Site	Rhizomelic Recessive Type	Usual Dominant C-H Type
Appendicular skeleton	Symmetrical rhizomelic shortening of upper limb with or without lower limb involvement. Metaphyses grossly splayed and surrounded by stippling coarse of variable degree	Epiphyseal and para-physeal stippling (mild to severe). Asymmetrical shortening of bone may occur
Axial skeleton	Stippling usually absent or mild. Coronal cleft vertebral bodies invariably present	Vertebral and paravertebral stippling usually present and may be severe. Vertebral body anomalies may be present. Scoliosis develops later (variable)
Laryngeal and tracheal cartilage calcification	Not usually present	Frequently present

[a] Modified from B. G. Heselson et al.: *Clinical Radiology, 29:* 684, 1978.[11]

may be differentiated from that of achondroplasia by the lack of symmetry. In adults, residual epiphyseal abnormalities may be indistinguishable from the deformities of multiple epiphyseal dysplasia[24] (Fig. 1320). Cretins may show epiphyseal fragmentations, but they are much larger, and conform to the expected shape of the epiphyses.[25]

HOLT-ORAM SYNDROME

The Holt-Oram syndrome consists of malformations of the hands, atrial septal defects, and cardiac arrhythmia.[1] Considerable variations of the skeletal and cardiac malformations occur.[2-6]

Clinical Features. It is transmitted by an autosomal dominant mode.[2-6] Congenital heart disease occurs in almost all patients. There are usually combinations of atrial and ventricular septal defects, persistent left superior vena cava, and malformations of the great vessels.[7] Bizarre intermittent cardiac arrhythmias are common and brachycardia (50–60/min) is frequent.

Roentgenographic Features. The changes vary from minor, symmetrical, subtle deformity of the hands, upper extremities, and shoulders to severe malformations and hyperplasia of the entire appendicular skeleton.[7] The lower extremities are uninvolved. The thumbs may lie in the same plane as the fingers and the phalanges curve inward; a rudimentary middle thumb phalanx may be present (Fig. 1321). There are wide varieties of some abnormalities from a normal thumb to a "fingerized" hypoplastic thumb or absent thumb, a triphalangeal thumb, a long clumsy-appearing thumb, and a normal thumb (Figs. 1321 and 1322). Often associated is an atrial septal defect although the thumb abnormality may be present without the cardiac defect.[8]

The carpals may be slender and the first and second metacarpals elongated or hypoplastic. In fully expressed cases the clavicles are short, and some degree of hypoplasia may be found in all the long bones. The glenoid fossa is often shallow. Voluntary dislocation of the shoulder is common.[7] The radiocarpal and radioulnar joints may be irregular. Other skeletal lesions are high arch palate, cervical scoliosis, and pectus excavatum.[1,7] Symmetry of skeletal lesions is the rule, but, in severe cases, the left side may be more severely affected. Unilateral involvement is not reported.

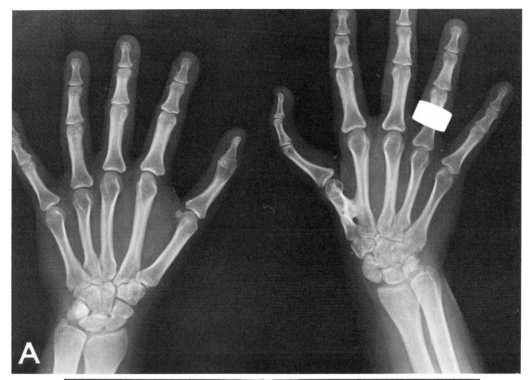
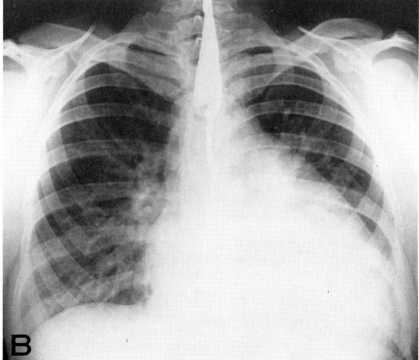

FIG. 1321. HOLT-ORAM SYNDROME

(A) Hands. A triphalangeal thumb and a synostosis between the right first and second metacarpals. The radial styloid is undeveloped and the articular surface of the radius is flattened. In the left hand, the interphalangeal joint is abnormal. (B) Large heart due to an intra-auricular septal defect. (Courtesy of Dr. Stanley Siegelman, Montefiore Hospital, New York, N.Y.)

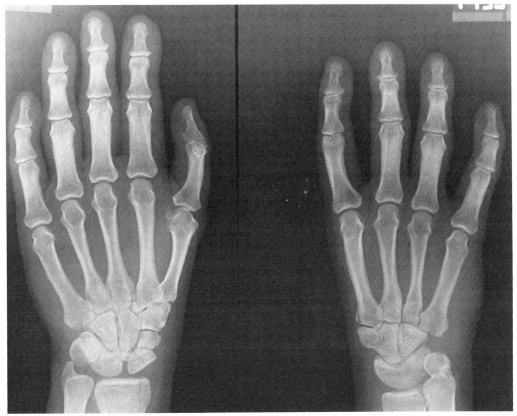

FIG. 1322. HOLT-ORAM SYNDROME
The thumb and related carpals are absent. The radial styloid is absent bilaterally. The right navicular is large.

METAPHYSEAL DYSOSTOSIS

At least five types of metaphyseal growth disturbances are recognized[1]: (1) Jansen type, (2) Schmid type, (3) Spahr type, (4) cartilage-hair hypoplasia, and (5) Pyle type.

McKusick[1] suggests that there are three types of metaphyseal dysostosis: (1) the rare Jansen type, with severe changes and no familial pattern; (2) the more common Schmid type, milder and with an autosomal dominant mode[2, 3]; and (3) the Spahr type,[4] with bone bowing, minor metaphyseal irregularities and an autosomal recessive mode.

Although most authorities agree that the Schmid type is a milder form of the Jansen type, clinical separation is justified until hereditary patterns are clearly elucidated.

Metaphyseal Dysostosis of Jansen[5]

This is a rare severe dysplasia due to excessive, proliferating hypertrophic cartilage in all metaphyses.

Clinical Features. There is no familial history in the few cases reported in the literature. The patients are severely dwarfed, and abnormalities may be detected at birth or soon thereafter. Mental and physical retardation is notable.

The facies are immature. Hypertelorism and exophthalmos are characteristic. Dwarfism is marked, especially in the lower extremities. Flexion contractures of the hips and knees cause a squat posture. The hands are clubbed and reach to the ankles in the erect position, producing a "monkeylike" stance. Muscles are atrophic and underdeveloped. The chest is narrowed superiorly and the ribs have a rosarylike appearance at the costochondral junctions.

Gram et al.[6] report hypercalcemia, hypophosphatemia, and elevated alkaline phosphatase. Cameron et al.[7] also found a high alkaline phosphatase but normal calcium and phosphorus in their cases. The life expectancy is unknown, but Holmann et al.[8] reported a patient 26 years old.

Roentgenographic Features. Early in infancy there is a cupping of the metaphyseal area of the long bones with irregularity of the margin of the metaphyses. The deformities progress until expansion of the metaphyseal areas appears, containing irregular stippled calcifications and radiolucencies due to calcified and uncalcified hypertrophied cartilage (Figs. 1323 and 1324). The cortex is intact. There is cupping of the metaphysis, with a notched metaphyseal center. The epiphyses are intact; the central portion of the epiphysis may protrude into the central notching of the metaphyses. The diaphyses are normal, except for a tendency toward bowing of the lower extremities. Cartilage formation may extend into the tibial diaphysis. The most marked changes are in both femoral and tibial ends, the proximal humerus and distal radius. The short tubular bones are similarly but less severely affected.

The pelvis is hypoplastic. There are large radiolucent areas intermingled with stippled calcified cartilage. The anterior rib ends present the same appearance as the long bone metaphyses. The skull is normal except for underdeveloped sinuses.

In the early stages, the differential diagnosis should consider renal osteodystrophy[9] and hypophosphatasia, but, as the condition progresses, the characteristic roentgen features become unmistakable.

Metaphyseal Dysostosis of Schmid

Metaphyseal dysostosis of Schmid is a milder form. The skeletal alterations are not present at birth, appearing soon after weight-bearing age, most often in the 3rd and 4th year.[10] There is an autosomal dominant mode.[2, 3] Dwarfism is mild. There is a bowing of the legs and a "waddling" gait.

Roentgenographic Features. After weight-bearing age, irregularity is noticeable in the first 0.5 cm of the metaphysis (Fig. 1325). There is a milder cupping of the metaphysis, and areas of radiolucency appear, intermingled with small dense areas of calcification. Diaphyseal and epiphyseal ossification is normal. Moderate bowing and minimal dwarfing occurs in the long bones of the lower extremities. Minimal dwarfing of all long bones appears in early childhood. Coxa vara and slipped epiphyses are common. The metaphyseal irregularity disappears with epiphyseal fusion.

Patients with vitamin D resistant rickets may present a similar appearance, but abnormal kidney function and response to vitamin D should aid in distinguishing the two. Patients with cartilage-hair hypoplasia may also show irregular metaphyses with bowing and dwarfism, but the fine, short, sparse, and brittle hair which always accompanies this syndrome distinguishes it from metaphyseal dysostosis.

Metaphyseal Dysostosis of Spahr

Metaphyseal dysostosis of Spahr differs from Schmid type by the autosomal recessive mode.[4] Roentgenographically, there is marked bowing of the legs with slight metaphyseal irregularities.

Cartilage-Hair Hypoplasia

In 1965, McKusick[1] described a type of dwarfism in the Amish of Lancaster County, Pennsylvania, which is distinguished from Ellis-van Creveld type and achondroplasia. It is a form of metaphyseal dysostosis with associated ectodermal abnormalities.

Clinical Features. It is probably transmitted by an autosomal recessive mode.[1] It is a short limbed variety, superficially resembling achondroplasia, but the head is normal. The hands and feet are short, pudgy, and loose jointed. The fingernails and toenails are short.

Inability to fully extend the elbows is frequent. Other less consistent features are rib beading, sternal deformity (pigeon breast), leg bowing, and ankle deformity. The hair is sparse, fine, and silky and breaks easily. The hair of the eyelids and eyebrows may also be sparse. Early baldness is a frequent feature. The hair is lighter in color than that of unaffected siblings. Congenital megacolon may occur and McKusick[1] believes susceptibility to viral infection is a feature of the syndrome.

Roentgenographic Features. The roentgen features are similar to metaphyseal dysostosis of Schmid. Metaphyseal scalloping, irregularity, sclerosis, and cystic changes occur in the tubular bones and the ribs. The entire metaphyseal width is involved, unlike other metaphyseal dysostoses that may involve a small portion, predominantly the medial aspect of the knees.

The epiphysis is normal and when epiphyseal closure occurs the characteristic features are lost. Ankle deformity is due to a short tibia and a longer fibula. Bone maturation is irregularly delayed. The carpal development is decidedly delayed, whereas the phalangeal epiphyses close early partially contributing to the brachydactyly.

The skull and spine are normal which distinguishes the condition from achondroplasia. Occasionally there is an increased height in the vertebral bodies (columnization). The pelvic outlet is compressed in the anteroposterior diameter, but the sacrosciatic notch is normal and

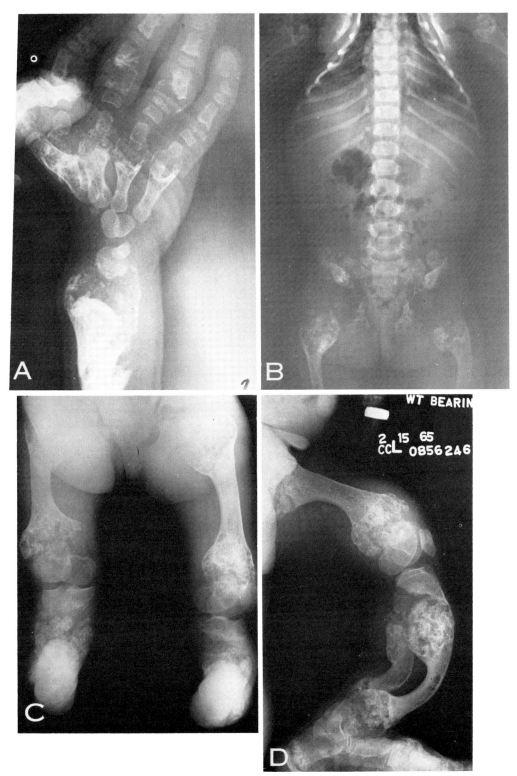

FIG. 1323. METAPHYSEAL DYSOSTOSIS OF JANSEN

Marked deformity of the metaphyseal area with expansion and stippled calcifications and radiolucencies due to calcified and uncalcified hypertrophied cartilage. Tibial and fibula bowing. The upper thoracic cage is constricted and the lower is flared. The pelvis is hypoplastic. (A) The hand. (B) Thorax, abdomen, and pelvis. (C) Anteroposterior projection of both lower extremities. (D) Lateral projection of left lower extremity. (Courtesy of Dr. Howard Steinbach, Moffitt Hospital, San Francisco, Calif.)

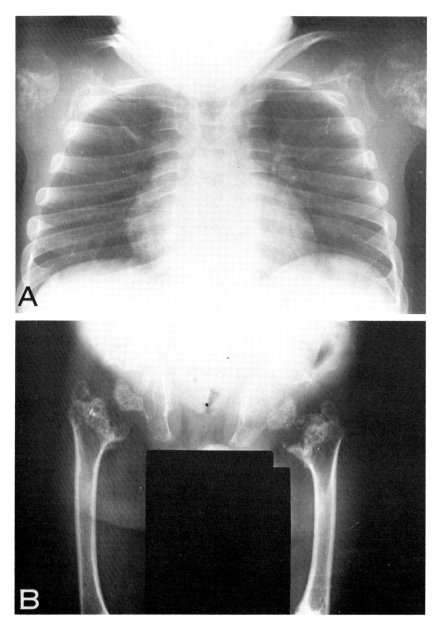

Fig. 1324. Metaphyseal Dysostosis of Jansen

A milder case than the preceding figure. The abnormalities are limited to the metaphyseal regions where there is irregular calcification and radiolucency. (*A*) The chest. The ribs are horizontal and there are calcifications in both humeral metaphyses. (*B*) Pelvis. Marked deformity of both femoral necks with irregular calcifications and ossification. Irregularity of the epiphyses is secondary to the metaphyseal change. (*C*) The hand. Irregular calcifications in the metaphyses of the radius and ulna. Shortening of the ulna and bowing of the radius. The tubular bones of the hands are small but the metaphyseal areas are relatively normal. (*D*) Lateral lumbar spine reveals calcifications in the end plates of the lumbar vertebrae. (*E*) Lower extremities. Irregular calcifications in all the metaphyseal areas of the long bones. The epiphyses are normal. The contour of the bone is normal. (Courtesy of Dr. Robert M. Peck, Tripler General Hospital, Hawaii.)

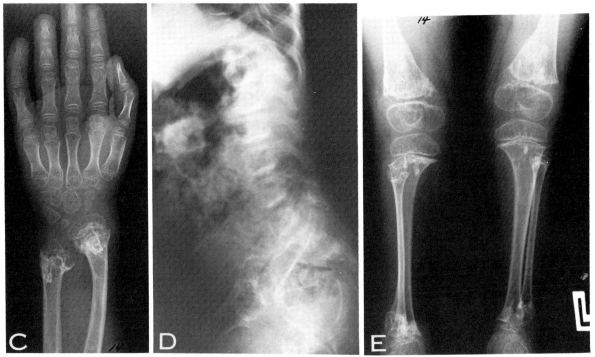

Fig. 1324 (C-E)

there is no shortening of the caudal segments as in achondroplasia.

Lymphopenic Agammaglobulinemia

Lymphopenic agammaglobulinemia may be associated with a dwarfism similar to cartilage-hair hypoplasia.[1-6] It is transmitted by an autosomal recessive mode.[1] Ectodermal abnormalities are variable and consist of absence of hair and eyebrows, icthyosiform skin lesions, erythroderma and cutis laxa.

Roentgenographic Features. The roentgenographic features are different than in achondroplasia. There is a short limbed dwarfism similar to achondroplasia; however, the pelvis is short but the iliac crest flares, unlike achondroplasia. Also the sacrosciatic notches are larger than in achondroplasia. The interpediculate spaces may be narrowed, but not as severely as in achondroplasia. The frontal bones are prominent but the nasal bridge is not depressed. The bones of the hands are short but not trident as in achondroplasia.

Other short limbed dwarf conditions manifested in the newborn period which clinically might be confused with the cases under discussion would include: hypochondroplasia,[7] diastrophic dwarfism,[8] chondrodystrophia calcificans congenita (stippled epiphyses),[9] chondroectodermal dysplasia (Ellis-van Creveld syndrome),[10] thoracic pelvic-phalangeal dystrophy (asphyxiating thoraci dystrophy),[11] mesomelic dwarfism of the hypoplastic ulna-fibula-mandible type,[12] cartilage-hair hypoplasia,[13] spondyloepiphyseal dysplasia congenita,[14] thanatophoric dwarfism,[15] achondrogenesis,[15] metatrophic dwarfism,[16] and osteogenesis imperfecta.[17]

Pyle Disease

Synonyms: familial metaphyseal dysplasia, metaphyseal dysplasia.

Pyle disease is a rare congenital and familial condition, characterized by symmetrical splaying of the ends of the long bones with normal midshafts. Pyle[1] described the first case in 1931; Hermel et al.[2] and Bakwin and Krida[3] described it in siblings.

Clinical Features. There are few clinical complaints. The patients are somewhat taller than average for their age, and often exhibit genu valgum. The enlargement of the tubular bones can be detected by palpation, but there is no pain or tenderness.

Roentgenographic Features. The long bones exhibit a splaying or spreading of the ends caused by lack of modeling. This widening is

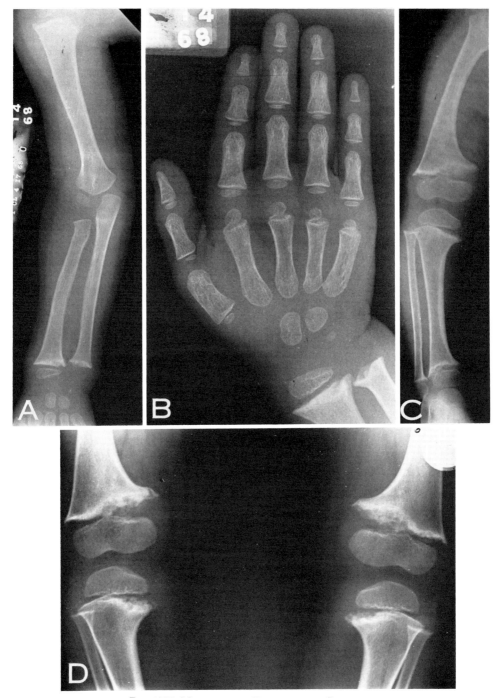

Fig. 1325. Metaphyseal Dysostosis of Schmid

Metaphyseal irregularities are present in the first 0.5 cm of the metaphyses. The epiphyses are normal. A milder form of metaphyseal dysostosis than in the previous two figures. The irregularities tend to disappear with epiphyseal fusion (A) Upper extremity, (B) hand, (C) lower extremity, and (D) knees. (Courtesy of Dr. Howard Steinbach, Moffitt Hospital, San Francisco, Calif.)

most striking at the distal end of the femur, radius, and ulna, and the proximal end of the humerus. Both ends of the tibia are equally affected. The cortex within the splaying is thin; elsewhere the bone is of normal diameter and normal cortical thickening (Figs. 1326 and 1327).

The hands may be involved most markedly in the proximal phalanges and distal metacarpals.

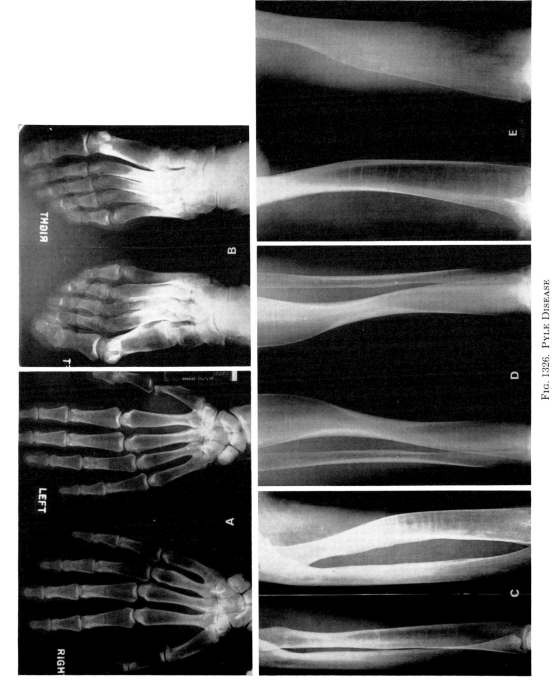

FIG. 1326. PYLE DISEASE.

Long bones exhibit splaying of the ends due to lack of modeling. The widening is most striking at the distal end of the femora, radii, and ulnae. The hands are also involved and show manifestations in the proximal phalanges and distal metacarpals. (A) Hands, (B) feet, (C) forearm, (D) legs, and (E) femurs. (Courtesy of M. B. Hermel, M.D.)

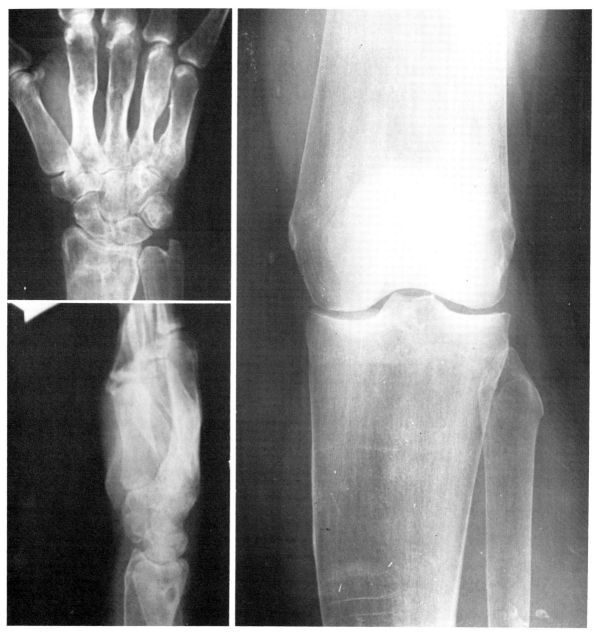

Fig. 1327. Pyle Disease

This is the sister of the patient seen in Figure 1326. The changes are similar and reveal the lack of modeling at the distal ends of the bones, particularly marked in the metacarpals, ulnae, and radii. The end of the femur, tibia, and fibula also show lack of modeling. (*Left*) Right hand, (*right*) right knee. (Courtesy of M. B. Hermel, M.D.)

Also recorded is splaying of the proximal ends of the clavicles, sternal ends of the ribs,[3] the mandible, rami of the ischium, and pubis. Hermel et al.[2] report retarded pneumatization of the mastoid and paranasal sinuses. Mori and Holt[4] believe that symmetrical hyperostosis of the calvaria and mandible, and hypertelorism are frequent. The vertebrae may be universally flattened, with increased density of the centra.[5]

DYSCHONDROSTEOSIS
(MADELUNG DEFORMITY)

Dyschondrosteosis is a hereditary mesomelic bone dysplasia characterized by bilateral Madelung wrist deformity.[1-6]

Clinical Features. It is transmitted by an autosomal dominant mode. There appears to be a 4:1 female to male predominance, but documented cases of males are rare.[6] Minor deformities of the male relatives of affected females are mentioned by several authors. Clinically, there is shortening of the forearms and legs in relation to the arms, thighs, hands, and feet (mesomelia). There is dorsal dislocation of the distal ulna, which is easily reducible but can not be maintained (Madelung deformity), and limited motion of the elbow and wrist.

Roentgenographic Features. Bilateral Madelung deformity is characteristic of this disease. In recognizing this condition, the fundamental criteria[6,7] are shortening of the radius; triangularization of the distal radial epiphysis, with the apex pointed medially; and a change in the inferior radial articular surface, with the carpal bones wedged between the deformed radius and protruding ulna, forming a triangle with the lunate at the apex (Fig. 1328).

Other changes of Madelung deformity may be present but, in mild cases, obscured by failure to obtain a true lateral view of the wrist[6] (Fig. 1329). These include the epiphysis of the distal radius faced ulnarly and palmarly; dislocation or subluxation of the distal radial ulnar articulation (Fig. 1330), with the distal ulna dorsal to the distal radius; and an arched curvature of the carpal bones that continues the arch of the dorsal bone of the radial diaphysis.

Some changes may only be noticed in the forearm roentgenogram when the hand is in pronation.[6] These are a lateral and dorsal bowing of the radius, most marked at the distal end; widened interosseous space due to lateral curvature of the radius; and, in the early adolescent, premature fusion in the ulnar half of the epiphyseal line of the distal radius.

Other components are less consistent. There may be an area of decreased bone density on the ulnar border of the radius, which extends for a short distance proximal to the fused epiphyseal line (Fig. 1330). Small bony excrescences may be noticed on the inferior ulnar border of the radius, and hypercondensation or trabeculations of the ulnar head may appear.

Patients with severe dyschondrosteosis may develop changes unrelated to Madelung deformity. The long bones may be thickened,[8] and coxa

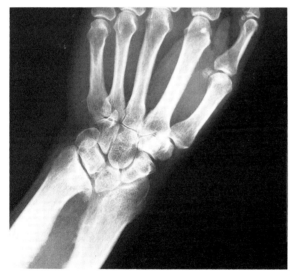

FIG. 1328. DYSCHONDROSTEOSIS (MADELUNG DEFORMITY)
Shortening of the radius, triangularization of the distal radial epiphysis with the apex pointed medially, angulation of the radial articular surface with the carpals wedged between deformed radius and the protruding ulna, and a defect on the medial side of the radius.

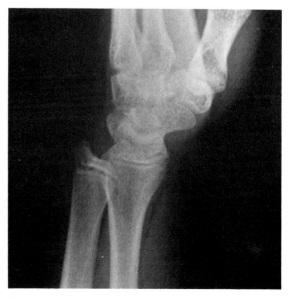

FIG. 1329. DYSCHONDROSTEOSIS (MADELUNG DEFORMITY)
A lateral projection reveals dorsal displacement of the ulna. No other deformities are present.

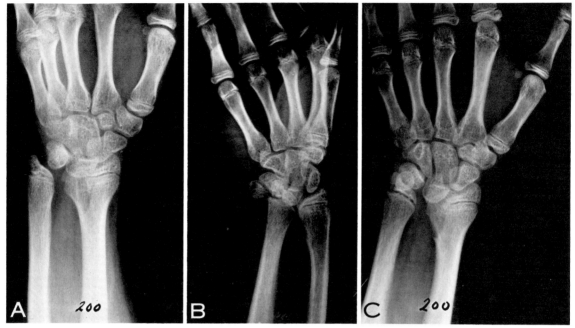

FIG. 1330. DYSCHONDROSTEOSIS (MADELUNG DEFORMITY)
(A) Subluxation of the ulna and a radiolucent area in the radial metaphysis. (B) Oblique projection reveals the metaphyseal defect to be cortical with very slight expansion of the inferior margin of the cortex. (C) Opposite oblique reveals the large defect in the radius and subluxation of the ulna.

valga[2] and cubitus valgus[4] may develop. The skull, spine, and pelvis are normal.

Although the wrist deformity is bilateral, asymmetrical changes occur; in trauma or infection, which may produce the same wrist deformity, the changes are unilateral. Multiple exostosis, Ollier disease, and multiple epiphyseal dysplasia may cause similar wrist and forearm deformities, but the skeletal survey easily distinguishes from dyschondrosteosis.[9] Mild thalidomide embryopathy[10] might cause similar changes, but the excessive mesomelic shortening or absence of the radius and fibula would not be seen in dyschondrosteosis.

Although Madelung disease is classically described in preadolescence, Langer believes they are the same condition and states "... The forearm deformity is primarily a result of unequal growth of the distal radial epiphyseal plate. For this reason, it may not become clinically manifested until the early adolescent growth spurt occurs."

ENCHONDROMATOSIS
(OLLIER DISEASE)

Synonyms: dyschondroplasia, multiple enchondromas, internal chondromatosis.

Enchondromatosis was first described and named by Ollier[1] in 1900. Dyschondroplasia is more descriptive, since it results from cartilage failing to undergo the normal process of enchondral bone formation. This produces rounded masses or columns of uncalcified cartilage within the metaphyses and diaphyses of certain bones which are invariably shortened. Solitary enchondromas are considered neoplasms[2] and not developmental anomalies. Enchondromatosis is often confused with multiple exostosis, although it can usually be distinguished roentgenographically; however, the two conditions may be found in the same individual.[3] When enchondromatosis and multiple cavernous hemangiomas occur together, the combination is called Maffucci disease.[4, 5]

Clinical Features. Heredity and familial influences play no part in enchondromatosis, whereas multiple exostosis commonly shows a hereditary tendency. Both sexes are affected, although it is slightly more common in men.[6]

The condition is detected in childhood usually through an arm or leg shortened by as much as

10 inches.[7] The hands and feet may be grossly deformed by the extreme hyperplasia of the ectopic cartilage within the small bones. Sometimes only half of a metaphysis is affected, leading to unequal growth of the epiphyses and deformity, such as genu valgum. The ulna is frequently shorter than the radius, causing curvature of the radius and radioulnar dislocation. Facial asymmetry has been reported[8] and Nielson[9] has described cranial nerve involvement.

Roentgenographic Features. Enchondromatosis is characterized by rounded masses or columnar streaks of decreased density, affecting the metaphysis and diaphysis of one or more tubular bones (Figs. 1331 and 1332). The more rapidly growing end of a bone is most often affected, but the other end may also be involved. The radiolucencies result from bone failing to replace cartilage during enchondral bone formation. Bone septa between the columns of radio-

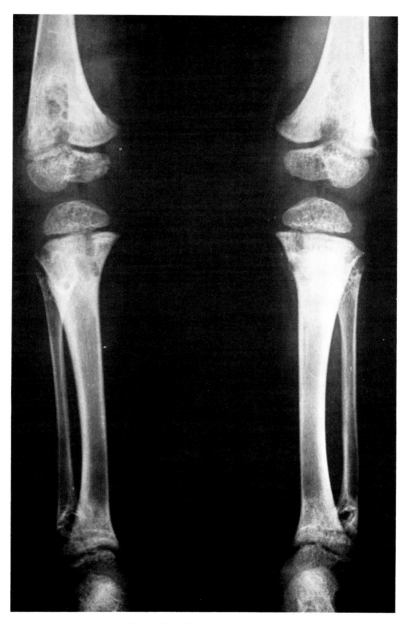

FIG. 1331. ENCHONDROMATOSIS
Multiple rounded radiolucencies at the ends of all long bones that represent cartilaginous rests. (Courtesy of John W. Hope, Children's Hospital, Philadelphia, Pa.)

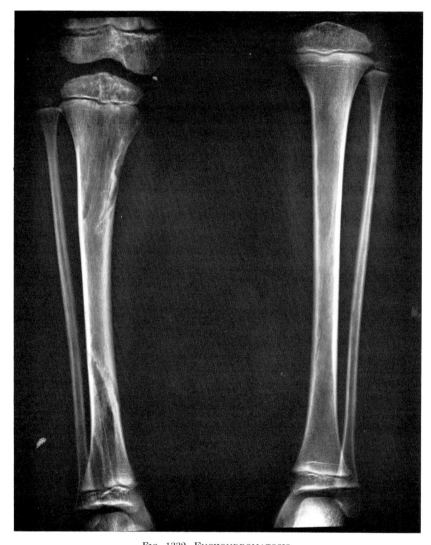

FIG. 1332. ENCHONDROMATOSIS
Disease is predominantly unilateral. Columnar streaks of decreased density in the right tibia. The tibia and fibula are dwarfed. (Courtesy of Herbert M. Stauffer, Temple University Hospital, Philadelphia, Pa.)

lucent cartilage rests cause a striated appearance (Fig. 1333), not to be confused with striated bone, in which dense vertical striation contains supervening normal bone. The radiolucencies extend for varying distances into the diaphyseal area, rarely involving the entire bone (Figs. 1334–1336). These changes are associated with dwarfing of the bones (Fig. 1331), and clublike deformity of the metaphyseal region due to lack of modeling (Figs. 1337 and 1338). Small or large columns of cartilage may extend into cortical bone and sometimes slightly outside the outline of the bone, where they may be mistaken for exostosis (Figs. 1337 and 1339). The bony spur points toward the joint, however, rather than away from it as in true exostosis.

Sometimes there is a central metaphyseal notch (Fig. 1331) but this may also indicate metaphyseal dysostosis, or trauma.

Early in the disease the cartilaginous areas are completely radiolucent. As the child grows older, he may show punctate calcifications which, eventually, become very dense (Figs. 1340 and 1341). Then they may fill in with bone and gradually disappear, although the ultimate appearance of the bones is rarely completely normal. The condition tends to be unilateral (Figs. 1342–1344) but there may be extensive bilateral disease, and

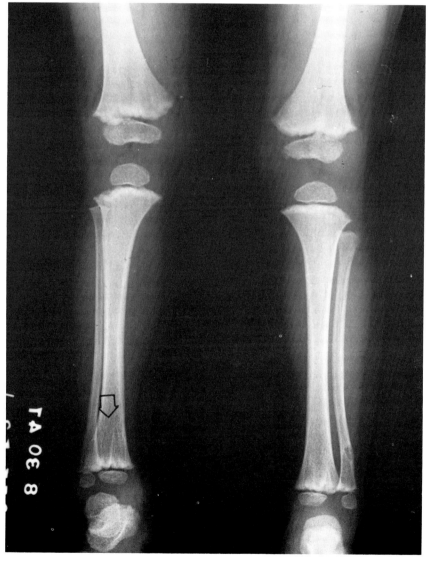

FIG. 1333. ENCHONDROMATOSIS

Columns of radiolucent cartilage rests are separated by bone septa (*arrow*) which give the metaphyseal end of the bone a striated appearance. (Courtesy of John W. Hope, Children's Hospital, Philadelphia, Pa.)

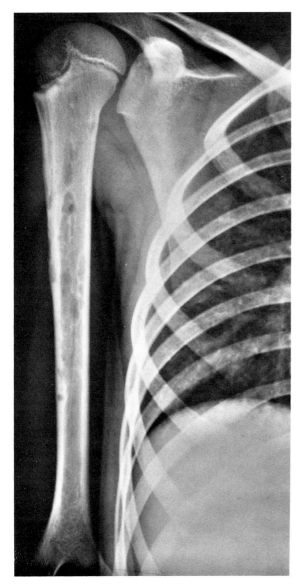

FIG. 1334. MULTIPLE ENCHONDROMATOSIS
Note that the radiolucencies do not extend throughout the entire bone. (Courtesy of Herbert M. Stauffer, Temple University Hospital, Philadelphia, Pa.)

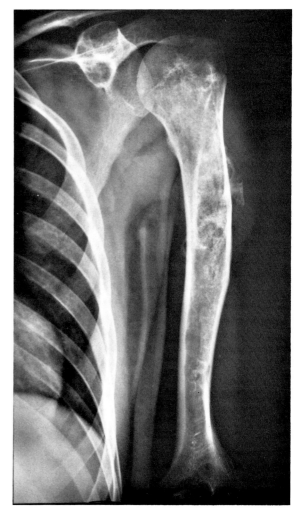

FIG. 1335. ENCHONDROMATOSIS
Irregular osteolytic areas extend throughout the shaft of the humerus surrounded by sclerotic margins. A small exostosis extends into the soft tissues at the level of the junction of the middle and upper thirds of the humerus. Cartilaginous rests in the glenoid of the scapula. (Courtesy of Herbert M. Stauffer, Temple University Hospital, Philadelphia, Pa.)

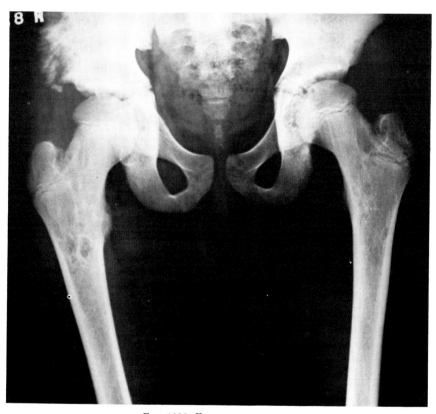

Fig. 1336. Enchondromatosis
Radiolucencies of both femurs and the right iliac bone. They do not extend throughout the entire shaft of the bone. (Courtesy of John W. Hope, Children's Hospital, Philadelphia, Pa.)

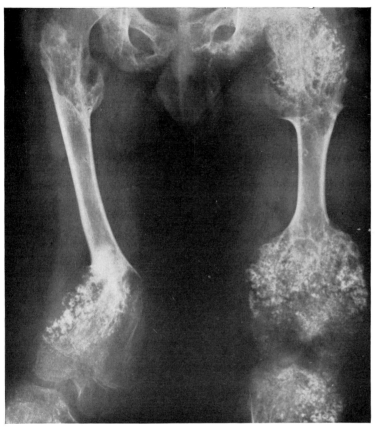

Fig. 1337. Enchondromatosis
Dwarfing of both femurs and clublike deformities of the metaphyseal regions due to the lack of modeling. Stippled calcifications are present in the cartilaginous masses.

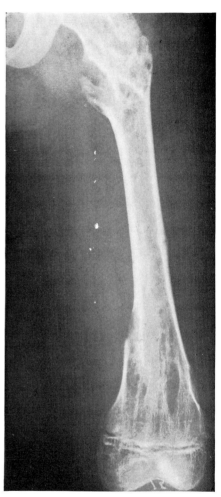

Fig. 1338. Enchondromatosis
Cartilaginous rests in the metaphyseal portions of the femur have caused a lack of modeling.

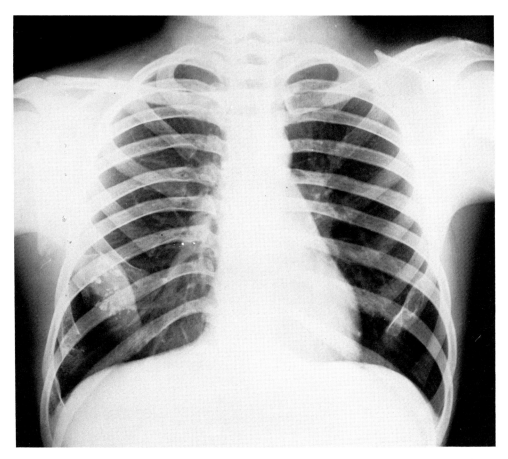

Fig. 1339. Enchondromatosis
Rib lesions might be mistaken for osteochondroma. (Courtesy of John W. Hope, Children's Hospital, Philadelphia, Pa.)

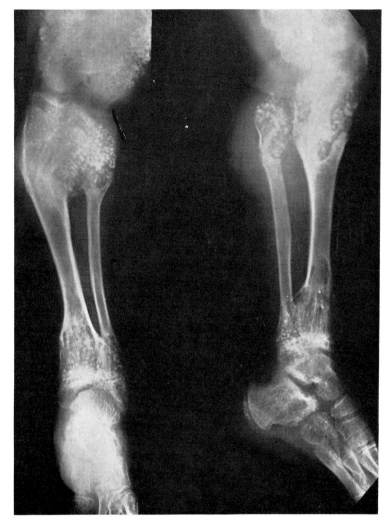

Fig. 1340. Enchondromatosis

Dwarfing bilaterally in the tibia and fibula. Clubbing at the metaphyseal regions and stippled calcifications in the cartilaginous rests.

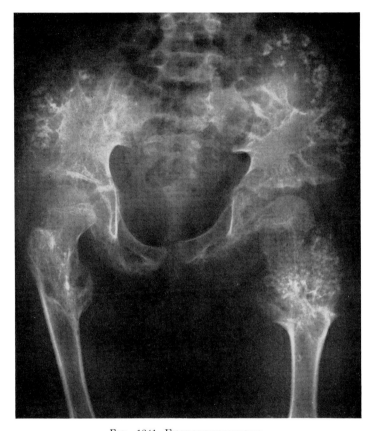
Fig. 1341. Enchondromatosis

Stippled calcification in cartilaginous rests. The calcification may not appear until later in the course of the disease. (Same case as seen in Figure 1340.)

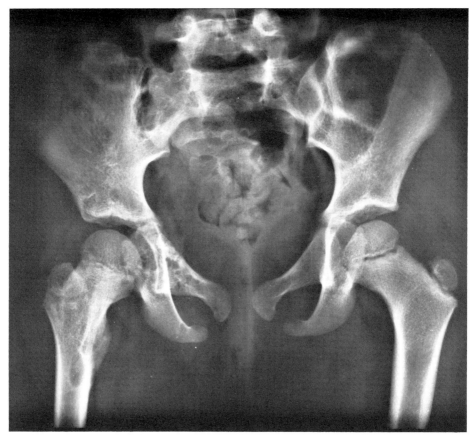

FIG. 1342. ENCHONDROMATOSIS

Lesions are predominantly unilateral. (Courtesy of Herbert M. Stauffer, Temple University Hospital, Philadelphia, Pa.)

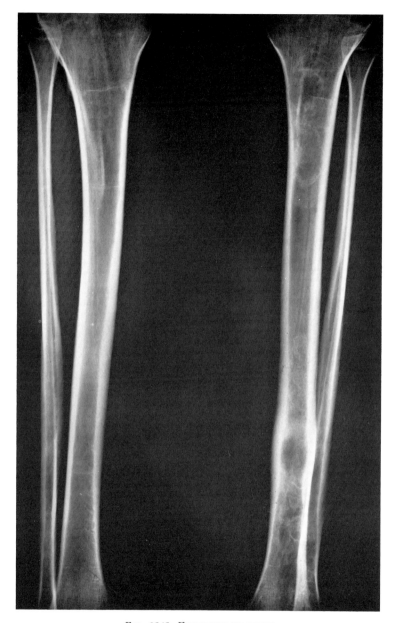

Fig. 1343. Enchondromatosis

Although the lesions are predominantly unilateral, some involvement of the other side is present. (Courtesy of Herbert M. Stauffer, Temple University Hospital, Philadelphia, Pa.)

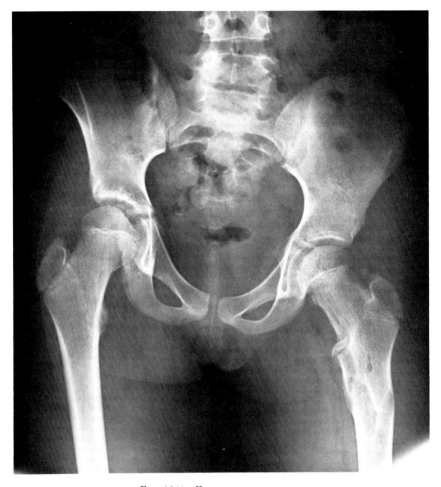

Fig. 1344. Enchrondromatosis
Tendency for unilateral involvement. (Courtesy of George T. Wohl, Philadelphia General Hospital, Philadelphia, Pa.)

sometimes all the long tubular bones and the bones in the hands and feet are affected.

The small bones of the hands and feet show round radiolucent areas of unreplaced cartilage (Figs. 1345 and 1346). In these bones there is a tendency for the cartilage to continue to proliferate and, at times, marked hyperplasia produces multiple deforming tumors that may break through the cortices (Fig. 1347). This sometimes appears in other bones, commonly the pelvic bones, where the bands or cartilage radiate, fanlike, from the center of the ilia to the crest (Fig. 1348). Malignancy occurs in enchondromatosis, but the incidence is difficult to evaluate; Jaffe[10] believes it to be about 50% (Fig. 1349). In children and young adults and the malignancy is usually osteosarcoma, whereas in older patients, chondrosarcoma and fibrosarcoma are the most common (Fig. 1350).

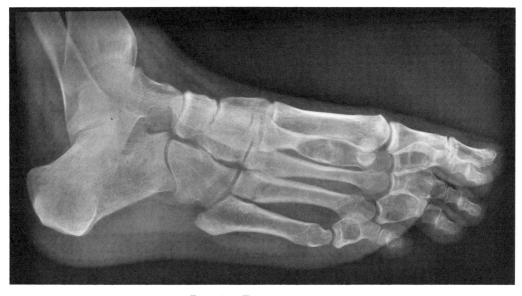

FIG. 1345. ENCHONDROMATOSIS
Small bones of the feet reveal radiolucent areas which represent unreplaced cartilage. Some calcification is present. (Courtesy of Herbert M. Stauffer, Temple University Hospital, Philadelphia, Pa.)

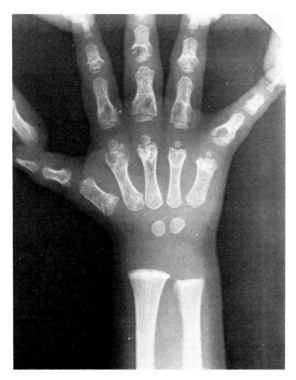

FIG. 1346. ENCHONDROMATOSIS
Small bones of the hand show multiple radiolucent areas representing unreplaced cartilage. (Courtesy of John W. Hope, Children's Hospital, Philadelphia, Pa.)

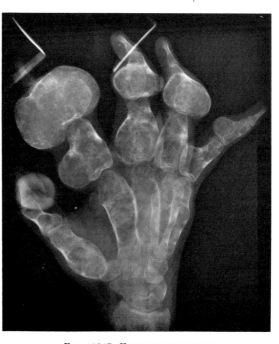

FIG. 1347. ENCHONDROMATOSIS
At times there is a unique tendency for the cartilage to continue to proliferate and multiple deforming tumors are formed. (Courtesy of Herbert M. Stauffer, Temple University Hospital, Philadelphia, Pa.)

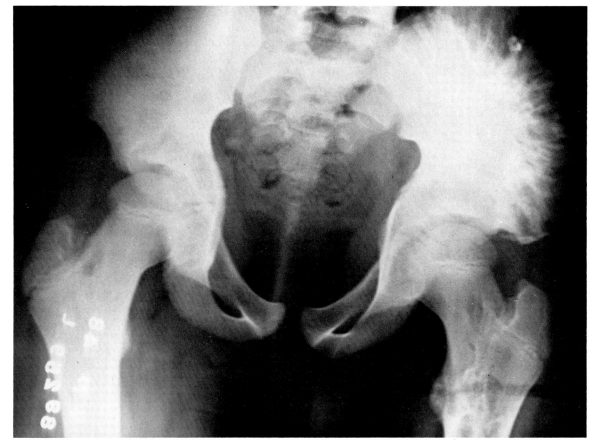

FIG. 1348. ENCHONDROMATOSIS
Pelvic bone involvement has produced bands of cartilage radiating fanlike from the center of the ilium to the crest. The left femur is similarly involved. (Courtesy of John W. Hope, Children's Hospital, Philadelphia, Pa.)

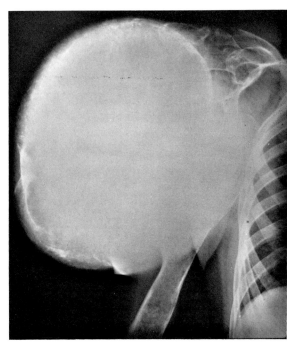

FIG. 1349. ENCHONDROMATOSIS WITH CHONDROSARCOMA
Large chondrosarcoma in the upper end of the humerus. (Courtesy of Herbert M. Stauffer, Temple University Hospital, Philadelphia, Pa.)

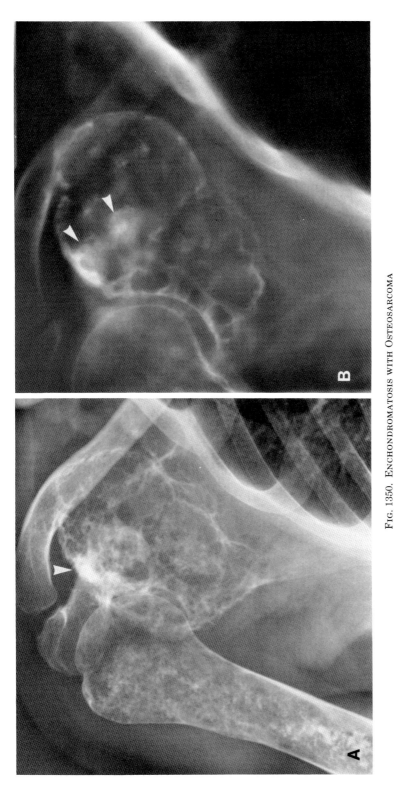

FIG. 1350. ENCHONDROMATOSIS WITH OSTEOSARCOMA

Extensive enchondromatosis in most of the bones; osteosarcoma in the right scapula. (A) Right shoulder. Extensive enchondromatosis in the humerus and scapula. Dense bone at the superior margin of the scapula (*arrow*) represents the osteosarcoma. The density suggests the presence of tumor osteoid. (B) Tomography. Better delineation of the osteoid density (*arrows*). (C) Left hand. Extensive enchondromatosis. (D) Left forearm. Enchondromatosis in the distal end of the radius with a small lesion in the midportion of the ulna. (E) Right forearm. Enchondromatosis in the radius. The ulna is spared. (F) Pelvis. Enchondromatosis in the ischial rami and both femurs. (G) Enchondromas in the proximal and distal ends of the tibia. (H) Enchondromatosis in both feet. (Courtesy of Dr. Beth Edeiken, M. D. Anderson Hospital and Tumor Institute, Houston, Tex.)

1260 DISEASES OF BONE

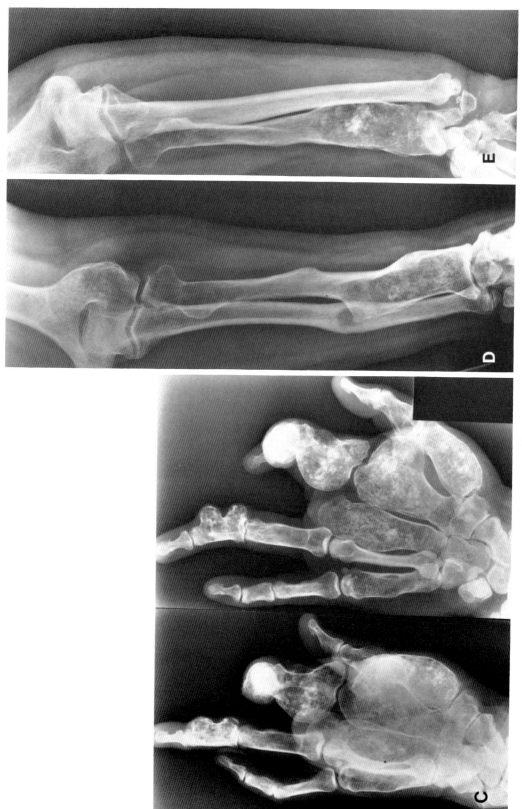

FIG. 1350 (C–E)

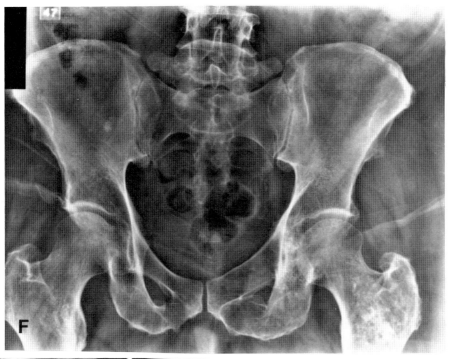

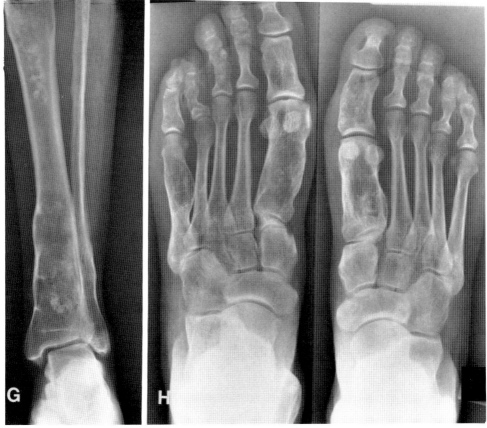

Fig. 1350 (F–H)

MAFFUCCI SYNDROME

This is a combination of enchondromatosis and multiple soft tissue hemangiomas that may have phleboliths (Fig. 1351). There may be no phleboliths in the newborn—the hemangiomas may disappear (Fig. 1352). It is asymmetric and hemangioma need not overlie bone lesions. It is not hereditary.[1]

The cartilaginous or vascular tumors may undergo malignant change.[2-5]

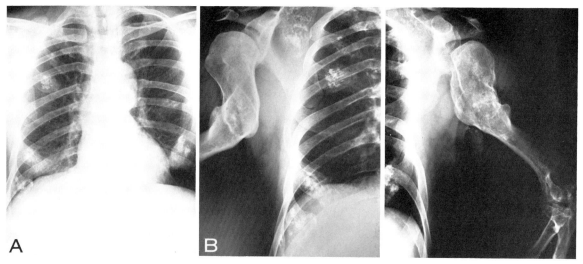

FIG. 1351. MAFFUCCI SYNDROME

(A) Chest. Multiple calcified enchondromas in the anterior ribs. (B) Both humeri have marked deformity due to enchondromatosis. (C) Right forearm and hand. Marked distortion of the bones with shortening of the ulna and curvature of the radius and ulna. Multiple defects are present in the small bones of the hand, due to enchondromatosis. A large soft tissue hemangioma with multiple phleboliths that extends from the elbow to the hand. (D) The opposite hand reveals small defects in the phalanges and distortion of the fourth metacarpal and the distal end of the ulna due to enchondromatosis. There was no hemangioma on this side. (Courtesy of Dr. Gary Romisher, Riverside, N.J.)

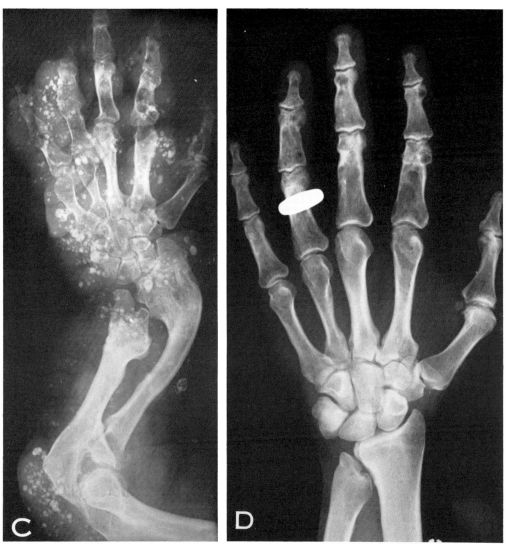

FIG. 1351 (*C* and *D*)

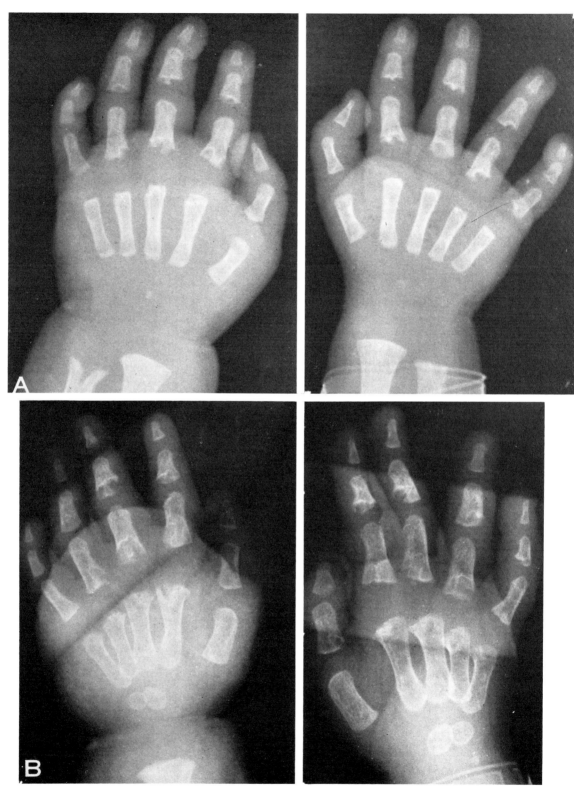

FIG. 1352. MAFFUCCI SYNDROME

(A) Newborn. Both hands have multiple defects at the ends of the small bones, particularly the proximal phalanges and the distal end of the ulna. A large soft tissue hemangioma without phleboliths on the left side. (B) At age 7 months the defects due to Ollier's disease are more prominent. The soft tissue hemangioma on the left has increased in size but no phleboliths are present. (C) Newborn pelvis. Small defects at the femoral ends. There is no evidence of hemangioma; (D) 7 months later there are definite abnormalities in the ends of the femurs and tibias due to enchondromatosis. (Courtesy of Dr. Donald Babbitt, Milwaukee, Wisc.)

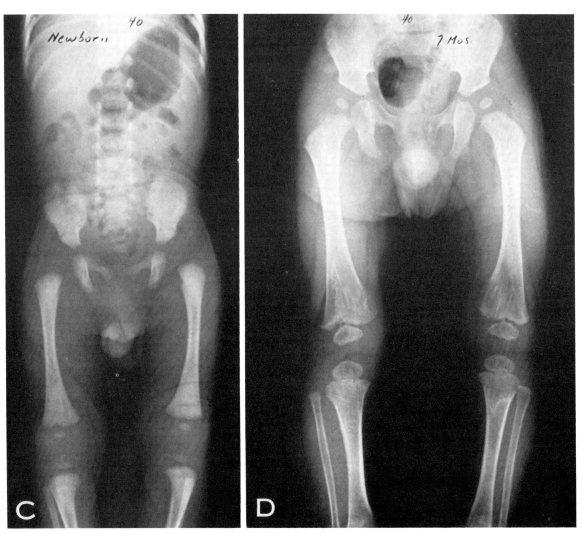

FIG. 1352 (*C* and *D*)

LAURENCE-MOON-BIEDL-BARDET SYNDROME

This syndrome includes polydactyly, retinitis pigmentosa, pseudo-Fröhlich type of obesity, genital hypoplasia, and mental retardation. Only 25% of the patients have the complete syndrome.[1, 2] Additional disturbances are as listed in Table 38.[3] Obesity and polydactyly are most common and suggest the syndrome.[4] It is probably transmitted by an autosomal recessive mode; and XXX chromosome is reported occasionally.[5]

Levy et al.[3] reported hip dysplasia with minimal subluxation and bilateral valgus deformity which may be associated with this syndrome.

Roentgenographic Features. Polydactyly of the feet and hands is nonsymmetrical and may be associated with syndactyly (Fig. 1353). Syndactyly may occur without polydactyly.

TABLE 38. ADDITIONAL DISTURBANCES DESCRIBED IN REPORTED CASES OF LAURENCE-MOON-BIEDL-BARDET SYNDROME

Syndactyly	Deafness
Dwarfism	Dental anomalies
Mongolian facies	Kidney anomalies
Congenital heart disease	Anal atresia
Microphthalmus	Diabetes insipidus
Cataracts	Valgus of knees
Convergent strabismus	Ataxia

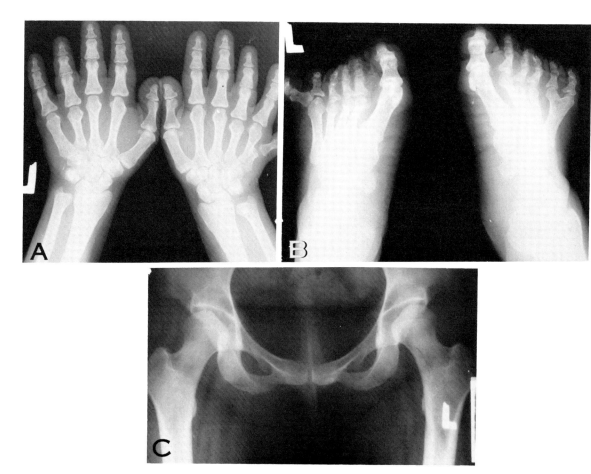

FIG. 1353. LAURENCE-MOON-BIEDL-BARDET SYNDROME

(A) Hands. A sixth digit projecting from bifid right fifth metacarpal. The left fourth and fifth and the right fourth metacarpals are short. (B) Feet. A thick right bifid metatarsal which projects digits five and six. The left sixth toe arises from the head of the fifth metatarsal. (C) Hips. The hips are not well seated in the acetabulum and there is a suggestion of subluxation on the left. Bilateral valgus deformity. (Courtesy of M. Levy, M. Lotom, and A. Fried: *Journal of Bone & Joint Surgery*, 52B: 318, 1970,[3] © British Editorial Society of Bone & Joint Surgery, London.)

SPONDYLOEPIPHYSEAL DYSPLASIAS
(EPIPHYSEAL DYSPLASIA, SPONDYLOEPIPHYSEAL DYSPLASIA, SPONDYLOEPIPHYSEAL METAPHYSEAL DYSPLASIA)

The spondyloepiphyseal dysplasias are chondrodystrophies with selective and combined involvement of the vertebrae and long bone epiphyses and metaphyses. The term is often used to cover any condition causing dwarfism and epiphyseal metaphyseal irregularities.

Morquio disease is often considered the prototype, but is best classified a mucopolysaccharidosis and used as an example of severe spondyloepiphyseal dysplasia. Ochronosis and juvenile rheumatoid arthritis may cause severe spondyloepiphyseal abnormalities, but they are, respectively, metabolic and collagen diseases, and appropriate laboratory tests and clinical evaluation identify them.

Some chondrodystrophies have typical radiologic and clinical characteristics that allow separate classification, e.g., metatrophic dwarfism, diastrophic dwarfism, chondroectodermal dysplasia, and stippled epiphyses.

The remaining ill defined spondyloepiphyseal dysplasias must be classifed by roentgenographic and clinical evaluation until more sophisticated laboratory diagnoses are developed. Most are discovered in the second year of life, and even though growth is practically normal in the first few months, they account for some of the most severe types of dwarfs.

Three principal groups are distinguished[1]: epiphyseal dysplasia, epiphyseal dysplasia with platyspondyly, and epiphyseal-metaphyseal dysplasia. There is considerable overlap in these groups.

Group I. Epiphyseal Dysplasia. TYPE I. The changes are limited to the long bone epiphyses and the vertebral plates. The vertebral heights are normal or only slightly decreased.

The epiphyses appear late and are irregular and small, and there may be fragmentation. The metaphyses are normal except for an infrequent juxtaepiphyseal irregularity. The limbs alone account for the moderately shortened stature. Secondary degenerative hip arthrosis occurs frequently.

TYPE II. *Proximal Limb Type.* The proximal epiphyses of the humeri and femora are most severely affected. The other epiphyses are normal or have only minor irregularities. The vertebral plates are irregular, usually only in the lower thoracic and upper lumbar vertebrae. Degenerative hip arthrosis usually occurs.

TYPE III. *Localized Type.* This is often characterized by bilateral isolated epiphyseal involvement of the hands and feet (peripheral dysostosis[2] and Thiemann disease[3]), elbows, hips, and knees. Late in the course, osteochondromas may appear.

Group II. Epiphyseal Dysplasia with Platyspondyly. TYPE I. *Spondyloepiphyseal Dysplasia Tarda.* The vertebral bodies are flattened and the trunk becomes obviously shortened after age 10. The pelvis is small, and the femoral capital epiphyses may be irregular and deformed. The other epiphyses are normal.

TYPE II. *Platyspondyly and Severe Irregularities of Proximal Epiphyses, with Minor Changes in Peripheral Epiphyses.* Congenital spondyloepiphyseal dysplasia[4] is the best example. The trunk is shortened at birth. The vertebrae are irregular, ovoid, and decreased in height. The proximal epiphyses of the humeri and femora are markedly deformed, although the other epiphyses are not seriously affected.

TYPE III. *Platyspondyly and Generalized Epiphyseal Abnormalities.* The vertebrae are flat, and generalized epiphyseal irregularities are severe. A hypoplastic lumbar vertebra is often present. Morquio disease is the best example, although it is better classified as a mucopolysaccharidosis. Spondylometaphyseal dysostosis[5] is another type, distinguished by the metaphyseal irregularity of the femoral neck and trochanters.

Group III. Epiphyseal-Metaphyseal Dysplasias. Epiphyseal and metaphyseal irregularity appears with or without spine involvement.

TYPE I. *Epiphyseal-Metaphyseal Deformities Quite Marked.* Micromelic dwarfism, short and thick diaphyses, and stubby hands and feet are the most apparent characteristics, and there may be isolated abnormality of the hands and feet (Silfverskiöld peripheral form[6] and Léri pleonosteosis).[7]

TYPE II. *Epiphyseal-Metaphyseal Dysplasia with Vertebral Involvement.* Micromelic dwarfism and the shortened trunk are usually present. Vertebral height is decreased and the plates irregular. The vertebral and peripheral manifestations are variable.

TYPE III. *Pseudoachondroplastic Form.* These patients look like achondroplastic dwarfs with rhizomelic limb shortening and normal trunk length. The skull is normal. The epiphyses are small and irregular, the metaphyses irregular and deformed. The vertebral bodies are irregular and deformed, biconvex with an anterior tongue in the infant, and a slightly wedged vertebra in the adult.

Spondyloepiphyseal Dysplasia Congenita

Spondyloepiphyseal dysplasia congenita is a heritable bone dysplasia which affects the spine and proximal long bone epiphyses at birth. It is easily mistaken for Morquio disease[1, 2] (Table 39).

Clinical Features. It is transmitted by an autosomal dominant mode.[2] Newborn infants are short, and talipes varus and cleft palate may occur. The face is flat[2] and the eyes wide spaced, with an upward slant of the palpebral fissures.[3-5]

The neck is short and the thorax barrelshaped. There is exaggeration of the thoracic kyphotic curve and lumbar lordotic curve. Knock-knee or bowleg deformities often appear. Scoliosis may develop in adolescence. The gait is waddling[6, 7] and muscle hypotonia may be noticed.

Myopia and/or retinal detachment are common.[1-4, 8, 9] The corneas are clear.[1, 7] The extent of dwarfism in the adult, due to short trunk and shortening of the proximal extremities, varies from 37 to 52 inches.[2] The hands and feet are normal.

Roentgenographic Features. The most striking features of all ages involve the spine, pelvis, and femoral heads[2] (Figs. 1354 and 1355). In the newborn, ossification is generally delayed and there is lack of ossification in the pubic bones, distal femoral and proximal tibial epiphyses, calcaneus, and talus. There is platyspondyly and in the lateral projection a trapezoid configuration of the centra due to a short posterior aspect can be seen. The iliac bones are broad at the base, with a lack of iliac flaring. The thorax is bell-shaped and the anterior rib ends are flared.

In the first year of life, the pubic bones remain underossified and the femoral heads do not ossify. The knee ossification centers appear late and irregularly. Platyspondyly continues and there is usually hypoplasia of one or several vertebrae at the thoracolumbar junction. The femora and tibiae are short, with metaphyseal irregularities appearing in the distal femora. During childhood, the platyspondyly persists (Fig. 1356), and there is a progressive dorsal lumbar lordosis.[2, 6] The acetabular roofs are horizontal but the fossa is deep. Underossification of the pubic bone diminishes but the wide cartilage remains wide. Ossification of the capital epiphyses remains retarded, and ossification occurring in multiple foci may produce a modeled appearance. Severe varus deformity of the femoral neck is usually noted. The trapezoid centra gradually assume a normal configuration. The intervertebral spaces become narrowed. The odontoid does not completely ossify. Shortening of the proximal long bones becomes particularly noticeable. The irregular and flattened aspect of the epiphyseal centers is most evident in the femoral capital epiphyses. Metaphyseal abnormalities may also appear most consistently and severely in the distal femora and proximal humeri. The metaphyseal ossification lines may be slightly irregular and convex. There may be gross irregularities with spur formations and splaying of the metaphyses.

The small tubular bones are usually normal, although metacarpal shortening may occur. The carpal and tarsal maturation is retarded. A steep anterior skull base is reported. The angle between the floor of the anterior fossa and clivus, normally 110–145°, may be as much as 165°.[2]

In adults there is severe platyspondyly and considerable narrowing of the intervertebral disk space, causing a marked short trunk dwarfism, contributed to by scoliosis and an increase in lumbar lordosis. The centra are irregular. The odontoid is still only partially ossified. Anterior hypoplasia of one or more vertebral centra persists, and the chest remains barrel-shaped. Severe coxa vara of the femoral neck continues. The femoral capital epiphyses are small and irregular, and they may remain unfused.

The long bones are short, with metaphyseal

TABLE 39. DIFFERENTIAL DIAGNOSIS OF SPONDYLOEPIPHYSEAL DYSPLASIA CONGENITA VERSUS MORQUIO DISEASE[a]

Differential	Spondyloepiphyseal Dysplasia Congenita	Morquio Disease
Clinical manifestation	Birth	End of first year
Flared ilia	Absent	Present
Deficient ossification of pubic bone	Present	Absent
Deficient ossification of superior acetabulum	Absent	Present
Acetabular angle	Small	Wide
Iliac angle	Small	Wide
Femoral neck	Varus	Valgus
Involvement of hands and feet	Minimal	Severe
Eye changes	Myopia	Corneal clouding
Keratosulfaturia	Absent	Present
Inheritance	Autosomal dominant	Autosomal recessive

[a] From J. W. Spranger and L. O. Langer, Jr.: *Radiology*, 94: 313, 1970.[2]

flaring, especially at the distal femora and both humeral ends. The articular surfaces are irregular. Genu valgum is common as are patella dislocations. The hands and feet may show minor flattening of the metatarsal and metacarpal heads. Irregularity of the proximal row of the carpals is frequent.

The major roentgen criteria of spondyloepiphyseal dysplasia congenita are summarized in Table 40. Spranger and Langer[2] summarized the frequency of roentgen features as shown in Table 41. The differential diagnosis includes achondroplasia, metatrophic dwarfism, Morquio disease, spondyloepiphyseal dysplasia tarda and diastrophic dwarfism.

Spondyloepiphyseal dysplasia is most easily confused with Morquio disease in both the child and the adult. Morquio disease, however, is transmitted by an autosomal recessive mode, and its corneal opacities and keratosulfaturia serve to distinguish it, as do the differences in the shape of the centra in the cervical, thoracic, and lumbar areas.[10] With Morquio disease, in early childhood, the flattening of the thoracic vertebrae is more marked, and in the lumbar region the centra have an oval configuration, later developing a defect in the anterior superior aspect, producing a beak. The disk spaces remain wide in Morquio disease, but become progressively narrower in spondyloepiphyseal dysplasia. There is no iliac flare and a horizontal acetabulum. The capital femoral epiphyses are relatively normal early and then flatten or disappear with erosion of the femoral neck. The changes in the proximal metatarsals in Morquio disease do not occur in spondyloepiphyseal dysplasia.[11] Finally, Langer and Carey[10] point out that coxa vara is never seen with Morquio disease, nor does dwarfism occur in the first year.

Spondyloepiphyseal dysplasia tarda appears at prepuberty and manifests a peculiar humplike deformity of the lumbar centra[12] with a deep narrow pelvic configuration which differentiates it from the congenital form.

Diastrophic dwarfism can be differentiated by the presence of clubfeet and the absence of severely retarded skeletal maturation. There is no platyspondyly. There is marked shortening and deformity of the bones of the hands and feet.

Metatrophic dwarfism is first manifested by marked streaklike underossifications of the vertebral centra, wide metaphyseal flaring, and trochanteric hyperplasia. Its progressively severe kyphoscoliosis and peculiar "battle-ax" pelvic configuration, with marked iliac flaring, distinguish it from spondyloepiphyseal dysplasia.[13]

In severe multiple epiphyseal dysplasia the spine may show minor changes and hypoplasia, but no generalized platyspondyly, although one or two anterior wedged vertebrae may be present. A severe form of spondyloepiphyseal dysplasia congenita, with marked dwarfism, has been called pseudoachondroplastic spondyloepiphyseal dysplasia. At first glance, this condition may resemble achondroplasia, but the changes in the pelvis and the lack of narrowing of the lower lumbar interpediculate spaces rule out this possibility.

Spondyloepiphyseal Dysplasia Tarda

Spondyloepiphyseal dysplasia tarda is an inherited epiphyseal dysplasia occurring only in males,[1] causing a mildly shortened stature and distinct radiologic features in the adult spine.[2]

Clinical Features. It is transmitted by a sex-linked recessive mode and appears in affected male offspring.[1] Approximately half of the daughters will be carriers. The height varies from 52 to 62 inches,[2,3] and the dwarfism is of a short trunk type.

Failure of normal growth becomes evident between the ages of 5 and 10 years. It is frequently associated with back pain. Premature hip osteoarthritis, resulting in pain and limitation of motion, occurs at some time after puberty.

Roentgenographic Features. A small pelvis with bilateral hip osteoarthritis in a young or middle-aged male is highly suggestive.[2] Langer[2] described a distinctive configuration in the adult lumbar centra, consisting of generalized flattening anteriorly, with a hump-shaped buildup of eburnated bone in the central and posterior portions of the superior and inferior plates. Bone is totally absent from the ringed apophyses. The disk spaces are narrowed. Platyspondyly extends through the thoracic and cervical regions, but the plate changes are not marked. Degenerative changes occur early, and mild kyphosis and scoliosis are frequent. In the infant and young child, no changes are reported. The earliest reported platyspondyly is at 5 years,[1] in a child with a roentgenogram age of 2½ years. Osteoarthritis of the hips usually becomes clinically apparent between the 3rd and 4th decades, but may appear at adolescence. It is characterized by marked joint space narrowing, cystic changes, spur formation, and productive new bone. Deformity of the femoral heads and necks results. The same type of change may occur in the shoulders. The other peripheral joints may show mild epiphyseal

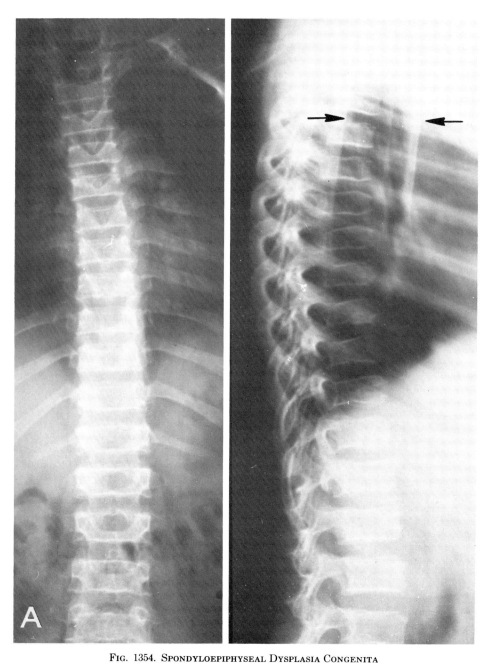

FIG. 1354. SPONDYLOEPIPHYSEAL DYSPLASIA CONGENITA

(A) Dorsal spine. Platyspondyly. (B) Hips and pelvis. Flattening of the femoral capital epiphyses and acetabular irregularities. (C) Open mouth view. The odontoid is aplastic (*arrows*). (Courtesy of Dr. Robert M. Peck, Tripler General Hospital, Hawaii.)

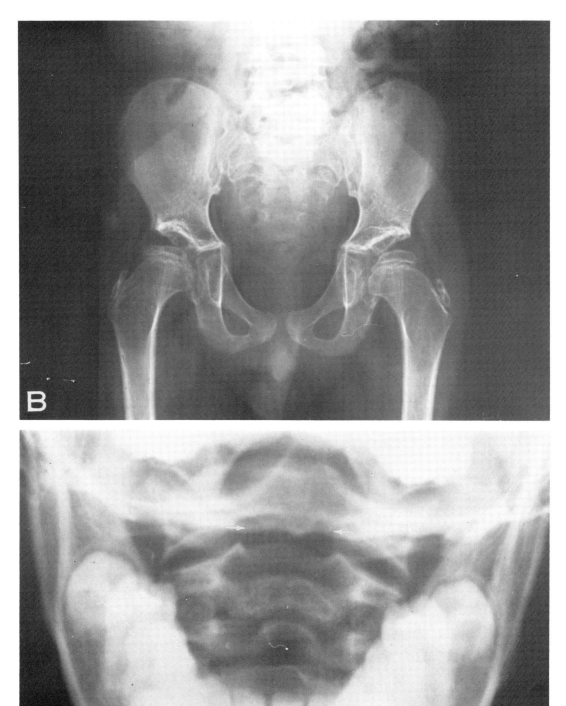

Fig. 1354 (B and C)

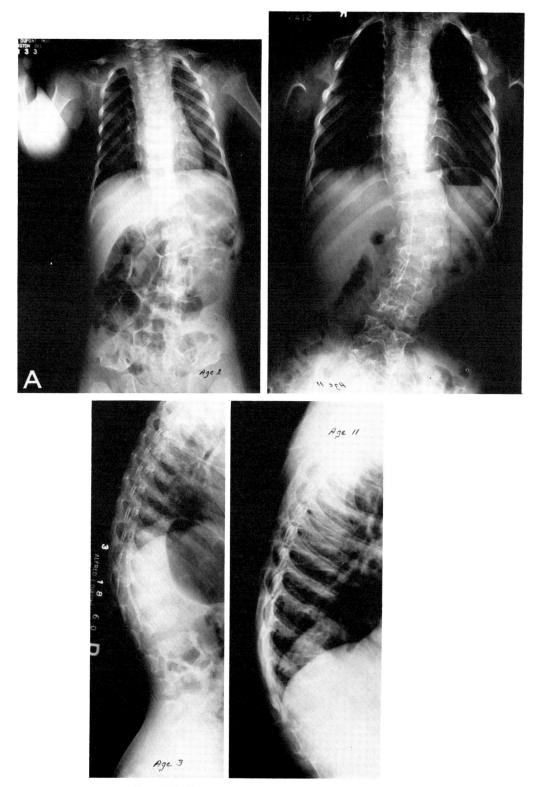

Fig. 1355. Spondyloepiphyseal Dysplasia Congenita

(A) Spine. Marked platyspondyly and scoliosis at age 3 that progresses. (*Upper left*) Anteroposterior view, age 3; (*upper right*) anteroposterior view, age 11; (*lower left*) lateral view, age 3; (*lower right*) lateral view, age 11. (B) Both hands. Marked distortion of the epiphyses at age 11. (C) Elbow at age 11. Marked distortion, irregularity, and flattening of the epiphyses. (D) Lower extremities. Progressive flattening and irregularity of the epiphyses of the knees. The femoral capital epiphyses are hypoplastic and irregular (*Left*) Age 2 (*center*), age 6, and (*right*) age 11.

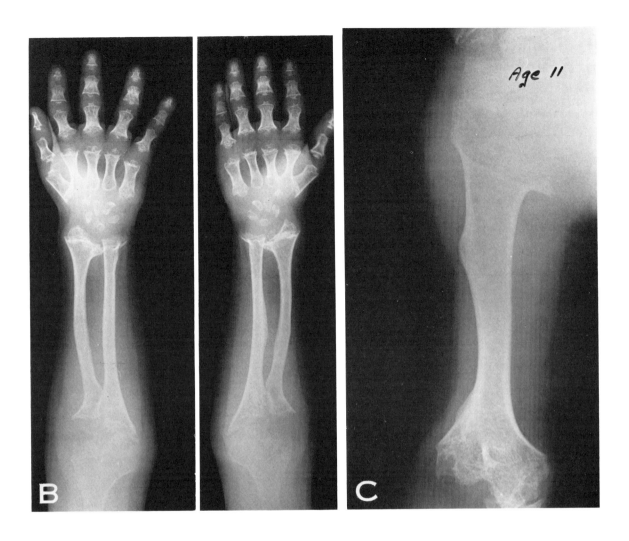
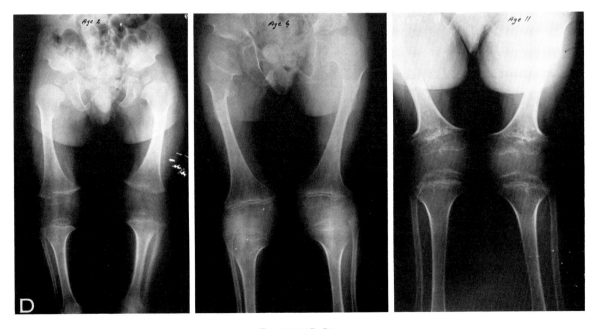

Fig. 1355 (*B–D*)

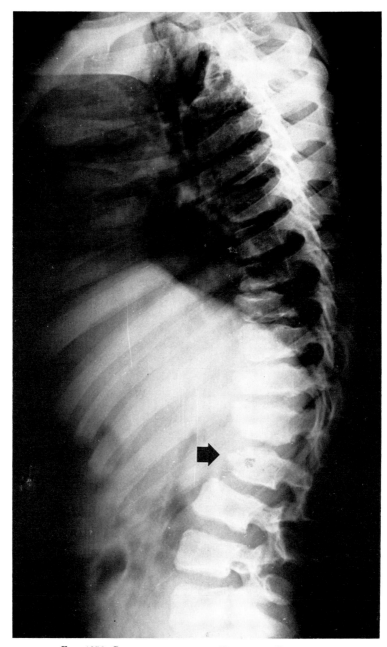

FIG. 1356. SPONDYLOEPIPHYSEAL DYSPLASIA CONGENITA

Universal vertebral plana with central beaking (arrow) of the hypoplastic and slightly displaced second lumbar vertebra. These changes are similar to those seen in Morquio's disease.

abnormality (Fig. 1357). The most constant findings in the peripheral joints are flattening of the articular surface of the ankles and the knees and shallowness of the intercondylar notch. Other findings include increase in anteroposterior and transverse diameter of the chest, small iliac wings, and relatively long pubic and ischial bones.[2]

Ochronosis may be suggested by the intervertebral disk space narrowing and the early arthritis. The humplike deformity of the vertebra may be superimposed on the intervertebral disk, and may resemble calcification, but close inspection and laminograms will exclude this possibility.

Morquio disease may be excluded by the absence of either severe platyspondyly or severe dysplasia of the peripheral epiphyses. Multiple epiphyseal dysplasia also presents more change in the peripheral joints and less in the spine than is seen in Morquio disease.

TABLE 40. MAJOR DIAGNOSTIC ROENTGENOGRAPHIC CRITERIA OF SPONDYLOEPIPHYSEAL DYSPLASIA CONGENITA[a]

A. *Infancy:*
   Retarded ossification of the skeleton
   Flattening, ovoid, or pear-shaped (immature) vertebral bodies
B. *Childhood:*
   Flattening, immaturity, and irregular ossification of vertebral bodies
   Odontoid hypoplasia
   Retarded pelvic ossification
   Low and broad iliac wings
   Horizontal, deep acetabular fossae with iliac angle approaching 90°
   Grossly retarded and abnormal ossification of the femoral head and neck; varus deformity
   Rhizomelic brachymelia
   Variably retarded and irregular epiphyseal and metaphyseal ossification of the long tubular bones
   Only minor abnormalities of the hands and feet
C. *Adulthood:*
   Severe spinal shortening with mild kyphoscoliosis and marked lumbar hyperlordosis
   Severe flattening and irregular outline of dorsal vertebral bodies
   Odontoid hypoplasia and lack of fusion with C2
   Marked coxa vara with high-riding femoral trochanter and normally located, deformed femoral head
   Relatively normal hands and feet

[a] From J. W. Spranger and L. O. Langer, Jr.: *Radiology*, 94: 313, 1970.[2]

TABLE 41. FREQUENCY OF ABNORMALITIES IN CASES BOTH PERSONAL AND LITERARY[a]

Roentgenographic Findings	Age in Years			
	0–1	1–6	7–16	Adults
Broad thorax	7/17	11/14	16/18	4/5
Odontoid hypoplasia	—	2/2	4/4	4/6
Dorsal vertebral bodies:				
Flattening	6/6	12/12	14/14	7/7
Anterior ossification defects	0/5	8/11	11/13	6/7
Pear-shaped	6/6	9/11	9/13	3/7
Lumbar vertebral bodies:				
Flattening	3/4	11/13	9/13	5/6
Anterior ossification defects	0/4	11/13	4/12	3/6
Pear-shaped	3/4	10/13	6/12	1/6
Thoracolumbar junction, anterior hypoplasia	0/4	6/11	6/15	3/6
Pelvis:				
Low, square iliac wing	5/10	19/19	20/22	0/8
Horizontal, deep acetabulum	8/10	19/19	23/23	0/8
Underossified os pubis	10/10	19/19	20/23	0/8
Underossified femoral head and neck with varus deformity	4/4[b]	19/19	23/23	8/8
Distal femur:				
Epiphyseal dysplasia	1/2[c]	11/13	13/17	13/17
Metaphyseal dysplasia	1/6	9/14	12/17	—
Proximal humerus:				
Epiphyseal dysplasia	—	8/11	7/8	2/5
Metaphyseal dysplasia	2/3	12/12	8/8	—
Wrist:				
Epiphyseal dysplasia	—	8/12	14/16	1/5
Metaphyseal dysplasia	2/4	8/13	8/16	—
Hand:				
Normal tubular bones	2/2	9/12[d]	15/15	3/5[d]
Multiple accessory epiphyses	—	7/12	13/15	—

[a] From J. W. Spranger and L. O. Langer, Jr.: *Radiology*, 94: 313, 1970.[2]
[b] Over 6 months of age.
[c] When epiphyseal ossification center is present.
[d] Mild metacarpal shortening and/or epiphyseal flattening present in 3 cases, ages 1–6 years, and in 2 adults.

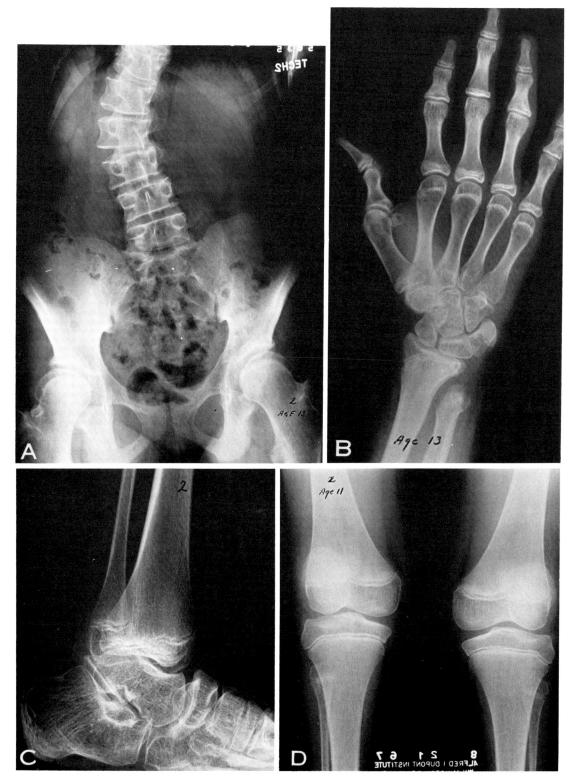

FIG. 1357. SPONDYLOEPIPHYSEAL DYSPLASIA TARDA

The epiphyseal dysplasia became evident after age 5. Flattening and irregularities of multiple epiphyses. Shortening of the long bones. Platyspondyly is moderate and there is a kyphoscoliosis. (*A*) Lumbar spine and pelvis at age 13. Flattening of the femoral capital epiphyses. Platyspondyly and scoliosis. (*B*) Hand at age 13. Shortening of the long bones and epiphyseal irregularity. Shortening of the ulna. (*C*) Ankle. Flattening of the distal tibial epiphyses and minor irregularities of the tarsals. (*D*) Knees at age 11. Flattening of the epiphyses.

EPIPHYSEAL DYSPLASIA MULTIPLEX

Synonyms: epiphyseal dysostosis,[1] multiple epiphyseal dysplasia, hereditary enchondral dysostosis.

Epiphyseal dysplasia multiplex is characterized by multiple epiphyseal irregularities without sclerosis. Fairbank[2-4] considers it distinct from dysplasia epiphyseal punctata. The cases reported as examples of multiple osteochondritis probably represent epiphyseal dysplasia multiplex.[5]

Clinical Features. Hereditary influences are evidence in more than half of the reported cases.[6-9] Some degree of dwarfism is present; the hands are short, the fingers blunt. Flexion deformities of many joints occur. The prognosis is excellent except for early degenerative arthritis of the hips and knees and mild dwarfism.[10]

Roentgenographic Features. The distinctive roentgenographic changes are bilateral symmetrical epiphyseal irregularity and hypoplasia without sclerosis. The epiphyses are mottled by irregular mineralization. The growth of the epiphysis may be delayed, but the maturation of epiphyseal fusion is normal. Fraying and irregular mineralization of the metaphysis occur rarely.

The large joints of the lower extremity are the most severely affected, and the shoulders, wrists, and elbows the least (Figs. 1358 and 1359). The tarsal and carpal bones may be hypoplastic. The small bones of the hands and feet are short and broad, but the relative length is maintained (Figs. 1360 and 1361). The changes are at a maximum in the 10th and 12th years.[11]

The lateral portion of the distal tibial epiphysis may be deficient, producing a wedge-shaped epiphysis and slanting of the tibiotalar joint[4] (Figs. 1362 and 1363). About half the cases will show this tibiotalar slant,[9] making it of diagnostic importance, although it is present in other conditions, such as hemophilia, and is not a pathogno-

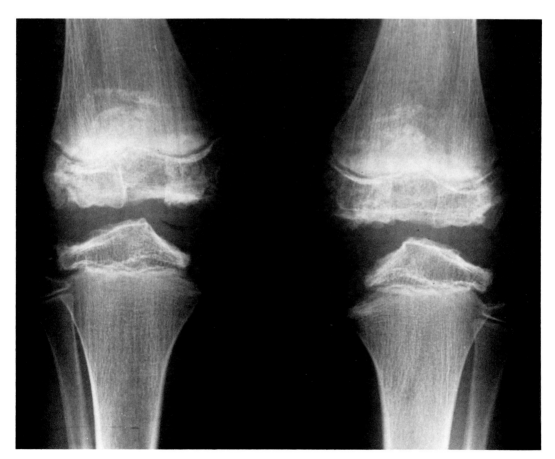

FIG. 1358. EPIPHYSEAL DYSPLASIA MULTIPLEX
Large lower extremity joints are most severely affected. Irregularity of the epiphyses. The metaphyseal areas are relatively unaffected. (Same case as seen in Figures 1359–1361.) (Courtesy of Paul K. Berg, Doctors Hospital, Staten Island, N.Y.)

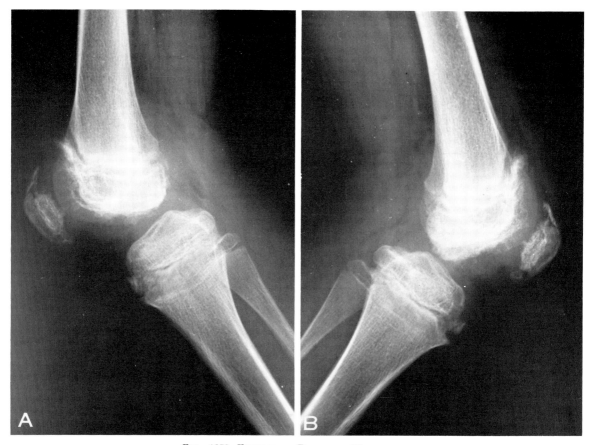

FIG. 1359. EPIPHYSEAL DYSPLASIA MULTIPLEX
(A) Right elbow. (Same case as seen in Figures 1358, 1360, and 1361.) (B) Left elbow. (Courtesy of Paul K. Berg, Doctors Hospital, Staten Island, N.Y.)

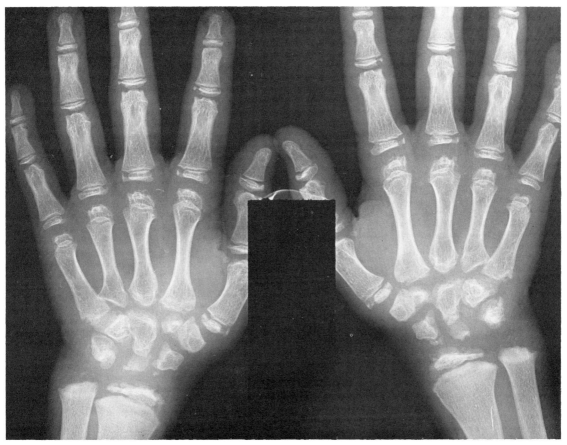

FIG. 1360. EPIPHYSEAL DYSPLASIA MULTIPLEX
Small bones of the hand and feet are short and broad, but the relative length is maintained. The epiphyses are irregular. (Same case as seen in Figures 1358, 1359 and 1361.) (Courtesy of Paul K. Berg, Doctors Hospital, Staten Island, N.Y.)

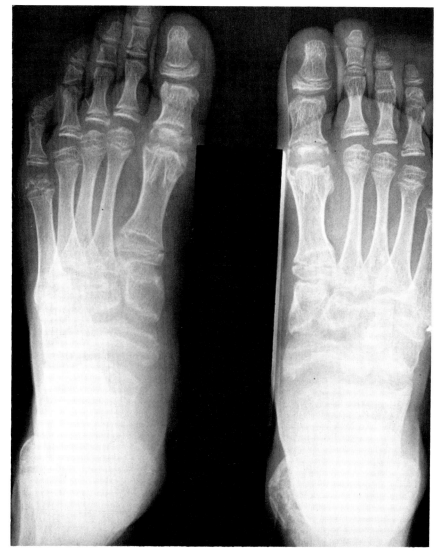

Fig. 1361. Epiphyseal Dysplasia Multiplex
(Same case as seen in Figures 1358–1360.) (Courtesy of Paul K. Berg, Doctors Hospital, Staten Island, N.Y.)

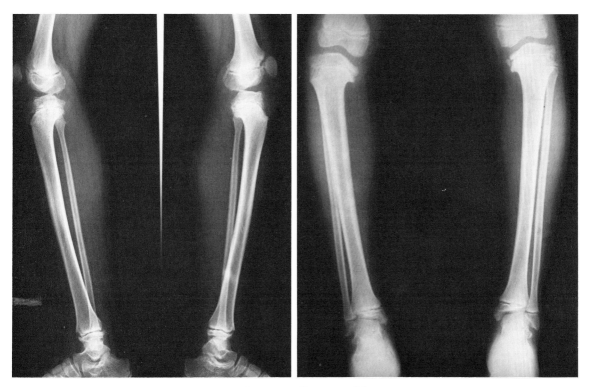

FIG. 1362. EPIPHYSEAL DYSPLASIA MULTIPLEX

Lateral portion of the distal tibial epiphysis is deficient, producing a wedgeshaped epiphysis and resulting in slanting of the tibiotalar joint. Changes in the proximal tibial epiphysis have produced a genu valgum deformity. (*Left*) Lateral views. (*Right*) Anteroposterior view.

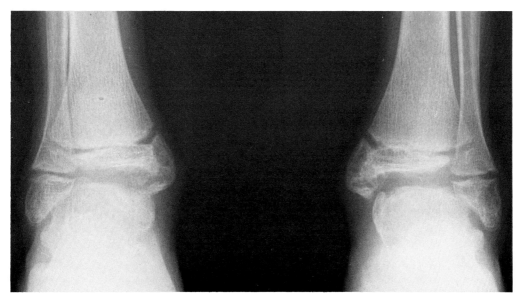

FIG. 1363. EPIPHYSEAL DYSPLASIA MULTIPLEX
(Courtesy of Paul K. Berg, Doctors Hospital, Staten Island, N.Y.)

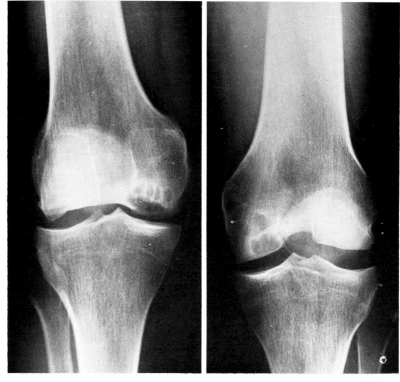

FIG. 1364. EPIPHYSEAL DYSPLASIA MULTIPLEX
There is a tendency for definite and complete improvement in the adult, although the contour of the epiphyseal areas may remain permanently irregular. (*Left*) Right knee, (*right*) left knee. (Courtesy of Paul K. Berg, Doctors Hospital, Staten Island, N.Y.)

monic sign.[10, 12] Vertebral changes, resembling Scheuermann's disease, may occur.[5] Jansen[1] has reported platyspondyly resembling Morquio disease.

There is a tendency toward definite and complete improvement in the adult, although the contour of the epiphyseal areas may be permanently irregular (Fig. 1364). The bones of the hands and feet will remain short and broad, and osteoarthrosis of the hips and knees at tibiotalar slant persists.

Epiphyseal dysplasia multiplex may mimic Legg-Perthes disease, but the involvement of multiple epiphyses and usual lack of metaphyseal change in the former serves to differentiate the two. Dysplasia epiphyseal punctata produces more marked change in the epiphyses, which appear to ossify from multiple centers, and their increased density is not seen in epiphyseal dysplasia multiplex. Morquio disease shows similar epiphyseal changes but its acetabular and vertebral involvement is distinctive.

Osteopoikilosis and osteopetrosis also may cause epiphyseal changes, but their increased density and the diaphyseal changes distinguish them. Fraub[13] has reported mottled epiphyses in association with pituitary gigantism, which may closely resemble epiphyseal dysplasia multiplex. Cretins show similar changes in the epiphyses, but their severely delayed bony maturation sets them apart. The epiphyseal deformities in adults due to stippling may be indistinct.[13]

PERIPHERAL DYSOSTOSIS

Peripheral dysostosis is a chondrodysplasia predominately manifested in the hands and feet.[1-7]

Clinical Features. There is a hereditary factor,[5] but the etiology is unknown. Mental retardation is not reported and there are no physical abnormalities other than the small hands and feet, although scalp alopecia is reported.[5] The individuals are short in stature, usually less than 65 inches.[8]

Roentgenographic Features. Short metacarpals, metatarsals, and phalanges, possibly the result of premature epiphyseal closure,[9] are the hallmark of the most severe form (Figs. 1365 and 1366). At times there is broadening of these bones which is usually more apparent than real. Shortening of one or more bones may occur in the milder forms, and in the severest type epiphyseal irregularities occur in the wrist and ankle. Enchondral bone disturbance is also reported in the ulna, patella, and the distal ends of the tibia and fibula.[1]

Singleton[9] reported long bones showing alteration of the modeling, similar to Pyle disease, and slenderness due to overconstriction and overtubulation.

The metacarpal and metatarsal anomalies of other disorders must be segregated from those of peripheral dysostosis. The following criteria for peripheral dysostosis are suggested[8]: (1) short but not deformed metacarpal and/or metatarsal, (2) stature less than 65 inches, (3) the skeletal survey otherwise normal, and (4) blood chemistry analyses within normal range.

Coned epiphyses are common, but, because they occur in a multitude of conditions, are not diagnostic of peripheral dysostosis (Table 42). The differential diagnosis from other disorders is shown in Table 43.

Since the deformities are usually limited to the hands and feet, there is no problem in differentiation from chondroectodermal dysplasia, achondroplasia, mucopolysaccharidosis, cleidocranial dysostosis, and multiple epiphyseal dys-

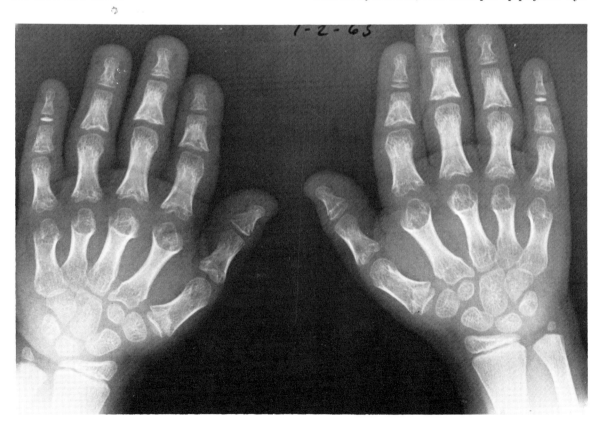

FIG. 1365. PERIPHERAL DYSOSTOSIS
The patient was short statured. Shortening of all the tubular bones. Cone-shaped metaphyses are present in the proximal and distal phalanges and proximal metacarpals. (Courtesy of Dr. Howard Steinbach, Moffitt Hospital, San Francisco, Calif.)

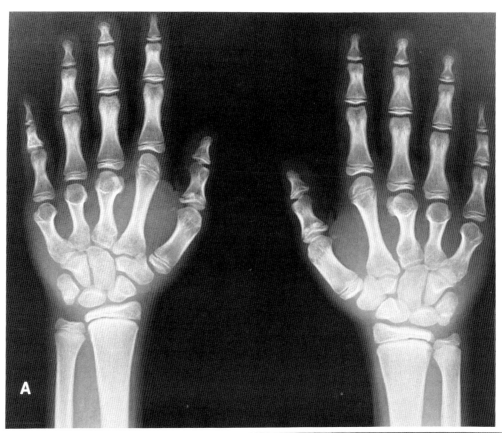

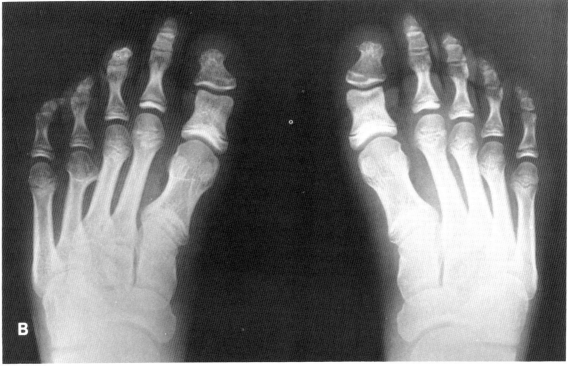

Fig. 1366. Peripheral Dysostosis

(A) Shortening of all tubular bones of the hand, particularly the first, fourth and fifth metacarpals, due to early epiphyseal closure. (B) Shortening of all the tubular bones of the foot, particularly the fourth metatarsal on the left, due to early epiphyseal closure. (C) Anteroposterior view revealing normal spine in this short-statured individual. (D) Lateral projection of spine.

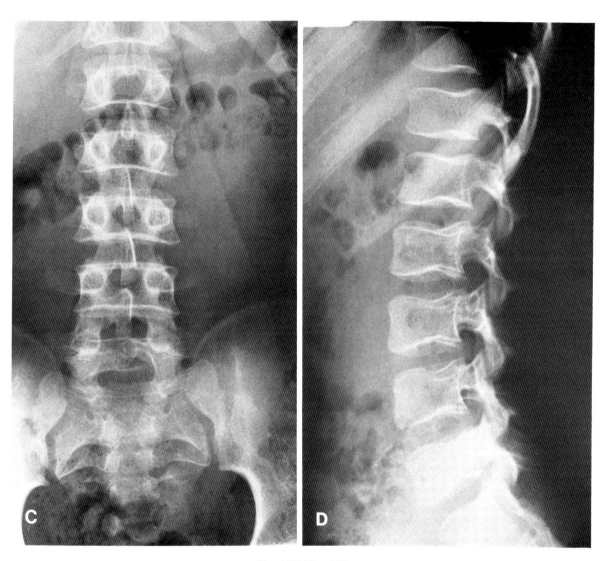

Fig. 1366 (*C* and *D*)

TABLE 42. DISEASES MANIFESTING PHALANGEAL CONE-EPIPHYSES[a]

No recognizable cause	Apert acrocephalosyndactyly
Cleidocranial dysostosis	Phalangeal gigantism
Chondroectodermal dysplasia	Marchesani syndrome
Trichorhinophalangeal syndrome	Pseudohypoparathyroidism
Peripheral dysostosis	Achondroplasia
Osteopetrosis	Multiple exostosis
Dysplasia epiphysialis hemimelica	Sickle cell dactylitis
	Kaschim-Beck disease
Freiberg infraction	Thiemann disease
Prader-Wili syndrome	Otopalatodigital syndrome
Cornelia deLange syndrome	Metaphyseal chondrodysplasia (type McKusick)
Asphyxiating thoracic dysplasia	Diastrophic dwarfism

[a] As shown by the number of diseases associated with coneshaped epiphyses, their occurrence alone is not diagnostic. Newcomb and Keats provide evidence that complete clinical and roentgen studies of families are necessary to exclude peripheral dysostosis as a diagnosis when cone epiphyses are present. (Modified from D. S. Newcomb and T. E. Keats: *American Journal of Roentgenology*, 106: 178, 1969.[8])

TABLE 43. DISORDERS CONSIDERED IN DIFFERENTIAL DIAGNOSIS OF PERIPHERAL DYSOSTOSIS[a]

Brachydactyly Type A-1[10, 11]	Silver dwarfism[14]
Brachydactyly Type B[10, 11]	Biemond's syndrome I[11, 15]
Brachydactyly Type C[10, 11]	Pseudohypoparathyroidism[16, 17]
Brachydactyly Type D[10, 11]	Pseudopseudohypoparathyroidism[17-19]
Brachydactyly Type E[10, 11]	
Brachydactyly Type E with renal anomalies	Gonadal dysgenesis[18, 19]
	Kleinfelter syndrome[19, 20]
Trichorhinophalangeal syndrome[12, 13]	Marchesani syndrome[21]
	Orodigitofacial syndrome[22]
Tabatznick's syndrome[11]	Multiple basal cell nevi-syndrome[8]
Myotonia dystrophica[8]	

[a] This is selected group of disorders that have roentgenographic findings primarily, but not necessarily exclusively, confined to the hand. It is not all-inclusive. (From D. S. Newcomb and T. E. Keats: *American Journal of Roentgenology*, 106: 178, 1969[8]).

plasia. Peripheral dysostosis is also easy to distinguish from Thiemann disease and Dietrick disease, since the former presents as dense and fragmented epiphyses of the fingers and toes, and the latter as ischemic necrosis of the metacarpal epiphyses.[9]

DYSPLASIA EPIPHYSIALIS HEMIMELICA

Synonym: tarsoepiphyseal aclasis.[2]

Eccentric cartilaginous overgrowth of one limb is the hallmark of this condition. It is probably due to a congenital error in development that, for a brief period and in a small area, affects either the preaxial or postaxial portion of the apical cap of a single limb bud.[2]

Roentgenographic Features. Eccentric cartilaginous overgrowth of one or more epiphyses of an upper or lower limb results in asymmetrical enlargement of the epiphyseal center. It is restricted to one-half of the epiphysis, either the medial or lateral half (Figs. 1367 and 1368). There may be an irregular enlargement or many irregular ossification centers, separate from the main epiphysis and so closely packed they seem to be a single bone mass. As the bone matures, the masses fuse and may attach themselves to the epiphysis, or they may be separate from or break from the main epiphysis. Varus and valgus deformities may result if they impinge on an articular surface. Bone shortening is rare.

The distal femur, distal tibia, and astragalus are most often involved,[1-7] but the small bones of the hands and feet may be affected. Carpal or tarsal bone involvement may manifest as premature appearance of the centers, which are irregularly enlarged.[2, 3, 7] The metacarpals and metatarsals may increase in length.[3]

Dysplasia epiphysialis hemimelica has been reported in two generations of the same family who had other combinations of cartilaginous tumors including intracapsular chondroma, extraskeletal osteochondroma, and typical osteochondroma.[8]

Epiphyseal osteochondromas[9] are probably a part of the same entity. Its asymmetrical epiphyseal involvement, its confinement to a single limb, and the lack of dwarfism distinguish it from multiple epiphyseal dysplasia and punctate epiphyses. Cretinoid epiphyseal dysgenesis may show spotty calcifications, but growth is normal.

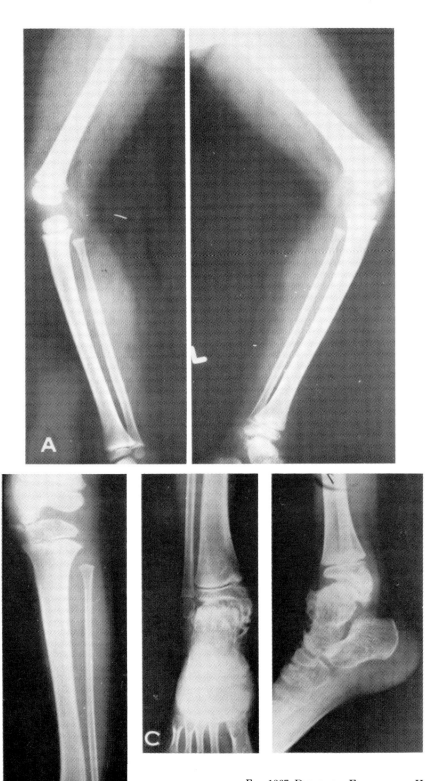

FIG. 1367. DYSPLASIA EPIPHYSIALIS HEMIMELICA
(A) Lateral projections of both legs reveal abnormalities of the distal femoral epiphyses and distal tibial epiphyses (*arrows*). Irregular enlargement of the epiphyseal centers and the distal femoral epiphysis extends into the knee joint. Secondary changes are present in the distal tibial metaphysis and the calcaneus. (B) Anteroposterior projection of the leg reveals the abnormality limited to the medial aspect of the epiphyses. (C) The enlargement of the distal tibial epiphysis is limited to the medial and posterior aspects. (Courtesy of Dr. Robert M. Peck, Tripler General Hospital, Hawaii.)

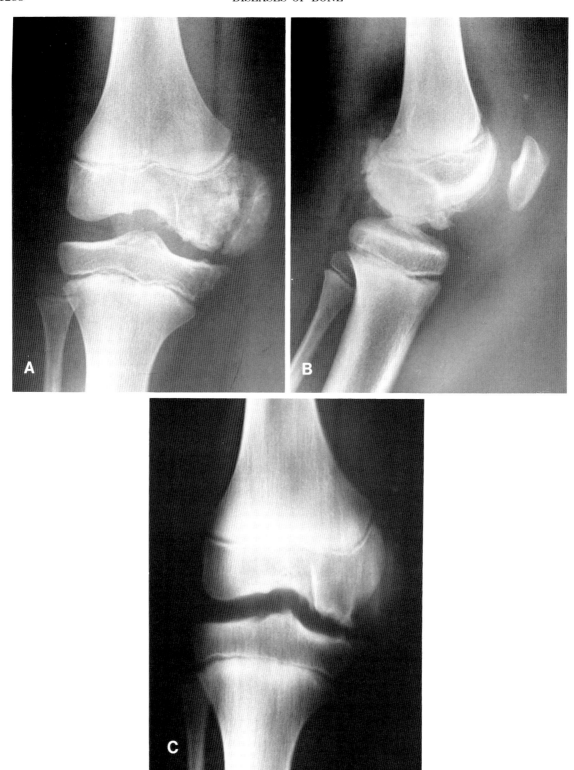

FIG. 1368. DYSPLASIA EPIPHYSIALIS HEMIMELICA

(A) Anteroposterior view. Large osteochondroma of the medial condyle and irregularity of the medial aspect of the tibial plateau. (B) Lateral view. New bone formation on the medial condyle. (C) Tomography. Irregularity of the joint surface and the osteochondroma formation.

OSTEOGENESIS IMPERFECTA

Osteogenesis imperfecta is a generalized disorder of connective tissue involving bone, sclera, inner ear, skin, ligaments, tendons, and fascia. Inherited as an autosomal dominant, the disease, clinically, is characterized by fragile bones, with or without the blue sclera and deafness.

A host of names have been applied to the disease, all concerned with the abnormal bones. Among these may be included "osteogenesis imperfecta,"[1] "mollities ossium,"[2] "fragilitas ossium,"[3] "osteopsathyrosis idiopathica,"[4] "osteogenesis imperfecta congenita" (OIC), and "osteogenesis imperfecta tarda" (OIT). Osteogenesis imperfecta tarda may be subdivided into the gravis (tarda-Type I) and levis varieties (tarda-Type II),[5] which may be based on the age of the child when fractures begin. Perhaps a more appropriate classification is the clinical variation of bowing: when there is bowing of the lower extremities the patient probably has tarda-Type I, and in those without bowing, tarda-Type II.[6] All the above represent the same fundamental defect varying in severity, time of onset, and points of involvement.

Clinical Features. The clinical features of osteogenesis imperfecta vary according to severity of the disease. The two main clinical types are osteogenesis imperfecta congenita, occurring early in life, and osteogenesis imperfecta tarda, occuring later.[7]

Osteogenesis imperfecta congenita is the more severe form of the disease. The main characteristics are its paper thin skull (caput membranaceum), and short, often deformed, extremities (micromelia). The skull is soft because of imperfect ossification. The short extremities, resulting from the bowing and deformity of multiple fractures, sometimes suggest achondroplasia. Death in utero or soon after birth is usually caused by intracranial hemorrhage.

Osteogenesis imperfecta tarda is a less severe form of the disease. Babies affected may exhibit no apparent abnormalities at birth. Fractures caused by trivial trauma may occur early or the bone fragility may not become evident until the child is weight-bearing, until puberty, or even late in adult life.

Blue sclerae are found in most patients with osteogenesis imperfecta but occasionally the sclerae are normal, even in unmistakable cases. Often the sclera immediately surrounding the cornea is not colored and appears as a white ring ("Saturn's ring").[8]

Deafness, clinically indistinguishable from other forms of otosclerosis, may appear during childhood or the teens. In women, it may be first observed during pregnancy.

The fragility of bones often seems to decrease as the child grows older. Female patients, particularly, tend to show improvement at puberty. This has led to the therapeutic use of estrogens, with some benefit.

Pathologic Considerations. The bone abnormality seems to be due to a failure in the deposition of normal collagen fibers in bone matrix. The collagen is immature, closely resembling the reticulum fibers found in fetal bones.[9] Enchondral bone growth proceeds normally up to the point of cartilaginous calcification, after which no true bone osteoid is laid down, particularly in the metaphysis, and the resultant brittle calcified cartilage fractures easily. Cortical bone is also abnormal and, for some reason, lacks a mature collagen matrix. Here, too, the collagen fibers laid down by the osteoblasts are disorganized, closely simulating fetal matrix. Stated differently, primary bone tissue is replaced rapidly by secondary bone, but, after birth, the secondary bone tissue is not formed normally.

According to the literature[8,10] osteogenesis imperfecta may be due to an enzyme deficiency resulting in abnormal connective tissue proteins. Some have reported fewer total osteoblasts in the condition, suggesting a relationship to enzymes. It is but fair to note that this decrease in osteoblasts has not been universally recorded.

Iliac biopsy specimens on 16 patients (8 severely affected, 7 moderately, and 1 mildly) showed an increased number of osteocytes and a decreased number of fractional areas of bone as compared to the control. The findings corresponded with the clinical severity of the disease. Patches of fiber bone were the only marked abnormalities found in the mild case, although all other parameters such as number of osteocytes, fractional areas of bone and loss of resorptive surface differed from the controls.[11]

Roentgenographic Features. The cardinal roentgen findings in osteogenesis imperfecta are fractures and thin, defective cortical bone. Because cortical thinning may be found in many forms of osteoporosis, the association of multiple fractures is required to establish the diagnosis (Fig. 1369). The roentgen diagnosis may be made in utero[12] (Fig. 1370).

Fairbank[13] divided the bone changes into three types: thick (Fig. 1371), slender and fragile (Fig. 1372), and cystic (Fig. 1373). These types are different expressions of the same disease process.[8]

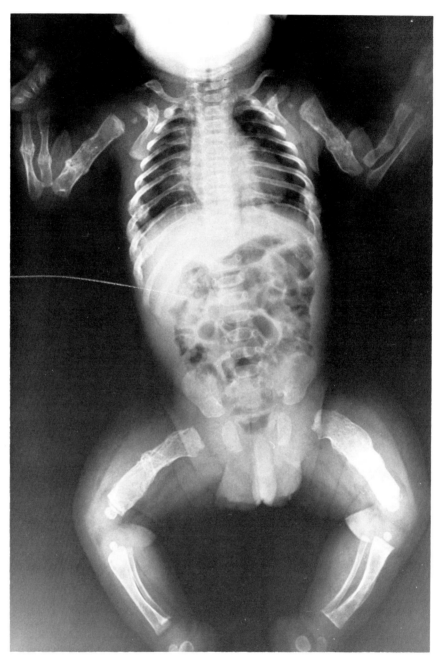

Fig. 1369. Osteogenesis Imperfecta

Cortical thinning, multiple fractures, and deossification characterize the roentgenographic changes of osteogenesis imperfecta. (Courtesy of John W. Hope, Children's Hospital, Philadelphia, Pa.)

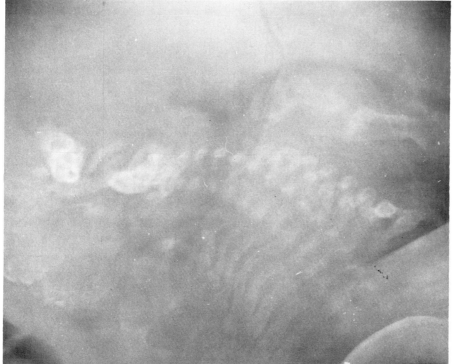 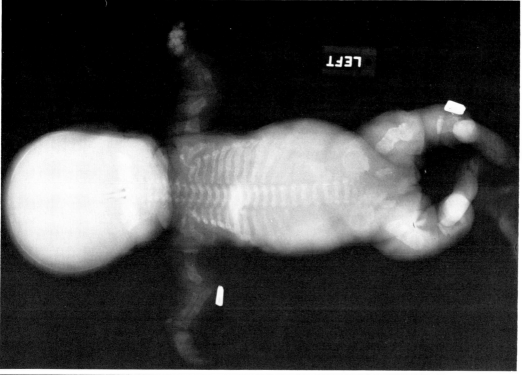

FIG. 1370. OSTEOGENESIS IMPERFECTA

(*Left*) In utero demonstration of osteogenesis imperfecta. (*Right*) Roentgenogram of same stillborn baby.

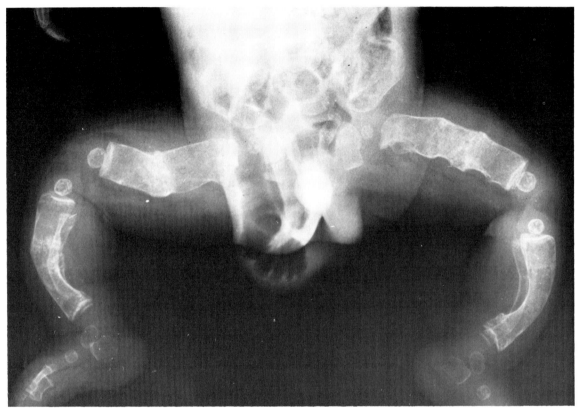

Fig. 1371. Osteogenesis Imperfecta

"Thick bone" type. Increase in the width of the bones with multiple fractures and bowing deformities. The cortices are thin. (Courtesy of Paul K. Berg, Doctors Hospital, Staten Island, N.Y.)

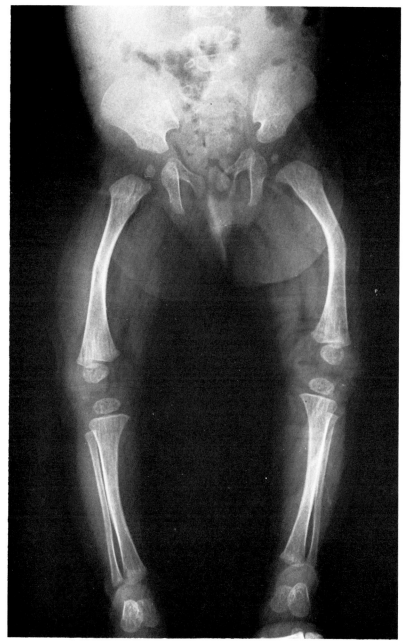

FIG. 1372. OSTEOGENESIS IMPERFECTA

"Slender bone" type. The bones are long and thin. There is bowing of several of the long bones due to fractures. A peculiar parallel deossification in the metaphyseal areas has been noted in several cases. (Courtesy of John W. Hope, Children's Hospital, Philadelphia, Pa.)

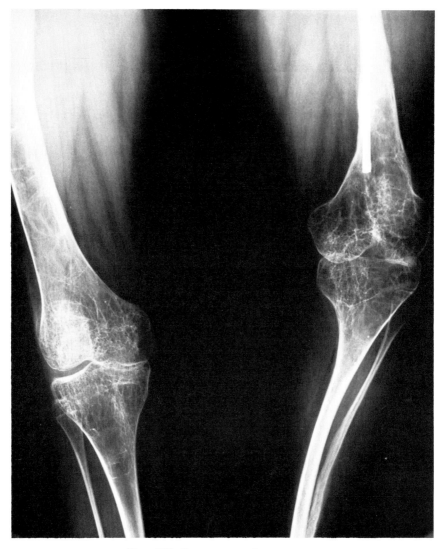

FIG. 1373. OSTEOGENESIS IMPERFECTA

"Cystic" type. The distal ends of both femurs show radiolucent areas as do the proximal ends of the tibiae. Due to marked deossification of disuse, although they resemble areas seen in the "cystic" type. (Courtesy of John W. Hope, Children's Hospital, Philadelphia, Pa.)

1. *Thick bone,* found usually in newborn infants, is a misnomer; rather it represents an increase in the width or diameter of the bone with bowing and deformity (Fig. 1374). The cortex is not thick; indeed, it is as paper-thin as in the other forms of osteogenesis imperfecta. Its density is greatly diminished, unless an abundance of callus has been formed as a result of fractures. The greater width, occurring usually in the proximal ends of the extremities (humeri and femora), gives the arms and legs the short, stubby appearance of achondroplasia, for which it may be mistaken.

2. *Slender fragile bone* is often found in utero and immediately after birth. Bone shafts are thin, and their cortex also thin and of decreased density. Bone trabeculae are poorly defined. The ends of the bone appear flared, because enchondral bone growth is normal up to the point of cartilaginous calcification. Deformities caused by fractures are common. Dense transverse lines of bone are, at times, noted at the junction of the metaphysis and epiphysis, due to microfractures of the brittle calcified cartilage. Their lack of symmetry differentiates them easily from growth lines. The bones of the calvaria are exceedingly thin, revealing the mosaic pattern of wormian bones which often persists throughout life (Fig. 1375) and is most marked along the lambdoidal suture. Skull fractures are rare. The tables are usually thin but may be abnormally thick. The vertebral bodies are decreased in density with the biconcave configuration of expanded intervertebral disks; multiple Schmorl nodules may be present (Fig. 1376). A general flattening is frequent. Sometimes they are coin-shaped, with anterior narrowing limited to one or two vertebrae.

3. The *cystic type* of osteogenesis imperfecta, according to Fairbank,[13] is rare. The condition starts at birth and is usually progressive. Multiple cystlike areas appear in the bones, particularly in the lower extremities. Fractures appear as the disease progresses, and eventually the long bones are markedly deformed by bowing. As in other types of osteogenesis imperfecta, the cortex is thin and the bone decreased in density.

In osteogenesis imperfecta fractures readily heal with normal or exuberant callus[14] (Figs. 1377 and 1378). Pseudoarthrosis occasionally develops, and this may be the first clinical manifestation of the disease. Sometimes the bone cortex is only slightly thinner than normal, passing as normal bone unless other findings create suspicion. These patients, too, not infrequently sustain fractures after trivial injuries (Fig. 1379).

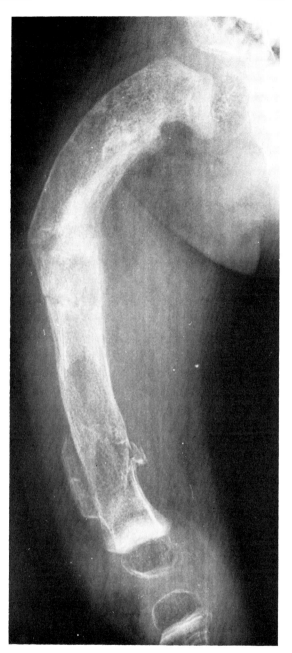

FIG. 1374. OSTEOGENESIS IMPERFECTA
"Thick bone" type; found usually in newborn infants. The width of the bone is wider rather than the cortex being thick. Bowing due to fractures.

The latter, plus the blue sclera, make the proper diagnosis easily apparent.

Multiple symmetrical fractures may occur with osteomalacia. In adults, there is little difficulty in differentiating from osteogenesis imper-

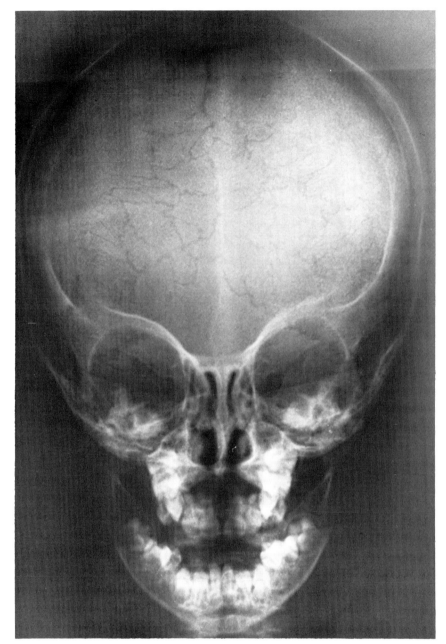

Fig. 1375. Osteogenesis Imperfecta
Mosaic pattern of wormian bones often persists throughout life. (Courtesy of John W. Hope, Children's Hospital, Philadelphia, Pa.)

fecta. Occasionally, idiopathic juvenile osteoporosis occurs just before puberty; this could be mistaken for osteogenesis imperfecta.

Multiple symmetrical fractures of bone may also develop in patients with elevated serum pyrophosphate levels, an apparent inhibitor of calcification.[15, 16] Although the serum pyrophosphate may be elevated in patients with osteogenesis imperfecta[17] and some other metabolic disorders, the difference is clinically evident.

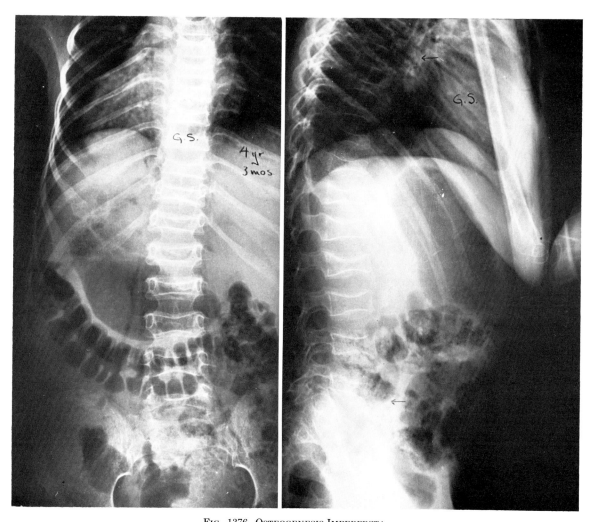

Fig. 1376. Osteogenesis Imperfecta
Vertebral bodies are decreased in density with biconcave configuration and expanded intervertebral disks. (*Left*) Anteroposterior view, (*right*) lateral view. (Courtesy of John W. Hope, Children's Hospital, Philadelphia, Pa.)

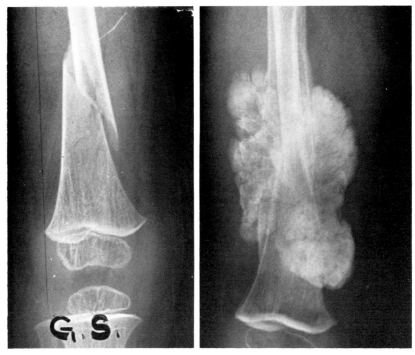

FIG. 1377. OSTEOGENESIS IMPERFECTA

(*Left*) At time of injury, a spiral fracture of the distal end of the femur is noted; (*right*) 7 weeks later, exuberant callus formation is present. (Courtesy of John W. Hope, Children's Hospital, Philadelphia, Pa.)

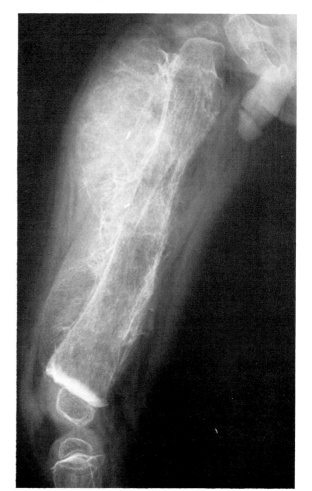

FIG. 1378. OSTEOGENESIS IMPERFECTA
Bone is wider than normal and multiple fractures are present. Exuberant callus formation. (Courtesy of John W. Hope, Children's Hospital, Philadelphia, Pa.)

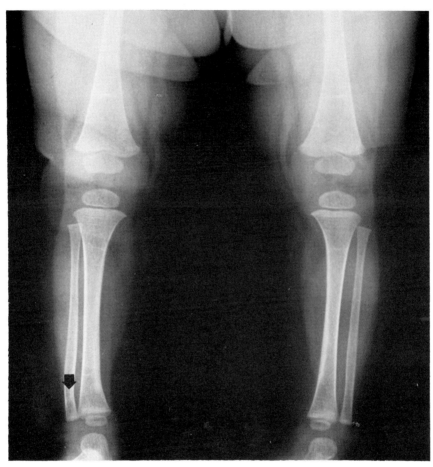

Fig. 1379. Osteogenesis Imperfecta
Sometimes the bone cortex is only slightly thinner than normal and may be overlooked unless other findings create suspicion. Minimal bowing of the distal end of the right fibula indicating a fracture (*arrow*). (Courtesy of John W. Hope, Children's Hospital, Philadelphia, Pa.)

ERYTHROGENESIS IMPERFECTA

Synonyms: congenital hypoplastic anemia,[2] pure red cell anemia, congenital regenerative anemia,[3] erythrophthisis,[4] idiopathic hypoplastic anemia,[5] chronic erythroblastopenia,[6] erythrodisgenetic anemia,[7] chronic erythrocytic hypoplasia,[8] primary red cell aplasia.[9]

Erythrogenesis imperfecta is an idiopathic, slowly progressive anemia, that begins in early infancy[2, 10] and is associated with congenital anomalies in one-third of the patients.[2, 11] Familial incidence is recorded.[2, 12]

Clinical Features. Erythrogenesis imperfecta usually begins within the first 3 months of infancy, but may commence as late as 4 years.[2] The median age is 1 month and the average age is 6 months. The characteristic bone marrow (hypoplasia or aplasia of erythroid cells) causes a normocytic, normochromic anemia with a normal number of platelets and leukocytes. Many patients were premature or small babies.[2]

Altman and Miller[13] describe urinary excretion of anthranilic acid, a breakdown product of tryptophan, and suggest the etiology is a disturbance of tryptophan metabolism. Minagi and Steinbach[11] reviewed 74 cases and found congenital anomalies in one-third (Table 44). The most frequent anomalies are retarded growth, mental retardation, congenital heart disease, and skeletal and renal anomalies. Blue sclera was noted in 2 cases. The prognosis is guarded, although periodic transfusions and steroid therapy may prolong life for years. Spontaneous remissions occur, especially after puberty.

Roentgenographic Features. Numerous metaphyseal dense growth arrest lines are present, presumably due to poor nutrition periodically improved by transfusion. Osteoporosis and dorsal compression fractures may occur. Retarded bone age has been noted. A supernumerary thumb and an extra thumb phalanx are described. Clinodactyly of the fifth finger may occur due to a hypoplastic middle phalanx (Fig. 1380).

Renal anomalies reported are bilateral double ureters, with congenital hydronephrosis, congenital dysplasia of one kidney, and unilateral absence of kidney and ureter.[11] These anomalies in a newborn may be confused with Fanconi's anemia, but clinical features and earlier age of onset of hematologic deficiency aid in identifying erythrogenesis imperfecta.

TABLE 44. Congenital Anomalies with Erythrogenesis Imperfecta Reported in 74 Patients[a]

Retarded growth	10
Mental retardation	4
Congenital heart disease	3
Skeletal anomaly	3
Congenital defect of osseous development (1)	
Extra phalanx of one thumb (1)	
Extra thumb (1)	
Renal anomaly	3
Bilateral double ureters with congenital hydronephrosis (1)	
"Congenital dysplasia" of one kidney (1)	
Absent left kidney and ureters (1)	
Webbed neck	2
Blue sclerae	2
Malformed ear, inguinal hernia, cleft palate, ptosis	1 of each

[a] From H. Minagi and H. Steinbach: *American Journal of Roentgenology*, 97: 100, 1968.[11]

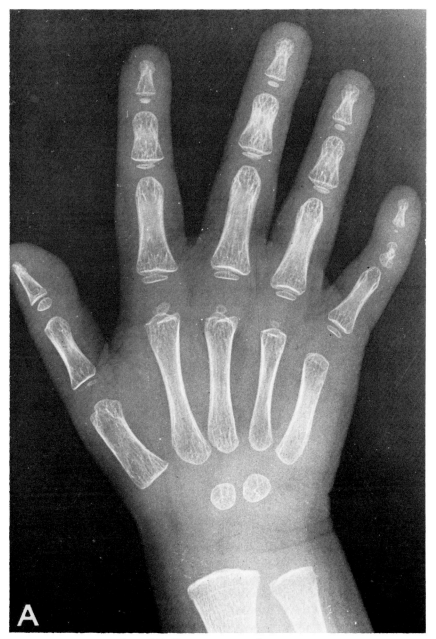

Fig. 1380. Erythrogenesis Imperfecta

(A) Cortical thinning and trabecular prominence due to osteoporosis. The fifth middle phalanx is hypoplastic. Growth lines in the radius and ulna. (B and C) Deossification and anterior notching in the lower dorsal vertebrae. (Courtesy of Dr. Howard Steinbach, Moffitt Hospital, San Francisco, Calif.)

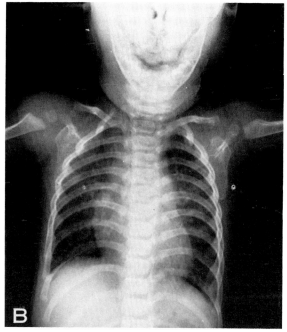
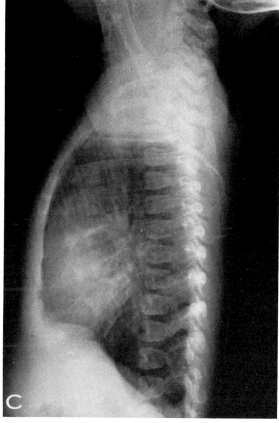

Fig. 1380 (B and C)

FANCONI ANEMIA

Fanconi anemia is a severe hypoplastic anemia with pancytopenia, often associated with brown skin pigmentation and multiple skeletal and urogenital anomalies (Table 45).

Smith[1] in 1919 first reported the syndrome of aplastic anemia and skin pigmentation. Fanconi[2] in 1927 added microcephaly with mental retardation and short stature. Uehlinger[3] in 1929 reported multiple skeletal, and urologic anomalies associated with the syndrome. Subsequently, over 100 cases are recorded[4-7] establishing the distinctive entity.

Clinical Features. Fanconi anemia is probably transmitted as an autosomal recessive.[8] There is a male predilection, and the disease is usually fatal within 5 years after the onset of anemia. Usually, characteristic congenital deformities are discovered before onset of blood dyscrasia[5] but thrombocytopenia may be present.[2] Either anemia[9] or congenital anomalies[10] may be the sole expression of the syndrome. The patients' families show a high incidence of leukemia.[11]

Excellent literature reviews tabulate the congenital anomalies in Fanconi syndrome, expecially those recorded by Minagi and Steinbach,[5] and McDonald and Goldschmidt.[9]

Skin pigmentation, due to melanin deposits,[12] is the most consistent abnormality, found in 74% of the patients. This tends to affect the skin of the trunk, axilla, groin, and neck. The pigmentation usually occurs before hematologic manifestations, although it may appear after anemia is evident.

The hematologic aspects are usually delayed, but can occur at any time between 17 months and 22 years of age. Frequently an acute febrile illness ushers in the clinical manifestations which are usually related to anemia, and include pallor, fatigue, and palpitations. Pancytopenia is most often present when symptoms commence. Anemia, thrombocytopenia, or leukopenia may occur alone, or in any combination. The anemia is usually of the macrocytic hyperchromic type.[5] Bone marrow hypoplasia may occur without peripheral blood changes.

TABLE 45. CONGENITAL ANOMALIES WITH FANCONI
ANEMIA REPORTED IN 68 PATIENTS[a]

Anomaly	No.	Percentage
Abnormal pigmentation	50	74
Skeletal anomalies	46	68
Retarded growth	37	54
Hypogonadism	27	40
Small head	19	28
Renal anomalies	20	29
Microphthalmia	13	19
Strabismus	11	16
Hyperreflexia	11	16
Undescended testis	7	16[b]
Cardiovascular anomalies	5	7
Deafness	4	6
Anomaly of external ears	3	4
Adrenal absent or hypoplastic	3	4
Inguinal hernia	3	4
Obesity	3	4
Hypospadias	2	5[b]
Hydrocephalus	2	3
Gynecomastia	2	3
Septate bladder, blue sclerae, anal stenosis, sacrococcygeal sinus, Meckel diverticulum, and aqueductal stenosis with absent septum pellucidum	1 of each	

[a] Modified from H. Minagi and H. Steinbach: *American Journal of Roentgenology,* 97: 100, 1968.[11]

[b] Percentage of 43 males.

Bleeding tendencies may include subcutaneous hemorrhage, epistaxis, and gastrointestinal bleeding. The malformations associated with Fanconi anemia are listed in Table 45. The skeletal anomalies are described below.

Roentgenographic Features. Anomalies of the radial component of the upper extremity are most distinctive and strongly suggest the diagnosis. Deformity of the phalanges of the thumb, the radial side of the forearm, and wrist are commonly associated anomalies (Table 46). Absence or hypoplasia of both thumb phalanges is most common (Fig. 1381). Usually both phalanges are hypoplastic, but occasionally the distal phalanx alone. Hypoplasia of the second and fifth proximal phalanges is recorded.[5, 13, 14] The first metacarpal may be hypoplastic with normal phalanges[15] (Fig. 1382). With absence of the phalanges, aplasia of the first metacarpal is the rule. Occasionally a supernumerary thumb is present.[16]

Aplasia of the radius may be unilateral or bilateral and is associated with absence of the corresponding thumb.[5] Either end of the radius may be aplastic or hyperplastic.[6] Absent or hypoplastic carpals (navicular and great multangular) occur and delayed carpal ossification is frequent.[6] Dwarfism is slight to moderate and minimal microcephaly may be detected.[5, 6] Renal anomalies are present in approximately 25% of these patients and include aplasia, ectopia, and horseshoe kidney.[4] The early differential diagnosis of Fanconi anemia depends on this association of rare anomalies. The anemic patient without typical anomalies cannot be diagnosed roentgenographically.

Isolated aplasia or hypoplasia of the radial components of the upper extremity may occur.[17-19] Whenever radial component anomalies occur, Fanconi anemia is suspected, but, if they are isolated, only subsequent developments will establish the diagnosis. If the radial component anomaly is not isolated, other congenital anomalies may distinguish the condition from Fanconi anemia, for many do not occur with it, including congenital heart disease, harelip, cleft palate, hydrocephalus, kyphosis, scoliosis, torticollis, rib deformity, hernia, and lung aplasia.[2, 18]

Kirkpatrick et al.[20] described 5 patients with a complex of anomalies associated with tracheoesophageal fistula and esophageal atresia, 3 of them with thumb hypoplasia. Juhl[6] refers to a similar case. However, this syndrome is easily differentiated by the tracheal and esophageal anomalies, not reported in Fanconi anemia.

Radial component abnormality of the upper extremity occurs in the Holt-Oram syndrome, but the associated anomalies include congenital heart disease, dislocation of the great multangular, carpal fusion, hypoplasia of the long bones, and pectus excavatum, distinguishing it from Fanconi anemia. Hypoplastic thumbs are described with trisomy 18[21] but this entity is easily differentiated.

Erythrogenesis imperfecta, described above, is a slowly progressive congenital hypoplastic anemia beginning in early infancy,[22, 23] which may be confused with Fanconi anemia, but associated anomalies are not as frequent or severe.[24]

TABLE 46. SKELETAL DEFORMITIES IN 44 PATIENTS WITH
FANCONI ANEMIA[a]

Deformity	Incidence
Absent, hypoplastic, or supernumerary thumb	34
Hypoplastic or absent radius	9
Congenital hip dislocation	4
Webbing, second and third toes	4
Flat feet	2
Klippel-Feil deformity	2
Club foot	1
Extra terminal phalanges of third and fourth toes	1
Sprengel deformity	1

[a] From H. Minagi and H. Steinbach: *American Journal of Roentgenology,* 97: 100, 1968.[11]

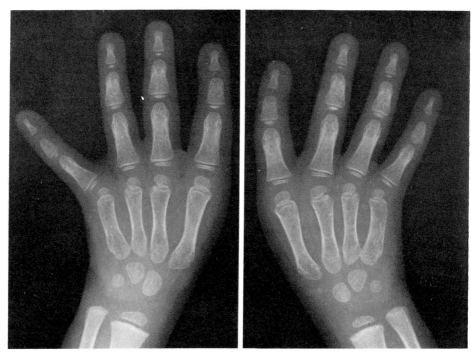

Fig. 1381. Fanconi Anemia
Absence of both thumbs.

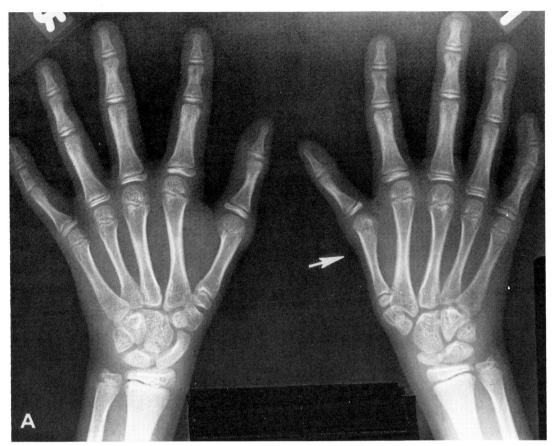

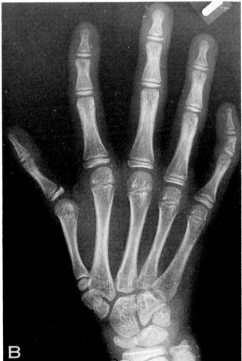

Fig. 1382. Fanconi Anemia
(A) Hypoplasia of the left metacarpal (*arrow*) and abduction of the thumb. Clinodactyly of both fifth fingers due to hypoplasia of the middle phalanx. (B) Left hand.

RADIAL APLASIA-THROMBOCYTOPENIA

Radial aplasia with thrombocytopenia is a rare disorder which must be distinguished from Fanconi pancytopenia.

Clinical Features. Aplasia or hypoplasia of the radius and thrombocytopenia are present at birth; transmission is by an autosomal recessive mode. There may or may not be cardiac defects, and anemia is frequent and out of proportion to apparent blood loss.[1,2]

Death occurs in early infancy in about 50% of the patients, as a result of hemorrhage, but the hematologic disorder improves with advancing age, and with early adequate treatment fatality is less likely.

Roentgenographic Features. Radial aplasia or hypoplasia (Fig. 1383) is usually bilateral and often associated with ulnar hypoplasia and defects of the hands, legs, and/or feet. Brachydactyly, syndactyly, short humerus, hypoplastic shoulder girdle, dislocation of the hip, and talipes deformity are reported.[3]

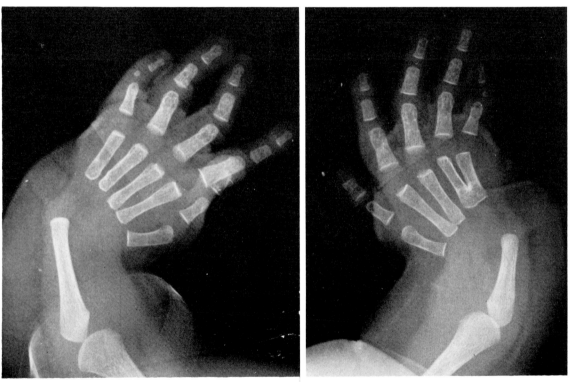

FIG. 1383. RADIAL APLASIA THROMBOCYTOPENIA
There is bilateral radial aplasia. The ulna is hypoplastic. There is hypoplasia of the fifth middle phalanges.

MELNICK-NEEDLES DYSPLASIA

This is a rare hereditary dysplasia which involves most of the bones.[1] It is probably transmitted by an autosomal recessive mode.

Roentgenographic Features. S-shaped tibial and radial bowing is the hallmark of the disease (Figs. 1384 and 1385). There is flaring of the long bone metaphyses, shortening and cortical irregularity, and coxa valga. The skull reveals delay in closure of the anterior fontanel and sclerosis of the base and mastoids. There is micrognathia, overbite, and tooth malalignment.

The ribs show cortical irregularity and a ribbon appearance. The vertebrae are increased in height and the lumbar disk spaces are narrowed. Excessive concavity of the anterior vertebral bodies produces a double beak appearance in the upper dorsal and lumbar spine. The iliac crests are flared, the obturator foramina narrowed and the ischii tapered. Sternal ossification is delayed. The clavicles may have cortical irregularity and flaring.

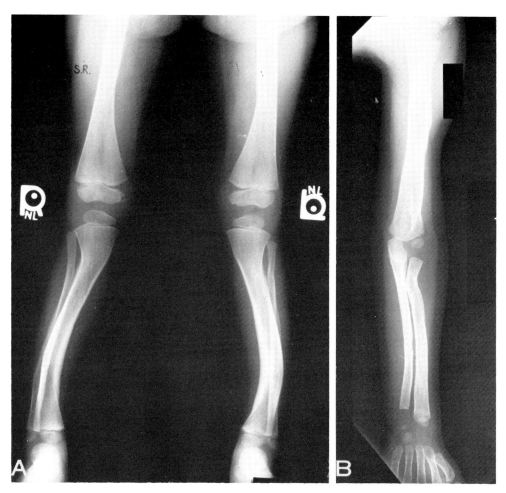

FIG. 1384. MELNICK-NEEDLES DYSPLASIA
(A) S-shaped bowing of the tibiae. (B) Angulation of the upper end of the radius and radial bowing of the ulna.

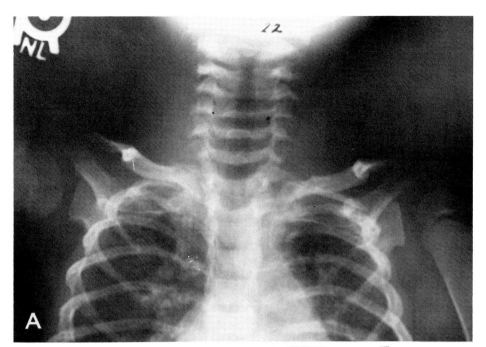
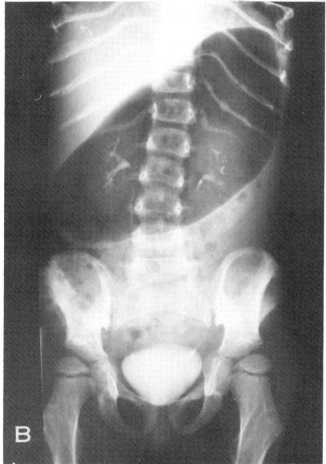

Fig. 1385. Melnick-Needles Dysplasia

(A) Ribbonlike irregularity of the ribs and deficient ossification of the distal ends of the clavicles. (B) Ribbonlike deformity of the ribs and irregular ossification of the pelvis and femora. Opaque material in the urinary tract. (C) Angulation and bowing of the upper end of the radius. (Courtesy of Dr. Robert M. Peck, Tripler General Hospital, Hawaii.)

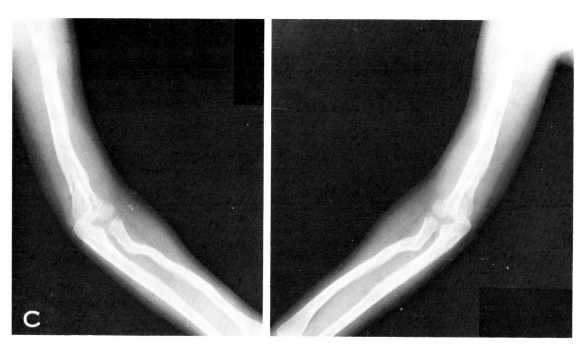

Fig. 1385(C)

PACHYDERMOPERIOSTOSIS

Synonyms: osteodermatopathia, hypertrophicans,[1] Touraine-Solente-Golé syndrome.[2]

Clinical Features. Onset is usually in adolescence, but may begin at any time from 3 to 38 years of age.[3] It consists of cylindrical skin thickening of the forearms and legs, spadelike enlargement of the hands and feet, terminal phalangeal clubbing, and large skin folds of the face and scalp. It progresses during the late 20s and 30s, and then stabilizes. It is more commmon in males, and thought to be transmitted by an autosomal dominant mode.[4, 5]

Roentgenographic Features. Periosteal reaction is most striking in the phalanges of the hands and feet and in the distal long bones. It is indistinguishable from other forms of periosteal reactions, which include pulmonary osteoarthropathy and thyroid acropachy (Fig. 1386). Acro-osteolysis may occur.[6–9] A thickened skin should suggest the diagnosis. Thyroid acropachy may be confused with it, but the history serves to distinguish the two.

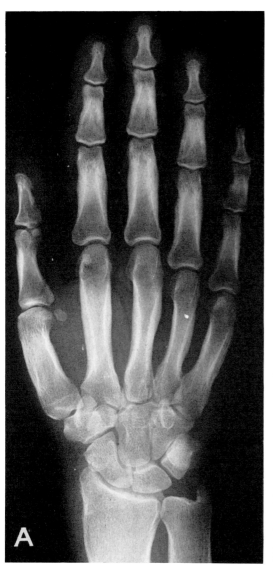
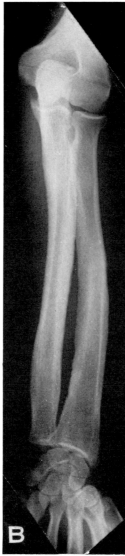
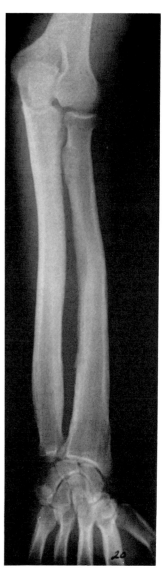

FIG. 1386. PACHYDERMOPERIOSTOSIS

Periosteal reaction has caused thickening of most of the bones. (*A*) Hand. Thickening is most noticeable in the metacarpals and proximal phalanges. This may be confused with thyroacropachy but the history serves to distinguish the two. (*B*) Thickening of the radius and ulna due to periosteal reaction. (*C*) Both legs. Periosteal thickening is evident. (Courtesy of Dr. Howard Steinbach, Moffitt Hospital, San Francisco, Calif.)

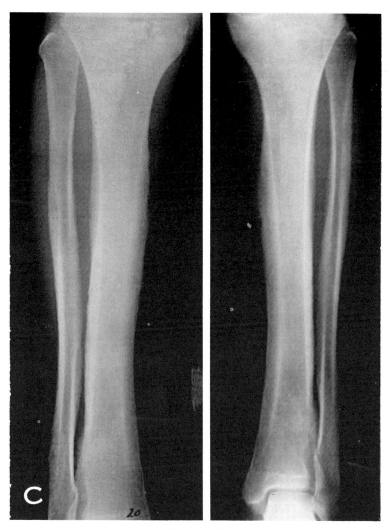

Fig. 1386(C)

RUBINSTEIN-TAYBI SYNDROME

The Rubinstein-Taybi[1] syndrome produces small statured individuals with mental retardation, broad thumbs and big toes, and other skeletal anomalies.[2]

Clinical Features. There is no familial history of chromosomal abnormalities, and the only consistent laboratory finding is a lack of B and E antigens.[2] Many anomalies are described, but Coffin[2] reports the common features as follows: (1) retarded mental and somatic development, (2) characteristic facies, (3) wide thumbs and first toes, (4) hypotonic muscles and joints, (5) peculiar stance and gait, and (6) severe swallowing difficulty in early infancy.

The clinical abnormalities are present at birth. The characteristic facies are: narrow and prominent forehead, nose beaklike due to a downward curved prolongation of the septum, palpebral fissures slanted downward, and ears positioned low. Hirsutism, scoliosis, undescended testes, angulated penis, and congenital heart disease are reported.[1, 2]

Roentgenographic Features. The broad thumbs and first toes are most prominent at the distal phalanges (Fig. 1387). The distal thumb phalanx may be deviated laterally and the big toe medially. Partial phalangeal duplication and delayed skeletal maturation is reported.[2] The

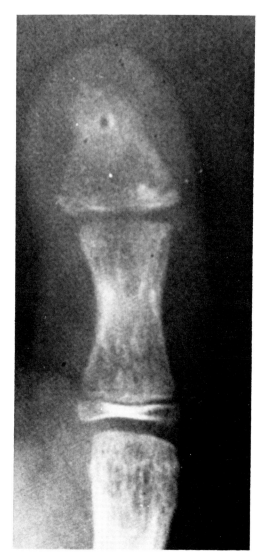

FIG. 1387. RUBINSTEIN-TAYBI SYNDROME
Roentgenogram of the thumb shows the broad distal phalanx. (Courtesy of Dr. Howard Steinbach, Moffitt Hospital, San Francisco, Calif.)

nasal septum is elongated and convex anteriorly.

The differential diagnosis includes other conditions which resemble some features of this syndrome. They are: acrocephalosyndactyly, certain forms of autosomal trisomy, Hallermann-Streiff syndrome, mongolism, birdheaded dwarfism, Treacher Collins syndrome,[3,4] and mandibulofacial dysostosis.[5] The Rubinstein-Taybi syndrome is distinguished by the broad thumbs and first toes.

NEVOID BASAL CELL CARCINOMA SYNDROME

Nevoid basal cell carcinoma syndrome[1] consists of multiple cutaneous basal cell carcinomas associated with jaw cysts, ectopic calcifications, and skeletal anomalies.[2-5]

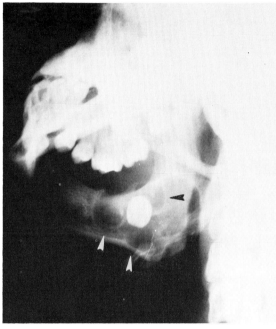

FIG. 1388. NEVOID BASAL CELL CARCINOMA
Multiple mandibular cysts. (Courtesy of Dr. Gerald Dodd, M. D. Anderson Hospital, Houston, Tex.)

Clinical Features. Nevoid basal cell carcinoma is inherited as an autosomal dominant trait.[6-10] The hallmark of the disease is the presence of multiple nevoid basal cell carcinomas, most commonly on the face, nose, mouth, chest, and back. The palms and soles are uncommon sites. They are detected in patients at an average age of 19 years.[9] They may develop before puberty and tend to be clinically benign. After puberty most lesions are aggressive and may metastasize. Dodd[10] reports the infrequency of nevoid basal cell carcinomas in young patients, increasing linearly to 100% in patients over 30.

Pitlike defects of the palms and soles may be found in the majority of patients. They have been noted as early as 5 years and in 97% of patients over 40 years.[10] They are due to maturation defects of the basal cell component of the skin.

Other less common skin lesions include epithelial cysts of the fingers, fibromas, neurofibromas, comedolike lesions, and chalazia.[11] Eye problems include extropia, exophoria,[9] congenital blindness, cataracts, and glaucoma.[12] A high incidence of medulloblastoma occurs in children with this syndrome.[10, 12, 13]

Roentgenographic Features. Multiple mandibular and maxillary cysts are frequent and usually detected at age 15[9] (Figs. 1388 and 1389). Small "cystlike" lesions may occur in the long

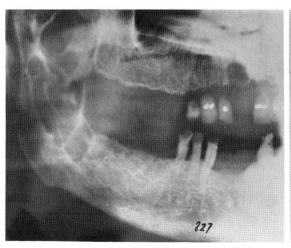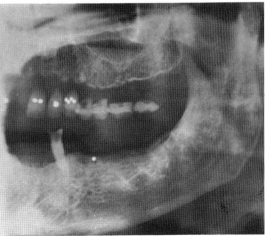

FIG. 1389. NEVOID BASAL CELL CARCINOMA
Multiple cysts in the right mandible. (Courtesy of Dr. Cesar Pedrosa, Madrid, Spain.)

bones (Fig. 1390). They vary from a few millimeters to several centimeters and may cause ectopic dentition. The maxillary cysts may protrude into the maxillary sinuses (Fig. 1392).

Rib anomalies involving the upper five ribs are common and may be bifid, fused, or dysplastic (Fig. 1391). The anterior rib ends are frequently flattened, broad, and bifid (Figs. 1391 and 1393).

Sprengel deformity, pectus excavatum, and pectus carinatum are reported.[10] Vertebral anomalies are most common in the cervical and upper dorsal regions. Spina bifida and scoliosis (Figs. 1394 and 1395) of the cervical and upper thoracic vertebrae are common anomalies. Hemivertebra and fused vertebrae occur less frequently (Fig. 1396).

Ectopic calcifications of the falx and tentorium is reported in 81% in Dodd's series as compared to 6-7% in the general population (Fig. 1397).[10, 14] The calcifications are more extensive than normal and appear lamellated. Bony bridging of the sella occurs (Fig. 1398). Calcification of the petroclinoid ligament is frequent. Calcifications are also reported in the subcutaneous tissues[10, 15, 16] (Fig. 1399), ovaries, sacrotuberous ligaments (Fig. 1400), and mesentery.[10, 17] Changes reported in the skull include hypertelorism, frontal and parietal bossing, and congenital hydrocephalus.[9, 10]

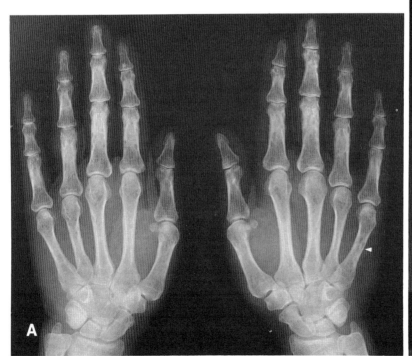
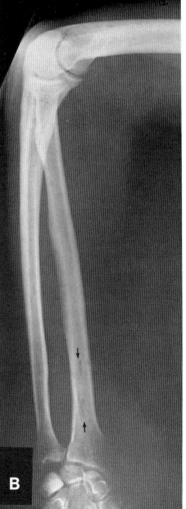

FIG. 1390. NEVOID BASAL CELL CARCINOMA
(A) Small radiolucencies in the fifth metacarpal (arrow). (B) Small radiolucencies in the radius (arrows).

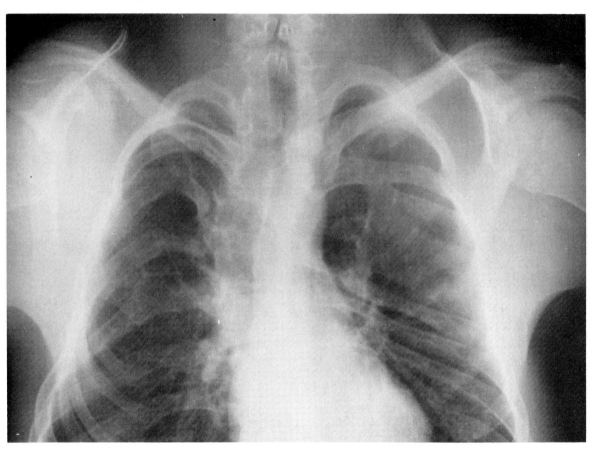

FIG. 1391. NEVOID BASAL CELL CARCINOMA
Sprengel deformity of the left scapula. Multiple anomalies of the upper rib cage. (Courtesy of Dr. Gerald Dodd, M. D. Anderson Hospital, Houston, Tex.)

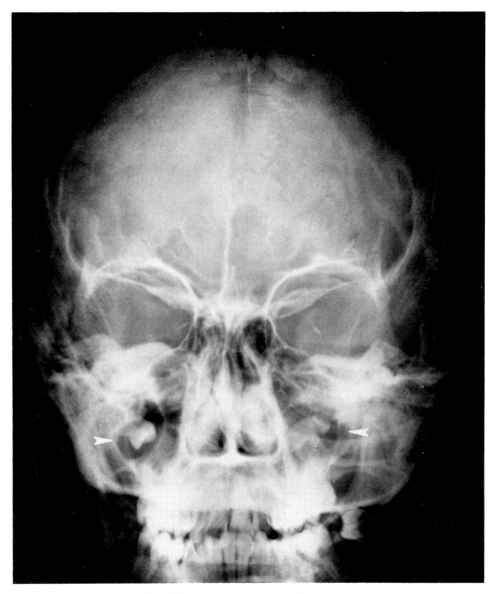

FIG. 1392. NEVOID BASAL CELL CARCINOMA
Bilateral dentigerous cysts (*arrows*).

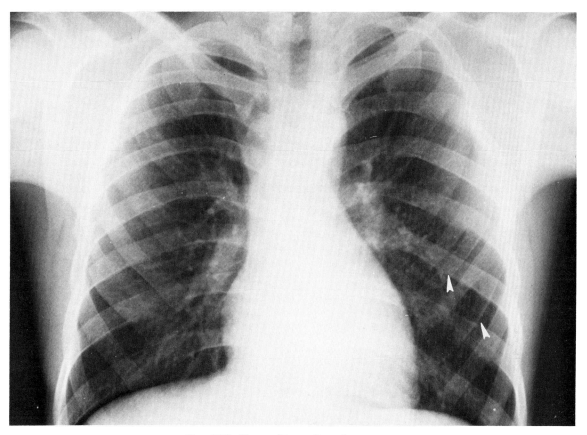

Fig. 1393. Nevoid Basal Cell Carcinoma
Two bifid ribs (*arrows*) on the left and a broad right fifth rib.

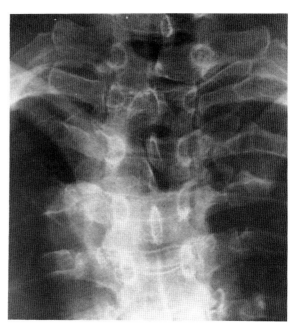

FIG. 1394. NEVOID BASAL CELL CARCINOMA
Bifid spinous processes of the first, second and third thoracic vertebrae. (Courtesy of Dr. Cesar Pedrosa, Madrid, Spain.)

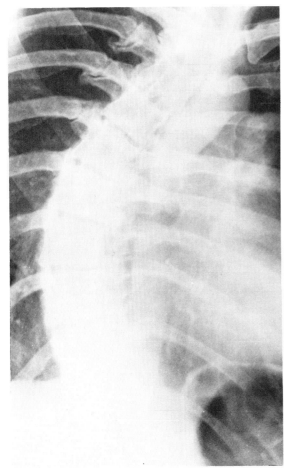

FIG. 1395. NEVOID BASAL CELL CARCINOMA
Scoliosis of the dorsal spine. (Courtesy of Dr. Gerald Dodd, M. D. Anderson Hospital, Houston, Tex.)

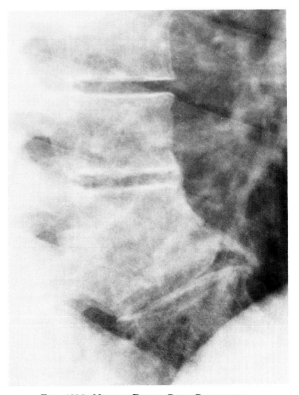

FIG. 1396. NEVOID BASAL CELL CARCINOMA
Block lower dorsal vertebra.

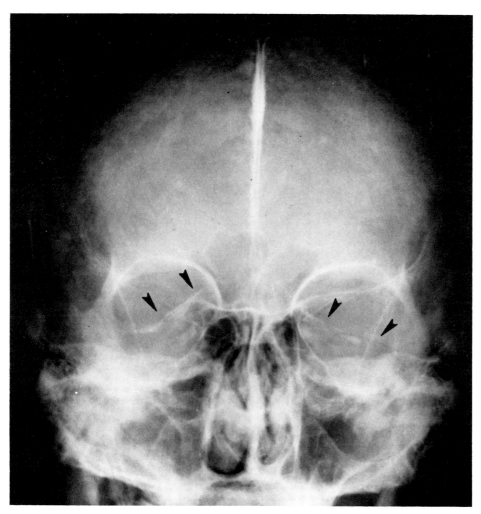

FIG. 1397. NEVOID BASAL CELL CARCINOMA
Ectopic calcification of the falx and tentorium (*arrows*).

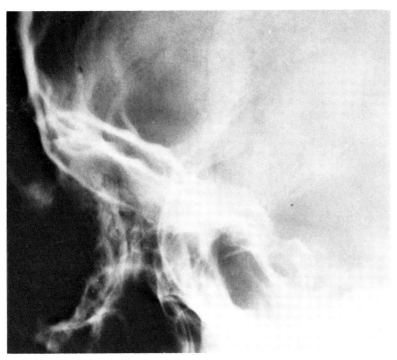

FIG. 1398. NEVOID BASAL CELL CARCINOMA
Bony bridging of the sella turcica.

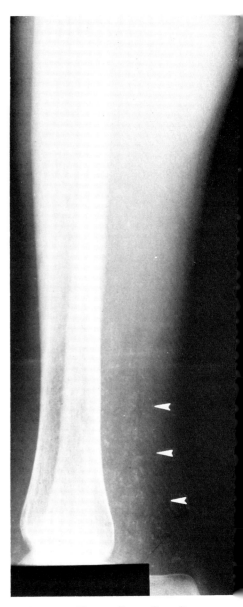

FIG. 1399. NEVOID BASAL CELL CARCINOMA
Calcifications in the soft tissues (*arrows*).

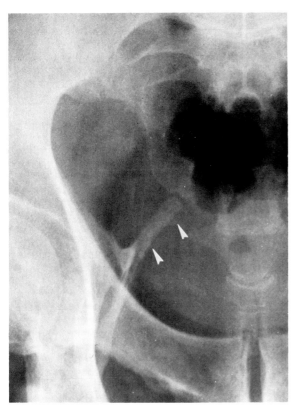

Fig. 1400. Nevoid Basal Cell Carcinoma
Calcification of the sacrotuberous ligament (*arrows*).

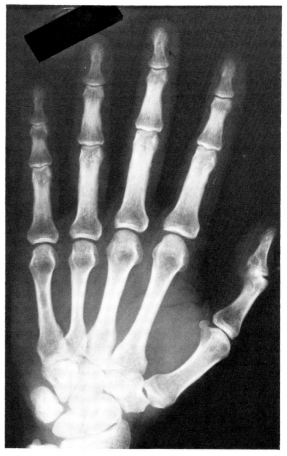

Fig. 1401. Nevoid Basal Cell Carcinoma
Shortening of the fourth and fifth metacarpals and a small cyst in the fifth metacarpal.

CORNELIA DE LANGE SYNDROME

Synonyms: typus degenerativus amstelodamensis, Amsterdam dwarfism.

Clinical Features. There is usually a normal prenatal history, although feeble fetal movements have been noted. Neonatal difficulties include cyanosis, apneic episodes, difficult feeding, vomiting, and recurrent infection. The etiology is unknown, but it is probably due to an undetected chromosomal abnormality, either a mutation[1-3] or an autosomal recessive mode.[4]

The de Lange syndrome denotes a peculiar form of mental retardation (intelligence quotient usually below 50) with characteristic facies: bushy eyebrows meeting in the midline, long curved eyelashes, high forehead, small nose with depressed bridge, nostrils tilted upward, excessive distance between the nose and upper lip, and small chin. The lips are thin and curved downward. The eyes may have an antimongoloid slant. The ears have a low insertion. The palate is arched, the teeth irregular, and the mandible underdeveloped. Other features include a short neck and hirsutism. Hypoplastic dermal ridges of the hands and feet and hypoplastic genitalia are common. The cry is often feeble, low pitched, and growling. The upper extremities are often short, with the thumb placed more proximally than is normal, and there is limited extension of the elbow. There may be partial webbing of the second and third toes. Congenital cardiac malformations are common. Growth failure often occurs.

Roentgenographic Features. The skull is small and brachycephalic. The cervical spine is small but there is no fusion.[5] The thumbs are often proximally placed and short, due to a hypoplastic first metacarpal (Fig. 1402).[6,7] There may be clinodactyly of the fifth finger. The distal phalanges are small and hypoplasia of the long bones is frequent (Fig. 1403), varying from phocomelia and micromelia to hypoplasia of a single bone. One or more of the forearm bones may be absent. The radii may be short and bowed, with resultant elbow dislocation (Fig. 1404). The distal articular surfaces of the radius and ulna may be angulated toward the midline. Involvement is usually more marked in the upper extremity. Retarded bone maturation is common, and modeling errors include broad humeral and femoral metaphyses and narrowed diaphyses in the lower extremities. Isolated bony anomalies, such as dislocated hips, short clavicles, and thin ribs may occur.

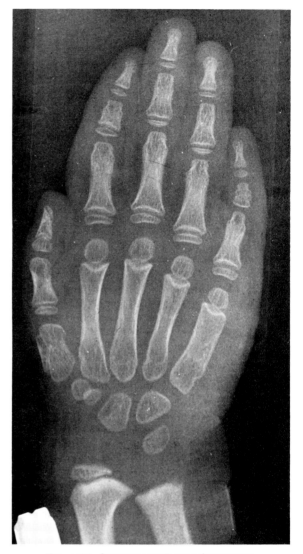

FIG. 1402. CORNELIA DE LANGE SYNDROME
Short first metacarpal and the thumb is proximally placed. Clinodactyly of the fifth finger due to hypoplasia of the middle phalanx.

No single radiographic finding is characteristic, but the combination is distinctive. Although the full syndrome is easily recognized, it must be differentiated from mucopolysaccharidosis, Turner syndrome, cretinism, Rubinstein-Taybi syndrome, and Trisomy 18. It most resembles Rubinstein-Taybi syndrome but the presence of broad thumbs, broad big toes, and the characteristic facies aids in differentiation.

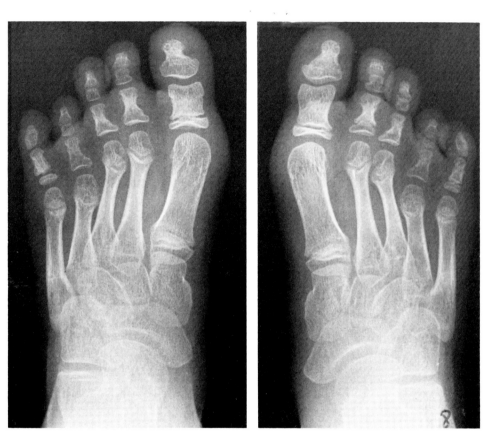

FIG. 1403. CORNELIA DE LANGE SYNDROME
All phalanges are short. Shortening of the fourth and fifth metatarsals.

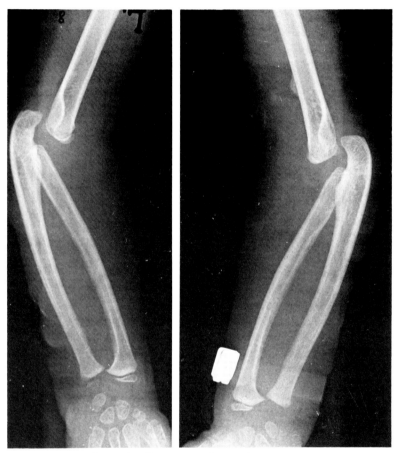

Fig. 1404. Cornelia de Lange Syndrome
Bilateral elbow dislocation.

ROTHMUND SYNDROME

Rothmund syndrome[1] is a hereditary ectodermal dysplasia similar to Werner syndrome (adult progeria).

Clinical Features. Characteristic skin changes develop between 3 months and 3 years of age[1,2] and consist of generalized skin atrophy, frequently mistaken for scleroderma. Subsequent impairment of joint motility results from tightening of the skin over the joints. The skin becomes smooth, thin, and transparent. Telangiectasia occurs over the entire body, most marked in the thighs, legs, and feet.[3] Cataracts usually develop early, but may develop later or not at all. Underdevelopment of the secondary sex characteristics is usual, but may be absent.

Roentgenographic Features. Resorption or absence of the terminal phalangeal tufts is a characteristic feature (Fig. 1405). Soft tissue calcifications are described but they are not as extensive as in scleroderma and do not occur in the tuft areas. The soft tissues are atrophic, and flaring of the iliac wings is reported.[3]

Differential diagnosis should include neurotrophic changes of all types, indifference to pain, scleroderma, hyperparathyroidism, thromboangiitis obliterans, hunger osteopathy, endocrine deficiency, allergic reactions, and pycnodysostosis.[4] Clinical and histochemical studies aid in the differentiation.

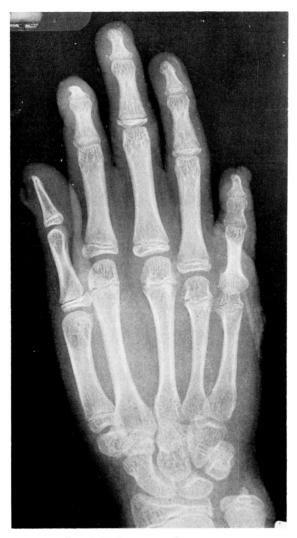

FIG. 1405. ROTHMUND SYNDROME
Resorption of the first, second, and third distal phalanges; a thumbnail abnormality is present. (Courtesy of Dr. Howard Steinbach, Moffitt Hospital, San Francisco, Calif.)

PROGRESSIVE MYOSITIS OSSIFICANS

Synonyms: ossificans progressive,[2] fibrositis ossificans progressiva,[1-4] fibrocellulitis.[5]

Myositis ossificans progressiva is an inflammatorylike disease of the fibrous connective tissue, marked by progressive ossification of the voluntary muscles, tendons, and fascia, and associated with congenital digital anomalies. It is hereditary and although early reports suggest a male predominance,[6,7] there are no sex-linked characteristics.

Clinical Features. Fibrositis of fascia and tendons usually appears within the 1st decade, and may be present at birth.[1,3,5,9] Onset rarely occurs after 20 years of age.[10-13]

Fibrositis presents initially as a subcutaneous swelling on the neck, back, and/or extremities, sometimes related to minor trauma. The swellings may be blood filled and tend to ulcerate and drain. Others come and go in a few days, and the patient may experience pyrexia and joint pain simulating rheumatic fever. There is a tendency for exacerbation and remission. Eventually, the muscles of the back and proximal extremities, including the knees and elbows, become rigid and the patients become severely disabled. A "wry neck" deformity often occurs, due to sternocleidomastoid muscle involvement, and jaw trismus results from involvement of the masticatory muscles. There are no reports of tongue, larynx, diaphragm, or sphincter involvement.[1]

Abnormal electrocardiogram is reported[8] possibly the result of myocardial fibrodysplasia. The usual causes of death are respiratory failure, due to involvement of the thoracic muscles, and inanition, due to masseter or temporal muscle involvement. Despite severe incapacitation, patients may survive for 60 years.[14]

Roentgenographic Features. The roentgenologic features fall into two categories, digital anomalies and ectopic ossification.

DIGITAL ANOMALIES. Digital anomalies may be recognized at birth before ectopic ossification becomes manifest. The most common anomalies are microdactyly of the big toes (90%) and thumbs (50%). There may be phalangeal shortening. Digital anomalies without myositis ossificans are reported in relatives of patients, indicating the hereditary nature.[1,15-23]

The microdactyly is due to phalangeal anomalies. The phalanges may be absent. Usually only one large phalanx is present in the big toe (Fig. 1406) or there may be a synostosis between the two phalanges with shortening. Microdactyly of this type is probably unique for this condition.[24]

Exostosis may develop, usually replacing the sesamoid bones. Hallux valgus is almost always present.[13]

Thumb anomalies are also due to phalangeal shortening, although the metacarpals may be short also (Fig. 1407). Synostosis of the metacarpal and proximal phalanx may occur, but synostosis of the phalanges of the thumb is uncommon. All fingers may be shortened[14] and radial curvature of the fifth finger is reported.[1] Other skeletal anomalies are rare, but broad femoral necks[25,26] and large epiphyseal ossification centers[1] are reported.

ECTOPIC OSSIFICATION. Although in most patients ossification does not appear in the early stages of the disease, it has been known to develop in fetal life.[1,3,5,9] Initially, a rounded or linear ossification occurs in the soft tissues, usually in the neck or shoulders (Figs. 1408–1410). The hips and proximal extremities and the dorsal aspect of the trunk are often involved. Eventually, columns and plates of bones replace the tendons, fascia, and ligaments in these areas

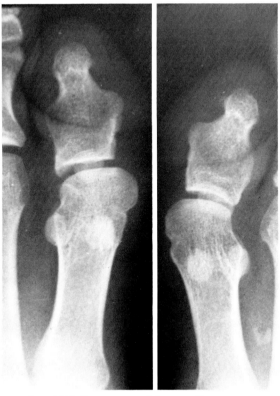

FIG. 1406. PROGRESSIVE MYOSITIS OSSIFICANS
One large phalanx of the big toe. (Courtesy of Dr. Howard Steinbach, Moffitt Hospital, San Francisco, Calif.)

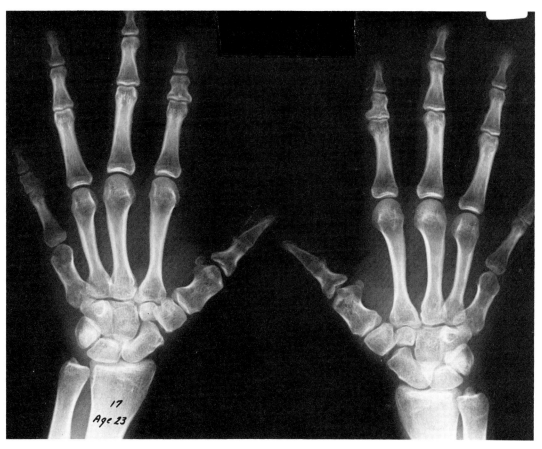

FIG. 1407. PROGRESSIVE MYOSITIS OSSIFICANS

The first and fifth metacarpals are short. Synostosis between the thumb phalanges. Hypoplasia of the middle phalanges of the second fingers. (Courtesy of Dr. Howard Steinbach, Moffitt Hospital, San Francisco, Calif.)

(Figs. 1408–1411). Ossifications commonly occur in the palmar and plantar fascia and at other areas of fibrous attachments, i.e., the occipital area of the skull or the anterior portion of the calcaneus.[1]

At birth, the cervical spine is normal, but some patients develop a progressive fusion of the centra of the vertebrae (Fig. 1412). The entire spine may fuse to such a degree that in the final stage it appears unsegmented.

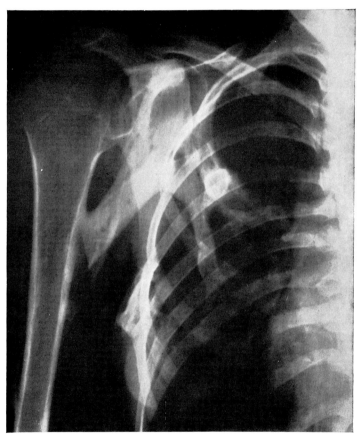

FIG. 1408. MYOSITIS OSSIFICANS PROGRESSIVA
Heterotopic bone lies in the axis of affected muscles and is attached to the skeleton in places.

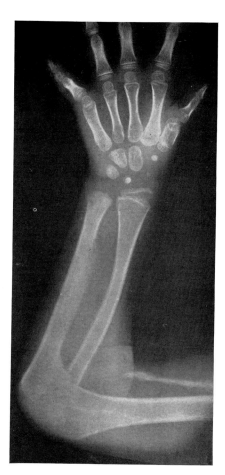

FIG. 1409. MYOSITIS OSSIFICANS PROGRESSIVA
Heterotopic bone is visible in the soft tissues anterior to the lower part of the humerus. Note the abnormally short first metacarpal and middle phalanx of the fifth finger, indicating the genetic origin of the disease.

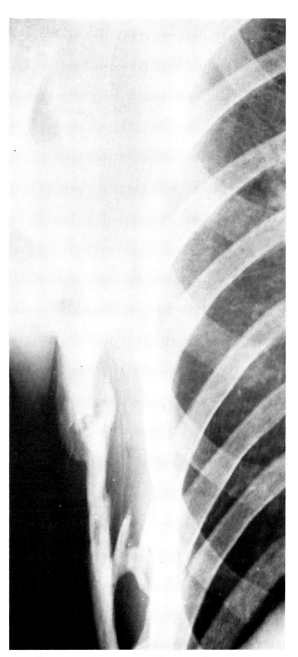

FIG. 1410. PROGRESSIVE MYOSITIS OSSIFICANS
Well developed ossifications in the lateral thoracic wall soft tissues.

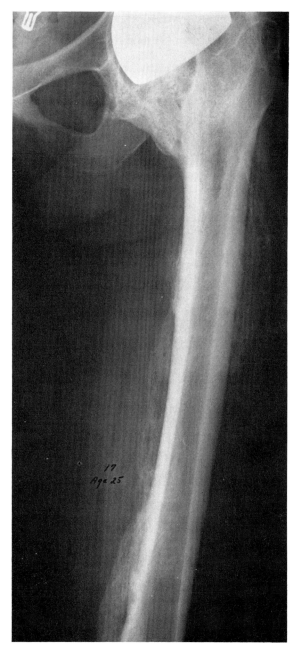

FIG. 1411. PROGRESSIVE MYOSITIS OSSIFICANS
Soft tissue calcifications in the thigh. A hip prosthesis is present.

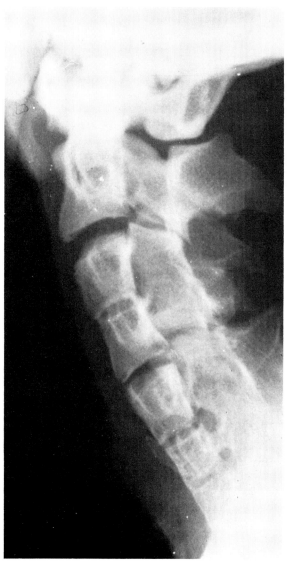

FIG. 1412. PROGRESSIVE MYOSITIS OSSIFICANS
Progressive fusion of the posterior spinal elements and partial fusion of the vertebral bodies.

MARFAN SYNDROME (ARACHNODACTYLY)

Arachnodactyly emphasizes the frequent association of skeletal abnormality with anomalous development of other body tissues.[1] This disorder is congenital and frequently inherited.[2] The most conspicuous feature in most patients is the elongation of the tubular bones. Muscular hypoplasia and hypotonicity are associated; joint hypermobility allows unusual feats of gymnastics. These conditions often lead to kyphoscoliosis and other postural deformities. Diminution of subcutaneous fat gives the patients an emaciated look. In approximately half of the cases there is dislocation of the ocular lenses, bilateral or unilateral.[3-5] Absence of the dilator muscle of the pupil is common, leading to contracted pupils unresponsive to mydriatics. In about one-third of the cases there is some type of congenital heart disease, most frequently interatrial septal defect. The major cardiovascular complications are aortic dilation,[6] dissecting aneurysm,[7] and mitral valvular anomalies.

Roentgenographic Features. In most instances the clinical diagnosis can be made without the aid of roentgenography. All that can be shown by roentgenologic methods is an unusual elongation of the tubular bones, especially of the phalanges, metacarpals, and metatarsals (Fig. 1413). The increase in bone length, unaccompanied by increased width, results in unusually slender bones. The cortices are thin and the trabecular structure is delicate. The other tubular bones are also elongated, so that the extremities are disproportionately long in relation to the trunk, which is not affected. Scoliosis is present in approximately 45%.[8]

Pectus excavatum is a common manifestation[5] and, when it is, hereditary Marfan syndrome should be suspected. Sinus of Valsalva aneurysms have been demonstrated angiocardiographically.[5, 6, 9-11] Mitral heart valvular disease with a floppy valve predominates in this condition. Aortic lesions occur later in the disease.[12] Absence of aortic calcification[13] with aneurysm is unusual with syphillis and aids in ruling it out. Homocystinuria may cause similar clinical and roentgenographic features. However, the abnormal urine and osteoporosis distinguishes it.

It is essential to distinguish Marfan syndrome from homocystinuria. No effective treatment is available for Marfan syndrome, but some patients with homocystinuria benefit from vitamin B_6 therapy, and in the future other types of homocystinuria may be correctable. Also, the risk of thrombosis exists in homocystinuria, and any surgical procedure accentuates this danger. A comparison of the main features of homocystinuria and Marfan syndrome is found in Table 49 under the discussion of homocystinuria.

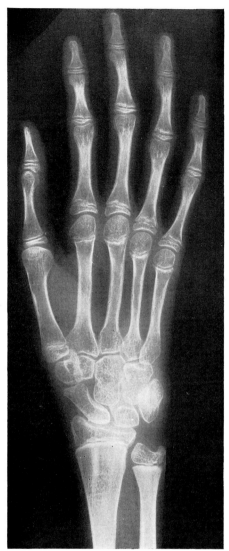

FIG. 1413. ARACHNODACTYLY
Elongation of all of the bones of the hand.

HOMOCYSTINURIA[1, 2]

Homocystinuria comprises a group of inborn errors of methionine metabolism transmitted by an autosomal recessive mode, which causes a syndrome similar to Marfan syndrome. A cross linkage can account for defects in both disorders due to similar but not identical defects, arising from two completely separate mechanisms, since they differ in mode of inheritance.[3]

There are three different enzymatic deficiencies identified[4]: cystathionine synthase, methyltetrahydrofolate-homocysteine methyltransferase, and methylenetetrahydrofolate reductase.[4-6]

The most common and best studied is the deficiency of cystathionine synthase, which is inherited by an autosomal recessive mode; however, at this locus there are at least two genetically heterogeneous diseases: pyridoxine responsive and pyridoxine resistant deficiencies.[4, 7-13]

Pathologic Features. The cystathionine synthase deficiency results in an accumulation of homocystine in the blood and its excretion in the urine, establishing the diagnosis. The urine may be examined chromatographically or electrophoretically to distinguish cystine from homocystine.[14] A simple urinary cyanide-nitroprusside screening test is virtually diagnostic. Enzymatic analysis of liver is also used in establishing the diagnosis.[15, 16] Evidence suggests that the metabolic errors causes a defect in the structure of collagen and/or elastin.[17]

A thrombotic tendency may be due to platelet "stickiness"[18] caused by homocystine.[19]

Clinical Features. There is considerable variation in incidence and severity of the abnormalities in homocystinuria, and many of them occur also in Marfan syndrome. Lenticular dislocations are frequent. Mild to moderate mental retardation occurs commonly.

A malar flush is observed in some patients.

There is a tendency to thrombose arteries and veins, and is the most serious threat to life. There may be a narrow, highly arched palate, crowding of the maxillary teeth and protrusion of the incisors.[3]

Sternal deformities (pectus excavatum or carintum) are common.[3]

Arachnadactyly and scoliosis occur less frequently than in Marfan syndrome.

Roentgenographic Features. There are no distinguishing features at birth. The genetic heterogeneity[6, 15] of the syndrome and the age of the patient will influence the roentgenographic features,[3] and explain some of the variations in the presence and severity of these features.

The severity of the enzymatic deficiency and its response to vitamin B_6 correlate quantitatively but not qualitatively with the roentgen features.[3] In children with ectopic lens skeletal involvement is more severe.[3]

The paranasal sinuses may be enlarged and suggest acromegaly; however, the pituitary fossa is normal.[20-22]

AXIAL SKELETON. Osteoporosis of the vertebrae is a frequent and striking feature. The bones appear less dense and are concave, particularly at the posterior portions of the vertebral bodies.[3] The vertebral bodies may be flattened or widened.[3] Osteoporosis is less frequent in children and in young adults.[4]

Scoliosis, usually mild, is present in the majority of patients.

Pectus excavatum or carinatum deformity of the sternum occurs in 75% of patients.

APPENDICULAR SKELETON. Osteoporosis, as evidenced by thinning of the cortices and diminished density, occurs frequently but is less common or severe than in the spine. There is usually widening of the metaphysis and enlargement of the ossification centers in the long bones, most evident at the knees and less obvious at the wrists, elbows and ankles. Multiple growth lines (Harris lines) occur frequently,[3] but may not be present in patients who are responsive to vitamin B_6 therapy.[4]

Genu valgum is frequent and coxa valga and coxa magnum occur. Perthes disease is sometimes associated, but it is not established whether this is a concurrent or related condition. The superior margin of the femoral neck may be straight or convex. Pes cavus or flat foot may occur.

Abnormalities in the upper limbs are not as severe or frequent. The humeral head may be enlarged. There may be a large capitate and hamate, and punctate ossification in the distal radial and ulnar epiphyseal cartilages.[23] Mild arachnodactyly may occur, especially in the toes.

Brill et al.[4] studied the features of 10 patients (6 children under 10 and 4 adults in the 3rd decade) in whom cystathionine synthase deficiency was established on the basis of enzymatic analysis of the liver.[15, 16] Some of these patients were completely or partially responsive to vitamin B_6. The incidence of roentgenographic features (Table 47),[4] and a summary of enzymatic, clinical and roentgenographic data are summarized in Table 48.[4]

TABLE 47. INCIDENCE OF POSITIVE ROENTGENOGRAPHIC FINDINGS IN 10 PATIENTS WITH HOMOCYSTINURIA DUE TO CYSTATHIONINE SYNTHASE DEFICIENCY[a]

	Adults	Children
A. *Spine*		
Biconcave vertebral bodies	2/4	2/6
Narrow upper thoracic disks	2/4	0/6
Osteoporosis	1/4	2/6
Scoliosis	3/4	1/6
B. *Chest*		
Pectus excavatum	2/4	0/6
Calcification in superior vena cava	1/4	0/6
C. *Skull and Face*		
Dural calcification	2/4	0/6
Wide diploic space	1/4	0/6
Prognathism	3/4	0/6
D. *Long Bones*		
Prominent growth arrest lines	2/4	5/6
Osteoporosis	1/4	1/6
Humerus varus	2/4	0/6
Bowed radius and ulna	2/4	0/6
Hemiatrophy	1/4	0/6
E. *Hands and Wrists*		
Metacarpal index >8.5	1/4	1/6
Metaphyseal spicules	0/4	5/6
Bone age	Adult	Within 1 yr of chronologic age
Large capitate and hamate	1/4	1/6
F. *Feet*		
Long talus	2/4	4/6
G. *Excretory Urogram*		
One kidney not visualized	1/4	0/6
Dense papillary blush	2/4	0/6

[a] From P. W. Brill et al.: *American Journal of Roentgenology*, 121: 45, 1974.[4]

TABLE 48. SUMMARY OF ENZYMATIC, CLINICAL AND ROENTGEN DATA IN 10 PATIENTS WITH CYSTATHIONINE SYNTHASE DEFICIENCY[a]

Patient	Age (yr)	Synthase Level (% of Normal) Response to Vitamin B_6	Main Clinical Features	Main Roentgen Features
C.H.	21	Synthase 15% B_6 responsive	Hemiparesis Thrombophlebitis Minor motor seizures Violent behavior	Hemiatrophy Prognathism Left kidney not visualized Scoliosis[b]
R.N.	21	Synthase 4% B_6 responsive	Slight mental deficiency Psychiatric disorder "Stroke" Lens dislocation	Biconcave vertebral bodies Narrow upper thoracic disks Humerus varus Bowed radius and ulna Prognathism Long talus IVP[c]: Dense papillary blush
L.M.	25	Synthase 0% B_6 unresponsive	Mental deficiency Thrombophlebitis Lens dislocation	Scoliosis Narrow upper thoracic disks Prominent growth arrest lines Calcified thrombus in superior vena cava Pectus excavatum Dural calcification Wide diploic space
M.S.	24	Synthase 0% B_6 unresponsive	Marfan-like appearance Lens dislocation Detached retina Thrombosis (testis) Learning disability	Biconcave vertebral bodies Osteoporosis Scoliosis Prominent growth arrest lines Humerus varus Bowed radius and ulna Pectus excavatum Prognathism Dural calcification Arachnodactyly (metacarpal index = 8.95) Large capitate and hamate Long talus IVP: Dense papillary blush
A.G.	5	Synthase 5% B_6 responsive	Normal clinically (sister of J.G.)	Growth arrest lines
J.G.	9	Synthase 5% B_6 responsive	Mental deficiency Behavior problem Lens dislocation (brother of A.G.)	Biconcave vertebral bodies Growth arrest lines Arachnodactyly (metacarpal index = 8.85)
R.C.	5	Synthase 3% B_6 partially responsive	Hyperactive behavior Myopia Treated with vitamin B_6 since age 12 mo (brother of M.C.)	Metaphyseal spicules Large capitate and hamate Long talus
M.C.	7	Synthase 0% B_6 partially responsive	Mental deficiency Hyperactive behavior Seizures Lens dislocation Retinal detachment (sister of R.C.)	Biconcave vertebral bodies Osteoporosis (spine) Growth arrest lines Metaphyseal spicules Long talus
J.O'R.	7	Synthase 0% B_6 unresponsive	Mental deficiency Lens dislocation (brother of A.O'R.)	Growth arrest lines Metaphyseal spicules Long talus
A.O'R.	9	Synthase 0% B_6 unresponsive	Lens dislocation (sister of J.O'R.)	Scoliosis Osteoporosis Metaphyseal spicules Growth arrest lines Long talus Metacarpal index = 8.5

[a] From P. W. Brill et al.: *American Journal of Roentgenology*, 121: 45, 1974.[4]

[b] This 21-year-old male patient had had a cerebral vascular accident in the neonatal period, resulting in left-sided hemiplegia. The scoliosis may have been secondary to the resultant hemiatrophy.

[c] IVP = intravenous pyelography.

TABLE 49. COMPARISON OF MAIN FEATURES OF HOMOCYSTINURIA AND MARFAN SYNDROME[a]

	Homocystinuria	Marfan Syndrome
Inheritance	Autosomal recessive	Autosomal dominant
Osteoporosis	Present	Absent
Spine	Biconcave vertebrae	Scoliosis common
Long bones	Wide metaphyses	Slender
Skin	Malar flush	Striae distensae
Epiphyses	Large	Normal
Arachnodactyly	Relatively uncommon	Very common
Feet	Pes cavus common	Long and flat
Coxa valga	Common	Common
Deformity of thorax	Very common	Common
Lens dislocation	Very common	Common
Arched palate	Very common	Common
Joint mobility	Normal or reduced	Normal or increased
Biochemical defect	Deficiency of cystathionine synthetase	None known
Urine	Homocystine present	No known abnormal substance excreted
Mental retardation	Common	Rare
Vascular disease	Thrombosis in medium sized arteries and veins	Dilatation and/or dissection of aorta. Thrombosis usually absent

[a] From R. O. Murray and H. G. Jacobson: *The Radiology of Skeletal Disorders,* Ed. 2, Vol. 2, Churchill Livingstone, London, 1977.[24]

It is important to distinguish Marfan syndrome from homocystinuria, for in the latter, one of the metabolic errors may be treated and it is likely that others may be treatable in the future. Since the risk of thrombosis exists in homocystinuria, any surgical procedure is dangerous. A comparison of the main features of homocystinuria and Marfan syndrome is presented in Table 49.[24]

EHLERS-DANLOS SYNDROME

Ehlers-Danlos syndrome is a disorder of the connective tissues with cutaneous, skeletal, ocular, and internal organ manifestations.

Clinical Features. It is inherited by an autosomal dominant mode,[1] and is present at birth, with skin velvety in appearance and feel.[1-6] The skin is hyperextensible, fragile, and brittle; minor trauma may cause large gaping wounds with little bleeding. Molluscoid pseudotumors[7,8] develop at pressure points, especially the heels, knees, and elbows. Small fatty cysts develop under the skin and when they calcify[9] are visible roentgenographically.[10]

Joint hyperextensibility is characteristic in childhood and becomes less marked with age. Some ocular changes reported are blue sclera,[6,11,12] microcornea with associated glaucoma,[12] microcornea and myopia,[13] keratoconus,[14] and extopia lentis,[14,15] pigmented retina, and retinal detachment.[16] Reports include diaphragmatic hernia,[1,17,18] ectasia of portions of the gastrointestinal[18] and respiratory tracts, spontaneous lung ruptures,[1,19] dissecting aortic aneurysm, and congenital malformations[1] Spontaneous rupture of large vessels[20] and multiple intercranial aneurysms[21] may occur.

Roentgenographic Features. The most characteristic change is multiple subcutaneous calcifications.[22,23] Ovoid, these vary in size from 2 to 10 mm. They are most frequent in the legs and resemble phlebolith calcification, but are more extensive. Hemarthrosis, especially in the knees, is common.[24] Flat feet and clubfoot are described.[16] Joints most often recurrently dislocated are the hip,[25-29] patella,[27,30-33] shoulder,[31,32] radius,[33] and clavicle.[34]

Other frequent manifestations include radial ulnar synostosis,[35] kyphoscoliosis,[27,36-41] spondylolisthesis,[1] and spina bifida occulta.[42] Ectopic bone formations may bridge the acetabulum and great trochanter,[43] probably secondary to hemorrhage. Roentgenographic features of the internal derangement may also be obvious, such as diaphragmatic hernia, ectasia of the gastrointestinal tract, and aneurysms of the great vessels.

Ehlers-Danlos syndrome may be combined with osteogenesis imperfecta. Blue sclera may occur with Ehlers-Danlos syndrome without osteogenesis imperfecta.

GENERALIZED CORTICAL HYPEROSTOSIS
(VAN BUCHEM DISEASE)

Van Buchem disease is a hereditary generalized sclerosis of the diaphysis with skull involvement.[1-3]

Clinical Features. It is transmitted either by an autosomal recessive[1] or an autosomal dominant mode.[4,5] It is asymptomatic except for paralysis of the facial nerve, observed in 3 of 7 patients, and auditory and ocular disturbance in 5 of 7.[3] These symptoms usually occur in the late teens. The sclerotic process is painless. The serum alkaline phosphatase is elevated; the serum calcium and phosphorous normal.

The etiology is unknown, but there is evidence of extensive new bone formation and probably normal resorptive mechanisms.[6]

Roentgenographic Features. There is symmetrical sclerosis of the skull, mandible, clavicles, ribs, and long bone diaphyses. The calvaria are thickened and sclerotic, with obliteration of the diploe, but the paranasal sinuses are well developed (Fig. 1414).

The diaphyseal sclerosis is accompanied by thickening of the endosteal surface of the cortex, which causes widening of the cortex but does not increase the diameter of the bone (Fig. 1415). The epiphyseal end of the bone is not affected. The ribs and clavicles are sclerotic and, late in the disease, their diameter is increased. The spinous processes are thickened and sclerotic but the bodies are only minimally affected (Fig. 1416). The pelvis, especially the acetabulum, is mildly sclerotic. Small bony excrescences may appear on the periosteal surface of the skull and other bones (Figs. 1417 and 1418).

Differential Diagnosis. The differential diagnosis includes: osteopetrosis, distinguishable in that it presents sclerosis of all bones and is not confined to the diaphysis. The skull is only slightly involved, and the serum alkaline phosphatase is normal or reduced. Generalized hyperostosis with pachydermia involves the entire long bone and produces considerable pain as well as characteristic skin changes. Hyperphosphatasia appears in infancy, with widened bones but decreased cortical density. Englemann disease usually involves the lower limbs and is rarely generalized. The skull is only slightly thickened. Pyle disease does not involve the mid diaphyses. Polyostotic fibrous dysplasia is rarely symmetrically generalized, and the paranasal sinuses are abnormal, with skull involvement.

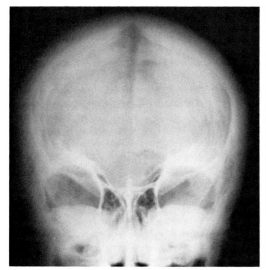 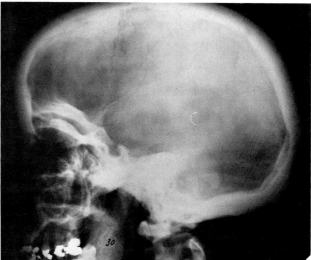

FIG. 1414. GENERALIZED CORTICAL HYPEROSTOSIS (VAN BUCHEM DISEASE)
Thickening and sclerosis of the calvaria with obliteration of the diploe. The paranasal sinuses are well developed. (*Left*) Posteroanterior projection, (*right*) lateral projection.

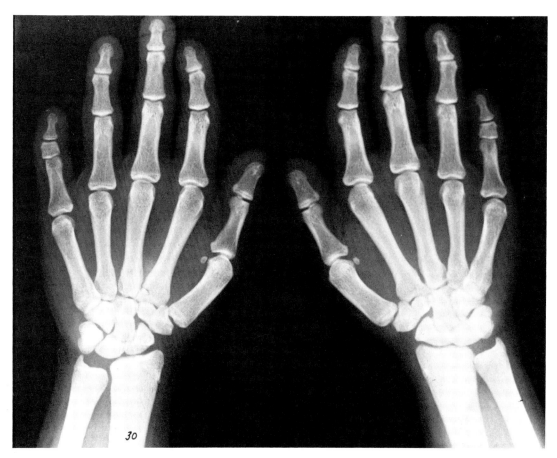

Fig. 1415. Generalized Cortical Hyperostosis (van Buchem Disease)
The bones of the hand have mild endosteal cortical thickening. The carpal bones are increased in density. The distal ends of the radii show lack of modeling and increased density.

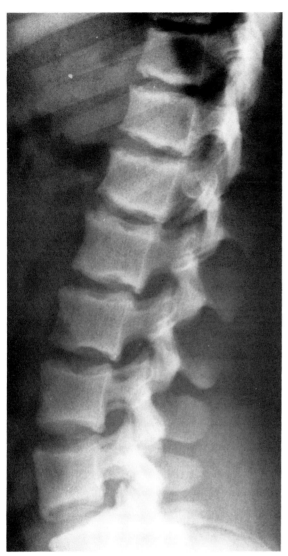

Fig. 1416. Generalized Cortical Hyperostosis (van Buchem Disease)

Increased density of the spinous processes and posterior elements of the lower lumbar spine. The vertebral end plates are irregular and increased in density.

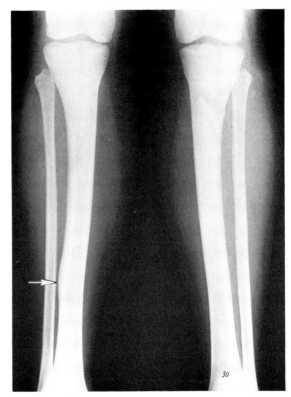

Fig. 1417. Generalized Cortical Hyperostosis (van Buchem Disease)

Increased density of the metaphyses and slight endosteal cortical thickening. A small bony excrescence is present on the periosteal surface of the right tibia (*arrow*).

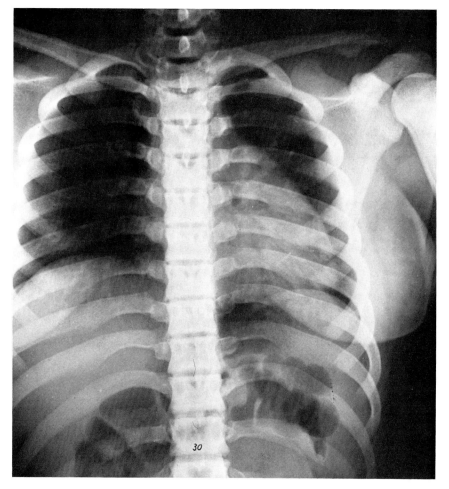

Fig. 1418. Generalized Cortical Hyperostosis (van Buchem Disease)
Increased density of the ribs and dorsal spine.

FOCAL DERMAL HYPOPLASIA (GOLTZ SYNDROME)

Synonyms: ectodermal and mesodermal dysplasia with osseous involvement,[1] congenital ectodermal dysplasia and mesodermal dysplasia,[2] combined mesoectodermal dysplasia,[3] and focal dermatophalangeal dysplasia.[4]

Focal dermal hypoplasia (FDH)[5] is a congenital dysplasia of ectodermal and mesodermal structures occurring almost exclusively in girls.[6, 7]

Clinical Features. It is possibly transmitted by X-linked dominant mode and is lethal in boys.[6, 8] Prematurity is common; physical and mental retardation is frequent.[9] The hallmark is universal superficial skin tumors, probably linear hamartomas[10] (nevoid neoplasms). Other skin lesions include atrophy, linear hyperpigmentation, and telangiectasia. Scarlike skin changes may be present at birth or occur later, preceded by blisters. Papillomas of the perorificial skin and oral mucosa and hypotrichosis are frequent, as are dystrophic nails. Colobomas, corneal opacity, and ocular hypertelorism are described.[9] Malformed ear and nasal cartilages occur. Stenosis of the external auditory canal, elongation of the tragus, neck webbing, and anal stenosis are reported.[9] A rudimentary tail is also reported, similar to that found in metatrophic dwarfism.[9, 11]

Roentgenographic Features. The most common features are asymmetric size, shape, and number of the teeth and appendicular bones, usually syndactyly of the third, fourth, and fifth digits and anodentia or malformed teeth. Fusion of the carpals and clinodactyly are also reported. Phalangeal tuft deformity may occur (Fig. 1419).

Osteoporosis and gracile long bones, tall lum-

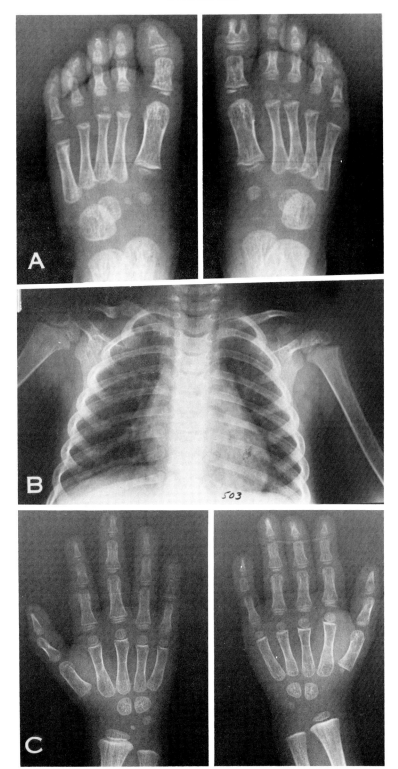

Fig. 1419. Focal Dermal Hypoplasia (Goltz Syndrome)
(A) Both feet. Clinodactyly of both fifth digits due to hypoplasia of the middle phalanx. Deformity of the first left distal phalanx. The bones have prominent trabeculations and thin cortices due to osteoporosis. (B) Chest. Right clavicular underdevelopment. (C) Hands. The left second, third, and fourth fingers are all approximately the same size. The fourth ray is longer than the second and third rays. Diffuse osteoporosis is present.

bar vertebrae and, coxa valga are probably due to disuse during the developmental period.[3, 9] Fused ribs, lack of segmentation, and clavicular underdevelopment (Fig. 1419) and aplasia occur and a "twisted rib" appearance, as with neurofibromatosis, has been described. Scoliosis and lack of vertebral segmentation are reported.

If the entire skeleton is not evaluated, there may be diagnostic confusion with neurofibromatosis, cleidocranial dysostosis, and metatrophic dwarfism. The multiple skeletal and dermatologic defects may also be confused with other conditions: incontinentia pigmenti, chondroectodermal dysplasia, oculoauriculovertebral dysplasia, Hallermann-Streiff syndrome, Möbius syndrome, Rothmund-Thomson syndrome, trisomy 13-15, and trisomy 17-18.[10, 12, 13]

ACROCEPHALOSYNDACTYLY AND ACROCEPHALOPOLYSYNDACTYLY

Temtamy and McKusick[1] divide acrocephalosyndactyly (ACS) into five types: Type I, Apert syndrome; Type II, Vogt cephalodactyly; Type III, acrocephalosyndactyly with asymmetry of skull and mild syndactyly; Type IV, Waardenburg type; and Type V, Pfeiffer type. Although they all have acrocephaly and syndactylism in common, they are probably due to different gene mutations, since there is no crossover of pedigree.[1]

In acrocephalopolysyndactyly, unlike Type V (Pfeiffer type), the polysyndactyly is a notable feature but large prominent great toes are present in both. Two types of acrocephalopolysyndactyly are described: Type I, Noack syndrome with an autosomal dominant inheritance, and Type II, Carpenter syndrome, which is recessive.[1] In contrast with Type V, the great toes in Noack syndrome prove radiologically to be two completely syndacteric great toes.[2] Carpenter syndrome is also distinguished by mental retardation, hypogenitalism, and more frequent cardiac abnormalities.[3]

Rubinstein-Taybi syndrome and Léri syndrome are also characterized by broad thumbs and great toes, but without acrocephaly or syndactyly. Léri syndrome also reveals ankylosis of the interphalangeal joints and hyperostosis or increased density of the long bones.

Acrocephalosyndactyly Type I (Apert Syndrome)

Acrocephalosyndactyly is a dysplasia characterized by a syndactyly of the hands and feet, and an increased vertical diameter of the skull (oxycephaly). Acrocephaly and oxycephaly may be used synonymously.[4] Until there is better delineation of hereditary factors and morphologic changes, acrocephalosyndactyly should be considered as a group of syndromes constituting a general category but separable into types.

Clinical Features. The disease is transmitted by an autosomal dominant mode,[5] although a mutation may account for sporadic cases.[6, 7] The parents tend to be older, and Blank[3] believes that the father's age (35–39) is the main factor.

The patient may be mentally retarded.[6, 8] The face is variably affected, usually with underdevelopment of the maxilla, approximated alveolar ridges, malformation of the teeth, and a high pointed arch of the palate. Prognathism is almost always present. A prominent vertical crest may be seen in the middle of the forehead. Bilateral exophthalmus is common. Increased intracranial pressure may occur in infancy, sometimes resulting in an enlarged sella. Variable soft tissue fusions of the fingers and toes are recorded.

Roentgenographic Features. Fusion of the distal portion of the fingers and toes is usually found, and may occur in the metatarsals; carpal fusions are also described (Fig. 1420). The middle phalanges may be absent; sometimes only one phalanx is present on each ray (Fig. 1421). Instead of fusions, in the older individual pseudoarthrosis may develop. Carpal and tarsal bones may be missing or supernumerary. The cervical spine may also fuse.

Schauerte and St. Aubin[9] describe the progressive pattern of the synostoses in the feet: calcaneus with cuboid, lateral cuneiform with third metatarsal, navicular with medial cuneiform, and phalanges at interphalangeal joints. The fusions of the phalanges of the hands and cervical spine are also progressive, which may account for their varying descriptions. In mature patients, fusions and pseudoarthroses are prominent, but in infants the fusions may not appear at all.

SKULL. The skull is oxycephalic. The occiput is flat and the facial bones are underdeveloped. The anterior fossa is V-shaped, due to elevation of the lateral margins of the lesser sphenoid wings. The sella may be enlarged. The paranasal sinuses are often underdeveloped. Hypertelorism may be present (Fig. 1422).

Oxycephaly is due to premature generalized craniostenosis, and may appear in syndromes not related to metabolic disease. Crouzon syndrome[10]

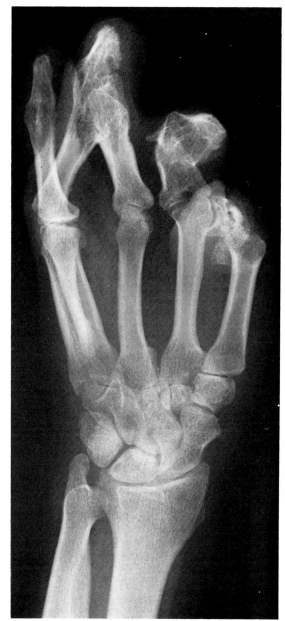

FIG. 1420. ACROCEPHALOSYNDACTYLY, TYPE I (APERT SYNDROME)
Fusion of the distal fingers and only one large phalanx for each finger.

tardation, coxa valga, genu valgum, pes varus, congenital heart disease, and abdominal hernias. The middle phalanges are triangular, producing ulnar deviation of the terminal phalanx of the third finger and radial deviation of the terminal phalanges of the fourth and fifth fingers.[12, 13]

Laurence-Moon-Biedl-Bardet syndrome (LMBB) is hereditary, and characterized by mild craniostenosis, mental retardation, hypogonadism, obesity, and polysyndactyly. Note that Apert, Carpenter, and LMBB syndromes share several common features.[11] All are hereditary and include some degree of mental retardation and polydactyly and/or syndactyly. Carpenter and LMBB syndromes include hypogonadism and obesity. Apert and Carpenter syndromes

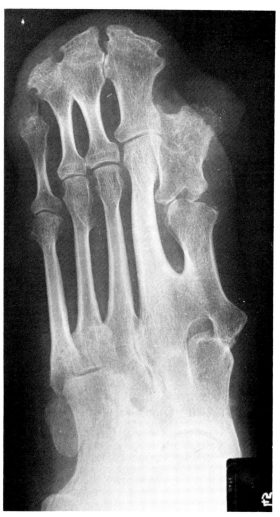

FIG. 1421. ACROCEPHALOSYNDACTYLY, TYPE I (APERT SYNDROME)
Fusions of the third and fourth toes and a synostosis between the second and third. Only one large phalanx is present on each toe. Fusion between the first and second metatarsals. Tarsal-tarsal and tarsometatarsal fusions.

(cranial-facial dysostosis) is a clinical complex characterized by a bilateral exophthalmus, strabismus, hypoplasia of the maxilla, prognathism, and a beak nose. It is probably the same as Apert syndrome without the syndactyly.[11]

The differential diagnosis should consider the following: Carpenter syndrome (acrocephalopolysyndactyly) is transmitted by an autosomal recessive mode[12] and is characterized by brachysyndactyly, hypogonadism, obesity, mental re-

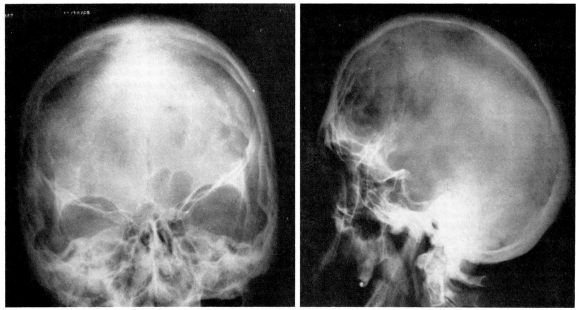

FIG. 1422. ACROCEPHALOSYNDACTYLY, TYPE I (APERT SYNDROME)

The skull is oxycephalic. The occiput is flat and the facial bones are underdeveloped. The anterior fossa is V-shaped due to elevation of the lateral margins of the lesser sphenoid wings. The sella is enlarged. (*Left*) Posteroanterior projection, (*right*) lateral projection.

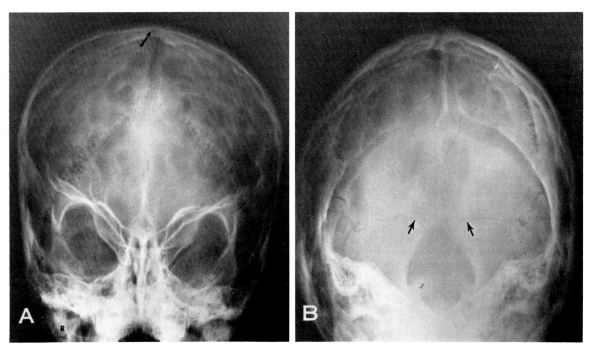

FIG. 1423. PFEIFFER SYNDROME

(*A*) Posteroanterior skull roentgenogram shows premature closure of the sagittal (*arrow*) and coronal sutures and increased digital markings. (*B*) Anteroposterior skull roentgenogram shows delay of closure of synchondroses (*arrows*) between exoccipital and supraoccipital portions of occipital bone. (*C*) Lateral skull roentgenogram shows closure of coronal suture, shallow anterior cranial fossa, increased height of cranial cavity, and flattening of nasal bridge. (Courtesy of J. M. Martsolf, J. B. Cracco, G. G. Carpenter, and A. E. O'Hara: *American Journal of Diseases of Children, 121:* 257, 1971,[22] © American Medical Association, Chicago, Ill.)

have acrocephaly; LMBB syndrome has a milder craniostenosis.[14]

Treacher Collins syndrome is occasionally associated with craniostenosis and characterized clinically[15-17] by low eyelids, notched between the middle and outer two-thirds, an antimongoloid slant of the eyes, depressed cheeks, a small jaw, auricular defects, stenosis or absence of the external auditory canals, middle and inner ear defects, and deafness.

Craniotelencephalic dysplasia[18] is a syndrome with frontal bone dysplasia, premature closure of the metopic, coronal, and sagittal cranial sutures, hypertelorism, ear deformity, and developmental neurologic retardation.

Acrocephalosyndactyly, Type II (Vogt Cephalodactyly)

This consists of syndactyly of the hands and feet, as in the Apert type, but with facial characteristics of Crouzon disease. Unlike Apert syndrome, the thumbs are free and large, and they differ from Type V (Pfeiffer type) in that they and the great toe consist of a single deformed phalanx. No hereditary transmission is reported.[1]

Acrocephalosyndactyly, Type III

This consists of mild acrocephaly, asymmetry of the skull, and partial soft tissue syndactyly of the hands and feet. It has been reported in two families involving two generations each.[1]

Acrocephalosyndactyly, Type IV (Waardenburg Type)

This is characterized by acrocephaly, asymmetry of the skull and orbits, strabismus, thin long pointed nose, and mild soft tissue syndactyly.[11]

Acrocephalosyndactyly, Type V (Pfeiffer Syndrome)

This consists of acrocephalosyndactyly associated with broad thumbs and great toes.[19-22] Other characteristics are flat nasal bridge, hypertelorism, slight antimongoloid slant of the palpebral fissures, high arched palate, varus deformities of the great toes, brachymesophalanx, trapezoidal first finger phalanx, broad first metatarsal, and normal intelligence[19] (Figs. 1423-1427). Martzell et al.[22] report other changes including a trapezoidal, normally placed first phalanx of the great toe, mild hearing loss, accessory epiphysis of the first and second metatarsals, and congenital pyloric stenosis. Transmission is by autosomal dominant mode.[22]

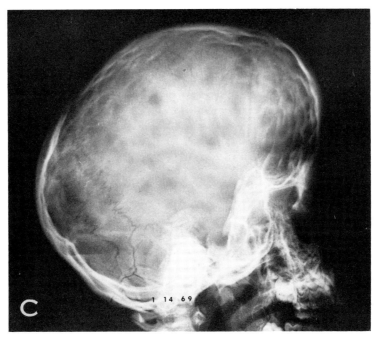

Fig. 1423 (C)

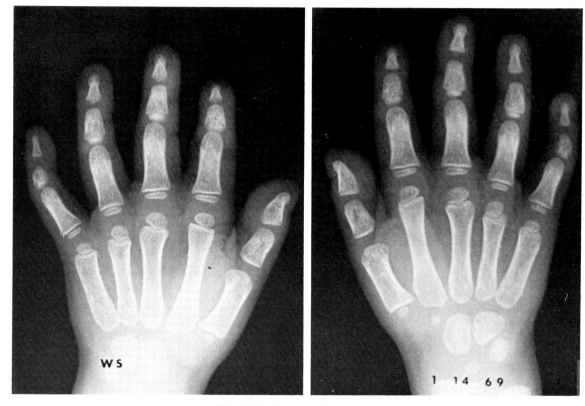

FIG. 1424. PFEIFFER SYNDROME
Hand roentgenograms show shortened middle phalanges, soft tissue hypertrophy of thumbs and partial membranous syndactyly between digits 2, 3, and 4 bilaterally. (Courtesy of J. T. Martsolf, J. B. Cracco, G. G. Carpenter, and A. E. O'Hara: *American Journal of Diseases of Children, 121:* 257, 1971,[22] © American Medical Association, Chicago, Ill.)

FIG. 1425. PFEIFFER SYNDROME
Trapezoid appearance of proximal phalanx of thumbs. (Courtesy of J. T. Martsolf, J. B. Cracco, G. G. Carpenter, and A. E. O'Hara: *American Journal of Diseases of Children, 121:* 257, 1971,[22] © American Medical Association, Chicago, Ill.)

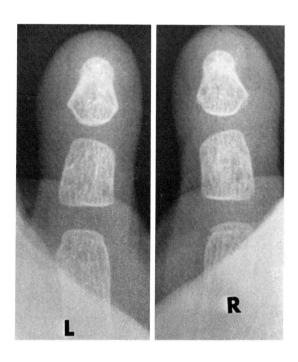

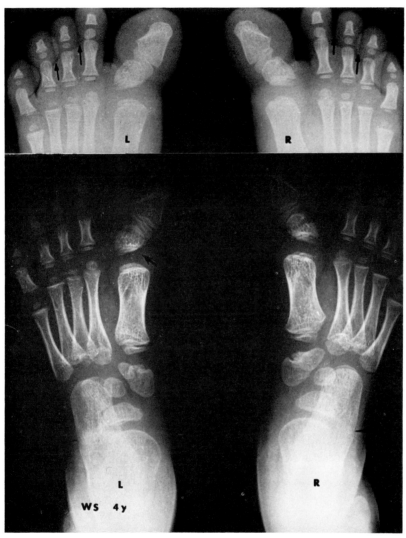

FIG. 1426. PFEIFFER SYNDROME
Roentgenograms of the feet show marked shortening of the middle phalanges, varus deformities of the great toes, trapezoid proximal phalanx of great toe (*top—arrows*), first metatarsal broadening, accessory epiphyses of the first and second metatarsals, partial membranous syndactyly of toes (*bottom—arrows*), and soft tissue hypertrophy of great toes bilaterally. (Courtesy of J. T. Martsolf, J. B. Cracco, G. G. Carpenter, and A. E. O'Hara: *American Journal of Diseases of Children, 121:* 257, 1971,[22] © American Medical Association, Chicago, Ill.)

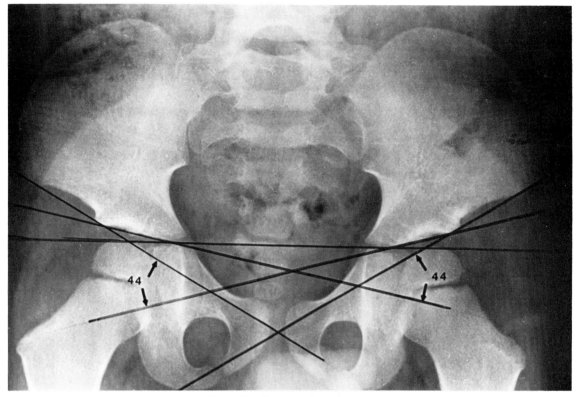

FIG. 1427. PFEIFFER SYNDROME

Roentgenogram of the pelvis shows a bilaterally decreased iliac angle and a 44 degree iliac index angle. (Courtesy of J. T. Martsolf, J. B. Cracco, G. G. Carpenter, and A. E. O'Hara: *American Journal of Diseases of Children, 121:* 257, 1971,[22] © American Medical Association, Chicago, Ill.)

OSTEOPETROSIS[1]

Synonyms: Albers-Schönberg disease, osteosclerosis, osteopetrosis generalisata, marble bones, osteosclerosis generalisata, chalk bones.

Osteopetrosis is a rare bone abnormality characterized by the persistence of calcified cartilage. In severe cases, encroachment on the marrow spaces causes severe anemia.

Clinical Features. Heredity plays a distinct role in osteopetrosis, which may be transmitted by either sex.[2] Consanguinity leads to a more severe form.[3] Lamy[4] believes there are two forms: (1) a benign and dominant, and (2) severe and recessive.

The clinical features include increased density of bones, anemia, and optic atrophy. Increased bone density results from the failure of primitive chondro-osteoid to resorb, so that it cannot be replaced by normal bone.[5] The bone is chalky and often fragile; multiple fractures sometimes occur.

Aplastic anemia is often present and may be associated with thrombocytopenia and leukopenia. Hepatomegaly, splenomegaly, and lymphadenopathy are frequent physical findings. There is no correlation between the severity of bone disease and the anemia; patients with sclerotic bone may have no anemia and an excellent prognosis. Optic atrophy, appearing in infants and children, results from skull deformity and other cranial nerve defects also occur. Except for infrequent serum calcium elevation, blood chemistry is normal. Prognosis is guarded. Massive hemorrhage and intercurrent infection are the usual cause of death. Sarcoma has been reported.

Pathologic Features. Bone changes are due to failure of resorptive mechanisms of calcified cartilage, which interferes with its normal replacement by mature bone. Formation of cartilage is normal up to the point of resorption of calcified cartilage. The columns of cartilage are disturbed at the edge of the zone of provisional calcification. Broad seams of osteoid form around the cartilage, creating a nest of osteoid with a cartilaginous nucleus. Normally the nucleus would resorb as soon as osteoid formed. Osteoid heaps up and encroaches on the marrow cavity. Modeling is prevented, resulting in flared metaphyses.

Transverse metaphyseal bands consist of immature bone alternating with bands of normal bone, which suggests that this can be an intermittent process. Osteopetrotic bone shows a disorganized structure resembling the mosaic pattern of Paget disease. Primary osteopetrosis is probably an endocrine disturbance related to the parafollicular cells of the thyroid which produce thyrocalcitonin.

Gruneberg[6, 7] described a mutant mouse (gray lethal) whose skeleton showed osteopetrosis. Barnicot[8] transplanted parathyroid tissue and bone of the "gray lethal" into normal sibling mice, and vice versa; bone erosion occurred in the normal mice but did not occur in the osteopetrotic mice. "Gray lethal" bone transplanted into the normal mice became osteoporotic, whereas normal bone transplanted into a "gray lethal" developed osteopetrosis. He concluded that the parathyroid gland and bone of the "gray lethal" were not inherently defective, but that an adverse environmental influence affected the bone remodeling, and parathyroid hormone was being metabolized at an abnormal rate.[8] Walker[9, 10] suspected that in the "gray lethal" osteoblastic activity was unresponsive to parathyroid hormone. He found that massive doses of parathyroid hormone caused osteoclastic activity within 24 hr, but thereafter subsided rapidly, followed by exuberant osteoblastic activity. This was correlated with the abundance of parafollicular cells, the source of thyrocalcitonin in the "gray lethal."[11] The parathyroid hormone stimulated the proliferation of parafollicular cells, which accounts for the delayed action in the bone. The parafollicular cells of the "gray lethal" are larger than normal, more oval, do not directly border on colloid, and are less basophilic. Therefore, Walker believed that "gray lethal" manifests an excessive action of thyrocalcitonin, and that this is the cause of osteopetrosis. Results of studies on other mutants with osteopetrosis are similar.[12]

Osteopetrosis has been produced in normal animals by maintaining a hypercalcemia-inducing regime.[13, 14] Raisz et al.[15] detected elevated plasma thyrocalcitonin in 5 patients with osteopetrosis. Osteopetrosis cannot be induced in animals that have been completely thyroidectomized, but this is a difficult state to achieve and if small remnants of thyroid gland tissue are present they will rapidly proliferate, and large populations of parafollicular cells will be produced.

Roentgenographic Features. Changes in density vary from none to extreme. The disease may become apparent in early childhood; if extreme bone density is present, it may be discovered in fetal bones in utero or any age group. The bone density appears amorphous, and transverse lines or longitudinal streaking may be identified. The bone appears structureless as its individual components (cortex, medullary cavity, epiphyseal plates, and trabecular pattern) are obliterated (Fig. 1428). Most or all bones may be involved (Figs. 1429–1431). The iliac bones may be affected early; long bone changes may not appear until later. Mandibular involvement is rare, and phalanges and calvaria exhibit less severe change.

The tubular bones lack modeling, which causes a flaring of the ends (Fig. 1432); their shafts may be widened. These bones may appear to have a

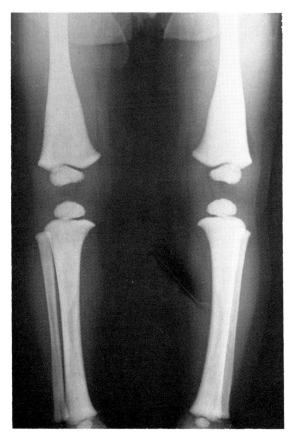

FIG. 1428. OSTEOPETROSIS
Extreme bone density and the bone appears to be without structure because of loss of individual components of cortex medullary cavity and trabecular pattern. Lack of modeling at the distal ends of the femurs.

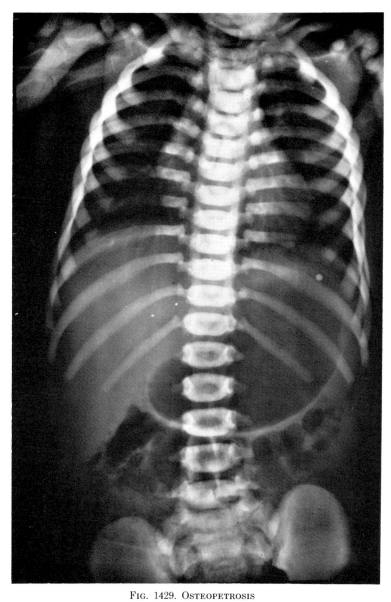

Fig. 1429. Osteopetrosis

Most of the bones show extreme increase in bone density. (Courtesy of Herbert M. Stauffer, Temple University Hospital, Philadelphia, Pa.)

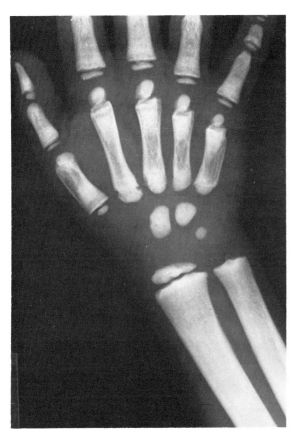

Fig. 1430. Osteopetrosis
Bone density is not extreme; however, bone within a bone in some of the metacarpals.

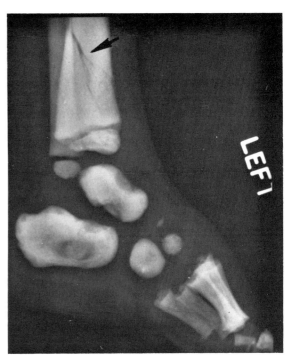

Fig. 1431. Osteopetrosis
Fracture to the distal end of the tibia. Extreme bone density is present with loss of normal trabecular pattern. (Courtesy of Herbert M. Stauffer, Temple University Hospital, Philadelphia, Pa.)

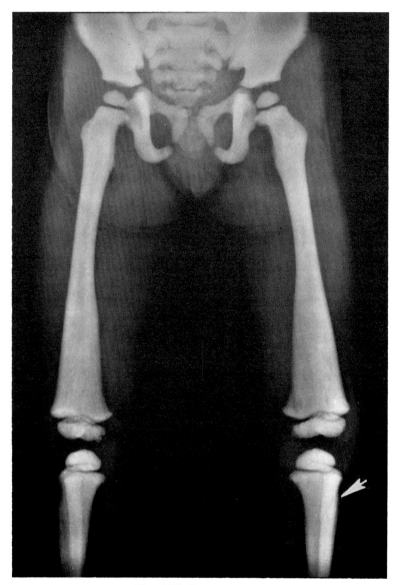

Fig. 1432. Osteopetrosis

Extreme bone density and a pathologic fracture (*arrow*) through the proximal end of the left fibula. Lack of modeling at the distal end of the femurs (Courtesy of Herbert M. Stauffer, Temple University Hospital, Philadelphia, Pa.)

miniature bone, inset in the host bone,[16] resembling a neonatal bone in size and shape. This is evidence of cessation of the sclerosing process near the time of birth. This "bone-in-bone" appearance may also be seen in the adult skeleton after Thorotrast administration in childhood. It is particularly prominent in the spine.[17-20]

Alternating dense and radiolucent metaphyseal transverse lines often occur,[22] and may be the only sign of disease. The epiphyses may consist of dense concentric circular zones. Longitudinal striations may be present in the metaphyses. Transverse bands are prominent in the phalanges. Iliac bones often present multiple, alternating, dense curved zones which parallel the crests and correspond to the transverse bands of tubular bones. Carpal and tarsal bones are as dense as other bones, and may show a dense center and radiolucent periphery, or normal center density with dense periphery, depending on the patient's age at onset. The centra of the vertebrae may be uniformly dense, but occasionally bands of normal bone in the center separate dense bands above and below. The skull base is also involved, although the calvaria is often spared. Increased density may not be marked; trabecular pattern, cortex, and medullary cavity may be distinguishable. Transverse bands and longitudinal metaphyseal striations are definite and must be differentiated from the results of heavy metal poisoning. Epiphyseal separation and fractures commonly occur.

Marked calcification of soft tissue and arteries indicates hypercalcemia. Osteopetrosis is generalized, unlike melorheostosis, which is limited to one extremity only. Heavy metal poisoning may be confused with the milder forms of osteopetrosis if the dense metaphyseal bands are obvious; however, a skeletal survey should eliminate this possibility.

DYSOSTEOSCLEROSIS

Dysosteosclerosis is a rare, presumably autosomal recessive bone dysplasia, with thickening and sclerosis of the base of the skull. This often impinges on the optic canals and causes blindness.[1-5] The ribs and vertebral bodies are dense, as in osteopetrosis. The metaphyses of the long bones are widened with a thin cortex. It is similar to metaphyseal dysplasia, but in young children there is a transverse band of metaphyseal density. The flattening of the vertebral bodies is also a distinctive finding, not a characteristic of osteopetrosis or metaphyseal dysplasia (Fig. 1433). There is dental hypoplasia, and the permanent teeth fail to erupt. Cutaneous and neurologic manifestations are inconsistent.[6,7]

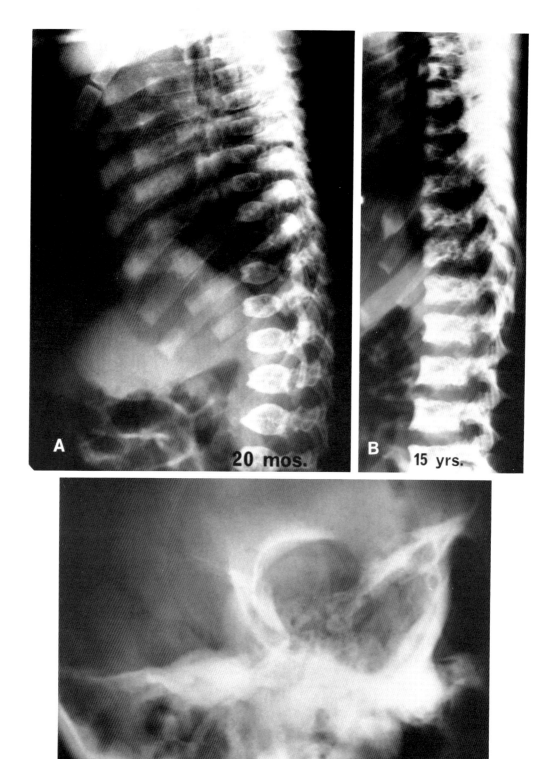

Fig. 1433. Dysosteosclerosis

(A) At 20 months, the thoracic bodies are slightly flattened, with anterior beaking of upper lumbar vertebrae. (B) At 15 years, the thoracic bodies are more flattened and more irregular in density. All bodies are concave anteriorly. (C) At 15 years, the hypertrophied bone narrows the optic canal. (D) The upper extremity at 20 months, 5½ years, and 15 years. Note the similarity to Pyle disease at 15 years. (E) The lower extremity at 20 months. The metaphyses are widened with dense metaphyseal bands. (*Middle*) At 5½ years, irregular densities and lack of modeling. (*Right*) At 15 years, the distal femur shows more advanced Erlenmeyer flask configuration with thin cortex reminiscent of Pyle disease. (F) Hands. (*Left*) At 3 years, tubular bones show similar but less severe abnormalities of modeling; (*right*) at 15 years, terminal tufts are relatively sclerotic. (G) Pelvis. (*Left*) At 15 years, x-shaped area of sclerosis in each lower ilium and wide metaphyses of upper femurs. (Courtesy of C. Stuart Houston, Saskatoon, Canada. Previously published by C. S. Houston, J. W. Gerrard, and E. J. Ives: *American Journal of Roentgenology, 130:* 988, 1978.[1])

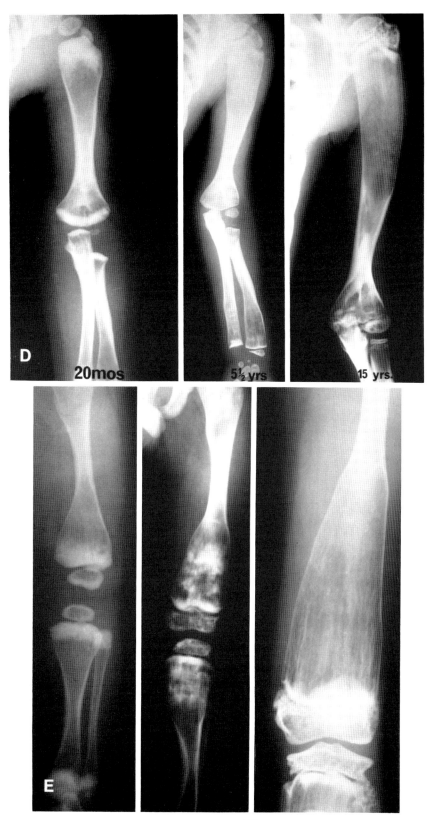

Fig. 1433 (D and E)

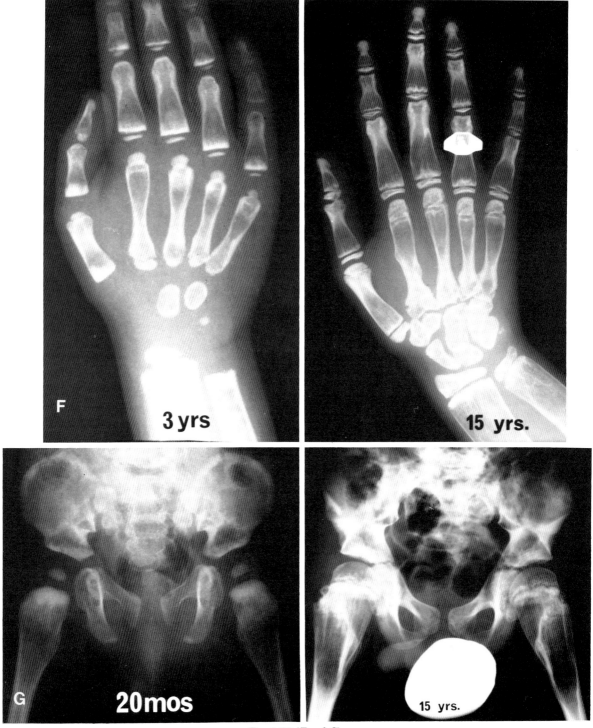

Fig. 1433 (F and G)

PYCNODYSOSTOSIS

This is a condensing bone disease with features similar to osteopetrosis and some resemblance to cleidocranial dysostosis. The characteristic features are increased bone density, predominantly in the skull and hands, dwarfism, and fragility.[1,2] It is transmitted by an autosomal recessive mode.

Pathologic Features. The precise nature of the basic defect of pycnodysostosis is unknown.

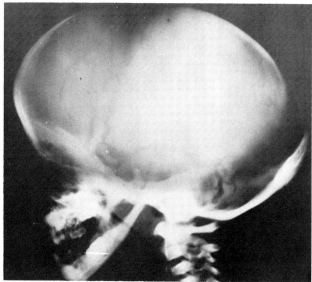
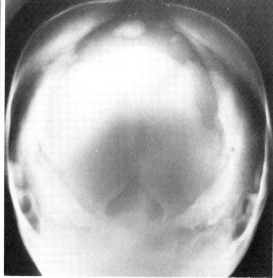

FIG. 1434. PYCNODYSOSTOSIS
Absence of fusion of the sutures and fontanels with remaining lakes of bone and dense base of skull, hypoplastic mandible and maxilla, and virtual disappearance of angle of mandible. (Previously published by Stanton E. Shuler in *Archives of Disease in Childhood*, 38: 620, 1963[1]; reprinted by permission of the author.)

Morphologically there is disorganization of bone structure at the level of lamella bone formation and of the osteum, although a chemical defect of the matrix components has not been determined. There are few osteoclasts and an excessive number of osteoblasts on the trabecular surface; this may be a physiologic response to an abnormal, or structurally inferior, matrix or a pathologic failure of the osteoclasts to initiate appropriate resorption and of bone surface cells to function. It is suggested that the cells do not respond normally to the demands of stress on the skeleton.[3]

Roentgenographic Features. Mandibular hypoplasia with mandibular angle loss results in craniofacial disproportion. Abnormalities in dentition are frequent, and there may be yellowish discoloration of the milk teeth.[4] The cranial sutures remain open; a wide anterior fontanel is usually present. Platybasia is frequent (Fig. 1434). There is nonpneumatization and hypoplasia of the paranasal sinuses.[5, 6] The long bones reveal a generalized increase in density, with cortical thickening but the medullary cavity is preserved. There is a tendency to multiple spontaneous fractures, less severe than in osteogenesis imperfecta (Fig. 1435). The distal phalanges are hypoplastic and the tufts may be absent (Fig. 1436). Hypoplasia may be present in the acromial end of the clavicle (Fig. 1437).

Vertebral segmentation may be lacking, especially at C1 and C2 and the lumbosacral junc-

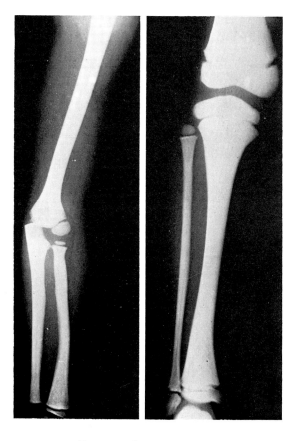

FIG. 1435. PYCNODYSOSTOSIS
Extreme density of the long bones. (*Left*) Upper extremity, (*right*) lower extremity. (Courtesy of Stanton E. Shuler.)

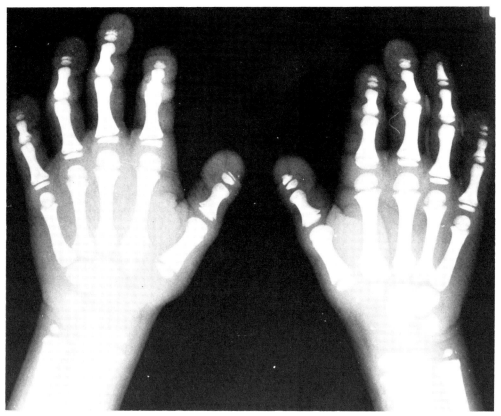

FIG. 1436. PYCNODYSOSTOSIS
Hypoplasia of the distal phalanges and increased density of the bones. (Previously published by Stanton E. Shuler in *Archives of Disease in Childhood, 38:* 620, 1963[1]; reprinted by permission of the author.)

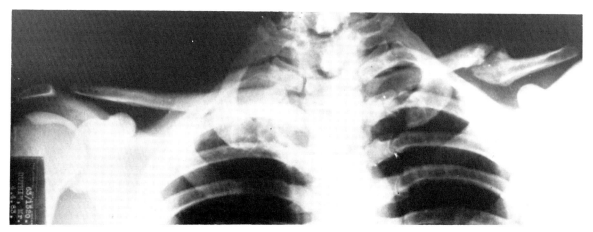

FIG. 1437. PYCNODYSOSTOSIS
Hypoplasia of the acromial end of the clavicles. Pseudoarthrosis of the left clavicle. (Previously published by Stanton E. Shuler in *Archives of Disease in Childhood, 38:* 620, 1963[1]; reprinted by permission of the author.)

TABLE 50. SALIENT FEATURES OF OSTEOPETROSIS, CLEIDOCRANIAL DYSOSTOSIS AND PYCNODYSOSTOSIS[a]

Features	Osteopetrosis	Cleidocranial Dysostosis	Pycnodysostosis
Base of skull	Dense	Normal or rarely dense	Dense
Cranial sutures	Normal	Normal	Open
Paranasal sinuses	Unaerated	Normal	Unaerated or closed
Mandible	Normal	Normal	Loss of angle
Clavicle	Present and normal	Absent or dysplastic	Present, sometimes dysplastic
Hands and feet	Normal	Normal	Aplastic tufts, short phalanges, overriding nails
Pelvis	Coxa vara	Normal	Coxa plana
Spontaneous fractures	Present	Absent	Present
Bone texture	Dense with obliteration of intramedullary canals	Normal	Dense without obliteration of intramedullary canals
Blood findings	Aplastic anemia	Normal	Normal
Genetic	Dominant; anemic type, recessive	Dominant	Recessive
Stature	Usually normal	Usually normal	Short

[a] From Z. Emami-Ahari et al.: *Journal of Bone and Joint Surgery, 51B*: 307, 1969.[10]

tion.[6] There may be coxa valga and shallow acetabulum with an increase in the acetabular angle of inclination. Spondylolisthesis and spondylolysis are described.[6–10]

The condition may be differentiated from osteopetrosis by the abnormalities in the skull, mandibular and distal phalangeal hypoplasia, and the lack of transverse metaphyseal bands of increased density, so characteristic of osteopetrosis. Anemia does not occur (Table 50).[10]

Pycnodysostosis may be distinguished from cleidocranial dysostosis by the dense bones, mandibular and terminal phalangeal hypoplasia, and the short stature.[10]

NEUROFIBROMATOSIS (VON RECKLINGHAUSEN DISEASE)

Neurofibromatosis is a congenital disease of the mesodermal and neuroectodermal elements. Smith[1] first described the disease, then, in 1882, von Recklinghausen[2] evaluated the histology and related the lesions to the nervous system. An excellent comprehensive review of this subject was presented by Dr. John F. Holt in the 1977 Edward B. D. Neuhauser Lecture.[3] The condition affects both the peripheral and central nervous systems, and is characterized by pigmented areas of the skin (cafe-au-lait), cutaneous fibromas (fibroma molluscum), and neurofibromas of the cranial and peripheral nerves. Sometimes the skeletal and endocrine systems are affected, so that the disease is not limited to the nervous system, although most of its significant changes do occur there.

The tumors of the peripheral nerves arise from the nerve sheaths and are usually discrete. A diffuse enlargement of a peripheral nerve is sometimes seen; disturbed bone growth often underlies it.

Clinical Features. Skin lesions are the most consistent finding. These are either pigment (cafe-au-lait spots) or fibromatous. The areas of pigmentation, which are the most common manifestation, have smooth outlines and vary in size from 1–2 mm to several centimeters in diameter. The cutaneous fibromas may be sessile or pedunculated, or they may appear as a large fold of skin, thickened and indurated, called "elephantiasis neuromatosa." These vary in size from pinhead dots to masses 5 cm or more in diameter. They may be few and single, or cover the body completely. They are usually asymptomatic, but may be painful and tender.

Neurofibromas of the cranial nerves may cause a variety of symptoms, depending on the nerve involved. Tumors of the spinal nerves may cause a multitude of motor symptoms, including paraplegia. The most serious and common complication is severe scoliosis, which may lead to paraplegia. The lesions may or may not progress. About 10% may undergo sarcomatous degeneration.[4,5] Hyperparathyroidism[6,7] and acromegaly also have been encountered in conjunction with neurofibromatosis.

Pathologic Considerations. The origin of the nerve tumors is unknown. Verocay,[8] Mason,[9] and Murray and Stout[10] have suggested that the cells arise from the Schwann syncytium. Tumors are found along the course of the peripheral and autonomic nerves, and the meninges. They may be discrete nodular lesions or may grow along the nerve, at times increasing the diameter of the entire nerve. Meningiomas are not uncommon and, occasionally, glial "tumors" appear, often near the floor of the third ventricle.[11] The neurofibroma may occur along the cranial or spinal nerves and erode the surrounding bone. The cafe-au-lait spots, caused by proliferation of melanoblasts, sometimes undergo malignant degeneration.

The skeletal system is involved in 80% of children with neurofibromatosis.[12] If macrocranium is included, 90% have skeletal abnormality.[3]

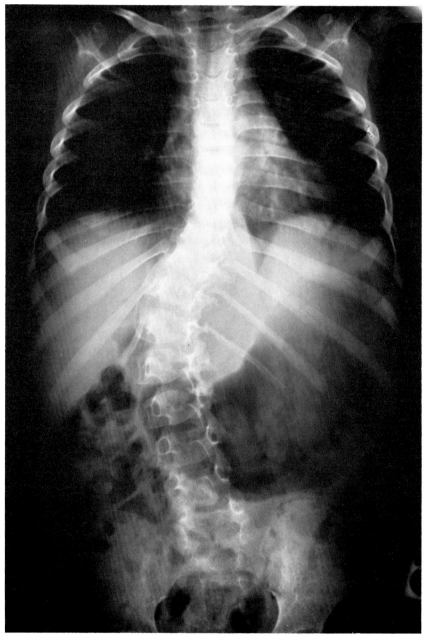

FIG. 1438. NEUROFIBROMATOSIS
Scoliosis occurs in about one-half of the patients. Its cause is unknown and it may not be associated with local neurofibroma. (Courtesy of John W. Hope, Children's Hospital, Philadelphia, Pa.)

Roentgenographic Features. The roentgenographic manifestations of neurofibromatosis include scoliosis, changes in bones due to erosion, abnormal growth, and pseudoarthrosis. Scoliosis occurs in about half of the patients. Its cause is unknown. It most commonly involves the lower thoracic spine (Fig. 1438). Usually there is sharp angulation; paraplegia is likely to occur. The scoliosis may be accompanied by kyphosis, particularly pronounced in the cervical region (Figs. 1439–1441). Local spinal neurofibromas, however, may not be present at all. Sometimes the scoliosis is secondary to a discrepancy in leg length, in which case the scoliotic deformity is less acute.

Collapse or wedging of vertebrae may occur, probably caused by deossification and replacement of the vertebral body by masses of spindle cell tumors.[13] Erosion in the posterior portion of the vertebral centrum or enlargement of spinal foramina may result from tumors arising from spinal nerves (Fig. 1440).

Cervical kyphosis is a frequent feature of neurofibromatosis combined with anterior and posterior scalloping of the vertebral bodies; it produces a vertebral body at the apex of the scoliosis which suggests the configuration of a golf tee.[3, 14] These changes are usually seen in adults, but are reported in younger individuals and may reach full proportions at an early age (Fig. 1447).[3, 15-17]

Erosion of the spinal elements are more often due to cysts or dilation of the spinal canal rather than to local neurofibroma[18] (Fig. 1442). Enlarged cervical foramina may be due to cervical perineural or extradural cysts rather than to a "dumbbell" neurofibroma.[19] Multiple or single foramina may be enlarged.

Intrathoracic meningoceles are most commonly seen in patients with neurofibromas and are usually asymptomatic.[18, 20-24] They produce

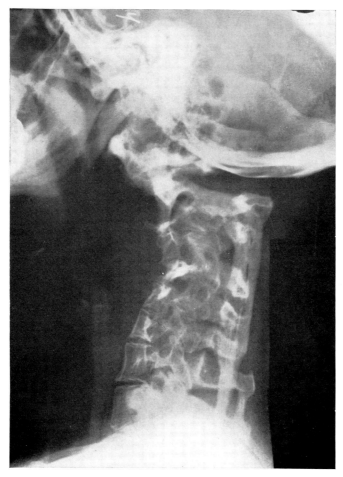

FIG. 1439. NEUROFIBROMATOSIS CAUSING DISTORTION OF BONE
Bone grafts in the cervical region to provide stability.

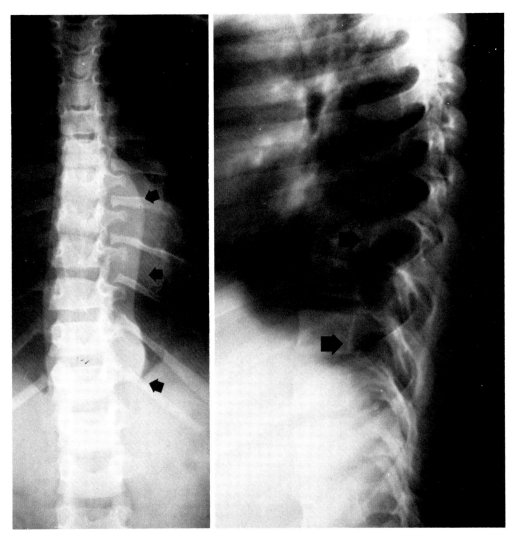

Fig. 1440. Neurofibromatosis

(*Left*) Multiple neurofibromas originating from spinal nerves produce a paravertebral mass (*arrows*). Rib changes are also noted. (*Right*) Erosion of the bone and the widening of the spinal canal (*arrows*). (Courtesy of John W. Hope, Children's Hospital, Philadelphia, Pa.)

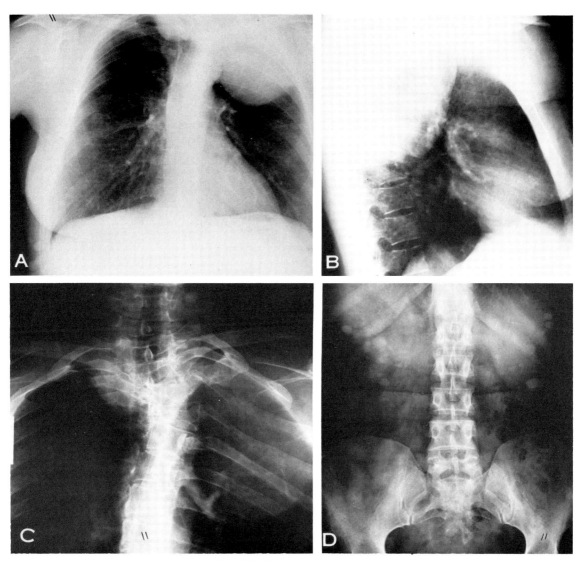

FIG. 1441. NEUROFIBROMATOSIS
(A) Chest roentgenogram reveals several large neurofibromas in the posterior mediastinum. (B) Lateral projection. (C) Roentgenogram of the upper dorsal spine reveals the neurofibroma and hemivertebra producing scoliosis of the upper dorsal spine. No local neurofibromas were associated. (D) Abdominal film reveals ovoid densities overlying the abdomen due to skin neurofibromas. (Courtesy of Dr. Gerald Murdock, Philadelphia, Pa.)

posterior mediastinal masses erroneously interpreted as neurofibromas. Myelography will show opaque filling and differentiate them from solid tumors.

Macrocranium and macroencephaly occur in most children with neurofibromatosis.[3, 12, 25-29] Normal pneumographic and arteriographic examinations and normal computer tomography suggest that the cause of the macrocranium is increase in the size of the brain.[3]

Changes in the skull, including the facial bones, are well documented. Frequently, a portion of the wall of the orbit or sphenoid bone are congenitally absent (Figs. 1443-1445).[30, 31] Neu-

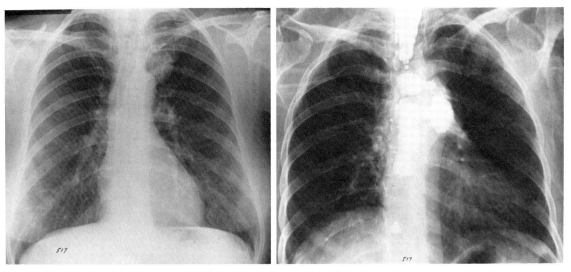

FIG. 1442. NEUROFIBROMATOSIS WITH ARACHNOID CYSTS

(*Left*) The survey chest reveals a density in the mediastinum thought to be a neurofibroma. (*Right*) The myelogram revealed the density to be due to arachnoid cysts. Several cysts are filled with opaque material. No local vertebral neurofibromas were present. Erosions of spinal elements are more often due to cysts or dilation of the spinal canal rather than local neurofibroma.

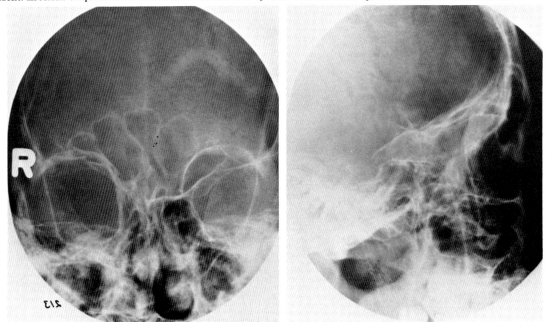

FIG. 1443. NEUROFIBROMATOSIS

Absence of the right sphenoid bone, which is not due to a neurofibroma. (Courtesy of Dr. Gerald D. Dodd, M. D. Anderson Hospital and Tumor Institute, Houston, Tex.)

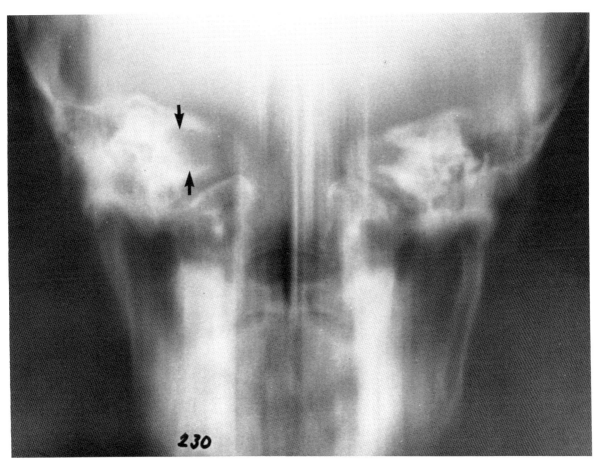

Fig. 1444. Neurofibromatosis
Neurofibromatosis with neurofibroma enlarging the auditory canal (*arrows*).

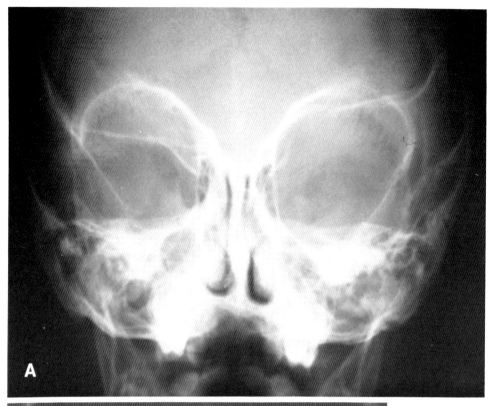
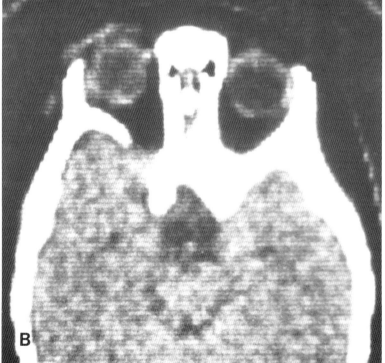

Fig. 1445. Neurofibromatosis
(A) Absence of the left sphenoid bone. (B) Computer tomography shows defect in the sphenoid bone. (Courtesy of Dr. John F. Holt, Ann Arbor, Mich.)

rofibromatous tissue is occasionally found adjacent to the osseous abnormality, but it is usually due to progressive mesodermal dysplasia. Calvarial defects encompassing or adjacent to the lambdoidal suture are frequent, and consist of rounded, oval or irregular bone lucencies involving the left lambdoidal suture, just posterior to the parietal mastoid and occipital mastoid sutures (Fig. 1446).[3, 32-37] The lesion is more common on the left side than the right by a ratio of 24 to 6.[38] Dysplastic lucencies of this sort may occur elsewhere in the calvarium; the most common site appears to be in the region of the sagittal suture posteriorly (Fig. 1446).[3, 39] Gliomas of the optic pathways are an integral part of neurofibromatosis,[40] and most of these lesions occur in children under the age of 12.[3, 41, 42] The radiologic diagnosis of optic gliomas is based primarily but not entirely on the demonstration of enlargement of one or both optic foramina. Computer tomography is probably the most accurate method in demonstrating enlarged optic canals and foramina.[3, 43] Acoustic neuromas are frequent (Fig. 1444), but rare in children.[3, 15, 44-47] The lesions are usually bilateral.[3, 15, 47] Dysplastic widening of an auditory canal may be confused with acoustic neuromas if the symptoms are not known.[3, 48] This may be accompanied by a combination of dysplastic widening of an auditory canal with enlargement of the contralateral optic foramen.[3, 49] Vertebral scalloping,[18, 19] or "gauge" defects, of the anterior and lateral aspects of the spine and the posterior aspect of the spine occur frequently. The posterior scalloping may be due to dural ectasia or neurofibromas.[50, 51] Anterior and lateral scalloping may be due to pressure from adjacent neurofibromas or a primary mesodermal dysplasia of bone (Fig. 1447).[52]

The periosteum is abnormal in some patients with neurofibromatosis.[53] The "gracile" bones and "twisted ribbon" rib appearance may be due to abnormal appositional growth. The response to trauma shows that the periosteum is easily stripped from the bone, which may result in large calcified subperiosteal hemorrhage (Figs. 1448 and 1449). Poor callus response and pseudoarthrosis may also result from the abnormal periosteum[54] (Fig. 1450).

Bones other than the vertebrae exhibit changes in fewer than 10% of the cases.[55] These changes are: (1) erosions from periosteal or adjacent soft tissue neurofibromas; (2) single or multiple "cystic" lesions within bone (Fig. 1451); (3) growth disturbances, including focal gigantism or dwarfism (Fig. 1452); (4) streaky increase in density of the tubular bone shaft; (5) softening and bowing of the ribs or long bones with or without multiple incomplete fractures; and (6) pseudoarthrosis, which may be a separate clinical entity. Soft tissue tumors may be evident and small exotoses may form adjacent to a soft tissue neurofibroma (Fig. 1453).

Erosions of the periosteal surface of the cortex may be due to periosteal neurofibromas or to tumors outside the bone (Fig. 1454). Intercostal tumors exemplify the latter; they erode the inferior margin of the rib, producing "notching," sometimes mistaken for coarctation of the aorta (Fig. 1455). The bone adjacent to the erosion may increase in density as much as several centimeters in either direction.

The "cystic" lesions in long bones are probably related to the osteolytic changes of vertebrae, representing deossification rather than replacement of bone by tumor. They may become large. Heublein et al.[56] reported a patient with a large excavation of the left ilium and complete resorption of the horizontal ramus of the right pubic bone. Holt and Wright[5] described multiple "cystic" lesions in the metaphyses of long bones in the mandible and vertebra, and have also seen these rarefied areas regress completely, and fill in with apparently normal bone. Histologically, most of these lesions prove to be nonosteogenic fibromas, and are usually much larger than those found without neurofibromatosis (Fig. 1456).

Disturbances in bone growth, especially of a tubular bone, is another manifestation of neurofibromatosis. The length of the bone is generally increased, and often accompanied by an increase in diameter. The hypertrophy may involve one or all the bones of an extremity, and hypertrophy of the soft tissues of the extremity keeps pace with it. When massive enlargement of the soft tissues exists, the condition is referred to as "elephantiasis neuromatosa."[57, 58] The bone overgrowth may be secondary to chronic hyperemia due to hemangiomatous and lymphangiomatous tissue. Elephantiasis neuromatosa must be differentiated from congenital hemihypertrophy; the two conditions may coexist.[16, 59-61] The affected bone usually corresponds to the distribution of an affected nerve. In some cases bone growth may be markedly retarded rather than accelerated. These bones, whether they are abnormally thick or slender, have a peculiar wavy appearance[62] (Figs. 1452 and 1457).

Long bones may be increased in density, irregularly thickened, and sometimes reveal linear, ridgelike densities at one end. Often they reflect no focal bone erosions to suggest adjacent soft tissue tumors. Softening of the bones with molding and deformity is especially common in the ribs[63] and tibia. The bowing may be extreme,

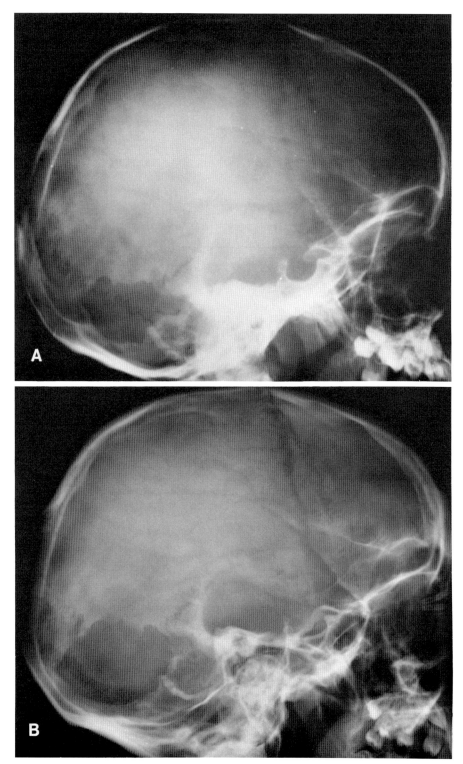

FIG. 1446. NEUROFIBROMATOSIS

Girl treated since infancy for severe facial neurofibromatosis. (A) At age five: Large left lambdoidal defect, the only visible abnormality. (B and C) At age 13, lambdoidal defect increased in size; a posterior sagittal defect appears. Dysplastic bone changes have enlarged the sella and almost completely destroyed the posterior two-thirds of skull base. (Courtesy of Dr. John F. Holt, Ann Arbor, Mich. Previously published in *American Journal of Roentgenology 130:* 615, 1978.[3])

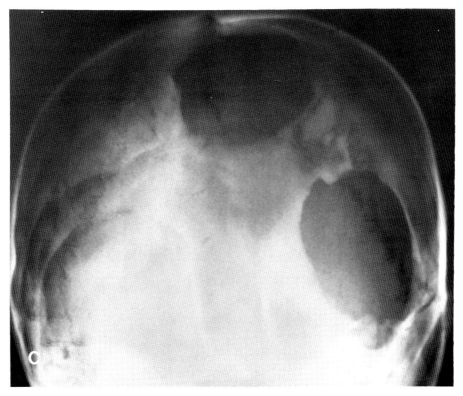

Fig. 1446(C)

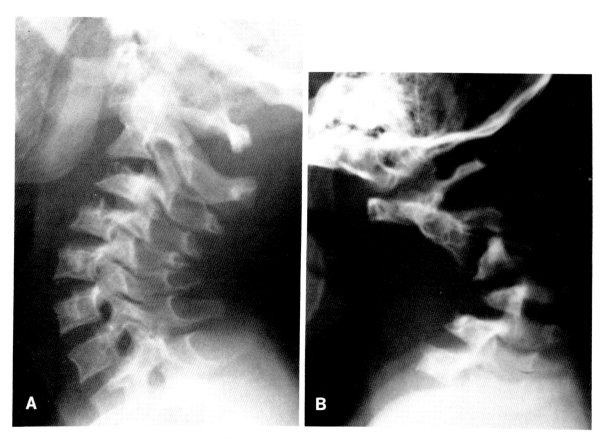

FIG. 1447. NEUROFIBROMATOSIS
Two patients with kyphosis. (A) Five-year-old boy with apex of upper cervical kyphosis at level of C3. Sharply pointed anterior aspect. Anterior and posterior scalloping of all vertebral bodies with "pointing" of vertebral margins, which at this age are normally rounded. (B) Six-year-old boy with more severe midcervical kyphosis. (Reprinted with permission from D. C. Harwood-Nash and C. R. Fitz: *Neuroradiology in Infants and Children*. C. V. Mosby Co., St. Louis, 1976.[15])

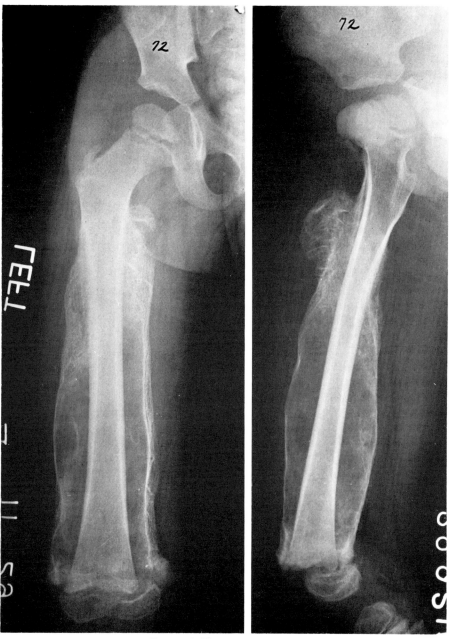

FIG. 1448. NEUROFIBROMATOSIS

Traumatic dislocation of the distal femoral epiphysis. The loosely attached periosteum is stripped from the bone and a large subperiosteal hematoma is calcifying. (*Left*) Posteroanterior projection, (*right*) lateral projection.

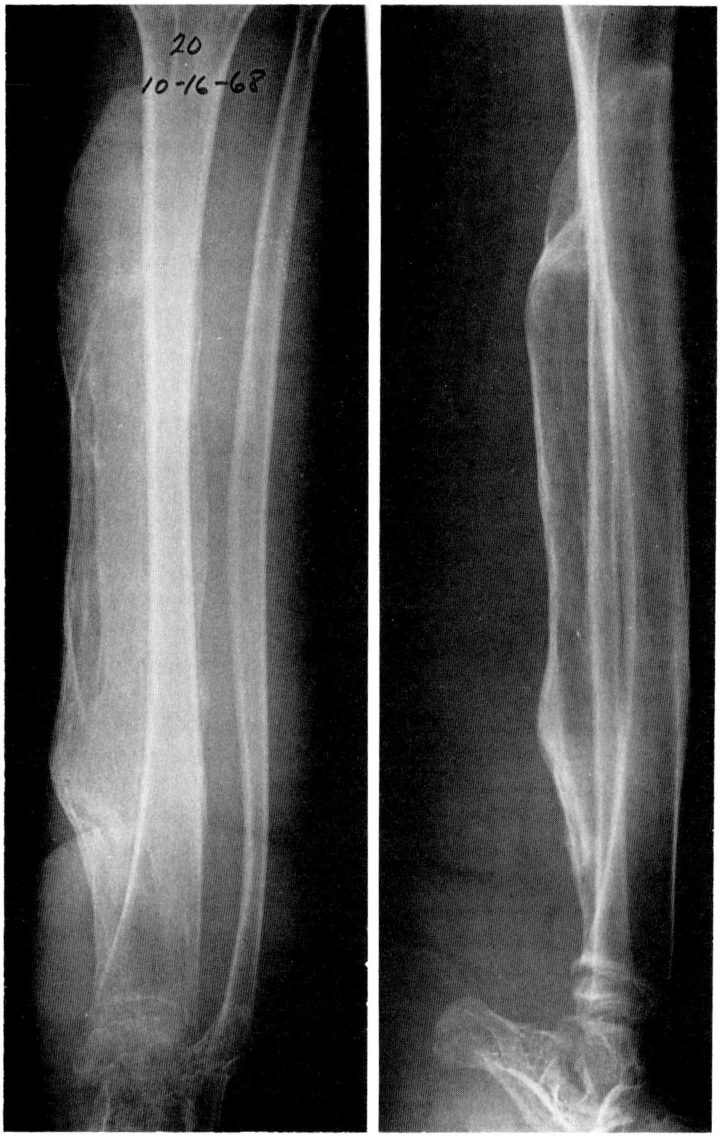

FIG. 1449. NEUROFIBROMATOSIS
The bones have a ribbonlike or gracile appearance. Three weeks after trauma a large calcified subperiosteal hemorrhage is present. The periosteum is loosely attached to the bone in some patients with neurofibromatosis which accounts for the gracile bones and the abnormal response to trauma.

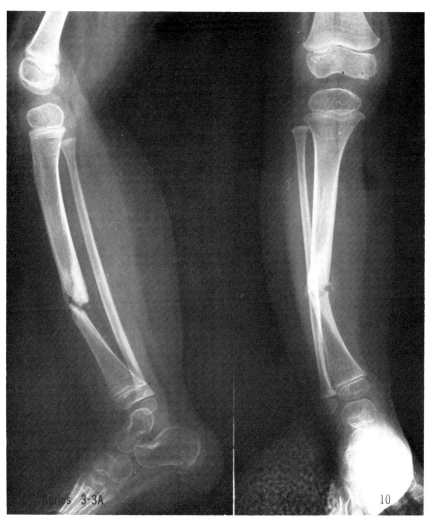

Fig. 1450. Pseudoarthrosis of Tibia

This may occur with neurofibromatosis or be an isolated entity. It is probably related to abnormal periosteum in patients with neurofibromatosis.

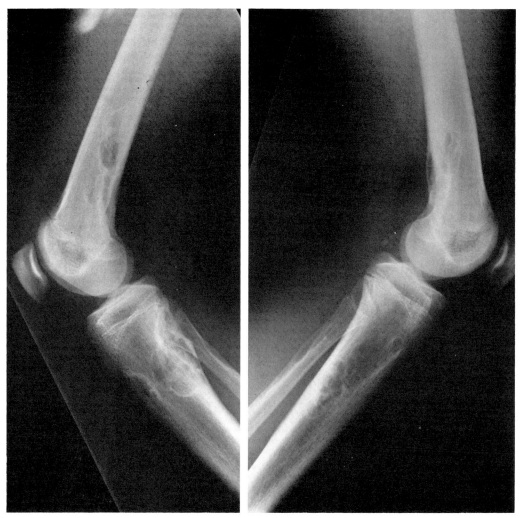

Fig. 1451. Neurofibromatosis

Multiple "cystic" lesions with sclerotic periphery are present. They occurred in multiple bones and were proved by biopsy to be nonosteogenic fibromas.

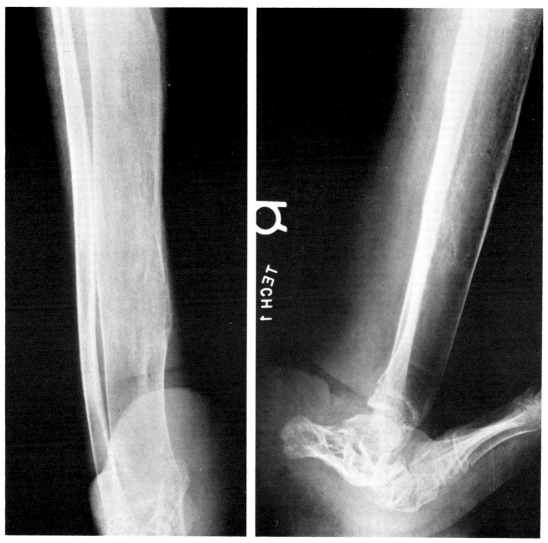

Fig. 1452. Neurofibromatosis
Focal gigantism of the tibia and deformity of the foot. Hypertrophy of the soft tissues. (*Left*) Anteroposterior projection, (*right*) lateral projection.

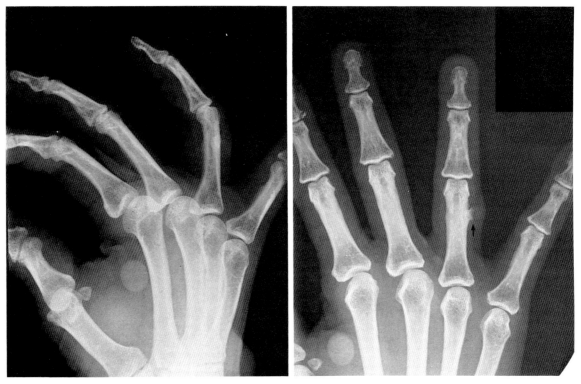

Fig. 1453. NEUROFIBROMA
Soft tissue tumor between the thumb and the index finger, with a small exostosis of the proximal phalanx of the fourth finger adjacent to a minor neurofibroma (*arrow*).

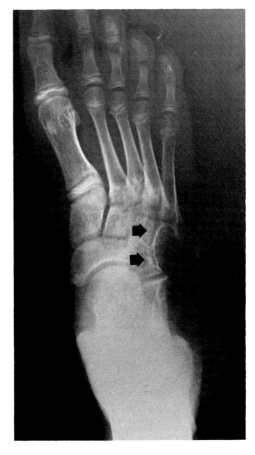

Fig. 1454. NEUROFIBROMATOSIS
Erosion of tarsal due to soft tissue neurofibroma (*arrows*). (Courtesy of John W. Hope, Children's Hospital, Philadelphia, Pa.)

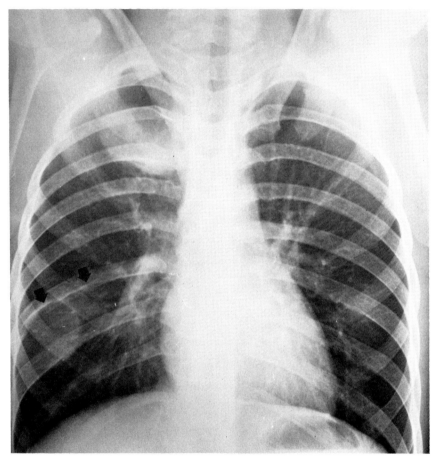

FIG. 1455. NEUROFIBROMATOSIS

Intercostal neurofibromas erode the inferior margin of the rib and may produce "notching" which may be mistaken for coarctation of the aorta (*arrows*). A large neurofibroma overlying the right upper lung. (Courtesy of John W. Hope, Children's Hospital, Philadelphia, Pa.)

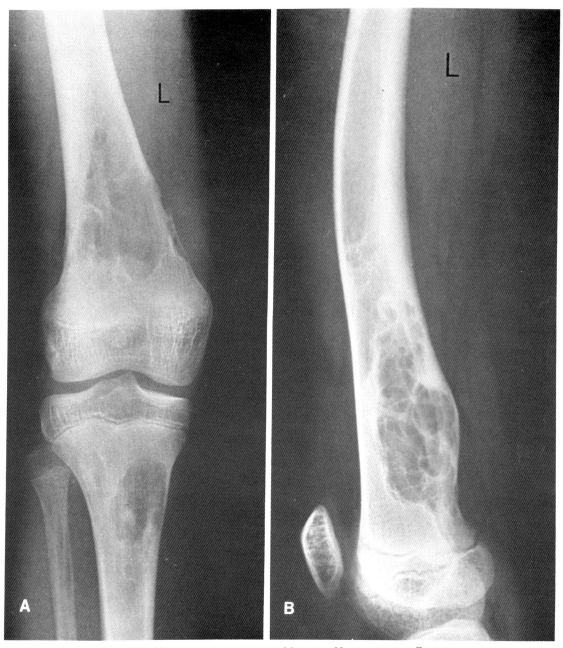

FIG. 1456. NEUROFIBROMATOSIS WITH MULTIPLE NONOSTEOGENIC FIBROMAS
(A) Large neurofibromas in the femur and tibia. Similar lesions were present in the other knee. (B) Lateral projection.

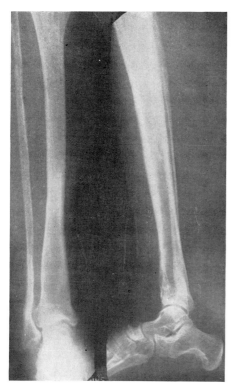

FIG. 1457. NEUROFIBROMATOSIS
Disturbance of tubular bone growth is usually manifested by an increase in the length and a peculiar wavy appearance.

perhaps accompanied by multiple incomplete fractures.[64] Some of these changes are due to osteomalacia, which may be associated with neurofibromatosis without dietary deficiencies (Fig. 1458).

Pseudoarthrosis is a complication of neurofibromatosis.[65, 66] Although it may occur as a separate clinical entity (Fig. 1450), one-half of the patients with pseudoarthrosis manifest other stigmata of neurofibromatosis.[13] Pseudoarthrosis begins with deossification of a weight-bearing bone, which may subsequently bow and then fracture. Because of poor callus formation, bone union does not occur and pseudoarthrosis ensues. These fractures were once attributed to intraosseous neurolemmoma; however, Holt and Wright[5] and Aegerter and Kirkpatrick[13] convincingly showed that fibro-osseous tissue adjacent to the fracture is the result and not the cause of the pseudoarthrosis.

Pseudofractures can occur at birth, most commonly in the lower two-thirds of the tibia. After fracture, the bones may seem to heal, but, with weight bearing, fractures recur and lead to eventual pseudoarthrosis. Congenital pseudoarthrosis can occur in the long bones of the forearm, and is usually associated with neurofibromatosis or fibrous dysplasia. It may also be idiopathic.[39, 67-77] Congenital anomalies of the skull, such as absence of the sphenoid bone, are not unusual, underscoring the fact that neurofibromatosis is not solely a disturbance of nerves. Spina bifida, lack of segmentation of the vertebral centra, congenital dislocation of the hip, clubfoot, and defects of the bones of the orbital structures have all been reported in conjunction with neurofibromatosis. Malignant degeneration is infrequent and occurs in the joints or soft tissues as neurofibrosarcoma (Fig. 1459).

Fibrosing alveolitis (interstitial lung disease) is reported.[78-83] Endocrine and metabolic lesions are comprehensively reviewed by Saxena[84]; the most common disorder is sexual precocity. Some of these children have large gliomas of the optic tract which may encroach upon the hypothalamus causing hypersecretion of growth hormone.[3, 85]

There have been some reports[86-88] of hyperparathyroidism in association with neurofibromatosis, and there may be a relationship between the two.

Neurofibromatosis and osteomalacia occur in combination and there is indication that it is more than a casual relationship.[63, 89-91] All reported cases were adults except for an 8-year-old boy,[3, 92] and it is believed that they represent examples of familial hypophosphatemic vitamin D resistant osteomalacia.[3]

The differential diagnosis in the usual case should not be difficult. Multiple fibromas of the skin are definitive. The bone changes may closely simulate fibrous dysplasia, but the cafe-au-lait spots of fibrous dysplasia are usually smoother and larger than those of neurofibromatosis.

Extensive hemangiomas of an extremity may result in gigantism, but the absence of cutaneous fibromas would distinguish it, as well as the frequent identification of phleboliths within the soft tissue masses. Rib notching may be confused with the rib notching of aortic coarctation or other congenital vascular disease.

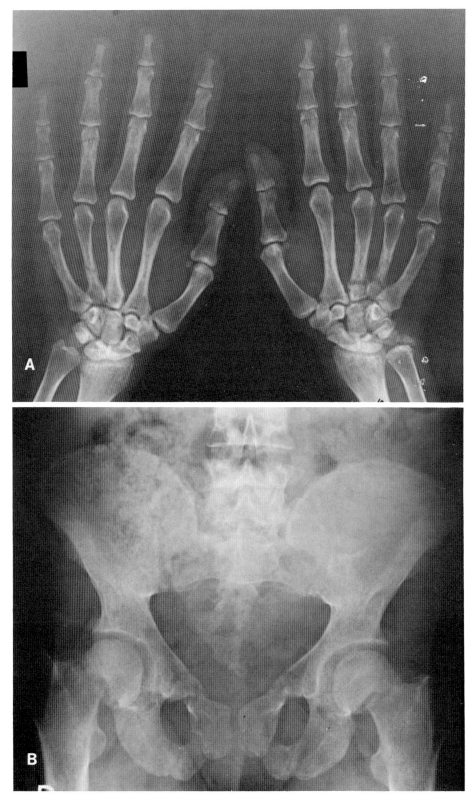

Fig. 1458. Neurofibromatosis with Osteomalacia

(A) Multiple fractures in both hands, particularly the fourth and fifth metacarpals and the ulnar styloid of the left hand, and fourth metacarpal of the right hand. (B) Pelvis: fractures of both femoral necks and multiple fractures of the pubic and ischial rami. (Courtesy of Dr. M. Bajogkli, Teheran, Iran.)

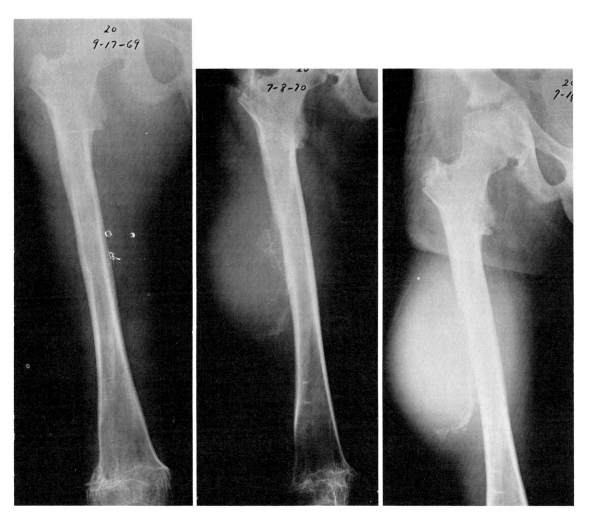

Fig. 1459. Neurofibroma and Neurofibrosarcoma

This patient had extensive neurofibromatosis with scoliosis and long bone deformities. (*Left*) In 1969 this patient began having pain in the mid-thigh. The bones show growth disturbance but no soft tissue mass or bone erosion. (*Center*) Six months later, there is a large soft tissue mass with calcification and erosion of the periosteal surface of the cortex. (*Right*) Soft tissue technique to show the large soft tissue mass and calcification.

MACRODYSTROPHIA LIPOMATOSA

Macrodystrophia lipomatosa is a rare congenital malformation characterized by an increase in the size of all elements and structures of a digit or digits.[1-4]

An overgrowth of fatty tissue is associated with localized gigantism, and many names have been used for it including: partial acromegaly, macrosomia, elephantiasis, megalodactyly, dactylomegaly, macrodactyly macroceir, and club finger.[1,2]

It is not hereditary.[2,3,5] Proposed causes include: lipomatous degeneration,[1] disturbed fetal circulation,[2] error in segmentation,[2] trophic influence of tumefied nerve,[1] and in utero disturbance of the growth limiting factor.[3] The distribution is usually in the area of the median nerve.[1] There is documented evidence of macrodactyly in patients with neurofibromatosis[6,7]; there may be a relationship between them.[6,8,9]

Roentgenographic Features. There is abnormality in both the soft tissues and osseous structures, as evidenced by an overgrowth most marked at the distal end of the digit and along its volar aspect. The asymmetric overgrowth causes a dorsal deviation of the affected parts, and most patients exhibit clinodactyly. Usually lucencies in the soft tissues indicate the dramatic overgrowth of fat.[4] The phalanges are long, broad, and often splayed at their distal ends (Fig. 1460). The articular surfaces may slant, and se-

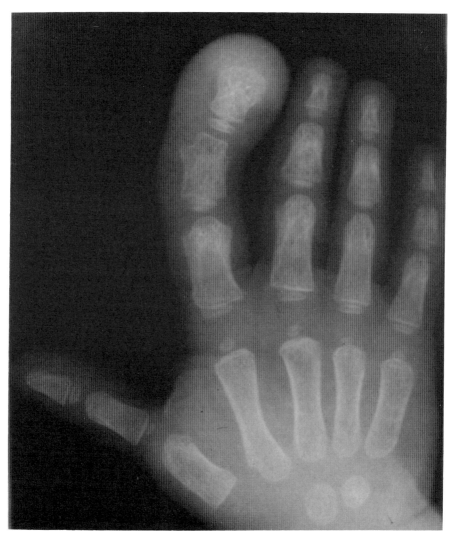

FIG. 1460. MACRODYSTROPHIA LIPOMATOSA
Overgrowth of fatty tissue with overgrowth of the soft tissues and bones of the index finger and ulnar and dorsal deviation of the finger. The phalanges are long, broad and splayed at their distal ends. (Courtesy of Dr. Amy Goldman, Hospital for Special Surgery, New York, N.Y.)

vere secondary degenerative joint disease occurs in childhood.[1,2,7] Syndactyly and polydactyly may be present.

The most difficult disease from which to distinguish it is neurofibromatosis, and, in fact, some patients also have neurofibromatosis. However, the macrodactyly in neurofibromatosis is usually the result of plexiform neurofibromas with hemangiomatous and lymphangiomatous element, combined with a mesodermal dysplasia. In neurofibromatosis, the enlarged digits are usually bilateral; there may not be contiguous involvement of an extremity, and the distal phalanges are not the most seriously affected. Also, the hemangiomatous elements of plexiform neurofibroma can cause premature fusion of the growth plates,[10] whereas growth in a digit caused by macrodystrophia lipomatosis stops when the patient reaches puberty. The enlarged osseous structures in neurofibromatosis may have a cortex and elongated sinuous appearance[10,11] due to periosteal abnormalities. Finally, the soft tissue lucencies associated with neurocutaneous manifestations in neurofibromatosis have not been reported with macrodactyly.[4]

ESSENTIAL OSTEOLYSIS

Essential osteolysis is a progressive, slow, bone-resorptive disease predominantly affecting the carpals, tarsals, metacarpals, metatarsals, and elbow. The etiology is unknown and there seem to be two types: (1) a hereditary type transmitted by an autosomal dominant mode[1,2] and (2) a nonfamilial type frequently associated with fatal neuropathy.[3-7]

The histologic features are proliferation and hyperplasia of the smooth muscle cells of the synovial arterioles, suggesting an inherent vascular abnormality.[1] The articular cartilages are totally replaced by fibrous tissue, and the spongy bone is progressively destroyed. The shaft may be spared but usually shows thinning and pointing. There are no inflammatory reactions.

Roentgenographic Features. Progressive osteolysis of the carpal and tarsal bones is the hallmark of the disease. These bones may be completely resorbed and the radius and ulna closely approximate the metacarpals (Figs. 1461 and 1463). The proximal ends of the metacarpals and metatarsals are markedly thinned and pointed; this may also occur in the phalanges. The radial articular surface becomes angulated and the distal end becomes pencil-thin. The remaining cortical bones are not deossified, although the large epiphyses of the humerus, femur, and tibia may show deossification without loss of outline. The contiguous metaphyses and shafts are normal. The proximal interphalangeal joints commonly reveal some narrowing with erosive changes but the interphalangeal and metatarsophalangeal joints, the distal interphalangeal joints, and the metacarpophalangeal joints are usually spared. The elbows commonly show the same type of destruction with lysis of bone and subluxation (Fig. 1463).

There may be bathyrocephalic depression of the base of the skull, particularly in the parietal and occipital areas, and asymmetric articular processes of the base of the atlas with platybasia.[8] Thoracic scoliosis is reported.[8]

Massive osteolysis (Gorham disease) may be distinguished by the local destruction of contiguous bones, not usually affecting the hands or feet. Bony hemangiomatosis may be the cause; it is easily differentiated from essential osteolysis.

Tabes, leprosy, syringomyelia, scleroderma, and Raynaud disease may all cause osteolysis, but their clinical features distinguish them. Four other conditions closely simulate the roentgenographic changes of essential osteolysis, but are also distinguished by their clinical features: (1) regional post-traumatic osteolysis[9,10]; (2) ulceromutilating acropathy, a condition of heredofamilial character typically involving the feet and accompanied by skin ulceration and sensory disturbances[11,12]; (3) mutilating forms of rheumatoid arthritis[12,13]; and (4) acrodinia mutilante, a nonhereditary entity with osteolysis of the phalanges, metacarpals and metatarsals (Fig. 1462).

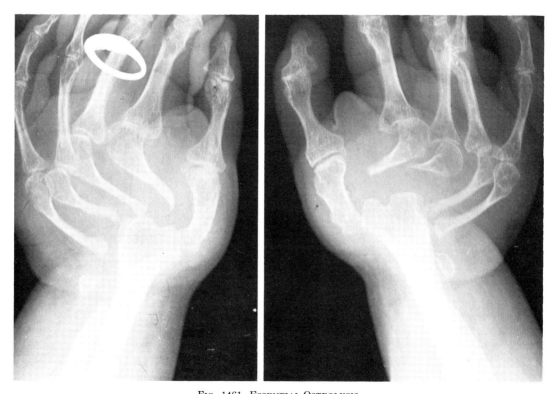

FIG. 1461. ESSENTIAL OSTEOLYSIS
Resorption of the proximal end of the metacarpals, the carpals and distal end of the ulna. (Courtesy of Dr. Robert M. Peck, Atlantic City, N.J.)

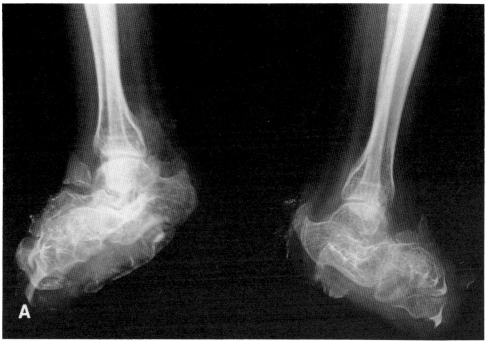
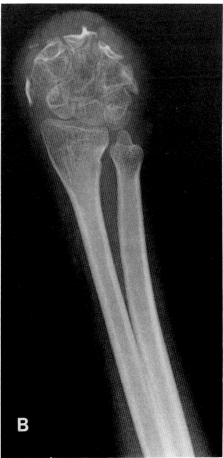

Fig. 1462. Acrodinia Multilente
Destruction of the phalanges, metacarpals and metatarsals. (A) Both feet. (B) Right hand. The left hand had similar changes. (Courtesy of Dr. Cesar Pedrosa, Madrid, Spain.)

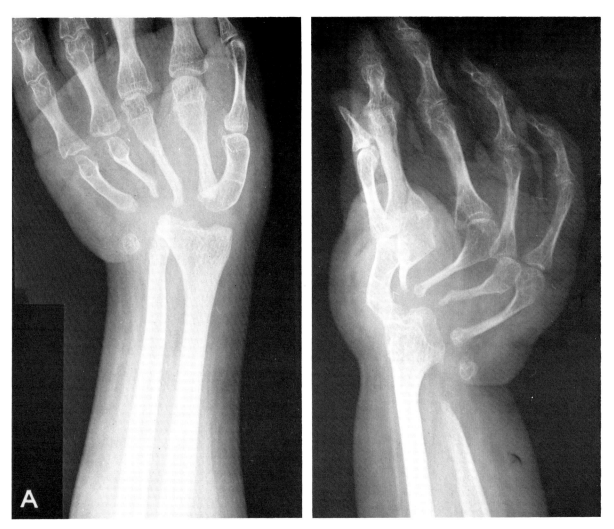

Fig. 1463. Essential Osteolysis

The son of the patient of Figure 1462. (A) Resorption of the carpals, proximal ends of the metacarpals, and the distal end of the ulna. (B) Resorption of the proximal end of the ulna and distal end of the humerus and dislocation.

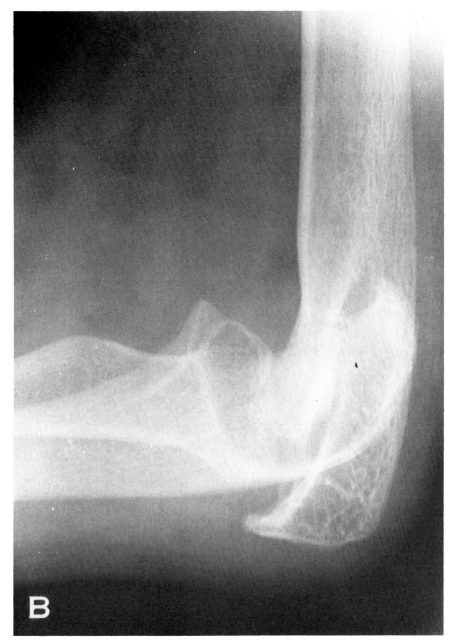

Fig. 1463(B)

ACRO-OSTEOLYSIS

Acro-osteolysis[1] is a condition that causes disintegration of bone, predominantly the phalanges, along with other features.

There are three types: familial, idiopathic, and acquired due to drugs and disease.

The *familial type* has a dominant inheritance mode and a 3-to-1 male predominance.[2, 3] It usually begins in a teenager with a swelling of the plantar aspect of the foot that gradually develops into a deep, wide ulcer. The underlying bone is involved and ejection of bone fragments occurs.[3] Although the ulcer may heal, recurrence is frequent. Osteolysis may occur without ulceration.[4] Destruction of the nails, sensory changes in the hands and feet, and joint hypermobility are common features.

The *idiopathic type* may be a mutant or a dominant trait in a homozygous form unrecognized in the parent. The fingernails usually remain intact, and sensory changes and plantar ulcers are rare.[1] It may be unilateral, and congenital malformation of the spine and feet may

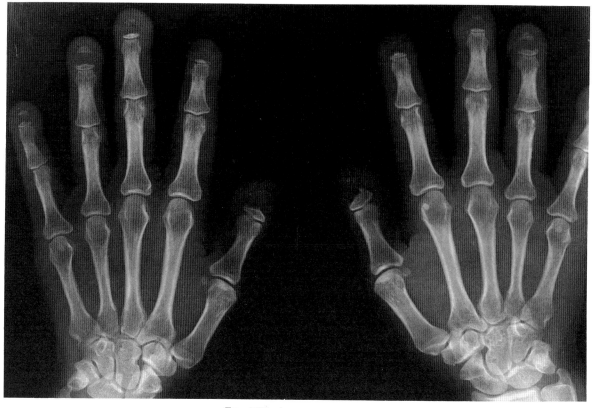

FIG. 1464. ACRO-OSTEOLYSIS
Destructive process of distal and middle phalanges.

be associated. Prognosis for life is good, but complete recovery is infrequent.

Acquired Acro-osteolysis may be caused by the following[5]: burns, chemical agents (polymerized vinylchloride), collagen disease (dermatomyositis and scleroderma), congenital insensitivity to pain, diabetes, Ehlers-Danlo syndrome, electric shock, epidermolysis bullosa dystrophica, frostbite, hyperparathyroidism, Kaposi sarcoma, leprosy, progeria, psoriasis, pyknodysostosis, Raynaud disease, Rothmund syndrome, sarcoidosis, syringomyelia, thromboangiitis obliterans (Buerger disease) Werner syndrome, and yaws.

Roentgenographic Features. The roentgen features of acro-osteolysis are similar in the hands and feet in all types. The destructive process, without periosteal reaction, usually involves the distal and middle phalanges, and begins in the teenager (Fig. 1464). The epiphyses resist osteolysis until late. The distal ends of the metacarpals and metatarsals may be affected.[6]

The familial and idiopathic forms may have the following abnormalities: compression fractures of the vertebrae with kyphosis, brittle bones with fractures, dolicocephaly with basilar impression and multiple wormian bones, and premature loss of teeth.

TUBEROUS SCLEROSIS

Synonyms: epiploia, Bourneville disease.[1]

Tuberous sclerosis is a rare hereditary mesodermal dysplasia, characterized by hamartomas most often in the brain, skin, bones, and kidneys, although any tissue or organ may be involved.[2-5] It is probably transmitted by an autosomal dominant mode.[6]

Clinical Features. The cardinal triad is adenoma sebaceum, mental deficiency, and epileptiform seizures. Mental deficiency with epilepsy is the most frequent clinical complaint. The clinical features may be absent or few,[7-11] with little change in the bones, kidneys, and other organs, and the diagnosis may be delayed until autopsy.

Deep pigmented skin nevi are usually the first lesions to be observed.[12] Other skin lesions are peau-chagrin or shagreen patches (rough skin) in the lumbar area, cafe-au-lait spots, cutaneous tags, fibromas, and subungual fibromas. Retinal whitish patches (phacomas) often appear. Spasticity and other signs of cerebral palsy are frequent.

Endocrine disturbances have been reported in some patients; they include abnormal pituitary adrenal function, thyroid disorders, abnormal responses to intravenous glucose tolerance tests, and high serum alkaline phosphate levels. These features would indicate that endocrine and metabolic dysfunction is a frequent component of tuberous sclerosis and may be evidenced by a variety of manifestations.[13]

It is usually discovered in children or teenagers, but its discovery may be delayed into adulthood if the skin changes are absent. The usual course is a variable progressive deterioration terminating in death from aspiration pneumonia, urinary tract infection, myocardial involvement, or neoplastic transformation to sarcoma and glioma.

Pathologic Features. Cerebral and visceral lesions are most frequent and occur in the brain, lungs, and kidney. Multiple focal ill defined masses of excessive neurologic cells occur in the brain, usually near the ventricle and basal ganglia, and may develop into large gliomas, usually in young adults, with minimal evidence of tuberous sclerosis elsewhere.[14]

The kidney usually contains single or multiple hamartomas,[15] mixtures of fat, smooth muscle, and blood vessels. The renal arteries are enlarged and tortuous, and aneurysms are frequently present. Rhabdomyomas and colonic adenomas are infrequent.

The lungs may have a dense fibrous and muscular stroma and diffuse cystic changes.[16] These changes are similar to those seen in pulmonary muscular hyperplasia.[17]

Roentgenographic Features. Changes occur in the skull, spine, long bones, kidneys, and lungs.[7, 9, 11, 18-24] Although individual roentgen findings are nonspecific, the combination of them should assure the diagnosis.

Calvarial sclerotic patches and/or thickening of the skull tables are frequent[25-29] and in most cases scattered intracranial calcifications appear in the basal ganglia (Fig. 1465).[9, 10, 29, 30] Cerebellar calcifications occur in 10–15%.[10, 29]

The long bones frequently develop undulating

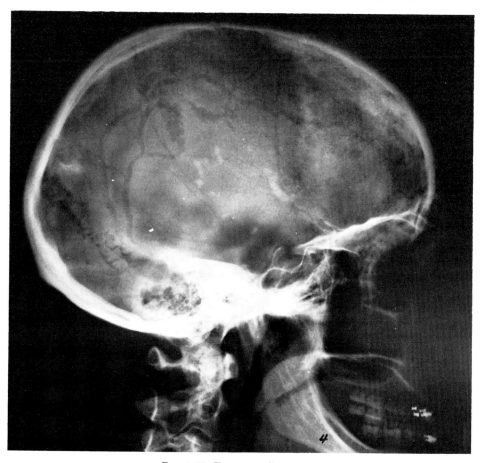

FIG. 1465. TUBEROUS SCLEROSIS
Skull. Calcifications in the basal ganglia.

periosteal new bone formation or periosteal "nodules," particularly on the metacarpals, metatarsals, and phalanges (Fig. 1466). Small cortical "cysts" with sclerotic rims appear on the phalanges. The trabecular pattern tends to be thickened (Figs. 1467 and 1468).[31, 32]

Diffuse sclerosis and sclerotic bone islands of the spine (Fig. 1469) and pelvis are reported,[7] most commonly appearing at the upper and lower margins of the vertebral bodies and pedicles and the pelvic brim.[29]

The frequent mass renal lesions, singular or multiple, may be confused with polycystic kidneys or carcinoma. Rare pulmonary features are a honeycomb lung pattern, most prominent in the lower lung fields, and a miliary nodular pattern. Spontaneous pneumothorax, hemoptysis, and dyspnea are complications commoner in the elderly.[29]

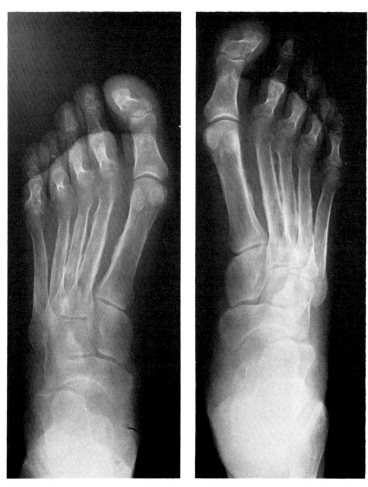

FIG. 1466. TUBEROUS SCLEROSIS
Undulating periosteal reaction and periosteal nodules of the metatarsals.

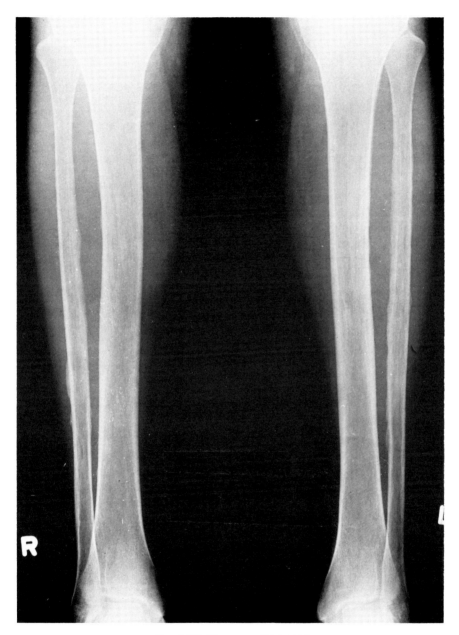

FIG. 1467. TUBEROUS SCLEROSIS
Undulating periosteal new bone formation or periosteal nodules (*arrows*).

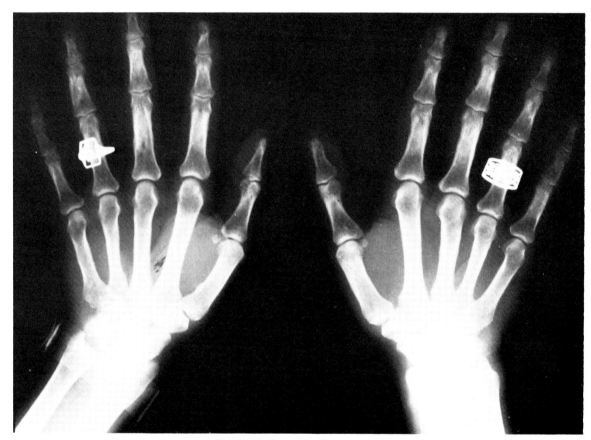

FIG. 1468. TUBEROUS SCLEROSIS
Disturbed bone growth particularly noticeable in the middle phalanges as evidenced by irregular ossification and small cystlike lesions.

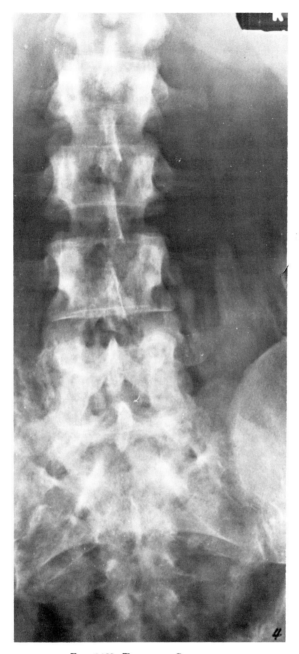

Fig. 1469. Tuberous Sclerosis

Small round densities are noted throughout the spine, most noticeable in the fourth and fifth vertebrae and the sacrum.

OSTEOPOIKILOSIS[1]

Synonym: osteopathia condensans dessiminata.[2]

Osteopoikilosis is a rare familial condition that produces ovoid or lenticular densities in cancellous bone. It may be related to melorheostosis and osteopathia striata.

Clinical Features. The condition is asymptomatic and never produces physical deformity. More males are affected than females; it has been found in fetal life and in patients over 60 years of age. Several cases have been associated with dermatofibrosis.[3-7]

Roentgenographic Features. Radiographic features are small foci of ovoid or lenticular opacifications, with the long axis of the opacities parallel to the long axis of the bone (Figs. 1470–1476). The densities vary in size from 2 mm to 2 cm and may show radiolucent centers. They occur in most metaphyses and epiphyses, and rarely extend to the midshaft. Occasionally they occur in the pelvis, not necessarily close to the joints or localized to metaphyses (Fig. 1472). They have been found in every bone, but are rare in the skull, ribs, vertebral centra, and mandible. They are shorter than the densities of osteopathia striata.

In the scapula and ilium, they appear to radiate from the glenoid fossa and acetabula. The opacities do not change in appearance after growth ceases. Holly[8] reports disappearance and reappearance in children and, occasionally, in adults. The disease may be found with melorheostosis or osteopathia striata. Its metaphyseal site differentiates osteopoikilosis from epiphyseal dysplasia, and the absence of diaphyseal involvement distinguishes it from melorheostosis.

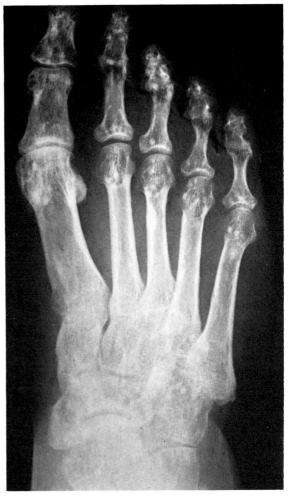

FIG. 1470. OSTEOPOIKILOSIS
Small oval densities are present in all bone predominately in the metaphyses.

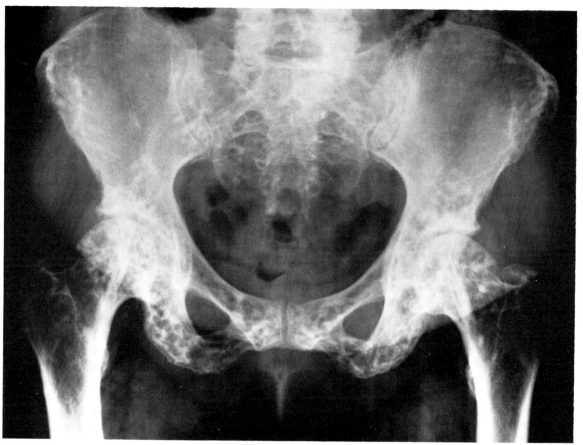

Fig. 1471. Osteopoikilosis
Oval bone densities in the bones of the pelvis.

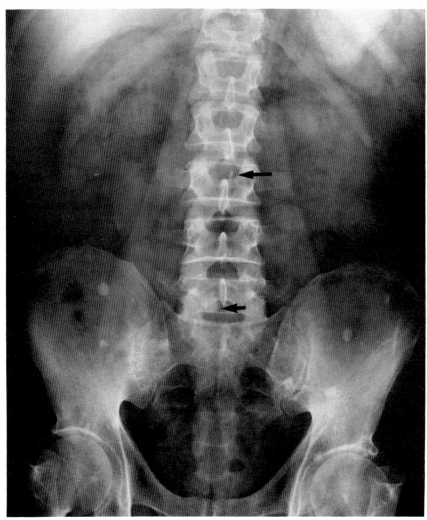

Fig. 1472. Osteopoikilosis
Scattered densities throughout the pelvis. One may be seen in the posterior elements of L3 and L5 (*arrows*).

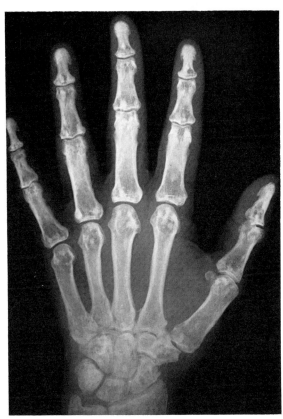

FIG. 1473. OSTEOPOIKILOSIS
Oval bone densities in all of the metaphyseal regions and carpals.

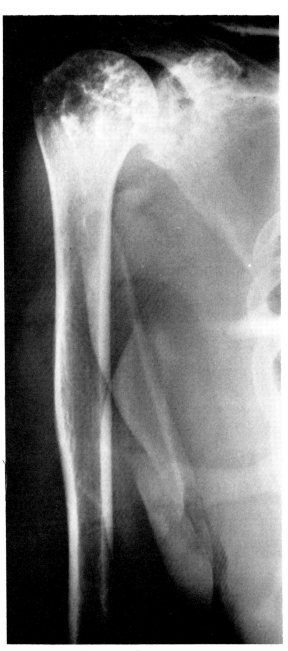

FIG. 1474. OSTEOPOIKILOSIS
Oval bone densities in the glenoid and humeral head. The humeral diaphysis is spared.

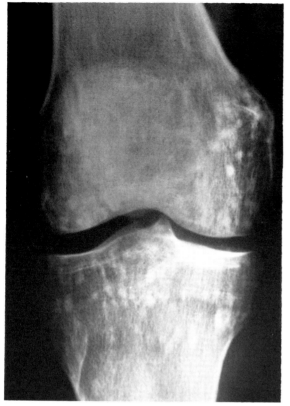

FIG. 1475. OSTEOPOIKILOSIS
Oval bone densities in the femur and tibia.

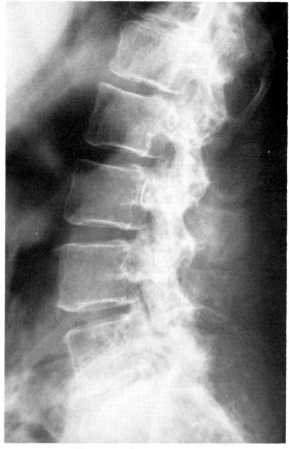

FIG. 1476. OSTEOPOIKILOSIS
Oval areas of increased density confined to the appendages of the vertebrae.

OSTEOPATHIA STRIATA

Osteopathia striata[1] is an unusual condition of the skeleton that may be related to osteopoikilosis and melorheostosis, and may occur with osteopetrosis.[2] It is characterized by dense longitudinal striations of the skeleton, particularly in metaphyseal areas.

Clinical Features. It is usually asymptomatic, although vague joint pains may occur.[3]

Roentgenographic Features. Osteopathia striata presents as elongated densities extending from the epiphyseal line into diaphyses. Usually every long bone is affected. Occasionally the streaks of density extend into an epiphysis. In the ilium the linear densities radiate from the acetabulum and fan out to the iliac crest in "sunburst fashion."

Osteopathia striata primarily affects metaphyseal portions of bone, which distinguishes it from osteopetrosis, melorheostosis, hereditary multiple diaphyseal sclerosis,[4] and progressive diaphyseal dysplasia (Engelmann disease). Occasionally, enchondromatosis shows streaking of metaphyses, but this is not as generalized as in osteopathia striata.

MELORHEOSTOSIS

Synonyms: Léri type of osteopetrosis, osteosi eburnizzani monomelica,[1] flowing hyperostosis.

Melorheostosis is a rare bone condition producing cortical thickening of the entire shaft of a tubular bone.

Clinical Features. The etiology is unknown and there is no sexual predominance. Although most cases are discovered between the ages of 5 and 20, it has been found in the neonatal period and in the elderly. Severe pain[2] is often the presenting symptom. As the result of cortical thickening, bone may encroach on nerves, blood vessels, or joints with limitation of motion or fusion of the joint. Thickening and fibrosis of the skin overlying bone may resemble scleroderma, and the soft tissue changes may precede the bone involvement.[3-13] Other cutaneous manifestations include tumors or malformations of blood vessels and lymphatics. The vascular lesions consist of hemangiomas, vascular nevi, glomus tumors, arteriovenous malformations and aneurysms. Lymphatic involvement causes lymphedema, trophedema and lymphangiectasia.[14]

Pathologic Features. The histologic features consist of areas of sclerotic bone, with alternating areas of mature and immature bone.[3] Concentric perivascular ossification has been reported, as has occasional fibrosis in the medullary cavity. The histologic features are not characteristic; only roentgen study confirms the diagnosis.

Roentgenographic Features. It presents as continuous or interrupted streaks or blotches of sclerotic bone along part or all of a tubular bone. These may encompass the entire cortex, or be limited to one side (Fig. 1477). It begins in the proximal end and extends distally, suggesting the flow of wax down a lighted candle.[15] In the beginning, the portions of the bone nearest the joints are spared. Eventually the dense bone extends into the epiphysis and may cross the joint (Fig. 1479), causing joint fusion. The thickened cortex may encroach on the medullary cavity, and tends to have a wavy pattern in children while being more linear in adults; it also has a tendency to protrude slightly beneath the periosteum[16] (Figs. 1480–1482).

Usually melorheostosis is limited to a single extremity with at least two bones involved, but multiple extremities can be involved (Fig. 1478).

The scapula and hemipelvis corresponding to the affected extremity may show small dense areas of bone, similar to those seen in osteopoikilosis (Fig. 1482). The small bones of the hand and wrist often contain small sclerotic islands. Soft tissue calcification and ossification may occur around the larger joints. The skull, spine, and ribs are rarely involved (Fig. 1483). Slow progression is the rule; Léri and Joanny[15] observed progression over a period of 7 years. The bone lesions do not regress.

The hyperostotic linear densities in bone have a peculiarly segmented distribution which does not correspond with the anatomical course of blood vessels or the mixed nerve roots of the limbs. They do, however, follow spinal sensory nerve sclerotomes in the majority of cases.[16,17] To account for its sclerotomal distribution it is proposed that melorheostosis may be the late result of a segmental sensory nerve lesion. Sclerotomal maps[17] are found in Figure 1478.

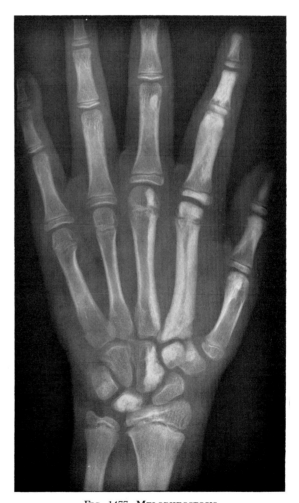

FIG. 1477. MELORHEOSTOSIS
Dense bone may encompass the entire cortex or it may be limited to one side. (Courtesy of Herbert M. Stauffer, Temple University Hospital, Philadelphia, Pa.)

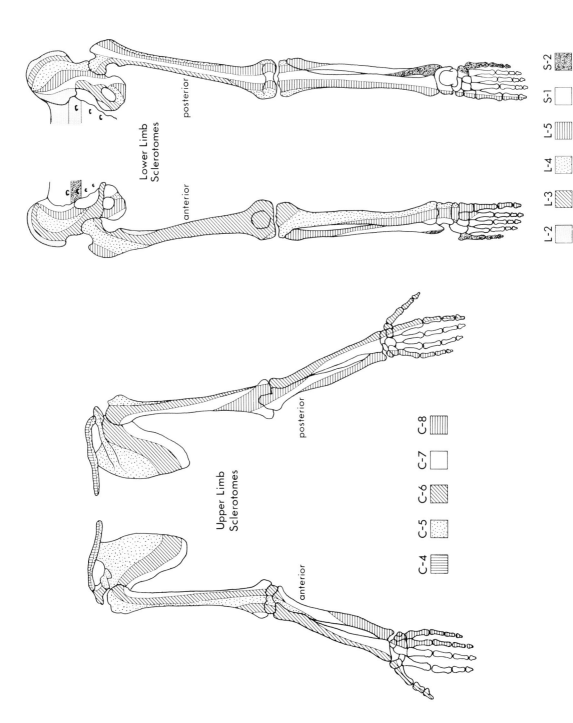

Fig. 1478. Sclerotomes (*Left*) Upper limb; (*right*) lower limb.

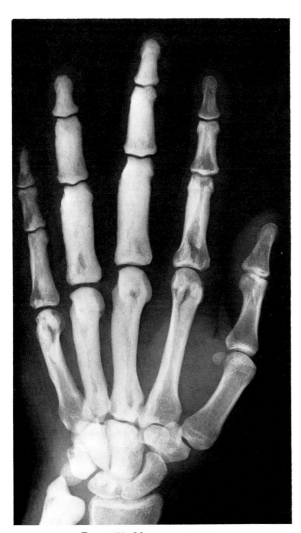

FIG. 1479. MELORHEOSTOSIS
(Courtesy of George T. Wohl, Philadelphia General Hospital, Philadelphia, Pa.)

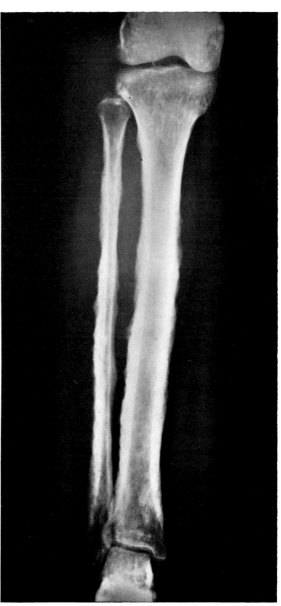

FIG. 1480. MELORHEOSTOSIS
Thickened cortex protrudes slightly beneath the periosteum. (Courtesy of Herbert M. Stauffer, Temple University Hospital, Philadelphia, Pa.)

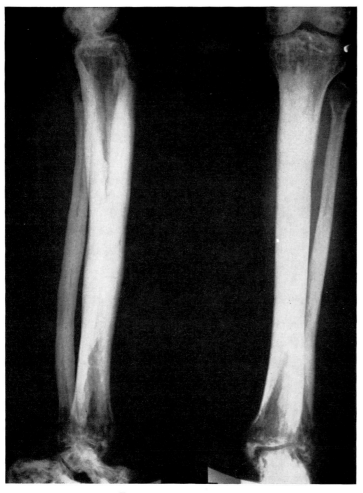

FIG. 1181. MELORHEOSTOSIS
Cortices are thick and dense. The bones are increased in transverse diameter and the medullary spaces are obliterated in places.

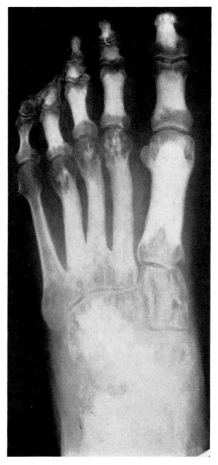

FIG. 1482. MELORHEOSTOSIS
Shafts of the long bones are affected but the ends of the bones are not. Small, rounded, and ovoid densities in the distal ends of the second, third, and fourth metatarsals suggest osteopoikilosis.

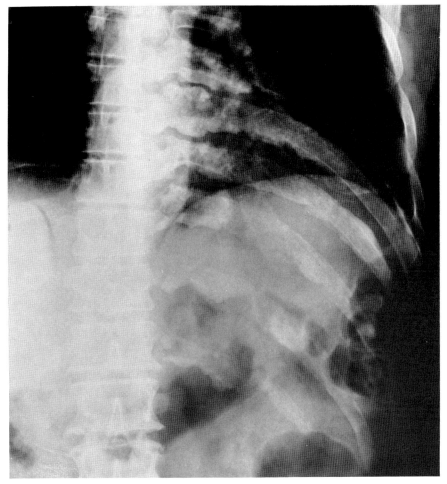

Fig. 1483. Melorheostosis
Hyperostotic lesions in the 7th through the 12th ribs, an unusual site for this condition.

Osteopoikilosis may be confused with melorheostosis if only the small bones of the hand or pelvis are evaluated. A skeletal survey will indicate the generalized nature of osteopoikilosis. Fibrosis dysplasia may also reveal diffuse sclerotic bone change, but the normal bone structure is not lost, and the bone is not as dense as in melorheostosis. Engelmann disease, which begins during infancy and childhood,[18] may be distinguishable from melorheostosis, which some consider a regional form of Engelmann disease. Hyperostosis may occur with neurofibromatosis, tuberous sclerosis and hemangiomas, and may mimic the appearance of melorheostosis.[19]

HEREDITARY MULTIPLE EXOSTOSIS

Synonyms: diaphyseal aclasis,[1] hereditary deforming chondroplasia,[2] external chondromatosis, multiple osteochondromas, cartilaginous exostosis, hereditary osteochondromatosis.

Multiple exostosis is a bone dysplasia in which osteochondromas affect many long bones and some flat bones. Stocks and Barrington[3] determined that hereditary factors were significant in 65% of a large series, usually, though not always, by male transmission. About half the offspring of an affected parent will show evidence of the condition. The exostoses are caused by disturbances of enchondral bone growth.

The etiology is unknown, but there are several possible explanations: (1) a small portion of the epiphyseal plate separates and sets up independent enchondral bone growth,[4] (2) the exostoses result from subperiosteal cartilaginous embryonal cell rests, (3) a defect in the periosteum near the epiphysis permits unrestrained growth of the epiphysis,[1] (4) impairment of osteolysis during modeling results in a broad metaphysis and formation of an epiphyseal line at right angles to the normal, and (5) groups of cells are displaced from the epiphysis in the periosteum and retain their chondrogenic potential.[5]

Clinical Features. Multiple exostosis is asymptomatic, except when it encroaches on contiguous soft tissue structures or becomes malignant. Tendons, blood vessels, and nerves may be impaired. Paraplegia occasionally occurs due to vertebral exostosis. The exostosis begins as a growth in childhood, and stops growing when the nearest epiphyseal center fuses. A sudden, painful growth spurt after long quiescence indicates malignancy. The incidence of malignancy is probably higher than that of solitary osteochondroma. When the bones reach maturity the exostoses stop growing and uptake of bone-seeking nucleides becomes normal. Where there is renewed growth the uptake of ^{99m}Tc-diphosphonate increases. The increased uptake indicates activity but not necessarily malignancy. Its measurement is more useful than radiographic bone survey and the periodic surveillance of adult patients with this disorder aid in discovering new growth.[6, 7]

Pathologic Considerations. The abnormalities of multiple exostosis consist of an ectopic cartilaginous rest in the metaphysis, and a defect in the periosteum at the site of the exostosis, responsible for the lack of modeling of the metaphysis. Histologic study reveals cancellous bone directly contiguous to the host bone and covered by a cartilaginous cap during the growth period. When growth ceases, a cap of hyaline cartilage forms. A bursa will often appear between the cap and the overlying soft tissues. The tissues of the exostosis are normal and are hamartomatous rather than neoplastic.[4]

Roentgenographic Features. Multiple exostoses vary in size and number, and any bone may be involved. There may be as many as 1,000 exostoses in the same patient,[3] although 10 or 12 are more usual. Most exostoses remain small and produce no symptoms.

The metaphysis of a long bone is frequently the seat of the exostosis, which grows out from the cortex close to the epiphyseal line, rarely from the epiphysis. As the bone grows, the distance from the epiphyseal line increases. The cortex and cancellous portion of the exostosis are directly contiguous to the host bone. On the epiphyseal side of the exostosis the cortex slopes gradually, and blends with adjacent epiphyseal

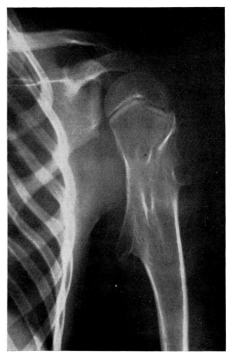

FIG. 1484. HEREDITARY MULTIPLE EXOSTOSES
Large exostosis on the medial aspect of the humerus shows the cortex and cancellous portion directly contiguous to the host bone. The cortex on the epiphyseal side slopes gradually, whereas on the diaphyseal side it joins the host bone almost at right angles. A small exostosis viewed en face appears as a radiolucency surrounded by a ringed zone of opacity.

bone, whereas on its diaphyseal side, it joins the host bone at right angles (Fig. 1484).

During the growth period the tip of the exostosis is cartilaginous and roentgenographically invisible; however, small punctate calcifications in cartilaginous caps sometimes occur. When viewed "en face" they appear as radiolucencies, due to the cortical defect, surrounded by a ring zone of opacity (cortex of the pedicle) (Fig. 1484). This may be mistaken for internal chondromatosis if other roentgenographic projections are not obtained.

The knee is involved most often (Fig. 1485), with the bone adjacent to the elbows also a common site. The radius is usually longer than the ulna and bowing occurs (Fig. 1486). The small bones of the hand and feet may have multiple exostoses (Figs. 1487-1489), usually small and rarely leading to impaired function. Spinal involvement, particularly the posterior elements may cause cervical cord compression.[3, 8-20] Occasionally an exostosis will show excessive growth, become irregular, and branch, indicating malignancy. The roentgenogram may reveal a fuzziness and irregularity of outline. Later a soft tissue mass appears, and finally bone destruction. Probably less than 5% of the patients will develop malignant degeneration (Fig. 1489).

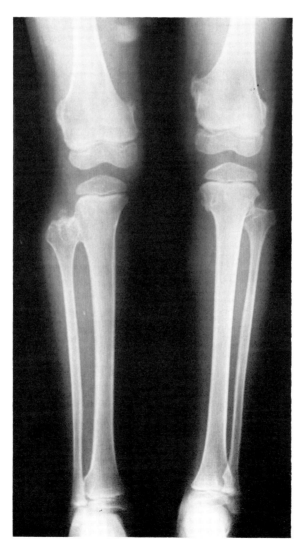

FIG. 1485. HEREDITARY MULTIPLE EXOSTOSES
Region of the knee is most often involved.

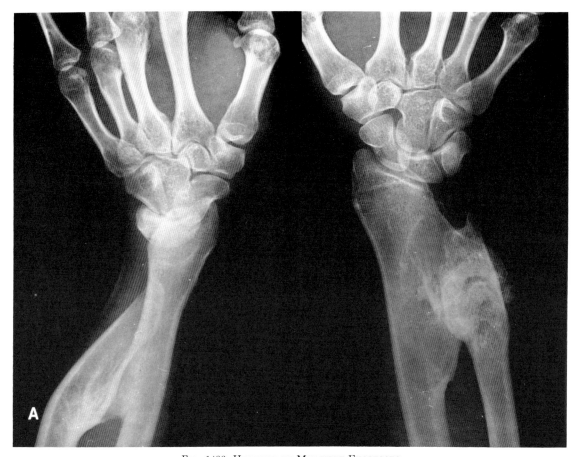

FIG. 1486. HEREDITARY MULTIPLE EXOSTOSES

(A) Both wrists. Shortening of the ulnas and exostoses. (B) Right knee, lateral projection. Multiple exostoses. (C) Left ankle. Multiple exostoses of the tibia and fibula and a large exostosis of the calcaneus.

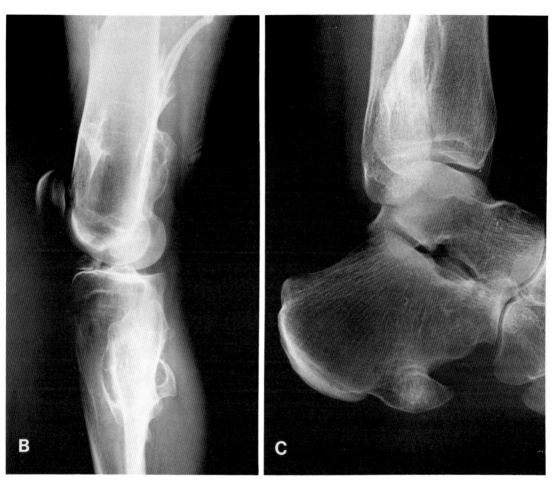

Fig. 1486 (*B* and *C*)

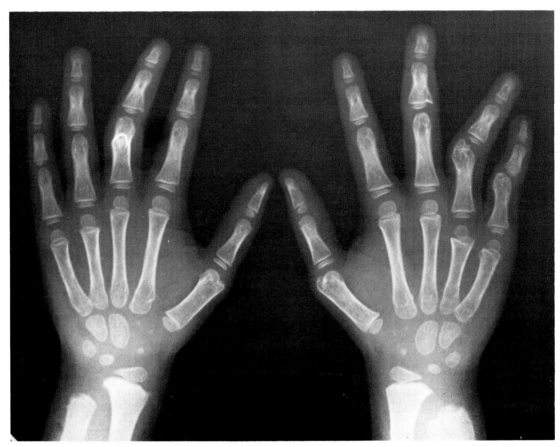

FIG. 1487. HEREDITARY MULTIPLE EXOSTOSES
Small bones of the hands and feet may show multiple exostoses. Usually they do not impair function.

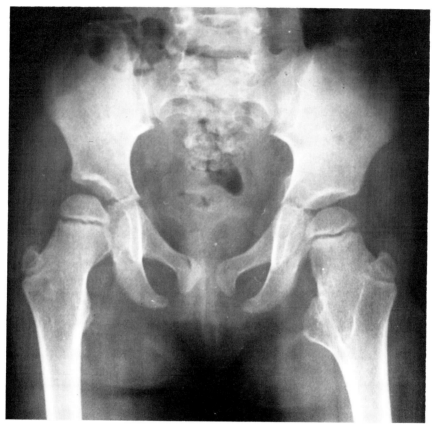

Fig. 1488. Hereditary Multiple Exostoses

1408 DISEASES OF BONE

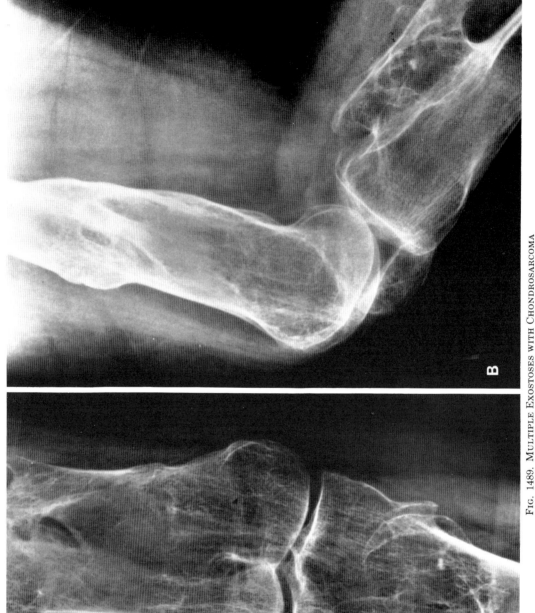

FIG. 1489. MULTIPLE EXOSTOSES WITH CHONDROSARCOMA

(A) Anteroposterior view of right knee. Multiple exostoses. (B) Lateral view of right knee. (C) Pelvis. Multiple exostoses. The exostoses on the left iliac crest with dense calcification were not malignant; that on the right iliac crest, with a large soft tissue mass and cartilaginous calcifications, proved to be chondrosarcoma. (D) Computer tomography reveals extension into the abdomen and destruction of the right transverse process. (Courtesy of Dr. Marvin Lindell, M. D. Anderson Hospital and Tumor Institute, Houston, Tex.)

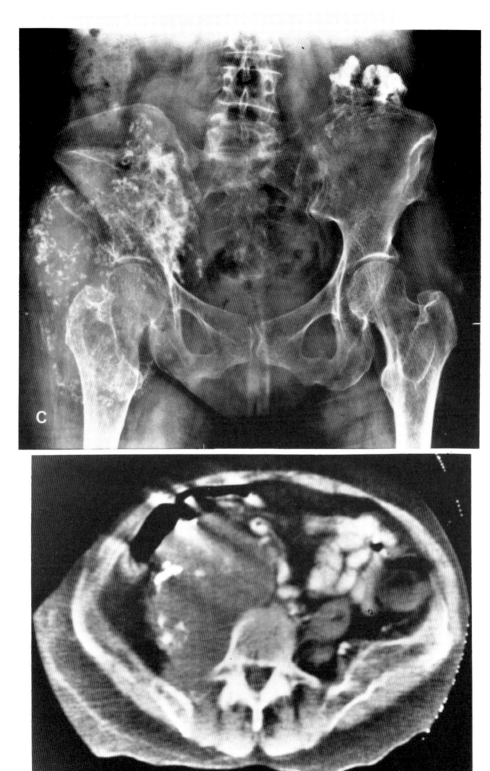

Fig. 1489 (*C* and *D*)

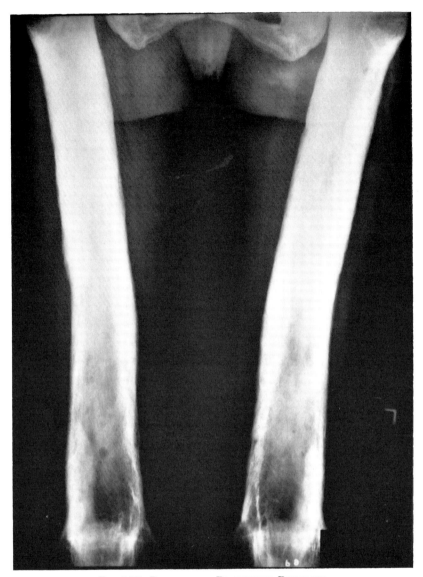

Fig. 1493. Progressive Diaphyseal Dysplasia

Thickening of both endosteal and periosteal portions of the cortex. Normal trabecular structure is lost. (Courtesy of Herbert M. Stauffer, Temple University Hospital, Philadelphia, Pa.)

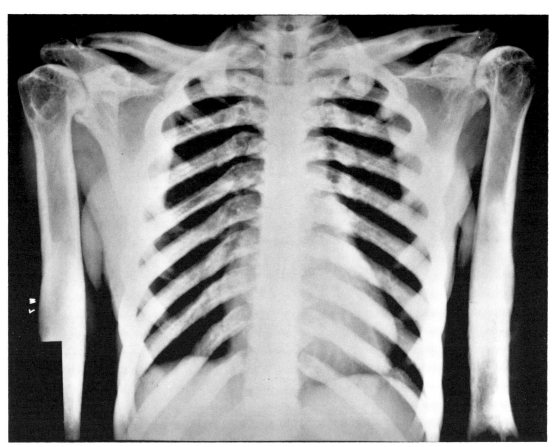

FIG. 1494. PROGRESSIVE DIAPHYSEAL DYSPLASIA

Increased density and thickening of the shafts of both humeri. The scapulae show increased density and prominent striations. The ribs and clavicles also show changes. (Same case as seen in Figures 1495 and 1496.) (Courtesy of Herbert M. Stauffer, Temple University Hospital, Philadelphia, Pa.)

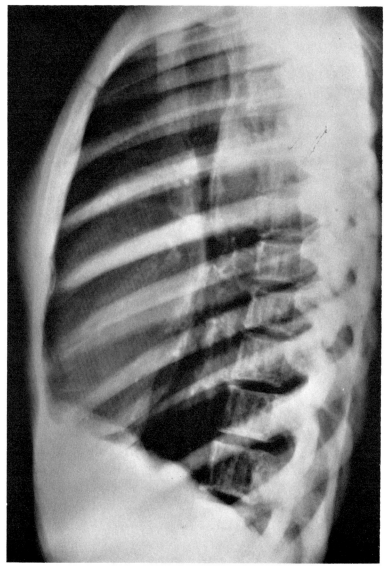

Fig. 1495. Progressive Diaphyseal Dysplasia
Vertebrae show increase in density and vertical bands. (Same case as seen in Figures 1494 and 1496). (Courtesy of Herbert M. Stauffer, Temple University Hospital, Philadelphia, Pa.)

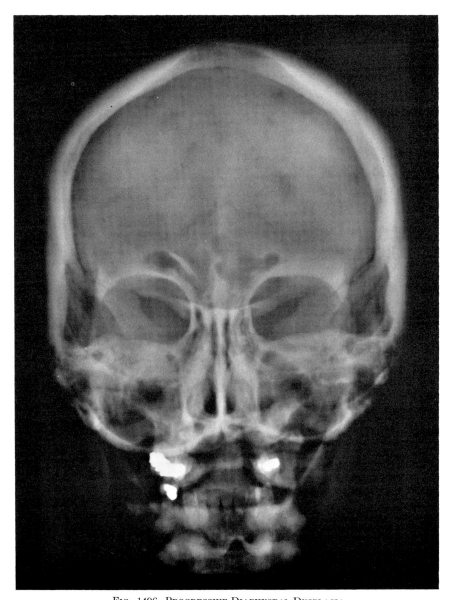

Fig. 1496. Progressive Diaphyseal Dysplasia
Tables of the skull are thickened and increased in density and normal trabecular structure is lost. (Same case as seen in Figures 1494 and 1495.) (Courtesy of Herbert M. Stauffer, Temple University Hospital, Philadelphia, Pa.)

CLEIDOCRANIAL DYSOSTOSIS (MUTATIONAL DYSOSTOSIS)

Cleidocranial dysostosis is an uncommon condition characterized by delayed or incomplete calvarial ossification and clavicular hypoplasia or aplasia. The long bones, pelvis, and spine may also show abnormalities of delayed ossification. Soule[1] reports that the first case was described by Cutter.[2] Marie and Sainton[3] supplied the name when they recognized it as a clinical entity. It is often familial. Occasionally, the cranial and clavicular abnormalities are lacking, and Rhinehart[4] suggested the name be changed to "mutational dysostosis."

Roentgenographic Features. The skull shows diminished or absent ossification early in infancy; in time, mineralization occurs and many wormian bones appear. Lack of ossification persists near the midline at the fontaneles and sutures, which appear widened (Figs. 1497 and 1498). Persistent metopic suture is common. The skull is brachycephalic, with prominent bossing. There is distinctive segmental calvarial thickening in the temporal and occipital bones.[5] The mandible is large and the mandibular suture closes late. Nasal bones may fail to ossify but without physical deformity. The palate is often narrow and high, and may be cleft. The paranasal sinuses are hypoplastic. Delayed or defective dentition is frequent.[6]

In 10% of the cases, the clavicle is absent, or it may show defective development in the inner, middle, or outer portion (Figs. 1499 and 1500). The developmental changes show a right clavicular selectivity (Fig. 1500).[5] The clavicle originates from three separate ossification centers, sternal, midportion, and acromial, any of which may show developmental abnormality. Rarely will the acromial end be present alone. The sternal and acromial portions may be normal, with

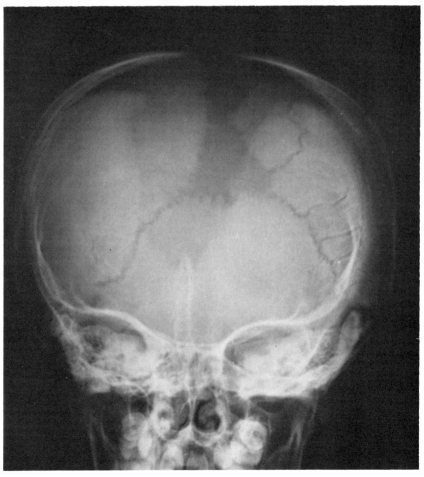

FIG. 1497. CLEIDOCRANIAL DYSOSTOSIS (MUTATIONAL DYSOSTOSIS)
The fontanels are open, the sutures widened, and wormian bones are present in this 3-year-old child.

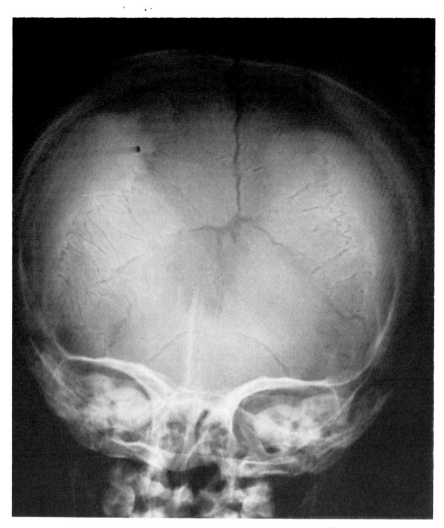

Fig. 1498. Cleidocranial Dysostosis (Mutational Dysostosis)
Multiple wormian bones, open sutures, and fontanels.

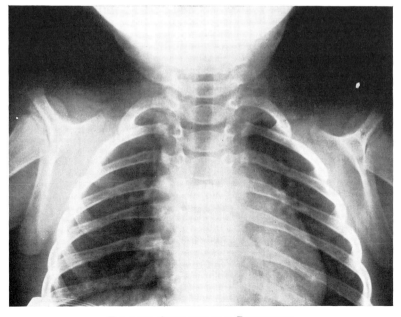

Fig. 1499. Cleidocranial Dysostosis
Clavicles are absent except for the medial ends, which are rudimentary.

a defective midportion, producing a pseudoarthrosis. Congenital pseudoarthrosis of the clavicle must be ruled out. It is a separate, isolated entity and unrelated to birth injury. It occurs in the newborn and is reported only in the right clavicle at the junction of the middle and outer thirds.[7] The scapulae are usually small and lie higher than normal. The radius is often short or completely absent.

Many other associated skeletal anomalies have been described. Delayed ossification is common in the bones surrounding the symphysis pubis, which simulates extrophy of the bladder (Figs. 1501 and 1502). Incomplete closure of vertebral neural arches, accessory epiphyses of hands, phalangeal hypoplasia, and anomalies of enchondral bone formation are described.[8] Femoral necks may be deformed or absent (Figs. 1502 and 1503); there may be lateral notching of the capital femoral epiphyses,[5] but the remainder of the femur and the tibia and fibula are usually normal.

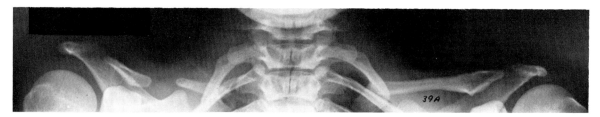

FIG. 1500. CLEIDOCRANIAL DYSOSTOSIS (MUTATIONAL DYSOSTOSIS)
There is right clavicular selectivity but in this instance bilateral involvement is present.

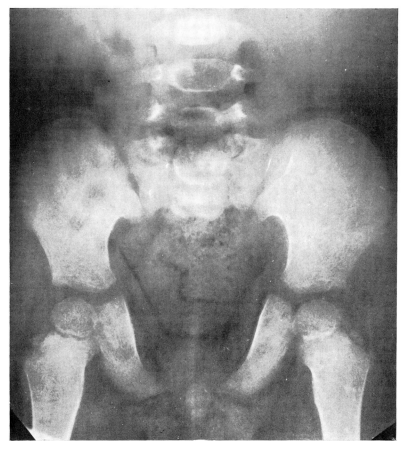

FIG. 1501. CLEIDOCRANIAL DYSOSTOSIS
Pubic bones are underdeveloped and the symphysis is greatly widened. This simulates the appearance of extrophy of the bladder.

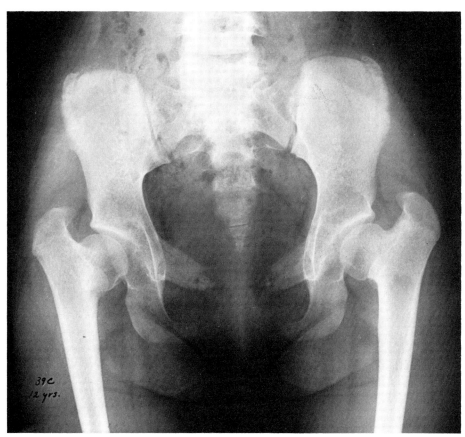

Fig. 1502. Cleidocranial Dysostosis (Mutational Dysostosis)
Defective ossification of the symphysis pubis. Bilateral coxa vara is present. The iliac wings are unflared.

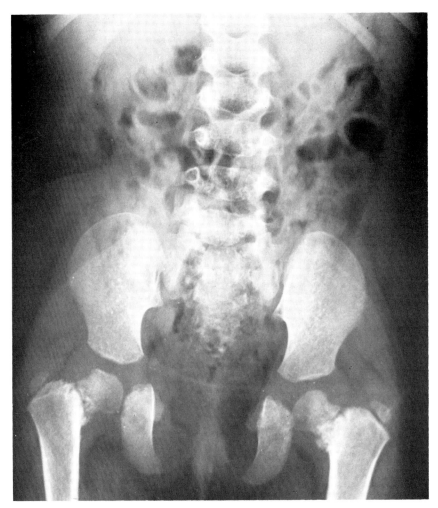

Fig. 1503. CLEIDOCRANIAL DYSOSTOSIS (MUTATIONAL DYSOSTOSIS)
Bilateral coxa vara and deficient ossification of the symphysis pubis.

ILIAC HORNS (FONG DISEASE,[1] FAMILIAL ONYCHO-OSTEODYSPLASIA)[2]

Iliac horns are found in combination with a variety of mesoectodermal anomalies.[3-5] The cutaneous anomalies are aplasia or hypoplasia of the thumb and index fingernails.

Roentgenographic Features. The characteristic feature is symmetrical posterior iliac "horns" formed by independent ossification centers (Fig. 1504).[6] Other skeletal anomalies include patellar and radial head hypoplasia (Fig. 1505), clinodactyly, and short fifth metacarpals.

Unilateral iliac horn (central posterior iliac process) occurs as an isolated anomaly of the pelvis and not associated with the syndrome of iliac horns.[7]

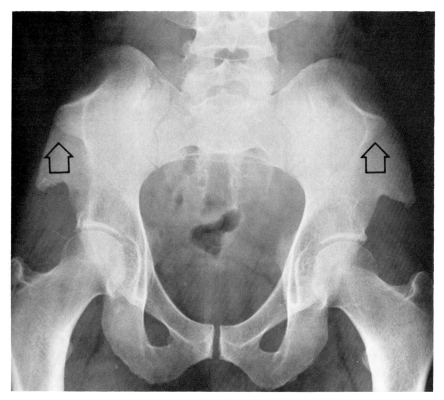

FIG. 1504. ILIAC HORNS
Bilateral iliac horns (*arrows*). Hypoplasia of the iliac bone. (Courtesy of George T. Wohl, Philadelphia General Hospital, Philadelphia, Pa.)

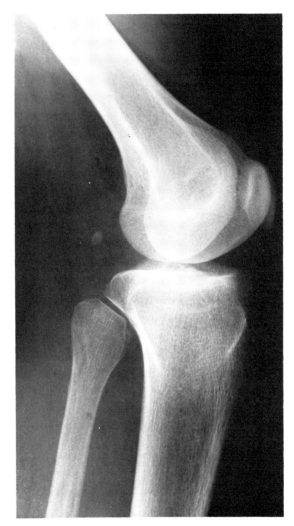

FIG. 1505. ILIAC HORNS (FONG DISEASE)
Hypoplasia of the inferior portion of the patella. (Same case as seen in Figure 1504. (Courtesy of George T. Wohl, Philadelphia General Hospital, Philadelphia, Pa.)

CONGENITAL INSENSITIVITY TO PAIN

Congenital insensitivity to pain is a rare condition and may be confused with other causes of pain insensitivity or indifference. MacMurray's[1] (1955) criteria are: (1) the defect must be present from birth and not acquired by disease or trauma, (2) there must be a general insensitivity to pain, and (3) there must be no general mental or physical retardation.

Other causes of pain insensitivity include: (1) sensory neuropathies,[2-6] (2) hysteria,[7] (3) syphilis,[8,9] (4) syringomyelia, (5) mental deficiency, and (6) organic brain disease. The typical history of painless injuries and burns is found with all types of pain insensitivity.

Roentgenographic Features. Roentgenographic features are most dramatic and include: (1) neuropathy (Fig. 1508), (2) fractures and dislocations (Figs. 1506, 1507, and 1509), and (3) longstanding infections (Fig. 1510). Neurotrophic joints are similar to those caused by syphilis, syringomyelia, or diabetes, and usually occur in weightbearing joints (Figs. 1511–1513).

Because fractures and dislocations go unnoticed, the lack of immobilization or protection result in bizarre deformity (Fig. 1512), and examination reveals gross displacement and considerable hemorrhage (Fig. 1514).

Osteomyelitis and septic arthritis may occur

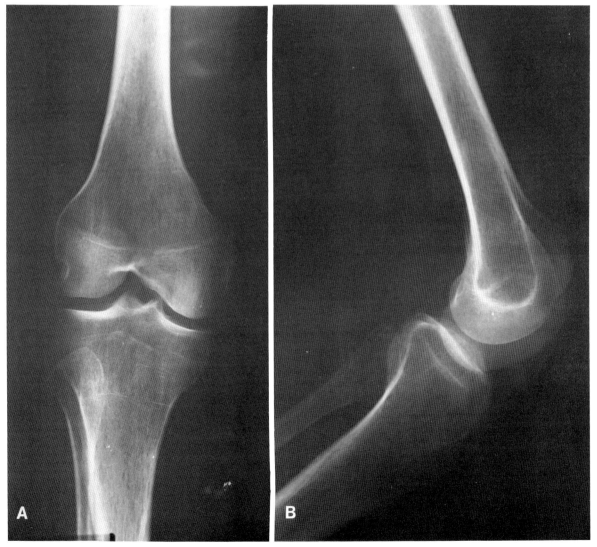

FIG. 1506. ILIAC HORNS (FONG DISEASE)
Absence of the patella. (A) Anteroposterior projection, (B) lateral projection.

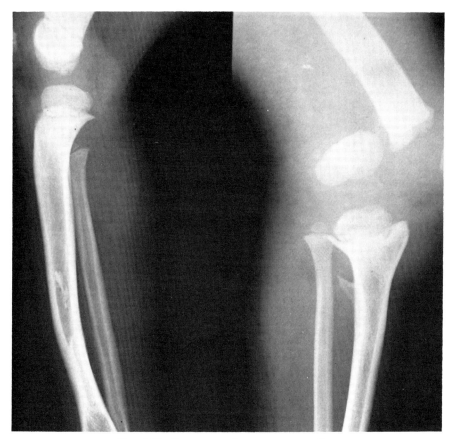

FIG. 1507. CONGENITAL INSENSITIVITY TO PAIN

Epiphyseal separation at the distal end of the femur was relatively asymptomatic and probably present for a long period of time. Considerable soft tissue swelling. Healed fractures in the shaft of the tibia. Irregularity at the proximal end of the tibia at the epiphyseal line represents probable infractions. Small exostosis on the upper portion of the tibia which may be the result of a previous fracture. (Courtesy of John A, Kirkpatrick, St. Christopher's Hospital, Philadelphia, Pa.)

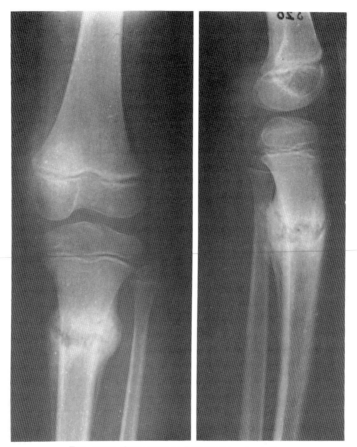

Fig. 1508. Congenital Insensitivity to Pain
Stress fracture of the upper end of the tibia. Considerable callus formation is present as the patient was using this limb.

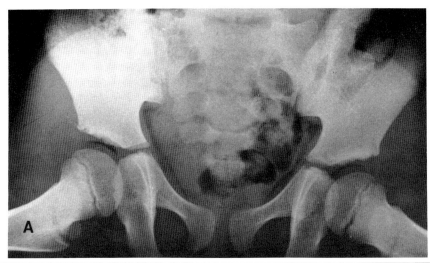
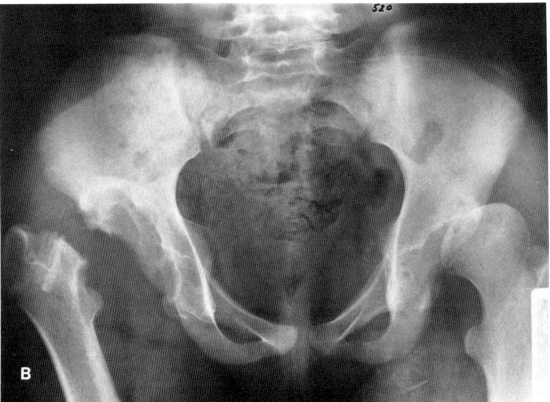

Fig. 1509. Congenital Insensitivity to Pain

(A) In 1950, the hips were normal. (B) Twelve years later, dislocation of the left hip with debris distal to the ischial spine. The right hip has resorption of the head and neck of the femur due to neuropathic joint.

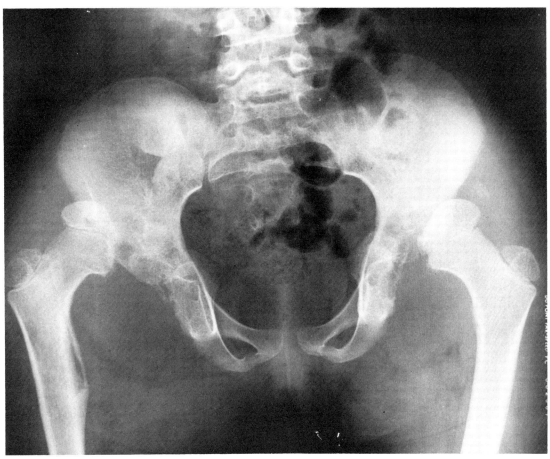

Fig. 1510. Congenital Insensitivity to Pain

Bilateral septic arthritis with dislocation. This patient had very few symptoms and this is the initial roentgenogram. The large abscesses had probably been present for several weeks. (Courtesy of John A. Kirkpatrick, St. Christopher's Hospital, Philadelphia, Pa.)

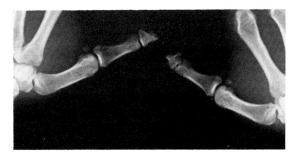

Fig. 1511. Congenital Insensitivity to Pain

Repeated trauma has caused resorption of both thumb tufts.

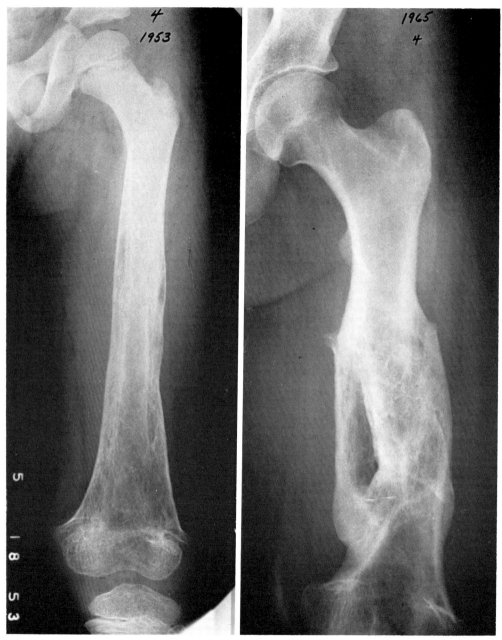

Fig. 1512. Congenital Insensitivity to Pain

(*Left*) Repeated trauma and infection has caused thickening of the right femur; (*right*) 12 years later there is marked deformity of the midportion of the femur due to repeated infections and trauma.

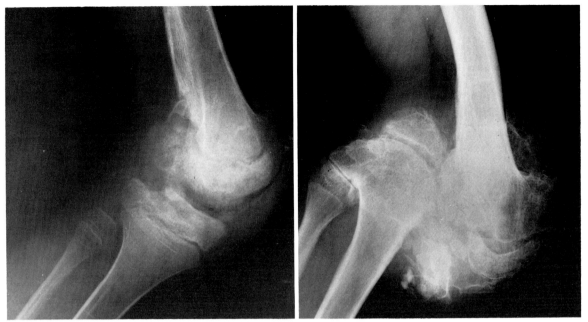

FIG. 1513. CONGENITAL INSENSITIVITY TO PAIN

(*Left*) Repeated trauma to the left knee has caused irregularity of the joint surfaces and periosteal reaction; (*right*) 10 years later, there is a Charcot joint.

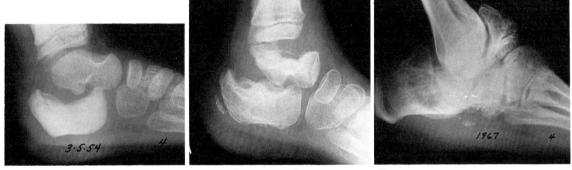

FIG. 1514. CONGENITAL INSENSITIVITY TO PAIN

(*Left*) Repeated trauma and infection has caused increase in deformity of the right calcaneus; (*center*) 3 years later there is deformity of the talus and calcaneus. The infection in the calcaneus has cleared but repeated trauma has caused further deformity; and (*right*) 13 years later, there is a Charcot of the talus, talocalcaneal, and tibiotalar joints. Marked deformity is present in the talus and cuboid and the anterior portion of the calcaneus.

and progress extensively before medical help is sought. Widespread destruction and joint subluxation and dislocation may appear. When Charcot joints, grossly displaced fractures, or extensive infection occur in children, some form of pain insensitivity should be suspected (Fig. 1513).

ANTERIOR TIBIAL BOWING
(TOXOPACHYOSTEOSE DIAPHYSAIRE TIBIO-PERONIERE)
(WEISMANN-NETTER SYNDROME)

Congenital painless, nonprogressive bilateral anterior leg bowing is unrelated to lues or rickets. Shortness of stature is a constant feature, often beginning in early childhood and becoming noticeable during puberty.[1] Less constant clinical features are mental retardation, goiter, and anemia.[1, 2] It may be familial or sporadic.

Roentgenographic Features. The tibias and fibulas are bowed anteriorly and usually bilaterally and symmetrically at the middiaphysis.

Medial bowing may also occur (Figs. 1515 and 1516). In contrast, bowing due to lues usually occurs in the upper end of the tibia, bowing due to rickets at the lower.

Thickening appears in the posterior tibial and fibular cortices, easily distinguished from anterior cortical thickening of the luetic sabre shin.[3] The trabecular pattern of the midportion of the bowed bones is distorted. Minor radioulnar bowing may occur. Kyphoscoliosis[3, 4] is reported. Extensive dural calcification may also occur (Fig. 1517).[2, 3]

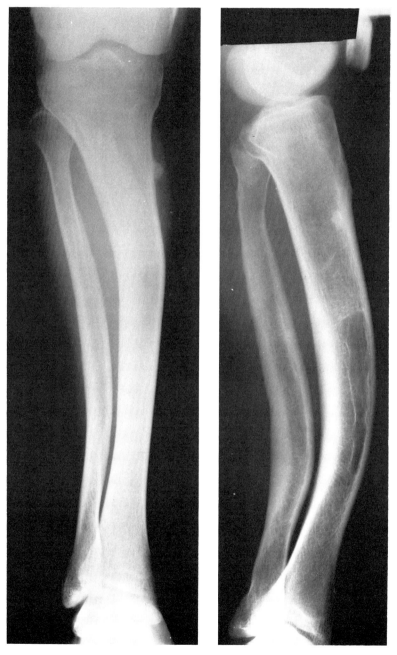

FIG. 1515. ANTERIOR TIBIAL BOWING

Anterior tibial and fibular bowing. A small exostosis is present on the upper medial aspect of the tibia. Thickening of the posterior cortices and disturbance of the trabecular pattern in the midportion of the tibia. (Courtesy of T. E. Keats and M. S. Alavi: *The American Journal of Roentgenology, 109:* 568, 1970,[3] © Charles C Thomas, Springfield, Ill.)

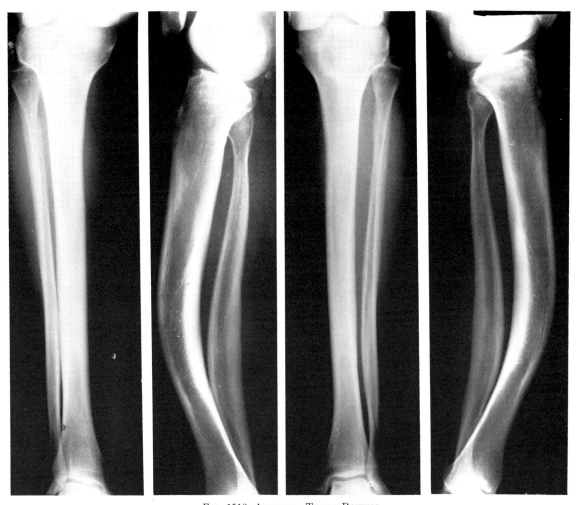

FIG. 1516. ANTERIOR TIBIAL BOWING

Bilateral anterior bowing of the tibias and fibulas. No medial bowing is present. (Courtesy of T. E. Keats and M. S. Alavi: *The American Journal of Roentgenology, 109:* 568, 1970,[3] © Charles C Thomas, Springfield, Ill.)

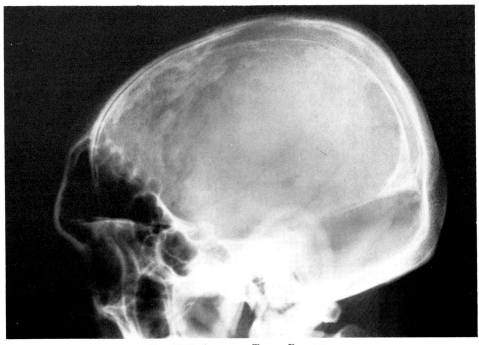

FIG. 1517. ANTERIOR TIBIAL BOWING

Extensive dural calcification is present. (Courtesy of T. E. Keats and M. S. Alavi: *The American Journal of Roentgenology, 109:* 568, 1970,[3] © Charles C Thomas, Springfield, Ill.)

AINHUM

Synonym: dactylolysis spontanea.

This dermatologic disorder of unknown etiology affects primarily the fifth toe, unilaterally or bilaterally, and sometimes other toes or fingers.[1] It is usually found in males in the 4th and 5th decades[2-4] and is most commonly found among Negroes in West Africa, but may be found in other areas, including the southern United States.[5]

Clinical Features. A deep soft tissue groove due to a hyperkeratotic band[3, 6] forms within the epidermis on the medial aspect, and extends to encompass the toe. As the band deepens over months or years it produces edema distally and eventually autoamputation.

Roentgenographic Features. The first sign is a soft tissue groove on the medial aspect of the distal portion of the fifth toe, near the interphalangeal joint. Bone resorption occurs on the medial aspect of the distal or middle phalanx and may progress until the bone is completely resorbed (Fig. 1518). The resorption is sharply demarcated. Soft tissue tumor occurs distal to the soft tissue contracture.

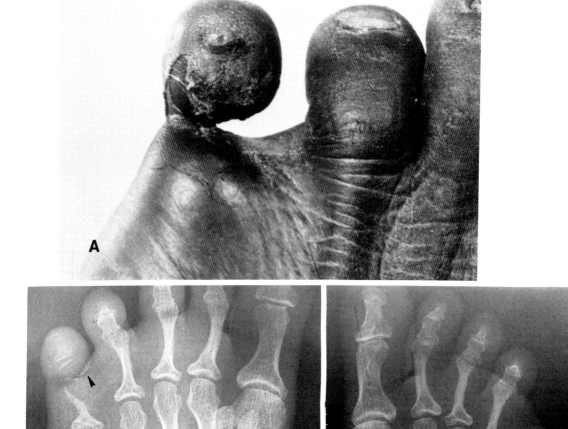

FIG. 1518. AINHUM

(A) Photograph of right foot reveals a soft tissue groove on the medial aspect of the distal portion of the big toe, near the interphalangeal joint. (B) Anteroposterior roentgenogram at the same time as (A) shows bone resorption at the site of the soft tissue groove. (C) The opposite foot shows complete resorption and sharp demarcation. (Courtesy of Dr. Stanley Bohrer, Ibadan, Nigeria, Africa.)

GARDNER SYNDROME

This syndrome of bony tumors (osteomas), soft tissue tumors, and colonic polyps,[1] is transmitted by an autosomal dominant mode.[2-4]

Roentgenographic Features. The skeletal abnormalities vary from cortical thickening to large osteomas that may be present on any bone.[5] The tubular bones most frequently have endosteal cortical thickening localized to one or several bones. There may be solid periosteal cortical thickening (Fig. 1519). Bone excrescences (osteomas) or exostoses may protrude from the periosteal surfaces (Fig. 1520). The superior aspects of the ribs may have a wavy cortical thickening.[5] The carpals and pelvis may have cortical thickening or exostoses.

Osteomas of the paranasal sinuses and outer skull table are frequent. Mandibular osteomas usually occur at the angle and are frequently lobular and highly characteristic (Fig. 1521). Wavy cortical thickening may be present in the body of the mandible. The skeletal lesion may precede the appearance of intestinal polyposis and, therefore, is an early sign. Patients with multiple osteomas or bone abnormalities should be carefully observed for the appearance of colonic polyps. Polyps also occur in the stomach, the duodenum, the ampulla of Vater, and in the remainder of the small intestine. The high incidence of carcinoma is in the duodenum and the ampulla of Vater.[6-9]

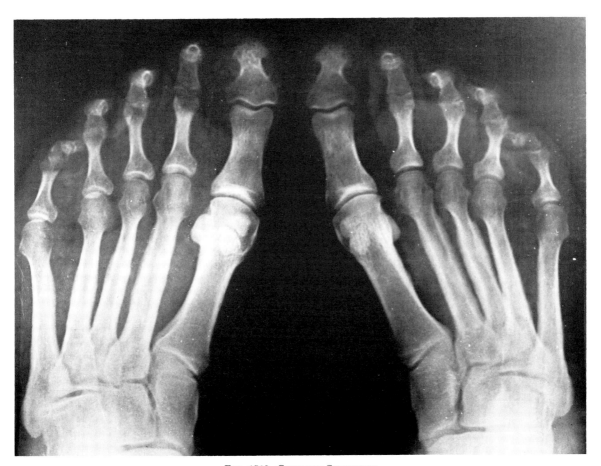

FIG. 1519. GARDNER SYNDROME
Solid periosteal cortical thickening of the metatarsals. (Courtesy of May Cliff, Temple University Hospital, Philadelphia, Pa.)

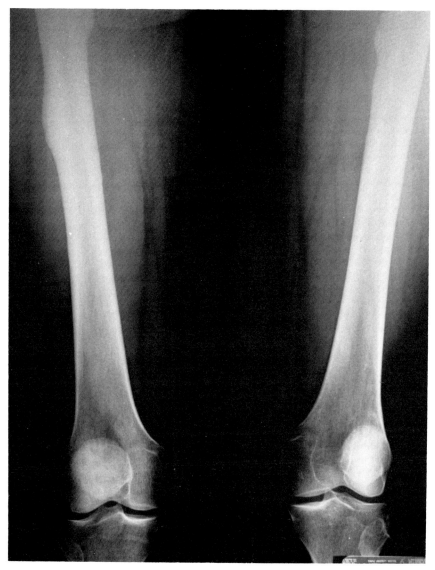

Fig. 1520. Gardner Syndrome
Bony excrescences are present in both femurs. (Courtesy of May Cliff, Temple University Hospital, Philadelphia, Pa.)

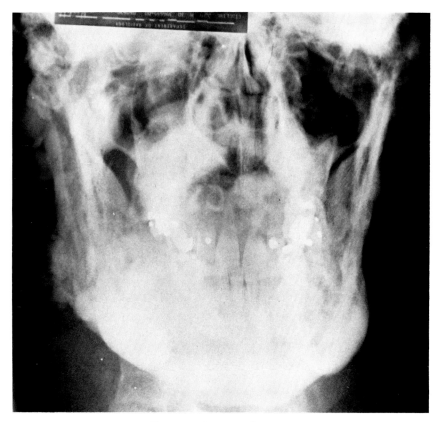

Fig. 1521. Gardner Syndrome
Multiple lobular osteomas of the mandible. (Courtesy of May Cliff, Temple University Hospital, Philadelphia, Pa.)

LARSEN SYNDROME

Larsen syndrome is a condition of congenital dislocation of multiple joints, accompanied by characteristic facies: prominent forehead, widely spaced eyes, and a depressed nasal bridge that gives the face a flattened appearance. Other findings include cleft palate, cleft uvula, cylindrical fingers with relatively short metacarpals, and abnormal segmentation of cervical and thoracic vertebrae. Mental development is normal.[1-3] Karyotype patterns have been normal and it is suggested that this is a generalized mesenchymal disorder of unknown etiology.[3]

Roentgenographic Features. Congenital dislocation of multiple joints most frequently affects the shoulders and the hips[4] (Figs. 1522 and 1523), and calcaneal valgus foot deformity is a usual finding. Bilateral valgus deformity of the hind foot[1] is also frequent. The knees often show dislocation, but may be normal.[1-3] A bipartite calcaneus with the posterior elements smaller than the distal is almost pathognomonic of Larsen disease.

Patients may show multiple congenital dislocations without the associated findings such as flattened facies or cleft palate.[5-16] However, at least some of the other characteristics must be present before the diagnosis is established.

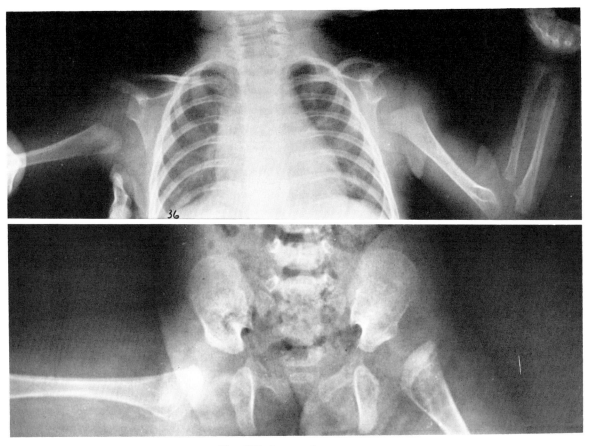

FIG. 1522. LARSEN SYNDROME
Dislocation of both shoulders and the left hip.

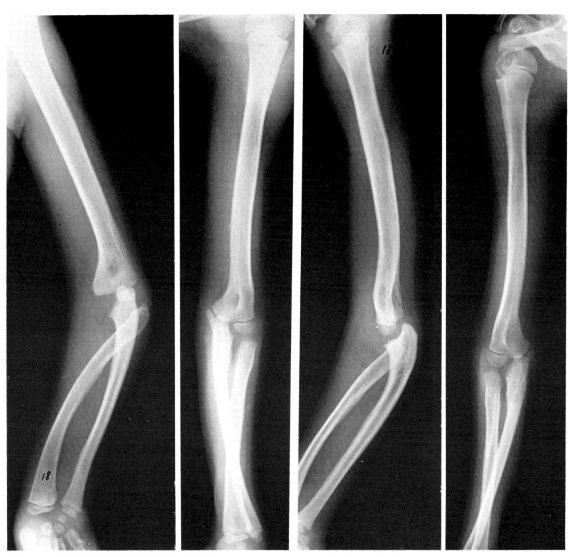

Fig. 1523. Larsen Syndrome
Dislocation of both elbows and both radial ulnar joints.

PROGERIA

Progeria or premature aging is a rare disease characterized by absence of hair, atrophic skin, and failure to grow at a normal rate.

Clinical Features. In the first 2 years the child develops degeneration of hair follicles, smooth tongue, hypoplastic nails, loss of subcutaneous fat, stiffness of joints and deficient growth.[1,2] Atherosclerosis may develop as early as 5 years.

The differential diagnosis of progeria includes other conditions of abiotrophic nature (the late onset of degenerative changes in the previously normal individual).[3] A comparative summary of Cockayne syndrome, homocystinuria, progeria and bird-headed dwarfs appears in Table 51.[4]

Roentgenographic Features. There is skeletal hypoplasia, most obvious in the facial bones, mandible, and hands. The vertebral bodies are oval and anterior rib notching persists. Scoliosis may develop. The tubular bone, ribs, and calvaria are thin. Fontanel ossification is delayed. The distal phalanges and clavicles are hypoplastic, probably due to bone resorption (Figs. 1524 and 1525).[5]

Some features are invariably present, and others are frequently present[6,7] (Table 52). The radiological features are summarized in Table 53.[7]

TABLE 51. COMPARATIVE SUMMARY OF COCKAYNE SYNDROME, HOMOCYSTINURIA, PROGERIA AND BIRD-HEADED DWARFS[a]

	Cockayne Syndrome	Homocystinuria	Progeria	Bird-headed Dwarf, Seckel Syndrome
Clinical Onset	During second year	After 1 yr	After 1 yr	Prenatal—very low birth weight
Dwarfism	Truncal	None	Truncal	Proportionate
Dolichostenomelia (long, thin extemities)	Present	Present	Present	Present
Joint abnormalities	Large epiphyses in *tarsal* centers and in *knees, elbows,* and *ankles* with flaring of the associated metaphyses producing joint contractures	Large epiphyses in distal *carpal* centers and in *knees,* and *ankles* with flaring of the metaphyses producing joint contractures	Capitellum sometimes large, periarticular fibrosis causing stiff joints	Dislocated radial head, congenital hip dislocation, absent patella
Osteoporosis	Present	Present	Variable	Present
Scoliosis	Present	Present	Variable	Present
Eye changes	Cataracts, retinal degeneration	Dislocation of lens, cataracts, retinal degeneration and detachment	Normal	Hypertelorism with ptosis of lids
Thoracic deformity	Pectus carinatum	Pectus excavatum and pectus carinatum	None	Defect in sternum
Skin	Photosensitivity, fine, fair hair	Malar flush, fine, fair hair	Thin skin, alopecia Brittle nails	Sparse hair
Mental Retardation	Always	Frequent, often with psychosis	Absent	Frequent
Inheritance	Autosomal recessive	Autosomal recessive	Unknown	Autosomal recessive
Facies	Senile	Normal	Senile	Bird-headed, prominent nose
Unusual manifestations	Intracranial calcifications	Thromboemboli	Premature arteriosclerosis, thin calvarium and open fontanelles	Absence of epiphyses of phalanges and metacarpals
Teeth	Carious	Crowded, small	Delay in eruption of deciduous and permanent teeth	Hypoplasia of enamel, partial anodontia

[a] From W. Riggs and J. Seibert: *American Journal of Roentgenology,* 116: 623–633, 1972.[4]

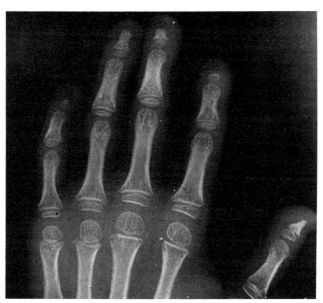
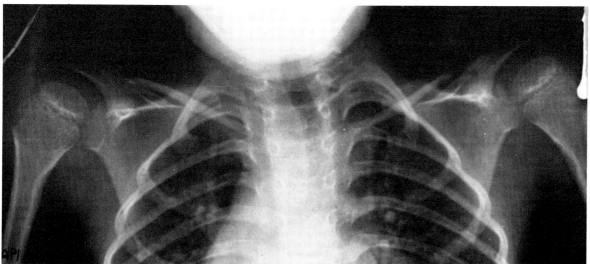

FIG. 1524. PROGERIA
The distal phalanges and clavicles are hypoplastic, probably due to bone resorption. (*Top*) Hand, (*bottom*) clavicles. (Courtesy of Dr. Howard Steinbach, Moffitt Hospital, San Francisco, Calif.)

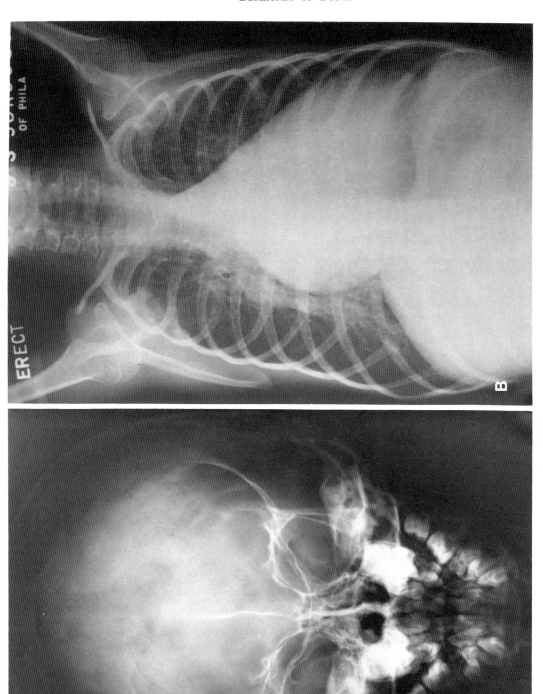

Fig. 1525. PROGERIA

(A) Skull shows hypoplasia of the facial bones and mandible. (B) Chest shows hypoplasia of the clavicles and enlarged heart. (C) Lower extremities show hypoplasia of the fibula. (D) Pelvis has a coxa valgum deformity from lack of weight-bearing.

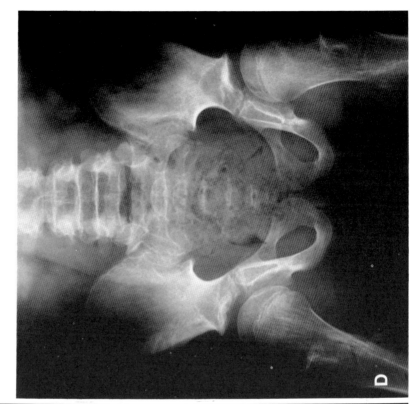

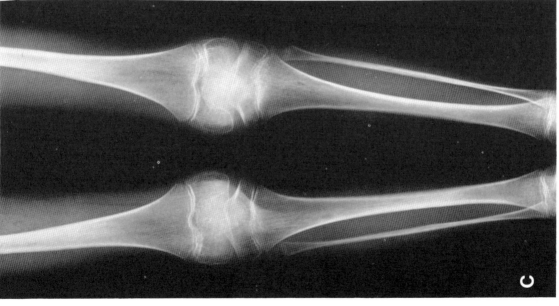

FIG. 1525 (*C* and *D*)

TABLE 52. Progeria Syndrome[a]

A. Characteristics Invariably Present
Short stature
Weight decreased for height
Sexual maturation absent
Subcutaneous fat diminished
Craniofacial disproportion
Clavicles short and dystrophic
"Horse-riding" stance
Gait wide based, shuffling
Coxa valga; limbs thin
Joints prominent and stiff
Micrognathia
Scalp veins prominent; alopecia
Eyes prominent
"Plucked-bird" appearance
Dentition delayed and abnormal
Thorax pyriform
B. Characteristics Frequently Present
Skin—thin, taut, wrinkled, "sclerodermatous"; brown spotted, prominent superficial veins, decreased sweating
Hypotrichosis, generalized
Eyebrows absent, eyelashes absent
Ears—protrude, lobes absent
Voice—thin, high pitched
Nails—dystrophic
Terminal phalanges radiolucent
Anterior fontanelle patent
Nose—glyphic, beaked
Nasal circumoral cyanosis
Lips thin

[a] From F. L. Debusk: *Clinical Pediatrics,* 10: 62, 1971.[6]

TABLE 53. Radiological Features of Progeria[a]

Skull	Craniofacial disproportion
	Micrognathia
	Re-widening of sutures
Hands and feet	Resorption of tips of terminal phalanges
Clavicles	Progressive resorption
Hips	Coxa vara
Long bones	Bizarre attenuation

[a] From P. P. Franklyn: *Clinical Radiology,* 27: 327, 1976.[7]

FAMILIAL OSTEODYSPLASIA

Familial osteodysplasia is a rare condition which is transmitted by an autosomal recessive mode. Although it is a general disorder, the most characteristic abnormalities are found in the facial bones.[1, 2]

Clinical Features. The striking facial features consist of prominent brows, flattening of the nasal bridge, prominent earlobes, dental malocclusion, and flattening of the mid-face and malar eminences.[1]

In one series,[1] all 4 patients of the same family were hyperuricemic but not hyperuricosuric.

Roentgenographic Features. The maxilla and zygomatic bones, including the malar portion and the arch, are small. Calvarial thinning and brachiocephaly are present as is a pointed configuration of the mastoids and hypoplasia of the petrous portion of the temporal bone.[1]

Thoracic scoliosis is frequent and the spinous processes of the cervical vertebrae are pointed.

The proximal phalanx of the second, third, fourth and fifth toes are relatively longer, and the middle phalanges of the second, third and fourth toes are relatively shorter.[1]

There is bilateral thinning of the superior pubic rami, and thickening of the cortices of some long bones, with encroachment on the medullary cavity.

REFERENCES

Approach to Patterns of Malformation and Introduction to Hereditary Dysplasias

1. Smith D. W.: Recognizable patterns of human malformation. In *Vol. VII: Major Problems in Clinical Pediatrics.* W. B. Saunders, Philadelphia, 1970.
2. Maroteaux, P.: Nomenclature internationale des maladies osseuses constitutionnelle. Ann Radiol, *13:* 455, 1970.
3. McKusick, V., and Scott, C. I.: A nomenclature for constitutional disorders of bone. J. Bone Joint Surg, *53A:* 978, 1971.
4. Special Report: International nomenclature of constitutional diseases of bone. AJR, *131:* 352, 1978.
5. Quelce-Salgado, A.: A new type of dwarfism with various bone aplasias and hypoplasias of the extremities. Acta Genet Stat Med, *14:* 63, 1964.
6. Langer, L. O., Jr., Spranger, J. W., Greinacher, I., and Herdman, R. C.: Thanatophoric dwarfism; a condition confused with achondroplasia in the neonate, with brief comments on achondrogenesis and homozygous achondroplasia. Radiology, *92:* 285, 1969.
7. Spranger, J. W., Opitz, J. M., and Bidder, U.: Heterogeneity of chondrodysplasia punctata. In *The Clinical Delineation of Birth Defects, XII: Skin, Hair and Nails.* Williams & Wilkins, Baltimore, 1971.
8. Langer, L. O., Jr., Baumann, P. A., and Gorlin, R. J.: Achondroplasia. AJR, *100:* 12, 1967.
9. Langer, L. O., Jr.: Diastrophic dwarfism in early infancy. AJR, *93:* 399, 1965.
10. LaRose, J. H., and Gay, B. B., Jr.: Metatrophic dwarfism. AJR, *106:* 156, 1969.
11. McKusick, V. A., Egeland, J. A., Eldridge, R., and Krusen, D. E.: Dwarfism in the Amish. I. The Ellisvan Creveld syndrome. Bull Johns Hopkins Hosp, *115:* 306, 1964.
12. Langer, L. O., Jr.: Thoracic-pelvic-phalangeal dystrophy; asphyxiating thoracic dystrophy of the newborn, infantile thoracic dystrophy. Radiology, *91:* 447, 1968.
13. Spranger, J. W., and Langer, L. O., Jr.: Spondyloepiphyseal dysplasia congenita. Radiology, *94:* 313, 1970.
14. Solonen, K. A., and Sulamaa, M.: Nievergelt syndrome and its treatment; a case report. Ann Chir Gynaecol Fenn, *47:* 142, 1958.
15. Langer, L. O., Jr.: Mesomelic dwarfism of the hypoplastic ulna, fibula, mandible type. Radiology, *89:* 654, 1967.
16. Forland, M.: Cleidocranial dysostosis; a review of the syndrome and report of a sporadic case, with hereditary transmission. Am J Med, *33:* 792, 1962.
17. Beals, R. K.: Hypochondroplasia; a report of five kindreds. J Bone Joint Surg, *51A:* 728, 1969.
18. Felman, A. H., and Kirkpatrick, J. A., Jr.: Madelung's deformity; observations in 17 patients. Radiology, *93:* 1037, 1969.
19. Lenz, W. D., and Holt, J. F.: Discussion. In *The Clinical Delineation of Birth Defects, IV: Skeletal Dysplasias,* pp. 71-75. Williams & Wilkins, Baltimore, 1969.
20. Rosenbloom, A. L., and Smith, D. W.: The natural history of metaphyseal dysostosis. J Pediatr, *66:* 857, 1965.
21. McKusick, V. A., Eldridge, R., Hostetler, J. A., Ruangwit, U., and Egeland, J. A.: Dwarfism in the Amish. II. Cartilage-hair hypoplasia. Bull Johns Hopkins Hosp, *116:* 285, 1965.
22. Shmerling, D. H., Prader, A., Hitzig, W. H., Giedion, A., Hardon, B., and Kuhni, M.: The syndrome of exocrine pancreatic insufficiency, neutropenia, metaphyseal dysostosis and dwarfism. Helv Paediatr Acta, *24:* 547, 1969.
23. Gatti, R. A., Platt, N., Pomerance, H. H., Hong, R., Langer, L. O., Jr., Kay, H. E. M., and Good, R. A.: Hereditary lymphopenic agammaglobulinemia associated with a distinctive form of short-limbed dwarfism and ectodermal dysplasia. J Pediatr, *75:* 675, 1969.
24. Kozlowski, K., Maroteaux, P., and Spranger, J.: La dysostose spondylo-metaphysaire. Presse Med, *75:* 2769, 1967.
25. Fairbank, T.: Dysplasia epiphysialis multiplex. Br J Surg, *34:* 225, 1947.
26. Tips, R. L., and Lynch, H. T.: Malignant congenital osteopetrosis resulting from a consanguineous marriage. Acta Paediatr, *51:* 585, 1962.
27. Stickler, G. B., Belau, P. G., Farrell, F. J., Jones, J. D., Steinberg, A. G., and Ward, L. E.: Hereditary progressive arthroophthalmopathy. Mayo Clinic Proc, *40:* 433, 1965.
28. Hall, J. G., and Dorst, J. P.: Four types of pseudoachondroplastic spondyloepiphyseal dysplasia (SED). In *The Clinical Delineation of Birth Defects, IV: Skeletal Dysplasias,* pp. 242-259. Williams & Wilkins, Baltimore, 1969.
29. Langer, L. O., Jr.: Spondyloepiphyseal dysplasia tarda; hereditary chondrodysplasia with characteristic vertebral configuration in the adult. Radiology, *82:* 833, 1964.
30. Giedion, A.: Das tricho-rhino-phalangeale syndrom. Helv Paediatr Acta, *21:* 475, 1966.
31. Bohme, A.: Kasuistischer beitrag zur thiemannschen Epiphysenerkrankung. Z Gesamte Inn Med, *18:* 491, 1963.
32. Allison, A. C., and Blumberg, B. S.: Familial osteoarthropathy of the fingers. J Bone Joint Surg, *40B:* 538, 1958.
33. Brailsford, J. F.: Chondro-osteo-dystrophy; roentgenographic and clinical features of a child with dislocation of vertebrae. Am J Surg, *7:* 404, 1929.
34. Rubin, P.: *Dynamic Classification of Bone Dysplasias.* Year Book, Chicago, 1964.
35. Solomon, L.: Hereditary multiple exostosis. J Bone Joint Surg, *45B:* 292, 1963.
36. Anderson, I. F.: Maffucci's syndrome; report of a case with review of the literature. S Afr Med J, *39:* 1066, 1965.
37. Caffey, J. P.: *Pediatric X-ray Diagnosis,* Ed. 5. Year Book, Chicago, 1967.
38. Albright, F., Butler, A. M., Hampton, A. O., and Smith, P.: Syndrome characterized by osteitis fibrosa disseminata, areas of pigmentation and endocrine dysfunction, with precocious puberty in females; report of 5 cases. N Engl J Med, *216:* 727, 1937.
39. Jones, W. A.: Cherubism; a thumbnail sketch of its diagnosis and a conservative method of treatment Oral Surg, *20:* 648, 1965.
40. Schnitka, T. K., Asp, D. M., and Horner, R. H.: Congenital generalized fibromatosis. Cancer, *11:* 627, 1958.
41. Remigio, P. A., and Grinvalsky, H. T.: Osteogenesis imperfecta congenita; association with conspicuous extraskeletal connective tissue dysplasia. Am J Dis Child *119:* 524, 1970.
42. McKusick, V. A.: Osteogenesis imperfecta. In *Heritable Disorders of Connective Tissue,* Ed. 3, pp. 230-270. C. V. Mosby, St. Louis, 1966.
43. Dent, C. E., and Friedman, M.: Idiopathic juvenile osteoporosis. Q J Med, *34:* 177, 1965.

44. Elsbach, L.: Bilateral hereditary micro-epiphyseal dysplasia of the hips. J Bone Joint Surg, *41B:* 514, 1959.
45. Johnston, C. C., Jr., Lavy, N., Lord, T., Vellios, F., Merritt, A. D., and Deiss, W. P., Jr.: Osteopetrosis; a clinical, genetic, metabolic, and morphologic study of the dominantly inherited benign form. Medicine, *47:* 149, 1968.
46. Andren, L., Dymling, J. F., Hogeman, K. E., and Wendeberg, B.: Osteopetrosis acro-osteolytica; a syndrome of osteopetrosis, acro-osteolysis and open sutures of the skull. Acta Chir Scand, *124:* 496, 1962.
47. Berlin, R., Hedensio, B., Lilia, B., and Linder, L.: Osteopoikilosis; a clinical and genetic study. Acta Med Scand, *181:* 305, 1967.
48. Campbell, C. P., Papademetroiu, T., and Bonfiglio, M.: Melorheostosis; a report of the clinical, roentgenographic and pathological findings in 14 cases. J Bone Joint Surg, *50A:* 1281, 1968.
49. Girdany, B. R.: Engelmann's disease (progressive diaphyseal dysplasia); a nonprogressive familial form of muscular dystrophy with characteristic bone changes. Clin Orthop, *14:* 102, 1959.
50. Halliday, J.: A rare case of bone dystrophy. Br J Surg, *37:* 52, 1949.
51. Van Buchem, F. S. P., Hadders, H. N., Hanse, J. F., and Woldring, M. G.: Hyperostosis corticalis generalisata; report of 7 cases. Am J Med, *33:* 387, 1962.
52. Caffey, J.: Congenital stenosis of medullary spaces in tubular bones and calvaria in two proportionate dwarfs—mother and son; coupled with transitory hypocalcemic tetany. AJR, *100:* 1, 1967.
53. Rimoin, D. L.: Pachydermoperiostosis (idiopathic clubbing and periostosis); genetic and physiologic considerations. N Engl J Med, *272:* 923, 1965.
54. Melnick, J. C., and Needles, C. F.: An undiagnosed bone dysplasia; a two family study of four generations and three generations. AJR, *97:* 39, 1966.
55. Gorlin, R. J., and Cohen, M. M., Jr.: Frontometaphyseal dysplasia; a new syndrome. Am J Dis Child, *118:* 487, 1969.
56. Holt, J. F.: The evolution of cranio-metaphyseal dysplasia. Ann Radiol, *9:* 209, 1966.
57. Millard, D. R., Jr., Maisels, D. O., Batstone, J. H. F., and Yates, B. W.: Craniofacial surgery in craniometaphyseal dysplasia. Am J Surg, *113:* 615, 1967.
58. Gorlin, R. J., Spranger, J., and Koszalka, M. F.: Genetic craniotubular bone dysplasias and hyperostoses; a critical analysis. In *The Clinical Delineation of Birth Defects, IV: Skeletal Dysplasias*, pp. 79–95. Williams & Wilkins, Baltimore, 1969.
59. Bakwin, H., Golden, A., and Fos, S.: Familial osteoectasia with macrocranium. AJR, *91:* 609, 1964.
60. Spranger, J., Albrecht, C., Rohwedder, H.-J., and Wiedemann, H. R.: Die Dyosteosklerose. Fortschr Roentgenstr, *109:* 504, 1968.
61. Shillito, J., Jr., and Matson, D. D.: Craniosynostosis; a review of 519 surgical patients. Pediatrics, *41:* 829, 1968.
62. Cross, H. E., and Opitz, J. M.: Craniosynostosis in the Amish. J Pediatr, *75:* 1037, 1969.
63. Vulliamy, D. G., and Normandale, P. A.: Cranio-facial dysostosis in a Dorset family. Arch Dis Child, *41:* 375, 1966.
64. Schauerte, E. W., and St. Aubin, P. M.: Progressive synosteosis in Apert's syndrome (acrocephalosyndactyly); with a description of roentgenographic changes in the feet. AJR, *97:* 67, 1966.
65. Temtamy, S. A.: Carpenter's syndrome; acrocephalopolysyndactyly, an autosomal recessive syndrome. J Pediatr, *69:* 111, 1966.
66. Rovin, S., Dachi, S. F., Borenstein, D. B., and Cotter, W. B.: Mandibulofacial dysostosis; a familial study of five generations. J Pediatr, *65:* 215, 1964.
67. Smith, J. L., and Stowe, F. R.: The Pierre Robin syndrome (glossoptosis, micrognathia, cleft palate); a review of 39 cases with emphasis on associated ocular lesions. Pediatrics, *27:* 128, 1961.
68. Falls, H. F., and Schull, W. J.: Hallermann-Streiff syndrome; a dyscephaly with congenital cataracts and hypotrichosis. Arch Ophthalmol, *63:* 409, 1960.
69. Rater, C.J., Selke, A. C., and Van Epps, E. F.: Basal cell nevus syndrome. AJR, *103:* 589, 1968.
70. Gunderson, C. H., Greenspan, R. H., Glasser, G. H., and Lubs, H. A.: The Klippel-Feil syndrome; genetic and clinical re-evaluation of cervical fusion. Medicine, *46:* 491, 1967.
71. Kirkham, T. H.: Cervico-oculo-acusticus syndrome with pseudo-papilloedema. Arch Dis Child, *44:* 504, 1969.
72. Schwarzweller, F.: Des angeborene schulterblatthochstand der wirbelsaule. Z Menschl Vereb Konstitut, *20:* 341, 1937.
73. Norum, R. A.: Costovertebral anomalies with apparent recessive inheritance. In *The Clinical Delineation of Birth Defects, IV: Skeletal Dysplasias*, pp. 326–329. Williams & Wilkins, Baltimore, 1969.
74. Rimoin, D. L., Fletcher, B. D., and McKusick, V. A.: Spondylocostal dysplasia; a dominantly inherited form of short-trunked dwarfism. Am J Med, *45:* 948, 1968.
75. Weyers, H., and Thier, C. J.: Malformations mandibulofaciales et delimitation d'un syndrome oculo-vertebral. J Genet Hum, *7:* 143–173, 1958.
76. Lucas, G. L., and Opitz, J. M.: The nail-patella syndrome; clinical and genetic aspects of 5 kindreds with 38 affected family members. J Pediatr, *68:* 273, 1966.
77. Bergsma, D. (ed.): *The Clinical Delineation of Birth Defects, III: Limb Malformations*. Williams & Wilkins, Baltimore, 1969.
78. Nevin, N. C., Dodge, J. A., and Kernohan, D. C.: Aglossia-adactylia syndrome. Oral Surg, *29:* 443, 1970.
79. Hansen, O. H., and Anderson, N. O.: Congenital radioulnar synostosis; report of 37 cases. Acta Orthop Scand, *41:* 225, 1970.
80. Temtamy, S., and McKusick, V. A.: Synopsis of hand malformations with particular emphasis on genetic factors. In *The Clinical Delineation of Birth Defects, III: Limb Malformations*, pp. 125–184. Williams & Wilkins, Baltimore, 1969.
81. Strasburger, A. F., Hawkins, M. R., Eldridge, R., Hargrave, R. L., and McKusick, V. A.: Symphalangism; genetic and clinical aspects. Bull Johns Hopkins Hosp, *117:* 108, 1965.
82. Poznanski, A. K., Gall, J. C. Jr., and Stern, A. M.: Skeletal manifestations of the Holt-Oram syndrome. Radiology, *94:* 45, 1970.
83. Lamy, M., and Maroteaux, P.: Acro-osteolyse dominante. Arch Fr Pediatr, *18:* 693, 1961.
84. Torg, J. S., and Steel, H. H.: Essential osteolysis with nephropathy; a review of the literature and case report of an unusual syndrome. J Bone Joint Surg, *50A:* 1629, 1968.
85. Torg, J. S., DiGeorge, A. M., Kirkpatrick, J. A., Jr., and Trujillo, M. M.: Hereditary multicentric osteolysis with recessive transmission; a new syndrome. J Pediatr, *75:* 243, 1969.
86. Williams, T. F., Winters, R. W., and Burnett, C. H.: A genetic study of familial hypophosphatemia and vitamin D resistant rickets. In *The Metabolic Basis of Inherited Disease*, Ed. 2, pp. 1179–1199. McGraw-Hill, New York, 1966.
87. Bianchine, J. W., Stambler, A. A., and Harrison, H. E.: Familial hypophosphatemic rickets showing autosomal

dominant inheritance. In *The Clinical Delineation of Birth Defects, X: Endocrine System*, pp. 287-295. Williams & Wilkins, Baltimore, 1971.
88. Dent, C. E., Friedman, M., and Watson, Lyal: Hereditary pseudovitamin D deficiency rickets ("hereditare pseudo-mangelrachitis"). J Bone Joint Surg, *50B:* 708, 1968.
89. McCance, R. A.: Osteomalacia with Looser's nodes (Milkman's syndrome) due to a raised resistance to vitamin D acquired about the age of 15 years. Q J Med, *16:* 33, 1947.
90. Bartter, F. C.: Hypophosphatasia. In *The Metabolic Basis of Inherited Disease*, Ed. 2, pp. 1015-1023. McGraw-Hill, New York, 1966.
91. Silverman, J. L.: Apparent dominant inheritance of hypophosphatasia. Arch Intern Med, *110:* 191, 1962.
92. Mann, J. B., Alterman, S., and Hills, A. G.: Albright's hereditary osteodystrophy comprising pseudohypoparathyroidism and pseudo-pseudohypoparathyroidism; with a report of 2 cases representing the complete syndrome occurring in successive generations. Ann Intern Med, *56:* 315, 1962.
93. McKusick, V. A.: The Marfan syndrome. In *Heritable Disorders of Connective Tissue*, Ed. 3, pp. 38-149. C. V. Mosby, St. Louis, 1966.
94. Durand, P., Borrone, C., and Della Cella, G.: Fucosidosis. J Pediatr, *75:* 665, 1969.
95. Austin, J. H.: Mental retardation. Metachromatic leucodystrophy. (sulfatide lipidosis, metachromatic leucoencephalopathy). In *Medical Aspects of Mental Retardation*, pp. 768-812. Charles C Thomas, Springfield, Ill., 1965.
96. Thieffry, S., Lyon, G., and Maroteaux, P.: Encephalopathie metabolique assocant mucopolysaccharidose et une sulfatidose. Arch Fr Pediatr, *24:* 425, 1967.
97. Gorlin, R. J., Meskin, L. H., and St. Geme, J. W.: Oculodentodigital dysplasia. J Pediatr, *63:* 69, 1963.
98. Ammann, F.: Investigations cliniques et genetiques sur le syndrome de Bardet-Biedl en Suisse. J Genet Hum, *18*(suppl): 1, 1970.
99. Klein, D., and Ammann, F.: The syndrome of Laurence-Moon-Bardet-Beidl and allied diseases in Switzerland; clinical, genetic and epidemiological studies. J Neurol Sci, *9:* 479, 1969.
100. Gorlin, R. J., Sedano, H. O., and Cervenka, J.: Popliteal pterygium syndrome; a syndrome comprising cleft lip-palate, popliteal and intercrural pterygia, digital and genital anomalies. Pediatrics, *41:* 503, 1968.
101. Walker, J. C., Jr., Meljer, R., and Aranda, D.: Syndactylism with deformity of the pectoralis muscle—Poland's syndrome. J Pediatr Surg, *4:* 569, 1969.
102. Rubinstein, J. H.: The broad thumbs syndrome—progress report 1968. In *The Clinical Delineation of Birth Defects, II: Malformation Syndromes*, pp. 25-41. Williams & Wilkins, Baltimore, 1969.
103. Juhl, J. H., Wesenberg, R. L., and Gwinn, J. L.: Roentgenographic findings in Fanconi's anemia. Radiology, *89:* 646, 1967.
104. Hall, J. G., Levin, Jack, Kuhn, J. P., Ottenheimer, E. J., van Berkum, K. A. P., and McKusick, V. A.: Thrombocytopenia with absent radius (TAR). Medicine, *48:* 411, 1969.
105. Gorlin, R. J., and Pindborg, J. J.: Orodigitofacial dysostosis. In *Syndromes of the Head and Neck*, pp. 438-446. McGraw-Hill, New York, 1964.
106. Rimoin, D. L., and Edgerton, M. T.: Genetic and clinical heterogeneity in the oral-facial-digital syndromes. J Pediatr, *71:* 94, 1967.
107. Black, J.: Low birth weight dwarfism. Arch Dis Child, *36:* 633, 1961.
108. Warkany, J., Monroe, B. B., and Sutherland, B. S.: Intrauterine growth retardation. Am J Dis Child, *102:* 249, 1961.
109. Pashayan, H., Whelan, D., Guttman, S., and Fraser, F. C.: Variability of the de Lange syndrome; report of 3 cases and genetic analysis of 54 families. J Pediatr, *75:* 853, 1969.
110. McKusick, V. A., Mahloudji, M., Abbott, M. H., Lindenberg, Richard, and Kepas, D.: Seckel's birdheaded dwarfism. N Engl J Med, *277:* 279, 1967.
111. Summitt, R. L., and Favara, B. E.: Leprechaunism (Donohue's syndrome); a case report. J Pediatr, *74:* 601, 1969.
112. Holden, J. D.: The Russell-Silver's dwarf. Dev Med Child Neurol, *9:* 457, 1967.
113. Macleod, W.: Progeria. Br J Radiol, *39:* 224, 1966.
114. MacDonald, W. B., Fitch, K. D., and Lewis, I. C.: Cockayne's syndrome; an heredofamilial disorder of growth and development. Pediatrics, *25:* 997, 1960.
115. Bloom, D.: The syndrome of congenital telangiectatic erythema and stunted growth. J Pediatr, *68:* 103, 1966.
116. Brocher, J. E., Klein, D., Bamatter, F., Franceschetti, A., and Boreux, G.: Rontgenologische befunde bei geroderma osteodysplastica hereditaria. Fortschr Geb Rontgenstr Nuklearmed, *109:* 185, 1968.
117. Scott, C. I.: Weill-Marchesani syndrome. In *The Clinical Delineation of Birth Defects, II: Malformation Syndromes*, pp. 238-240. Williams & Wilkins, Baltimore, 1969.
118. Kenny, F. M., Aceto, T., Jr., Purisch, M., Harrison, H. E., Harrison, H. C. and Blizzard, R. M.: Metabolic studies in a patient with idiopathic hypercalcemia of infancy. J Pediatr, *62:* 531, 1963.
119. Spranger, J. W., and Wiedemann, H. R.: The genetic mucolipidoses; diagnosis and differential diagnosis. Humangenetik *9:* 113, 1970.
120. Spranger, J., Wiedmann, H. R., Tolksdorf, M., Graucob, E., and Caesar, R.: Lipomucopolysaccharidose. Z Kinderheilkd, *103:* 285, 1968.
121. Leroy, J. G., DeMars, R. I., and Opitz, J. M.: I-Cell disease. In *The Clinical Delineation of Birth Defects, IV: Skeletal Dysplasias*, pp. 174-185. Williams & Wilkins, Baltimore, 1969.
122. Scott, C. I., Jr., and Grossman, M. S.: Pseudo-Hurler polydystrophy. In *The Clinical Delineation of Birth Defects, IV: Skeletal Dysplasias*, pp. 349-355. Williams & Wilkins, Baltimore, 1969.
123. Ockerman, P. A.: A generalized storage disorder resembling Hurler's syndrome. Lancet, *2:* 239, 1967.

Clinodactyly, Camptodactyly, and Crooked Fingers

1. Poznanski, A. K., Pratt, G. B., Mason, G., and Weiss, L.: Clinodactyly, camptodactyly, Kirner's deformity, and other crooked fingers. Radiology, *93:* 573, 1969.
2. Ashley, L. M.: Inheritance of streblomicrodactyly. J Hered, *38:* 93, 1947.
3. Dutta, P.: Inheritance of the radially curved little finger. Acta Genet, *15:* 70, 1965.
4. Hersh, A. H., DeMarinis, F., and Stecher, R. M.: On inheritance and development of clinodactyly. Am J Hum Genet, *5:* 257, 1953.
5. Garn, S. M., Fels, S. L., Rohman, C. G., and Lee, M.: Representative digital metacarpal and carpal abnormalities in Guatemala and El Salvador. Progress report 66-2, Part II on PH-43-65-1006 with Advanced Projects Agency. Monitored by the National Section, Office of International Research, National Institutes of Health, under ARPA Order 580, June 1966.

6. Hefner, R. A.: Inheritance of crooked little fingers (streblomicrodactyly). J Hered, 20: 395, 1929.
7. Curratino, G., and Waldman, I.: Camptodactyly. Am J Roentgenol Radium Ther Nucl Med, 92: 1312, 1964.
8. Hefner, R. A.: Inherited abnormalities of the fingers. J Hered, 15: 481, 1924.
9. Smith, R. J., and Kaplan, E. B.: Camptodactyly and similar atraumatic flexion deformities of the proximal interphalangeal joints of the fingers; study of 31 cases. J Bone Joint Surg, 50A: 1187, 1968.
10. Stoddard, S. E.: Nomenclature of hereditary crooked fingers; streblomicrodactyly and camptodactyly—are they synonyms? J Hered, 30: 511, 1939.
11. Welch, J. P., and Temtamy, S. A.: Hereditary contractures of the fingers (camptodactyly). J Med Genet, 3: 104, 1966.
12. Littman, A., Yates, J. W., and Treger, A.: Camptodactyly—a kindred study. JAMA, 206: 1565, 1968.
13. Hoefnagel, D., and Gerald, P. S.: Herditary brachydactyly. Ann Hum Genet, 29: 377, 1966.
14. Pol, P.: Brachydaktylie-klinodaktylie-hyperphalangie und ihre grundlagen. Arch Pathol Anat Klin Med, 229: 388, 1921.
15. Robinson, G. C., Wood, B. J., Miller, J. R., and Baillie, J.: Hereditary brachydactyly and hip disease; unusual radiological and dermatoglyphic findings in a kindred. J Pediatr, 72: 539, 1968.
16. Hefke, H. W.: Roentgenologic study of anomalies of hands in 100 cases of mongolism. Am J Dis Child, 60: 1319, 1940.
17. Oster, J.: Mongolism; a clinicogenealogical investigation comparing 526 mongols living in Seeland and neighboring islands in Denmark. Danish Scientific Press Ltd., Copenhagen, 1953.
18. Smith, T. T.: Peculiarity in shape of hand in idiots of mongol type. Pediatrics, 2: 315, 1896.
19. Moseley, J. E., Moloshik, R. E., and Freiberger, R. H.: Silver syndrome: congenital asymmetry, short stature and variations in sexual development, roentgen features. AJR, 97: 74, 1966.
20. Silver, H. K.: Asymmetry, short stature, and variations in sexual development. Am J Dis Child, 107: 495, 1964.
21. Gerald, B., and Umansky, R.: Cornelia de Lange syndrome; radiographic findings. Radiology, 88: 96, 1967.
22. Lee, F. A.: Generalized overconstriction of long bones and unilateral Kirner's deformity in a de Lange dwarf. Am J Dis Child, 116: 599, 1968.
23. McArthur, R. G., and Edwards, J. H.: de Lange syndrome; report of 20 cases. Can Med Assoc J, 96: 1185, 1967.
24. Silver, H. K.: The de Lange syndrome; typus amstelodamensis. Am J Dis Child, 108: 523, 1964.
25. Houston, C. S.: Roentgen findings in the XXXXY chromosome anomaly. J Can Assoc Radiol, 18: 258, 1967.
26. Doege, T. C., Thuline, H. C., Priest, J. H., Norby, D. E., and Bryant, J. S.: Studies of a family with the oral-facial-digital syndrome. N Engl J Med, 271: 1073, 1964.
27. Dudding, B. A., Corlin, R. J., and Langer, L. O., Jr.: Otopalato-digital syndrome; a new symptom-complex consisting of deafness, dwarfism, cleft palate, characteristic facies and a generalized bone dysplasia. Am J Dis Child, 113: 214, 1967.
28. Corlin, R. J., Meskin, L. H., and St. Geme, J. W.: Oculodento-digital dysplasia. J Pediatr, 63: 69, 1963.
29. Corlin, R. J., and Psaume, J.: Orodigitofacial dysostosis—a new syndrome; study of 22 cases. J Pediatr, 61: 520, 1962.
30. Langer, L. O., Jr.: Roentgenographic features of the otopalato-digital (OPD) syndrome. AJR, 100: 63, 1967.
31. Rajic, D. S., and de Veber, L. L.: Hereditary oculodentoosseous dysplasia. Ann Radiol, 9: 224, 1966.
32. Kosowicz, J.: Roentgen appearance of the hand and wrist in gonadal dysgenesis. AJR, 93: 354, 1965.
33. Gonadal dysgenesis; clinical and roentgenologic manifestations. AJR, 87: 1116, 1962.
34. Gellis, S. S., and Feingold, M.: Male Turner's syndrome. Am J Dis Child, 112: 63, 1966.
35. Nora, J. J., and Sinha, A. K.: Direct familial transmission of Turner phenotype. Am J Dis Child, 116: 343, 1968.
36. Gall, J. C., Jr., Stern, A. M., Cohen, M. M., Adams, M. S., and Davidson, R. T.: Holt-Oram syndrome; clinical and genetic study of a large family. Am J Hum Genet, 18: 187, 1966.

Hereditary Bowing of Terminal Phalanges of Fifth Fingers

1. Kirner, J.: Doppelseitige verkrummungen des kleinfingerendgliedes als selbstandiges krankheitsbild. Fortschr Geb Roentgenstr, 36: 804, 1927.
2. McKusick, V. A.: *Mendelian Inheritance in Man: Catalogs of Autosomal Dominant, Autosomal Recessive, and X-linked Phenotypes*, Ed. 2 Johns Hopkins Press, Baltimore, 1968.
3. Dykes, R. G.: Kirner's deformity of the little finger. J Bone Joint Surg, 60B: 58, 1978.
4. Wilson, J. N.: Dystrophy of fifth finger; report on 4 cases. J Bone Joint Surg, 34B: 236, 1952.
5. Blank, E., and Girdany, B. R.: Symmetric bowing of the terminal phalanges of the fifth fingers in a family (Kirner's deformity). Radiology, 93: 367, 1965.
6. Moseley, J. E., Moloshok, R. E., and Freiberger, R. H.: Silver syndrome; congenital asymmetry, short stature and variations in sexual development, roentgen features. AJR, 97: 74, 1966.
7. Lee, F. A.: Generalized overconstriction of long bones and unilateral Kirner's deformity in a de Lange dwarf. Am J Dis Child, 116: 599, 1968.

Achondroplasia

1. Parrot, M. J.: Sur la malformation achondroplasique et le dieu Ptah. Bull Soc Anthropol Paris, 1 (3rd ser.): 296, 1878.
2. Kaufmann, E.: *Untersuchungen uber die Sogenannte foetale Rachitis (Chondrodystrophy Foetalis)*. Georg Reimer, Berlin, 1892.
3. Phemister, D. B.: General pathology of bone in children. In *Abt's Pediatrics*, Ed. 5, p. 40. W. B. Saunders, Philadelphia, 1924.
4. Lamy, M. E.: Hereditary disorders of bones—an overview. In *The Clinical Delineation of Birth Defects, IV: Skeletal Dysplasias*, pp.8–16 Williams & Wilkins, Baltimore, 1969.
5. Bailey, J. A., II.: Elbow and other upper limb deformities in achondroplasia. Clin Orthop, 80: 75, 1971.
6. Dandy, W. E.: Hydrocephalus in chondrodystrophy. Bull Johns Hopkins Hosp, 32: 1, 1921.
7. Marie, P.: L'achondroplasie dans l'adolescent et l'age adult. Presse Med, 8: 17, 1900.
8. Rimoin, D. L., et al.: Endochondral ossification in achondroplastic dwarfism. N Engl J Med, 283: 728, 1970.
9. Cocchi, U.: In *Roentgen Diagnosis*, Vol. 1, H. R. Schinz (ed.). Grune & Stratton, New York, 1951.
10. Fairbank, H. A. T.: In *Diseases of Children*, Ed. 3, p. 865, A. E. Garrod, F. E. Batten and H. Thursfield (eds.). Williams & Wilkins, Baltimore, 1934.
11. Fairbank, T.: *An Atlas of General Affections of the Skeleton*. Williams & Wilkins, Baltimore, 1951.

12. Caffey, J.: Achondroplasia of the pelvis and lumbosacral spine; some radiographic features. AJR, *80;* 458, 1958.
13. Jeune, M., Carron, R., Beraud, C., and Loaec, Y.: Polychondrodystrophie avec blocage thoracique d'evolution fatale. Pediatrie, *9:* 390, 1954.
14. Jeune, M., Beraud, C., and Carron, R.: Dystrophic thoracique asphyxiante de caractere familial. Arch Fr Pediatr, *12:* 886, 1955.
15. Kaufmann, H. J.: *Rontgenbefunde am kindlichen Becken bei angeborenen Skelettaffektionen und chromosomalen Abberationen.* Georg Thieme Verlag, Stuttgart, 1964.
16. Maroteaux, P., and Savart, P.: La dystrophic thoracique asphyxiante; etude radiologique et rapports avec le syndrome d'Ellis et van Creveld. Ann Radiol, *7:* 332, 1964.
17. Pirnar, T., and Neuhauser, E. B. D.: Asphyxiating thoracic dystrophy of the newborn. AJR, *98:* 358, 1966.
18. Aegerter, E., and Kirkpatrick, J. A.: *Orthopedic Diseases.* W. B. Saunders, Philadelphia, 1964.
19. Silverman, F. N., and Brunner, S.: Errors in the diagnosis of achondroplasia. Acta Radiol, *6:* 305, 1967.
20. Rubin, P.: Achondroplasia versus pseudoachondroplasia. Radiol Clin North Am, *1:* 621, 1963.
21. Silverman, F. N.: A differential diagnosis of achondroplasia. Radiol Clin North Am, *6:* 223, 1968.
22. Lamy, M., and Maroteaux, P.: Le nanisme diastrophique. Presse Med, *68:* 1977, 1960.
23. Taybi, H.: Diastrophic dwarfism. Radiology, *80:* 1, 1963.
24. McKusick, V. A., Eldrige, R., and Hostetler, J. A.: Dwarfism in the Amish. II. Cartilage-hair hypoplasia. Bull Johns Hopkins Hosp., *116:* 285, 1965.
25. Lamy, M., and Maroteaux, P.: *Les chondrodystrophies genotypiques.* L'Expansion Scientifique Francaise, Paris, 1960.
26. Maroteaux, P., and Lamy, M.: Le diagnostic des namismes chondrodystrophiques chez les nouveauxnes. Arch Fr Pediatr, *25:* 241, 1968.
27. Ravenna, F.: Achondroplasie et chondrohypoplasie, contribution clinique. N Iconogr Salpertriere, *26:* 157, 1913.
28. Léri, A. and Linossier, M.: Hypochondroplasie hereditaire. Bull Soc Med Hosp Paris, *48:* 1780, 1924.
29. Kozlowski, K., and Zychowicz, C.: Hypochondroplasie (ein weitere Beitrag). Fortschr Geb Roentgenstr Nuklearmed *101:* 531, 1964.
30. Rubin, P.: *Dynamic Classification of Bone Dysplasias.* Year Book, Chicago, 1964.
31. Maroteaux, P., Spranger, J., and Wiedemann, H.: Der metatropische Zwergwuchs. Arch Kinderheilk, *173:* 211, 1966.
32. Lachman, R. S., Rimoin, D. L., Hollister, D. W., Dorst, J. P., Siggers, D. C., McAlister, W., Kaufman, R. L., and Langer, L. O.: The Kniest syndrome. AJR, *123:* 805, 1975.
33. Kniest, W.: Zur Abgrenzung der Dysostosis enchondralis von der Chondrodystrophie. Z Kinderheilkd, *70:* 633, 1952.
34. Maroteaux, P., and Spranger, J.: La maladie de Kniest. Arch Fr Pediatr, *30:* 735, 1973.
35. Rimoin, D. L., Hollister, D. W., Silberberg, R., Lachman, R. S., McAlister, W., and Kaufman, R.: Kniest (Swiss cheese cartilage) syndrome; clinical, radiographic, histologic and ultrastructural studies. Clin Res, *21:* 296, 1973.
36. Siggers, D., Rimoin, D. L., Kaufman, R. L., Dorst, J. P., McKusick, V. A., Doty, S. B., Williams, B. R., Hollister, D. W., Silberberg, R., and Cranley, R. E.: Kniest syndrome in skeletal dysplasias. Birth Defects, *10:* 193, 1974.
37. Larose, J. H., and Gay, B. B., Jr.: Metatrophic dwarfism. AJR, *106:* 156, 1969.
38. Rimoin, D. L., Hollister, D. W., Siggers, D., Silberberg, R., Lachman, R. S., McAlister, W., Kaufman, R., McKusick, V. A., and Dorst, J.: Clinical, radiographic, histologic and ultrastructural definition of Kniest syndrome. Pediatr Res, *7:* 348, 1973.
39. Rimoin, D. L., Hollister, D. W., Lachman, R. S., Kaufman, R. L., McAlister, W. H., Rosenthal, R. E., and Hughes, G. N. F.: Histological studies in chondrodystrophies. Proc 1972 Birth Defects Original Article Series (to be published).
40. Rimoin, D. L.: Histopathology and ultrastructure of cartilage in chondrodystrophies in skeletal dysplasia. Birth Defects, *10:* 1, 1974.
41. McKusick, V., Egeland, J. A., Eldrige, R., and Krusen, D. E.: Dwarfism in the Amish. I. The Ellisvan Creveld syndrome. Bull Johns Hopkins Hosp, *115:* 306, 1964.
42. Kaufmann, H. J.: The pelvis in the Ellis-van Creveld syndrome. Ann Radiol, *8:* 146, 1965.
43. Hanissian, A. S., Riggs, W. W., Jr., and Thomas, D. A.: Infantile thoracic dystrophy—a variant of Ellisvan Creveld syndrome. J Pediatr, *71:* 855, 1967.
44. Conradi, E.: Vorzeitiges Auftreten von Knochen und eigenartigen Verkalkungskernen bei chondrodystrophia fotalis hypoplastica. Histologische un Rontgenuntersuchungen. Jahrb Kinderheilkd, *80:* 16, 1914.
45. Silverman, F. N.: Dysplasies epiphysaires; entite proteiforme. Ann Radiol, *4:* 833, 1961.
46. Jansen, M.: Uber atypische condrodystrophie (Achondroplasie) und uber eine noch nicht beschriebene angeborene Wachstumstorung des Knochensystems: Metaphysare dysostosis. Z Orthop Chir, *61:* 253, 1934.
47. Schmid, F.: Beitrag zur dysostosis enchondralis metaphysaria. Monatsschr Kinderheilkd, *97:* 393, 1949.
48. Lenk, R.: Hereditary metaphyseal dysostosis. AJR, *76:* 569, 1956.
49. Maroteaux, P., Lamy, M., and Robert, J. M.: Le nanisme thanatophore. Presse Med, *75:* 2519, 1967.

Thanatophoric Dwarfism

1. Maroteaux, P., Lamy, M., and Robert, J. M.: Le nanisme thanatophore. Presse Med, *75:* 2519, 1967.
2. Langer, L. O., Jr., Spranger, J. W., Greinacher, I., and Herdman, R. C.: Thanatophoric dwarfism. Radiology *92:* 285, 1969.
3. Keats, T. E., Riddervold, H. O., and Michaelis, L. L.: Thanatophoric dwarfism. AJR, *108:* 473, 1970.
4. Cremin, B. J., and Shaff, M. I.: Ultrasonic diagnosis of thanatophoric dwarfism in utero. Radiology, *124:* 479, 1977.
5. Kaufmann, J. H.: In utero diagnosis of skeletal dysplasias. Significance of prenatal diagnosis of severe forms incompatible with life. Presented at the 13th meeting of the European Society of Pediatric Radiology, Stockholm, May 19–22, 1976.
6. Cronberg, N. E.: A case of chondrodystrophia foetalis, diagnosed by x-ray examination before delivery. Acta Obstet Gynaecol Scand, *13:* 275, 1933.
7. Wichtl, O.: Zur pranatalen diagnose der chondrodystrophie (achondroplasie). Fortschr Geb Rontgenstr Nuklearmed, *103:* 114, 1965.
8. Zellweger, H., and Taylor B.: Genetic aspects of achondroplasia. J Lancet, *85:* 8, 1965.
9. Holtermueller, K., and Wiedemann, H. R.: Kleeblattschädel syndrome. Med Monatsschr, *14:* 439, 1960.
10. Angle, C. R., McIntyre, M. S., and Moore, R. C.: Cloverleaf skull; kleeblattschadel-deformity syndrome. Am J Dis Child, *114:* 198, 1967.
11. Bloomfield, J. A.: Cloverleaf skull and thanatophoric dwarfism. Australas Radiol, *14:* 429, 1970.

12. Liebaldt, G.: "Das Kleeblatt"-Schadel-Syndrom, al: Beitrag zur formalen Genese der Entwicklungsstorungen des Schadeldaches. Ergeb Allgem Pathol, 45: 23, 1964.
13. Partington, M. W., Gonzales-Crussi, F., Khakee, S. G., et al.: Cloverleaf skull and thanatophoric dwarfism. Report of four cases, two in the same sibship. Arch Dis Child, 46: 656, 1971.
14. Young, R. S., Pochaczevsky, R., Leonidas, J. C., Wexler, I. B., and Ratner, H.: Thanatophoric dwarfism and cloverleaf skull ("Kleeblattschadel"). Radiology, 106: 401, 1973.

Familial Asphyxiating Thoracic Dystrophy

1. Jeune, M., Carron, R., Beraud, C., and Loaec, Y.: Polychondrodystrophie avec blocage thoracique d'evolution fatale. Pediatrie, 9: 390, 1954.
2. Jeune, M., Beraud, C., and Carron, R.: Dystrophie thoracique asphyxiante de caractere familial. Arch Fr Pediatr, 12: 886, 1955.
3. Langer, L. O., Jr.: The thoracic-pelvic-phalangeal dystrophy. In *The Clinical Delineation of Birth Defects, IV: Skeletal Dysplasias*, pp.55-64 Williams & Wilkins, Baltimore, 1969.
4. Giedion, A.: Das tricho-rhino-phalangeal syndrom. Helv Paediatr Acta, 21: 475, 1966.

Diastrophic Dwarfism

1. Rubin, P. *Dynamic Classification of Bone Dysplasias.* Year Book, Chicago, 1969.
2. Lamy, M., and Maroteaux, P.: Le nanisme diastrophique. Presse Med, 68: 1977, 1960.
3. Ford, N., Silverman, F. N., and Kozlowski, K.: Spondyloepiphyseal dysplasia (pseudo-achondroplastic type). AJR, 86: 462, 1961.
4. Kaplan, M., Sauvegrain, J., Hayem, F., Drapeau, P., Maugey, F., and Boulle, J.: Etude d'um nouveau cas de nanisme diastrophique. Arch Pediatr, 18: 981, 1961.
5. Langer, L. O., Jr.: Diastrophique dwarfism in early infancy. AJR, 93: 399, 1965.
6. Mouledous, P.: un nouveau cas de nanisme diastrophique. Toulouse Med, 63: 617, 1962.
7. Niemann, N., Pierson, M., Manciaux, M., and Sapelier, J.: A propos d'un nouveau cas de nanisme diastrophique. Arch Fr Pediatr, 21: 957, 1964.
8. Paul, S. S., Rao, P. L., Mullick, P., and Saigal, S.: Diastrophic dwarfism; a little known disease entity. Clin Pediatr, 4: 95, 1965.
9. Salle, B., Picot, C., Vauzelle, J. L., Deffrenne, P., Monnet, P., Francois, R., and Robert, J. M.: Le nanisme diastrophique; a propos de trois observations chez le nouveau-ne. Pediatrie, 21: 311, 1966.
10. Stover, C. N., Hayes, J. T., and Holt, J. F.: Diastrophic dwarfism. AJR, 89: 914, 1963.
11. Taybi, H.: Diastrophic dwarfism. Radiology, 80: 1, 1963.
12. Amuzo, S. J.: Diastrophic dwarfism. J Bone Joint Surg, 50; 113, 1968.
13. Wilson, D. W., et al.: Diastrophic dwarfism. Arch Dis Child, 44: 48, 1969.
14. Motta, C., and Pizzetti, M.: Su di un raro caso di osteochondrodistrofia del accrescimento con note acondroplastiche associato ad iperelasticita cutanea e lassita legamentosa. Ortop Traumatol, 26: 269, 1958.
15. Herring, J. A.: The spinal disorders in diastrophic dwarfism. J Bone Joint Surg, 60A: 177, 1978.

Hypochondroplasia

1. Revenna, F.: Achondroplasie et chondrophypoplasie; contribution clinique. N Iconogr Salpertiriere, 26: 157, 1913.
2. Léri, A., and Linossier, M.: Hypochondroplasie hereditaire. Bull Mem Soc Med Hop Paris, 48: 1780, 1924.
3. Kozlowski, K.: Hypochondroplasia. Pol Rev Radiol Nucl Med, 29: 450, 1965.
4. Book, J. A.: A clinical and genetical study of disturbed skeletal growth (chondrohypoplasia). Hereditas, 36: 161, 1950.
5. Langer, L. O., Jr.: Mesomelic dwarfism of the hypoplastic ulna, fibula, mandible type. Radiology, 89: 654, 1967.
6. Walker, B. A., Murdoch, L., McKusick, V. A., Langer, L. O., and Beals, R. K.: Hypochondroplasia. Am J Dis Child, 122: 95, 1971.
7. Murdoch, J. L., Walker, B. A., and J. G. Hall, et al.: Achondroplasia; a genetic and statistical survey. Ann Hum Genet, 33: 227, 1970.
8. Penrose, L. S.: Parental age and mutation. Lancet, 2: 312, 1955.
9. Beals, R. K.: Hypochondroplasia; a report of five kindreds. J Bone Joint Surg, 51A: 728, 1969.
10. Maroteaux, P., and Lamy, M.: Achondroplasia in man and animals. Clin Orthop, 33: 91, 1964.
11. Rubin, P.: *Dynamic Classification of Bone Dysplasias.* Year Book, Chicago, 1964.
12. Langer, L. O., Spranger, J. W. et al.: Thanatophoric dwarfism. Radiology, 92: 285, 1969.

Mesomelic Dwarfism

1. Blockey, N. J., and Lawrie, J. H.: Unusual symmetrical distal limb deformity in siblings. J Bone Joint Surg, 45B: 745, 1963.
2. Langer, L. O., Jr.: Mesomelic dwarfism of the hypoplastic ulna, fibula, mandible type. Radiology, 89: 654, 1967.
3. Langer, L. O., Jr.: Dyschondrosteosis, a hereditable bone dysplasia with characteristic roentgenographic features. AJR, 95: 178, 1965.
4. Lamy, M., and Bienenfeld, C.: La Dyschondrosteose. In *Symposium on International Medical Genetics*, pp. 153-164. Gregorio Mendel, Rome, 1954.
5. Nievergelt, K.: Positiver vaterschaftsnachweis auf grund erblicher Missbildungen der Extremitaten. Arch Julius Klaus-Stift Verebungsforsch, 19: 157, 1944.
6. Solonen, K. A., and Sulamaa, M.: Nievergelt syndrome and its treatment; a case report. Ann Chir Gynaecol Fenn, 47: 142, 1958.
7. McKusick, V. A., Egeland, J. A., Eldridge, R., and Krusen, D. E.: Dwarfism in the Amish. I. The Ellis-van Creveld syndrome. Bull Johns Hopkins Hosp, 115: 306, 1964.
8. Caffey, J.: Achondroplasia of pelvis and lumbosacral spine; some roentgenographic features. AJR, 80: 449, 1958.
9. Langer, L. O., Jr., Baumann, P. A., and Gorlin, R. J.; Achondroplasia. AJR, 100: 12, 1967.

Achondrogenesis

1. Freire-Maia, N., and Lenz, W. D.: Discussion (Differential diagnosis between thanatophoric dwarfism and achondrogenesis). In *The Clinical Delineation of Birth Defects, IV: Skeletal Dysplasias*, pp. 14. Williams & Wilkins, Baltimore, 1969.
2. Parenti, G. C.: La anosteogenesi (un varieta dell osteogenesi imperfecta). Pathologica, 28: 447, 1936.
3. Fraccaro, M.: Contributo allo studio delle malattie del mesenchima osteopoietico; l'acondrogenesi. Folia Hered Pathol, 1: 190, 1952.
4. Grebe, H.: Die achondrogenesis; ein einfach rezessives erbnerkmal. Folia Hered Pthol, 2: 23, 1952.
5. Grebe, H.: *Chondrodysplasie*, pp. 300-303. Gregorio Mandel, Rome, 1955.

6. Quelce-Salgado, A.: A new type of dwarfism with various bone aplasias and hypoplasias of the extremities. Acta Genet, 14: 63, 1964.
7. Langer, L. O., Spranger, J. W., Greinacher, I., and Herdman, R. C.: Thanatophoric dwarfism; a condition confused with achondroplasia in the neonate, with brief comments on achondrogenesis and homozygous achondroplasia. Radiology, 92: 285, 1969.

Ellis-van Creveld Syndrome

1. McIntosh, R.: *Diseases of Infancy and Childhood*, Ed. 10, p. 362. Holt & Howland, New York, 1933.
2. Ellis, R. W. B., and Creveld, S. van: A syndrome characterized by ectodermal dysplasia, polydactyly, chondrodysplasia and congenital morbus cordis. Arch Dis Child, 15: 65, 1940.
3. Metrakos, J. D., and Fraser, F. C.: Evidence for a hereditary factor in chondro-ectodermal dysplasia (Ellis-van Creveld syndrome). Am J Hum Genet, 6: 260, 1954.
4. Dayer, L.: Le syndrome d'Ellis-van Crevald. Une forme de dysplasie chondroectodermique. Discussion des limites de l'affection. Bibl Paediatr. (Also Suppl. to Annales Paediatrica Karger) Bale-New York, 1960.
5. Ellis, R. W. B., and Andrew, J. D.: Chondroectodermal dysplasia. J Bone Joint Surg, 44B: 626, 1962.
6. McKusick, V. A., Egeland, J. A., Eldridge, R., and Krusen, D. E.: Dwarfism in the Amish. I. The Ellis-van Creveld syndrome. Bull Johns Hopkins Hosp, 115: 306, 1964.
7. Bodenhoff, J., and Gorlin, R.: Natal and neonatal teeth; folklore and fact. Pediatrics, 32: 1087, 1963.
8. Tubbs, F. E., Crevasse, L., and Green, L. R., Jr.: Congenital heart disease in an adult with the Ellis-van Creveld syndrome. N Engl J Med, 57: 829, 1962.
9. Gitkis, F. L.: Single atrium and the Ellis-van Creveld syndrome. Report of 2 cases in siblings. Acta Paediatr, 47: 142, 1958.
10. Keizer, D. R. R., and Schilder, J. H.: Ectodermal dysplasia, achondroplasia, and congenital morbus cordis. Am J Dis Child, 82: 345, 1951.
11. Caffey, J.: Chondroectodermal dysplasia (Ellis-van Creveld disease); report of 3 cases. AJR, 68: 875, 1952.
12. Feingold, M.: Ellis-van Creveld syndrome. Clin Pediatr, 5: 431, 1966.
13. Darling, R. C., Rothney, E. W., and Craig, M.D.: Total pulmonary venous drainage into the right side of the heart. Lab Invest, 6: 44, 1957.
14. Husson, G. S., and Parkman, P.: Chondroectodermal dysplasia (Ellis-van Creveld syndrome) with a complex cardiac malformation. Pediatrics, 28: 285, 1961.
15. Smith, H. L., and Hand, A. M.: Chondroectodermal dysplasia (Ellis-van Creveld syndrome). Pediatrics, 21: 298, 1958.

Chondrodystrophia Calcificans Congenita

1. Hunermann, C.: Chondrodystrophia calcificans congenital als abortive form der chondrodystrophie. Z Kinderheilkd, 51: 1, 1931.
2. Bateman, D.: Two cases, and specimens from third case, of punctate epiphyseal dysplasia. Proc R Soc Med, 29: 745, 1936.
3. Fairbank, H. A. T.: Generalized diseases of skeleton. Proc R Soc Med, 28: 1611, 1935.
4. Conradi, E.: Vorzeitiges Auftreten von Knochen-und eigenartigen Verkalkungskernen bei chondrodystrophia fotalis hypoplastica. Histologische und Rontgenuntersuchungen. Jahrb Kinderheilkd, 80: 86, 1914.
5. Fraser, F. C., and Scriver, J. B.: Hereditary factor in chondrodystrophia calcificans congenita. N Engl J Med, 250: 272, 1954.
6. Maitland, D. G.: Punctate epiphyseal dysplasia occurring in 2 members of same family. Br J Radiol, 12: 91, 1939.
7. Raap, G.: Chondrodystrophia calcificans congenita. AJR, 49: 77, 1943.
8. Fairbank, T.: *An Atlas of General Affectations of the Skeleton*, pp. 102–105. Williams & Wilkins, Baltimore, 1951.
9. Brogdon, B. G., and Crow, N. E.: Chondrodystrophia calcificans congenita. AJR, 80: 443, 1958.
10. Cocchi, U.: *Roentgen Diagnosis, Vol. 1*, H. R. Schinz (ed.). Grune & Stratton, New York, 1951.
11. Heselson, N. G., Cremin, B. J., and Beighton, P.: Lethal chondrodysplasia punctata. Clin Radiol, 29: 679, 1978.
12. Spranger, J. W., Langer, L. O., and Wiedemann, H. R.: Bone Dysplasias: An Atlas of *Constitutional Disorders of Skeletal Development*, pp. 3–9. W. B. Saunders, Philadelphia, 1974.
13. Sheffield, L. J., Danks, D. M., Mayne, V., and Hutchinson, L. A.: Chondrodysplasia punctata—23 cases of a mild and relatively common variety. J Pediatr, 86: 916, 1976.
14. Happle, R., Matthias, H-H., and Macher, E.: Sex linked chondrodysplasia punctata? Clin Genet, 11: 73, 1977.
15. Cremin, B. J., and Beighton, P.: Dwarfism in the newborn; the nomenclature and genetic significance. Br J Radiol, 47: 77, 1974.
16. Kaufmann, H. J., Mahoubi, S., Sprackman, T. J., Capitano, M. A., and Kirkpatrick, J.: Tracheal stenosis as a complication of chondrodysplasia punctata. Ann Radiol, 1: 203, 1976.
17. Afshani, E., and Girdany, B. R.: Atlanto-axial dislocation in chondrodysplasia punctata. Report of the findings in two brothers. Radiology, 102: 399, 1972.
18. Becker, M.: Personal communication, 1979.
19. Harris, H. A.: *Bone Growth in Health and Disease*. Oxford University Press, London, 1933.
20. Meschen, I.: *Roentgen Signs in Clinical Diagnosis*, pp. 165–166. W. B. Saunders, Philadelphia, 1956.
21. Hassler, E., and Schallock, G.: Chondrodystrophia calcificans. Monatschr Kinderheilkd, 82: 133, 1940.
22. Jorup, S.: Fall von chondrodystrophia congenita calcificans. Acta Radiol, 25: 580, 1944.
23. Hilliard, C.: Case of chondro-osseous dystrophy with punctate epiphyseal dysplasia. Br J Radiol, 16: 144, 1943.
24. Silverman, F. N.: Discussion on the relation between stippled epiphyses and the multiplex form of epiphyseal dysplasia. In *The Clinical Delineation of Birth Defects, IV: Skeletal Dysplasias*, pp. 68–70. Williams & Wilkins, Baltimore, 1969.
25. Reilly, W. A., and Smyth, F. S.: Stippled epiphyses with congenital hypothyroidism (cretinoid epiphyseal dysgenesis). AJR, 40: 675, 1938.

Holt-Oram Syndrome

1. Graham, J. B.: Hereditary chronic kidney disease; an alternative to partial sex linkage in the Utah kindred. Am J Hum Genet, 11: 333, 1959.
2. Holmes, L. B.: Congenital heart disease and upper extremity deformities. N Engl J Med, 272: 437, 1965.
3. Lewis, K. B., Bruce, R. S., and Motulsky, A. G.: Upper limb cardiovascular syndrome; an autosomal dominant genetic effect on embryogenesis, Circulation, 30 (supp 111): 113 (abstr.), 1964.
4. McKusick, V. A.: *Medical Genetics*, p. 426. C. V. Mosby, St. Louis, 1958–1960.
5. Pruzanski, W.: Familial congenital malformation of the heart and upper limbs; a syndrome of Holt-Oram. Cardiologia, 45: 21, 1964.
6. Zetterqvist, P.: The syndrome of familial atrial septal

defect, heart arrhythmia, and hand malformation (Holt-Oram) in mother and son. Acta Paediatr Scand, 52: 115, 1963.
7. Gall, J. C., Jr., Stern, A. M., Cohen, M., Adams, M., and Davidson, R. T.: Holt-Oran syndrome; clinical and genetic study of a large family. Am J Hum Genet, 18: 187, 1966.
8. Silverman, et al.: Holt-Oram syndrome—the long and the short of it. Am J Cardiol, 25: 11, 1970.

Metaphyseal Dysostosis

1. McKusich, V.: *Heritable Disorders of Connective Tissue*, Ed. 3. C. V. Mosby, St. Louis, 1966.
2. Stickler, G., Maher, F., Hunt, J., Burke, E., and Rosevear, J.: Familial bone disease resembling rickets (hereditary metaphyseal dysostosis). Pediatrics, 29: 996, 1962.
3. Miller, S., and Paul, L.: Roentgen observations in familial metaphyseal dysostosis. Radiology, 83: 665, 1964.
4. Spahr, A., and Spahr-Hartmann, I.: Dysostose metaphysaire familiale; etude de 4 cas dans une fratrie. Helv Paediatr Acta, 16: 836, 1961.
5. Jansen, M.: Uber atypische chondrodystrophie (achondroplasie) und uber eine noch nich beschriebene angeborene Wachstumsstorung des Knochensystems; Metaphysare dysostosis. Z Orthop Chir, 61: 255, 1934.
6. Gram, P. B., Fleming, J. L., Frame, B., and Fine, G.: Metaphyseal chondrodysplasia of Jansen. J Bone Joint Surg, 41A: 951, 1959.
7. Cameron, J. A. P., Young, W. B., and Sissons, H. A.: Metaphyseal dysostosis; report of a case. J Bone Joint Surg, 36B: 622, 1954.
8. Holmann, G., Hackenbroch, M., and Lindemann, K.: *Handbuch der Orthopadie, Vol. 1*. Georg Thieme Verlag, Stuttgart, 1957.
9. Muller, G. M., and Sissons, H. A.: A case of renal rickets simulating metaphyseal dysostosis. J Bone Joint Surg, 33B: 231, 1951.
10. Schmid, F.: Beitrag zur dysostosis enchodralis metaphysaria. Monatsschr Kinderheilkd, 97: 393, 1949.

Lymphopenic Agammaglobulinemia

1. Gatti, R. A., et al.: Hereditary lymphopenic agammaglobulinemia associated with a distinctive form of short-limbed dwarfism and ectodermal dysplasia. J Pediatr, 75: 675, 684, 1969.
2. McKusick, V. A., and Cross, H. E.: Ataxiatelangiectasia and Swiss-type agammaglobulinemia. JAMA, 195: 739, 1966.
3. Davis, J. A.: A case of Swiss-type agammaglobulinemia and achondroplasia. Br Med J, 2: 1371, 1966.
4. Fulginiti, V. A., Hathaway, W. E., Pearlman, D. S., and Kempe, C. H.: Agammaglobulinemia and achondroplasia. Br Med J, 2: 242, 1967.
5. Fulginiti, V. A., et al: Progressive vaccinia in immunologically deficient individuals. In *Immunologic Deficiency Diseases in Man*, D. Bergsma and R. A. Good (eds.), pp. 129-151. Williams & Wilkins, Baltimore, 1968.
6. Alexander, W. J., and Dunbar, J. S.: Unusual bone changes in thymic alymphoplasia. Ann Radiol, 11: 389, 1968.
7. Kozlowski, K.: Hypochondroplasia. Rev Radiol, 29: 450, 1965.
8. Langer, L. O.: Diastrophic dwarfism in early infancy. AJR, 93: 399, 1965.
9. Fraser, F. C., and Scrurer, T. B.: A hereditary factor in chondrodystrophic calcificans congenita. N Engl Med J, 25: 272, 1954.
10. McKusick, V. A., Egeland, J. A., Eldridge, R. and Krusen, D. E.: Dwarfism in the Amish. I. The Ellis-van Creveld syndrome. Bull Johns Hopkins Hosp, 115: 306, 1964.
11. Langer, L. O.: Thoracic-pelvic-phalangeal dystrophy; asphyxiating thoracic dystrophy of the newborn, infantile thoracic dystrophy. Radiology, 91: 3, 1968.
12. Langer, L. O.: Mesomelic dwarfism of the hypoplastic ulna, fibula, mandible type. Radiology, 89: 654, 1967.
13. McKusick, V. A., Eldridge, R., Hostetler, J. A., Ruangivit, V., and Egeland, J. A.: Dwarfism in the Amish. II. Cartilage-hair hypoplasia. Bull Johns Hopkins Hosp, 116: 285, 1965.
14. Spranger, J., and Wiedemann, H. R.: Dysplasia spondyloepiphysaria congenita. Helv Paediatr Acta, 21: 598, 1966.
15. langer, L. O., Spranger, J. W., Herdman, R. C., and Greinacher, I.: Thanatophoric dwarfism; a condition confused with achondroplasia in the neonate with brief comments on achondrogenesis and homozygous achondroplasia. Radiology, 92: 285, 1969.
16. Maroteaux, P., Spranger, J., and Wiedemann, H. R.: Der metatrophische Zwergwuchs. Arch Kinderheilkd, 173: 211, 1966.
17. Fairbank, T.: *Atlas of General Affection of the Skeleton*. Williams & Wilkins, Baltimore, 1951.

Pyle Disease

1. Pyle, E.: A case of unusual bone development. J Bone Joint Surg, 13A: 874, 1931.
2. Hermel, M. B., Gershon-Cohen, J., and Jones, D. T.: Familial metaphyseal dysplasia. AJR, 70: 413, 1953.
3. Bakwin, H., and Krida, R.: Familial metaphyseal dysplasia. Am J Dis Child, 53: 1521, 1937.
4. Mori, P. A., and Holt, J. F.: Cranial manifestations of familial metaphyseal dysplasia. Radiology, 66: 335, 1956.
5. Caffey, J.: *Pediatric X-ray Diagnosis*. Year Book, Chicago, 1961.

Dyschondrosteosis

1. Léri, A., and Weill, J.: Une affection congenitale et symmetrique du developpement osseux; la dyschondrosteose. Bull Mem Soc Med Hop Paris, 53: 1491, 1929.
2. Léri, A., Flandin, C., and Arnaudet, L.: Dyschondrosteose; presentation d'un nouveau cas. Bull Mem Soc Med Hop Paris, 54: 385, 1930.
3. Gareiso, A., Pellarano, J. C., and Schere, S.: Sobre un caso de dischondrosteosis. Rev Med Lat Am, 22: 1937.
4. Kaplan, M., Guy, E., and Cantagrel, A.: Dyschondrosteose familiale. Presse Med. 59: 1723, 1951.
5. Brocher, J. E., and Klein, D.: Kie dyschondrosteose. Fortschr Geb Rontgenstr Nuklearmed, 96: 496, 1962.
6. Langer, L. O.: Dyschondrosteosis, a hereditable bone dysplasia with characteristic roentgenographic features. AJR, 95: 178, 1965.
7. Dannenberg, M., Anton, J. I., and Spiegel, M. B.: Madelung's deformity; consideration of its roentgen diagnostic criteria. AJR, 42: 671, 1939.
8. Bertolotti, M.: Nanisme familial par aplasie chondrale systmatisee; mesomelie et brachymelie metapodiale symetrique. Presse Med, 21, 165, 1913.
9. Lamy, M., and Maroteaux, P.: Les chondystrophies genotypiques. L'Expansion Scientific Francaise: 33-39, 1960.
10. Lamy, M., and Bienefeld, Ch.: La dyschondrosteose. Acta Genet, 3: 153, 1954.

Enchondromatosis

1. Ollier, L.: De la dyschondroplasie. Bull Soc Chir Lyon, 3: 22, 1900.

2. Willis, R. A.: *Pathology of Tumours*, p. 673. Butterworth, London, 1948.
3. Voorhoeve, N.: L'image radiologique non encore decrite d'une anomalie du squelette; ses rapports avec la dyschondroplasie et l'osteopathia condensans disseminata. Acta Radiol, *3:* 407, 1924.
4. Maffucci, A.: Di un caso encondroma ed angioma multiplo; contribuzione alla genesi embrionale dei tumori. Movimento Medico-Chirurgico, 2s, *3:* 399, 1881.
5. Bean, W.: Dyschondroplasia and hemangiomata (Maffucci's syndrome). Arch Intern Med, *102:* 544, 1958.
6. Stocks, P., and Barrington, A.: Hereditary disorders of bone development. I. Diaphysial aclasis (multiple exostoses), multiple enchondromata, cleido-cranial dysostosis. In *Treasury of Human Inheritance, Vol. 3.* Cambridge University Press, London, 1925.
7. Fairbank, T.: *An Atlas of General Affections of the Skeleton.* Williams & Wilkins, Baltimore, 1951.
8. Hunter, D., and Wiles, P.: Dyschondroplasia (Ollier's disease): with report of case. Br J Surg, *22:* 507, 1935.
9. Nielson, J. L., Jr.: Ollier's disease; report of the first case with involvement of the optic nerve. Bull Los Angeles Neurol Soc, *6:* 104, 1941.
10. Jaffe, H. L.: *Tumors and Tumorous Conditions of the Bones and Joints.* Lea & Febiger, Philadelphia, 1958.

Maffucci Syndrome

1. Maffucci, A.: Di un caso encondroma ed angioma multiplo; contribuzione alla genesi embrionale dei tumori. Movimento Medico-Chirurgico, 2s, *3:* 399, 1881.
2. Cook, P. L., and Evans, P. G.: Chondrosarcoma of the skull in Maffucci's syndrome. Br J Radiol, *50:* 833, 1977.
3. Lewis, J., and Ketcham, A. S.: Maffucci's syndrome. Functional and neoplastic significance. J Bone Joint Surg, *55A:* 1465, 1973.
4. Boinet, M. E.: Enchondrose rachitiforme. Arch Gen Med, *194:* 2689, 1904.
5. Strang, C., and Rannie, I.: Dyschondroplasia with haemangiomata (Maffucci's syndrome). J Bone Joint Surg, *32B:* 376, 1950.

Laurence-Moon-Biedl-Bardet Syndrome

1. Reilly, W. A., and Lisser, H.: Laurence-Moon-Biedl syndrome. Endocrinology, *16:* 337, 1932.
2. Warkany, J., and Weaver, T. S.: Heredofamilial deviation. II. Enlarged parietal foramens combined with obesity, hypogenitalism, microphthalmos and mental retardation. Am J Dis Child, *60:* 1147, 1940.
3. Levy, M., Lotom, M., and Fried, A.: The Laurence-Moon-Biedel-Bardt syndrome; report of 3 cases in a Jewish Yemenite family. J Bone Joint Surg, *52B:* 318, 1970.
4. Lisser, H., and Escamilla, R. F.: *Atlas of Clinical Endocrinology*, p. 417. C. V. Mosby, St. Louis, 1957.
5. Bowen, P., Ferguson-Smith, M. A., Mosier, D., Lee, C. S. N., and Butler, H. G.: The Laurence-Moon syndrome; association with hypogonadotrophic hypogonadism and sex-chromosome aneuploidy. Arch Intern Med, *116:* 598, 1965.

Introduction to Spondyloepiphyseal Dysplasias

1. Maroteaux, P.: Spondyloepiphyseal dysplasias and metatropic dwarfism. In *The Clinical Delineation of Birth Defects, IV:* Skeletal Dysplasias, pp. 35–44. Williams & Wilkins, Baltimore, 1969.
2. Brailsford, J. F.: *The Radiology of Bones and Joints*, Vol. 1, Ed. 4. Williams & Wilkins, Baltimore, 1948.
3. Thiemann, H.: Juvenile epiphysenstorungen, idiopatische erkrankung der epiphysenknorpel der fingerphalangen. Fortschr Geb Roentgenstr, *14:* 79, 1909.
4. Spranger, J., and Wiedemann, H. R.: Dysplasia spondyloepiphysaria congenita. Helv Paediatr Acta, *21:* 598, 1966.
5. Kozlowski, K., Maroteaux, P., and Spranger, J.: La dysostose spondylo-metaphysaire. Presse Med, *75:* 2769, 1967.
6. Silfverskiöld, N.: Sur la question de l'achondroplasie atypique et de sa forme peripherique. Acta Radiol Scand, *5:* 223, 1926.
7. Léri, A.: Dystrophie osseuse généralisée congénitale et héréditaire; la pléonostéose familiale. Presse Med, *30:* 13, 1922.

Spondyloepiphyseal Dysplasia Congenita

1. Spranger, J., and Wiedemann, H. R.: Dysplasia spondyloepiphysaria congenita. Helv Paediatr Acta, *21:* 598, 1966.
2. Spranger, J., and Langer, L. O., Jr.: Spondyloepiphyseal dysplasia congenita. Radiology, *94:* 313, 1970.
3. Bach, C., et al.: Dysplasie spondyloepiphysaire congenitale avec anomalies multiples. Arch Fr Pediatr, *24:* 23, 1967.
4. Fraser, G. R., and Friedmann, A. J.: *The Causes of Blindness in Childhood.* Johns Hopkins Press, Baltimore, 1967.
5. Uhlig, H.: Dysostosis enchondralis-typ bartenwerfer. Arch Kinderheilkd, *148:* 22, 1954.
6. Barbe, P.: Un cas de maladie de Morquio. Arch Fr Pediatr, *12:* 202, 1955.
7. Pisani, G., and Schrurmacher, E.: Le osteochondrodistrofie sistemiche dellaccrescimento; a propositio di un caso disostosi encondrale politopa tipo Morquio. Minerva Med, *48:* 2028, 1957.
8. Kozlowski, K., Bittel, D., and Budzinska, A.: Spondyloepiphyseal dysplasia congenita. Ann Radiol, *11:* 367, 1968.
9. Roaf, R., Longmore, J. B., and Forrester, R. M.: A childhood syndrome of bone dysplasia, retinal detachment and deafness. Dev Med Child Neurol, *9:* 464, 1967.
10. Langer, L. O., Jr., and Carey, L. S.: The roentgenographic features of the KS mucopolysaccharidosis of Morquio (Morquio-Brailsford's disease). AJR, *97:* 1, 1966.
11. Brailsford, J. F.: *The Radiology of Bones and Joints*, Ed. 5. Williams & Wilkins, Baltimore, 1953.
12. Langer, L. O., Jr.: Spondyloepiphyseal dysplasia tarda; hereditary chondrodysplasia with characteristic vertebral configuration in the adult. Radiology, *82:* 833, 1964.
13. Maroteaux, P., Spranger, J., and Wiedemann, H. R.: Der metatropische Zwergwuchs. Arch Kinderheilkd, *173:* 211, 1966.

Spondyloepiphyseal Dysplasia Tarda

1. Maroteaux, P., Lamy, M., and Bernard.: La dysplasie spondyloepiphysaire tradive; description clinique et radiologique. Presse Med, *65:* 1205, 1957.
2. Langer, L. O., Jr.: Spondyloepiphyseal dysplasia tarda. Radiology, *82:* 833, 1964.
3. Poker, N., Finby, N., and Archibald, R.: Spondyloepiphysial dysplasia tarda. Radiology, *85:* 474, 1965.

Epiphyseal Dysplasia Multiplex

1. Jansen, M.: Uber atypische chondrodystrophie (achondroplasie) und uber eine noch nicht beschriebene angeborene Wach-stumsstorung des Knochensystems; Metaphysare dysostosis. Z Orthop Chir, *61:* 253, 1934.
2. Fairbank, H. A. T.: Generalized diseases of skeleton. Proc R Soc Med, *28:* 1611, 1935.

3. Fairbank, H. A. T.: Dysplasia epiphysealis multiplex. Proc R Soc Med, 39: 315, 1946.
4. Fairbank, H. A. T.: Dysplasia epiphysialis multiplex. Br J Surg, 34: 225, 1947.
5. Fairbank, T.: *An Atlas of General Affections of the Skeleton.* pp. 91–101. Williams & Wilkins, Baltimore, 1951.
6. Maudsley, R. H.: Dysplasia epiphysialis multiplex, report of 14 cases in 3 families. J Bone Joint Surg, 37B: 228, 1955.
7. Shephard, E.: Multiple epiphysial dysplasia. J Bone Joint Surg 38B: 458, 1956.
8. Waugh, W.: Dysplasia epiphysialis multiplex in 3 sisters. J Bone Joint Surg, 34B: 82, 1952.
9. Leeds, N. E.: Epiphysial dysplasia multiplex. AJR, 84: 506, 1960.
10. Jackson, W. P. U., Hanelin, J., and Albright, F.: Metaphyseal dysplasia, epiphyseal dysplasia; diaphyseal dysplasia, and related conditions; multiple epiphyseal dysplasia; its relation to other disorders of epiphyseal development. Arch Intern Med, 94: 886, 1954.
11. Lane, J. W.: Roentgenographic manifestations of the cartilaginous dysplasias. Am J Med Sci, 240 (N.S.): 138/636, 1960.
12. Edeiken, J., et al.: To be published.
13. Silverman, F. N.: Discussion on the relation between stippled epiphyses and the multiplex form of epiphyseal dysplasia. In *The Clinical Delineation of Birth Defects, IV: Skeletal Dysplasias,* p. 68. Williams & Wilkins, Baltimore, 1969.

Peripheral Dysostosis

1. Brailsford, J. F.: *The Radiology of Bones and Joints,* Ed. 4. Williams & Wilkins, Baltimore, 1948.
2. Graziansky, W.: Die Kaschin-Becksche Krankheit im Rontgenbild. Fortschr Geb Rontgenstr, 50: 367, 1934.
3. Kohler, A.: *Borderlands of the Normal and Early Pathology in Skeletal Roentgenology,* T. Case (ed). Grune & Stratton, New York, 1956.
4. Laurent, Y., and Brombart, M.: Variation tres rare de l' ossification des phalanges des orteils. J Belge Radiol, 36: 102, 1953.
5. Liess, G.: A formige epiphysen an Handen und Fussen (periphere dysostosen). Fortschr Geb Rontgenstr Nuklearmed, 81: 178, 1954.
6. Razelli, A.: Eine seltene Ossifikationsanomalie an den Grundphalangen der Zehen (Zapfenepiphysen). Fortschr Geb Rontgenstr Nuklearmed, 76: 261, 1952.
7. Silfverskiöld, N.: Sur la question de l'achrondroplasie atypique et de sa forme peripherique. Acta Radiol, 5: 223, 1926.
8. Newcombe, D. S., and Keats, T. E.: Roentgenographic manifestations of hereditary peripheral dysostosis. AJR, 106: 178, 1969.
9. Singleton, E. B., Daeschner, W., and Teng, C. T.: Peripheral dysostosis. AJR, 84: 499, 1960.
10. Bell, J.: On brachydactyly and symphalangism. In *Treasury of Human Inheritance, Vol. 5.* Cambridge University Press, London, 1951.
11. Tetamy, S.: Genetic factors in hand malformations. Ph.D. Thesis, The Johns Hopkins University, 1966.
12. Giedion, A.: Das tricho-rhino-phalangeal syndrome. Helv Paediatr Acta, 21: 475, 1966.
13. Beals, R. K.: Tricho-rhino-phalangeal dysplasia. Report of a kindred. J Bone Joint Surg, 55A: 821, 1973.
14. Silver, H. K.: Asymmetry, short stature, and variations in sexual development; syndrome of congenital malformations. Am J Dis Child, 107: 495, 1964.
15. Biedmond, A.: Brachydactylie, nystagmus en cerebellaire ataxie als familiar syndroom. Ned Tijdschr Geneeskd, 78: 1423, 1934.
16. Albright, F., Bernett, C. H., Smith, P. H., and Parson, W.: Pseudo-hypoparathyroidism; example of Seabright-Bantam syndrome; report of 3 cases. Endocrinology, 30: 922, 1942.
17. Steinbach, H. L., and Young, D. A.: Roentgen appearance of pseudohypoparathyroidism (PH) and pseudo-pseudohypoparathyroidism (PPH); differentiation from other syndromes associated with short metacarpals, metatarsals, and phalanges. AJR, 97: 49, 1966.
18. Albright, F., Forbes, A. P., and Henneman, P. H.: Pseudo-pseudohypoparathyroidism. Trans Assoc Am Physicians, 65: 337, 1952.
19. Arkless, R., and Graham, C. B.: Unusual case of brachydactyly; peripheral dysostosis? pseudo-pseudohyperparathyroidism? cone epiphyses? AJR, 99: 724, 1967.
20. Stewart, J. S.: S. Klinefelter's syndrome; clinical and hormonal aspects. Q J Med, 28: 561, 1959.
21. Zabriskie, J., and Resiman, M.: Marchesani syndrome. J Pediatr, 52: 158, 1958.
22. Gorlin, R. J., and Psaume, J.: Orodigitofacial dysostosis; new syndrome. J. Pediatr, 61: 520, 1962.

Dysplasia Epiphysialis Hemimelica

1. Trevor, D.: Tarso-epiphyseal aclasis; congenital error of epiphyseal development. J Bone Joint Surg, 32B: 204, 1950.
2. Fairbank, T. J.: Dysplasia epiphysialis hemimelica (tarso-epiphysial aclasis). J Bone Joint Surg 38B: 237, 1956.
3. Keats, T. E.: Dysplasia epiphysialis hemimelica (tarsoepiphyseal aclasis). Radiology, 68: 558, 1957.
4. Mouchet, A., and Belot, L.: La tarsomegalie. J Radiol Electrol, 10: 289, 1926.
5. Ingelmans, P., and Lacheretz, M.: A propos d'un cas de chondrodystrophie epiphysaire. Rev Chir Orthop, 39: 242, 1953.
6. Donaldson, J. S., Sankey, H. H., Girdany, B. R., and W. F.: Osteochondroma of the distal femoral epiphysis. J Pediatr, 43: 212, 1953.
7. D'Angio, G. J., Ritvo, M., and Ulin, R.: Clinical and roentgen manifestations of tarso-epiphyseal aclasis; review of the manifestations of tarso-epiphyseal aclasis; review of the literature and report of an additional case. AJR, 74: 1068, 1955.
8. Hensinger, R. N., Cowell, H. R., Ramsey, P. L., and Leopold, R. G.: Familial dysplasia epiphysealis hemimelica, associated with chondromas and osteochondromas. J Bone Joint Surg, 56A: 1513, 1974.
9. Caffey, J.: *Pediatric X-ray Diagnosis,* Ed. 3. Year Book, Chicago, 1956.

Osteogenesis Imperfecta

1. Vrolik, W.: Tabulae ad illustrandam embryogenesin hominis et mammalium, tam naturalem quam abnormem. Amstelodami, G. M M. P. Londonck, 1843.
2. Ormerod, E. L.: An account of a case of mollities ossium. Br Med J, 2: 735, 1859.
3. Gurlt, E.: *Handbuch der Lehre von den Knochenbruchen,* Bd. I, pp. 147–154. Berlin, 1862–1865.
4. Lobstein, J. G. C. F. M.: *Lehrbuch der Pathologischen Anatomie,* Bd. II, p. 179. Stuttgart, 1835.
5. King, J. D., and Bobechko, W. P.: Osteogenesis imperfecta. An orthopaedic description and surgical review. J Bone Joint Surg, 53B: 72, 1971.
6. Falvo, K. A., Root, L., and Bullough, P. G.: Osteogenesis imperfecta; clinical evaluation and management. J Bone

Joint Surg, *56A:* 783, 1974.
7. Seedorff, K. S.: Osteogenesis imperfecta; a study of clinical features and hereditary based on 55 Danish families comprising 180 affected persons. Ejnar Munksgaard, Copenhagen, 1949.
8. McKusick, V. A.: *Heritable Disorders of Connective Tissue,* Ed. 2. C. V. Mosby, St. Louis, 1960.
9. Follis, R. H., Jr.: Histochemical studies on cartilage and bone. III. Osteogenesis imperfecta. Bull Johns Hopkins Hosp, *93:* 386, 1953.
10. Giordano, A.: Hereditary disease of the osteocartilaginous system; comparative morphological basis. Acta Genet, *7:* 155, 1957.
11. Falvo, K. A., and Bullough, P. G.: Osteogenesis imperfecta; a histometric analysis. J Bone Joint Surg, *55A:* 275, 1973.
12. Danelius, G.: Osteogenesis imperfecta intrauterin diagnostiziert. Arch Gynaekol, *154:* 160, 1933.
13. Fairbank, H. A. T.: *An Atlas of General Affections of the Skeleton.* Williams & Wilkins, Baltimore, 1951.
14. Aegerter, E., and Kirkpatrick, J. A., Jr.: *Orthopedic Diseases.* W. B. Saunders, Philadelphia, 1964.
15. Fulkerson, J. P., and Ozonoff, M. B.: Multiple symmetrical fractures of bone of unresolved etiology. AJR, *129:* 313, 1977.
16. Fleisch, H.: Role of nucleation and inhibition of calcification. Clin Orthop, *32:* 170, 1964.
17. Solomons, C. C., and Styner, J.: Osteogenesis imperfecta; effect of magnesium administration on pyrophosphate metabolism. Calcif Tissue Res, *3:* 318, 1969.

Erythrogenesis Imperfecta

1. Cathie, I. A. B.: Erythrogenesis imperfecta. Arch Dis Child, *25:* 313, 1950.
2. Diamond, L. K., and Blackfan, K. D.: Hypoplastic anemia. Am J Dis Child, *56:* 464, 1938.
3. Stransky, E.: Clinical hematology of infants; a regenerative anemia in young children. Z Kinderheilkd, *39:* 553, 1925.
4. Lescher, F. G., and Hubble, D.: A correlation of certain blood diseases with the hypothesis of bone-marrow deficiency or hypoplasia. Q J Med, *1:* 425, 1932.
5. Rinvik, R.: Two cases of idiopathic hypoplastic anemia in infants. Acta Paediatr (Upps) *28:* 304, 1941.
6. Hansen, H. G.: Uber die essentielle erythroblastopenie. Acta Haematol, *6:* 334, 1951.
7. Martoni, L., and Palesi, S.: Un caso di anemia tipo Blackfan e. Diamond. Clin Pediatr (Bologna), *34:* 753, 1952.
8. Seaman, A. J.: Cobalt and cortisone therapy of chronic erythrocytic hypoplasia. Am J Med, *13:* 99, 1952.
9. Donnelly, M.: A case of primary red-cell aplasia. Br Med J, *1:* 438, 1953.
10. Josephs, H. W.: Anaemia of infancy and early childhood. Medicine, *15:* 307, 1936.
11. Minagi, H., and Steinbach, H.: Roentgen appearance of anomalies associated with hypoplastic anemias of childhood; Fanconi's anemia and congenital hypoplastic anemia (erythrogenesis imperfecta). AJR, *97:* 100, 1968.
12. Burgert, E. O., Jr., Kennedy, R. L. J., and Pease, G. L.: Congenital hypoplastic anemia. Pediatrics, *12:* 218, 1954.
13. Altman, K. I., and Miller, G.: Disturbance of tryptophan metabolism in congenital hypoplastic anaemia. Nature, *172:* 868, 1953.

Fanconi Anemia

1. Smith, L. W.: Report on unusual case of aplastic anemia. Am J Dis Child, *17:* 174, 1919.
2. Fanconi, G.: Familiare infantile perniziosaartige anamie (perniziöses Blutbild und Konstitution). Jahrb Kinderheilkd, *117:* 257, 1927.
3. Uehlinger, E.: Konstitutionelle infantile (perniziosaartige) anemia. Klin Wochenschr, *8:* 1501, 1929.
4. Nilsson, L. R.: Chronic pancytopenia with multiple congenital abnormalities (Fanconi's anemia). Acta Paediatr, *49:* 518, 1960.
5. Minagi, H., and Steinbach, H. L.: Roentgen appearance of anomalies associated with hypoplastic anemias of childhood; Fanconi's anemia and congenital hypoplastic anemia (erythrogenesis imperfecta). AJR, *97:* 100, 1967.
6. Juhl, J. H., Wesenberg, R. L., and Gwinn, J. L.: Roentgenographic findings in Fanconi's anemia. Radiology, *89:* 646, 1967.
7. Gmyrek, D., Witkowski, R., Sylim-Rapoport, I., and Jacobasch, G.: Chromosomal aberrations and abnormalities of red-cell metabolism in a case of Fanconi's anaemia before and after development of leukaemia. Ger Med Mon *13:* 105, 1968.
8. Beautyman, W.: Case of Fanconi's anaemia. Arch Dis Child, *26:* 238, 1951.
9. McDonald, R., and Goldschmidt, B.: Pancytopenia with congenital defects (Fanconi's anemia). Arch Dis Child, *35:* 367, 1960.
10. Kunz, H. W.: Hypoplastic anemia with multiple congenital defects (Fanconi syndrome). Pediatrics, *10:* 286, 1952.
11. Garriga, S., and Crosby, W. H.: Incidence of leukemia in families of patients with hypoplasia of marrow. Blood, *14:* 1008, 1959.
12. Kessel, I., and Cohen, H.: Case of Fanconi's anaemia. S Afr Med J, *27:* 883, 1953.
13. Delage, J. M.: Pancytopenie familiale. Laval Med, *21:* 334, 1956.
14. Imerslund, O.: Hypoplastic anemia with multiple malformations (Fanconi syndrome). Nord Med, *50:* 1301, 1953.
15. Sjolin, S., and Wranne, L.: Erythropoietic dysfunction in case of Fanconi's anaemia. Acta Haematol, *28:* 230, 1962.
16. Estren, S., Suess, J. F., and Dameshek, W.: Congenital hypoplastic anemia associated with multiple developmental defects (Fanconi syndrome); report of case. Blood, *2:* 85, 1947.
17. Entin, M. A.: Reconstruction of congenital aplasia of radial component. Surg Clin North Am, *44:* 1091, 1964.
18. Riordan, D. C.: Congenital absence of the radius. J Bone Joint Surg, *37A:* 1129, 1955.
19. Birch-Jensen, A.: Congenital deformities of the upper extremities. Andelsbogtryckeriet, Odense, Denmark, 1949.
20. Kirkpatrick, J. A., Wagner, M. L., and Pilling, G. P.: A complex of anomalies associated with tracheo-esophageal fistula and esophageal atresia. AJR, *95:* 208, 1965.
21. Voorhess, M. L., Aspillaga, M. J., and Gardner, L. I.: Trisomy 18 syndrome with absent radius, varus deformity of the hand and rudimentary thumb; report of a case. J Pediatr, *65:* 130, 1964.
22. Diamond, L. K., and Blackfan, K. D.: Hypoplastic anemia. Am J Dis Child, *56:* 464, 1938.
23. Josepha, H. W.: Anaemia of infancy and early childhood. Medicine, *15:* 307, 1936.
24. Diamond, L. K., Allen, D. M., and Magill, F. B.: Congenital (erythroid) hypoplastic anemia; 25-year study. Am J Dis Child, *102:* 403, 1961.

Radial Aplasia Thrombocytopenia

1. Gross, H., Groh, C., and Weippl, G.: Congenitale hypoplastische Thrombopenie mit Radialaplasie. Neue Osterr Z Kinderheilkd *1:* 574, 1956.

2. Shaw, S., and Oliver, R. A. M.: Congenital hypoplastic thrombocytopenia with skeletal deformities in siblings. Blood, 14: 374, 1956.
3. Smith, D. W.: *Recognizable Patterns of Human Malformation, Vol. VII.* W. B. Saunders, Philadelphia, 1970.

Melnick-Needles Dysplasia

1. Melnick, J. C., and Needles, C. F.: An undiagnosed bone dysplasia; a 2-family study of 4 generations and 3 generations. AJR, 97: 39, 1966.

Pachydermoperiostosis

1. Tornblom, M., Malers, E., and Wollenius, G.: Osteodermatopathia hypertrophicans. Acta Med Scand, 164: 325, 1959.
2. Touraine, A., Solente, G., and Golé, L.: Un syndrome osteodermopathique; la pachydermie plicaturee avec pachyperiostose; des extremities. Presse Med, 43: 1820, 1935.
3. Vogl, A., and Goldfisher, S.: Pachydermoperiostosis; primary or idiopathic hypertrophic osteoarthropathy. Am J Med, 33: 166, 1962.
4. Fried, B. M.: Chronic pulmonary osteoarthropathy; dyspituitarism as probable cause. Arch Intern Med, 72: 565, 1943.
5. Hambrick, G. W., Jr., and Carter, M.: Pachydermoperiostosis. Arch Dermatol, 94: 594, 1966.
6. Guyer, P. B., Brunton, F. J., and Wren, M. W. G.: Pachydermoperiostosis with acro-osteolysis. J Bone Joint Surg, 60B: 219, 1978.
7. Kreel, L.: *Outline of Radiology*, pp. 312 and 363-364. William Heinemann, London, 1971.
8. Taybi, H.: Touraine-Solente-Gole syndrome. In *Radiology of Syndromes*, p. 264. Year Book, Chicago, 1975.
9. Weens, H. S., and Brown, C. E.: Atrophy of terminal phalanges in clubbing and hypertrophic osteoarthropathy. Radiology, 45: 27, 1945.

Rubinstein-Taybi Syndrome

1. Rubinstein, J. H., and Taybi, H.: Broad thumbs and toes and facial anomalies. Am J Dis Child, 105: 588, 1963.
2. Coffin, G. S.: Brachydactyly, peculiar facies and mental retardation. Am J Dis Child, 108: 351, 1964.
3. Berry, G. A.: Note on congenital defect (coloboma) of lower lid. R Lond Ophthalmol Hosp Rep, 12: 255, 1889.
4. Treacher Collins, E.: 8. Case with symmetrical congenital notches in outer part of each lower lid and defective development of malar bones; 9. Case with symmetrical congenital notches in outer part of each lower lid and defective development of malar bones. Trans Ophthalmol Soc UK, 20: 190, 1900.
5. Seckel, H. P.: *Bird-Headed Dwarfs.* Charles C Thomas, Springfield, Ill., 1960.

Nevoid Basal Cell Carcinoma Syndrome

1. Nomaland, R.: Multiple basal cell epitheliomas originating from congenital pigmented basal cell nevi. Arch Dermatol, 25: 1002, 1932.
2. Binkley, G. W., and Johnson, H. H., Jr.: Epithelioma adenoides cysticum; basal cell nevi, agenesis of corpus callosum and dental cysts. Arch Dermatol, 63: 73, 1951.
3. Straith, F. E.: Hereditary epidermoid cyst of the jaws. Am J Orthod Oral Surg, 25: 673, 1939.
4. Gorlin, R. L., and Goltz, R. W.: Multiple nevoid basal-cell epithelioma, jaw cysts, and bifid rib. N Engl J Med, 262: 908, 1960.
5. Howell, J. B., and Caro, M. R.: The basal cell nevus. Arch Dermatol, 79: 67, 1959.
6. Anderson, D. E.: Linkage analysis of the nevoid basal cell carcinoma syndrome. Ann Hum Genet, 32: 113, 1968.
7. Anderson, D. E., and Cook, W. A.: Jaw cysts and the basal cell nevus syndrome. J Oral Surg, 24: 15, 1966.
8. Anderson, D. E., McClendon, J. L., and Howell, J. B.: Genetics and skin tumors, with special reference to basal cell nevi. In *Tumors of the Skin.* Year Book, Chicago, 1963.
9. Anderson, D. E., Taylor, W. B., Falls, H. F., and Davidson, R. T.: The nevoid basal cell carcinoma syndrome. Am J Hum Genet, 19: 12, 1967.
10. Dodd, G. D.: Nevoid basal cell carcinoma syndrome; analysis of 90 examined patients. To be published.
11. Howel, J. B., Anderson, D. E., and McClendon, J. L.: The basal cell nevus syndrome. JAMA, 190: 274, 1964.
12. Gorlin, R. J., Vockers, R. A., Kelln, E., and Williamson, J. J.: The multiple basal cell nevi syndrome. Cancer, 18: 89, 1965.
13. Graham, J. K., McJimsey, B. A., and Hardin, J. C.: Nevoid basal cell carcinoma syndrome. Arch Otolaryngol, 87: 72, 1968.
14. Dyke, C. G.: Indirect signs of brain tumor as noted in the routine roentgen examinations: displacement of pineal shadow; survey of 3,000 consecutive skull examinations. AJR, 23: 598, 1930.
15. Dodd, G. D., Anderson, E. D., and Jing, B.: The nevoid basal cell carcinoma syndrome. Proceedings of the Fifty-fourth Scientific Assembly and Annual Meeting of the Radiological Society of North America, p. 266, 1968.
16. Murphy, K. J.: Subcutaneous calcification in the nevoid basal-cell carcinoma syndrome; response to parathyroid hormone and relationship to pseudohypoparathyroidism. Clin Radiol, 20: 287, 1969.
17. Clendenning, W. E., Herdt, J. R., and Block, J. B.: Ovarian fibromas and mesenteric cysts; their association with hereditary basal cell cancer of the skin. Am J Obstet Gynecol, 87: 1008, 1963.

Cornelia de Lange Syndrome

1. Ptacek, L. J., Opitz, J. M., Smith, D. W., Gerritsen, T., and Waisman, H. A.: The Cornelia de Lange syndrome. J Pediatr, 63: 1000, 1963.
2. Lenz, W.: Anomalien des Wachstums und der Korperform. In *Humangenetik, Vol. 2,* P. E. Becker (ed.) G. Thieme Verlag, Stuttgart, 1964.
3. Falek, A., Schmidt, R., and Jervis, G. A.: Familial de Lange syndrome with chromosome abnormalities. Pediatrics, 37: 92, 1966.
4. Opitz, J. M., Segal, A. T., Lehrke, R., and Nadler, H.: Brachmann/de Lange syndrome. Lancet, 2: 1019, 1964.
5. Gerald, B., and Umansky, R.: The Cornelia de Lange syndrome; radiographic findings. Radiology, 88: 96, 1967.
6. Verger, P., Martin, C., and Mortureaux, Y.: Typus amstelodamensis (C. de Lange); trois observations nouvelles. Arch Fr Pediatr, 22: 91, 1965.
7. Borghi, A., Cavina, C., Maiello, M., Taddei, L., and Bigozzi, U.: Typus degenerativus amstelodamensis (syndrome de Cornelia de Lange); considerations cliniques et radiologiques sur trois observations, et quelques remarques sur l'evolution de la "maladie" a l'age post-pubertaire. Presse Med, 72: 3373, 1964.

Rothmund Syndrome

1. Thannhauser, S. J.: Werner's syndrome (Progeria of the adult) Rothmund's syndrome; two types of closely related

heredofamilial atrophic juvenile cataracts and endocrine features; a critical study with 5 new cases. Ann Intern Med, 23: 559, 1945.
2. Merz, E. H., Tausk, K., and Dukes, E.: Meso-ectodermal dysplasia and its variants, with particular reference to the Rothmund-Werner syndrome. Am J Ophthalmol, 55: 488, 1963.
3. Maurer, R. M., and Langford, O. L.: Rothmund's syndrome. Radiology, 89: 706, 1967.
4. McAfee, J. G., and Donner, M. W.: Differential diagnosis of radiographic changes in the hands. Am J Med Sci, 245: 592, 1963.

Progressive Myositis Ossificans

1. McKusick, V. A.: *Heritable Disorders of Connective Tissue*, Ed. 3. C. V. Mosby, St. Louis, 1966.
2. Greig, D. M.: *Clinical Observations of the Surgical Pathology of Bone*, p. 170. Oliver and Boyd, Edinburgh, 1931.
3. Mair, W. F.: Myositis ossificans progressiva. Edinburgh Med J, 39: 13/69, 1932.
4. Vastine, J. H., II, Vastine, M. F., and Arango, O.: Myositis ossificans progressiva in homozygotic twins. AJR, 59: 204, 1948.
5. Rosenstirn, J.: A contribution to the study of myositis ossificans progressiva. Ann Surg, 68: 485/591, 1918.
6. Helferich, H.: Ein fall von sogenannter myositis ossificans progressive. Med Klin, 25: 1661, 1929.
7. Rolleston, H. D.: Progressive myositis ossificans with references to other developmental diseases of the mesoblast. Clin J, 17: 209, 1901.
8. Lutwak, L.: Myositis ossificans progressiva; mineral, metabolic and radioactive calcium studies of the effects of hormones. Am J Med, 37: 269, 1964.
9. Hutchinson, J.: Reports of hospital practice. Med Times Gaz, 1: March 31, 1860.
10. Gruber, G. B.: Anmerkungen zur frage der weichteilverknocherung besonders der myopathia osteoplastica. Virchows Arch, 260: 457, 1926.
11. Frejka, B.: Heterotopic ossification and myositis ossificans progressiva. J Bone Joint Surg, 11A: 157, 1929.
12. Hirsch, F., and Low-Beer, A.: Ueber einen fall von myositis ossificans progressiva. Med Klin, 25: 1661, 1929.
13. Fletcher, E., and Moss, M. S.: Myositis ossificans progressiva. Ann Rheum Dis, 24: 267, 1965.
14. Fairbank, H. A. T.: *An Atlas of General Affections of the Skeleton*. Williams & Wilkins, Baltimore, 1954.
15. Painter, C. F., and Clark, J. D.: Myositis ossificans. Am J Orthop Surg, 6: 626, 1908-9.
16. Koontz, A. R.: Myositis ossificans progressiva. Am J Med Sci, 174: 406, 1927.
17. Tokura, N.: Letter to the editor Tokyo Izi Sinsi, 1498, 1930.
18. Uehlinger, E.: Myositis ossificans progressiva. Ergeb Med Strahlenforsch, 7: 175, 1936.
19. van Creveld, S., and Soeters, J. M.: Myositis ossificans progressiva. Am J Dis Child, 62: 1000, 1941.
20. Riley, H. D., Jr., and Christie, A.: Myositis ossificans progressiva. Pediatrics, 8: 753, 1951.
21. Knorre, G. V.: Uber die myositis ossificans progressiva. Z Menschl Vererb Konstitutions-lehre, 33: 85, 1955.
22. Eaton, W. L., Conkling, W. S., and Daeschner, C. W.: Early myositis ossificans progressiva occurring in homozygotic twins; a clinical and pathological study. J. Pediatr, 50: 591, 1957.
23. Stonham, C.: Myositis ossificans. Lancet, 2: 1481, 1892.
24. Pol, P.: Brachydaktylie-klinodaktylie-hyperphalangie und ihre grundlagen. Arch Pathol Anat Klin Med, 229: 388, 1921.
25. Griffith, G.: Progressive myositis ossificans; report of a case. Arch Dis Child, 24: 71, 1949.
26. Vastine, J. H., II, Vastine, M. F., and Arango, O.: Genetic influence of osseous development with particular reference to the deposition of calcium in the costal cartilages. AJR, 59: 213, 1948.

Marfan Syndrome

1. Olcott, C. T.: Arachnodactyly (Marfan's syndrome) with severe anemia. Am J Dis Child, 60: 660, 1940.
2. Parker, A. S., Jr., and Hare, H. F.: Arachnodactyly. Radiology, 45: 220, 1945.
3. Burch, F. E.: Association of ectopia lentis with arachnodactyly. Arch Opthalmol, 15: 645, 1936.
4. Etter, L., and Glover, L. P.: Arachnodactyly complicated by dislocation of lens and death from rupture of dissecting aneurysm of aorta. JAMA, 123: 88, 1943.
5. McKusick, V. A.: *Heritable Disorders of Connective Tissue*. C. V. Mosby, St. Louis, 1966.
6. Baer, R. W., Taussig, H. B., and Oppewheimer, E. H.: Congenital aneurysmal dilatation of the aorta associated with arachnodactyly. Bull Johns Hopkins Hosp, 72: 309, 1943.
7. Piper, R. K., and Irvine-Jones, E.: Arachnodactylia and its association with congenital heart disease. Am J Dis Child, 31: 832, 1926.
8. Robins, P. R., Moe, J. H., and Winter, R. B.: Scoliosis in Marfan's syndrome. Its characteristics and results of treatment in thirty-five patients. J Bone Joint Surg, 57A: 358, 1975.
9. Gore, I.: The pathogenesis of dissecting aneurysm of the aorta. Arch Pathol, 53: 142, 1952.
10. Feldman, L., Friedlander, J., Dillon, R., and Wallyn, R.: Aneurysm of right sinus of Valsalva with rupture into right atrium and into right ventricle. Am Heart J, 51: 314, 1956.
11. Steinberg, I., and Geller, W.: Aneurysmal dilatation of aortic sinuses in arachnodactyly; diagnosis during life in 3 cases. Ann Intern Med, 43: 120, 1955.
12. Murray, R. O., and Jacobson, H. G.: *The Radiology of Skeletal Disorders*, Ed. 2, Vol 2, p. 1105. Livingstone, Edinburgh, 1977.
13. McCann, J. S., and Porter, D. C.: Calcification of the aorta as an aid to the diagnosis of syphilis. Br Med J, 1: 826, 1956.

Homocystinuria

1. Field, C. M. B., Carson, N. A. J., Cusworth, D. C., Dent, C. E., and Neill, D. W.: Homocystinuria, a new disorder of metabolism. Abstracts of the 10th International Congress of Pediatrics, p. 274, Lisbon, 1962.
2. Gerritsen, T., and Waisman, H. A.: Homocystinuria, an error in the metabolism of methionine. Pediatrics, 33: 413, 1964.
3. Brenton, D. P., and Dow, C. J.: Homocystinuria and Marfan's syndrome. A comparison. J Bone Joint Surg, 54B: 277, 1972.
4. Brill, P. W., Mitty, H. A., and Gaull, G. E.: Homocystinuria due to cystathionine synthase deficiency: clinical-roentgenologic correlations. AJR, 121: 45, 1974.
5. Carson, N. A. J., and Raine, D. N.: *Inherited Disorders of Sulphur Metabolism*. J. A. Churchill, London, 1971.
6. Uhlendorf, B. W., Conerly, E. B., and Mudd, S. H.: Homocystinuria: studies in tissue culture. Pediatr Res, 7: 645, 1973.
7. Brenton, D. P., Cusworth, D. C., Dent, C. E., and Jones, E. E.: Homocystinuria; clinical and dietary studies. Q J Med, 35: 325, 1966.

8. Brett, E. M.: Homocystinuria with epilepsy. Proc Roy Soc Med, *59:* 484, 1966.
9. Carson, N. A. J., Dent, C. E., Field, C. M. B., and Gaull, G. E.: Homocystinuria: clinical and pathological review of ten cases. J Pediatr, *66:* 565, 1965.
10. Gaull, G. E.: Homocystinuria, vitamin B_6 and folate: metabolic interrelationships and clinical significance. J Pediatr, *81:* 1014, 1972.
11. McKusick, V. A.: *Heritable Disorders of Connective Tissue,* Ed. 4, pp. 224–281. C. V. Mosby, St. Louis, 1972.
12. Morrow, G., and Barness, L. A.: Combined vitamin responsiveness in homocystinuria. J Pediatr, *81:* 946, 1972.
13. Schimke, R. N., McKusick, V. A., Huang, T., and Pollack, A. D.: Homocystinuria; studies of 20 families with 38 affected members. JAMA, *193:* 711, 1965.
14. Cusworth, D. C., and Gattereau, A.: Inhibition of renal tubular reabsorption of homocystine by Lysine and Arginine. Lancet, *2:* 916, 1968.
15. Gaull, G. E., Sturman, J. A., and Shaffner, F.: Homocystinuria due to cystathionine synthase deficiency: enzymatic and ultrastructural studies (to be published).
16. Gaull, G. E., Sturman, J. A., and Rassin, D. K.: Enzymatic and metabolic studies of homocystinuria; effects of pyridoxine. Neuropaediatrie, *1:* 199, 1969.
17. Gibson, J. B., Carson, N. A. J., and Neill, D. W.: Pathologic findings in homocystinuria. J Clin Pathol, *17:* 427, 1964.
18. McDonald, L., Bray, C., Field, C., Love, F., and Davies, B.: Homocystinuria. thrombosis and the blood-platelets. Lancet, *1:* 745, 1964.
19. Barber, G. W., and Spaeth, G. L.: The successful treatment of homocystinuria with pyridoxine. J Pediatr, *75:* 463, 1969.
20. Brenton, D. P., Cusworth, D. C., and Gaull, G. E.: Homocystinuria. Metabolic studies on 3 patients. J Pediatr, *67:* 58, 1965.
21. Smith, S. W.: Roentgen findings in homocystinuria. AJR, *100:* 147, 1967.
22. MacCarthy, J. M. T., and Carey, M. C.: Bone changes in homocystinuria. Clin Radiol, *19:* 128, 1968.
23. Morreels, C. L., Fletcher, B. D., Weilbaecher, R. G., and Dorst, J. P.: The roentgenographic features of homocystinuria. Radiology, *90:* 1150, 1968.
24. Murray, R. O., and Jacobson, H. G.: *The Radiology of Skeletal Disorders,* Vol. 2, Ed. 2, pp. 1105. Livingstone, Edinburgh, 1977.

Ehlers-Danlos Syndrome

1. McKusick, V. A.: *Heritable Disorders of Connective Tissue,* Ed. 3. C. V. Mosby, St. Louis, 1966.
2. Benjamin, B., and Weiner, H.: Syndrome of cutaneous fragility and hyperelasticity and articular hyperlaxity. Am J Dis Child, *65:* 247, 1943.
3. Burrows, A., and Turnbull, H. M.: Cutis hyperelastica (Ehlers-Danlos syndrome). Br J Dermatol, *50:* 648, 1938.
4. King-Louis, F. J.: Two cases of Ehlers-Danlos syndrome. Proc R Soc Med, *39:* 135, 1946.
5. Ringrose, E. J., Nowlan, F. B., and Perry, H.: Ehlers-Danlos syndrome; report of a case. Arch Dermatol Syphliol, *62:* 443, 1950.
6. Rossi, E., and Angst, H.: Das Danlos-Ehlers syndrome. Helv Paediatr Acta, *6:* 245, 1951.
7. Pautrier, M.: Note histologique sur un cas de cutis elastica, avec pesudotumereurs aux genoux et aux coudes, presente par M. Danlos. Bull Soc Fr Dermatol Syphiligr, *19:* 72, 1908.
8. Poumeau-Delille, G., and Soulie, P.: Un cas d'hyperlaxite cutanee et articularie avecicatrices atrophiques et pseudotumeurs molluscoides (syndrome d'Ehlers-Danlos). Bull Soc Med Hop Paris, *50:* 593, 1934.
9. Weber, F. P., and Aitken, J. K.: Nature of the subcutaneous spherules in some cases of Ehlers-Danlos syndrome. Lancet, *1:* 198, 1938.
10. Lapayowker, M. S.: Cutis hyperelastica; the Ehlers-Danlos syndrome. AJR, *84:* 232, 1960.
11. Bossu, A., and Lambrechts: Manifestations oculaires du syndrome d'Ehlers-Danlos. Bull Soc Fr Ophtalmol, *1:* 211, 1953.
12. Durham, D. G.: Cutis hyperelastica (Ehlers-Danlos syndrome) with blue scleras, microcornea, and glaucoma. Arch Ophthalmol, *49:* 220, 1953.
13. Schaper, G.: Familiares Vorkommen von Ehlers-Danlos syndrome ein Beitrag zur Klinik und Pathogenese. Z Kinderheilkd, *70:* 504, 1952.
14. Thomas, C., Cordier, J., and Algan, B.: Les alterations oculaires de la maladie d'Ehlers-Danlos. Arch Ophthalmol, *14:* 691, 1954.
15. Thomas, C., Cordier, J., and Algan, B.: Une etiologie nouvelle du syndrome de luxation spontanee des cristallins: la maladie d'Ehlers-Danlos. Bull Soc Belge Ophtalmol, *100:* 375, 1952.
16. Bonnet, P.: Les manifestations oculaires de la maladie d'Ehlers-Danlos. Bull Soc Fr Ophtalmol, *1:* 211, 1953.
17. Brombart, M., Coupatez, G., and Laurent, Y.: Contribution a l'etude de l'etiologie de la hernie hiatale et al diverticulose du tube digestif; un case de maladie d'Ehlers-Danlos associee a une hernia hiatale, un deverticule d l'estomac, un diverticule duodenal, une diverticulose colique et une anemie sideropenique. Arch Mal Appar Dig, *41:* 413, 1952.
18. Mounier-Kuhn, P., and Meyer, L.: Mega-organes (oesophage, trachee, colon), syndromes de Mickulica et d'Ehlers-Danlos chez une heredo-syphilitique. Bull Soc Med Hop Lyon, November 9, 1943.
19. Packer, B. D., and Blades, J. F.: Dermatorrhexia, a case report; the so-called Ehlers-Danlos syndrome. Va Med Mon, *81:* 21, 1954.
20. McFarland, W., and Fuller, D. E.: Mortality in Ehlers-Danlos syndrome due to spontaneous rupture of large arteries. N Engl J Med, *271:* 1309, 1964.
21. Rubenstein, M., and Cohen, N. H.: Ehlers-Danlos syndrome associated with multiple intracranial aneurysms. Neurology, *14:* 125, 1964.
22. Bolam, R. M.: A case of Ehlers-Danlos syndrome. Br J Dermatol, *50:* 174, 1938.
23. Holt, J. R.: The Ehlers-Danlos syndrome. AJR, *55:* 420, 1946.
24. Murray, J. E., and Tyars, M. D.: A case of Ehlers-Danlos syndrome. Br Med J, *1:* 974, 1940.
25. Carter, C. O., and Wilkinson, J.: Persistent joint laxity and congenital dislocation of the hip. J Bone Joint Surg, *46B:* 40, 1964.
26. Danlos, M.: Un cas cutis laxa avec tumeurs par contusion chronique des coudes et des genoux (xanthome juvenile pseudo-diabetique de M. M. Hallopeau et Mace de Lepinay). Bull Soc Fr Dermatol Syphiligr, *19:* 70, 1908.
27. Noto, P.: Cited by Pelbois and Rollier.
28. Pascher, F.: Ehlers-Danlos syndrome. Arch Dermatol Syphilol, *67:* 214, 1953.
29. Weill, J., and Martineau, J.: A propos d'un cas de maladie d'Ehlers-Danlos; etude anatomoclinique et biologique. Bull Soc Fr Dermatol Syphiligr, *44:* 99, 1937.
30. Carter, C. O., and Sweetman, R.: Familial joint laxity and recurrent dislocation of the patella. J Bone Joint Surg, *40B:* 664, 1958.
31. Carter, C. O., and Sweetman, R.: Recurrent dislocation of the patella and of the shoulder. J Bone Joint Surg,

42B: 721, 1970.
32. DuBois, Cutis laxa (abstract). Zentralbl Haut Geschlechtskr, 35: 52, 1931.
33. Key, J. A.: Hypermobility of joints as a sex-linked hereditary characteristic. JAMA, 88: 1710, 1927.
34. Gould, G. M., and Pyle, W. L.: *Anomalies and Curiosities of Medicine*, p. 217. W. B. Saunders, Philadelphia, 1897.
35. Coventry, M. B.: Some skeletal changes in the Ehlers-Danlos syndrome. J Bone Joint Surg, 43A: 855, 1961.
36. Debre, R., and Semalaigne, G.: A propos de la maladie d'Ehlers chez le nourrisson. Bull Soc Med Hop Paris, 57: 849, 1936.
37. Rollhauser, H.: Die Zugfestigkeit de menschlichen Haut. Gegenbaur Morph Jahrb, 90: 249, 1950.
38. Shapiro, S. K.: A case of Meekrin-Ehlers-Danlos syndrome with neurologic manifestations. J Nerv Ment Dis, 115: 64, 1952.
39. Sutro, C. J.: Hypermobility of bones due to "overlengthened" capsular and ligamentous tissues; case for recurrent intraarticular effusions. Surgery, 21: 67, 1947.
40. Turkington, R. W., and Grude, H. E.: Ehlers-Danlos syndrome and multiple neurofibromatosis. Ann Intern Med, 61: 549, 1964.
41. Wenzel, H. G.: Untersuchungen uber die Dehnbarkeit und Zerreissbarkeit der Haut. Zentralbl Allg Pathol, 85: 117, 1949.
42. Kirk, E., and Kvorning, S. A.: Quantitative measurements of the elastic properties of the skin and subcutaneous tissue in young adults and old individuals. J Gerontol, 4: 273, 1949.
43. Katz, I., and Steiner, K.: Ehlers-Danlos syndrome with ectopic bone formation. Radiology, 65: 352, 1955.

Generalized Cortical Hyperostosis

1. van Buchem, F. S. P., Hadders, H. N., and Ubbens, R.: An uncommon familial systemic disease of the skeleton; hyperostosis corticalis generalisata familiaris. Acta Radiol, 44: 109, 1955.
2. van Buchem, F. S. P., and Hadders, H. N.: Hyperostosis corticalis generalisata. Schweiz Med Wochenschr, 87: 231, 1957.
3. van Buchem, F. S. P., Hadders, H. N., Hansen, J. F., and Woldring, M. G.: Hyperostosis corticalis generalisata. Am J Med, 33: 1962.
4. Owen, R. H.: Van Buchem's disease (hyperostosis corticalis generalisata). Br J Radiol, 49: 126, 1976.
5. Maroteaux, P., Fontaine, G., and Scharfman, W.: L'hyperostose corticole generalisee. Arch Fr Pediatr, 28: 685, 1971.
6. Hinkel, C. L., and Beiler, D. D.: Osteopetrosis in adults. AJR, 74: 46, 1955.

Focal Dermal Hypoplasia

1. Cole, H. N., Driver, J. R., Giffen, H. K., Norris, C. B., and Strand, G.: Ectodermal and mesodermal dysplasia with osseous involvement. Arch Dermatol Syphilol, 44: 773, 1941.
2. Freeman, C. D., Jr.: Congenital ectodermal dysplasia and mesodermal dysplasia. Arch Dermatol Syphilol, 71: 667, 1955.
3. Holden, J. D., Akers, W. A.: Goltz's syndrome; focal dermal hypoplasia; combined mesoectodermal dysplasia. Am J Dis Child, 114: 292, 1967.
4. Warburg, M.: Focal dermato-phalangeal dysplasia. Acta Ophthalmol, 46: 137, 1968.
5. Gorlin, R. J., Meskin, L. H., Peterson, W. C., Jr., and Goltz, R. W.: Focal dermal hypoplasia syndrome. Acta Derm Venereol, 43: 421, 1963.
6. Goltz, R. W., Peterson, W. C., Gorlin, R. J., and Ravits, H. G.: Focal dermal hypoplasia. Arch Dermatol Syphilol, 86: 708, 1962.
7. Jessner, M.: Naeviforme, poililodermie-artige Hautveranderungen mit Missbildungen (Schwimmhautbildungen an den Fingern, Papillome an Anus). Zentralbl Haut Geschlechtskr, 27: 468, 1928.
8. Martin-Scott, I.: Congenital focal dermal hypoplasia. Br J Dermatol, 77: 60, 1965.
9. Ginsburg, L. D., Sedano, H. O., and Gorlin, R. J.: Focal dermal hypoplasia syndrome. AJR, 110: 1970.
10. Howell, J. B.: Nevus angiolipomatosus vs. focal dermal hypoplasia. Arch Dermatol Syphilol, 92: 238, 1965.
11. Silverman, F. N.: Differential diagnosis of achondroplasia. Radiol Clin North Am, 6: 223, 1968.
12. Gorlin, R. J., and Pindborg, J. J.: *Syndromes of the Head and Neck*, p. 580. McGraw-Hill, New York, 1964.
13. James, A. E., Jr., Belcourt, C. L., Atkins, L., and Janower, M. L.: Trisomy 18. Radiology, 92: 37, 1969.

Acrocephalosyndactyly and Acrocephalopolysyndactyly

1. Temtamy, S., and McKusick, V. A.: Synopsis of hand malformations with particular emphasis on genetic factors. In *The Clinical Delineation of Birth Defects, III: Limb Malformations*, pp. 125–184. Williams & Wilkins, Baltimore, 1969.
2. Noack, M.: Beitrag zum Krankheitsbild der Akrozephalosyndaktylie (Apert). Arch Kinderheilkd, 160: 168, 1959.
3. Temtamy, S. A.: Carpenter's syndrome; acrocephalopolysyndactyly, an autosomal recessive syndrome. J Pediatr, 69: 111, 1966.
4. Crome, L.: Acrocephalosyndactyly (Apert's syndrome). J Ment Sci, 107: 459, 1961.
5. Penrose, L. S.: Acrocephalosyndactyly (Apert's syndrome). Sci Mon NY, 52: 359, 1941.
6. Blank, C. E.: Apert's syndrome (a type of acrocephalosyndactyly); observations in a British series of 39 cases. Ann Hum Genet, 24: 151, 1960.
7. Buchanan, R. C.: Acrocephalosyndactyly or Apert's syndrome. Br J Plast Surg, 21: 406, 1968.
8. Gunther, H.: Turmschadel als Konstitutionsanomalie und als kliousches Symptom. Ergeb Inn Med Kinderheilk, 40: 40, 1931.
9. Schauerte, E. W., and St. Aubin, P. M.: Progressive synostosis in Apert's syndrome (acrocephalosyndactyly). AJR, 97: 67, 1966.
10. Rubin, P.: *Dynamic Classification of Bone Dysplasias*. Year Book, Chicago, 1964.
11. Ferriman, D.: *Acrocephaly Acrocephalosyndactyly*. Oxford University Press, London, 1941.
12. Temtamy, S. A.: Carpenter's syndrome, acrocephalopolysyndactyly; autosomal recessive syndrome. J Pediatr, 69: 111, 1966.
13. Andersson, H., and Gomes, S. P.: Craniosynostosis, review of literature and indications for surgery. Acta Paediatr Scand, 57: 47, 1968.
14. Duggan, C. A., Keener, E. B., and Gay, B. B.: Secondary craniosynostosis. AJR, 109: 277, 1970.
15. Caffey, J.: *Pediatric X-ray Diagnosis: A Textbook for Students and Practitioners of Pediatrics, Surgery and Radiology*, Ed. 5. Year Book, Chicago, 1967.
16. Dunn, F. H.: Nonfamilial and nonhereditary craniofacial dysostosis; variant of Crouzon's disease. AJR, 84: 472, 1960.
17. Stovin, J. J., Lyon, J. A., Jr., and Clemmens, R. L.: Mandibulofacial dysostosis. Radiology, 74: 225, 1960.
18. Jabbour, J. T., and Taybi, H.: Craniotelencephalic dysplasia. Am J Dis Child, 108: 627, 1964.

19. Pfeiffer, R. A.: Dominant erbliche acrocephalosyndaktylie. Z Kinderheilkd, *90:* 301, 1964.
20. Pfeiffer, R. A.: Associated deformities of the head and hands. In *The Clinical Delineation of Birth Defects, III: Limb Malformations,* pp. 18–34. Williams & Wilkins, Baltimore, 1969.
21. Asnes, R. S., and Morehead, C. D.: Pfeiffer syndrome. In *The Clinical Delineation of Birth Defects, III: Limb Malformations,* pp. 198–202. Williams & Wilkins, Baltimore, 1969.
22. Martsolf, J. T., Cracco, J. B., Carpenter, G. G., and O'Hara, A. E.: Pfeiffer syndrome. Am J Dis Child, *121:* 257, 1971.

Osteopetrosis

1. Karshner, R. G.: Osteopetrosis. AJR, *16:* 405, 1926.
2. McPeak, C. N.: Osteopetrosis; report of 8 cases occurring in 3 generations of 1 family. AJR, *36:* 816, 1936.
3. Nussey, A. M.: Osteopetrosis. Arch Dis Child, *13:* 161, 1938.
4. Lamy, M. E.: Hereditary disorders of bones—an overview. In *The Clinical Delineation of Birth Defects, IV: Skeletal Dysplasias,* pp. 8–13. Williams & Wilkins, Baltimore, 1969.
5. Zawisch, C.: Marble bone disease; a study of osteogenesis. Arch Pathol, *43:* 55, 1947.
6. Gruneberg, H.: A new sublethal colour mutation in the house mouse. Proc R Soc, *118:* 321, 1935.
7. Gruneberg, H.: Grey-lethal, a new mutation in the house mouse. J Heredity, *27:* 105, 1936.
8. Barnicot, N. A.: Studies on the factors involved in bone absorption. I. Effect of subcutaneous transplantation of bone of the grey-lethal house mouse into normal hosts and of normal bone into grey-lethal hosts. Am J Anat, *68:* 497 (plates 528–31), 1941.
9. Walker, D. G.: Counteraction to parathyroid therapy and osteopetrotic mice as revealed in the plasma level and ability to incorporate 3-H-proline into bone. Endocrinology, *79:* 836, 1966.
10. Walker, D. G.: Elevated bone collagenolytic activity and hyperplasia of parafollicular light cells of the thyroid gland in parathormone-treated grey-lethal mice. Z Zellforsch Mikrosk Anat, *72:* 100, 1966.
11. Harkis, G. K., Williams, G. A., and Tennenouse, A.: Thyrocalcitonin; cytological localization by immunofluorescence. Science, *152:* 73, 1966.
12. Marks, S. C., and Walker, G. D.: The role of the parafollicular cell of the thyroid gland in the pathogenesis of congenital osteopetrosis in mice. Am J Anat, *126:* 299, 1969.
13. Marks, S. C.: The parafollicular cell of the thyroid gland as the source of an osteoblast-stimulating factor; evidence from experimentally osteopetrotic mice. J Bone Joint Surg, *51A:* 875, 1969.
14. Walker, D. G.: Osteopetrosis. In *The Clinical Delineation of Birth Defects, IV. Skeletal Dysplasias,* p. 308. Williams & Wilkins, Baltimore, 1969.
15. Raisz, L. G., Au, W. Y. W., Friedman, J., and Niemann, I.: Thyrocalcitonin and bone resorption. Am J Med, *43:* 684, 1967.
16. Caffey, J.: *Pediatric X-ray Diagnosis.* Year Book, Chicago, 1961.
17. Teplick, J. G., Head, G. L., Kricun, M. E., and Haskin, M. E.: Ghost infantile vertebrae and hemipelves within adult skeleton from Thorotrast administration in childhood. Radiology, *129:* 657, 1978.
18. Janower, M. L., Miettinen, O. S., and Flynn, M. J.: Effects of long-term thorotrast exposure. Radiology, *103:* 13, 1972.
19. Symposium on distribution, retention and late effects of thorium dioxide. Ann NY Acad Sci, *145:* 523, 1967.
20. Looney, W. B.: An investigation of the late clinical findings following thorotrast (thorium dioxide) administration. AJR, *83:* 163, 1960.
21. Looney, W. B.: Late clinical changes following the internal deposition of radioactive materials. Ann Intern Med, *42:* 378, 1955.
22. Sear, H. R.: A case of Albers-Schonberg's disease. Br J Surg, *14:* 657, 1927.

Dysosteosclerosis

1. Houston, C. S., Gerrard, J. W., and Ives, E. J.: Dysosteosclerosis. AJR, *130:* 988, 1978.
2. Spranger, J. W., Langer, L. O., and Wiedemann, H. R.: *Bone Dysplasias. An Atlas of Constitutional Disorders of Skeletal Development.* W. B. Saunders, Philadelphia, 1974.
3. Spranger, J., Albrecht, C., Rohwedder, H. J., and Wiedemann, H. R.: Die Dysosteosklerose eine Sonderform der generalisierten Osteosklerose. Fortschr Geb Roentgenstr Nuklearmed, *109:* 504, 1968.
4. Utz, V. W.: Manifestation der Dysosterosklerose im Kieferbereich. Dtsch Zahnaerztl Z, *25:* 48, 1970.
5. Stehr, L.: Pathogenese und Klinik der Osteosklerosen. Arch Orthop Unfallchir, *41:* 156, 1941.
6. Roy, C., Maroteaux, P., Kremp, L., Courtecuisse, W., and Alagille, D.: Un nouveau syndrome osseux avec anomalies curanees et troubles neurologiques. Arch Fr Pediatr, *25:* 893, 1968.
7. Leisti, J., Kaitila, I., Lachman, R. S., Asch, M. D., and Rimoin, D. L.: Dysosteosclerosis (case report). Birth Defects, *11:* 349, 1975.

Pycnodysostosis

1. Shuler, S. E.: Pycnodysostosis. Arch Dis Child, *38:* 620, 1963.
2. Maroteaux, P., and Lamy, M.: The malady of Toulouse-Lautrec. JAMA, *191:* 111, 1965.
3. Meredith, S. C., Simon, M. A., Laros, G. S., and Jackson, M. A.: Pycnodysostosis. A clinical, pathologic and ultramicroscopic study of a case. J Bone Joint Surg, *60A:* 1122, 1978.
4. Muthukrishnan, N., and Shetty, M. V. K.: Pycnodysostosis. AJR, *114:* 247, 1972.
5. Dusenberry, James F., Jr., and Kane, John J.: Pycnodysostosis; report of 3 new cases. AJR, *99:* 717, 1967.
6. Elmore, S. M.: Pycnodysostosis; a review. J Bone Joint Surg, *49A:* 153, 1967.
7. Elmore, S. M., Nance, W. E., and McGee, B. J.: Pycnodysostosis with a familial chromosome anomaly. Am J Med, *40:* 273, 1966.
8. Grepl, J.: Albers-Schonbergsche osteopetrosis mit weniger haufigen Knochenveranderungen. Forschr Geb Rontgenstr Nuklearmed, *83:* 229, 1955.
9. Sjolin, K. E.: Gargoylism; forme fruste. Acta Paediatr, *40:* 165, 1951.
10. Emami-Ahari, Z., Zarabi, M., and Javid, B.: Pycnodysostosis. J Bone Joint Surg, *51B:* 307, 1969.

Neurofibromatosis

1. Smith, R.: *Treatise on the Pathology, Diagnosis and Treatment of Neuroma.* Hodges Smith, Dublin, 1849.
2. von Recklinghausen, F.: Ueber die multiplen Fibrome der Haut und ihre Beziehung zu den multiplen Neuromen. A. Hirschwald (in Virchow's Festschrift), Berlin, 1882.
3. Holt, J. F.: Neurofibromatosis in children. AJR, *130:* 615, 1978.

4. Hosoi, K.: Multiple neurofibromatosis (von Recklinghausen's disease), with special reference to malignant transformation. Arch Surg, 22: 258, 1931.
5. Holt, J. F., and Wright, E. M.: Radiologic features of neurofibromatosis. Radiology, 51: 647, 1948.
6. Stalmann, A.: Nerven-, Haut- und Knochenveran derungen bei der Neurofibromatosis Recklinghausen und ihre entstehungsges-chichtlichen Zusammenhaunge. Virchows Arch Pathol Anat Physiol Klin Med, 288: 96, 1933.
7. Cohen, R., and Douady, D.: Coexistence des deux maladies de Recklinghausen chez un sujet; leurs liens nosologiques. Presse Med, 44: 2063, 1963.
8. Verocay, J.: Quoted by H. W. Cushing: *Tumors of the Nervus Acousticus and the Syndrome of the Cerebellopontile Angle.* W. B. Saunders, Philadelphia, 1917.
9. Masson, P.: Experimental and spontaneous schwannomas (peripheral gliomas). Am J Pathol, 8: 367, 1932.
10. Murray, M. R., and Stout, A. P.: Schwann cell versus fibroblast as the origin of the specific nerve sheath tumor. Am J Pathol, 16: 41, 1940.
11. Aegerter, E., and Smith, L. W.: Case of diffuse neurofibromatosis involving cranial peripheral and sympathetic nerves, accompanied by tumor of hypothalamus. Am J Cancer, 31: 212, 1937.
12. Holt, J. F., and Kuhns, L. R.: Macrocranium and macroencephaly in neurofibromatosis. Skeletal Radiol, 1: 25, 1976.
13. Aegerter, E., and Kirkpatrick, J. A., Jr.: *Orthopedic Diseases.* W. B. Saunders, Philadelphia, 1964.
14. Heard, G. E., Holt, J. F., and Naylor, B.: Cervical vertebral deformity in von Recklinghausen's disease of the nervous system. A review with necropsy findings. J Bone Joint Surg, 44B: 880, 1962.
15. Harwood-Nash, D. C. and Fitz, C. R.: *Neuroradiology in Infants and Children.* C. V. Mosby, St. Louis, 1976.
16. Caffey, J.: *Pediatric X-ray Diagnosis,* Ed. 6. Year Book, Chicago, 1972.
17. Murray, R. O., and Jacobson, H. G.: *The Radiology of Skeletal Disorders,* Ed. 2. Livingstone, Edinburgh, 1977.
18. Salerno, N., and Edeiken, J.: Vertebral scalloping in neurofibromatosis. Radiology, 97: 509, 1970.
19. Edeiken, J., Zervas, N. T., and Clearfield, R.: Cervical perineural or extradural cysts. Clin Orthop, 44: 1966.
20. Hillenius, L.: Intrathoracic meningocele. Acta Med Scand, 163: 15, 1959.
21. Kent, E. M., Blades, B., Valle, A. R., and Graham, E. A.: Intra-thoracic neurogenic tumors. J Thorac Surg, 13: 116, 1944.
22. Loop, J. W., Akeson, W. H., and Clawson, D. K.: Acquired thoracic abnormalities in neurofibromatosis. AJR, 93: 416, 1965.
23. Meszaros, W. T., Guzzo, F., and Schorsch, H.: Neurofibromatosis. AJR, 98: 557, 1966.
24. Sengpiel, G. W., Ruzicka, F. F., and Lodmell, E. A.: Lateral intrathoracic meningocele. Radiology, 50: 515, 1968.
25. DeMyer, W.: Megalencephaly in children. Clinical symptoms, genetic patterns, and differential diagnosis from other causes of megalocephaly. Neurology, 22: 634, 1972.
26. Norman, M. E.: Neurofibromatosis in a family. Am J Dis Child, 123: 159, 1972.
27. Weichert, K. A., Dine, M. S., Benton, C., and Silverman, F. N.: Macrocranium and neurofibromatosis. Radiology, 107: 163, 1973.
28. Haas, L. L.: Roentgenological skull measurements and their diagnostic applications. AJR, 67: 197, 1952.
29. Gordon, I. R. S.: Measurement of cranial capacity in children. Br J Radiol, 39: 377, 1966.
30. LeWald, L. T.: Congenital absence of the superior orbital wall associated with pulsating exophthalmos: report of four cases. AJR, 30: 756, 1933.
31. Binet, E. F., Kieffer, S. A., Martin, S. H., and Peterson, H. O.: Orbital dysplasia in neurofibromatosis. Radiology, 93: 829, 1969.
32. Joffe, N.: Calvarial bone defects involving the lambdoidal suture in neurofibromatosis. Br J Radiol, 38: 23, 1965.
33. Davidson, K. C.: Cranial and intracranial lesions in neurofibromatosis. AJR, 98: 550, 1966.
34. Saha, M. M., Agarwal, K. N., and Bhardwaj, O. P.: Calvarial bone defects in neurofibromatosis. A case report. AJR, 105: 319, 1969.
35. Eickhoff, U., and Fischer, W.: Eine seltene kombinierte Dysplasie bei der Neurofibromatosis Recklinghausen. Forschr Geb Roentgenstr Nuklearmed, 116: 776, 1972.
36. Weber, K., and Grauthoff, H. J.: Lambdanaht-Defekt typische Veranderung der Neurofibromatosis Rechlinghausen. Fortschr Geb Roentgenstr Nuklearmed, 118: 230, 1973.
37. Handa, J., Koyama, T., Shimizu, Y., and Yoneda, S.: Skull defect involving the lambdoid suture in neurofibromatosis. Surg Neurol, 3: 119, 1975.
38. Holt, J. F.: Personal communication, 1978.
39. Hunt, J. C., and Pugh, D. G.: Skeletal lesions in neurofibromatosis. Radiology, 76: 1, 1961.
40. Wolter, E.: Neuropathy of the eye and adnexa. In *Pathology of the Nervous System,* pp. 674–675, J. Minckler (ed.). McGraw-Hill, New York, 1968.
41. Fowler, F. D., and Matson, D. D.: Gliomas of the optic pathways in childhood. J Neurosurg, 14: 515, 1957.
42. Harwood-Nash, D. C.: Optic gliomas and pediatric neuroradiology. Radiol Clin North Am, 10: 83, 1972.
43. Harwood-Nash, D. C.: Axial tomography of the optic canals in children. Radiology, 96: 367, 1970.
44. Young, D. F., Eldridge, R., Nager, G. T., Deland, F. H., and McNew, J.: Hereditary bilateral acoustic neuroma (central neurofibromatosis). Birth Defects 7: 73, 1971.
45. Alliez, J., Masse, J. L., and Alliez, B.: Bilateral tumors of the acoustic nerve and Recklinghausen's disease observed in several generations. Considerations on hereditary acoustic nerve tumors. Rev Neurol (Paris), 131: 545, 1975.
46. Hitselberger, W. E., and Hughes, R. L.: Bilateral acoustic tumors and neurofibromatosis. Arch Otolaryngol, 88: 700, 1968.
47. Krause, C. J., and McCabe, B. F.: Acoustic neuroma in a 7-year-old girl. Report of a case. Arch Otolaryngol, 94: 359, 1971.
48. Hill, M. C., Oh, S. K., and Hodges, F. J.: Internal auditory canal enlargement in neurofibromatosis without acoustic neuroma. Radiology, 122: 730, 1977.
49. Sarwar, M., and Swischuk, L. E.: Bilateral internal auditory canal enlargement due to dural ectasia in neurofibromatosis. AJR, 129: 935, 1977.
50. Hagelstam, L.: On the deformities of the spine in multiple neurofibromatosis (von Recklinghausen). Acta Chir Scand, 93: 169, 1946.
51. Laws, J. W., and Pallis, C.: Spinal deformities in neurofibromatosis. J Bone Joint Surg, 45B: 674, 1963.
52. Casselman, E. S., and Mandell, G. A.: Vertebral scalloping in neurofibromatosis. Radiology, 131: 89, 1979.
53. Aegerter, E.: Personal communication, 1971.
54. Pitt, M. J., Mosher, J. F., and Edeiken, J.: Periosteum and abnormal bone in neurofibromatosis. Radiology, 103: 143, 1972.
55. Reuben, M. S.: von Recklinghausen's disease; review of the literature with report of case. Arch Pediatr, 51: 522, 1934.
56. Heublein, G. W., Pendergrass, E. P., and Widmann, B.

P.: Roentgenographic findings in the neurocutaneous syndromes. Radiology, 35: 701, 1940.
57. Moore, B. H.: Some orthopaedic relationships of neurofibromatosis. J Bone Joint Surg, 23A: 109, 1941.
58. Wescott, R. J., and Ackerman, L. V.: Elephantiasis neuromatosa. Arch Dermatol Syphiol, 55: 233, 1947.
59. Preston, J. M., Starshak, R. J., and Oechler, H. W.: Neurofibromatosis; unusual lymphangiographic findings. AJR, 132: 474, 1979.
60. Lantsov, V. P., and Kramorev, V. A.: State of the lymphatic system in neurofibromatosis (according to roentgenololymphographic data). Vestn Roentgenol Radiol, 2: 99, 1976.
61. Praharaj, K. C., Mohanta, K. D., Nanda, B. K., and Nanca, C. N.: Congenital hemihypertrophy with neurofibromatosis. Indian J Med Sci, 23: 146, 1969.
62. Pugh, D. G.: *Roentgenologic Diagnosis of Diseases of Bones.* Williams & Wilkins, Baltimore, 1954.
63. Gould, E. P.: The bone changes occurring in von Recklinghausen's disease. Q J Med, 11: 221, 1918.
64. Brailsford, J. F.: Discussion on generalized disease of bone in the adult. Proc R Soc Med, 41: 738, 1948.
65. Ducroquet, R.: A propos des pseudoarthroses et inflexions congenitales du tibia. Mem Acad Chir, 63: 863, 1937.
66. Barber, C. G.: Congenital bowing and pseudoarthrosis of the lower leg; manifestations of von Recklinghausen's neurofibromatosis. Gynecol Obstet, 69: 618, 1939.
67. Cleveland, R. H., Gilsanz, V., and Wilkinson, R. H.: Congenital pseudarthrosis of the radius. AJR, 130: 955, 1978.
68. Ferguson, A. B.: *Orthopedic Surgery in Infancy and Childhood,* Ed. 3. Williams & Wilkins, Baltimore, 1968.
69. Aegerter, E. R., and Kirkpatrick, J. A.: *Orthopedic Diseases,* Ed. 4. W. B. Saunders, Philadelphia, 1975.
70. Grunberg, L. A., and Schwartz, A.: Congenital pseudarthrosis of the distal radius. South Med J 68: 1053, 1975.
71. Sprague, B. L., and Brown, G. A.: Congenital pseudarthrosis of the radius. J Bone Joint Surg, 56A: 191, 1974.
72. Cobb, N.: Neurofibromatosis and pseudarthrosis of the ulna: a case report. J Bone Joint Surg, 50B: 146, 1968.
73. Shertzer, J. H., Bickel, W. H., and Stubbins, S. G.: Congenital pseudarthrosis of the ulna; report of two cases. Minn Med, 52: 1061, 1969.
74. Mollon, R. A. B., and Baird, D. S. C.: Pseudarthrosis of the radius. J R Coll Surg Edinb, 21: 376, 1976.
75. Richin, P. F., Kranik, A., Van Herpe, L., and Suffecool, S. L.: Congenital pseudarthrosis of both bones of the forearm. J Bone Joint Surg, 58A: 1032, 1976.
76. Moore, J. E.: Delayed autogenous bone graft in the treatment of congenital pseudarthrosis. J Bone Joint Surg, 31A: 23, 1949.
77. Madsen, E. T.: Congenital angulations and fractures of the extremities. Acta Orthop Scand, 25: 242, 1956.
78. Webb, W. R., and Goodman, P. C.: Fibrosing alveolitis in patients with neurofibromatosis. Radiology, 122: 289, 1977.
79. Davidson, L. M.: Neurofibromatosis with diffuse interstitial fibrosis and phaeochromocytoma. Br J Radiol, 40: 549, 1967.
80. Israel-Asselain, R., Chebat, J., Sors, C., et al.: Diffuse interstitial pulmonary fibrosis in a mother and son with von Recklinghausen's disease. Thorax, 20: 153, 1965.
81. Massaro, D., and Katz, S.: Fibrosing alveolitis; its occurrence, roentgenographic and pathologic features in von Recklinghausen's neurofibromatosis. Am Rev Respir Dis, 93: 934, 1966.
82. Massaro, D., Katz, S., Matthews, M. J., et al.: Von Recklinghausen's neurofibromatosis associated with cystic lung disease. Am J Med, 38: 233, 1965.
83. Patchefsky, A. S., Atkinson, W. G., Hock, W. S., et al.: Interstitial pulmonary fibrosis and von Recklinghausen's disease. An ultrastructural and immunofluorescent study. Chest, 64: 459, 1973.
84. Saxena, K. M.: Endocrine manifestations of neurofibromatosis in children. Am J Dis Child, 120: 265, 1970.
85. Costin, G., Fetterman, R. A., and Kogut, M. D.: Hypothalamic gigantism. J Pediatr, 83: 419, 1973.
86. Clark, D. H., Lumberton, N. C., and Mathews, W. R.: Simultaneous occurrence of von Recklinghausen's neurofibromatosis and osteitis fibrosa cystica. Report of a case showing both diseases in addition to liposarcoma. Surgery, 33: 434, 1953.
87. Foukas, M., and Skouteris, A.: Uber das Zusammentreffen von morbus Recklinghausen. Hyperparathyreoidismus und Graviditat. Zentralbl Gynaekol, 88: 999, 1966.
88. Daly, D., Kaye, M., and Estrada, R. L.: Neurofibromatosis and hyperparathyroidism—a new syndrome? Can Med Assoc J 103: 258–259, 1970.
89. Dent, C. E.: Metabolic forms of rickets (and osteomalacia). In *Inborn Errors of Calcium and Bone Metabolism,* pp. 124–149, H. Bickel, and J. Stern (eds.). University Park Press, Baltimore, 1976.
90. Saville, P. D., Nassim, J. R., Stevenson, F. H., Mulligan, L., and Carey, M.: Osteomalacia in von Recklinghausen's neurofibromatosis. Metabolic study of a case. Br J Med, 1: 1311, 1955.
91. Nordin, B. E. C., and Fraser, R.: Vitamin-D resistant osteomalacia associated with neurofibromatosis. Proc R Soc Med, 46: 302, 1953.
92. Balsan, S., Guivarch, J., Dartois, A. M., and Royer, P.: Rachitisme vitamine-resistant associe a une neurofibromatose probable chez un enfant. Arch Fr Pediatr, 24: 609, 1967.

Macrodystrophia Lipomatosa

1. Kelikian, H.: Macrodactyly. In *Congenital Deformities of the Hand and Forearm,* pp. 610–660, H. Kelikian (ed.). W. B. Saunders, Philadelphia, 1974.
2. Littler, W. J., Cramer, L. M., and Smith, J. W.: *Symposium on Reconstructive Hand Surgery,* pp. 218–220. C. V. Mosby, St. Louis, 1974.
3. Barsky, A. J.: Macrodactyly. J Bone Joint Surg, 49A: 1255, 1967.
4. Goldmany, A. B., and Kaye, J. J.: Macrodystrophia lipomatosa; radiographic diagnosis. AJR, 128: 101, 1977.
5. Minkowitz, S., and Minkowitz, F.: A morphological study of macrodactylism. A case report. J Pathol, 90: 323, 1965.
6. Moore, B. H.: Macrodactylism and associated peripheral nerve changes. J Bone Joint Surg, 26: 282, 1944.
7. Thorne, F. L., Posch, J. L., and Mladick, R. A.: Megalodactyly. Plast Reconstr Surg, 41: 232, 1968.
8. Tuli, S. M., Khanna, N. N., and Sinha, G. P.: Congenital macrodactyly. Br J Plast Surg, 22: 237, 1969.
9. Inglis, K.: Local gigantism (a manifestation of neurofibromatosis); its relation to general gigantism and to acromegaly. Illustrating the influence of intrinsic factors in disease when development is abnormal. Am J Pathol, 26: 1059, 1950.
10. Meszaros, W. T., Guzzo, F., and Schorsch, H.: Neurofibromatosis. AJR, 98: 557, 1966.
11. Pitt, M. J., Mosher, J. F., and Edeiken, J.: Abnormal periosteum and bone in neurofibromatosis. Radiology, 103: 143, 1972.

Essential Osteolysis

1. Shurfleff, D. B., Sparkes, R. S., Clawson, D. K., Guntheroth, W. G., and Mottet, N. K.: Hereditary osteolysis with hypertension and nephropathy. JAMA, 188: 363, 1964.

2. Thieffry, S., and Sorrel-Dejerine, J.: Forme speciale d' osteolyse essentielle hereditaire et familiale a stabilisation spontanee survenant das l'efance. Presse Med, 66: 1856, 1958.
3. Derot, M., Rathery, M., Rosselin, G., and Catellier, C.: Acroosteolyse du carpe, pied creux, scoliose et strabisme chez une jeune fille atteinte d'une insuffisance renale. Bull Mem Soc Med Hop Paris, 77: 223, 1961.
4. Mahoudeau, D., Dubrisay, J., Elissalde B., and Sraer, C.: Osteolyse essentielle et nephrite. Bull Mem Soc Med Hop, Paris, 77: 229, 1961.
5. Marie, Julien, Leveque, B., Lyon, G., Bebe, M., and Watchi, J. M.: Acro-osteolyse essentielle compliquee d' insuffisance renale d'evolution fatale. Presse Med, 71: 249, 1963.
6. Torg, J. S., and Steel, H. H.: Essential osteolysis with nephropathy; a review of the literature and case report of an unusual syndrome. J Bone Joint Surg, 50A: 1629, 1968.
7. Amin, P. H., and Evans, A. N. W.: Essential osteolysis of carpal and tarsal bones. Br J Radiol, 51: 539, 1978.
8. Lagier, R., and Rutushauser, E.: Osteoarticular change in a case of essential osteolysis; an anatomical and radiological study. J Bone Joint Surg, 47B: 339, 1965.
9. Mouchet, A.: Osteolyses post-traumatiques. Rev Rhum Mal Ostoartic, 10: 43, 1943.
10. Crasselt, C.: Die akroosteolyse; 2Teil, zur Differtialdiagnose der lokalisierten akroosteolyse und die atiologie des akroosteolysesyndroms. Z Orthop, 94: 33, 1961.
11. Thevenard, A.: L'acropathie ulcero-mutilante familiale. Acta Neurol Psychiatr Belg, 53: 1, 1953.
12. Lievre, J. A., and Gama, G.: L'acroosteolyse. Bull Mem Soc Med Hop Paris, 73: 109, 1957.
13. Eisenstadt, H. B., and Eggers, G. W. N.: Arthritis mutilans (Doigt, Main, Pied en Lorgnette). J Bone Joint Surg, 37A: 337, 1955.

Acro-osteolysis

1. Harnasch, H.: Acro-osteolysis, new disease picture. Fortschr Geb Roentgenstr, 72: 352, 1950.
2. Giaccai, L.: Familial and sporadic neurogenic acro-osteolysis. Acta Radiol, 38: 17, 1952.
3. Harms, I.: Familial acro-osteolysis. Fortschr Geb Roentgenstr, 80: 727, 1954.
4. Cheney, W. D.: Acro-osteolysis. AJR, 94: 595, 1965.
5. Steinbach, H. L., Gold, R. H., and Preger, L.: Acroosteolysis. In *Roentgen Appearance of the Hand in Diffuse Disease*. Year Book, Chicago, 1975.
6. Dorst, J. P., and McKusick, V. A.: Acro-osteolysis (Cheney's syndrome). Birth Defects, 5: 215, 1969.

Tuberous Sclerosis

1. Bourneville, D.: Contribution a l'etude de l'idiotie. Observation III: Scléreuse tubéreuse des circonvolution cérébales; idiotie et epilepsie hémiplégique. Arch Neurol 1: 81, 1880.
2. Cooper, J. R.: Brain tumors in hereditary multiple system hamartomatosis (tuberous sclerosis). J Neurosurg, 34: 194, 1971.
3. Moolten, S. E.: Hamartial nature of the tuberous sclerosis complex and its bearing on the tumor problem; report of a case with tumor anomaly of the kidney and adenoma sebaceum. Arch Intern Med, 69: 589, 1942.
4. Medley, B. E., McLeod, R. A., and Houser, W.: Tuberous sclerosis. Semin Roentgenol, 11: 35, 1976.
5. Herz, D. A., Liebeskind, A., Rosenthal, A., and Schechter, M. M.: Cerebral angiographic changes associated with tuberous sclerosis. Radiology, 115: 647, 1975.
6. Penrose, L. S.: Autosomal mutation and modification in man with special reference to mental defect. Ann Eugen, 7: 1, 1936.
7. Ashby, D. W., and Ramage, D.: Lesions of the vertebrae and innominate bones in tuberous sclerosis. Br J Radiol, 30: 274, 1957.
8. Brain, W. R.: *Brain's Disease of the Nervous System*. Oxford University Press, London, 1969.
9. Holt, J. F., and Dickerson, W. W.: The osseous lesions of tuberous sclerosis. Radiology, 58: 1, 1952.
10. Ross, A. T., and Dickerson, W. W.: Tuberous sclerosis. Arch Neurol Psychiatr, 50: 233, 1943.
11. Teplick, J. G.: Tuberous sclerosis; extensive roentgen findings without the usual clinical picture, a case report. Radiology, 93: 53, July, 1969.
12. Gold, A. P., and Freeman, J. M.: Depigmented nevi; the earliest signs of tuberous sclerosis. Pediatrics, 35: 1003, 1965.
13. Sareen, C. K., Ruvalcaba, R. H. A., Scotvold, M. J., Mahoney, C. P., and Kelley, V. C.: Tuberous sclerosis. Clinical, endocrine and metabolic studies. Am J Dis Child, 123: 34, 1972.
14. Globus, J. H., Strauss, I., and Selinsky, H.: Das Neurospongioblastoma, eine primere Gehirngeschwulst bei disseminierter Neurospongioblastose (tuberose sklerose). Z Gesamte Neurol Psychiatr, 140: 1, 1932.
15. Berg, G., and Nordenskjold, A.: Pulmonary alterations in tuberous sclerosis. Acta Med Scand, 125: 428, 1946.
16. Buhl, L.: Lungenentzundung, Tuberculose und Schwindsucht, p. 58. Munchen, 1872.
17. Rubenstein, L., Gutstein, W. H., and Lepow, H.: Pulmonary muscular hyperplasia (muscular cirrhosis of the lungs). Ann Intern Med, 42: 36, 1955.
18. Ackerman, A. J.: Pulmonary and osseous manifestations of tuberous sclerosis, with some remarks on their pathogenesis. AJR, 51: 315, 1944.
19. Budenz, G. C.: Tuberous sclerosis, a neurocutaneous syndrome. Report of a case. Radiology, 55: 522, 1950.
20. Crosett, A. D., Jr.: Roentgenographic findings in the renal lesion of tuberous sclerosis. AJR, 98: 739, 1966.
21. Gottlieb, J. S., and Lavine, G. R.: Tuberous sclerosis with unusual lesions of bone. Arch Neurol Psychiatr, 33: 379, 1935.
22. Khilnani, M. T., and Wolf, B. S.: Hamartolipoma of the kidney; clinical and roentgen features. AJR, 86: 830, 1961.
23. Milledge, R. D., Gerald, B. E., and Carter, W. J.: Pulmonary manifestations of tuberous sclerosis. AJR, 98: 734, 1966.
24. MacCarty, W. C., Jr., and Russell, D. G.: Tuberous sclerosis; report of a case with ependymoma. Radiology, 71: 833, 1958.
25. Dalsgaard-Nielsen, T.: Tuberous sclerosis with unusual roentgen picture. Nord Med Tidsskr, 10: 1541, 1935.
26. Dickerson, W. W.: Characteristic roentgenographic changes associated with tuberous sclerosis. Arch Neurol Psychiatr, 53: 199, 1945.
27. Kveim, A.: Uber Adenoma Sebaceum (Morbus Pringle), und seinen Platz im neurokutonen Syndrom—tuberose Gehirnsklerose—und dessen Beziehung zur v. Recklinghausenschen Krentheit. Acta Derm Venerol, 18: 637, 1937.
28. Hall, G. S.: Tuberose sclerosis, rheostosis, and neurofibromatosis. Q J Med, 9: 1, 1940.
29. Green, G. J.: The radiology of tuberose sclerosis. Clin Radiol, 19: 135, 1968.
30. Marcus, H.: Svenska Lak Salisk, Forh. p. 1412; cited by Dickerson, W. W., in Arch Neurol Psychiatr, 53: 199, 1945.
31. Whitaker, P. H.: Radiological manifestations in tuberose

sclerosis. Br J Radiol, 32: 152, 1959.
32. Harkins, T. D.: Radiological bone changes in tuberose sclerosis. Br J Radiol, 32: 157, 1959.

Osteopoikilosis

1. Stieda, A.: Ueber umschriebene Knochenverdichtungen im Bereich der Substantia spongiosa im rontgenbilde. Beitr Klin Chir, 45: 700, 1905.
2. Wachtel, H.: Ueber einen Fall von Osteopathia condensans disseminata. Fortschr Geb Rontgenstr, 27: 624, 1919-20.
3. Buschke, A., and Ollendorff, H.: Ein Fall von Dermatofibrosis lenticularis disseminata und Osteopathia condensans disseminata. Dermatol Wochenschr, 86: 257, 1928.
4. Windholz, F.: Verlaufsbeobachtungen bei osteopoikilie. Fortschr Geb Rontgenstr, 48: 720, 1933.
5. Svab, V.: A propos de l'osteopoecilie hereditaire. J Radiol Electrol, 16: 405, 1932.
6. Busch, K. F. B.: Familial disseminated osteosclerosis. Acta Radiol, 18: 693, 1937.
7. Lindbom, A.: Zwei neue falle mit streifenformiger osteopoikilie (Voorhoeve). Acta Radiol, 23: 296, 1942.
8. Holly, L. E.: Osteopoikilosis; 5-year study. AJR, 36: 512, 1936.

Osteopathia Striata

1. Voorhoeve, N.: L'image radiologique non encore decrite d'une anomalie de la squellette. Acta Radiol, 3: 407, 1924.
2. Hurt, R. L.: Osteopathia striata-Voorhoeve's disease; report of a case presenting the features of osteopathia striata and osteopetrosis. J Bone Joint Surg, 35B: 89, 1953.
3. Fairbank, H. A. T.: Osteopathia striata. J Bone Joint Surg, 32: 117, 1950.
4. Ribbing, S.: Hereditary, multiple, diaphyseal sclerosis. Acta Radiol, 31: 522, 1949.

Melorheostosis

1. Putti, V.: L'osteosi eberneizzante monomelica (una nuova sindrome osteopatica). Chir Organi Mov, 11: 335, 1927.
2. Franklin, E. L., and Matheson, I.: Melorheostosis; report on case with review of literature. Br J Radiol, 15: 185, 1942.
3. Fairbank, H. A. T.: *An Atlas of General Affections of the Skeleton*. Williams & Wilkins, Baltimore, 1951.
4. Soffa, D. J., Sire, D. J., and Dodson, J. H.: Melorheostosis with linear sclerodermatous skin changes. Radiology, 114: 577, 1975.
5. Campbell, C. J., Papademetriou, R., and Bonfiglio, M.: Melorheostosis. J Bone Joint Surg, 50A: 1281, 1968.
6. Clement, R., and Combes-Hamelle, A.: Melorheostose et sclerodermie en bandes. Osteopycnose et histopycnose. Bull Mem Soc Med Hop Paris, 58: 423, 1943.
7. Dillehunt, R. B., and Chuinard, E. G.: Melorheostosis Leri; a case report. J Bone Joint Surg, 18: 991, 1936.
8. Ernsting, G.: Weichteilveranderungen als diagnostisches Leit-symptom der Melorheostose. Z Orthop, 102: 126, 1966.
9. Gillespie, J. B., and Siegling, J. A.: Melorheostosis Leri. Am J Dis Child, 55: 1273, 1938.
10. Maroteaux, P., and Lamy, M.: Melorheostose, osteopoecilie et sclerodermie en bandes. Ann Pediatr (Paris), 8: 576, 1961.
11. Muller, S. A., and Henderson, E. D.: Melorheostosis with linear scleroderma. Arch Dermatol, 88: 142, 1963.
12. Thompson, N. M., Allen, C. E. L., and Andrews, G. S., et al.: Scleroderma and melorheostosis. Report of a case. J Bone Joint Surg, 33B: 430, 1951.
13. Wagers, L. T., Young, A. W., Jr., and Ryan, S. F.: Linear melorheostotic scleroderma. Br J Dermatol, 86: 297, 1972.
14. Morris, J. M., Samilson, R. L., and Corley, C. L.: Melorheostosis. Review of the literature and report of an interesting case with a nineteen year followup. J Bone Joint Surg, 45A: 1191, 1963.
15. Léri, A., and Joanny, J.: Une affection non decrite des os; hyperostose "en coulee" sur tout la longueur d'un membre ou melorheostose. Bull Mem Soc Med Hop Paris, 46: 1141, 1922.
16. Murray, R. O., and McCredie, J.: Melorheostosis and the sclerotomes; a radiological correlation. Skeletal Radiol, 4: 57, 1979.
17. Inman, V. I., and Saunders, J. B. deC.: Referred pain from skeletal structures. J Nerv Ment Dis, 99: 660, 1944.
18. Caffey, J.: *Pediatric X-ray Diagnosis*. Year Book, Chicago, 1961.
19. Beauvais, P., Faure, C., Montagne, J-P., Chigot, P. L., and Maroteaux, P.: Leri's melorheostosis: three pediatric cases and a review of the literature. Pediatr Radiol, 6: 153, 1977.

Hereditary Multiple Exostosis

1. Keith, A.: Studies on the anatomical changes which accompany certain growth-disorders of the human body. J Anat, 54: 101, 1920.
2. Ehrenfried, A.: Hereditary deforming chondrodysplasiamultiple cartilaginous exostosis. JAMA, 68: 502, 1917.
3. Stocks, P., and Barrington, A.: Hereditary disorders of bone development. I. Diaphyseal aclasis (multiple exostoses), multiple enchondromata, cleido-cranial dysostosis. In *Treasury of Human Inheritance*, Vol. 3. Cambridge University Press, London, 1925.
4. Aegerter, E., and Kirkpatrick, J. A.: *Orthopedic Diseases*. W. B. Saunders, Philadelphia, 1964.
5. Langenskiold, A.: Normal and pathological bone growth in the light of the development of cartilaginous foci in chondro-dysplasia. Acta Chir Scand, 95: 367, 1947.
6. Epstein, D. A., and Levin, E. J.: Bone scintigraphy in hereditary multiple exostoses. AJR, 130: 331, 1978.
7. Gilday, D. L., and Ash, J. M.: Benign bone tumors. Semin Nucl Med, 6: 33, 1976.
8. Madigan, R., Worrall, T., and McClain, E. I.: Cervical cord compression in hereditary multiple exostosis. J Bone Joint Surg, 56A: 401, 1974.
9. Adam, H., and Morin, P.: Compression medullaire par exostose au cours d'une maladie exostosante. Presse Med, 74: 660, 1966.
10. Cannon, J. F.: Hereditary multiple exostoses. Am J Hum Genet, 6: 419, 1954.
11. Carmel, P. W., and Crammer, F. J.: Cervical cord compression due to exostosis in a patient with hereditary multiple exostoses. J Neurosurg, 28: 500, 1968.
12. Decker, R. D., and Wei, W. C.: Thoracic cord compression from multiple hereditary exostoses associated with cerebellar astrocytoma. Case report. J Neurosurg, 30: 310, 1969.
13. Gokay, H., and Bucy, P. C.: Osteochondroma of the lumbar spine. Report of a case. J Neurosurg, 12: 72, 1955.
14. Goldenberg, D. B., Reinhoff, W. F., III, and Rao, P. S.: Osteochondroma with spinal cord compression. Report of a case. J Can Assoc Radiol, 19: 192, 1968.
15. Goncharova, R. F., Dogaeva, M. A., and Koraidi, L. S.: Spinal cord compression by exostosis. Vopr Neurokhir, 29: 60, 1965.
16. Larson, N. E., Dodge, Jr., H. W., Rushton, J. G., and Dahlin, D. C.: Hereditary multiple exostoses with com-

pression of the spinal cord. Proc Staff Meet Mayo Clin, 32: 728, 1957.
17. Ochsner, E. H., and Rothstein, T.: Multiple exostoses, including an exostosis within the spinal canal with surgical and neurological observations. Ann Surg, 46: 608, 1907.
18. Slepian, A., and Hamby, W. B.: Neurologic complications associated with hereditary deformity chondrodysplasia. Review of the literature and a report of two cases occurring in the same family. J Neurosurg, 8: 529, 1951.
19. Stanley, B.: *Illustrations of the Effects of an Injury of the Bone*, p. 29. Longham, Brown and Co., London, 1849.
20. Vinstein, A. L., and Franken, Jr., E. A.: Hereditary multiple exostoses. Report of a case with spinal cord compression. AJR, 112: 405, 1971.

Progressive Diaphyseal Dysplasia

1. Neuhauser, E. B. D., et al.: Progressive diaphyseal dysplasia. Radiology, 51: 11, 1948.
2. Engelmann, G.: Osteopathia hyperostotica sclerotisans multiplex infantilis. Fortschr Geb Rontgenstr, 39: 1101, 1929.
3. Ribbing, S.: Hereditary multiple diaphyseal sclerosis. Acta Radiol, 31: 522, 1949.
4. Caffey, J.: *Pediatric X-ray Diagnosis*. Year Book, Chicago, 1961.
5. Sear, H. R.: Engelmann's disease. Br J Radiol, 21: 236, 1948.
6. Bingold, A. C.: Engelmann's disease; osteopathia hyperostotica (sclerotisans) multiplex infantilis; progressive metaphyseal dysplasia. Br J Surg, 37: 266, 1950.
7. Singleton, E. B., et al.: Progressive diaphyseal dysplasia (Engelmann's disease). Radiology, 67: 233, 1956.

Cleidocranial Dysostosis

1. Soule, A. B., Jr.: Mutational dysostosis (cleidocranial dysostosis). J Bone Joint Surg, 28A: 81, 1946.
2. Cutter, E.: Descriptive Catalogue of the Warren Anatomical Museum (J. B. S. Jackson), 21, No. 217, Boston, 1870.
3. Marie, P., and Sainton, P.: Sur la dysostose cleidocranienne hereditaire. Rev Neurol, 6: 835, 1898.
4. Rhinehart, B. A.: Cleidocranial dysostosis (mutational dysostosis) with a case report. Radiology, 26: 741, 1936.
5. Jarvis, J. L., and Keats, T. E.: Cleidocranial dysostosis. A review of 40 new cases. AJR, 121: 5, 1974.
6. Anspach, W. E., and Huepel, R. C.: Familial cleidocranial dysostosis (cleidal dysostosis). Am J Dis Child, 58: 786, 1939.
7. Alldred, A. J.: Congenital pseudoarthrosis of the clavicle. J Bone Joint Surg, 45B: 312, 1963.
8. Caffey, J.: *Pediatric X-ray Diagnosis*. Year Book, Chicago, 1961.

Iliac Horns

1. Fong, E. E.: "Iliac horns" (symmetrical bilateral central posterior iliac processes); a case report. Radiology, 47: 517, 1946.
2. Zimmerman, C.: Iliac horns; a pathognomonic roentgen sign of familial onycho-osteodysplasia. AJR, 86: 478, 1961.
3. Hawkins, C. F., and Smith, O. E.: Renal dysplasia in a family with multiple hereditary abnormalities including iliac horns. Lancet, 1: 803, 1950.
4. Roeckrath, W.: Hereditary osteo-onycho-dysplasia. Fortschr Geb Roentgenstr Nuklearmed, 75: 700, 1951.
5. Thompson, E. A., Walker, E. T., and Weens, H. S.: Iliac horns (an osseous manifestation of hereditary arthrodysplasia associated with dystrophy of the fingernails). Radiology, 53: 88, 1949.
6. Caffey, J.: *Pediatric X-ray Diagnosis*. Year Book, Chicago, 1961.
7. Wasserman, D.: Unilateral iliac horn (central posterior iliac process). Case report. Radiology, 120: 562, 1976.

Congenital Insensitivity to Pain

1. McMurray, G. A.: Congenital insensitivity to pain and its implications for motivational theory. Can J Psychol, 9: 121, 1955.
2. Denny-Brown, D.: Hereditary sensory radicular neuropathy. J Neurol Neurosurg Psychiatry, 14: 237, 1951.
3. Mandell, A. J., and Smith, C. K.: Hereditary sensory radicular neuropathy. Neurology, 10: 627, 1960.
4. Munro, M.: Sensory radicular neuropathy in a deaf child. Br Med J, 1: 541, 1956.
5. Parks, H., and Staples, O. S.: Two cases of Morvan's syndrome of uncertain cause. Arch Intern Med, 75: 75, 1945.
6. Walker, C. H. M.: Sensory radicular neuropathy; report on 2 cases. Great Ormond St. J, 10: 72, 1955-56.
7. Burr, C. W.: Two cases of general anesthesia. Univ Med Mag Univ Pa, 13: 245, 1900-01.
8. Berkley, H. J.: The pathological findings in a case of general cutaneous and sensory anaesthesia without physical implications. Brain, 23: 111, 1900.
9. Carezzano, P.: Il sintomo del lombroso come carattere familiare. Arch Antropol Criminol Torino, 48: 827, 1928.

Anterior Tibia Bowing

1. Weismann-Netter, R., and Stuhl, L.: D'une ostéopathie congénitale éventuellement familiale. Presse Med, 62: 1618, 1954.
2. Heully, F., Gaucher, A., Gaucher, P., Laurent, J., and Vautrin, D.: Une curieuse association; maladie de Weismann-Netter et Stuhl, maladie de Biermer, goitre congenital. Presse Med, 75: 1577, 1967.
3. Keats, T. E., and Alavi, M. S.: Toxopachyostéose diaphysaire tibio-péronière (Weismann-Netter syndrome). AJR, 109: 568, 1970.
4. Larcan, A., Cayotte, L., Gaucher, A., and Bertheau, J. M.: La toxopachyosteose de Weismann-Netter. Ann Med Nancy, 2: 1724, 1963.

Ainhum

1. Earle, K. V.: Ainhum of the fingers; case from Sierra Leone. Trans R Soc Trop Med Hyg, 52: 570, 1958.
2. Browne, S. G.: Ainhum; clinical and etiological study of 83 cases. Ann Trop Med, 55: 314, 1961.
3. Kean, B. H., Tucker, H. A., and Miller, W. C.: Ainhum; clinical summary of 45 cases on Isthmus of Panama. Trans R Soc Trop Med Hyg, 39: 331, 1946.
4. Spinzig, E. W.: Ainhum; its occurrence in United States, with report of 3 cases. AJR, 42: 246, 1939.
5. Fetterman, L. E., Hardy, R., and Lehrer, H.: The clinicoroentgenologic features of ainhum. AJR, 100: 512, 1967.
6. Cole, G. J.: Ainhum; account of 54 patients with special reference to etiology and treatment. J Bone Joint Surg, 47B: 43, 1965.

Gardner Syndrome

1. Watne, A. L., Johnson, J. G., and Chang, C. H.: The challenge of Gardner's syndrome. Cancer, 19: 267, 1966.
2. Gardner, E. J.: A genetic and clinical study of intestinal polyposis, a predisposing factor for carcinoma of the colon and rectum. Am J Hum Genet, 3: 167, 1951.
3. Gardner, E. J., and Plenk, H. P.: Hereditary pattern for

multiple osteomas in a family group. Am J Hum Genet, 4: 31, 1952.
4. Gardner, E. J., and Richard, R. C.: Multiple cutaneous and subcutaneous lesions occurring simultaneously with hereditary polyposis and osteomatosis. Am J Hum Genet, 5: 139, 1953.
5. Chang, C. H. (J.), Piatt, E. D., Thomas, K. E., and Watne, A. L.: Bone abnormalities in Gardner's syndrome. AJR, 103: 645, 1968.
6. Schulman, A.: Gastric and small bowel polyps in Gardner's syndrome and familial polyposis coli. J Assoc Can Radiol, 27: 206, 1976.
7. Bussey, H. J. R.: Gastrointestinal polyposis. Gut, 11: 970, 1970.
8. Morson, B. C., and Dawson, I. M. P.: Familial adenomatous polyposis. In *Gastrointestinal Pathology*. Blackwell, Oxford, 1972.
9. Sachatello, C. R.: Familial polyposis of the colon—a four decade follow-up. Cancer, 28: 581, 1971.

Larsen Syndrome

1. Kaijser, R.: Ueber kongenitale Kniegelenksluxationen. Acta Orthop Scand, 6: 1, 1935.
2. Larsen, L. J., Schottstaedt, E. R., and Bost, F. C.: Multiple congenital dislocations associated with characteristic facial abnormality. J Pediatr, 37: 574, 1950.
3. Latta, R. J., Graham, C. B., Aase, J., Scham, S. M., and Smith, D. W.: Larsen's syndrome; a skeletal dysplasia with multiple joint dislocations and unusual facies. J Pediatr, 78: 291, 1971.
4. Azimi, F., Edeiken, J., and Macewen, G. D.: Larsen's syndrome. Congenital dislocation of multiple large joints of the extremities associated with an unusual flat facies. Australas Radiol 18: 333, 1974.
5. Chandler, F. A.: Congenital dislocation of hip. J Bone Joint Surg, 11A: 456, 1929.
6. Curtis, B. H., and Fisher, R. L.: Congenital hyperextension with anterior subluxation of the knee. J Bone Joint Surg, 11: 456, 1929.
7. Hass, J., and Hass, R.: Arthrochalasis multiplex congenita; congenital flaccidity of joints. J Bone Joint Surg, 40A: 663, 1958.
8. Ichiseki, H., and Kuno, M.: Case of multiple abnormalities with congenital shoulder dislocation, fusion of the humerus and radius and defect of frontal bone. Orthop Surg, 19: 740, 1968.
9. Kellog: Quoted by Weil, S.: Die angeborenen Erkrankungen der Schiltergegend und des Schultergelenkes. In *Handbuch der Orthopedie*, Ban III: 31. Georg Thieme Verlag, Stuttgart, 1959.
10. Koehler, A.: Mitteilung eines Falles angeborener Luxation der unteren Extremitaeten. Z Orthop Chir, 58: 401, 1933.
11. Legal: Quoted by Weil, S.: Die angeborenen Erkrankunger er Schultergegend und des Schultergelenkes. In *Handbuch der Orthopedie*, Band III: 32. Georg Thieme Verlag, Stuttgart, 1959.
12. McFarland, B. L.: Congenital dislocation of the knee. J Bone Joint Surg, 11A: 281, 1929.
13. Sorrel, E., and Grand-Lambling: Triple malformation congenitale de members inferieurs. Bull Soc Pediatr Paris, 30: 169, 1932.
14. Valentin, B.: Die kongenitale Schulterluxation; Bericht ueber drei Faelle in einer Familie. Z Orthop Chir, 55: 229, 1931.
15. Weil, S.: Die angeborenen Erkrankungen der Schultergegend und des Schultergelenkes. In *Handbuch der Orthopedie*, Band III, 31. Georg Thieme Verlag, Stuttgart, 1959.
16. Wolff, S.: Ueber einen Fall von kongenitaler Schulterluxation. Z Orthop Chir, 51: 199, 1929.

Progeria

1. Hutchinson, J.: Congenital absence of hair and mammary glands with atrophic condition of the skin and its appendages in a boy whose mother had been almost wholly bald from alopecia areata from the age of 6. Trans Med Chir Soc Edinburgh, 69: 473, 1886.
2. Gilford, H.: Progeria; a form of senilism. Practitioner, 73: 188, 1904.
3. Marie, J., Veveque, B., Hesse, J. C., and Buri, J.: Nanism avec retinite pigmentaire et surdite; syndrome de Cockayne. Sem Hop Paris, 34: 2808, 1958.
4. Riggs, W., and Seibert, J.: Cockayne's syndrome. Roentgen findings. AJR, 116: 623, 1972.
5. Wadia, R. S.: Progeria. In *The Clinical Delineation of Birth Defects, II: Malformation Syndromes*, pp. 247–249. Williams & Wilkins, Baltimore, 1969.
6. DeBusk, F. L. Progeria. Clin Pediatr, 10: 62, 1971.
7. Franklyn, P. P.: Progeria in siblings. Clin Radiol, 27: 327, 1976.

Familial Osteodysplasia

1. Buchignani, J. S., Cook, A. J., and Anderson, L. G.: Roentgenographic findings in familial osteodysplasia. AJR, 116: 602, 1972.
2. Anderson, L. G., Cook, A. J., Coccaro, P. J., and Bosma, J. F.: Familial osteodysplasia. JAMA, 220: 1687, 1972.

16

Mucopolysaccharidoses

The mucopolysaccharidoses are genetically determined disorders of mucopolysaccharide metabolism, resulting in a wide variety of skeletal, visceral, and mental abnormalities.

Mucopolysaccharides are macromolecules of uronic acids and hexosamine distributed through the ground substance of connective tissues. They are important components of cartilage, the cornea, the vascular walls, and subcutaneous tissues (Table 54). Chondroitin sulfate B, heparitin sulfate, and keratosulfate, to date, have been implicated in genetically determined disorders.

The normal urinary excretion of mucopolysaccharides is 5–15 mg per day, consisting of approximately 80% chondroitin sulfate A, 10% chondroitin sulfate B, and 10% heparitin sulfate.[1] Disturbed mucosaccharide metabolism may be detected by abnormally high percentages of mucopolysaccharides in the urine. Mucopolysaccharide excretion may be normally greater in infants and children,[2] but does not reach the levels present with mucopolysaccharidoses. Metachromic granules in circulating lymphocytes and bone marrow cells[3] (Reilly bodies) may be detected in some cases, and represent mucopolysaccharide inclusions.[4] However, there are cases which defy classification, because either there is no mucopolysacchariduria, or the identical mucopolysaccharide is implicated in different syndromes. Also, one mucopolysaccharide may be present in the urine, and another deposited in body tissues. Mucopolysacchariduria may be found in conditions other than mucopolysaccharidosis, including multiple exostosis[5] (hyaluronic acid and chondroitin sulfate A, B, and C), Marfan[6] syndrome (hyaluronic acid and chondroitin sulfate A and B), rheumatic disease, neoplasia, cardiac decompensation, arterial hypertension, hepatic cirrhosis, glomerulonephritis, diabetes mellitus, rheumatoid arthritis, and other collagen diseases.[7]

Hurler syndrome and Morquio syndrome, originally considered chondro-osteodystrophy, lipochondrodystrophy, or dysplasia, are known to result from genetically determined disorders of mucopolysaccharide metabolism. Previously classified by clinical and roentgenographic features, a wide spectrum of changes was noted and differentiation was difficult or impossible. Sophisticated biochemical analysis now identifies six distinct entities. The six are well known by eponyms and are numbered by McKusick[8]: MPS-I—Hurler (gargoylism, MPS-II—Hunter, MPS-III—Sanfilippo, MPS-IV—Morquio, MPS-V—Scheie, and MPS-VI—Maroteaux-Lamy.

MPS-I—HURLER SYNDROME (GARGOYLISM)

Synonyms: lipochondrodystrophy, gargoylism,[1] osteochondrodystrophy, dysostosis multiplex,[2] and Pfaundler-Hurler disease.

The Hurler syndrome is a rare congenital disturbance of mucopolysaccharide metabolism.[3] Lipoid accumulation in the central nervous system[4] and other organs[5] is a secondary manifestation. It is inherited as an autosomal recessive[6] and occurs in approximately 1 of 10,000 births.[7]

Clinical Features. Clinical features are rarely present at birth. They may commence in early infancy, but usually appear after the first year of life. Progressive mental and physical deterioration is the rule, and death results at a younger age than in other mucopolysaccharidoses. The physical features are large head with eyes wide apart, sunken bridge of the nose, everted lips, and protruding tongue. The teeth are widely separated and poorly formed, and the facies heavy and ugly. The abdomen is protuberant, the liver and spleen enlarged. The hands are trident, with stiff joints, and clawing develops early. There is dorsolumbar kyphosis. Corneal opacities develop after the first year and usually progress to blindness. Progressive mental retardation is the rule.

Growth disturbance results in dwarfism and restriction of motion of the joints, with flexion deformity of the knees and hips. Patients who live to maturity are mentally deficient, blind, deformed, dwarfed, and incapable of caring for themselves.

TABLE 54. GENETIC MUCOPOLYSACCHARIDOSES (MPS)

MPS	Mental	Skeletal	Dwarfism	Corneal	Genetic	Biochemical	Remarks
I. Hurler	+++ (early)	+++	+++	+++ (diffuse)	Autosomal recessive	Chondroitin sulfate B, 80%; heparitin 10%	Early corneal clouding and severe mental deterioration, grave prognosis
II. Hunter	++ (late)	++	++	+/0	X-linked recessive	Chondroitin sulfate B, 55%; heparitin sulfate	No corneal clouding; mild course
III. Sanfilippo	+++ (early)	+	+	+/0	Autosomal recessive	Heparitin sulfate	Mild somatic; severe mental deterioration
IV. Morquio	0/+	+++	+++	++ (late)	Autosomal recessive	Keratosulfate	Severe characteristic skeletal changes; usually unimpaired mentally
V. Scheie	0	+/0	0/+	+++ (peripheral)	Autosomal recessive	Chondroitin sulfate B	Aortic regurgitation; frequent psychosis; unimpaired mentally
VI. Maroteaux-Lamy	0	+++	+++	+++	Autosomal recessive	Chondroitin sulfate B	Early corneal clouding grave manifestation; mentally unimpaired

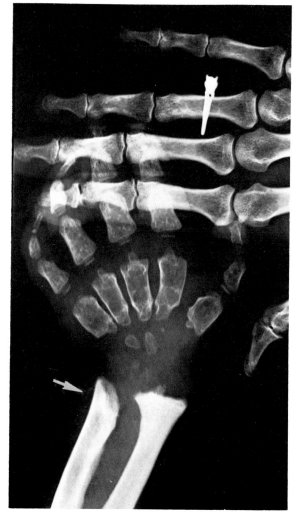

FIG. 1526. MPS-I (HURLER SYNDROME)
The proximal ends of the metacarpals are pointed, characteristic of the condition. The distal end of the ulna is tapered and the articular surface apposes the radius. (Courtesy of John W. Hope, Children's Hospital, Philadelphia, Pa.)

The diagnosis is confirmed biochemically by the presence of excessive mucopolysacchariduria of chondroitin sulfate B and heparitin sulfate, and by the presence of metachromic granules (Reilly bodies) in circulating white blood cells or bone marrow cells.

Roentgenographic Features. The skeletal changes at birth consist of thick periosteal cloaking of long bone diaphyses, generalized osteoporosis, and metaphyseal cupping.[8] However, these early changes regress and there may be a period when the bones appear normal.

Later, changes in tubular bones are most marked in the upper extremities and characteristically show swellings in the central portions and tapering of one or both ends. The latter is conspicuous at the distal ends of the humerus, radius, and ulna, and the proximal end of the metacarpals (Figs. 1526 and 1527). The tapered articular ends of the radius and ulna face each other. Enlargement of the diaphysis may result from cortical thickening, but the cortex often is thin, and enlargement of the shaft is due to dilation of the medullary canal. The ribs are wide and tapered at both ends (Fig. 1528). Deossification is constant and prominent in the hands and tubular bones. Lower extremities exhibit less dramatic change. The proximal femur is constricted but the remainder of the femur may be enlarged. Bone tapering is not frequent. Coxa valga or vara is common. The pelvis presents flaring of iliac wings and tapering of rami (Fig. 1528).

The pituitary fossa is almost always enlarged; its J-shaped (Fig. 1529) walls are well formed and not eroded. The enlargement may occur without hydrocephalus or pituitary tumor and is due to a disturbance of sphenoid bone growth.

Vertebral changes are distinctive and, except for dorsolumbar scoliosis, are not similar to those of Morquio syndrome. The centra are oval due to a convexity of upper and lower surfaces. The height of the centrum is either increased or normal, in contrast to the platyspondyly of Morquio

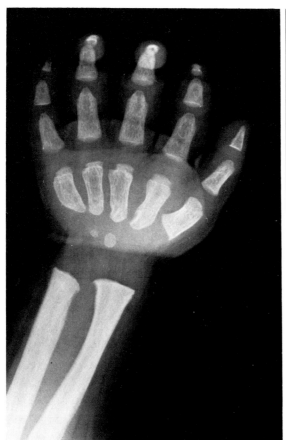
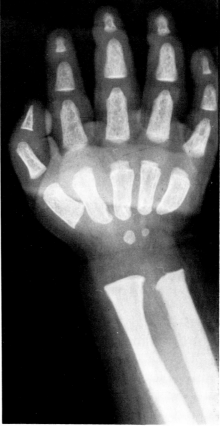

FIG. 1527. MPS-I (HURLER SYNDROME)
Deossification and tapering of the proximal end of the metacarpal. The distal ends of the radius and ulna show minimal changes. (Courtesy of John W. Hope, Children's Hospital, Philadelphia, Pa.)

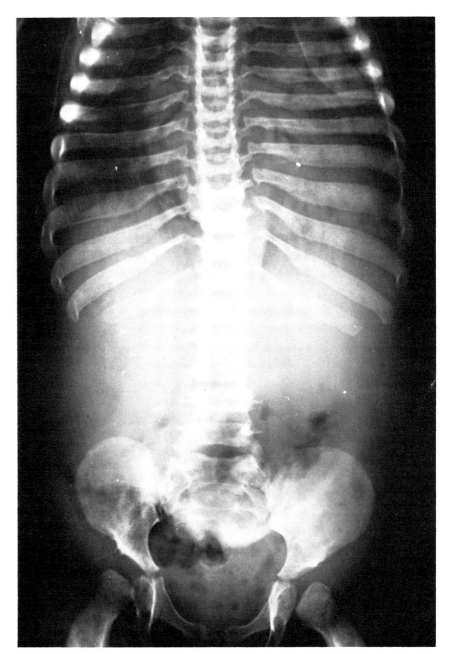

Fig. 1528. MPS-I (Hurler Syndrome)
The ribs are widened and tapered at the vertebral end. Tapering of the caudal portions of the iliac bones and minor changes in the hips. (Courtesy of John W. Hope, Children's Hospital, Philadelphia, Pa.)

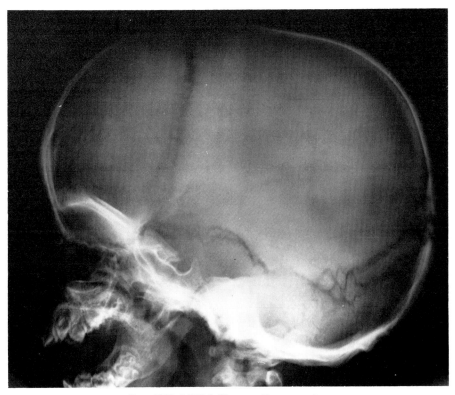

FIG. 1529. MPS-I (HURLER SYNDROME)

The pituitary fossa is elongated and J-shaped. The walls of the sella are well formed and there is elongation of the anterior clinoids. (Courtesy of John W. Hope, Children's Hospital, Philadelphia, Pa.)

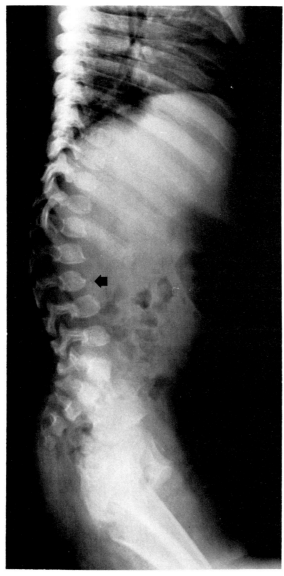

FIG. 1530. MPS-1 (HURLER SYNDROME)
Vertebral changes are distinctive and have oval centra with convexity of the inferior and superior surfaces. The height of the centra is increased. The third lumbar vertebra (*arrow*) is hypoplastic and displaced posteriorly. Anteroinferior beaking of the upper lumbar vertebrae. (Courtesy of John W. Hope, Children's Hospital, Philadelphia, Pa.)

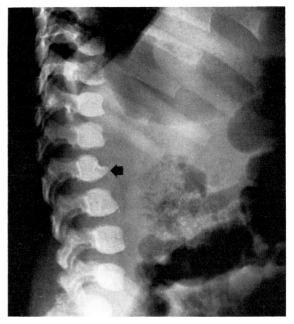

FIG. 1531. MPS-I (HURLER SYNDROME)
Prominent anterior beaking of the second lumbar vertebra (*arrow*) which is hypoplastic and displaced posteriorly. The oval vertebrae and convexity of the inferior and superior surfaces serve to differentiate it from Morquio's syndrome. (Courtesy of John W. Hope, Children's Hospital, Philadelphia, Pa.)

syndrome (Fig. 1530). The centrum of the second lumbar vertebra is usually hypoplastic and displaced posteriorly, and a "beak" projects from the anteroinferior aspect (Fig. 1531). Beaking may occur in several vertebrae adjacent to the small displaced vertebra. The abnormal vertebra may be the 12th dorsal or first lumbar.

Hurler syndrome may be confused with Morquio syndrome, but changes in the spine, skull, and long bones serve to distinguish the two. Morquio syndrome does not show sellar enlargement. The inferior beaking of the vertebral centra differs from the central beaking of Morquio syndrome. Tapering of long bones may be present in Morquio syndrome, but will not be marked. Acetabular irregularity is always seen in Morquio syndrome, but is not prominent in Hurler syndrome.

Dwarfism is common in Hurler syndrome, and may be mistaken for achondroplasia. Shortened extremities are not present at birth in Hurler syndrome. Hepatomegaly and splenomegaly do not occur in Morquio syndrome or in achondroplasia. Neurovisceral lipogranulomatosis and I-cell disease have similar roentgenographic features.

MPS-II—HUNTER SYNDROME

The Hunter syndrome is a mucopolysaccharidosis similar to Hurler syndrome but it is less severe clinically and is genetically distinct. It is inherited as an X-linked recessive trait and occurs in approximately 1 of 40,000 births.

Clinical Features. The clinical features are the same as seen in Hurler syndrome but less severe. The physical features are a large head with eyes wide apart, sunken bridge of nose, everted lips, and protruding tongue. The teeth are widely spaced and poorly formed and the facies are heavy and ugly. Stiff joints are frequent and sometimes lead to "clawhands." Dwarfism and hepatosplenomegaly are also common. Mental deterioration occurs but progresses more slowly than in Hurler syndrome. Lumbar gibbus, so common with Hurler syndrome, has not been noted. Progressive deafness is frequent.[1] There may be nodular skin lesions of the thorax and upper extremities.

Blindness may occur in older patients, due to an atypical retinitis pigmentosa.[1-3] Corneal clouding may be detectable by slit lamp examination, but does not progress to blindness. This serves to distinguish Hurler syndrome from Hunter syndrome in older patients. Patients with the Hunter syndrome frequently survive to age 30 or 40, and may live to age 60.[3]

There are metachromic granules in the white blood cells. Excessive urinary excretion of chondroitin sulfate B and heparitin sulfate is present both in Hurler and Hunter syndromes. Terry and Linker[4] reported a proportional difference of mucopolysacchariduria: chondroitin sulfate B is 80% with Hurler syndrome and 55% with Hunter syndrome. If this is true, early differentiation is possible.

Roentgenographic Features. The roentgenographic features of Hunter syndrome are similar to Hurler syndrome. The changes are less severe and tend to occur at a later age.

Joint involvement in adults may be the most severe complaint. Articular destruction and secondary degenerative changes occur (Fig. 1532). Dwarfism is usually mild as are bone abnormalities. Hypoplasia of the second and fifth middle phalanges occurs (Fig. 1532).

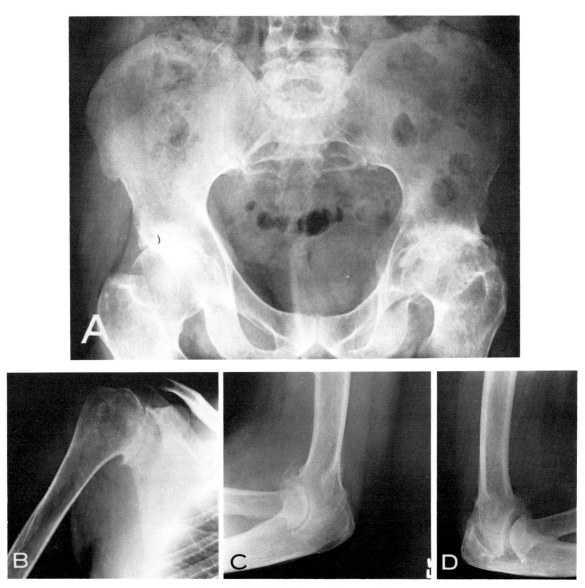

FIG. 1532. MPS-II (HUNTER SYNDROME)

The stigmata of Hunter syndrome are not as severe as MPS-I (Hurler syndrome). Most symptoms relate to joint destruction in the adult. (A) Marked narrowing of both hip joints with secondary degenerative changes. The iliac crests are not flared and there are minor irregularities of the vertebral plates. (B) Narrowing of the shoulder joint with secondary degenerative changes. (C) Narrowing of the elbow joint with secondary arthritic change and fusion of the humerus and ulna. (D) Joint space widening and secondary arthrotic change of the elbow. (E) The vertebrae are normal except for vertebral plate irregularities. (F) The ribs are spatulous and the vertebral ends are bulbous. (G). The hands are small and there is hypoplasia of the second and fifth middle phalanges. The proximal ends of the metacarpals are not pointed; however, the articular surfaces of the radius and ulna are angled toward each other. There is minor narrowing of the wrist joint.

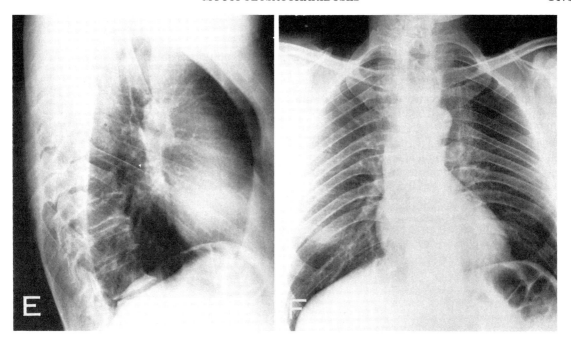
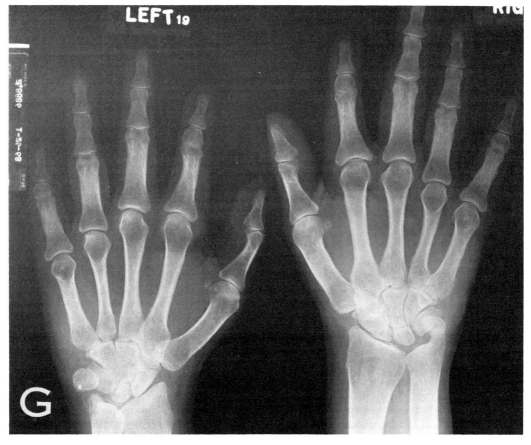

Fig. 1532 (E – G)

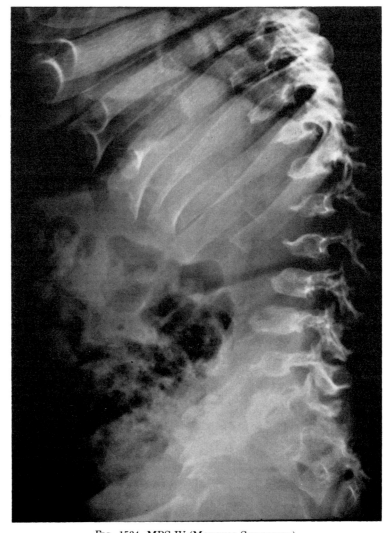

Fig. 1534. MPS-IV (Morquio Syndrome)
Universal vertebral plana with central beaking of multiple vertebrae. (Same case as Figures 1536, 1538, and 1541). (Courtesy of Herbert M. Stauffer, Temple University Hospital, Philadelphia, Pa.)

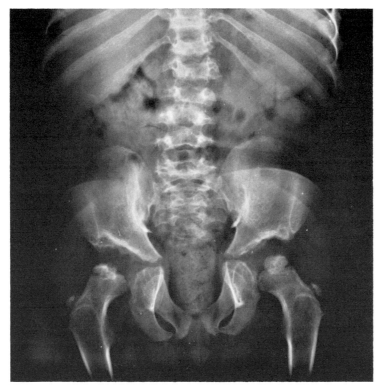

FIG. 1535. MPS-IV (MORQUIO SYNDROME)

The femoral capital epiphyses are flattened and the acetabula irregular. Pointing of the caudal ends of the iliac bones. Universal vertebral plana is evident. (Same case as Figures 1533 and 1540). (Courtesy of Herbert M. Stauffer, Temple University Hospital, Philadelphia, Pa.)

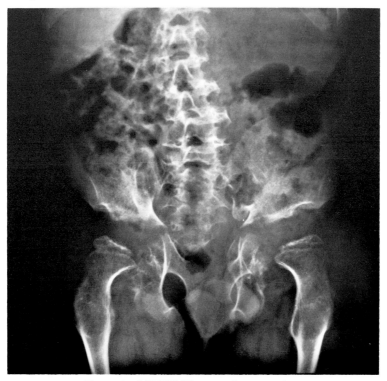

FIG. 1536. MPS-IV (MORQUIO SYNDROME)

Universal vertebral plana. Flattening of the femoral capital epiphyses and irregularity of the acetabula. Coxa valga is present. (Same case as Figures 1534, 1538, and 1541). (Courtesy of Herbert M. Stauffer, Temple University Hospital, Philadelphia, Pa.)

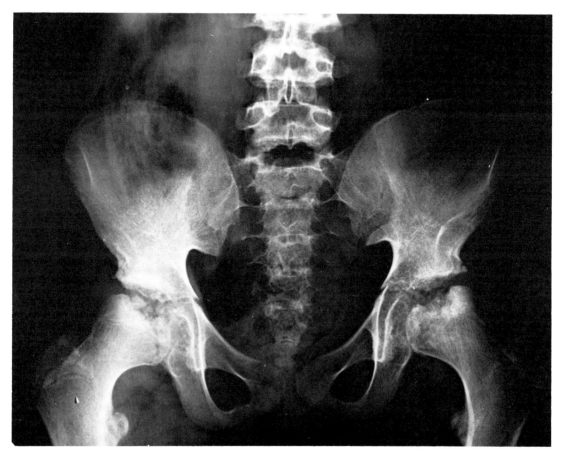

Fig. 1537. MPS-IV (Morquio Syndrome)
The femoral heads are irregular and fragmented, and acetabular irregularity is present.

(Fig. 1538). The whafts are short and thickened (Fig. 1539).

Morquio and Hurler syndromes may be confused, but changes in the spine help differentiate the two. Vertebral bodies are flattened in Morquio syndrome but are normal or heightened in Hurler syndrome. Also, in Hurler syndrome, the vertebral beak is inferior; in Morquio syndrome the beak is central. The carpal and tarsal bones may be irregular and tapering of the proximal metacarpals may occur (Figs. 1540 and 1541).

Multiple epiphyseal abnormalities simulate late dysplasia epiphysalis multiplex; however, the latter rarely affects the spine. Other conditions with which it may be confused are diastrophic dwarfism and spondyloepiphyseal dysplasia.

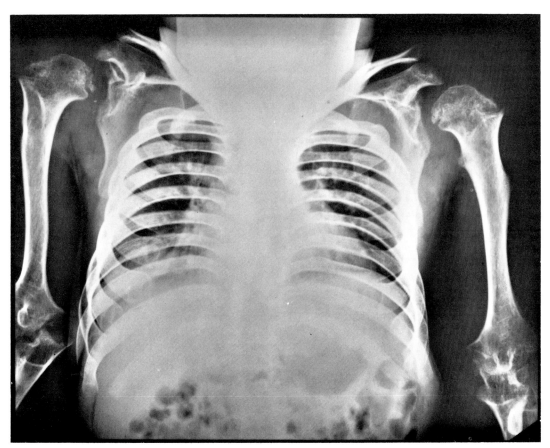

FIG. 1538. MPS-IV (MORQUIO SYNDROME)
The metaphyseal ends of the bones are widened and irregular and accommodate enlarged and irregular epiphyses. (Same cases as Figures 1534, 1536, and 1541). (Courtesy of Herbert M. Stauffer, Temple University Hospital, Philadelphia, Pa.)

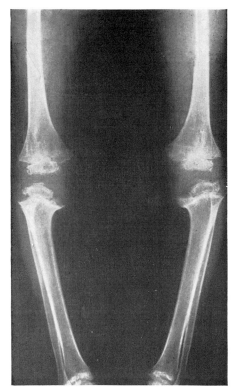

FIG. 1539. MPS-IV (MORQUIO SYNDROME)
Marked metaphyseal irregularity.

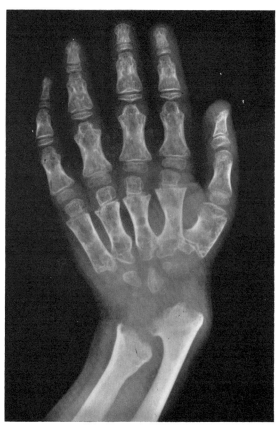

FIG. 1541. MPS-IV (MORQUIO SYNDROME)
The distal radial ulnar articulation is oblique and the articular surfaces are angled toward each other. The proximal metacarpals are tapered and there is brachydactyly. (Same case as Figures 1534, 1536, and 1538). (Courtesy of Herbert M. Stauffer, Temple University Hospital, Philadelphia, Pa.)

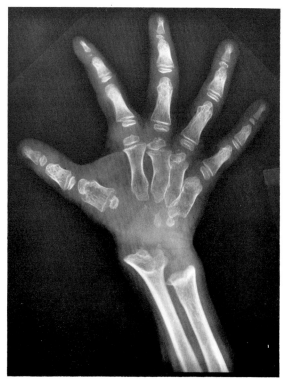

FIG. 1540. MPS-IV (MORQUIO SYNDROME)
Shortening of the ulna, flaring and irregularity of the distal ends of the radius and ulna. The metacarpals are pointed. (Same case as Figures 1533 and 1535). (Courtesy of Herbert M. Stauffer, Temple University Hospital, Philadelphia, Pa.)

MPS-V—SCHEIE SYNDROME

The Scheie syndrome, originally described as a variant of the Hurler syndrome,[1] is a mucopolysaccharidosis in which there is excessive urinary excretion of chondroitin sulfate B. It is probably inherited as an autosomal recessive trait.

Clinical Features. The "broad mouthed" and coarse facial characteristics of Hurler syndrome are present in this syndrome. The few patients reported are strikingly similar in appearance.[1,2] The head is large and eyes wide apart. The neck is short. The joints are stiff and the hands "clawed." Median nerve compression in the carpal tunnel is reported in the majority of cases.[2]

Dwarfism is not a striking feature in the few cases reported.[2] Aortic valve abnormalities are reported in all cases, and aortic regurgitation is the most common reflection. Corneal clouding is uniform early, and later becomes denser peripherally. Retinitis pigmentosis is also a feature.

The majority of patients have unimpaired intellect and most have above average intelligence; however, psychosis is a common feature in the few cases reported.[1,2] Too few cases are reported to estimate the prognosis. The patients seem to live longer than those with other mucopolysaccharidoses. One patient was seen as old as the late 40s, and several were in their mid-30s.

Metachromatic granules (Reilly bodies) are absent or ill defined. The presence of excessive urinary excretion of chondroitin sulfate B and the absence of heparitosulfaturia differentiate MPS-IV (Scheie syndrome) from other mucopolysaccharidoses, with the exception of MPS-VI (Maroteaux-Lamy syndrome). Although the latter has excessive urinary excretion of chondroitin sulfate B, skeletal changes and dwarfism are pronounced, and they serve to distinguish it.

Roentgenographic Features. The roentgenographic features are minimal. Dwarfism is uncommon.

MPS-VI—MAROTEAUX-LAMY SYNDROME

Synonym: polydystrophic dwarfism.

The Maroteaux-Lamy syndrome was originally described as polydystrophic dwarfism by Maroteaux et al.[1] and Maroteaux and Lamy.[2] It is a mucopolysaccharidosis in which the clinical characteristics are similar to those of the Hurler and Hunter syndromes, and the biochemical findings (excessive excretion of chondroitin sulfate B) are identical with those of the Scheie syndrome. Perhaps this syndrome is a severe form of the Scheie syndrome, but the marked osseous disturbances justify a separate classification.

Clinical Features. The striking clinical features are dwarfism and corneal clouding without mental impairment. The physical and facial characteristics are as described in the Hurler syndrome. Excessive urinary excretion of chondroitin sulfate B distinguishes MPS-IV from other mucopolysaccharidoses, with the exception of the Scheie syndrome.

Roentgenographic Features. The roentgenographic features are difficult to distinguish from those of Hurler syndrome (Figs. 1542–1545). The femoral capital epiphyses are frequently irregular and flattened (Figs. 1542 and 1545). The hand abnormalities are not as severe as those seen in Hurler syndrome (Fig. 1542).

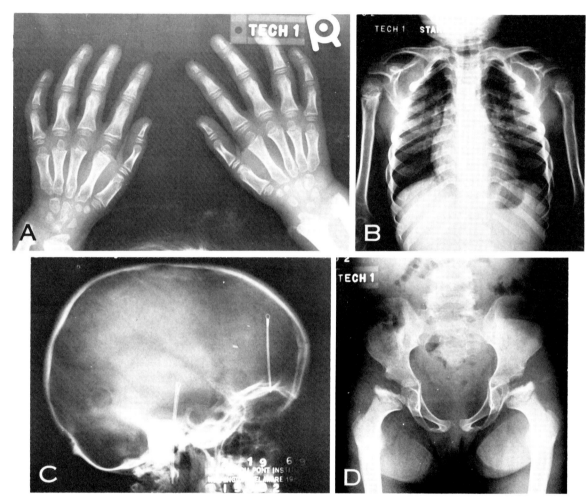

FIG. 1542. MPS-VI (MAROTEAUX-LAMY SYNDROME)

(A) The proximal ends of the metacarpals are slightly pointed. The articulating surface of the ulna is tilted toward the radius. (B) The ribs are spatulous. (C) J-shaped sella indistinguishable from MPS-I (Hurler syndrome). (D) The iliac bones do not flare and there is tapering. The femoral capital epiphyses are flattened and irregularly dense. The acetabula are relatively normal. The changes are more marked than MPS-I (Hurler syndrome) and serves to distinguish them.

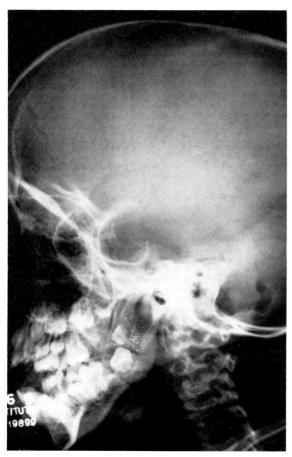

Fig. 1543. MPS-VI (Maroteaux-Lamy Syndrome)
The sella is enlarged. This is a sibling of Figure 1542.

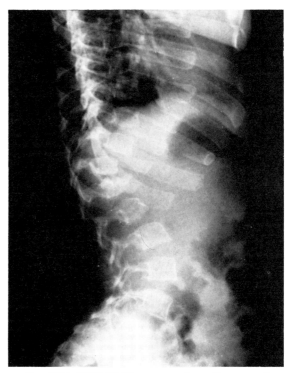

Fig. 1544. MPS-VI (Maroteaux-Lamy Syndrome)
This is a sibling of Figures 1542 and 1543. Hypoplasia of the twelfth thoracic vertebra and the first and second lumbar vertebrae with beaking. The findings are similar to MPS-I (Hurler syndrome).

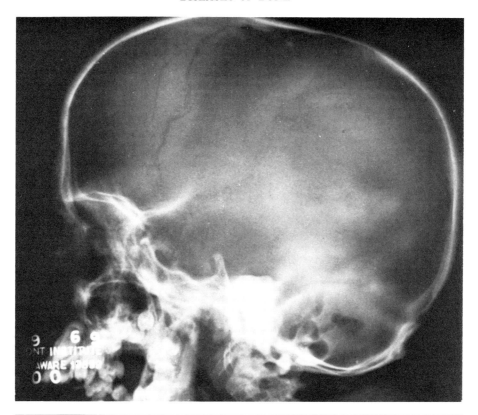

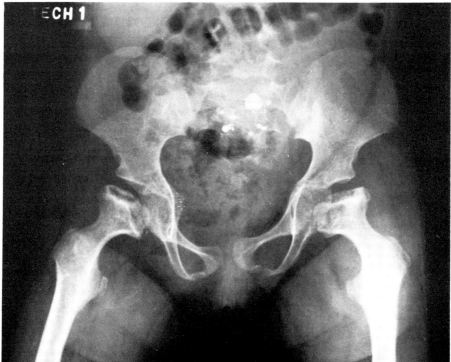

FIG. 1545. MPS-VI (MAROTEAUX-LAMY SYNDROME)

(*Top*) The patient is a sibling of Figures 1542–1544. J-shaped sella. (*Bottom*) Flattening and irregularity of the femoral capital epiphyses. The acetabula are slightly abnormal. Tapering of the iliac bones.

REFERENCES

Mucopolysaccharidoses: Introduction

1. Linker, A., and Terry, K. D.: Urinary acid mucopolysaccharides in normal man and in Hurler's syndrome. Proc Soc Exp Biol Med, *113:* 743, 1963.
2. Dorfman, A.: The Hurler syndrome: In *First International Conference on Congenital Malformations*, M. Fishbein (ed.). J. B. Lippincott, Philadelphia, 1962.
3. Reilly, W. A.: The granules in the leukocytes in gargoylism. Am J Dis Child, *62:* 489, 1941.
4. Mittwoch, U.: Inclusions of mucopolysaccharide in lymphocysts of patients with gargoylism. Nature, *191:* 1315, 1961.
5. Lorincz, A. E.: Heritable disorders of acid mucopolysaccharide metabolism in humans and in shorter dwarf cattle. Ann NY Acad Sci, *91:* 644, 1961.
6. Berenson, G. S., and Serra, M. T.: Polysaccharides in urine from patients with Marfan's syndrome. Fed Proc, *18:* 190, 1959.
7. King, J. S.: Acid mucopolysaccharides in urine. In *High Molecular Weight Substances in Human Urine*, J. S. King and W. H. Boyce (eds.). Charles C Thomas, Springfield, Ill., 1963.
8. McKusick, V. A.: *Heritable Disorders of Connective Tissue*, Ed. 3. C. V. Mosby, St. Louis, 1966.

I. Hurler Syndrome

1. Ellis, R. W. B., Sheldon, W., and Capon, N. B.: Gargoylism (chondro-osteo-dystrophy, corneal opacities, hepatosplenomegaly, and mental deficiency). Q J Med, *5:* 119, 1936.
2. Binswanger, E., and Ullrich, O.: Uber die "Dysostosis Multiplex" (Typus Hurler) und ihre Beziehungen zu anderen Konstitutionsanomalien. Z Kinderheilkd, *54:* 699, 1933.
3. Stanbury, J. B., Fredrickson, D. S., and Wyngaarden, J. B.: *The Metabolic Basis of Inherited Disease*, McGraw-Hill, New York, 1960.
4. Tuthill, C. R.: Juvenile amaurotic idiocy; marked adventitial growth associated with skeletal malformations and tuberculomas. Arch Neurol Psychiatr, *32:* 198, 1934.
5. Kressler, R. J., and Aegerter, E. E.: Hurler's syndrome (gargoylism); summary of literature and report of case with autopsy findings. J Pediatr, *12:* 579, 1938.
6. Lamy, M., Maroteaux, P., and Bader, J. P.: Etude genetique du gargoylisme. J Genet Hum, *6:* 156, 1957.
7. McKusick, V. A., and Milch, R. A.: The clinical behavior of genetic disease; selected aspects. Clin Orthop, *33:* 22, 1964.
8. Caffey, J.: *Pediatric X-ray Diagnosis*. Year Book, Chicago, 1961.

II. Hunter Syndrome

1. McKusick, V. A.: *Heritable Disorders of Connective Tissue*, Ed. 2. C. V. Mosby, St. Louis, 1960.
2. Hooper, J. M. D.: An unusual case of gargoylism. Guy's Hosp Rep, *101:* 222, 1952.
3. Beebe, R. T., and Formel, P. F.: Gargoylism; sex-linked transmission in nine males. Trans Am Clin Climatol Assoc, *66:* 199, 1954.
4. Terry, K., and Linker, A.: Distinction among 4 forms of Hurler's syndrome. Proc Soc Exp Biol Med, *115:* 394, 1964.

III. Sanfilippo Syndrome

1. Terry, K., and Linker, A.: Distinction among 4 forms of Hurler's syndrome. Proc Soc Exp Biol Med, *115:* 394, 1964.
2. Sanfilippo, S. J., Podosin, R., Langer, L. O., and Good, R. A.: Mental retardation associated with acid mucopolysacchariduria (heparitin sulfate type). J Pediatr, *63:* 837, 1963.

IV. Morquio Disease

1. Brailsford, J. E.: Chondro-osteo-dystrophy; roentgenographic and clinical features of a child with dislocation of vertebrae. Am J Surg, *7:* 404, 1929.
2. Morquio, L.: Sur une forme de dystrophie osswuse familiale. Bull Soc Pediatr Paris, *27:* 145, 1929.
3. Ullrich, O.: Die Pfaundler-Hurlersche Krankheit. Ergeb Inn Med Kinderheilkd, *63:* 929, 1943.
4. Weidmann, H. R.: Ausgedehnte und allgemeine erblich bedingte Bildungs- und Wachstumsfehler dis Knochengerustes. Monatsschr Kinderheilkd, *102:* 136, 1954.
5. Zellweger, H., Ponseti, I. V., Pedrini, V., Stamler, F. S., and von Noorder, G. K. Morquio-Ullrich's disease. J Pediatr, *59:* 549, 1961.
6. McKusick, V. A., et al.: The genetic mucopolysaccharidoses. Medicine, *44:* 445, 1965.
7. Jacobsen, A. W.: Hereditary osteochondrodystrophia deformans. JAMA, *133:* 121, 1939.
8. Smith, R., and McCort, J. J.: Osteochondrodystrophy (Morquio-Brailsford type); occurrence in 3 siblings. Calif Med, *88:* 55, 1959.
9. Lamy, M., and Maroteaux, P.: La nanisme disastrophique. Presse Med, *68:* 1977, 1960.
10. McKusick, V. A., and Milch, R. A.: The clinical behavior of genetic disease; selected aspects. Clin Orthop, *33:* 22, 1964.
11. Cameron, J. M., and Gardiner, T. B.: Atypical familial osteochondrodystrophy. Br J Radiol, *36:* 135, 1963.
12. Helwig-Larsen, H. G., and Morch, E. T.: Genetic aspects of osteochondrodystrophy. Acta Pathol Microbiol Scand, *22:* 335, 1945.
13. Linker, A., and Terry, K. D.: Urinary acid mucopolysaccharides in normal man and in Hurler's syndrome. Proc Soc Exp Biol Med, *113:* 743, 1963.
14. Rubin, P.: *Dynamic Classification of Bone Dysplasias*. Year Book, Chicago, 1964.
15. Bartman, J., Mandelbaum, I. M., and Gregoire, P. E.: Mucopolysaccharides of serum and urine in a case of Morquio's disease. Clin Biol, *8:* 250, 1963.
16. Maroteaux, P., and Lamy, M.: La maladie de Morquio. Presse Med, *71:* 2091, 1963.
17. Maroteaux, P., and Lamy, M.: Opacites corneennes et trouble metabolique dans la maladie de Morquio. Clin Biol, *6:* 481, 1961.
18. Pedrini, N., Lenuzzi, L., and Zambotti, V.: Isolation and identification of keratosulfate in urine of patients affected by Morquio-Ullrich disease. Proc Soc Exp Biol Med, *110:* 847, 1962.
19. Sorsby, A.: Hereditary affections of the retina and choroid. Acta Genet Med, *13:* 20, 1964.
20. Robins, M. M., Stevens, H. F., and Linker, A.: Morquio's disease; an abnormality of mucopolysaccharide metabolism. J Pediatr, *62:* 881, 1963.
21. Fairbank, T.: *An Atlas of General Affections of the Skeleton*. Williams & Wilkins, Baltimore, 1951.

22. von Noorden, G. K., Zellweger, H., and Poreseti, I. V.: Ocular findings in Morquio-Ullrich's disease. Arch Ophthalmol, *64:* 585, 1960.

V. Scheie Syndrome

1. Scheie, H. G., Hambrick, G. W., Jr., and Barnes, L. A.: A newly recognized forme fruste of Hurler's disease (gargoylism). Am J Ophthalmol, *53:* 753, 1962.

2. McKusick, V. A.: The genetic mucopolysaccharidoses. Medicine, *44:* 445, 1965.

VI. Maroteaux-Lamy Syndrome

1. Maroteaux, P., Leveque, B., Marie, J., and Lamy, M.: Une nouvelle dysostose avec elimination urinaire de condroitine-sulfate B. Presse Med, *71:* 1849, 1963.

2. Maroteaux, P., and Lamy, M.: Hurler's disease, Morquio's disease and related mucopolysaccharidoses. J Pediatr, *67:* 312, 1965.

17
Chromosomal Abnormalities

Chromosomal abnormalities are associated with certain characteristics of physical and mental diseases. The term "chromosomal abnormality" relates to morphologic change; abnormalities in the configuration of the individual chromosomes or addition or deletion of chromosomes. Eventually sophisticated studies will define biochemical changes within the chromosome. It is increasingly evident that some genetic defects show characteristic roentgen changes which are confirmed by karyotype analysis.

The diagnosis of syndromes involving the chromosome aberration is ultimately made on the basis of cytological studies. These studies are not done routinely; a chromosomal defect is often initially suggested by roentgen studies. Since clinical findings are often minimal, particularly in the newborn, radiographic studies can be of great value in screening the possible chromosomal abnormality, although the absence of x-ray findings does not exclude such abnormalities.

Chromosomal study became practical in 1952 when Hsu[1] introduced a new method of cell culture. In 1956, Tijio and Levan[2] proved that fetal cell cultures contain 46 chromosomes, rather than 48,[3] as was previously believed. The first clinical application of this knowledge to abnormality was in 1959, when LeJeune et al.[4] described a constant chromosome deviation in mongoloids.

The type and characteristics of each cell in the body is determined by a genetic material in chromosomes, located in cell nuclei. Somatic cells, which include all but germ cells, normally contain 22 pairs of autosomes and 2 sex chromosomes, or 46 chromosomes in all. Germ cells contain half of this complement; the number of these chromosomes is called the "haploid number." The number of chromosomes in the somatic cell is called the "diploid" (double). The female cell contains 2 X sex chromosomes; the male cell contains 1 X and 1 Y sex chromosome.

CELL DIVISION

Mitosis and meiosis are the two important mechanisms of cell division.[5]

Mitosis

Somatic cells reproduce by mitosis which consists of 5 phases: (1) *interphase*, when chromatin is not visible. When chromatin becomes visible, the (2) *prophase* has begun, and the chromatin appears as dense staining, double strands united at the centromere. Two centrioles migrate to the poles of the cells and act as attachments for the mitotic spindle. The (3) *metaphase* starts with the lining up of chromosomes in an "X" or "wishbone" configuration, each representing 2 chromatids (spiral filaments that make up chromosomes, and separate in cell division) of a single chromosome. The spindle apparatus is attached to the centromere. The (4) *anaphase* begins with complete separation of chromatids into 2 separate chromosomes, and the (5) *telephase* occurs when the entire cell has divided.

Meiosis

This is a process of reduction division by germ cells. Reduction of chromosomes is necessary because a germ cell, or gamete, must contain a haploid number of chromosomes to match the haploid number of gametes of the opposite sex at the time of fertilization. Two successive divisions occur during the process of meiosis: (1) the *reductional division* produces 2 daughter cells that contain a haploid number of chromosomes. So, in the testes the parent spermatocyte, with 22 pairs of chromosomes plus X and Y sex chromosomes, divides to form 22 autosomes with either an X or Y sex chromosome. In the ovary the primary oocyte produces a daughter cell with 22 autosomes plus an X sex chromosome. (2) There is a *second division* of chromosomes of daughter cells, so that in the process of meiosis 4 haploid gametes are produced from each parent cell, ensuring a consistent number of chromosomes in each generation.

During meiosis the chromosomes line up in pairs. Then, as division occurs, each pair separates and each migrates toward its new cells. This process of separation and migration is known as *disjunction*, which is the basic process in the formation of a haploid number of chromosomes in spermatozoon or ovum. Failure of proper disjunction is called *nondisjunction*, and is found in Turner and Klinefelter syndromes.

The mature ovum contains 22 autosomes plus an X chromosome, and the spermatozoon contains 22 autosomes plus an X or Y chromosome so that, at conception, union of the two form a fertilized ovum containing a diploid number of chromosomes (44 chromosomes and either 2 X chromosomes or 1 X and 1 Y chromosome). The 44 chromosomes plus 2 X chromosomes produce a female and 44 plus X and Y produce a male.

Normal Chromosome Morphology

Karyotype analysis consists of separating and identifying the chromosome pairs according to length and location of the centromere. In the Denver system, the chromosomes are arranged in groups, because each chromosome cannot be morphologically identified. These groups are labeled from A to G or from I to VII. Group A contains 3 sets (1–3); Group B contains 2 sets (4–5); Group C, 7 sets (6–12) and the sex chromosomes (XX); Group D, 3 sets (13–15); Group E, 3 sets (16–18); Group F, 2 sets (19–20); Group G, 2 sets (21–22) and Y chromosome. Individual pairs may sometimes be separated, especially 1, 2, 3, 16, and Y, but most frequently only the groups can be separated by the length of the chromosomes and the position of the centrum. X chromosomes fall into Group C, and Y chromosomes are Group G. So the normal female will have 16 chromosomes (XX) in Group C and 4 in Group G, and the male 15 in Group B (X) and 5 (Y) in Group G (Figs. 1546 and 1547).

Morphologic Chromosomal Alterations

Morphologic chromosomal alterations consist of abnormalities in number or configuration, or both. Morphologic changes are not necessarily

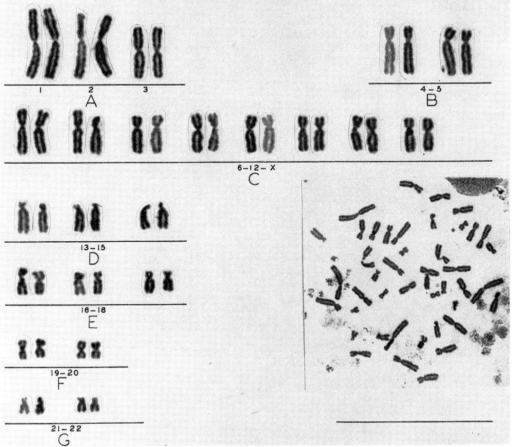

FIG. 1546. NORMAL FEMALE KARYOTYPE

In Group C there are 7 sets of autosomal chromosomes (6–12) plus 2 X chromosomes. Note that in Group G there are only 2 sets of chromosomes. Compare this to the normal male karyotype (Fig. 1547).

CHROMOSOMAL ABNORMALITIES 1489

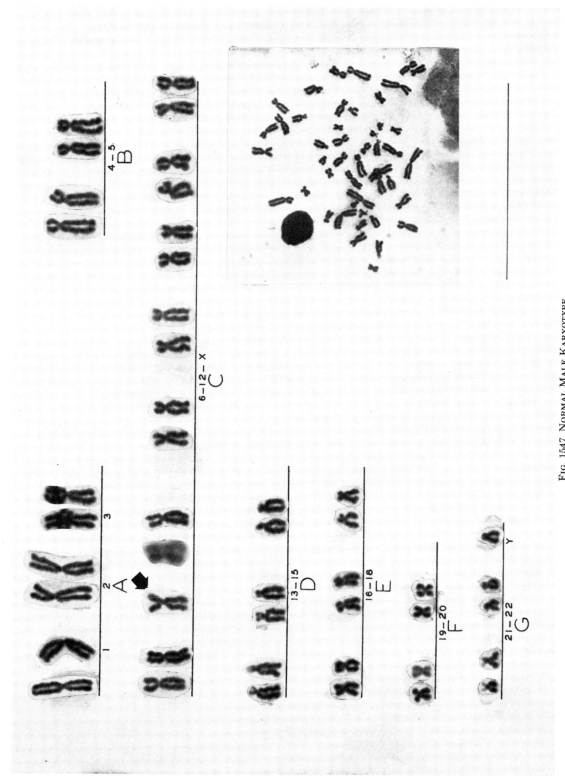

FIG. 1547. NORMAL MALE KARYOTYPE

There are 22 pairs of autosomes and XY sex chromosomes. The X chromosome is in Group C (*arrow*). The Y chromosome is in Group G. Compare this to normal female karyotype (Fig. 1546).

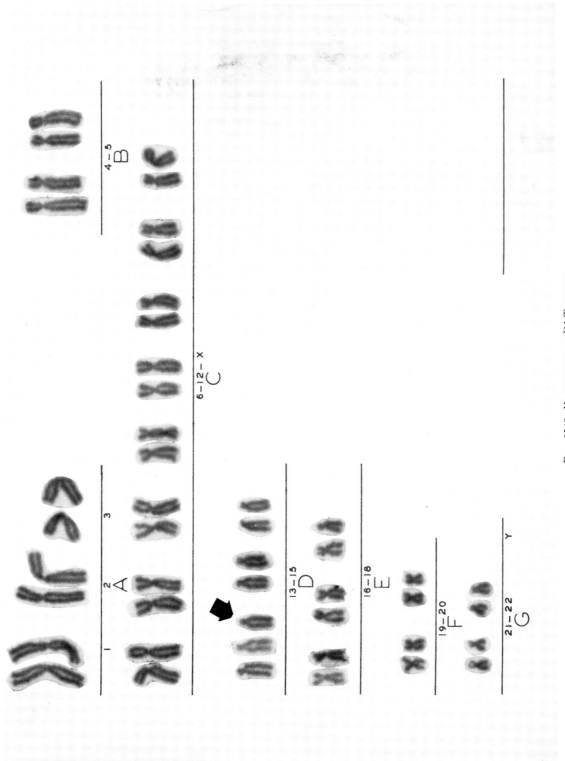

Fig. 1548. Karyotype of D1 Trisomy
Shows an extra chromosome in Group D (*arrow*). Compare this with normal female karyotype (Fig. 1546).

related to genetic aberrations. Hereditary and familial conditions are related to nucleic acid composition in chromatin (DNA) and do not alter the karyotype pattern.

Possible anomalies of karyotype pattern are numerous, but so far only a small number are identified: (1) *monosomy*: loss of an entire chromosome, which is almost always incompatible with life except when sex chromosomes are involved; (2) *trisomy*: the presence of an extra chromosome, which is compatible with life, in many instances (Fig. 1548); and (3) *grossly abnormal or deformed chromosomes*, most often seen in acquired disease such as leukemia. Loss or gain of a sex chromosome is compatible with life, but such persons are usually infertile.

The etiology of chromosomal defects is not understood, but certain conditions appear to have some bearing on them. Most chromosomal abnormalities are associated with older mothers; paternal age does not matter. This suggests that ova, which are present at birth, are affected as they grow older, whereas spermatozoa, being continually produced, have little chance to change because of age. Radiation,[6,7] chronic disease, genetic influences, and mutagenic viruses have all been implicated as causes of chromosomal defects. Translocation (the breaking of chromatids and reunion with the wrong fragments) and nondisjunction are the usual causes of chromosomal abnormalities.

Nuclear Chromatin and Sex Significance

Nuclear sexing refers to the presence of darkly staining chromatin material in the cell nuclei of females. Barr and Bertrum[8] first reported the presence of this darkly staining material in the neural tissues of cats, as well as its absence in male cats. In 1955, Moore and Barr[9] reported that 50–60% of the buccal mucosal cells of human females contain nuclear sex chromatin, and less than 2% of male cells. Davidson and Smith[10] found dark chromatin in polymorphonuclear leukocytes in females, ovoid spots connected to the nucleus by a fine strand that suggests a drumstick. It is generally accepted that the nuclear sex chromosome of buccal mucosal cells and the "drumstick" of leukocytes represent the same phenomenon. It is believed that these cell chromatin bodies are derived from a pyknotic X chromosome, and that all but one of the X chromosomes in a cell nucleus produce this substance; because there is only one X chromosome in the male it can produce no nuclear sex chromatin. In normal females, on the other hand, there are two X chromosomes; one X chromosome is in the cell and the other may become the dark staining chromatin. Supernumerary X chromosomes will be reflected by excess sex nuclear chromatin. If this hypothesis proves to be true, then the total X chromosomes can be estimated by adding 1 to the number of chromatin bodies in the nuclei of cells.

Gonadal dysgenesis refers to a group of syndromes, each of which is characterized by some degree of abnormal gonadal development. Two of these syndromes, Turner and Klinefelter, are related to abnormal sex chromosomal patterns, but not all these patients have abnormal chromosomal patterns, nor do all patients with abnormal patterns have these syndromes.

Pathogenesis

The basic abnormality of Turner and Klinefelter syndromes is in the meiotic phase of cell division of either parent, wherein sex chromosomes fail to separate from each other (nondisjunction). The autosomal chromosomes are normal. The resulting gamete of the male parent will contain either XY sex chromosomes or no sex chromosomes; the female XX chromosome or O chromosome. The possible sex chromosome constitutions are listed in Table 55 (Figs. 1549 and 1550). Table 55 shows that nondisjunction may lead to two possible patterns in the male and to four patterns in the female. The last pattern, YO, has not yet been found, indication that at least one X chromosome is necessary for successful fertilization. The remaining combinations (XXX, XXY, XO) are those seen in the "super female syndrome," Klinefelter syndrome, and Turner syndrome, respectively. Nondisjunction in either parent may result in Klinefelter or Turner syndrome, but only the female nondisjunction can lead to the super female syndrome.

TABLE 55. POSSIBLE SEX CHROMOSOME CONSTITUTIONS

Male	Female	Children	Syndrome
A. Nondisjunction in Male			
O	X	XO	Turner
XY	X	XXY	Klinefelter
B. Nondisjunction in Female			
X	XX	XXX	Super female
X	O	XO	Turner
Y	XX	XXY	Klinefelter
Y	O	YO	Nonfertilization

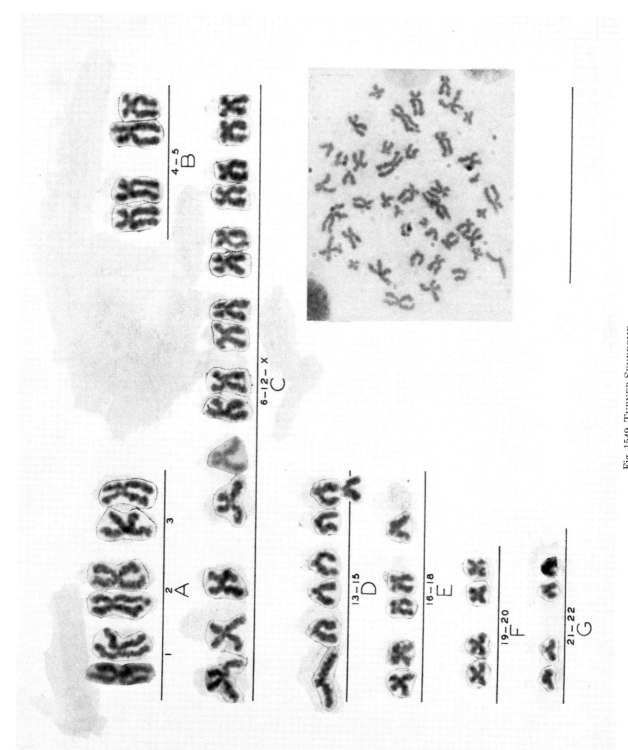

Fig. 1549. TURNER SYNDROME

Karyotype showing a single X chromosome in Group C and no Y chromosome in Group G. Compare this to normal male karyotype (Fig. 1547).

CHROMOSOMAL ABNORMALITIES 1493

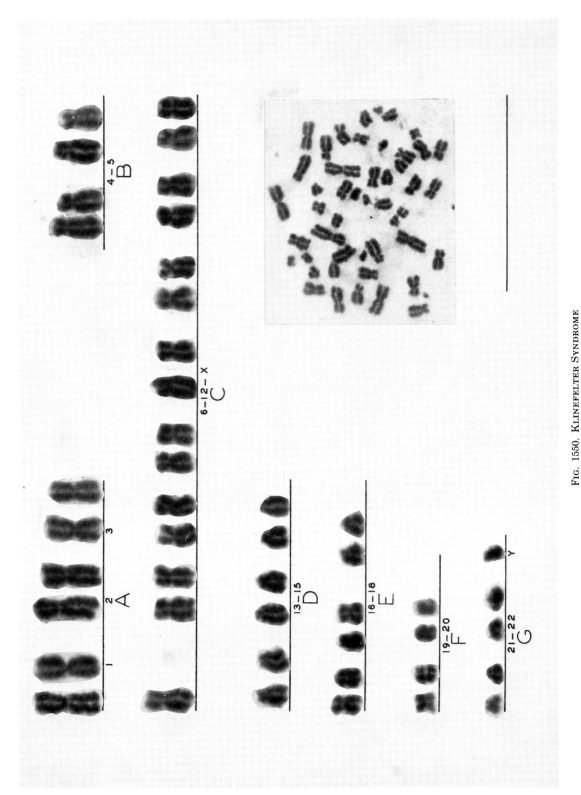

FIG. 1550. KLINEFELTER SYNDROME
Karyotype showing XX chromosome in Group C and Y chromosome in Group G.

TURNER SYNDROME

In 1938, Turner[1] described a syndrome of sexual infantilism, congenital webbed neck, and cubitus valgus. When the technique of nuclear sex determination was applied, it was found that most patients with Turner syndrome had a male chromatin pattern,[2] which led to the hypothesis that they would be deficient in a sex chromosome. When the methods of obtaining human karyotype patterns were improved, this turned out to be indeed the case.[3]

In the past, clinical findings similar to those in Turner syndrome were found in male patients and in some female patients with a normal karyotype pattern. These syndromes were called "male Turner syndrome" and "pseudo-Turner syndrome." It is now preferred that "Noonan syndrome" be the term for the latter conditions,[4-6] due to hereditable disease and usually with the normal karyotype pattern.

Clinical Features. This is the syndrome usually suspected in adolescence, since somatic stigmata are often absent or not apparent. The first overt signs are primary amenorrhea, absence of the prepubertal growth spurt, and lack of secondary sex characteristics. Short stature, increased carrying angle of the arms, webbed neck, mental deficiency, coarctation of the aorta, aortic stenosis and idiopathic hypertension are common manifestations. Other abnormalities include shield chest, low irregular nuchal hair line, high palate, osteoporosis, multiple pigmented nevi, and keloid formation.[7-13]

Most patients with Turner Syndrome have only one X chromosome (complete monosomy). Other chromosomal aberrations consist of partial monosomy (structurally altered second X chromosomes), mosaicism, involving XO and another sex karyotype, or an X with a severely deleted Y chromosome.

Roentgenographic Features. These are described by Kosowicz[14]; usually only a few changes are demonstrable. The maturation of the skeleton progresses normally until the age of 14 or 15, and all ossification centers appear at the normal age. Fusion of the epiphyses, however, is delayed until after the age of 20, (Figs. 1551 and 1552).

A decrease in density of the skeleton, particularly the carpal and tarsal bones and vertebrae, is sometimes apparent in children, and the rule in patients over 18 (Fig. 1553). In the hand and wrist, deossification is especially marked in the carpal bones. The metacarpal sign[15] is due to a relative shortening of the fourth metacarpal. Normally a line extending tangentially from the distal ends of the fourth and fifth metacarpals extends to the distal end of the third metacarpal head (Figs. 1554–1557). A positive metacarpal sign is indicated when this line passes through the third metacarpal. Keats and Burns[16] found that it is frequently present in patients with gonadal dysgenesis, who have either male or female sex chromatins; however, it may appear in normal patients or in patients with unrelated growth disorders such as pseudo-pseudohypoparathyroidism, melorheostosis, or ectodermal dysplasia. This sign, when correlated with other clinical and radiographic changes, is helpful but not diagnostic.

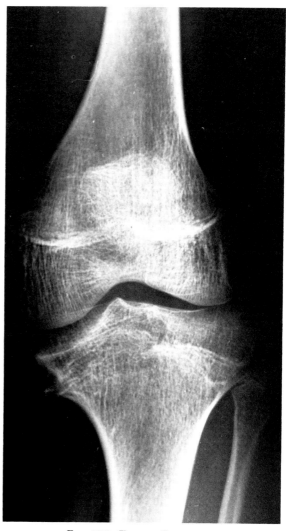

FIG. 1551. TURNER SYNDROME
A 19-year-old girl. Delayed fusion of the epiphyses and deossification of the bone. The knees are deformed much as in Blount disease. The medial tibial plateau is depressed.

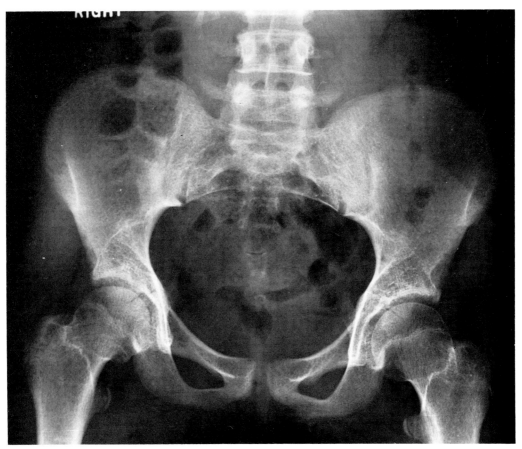

FIG. 1552. TURNER SYNDROME
Same patient as seen in Figure 1551. Delayed epiphyseal fusion and deossification.

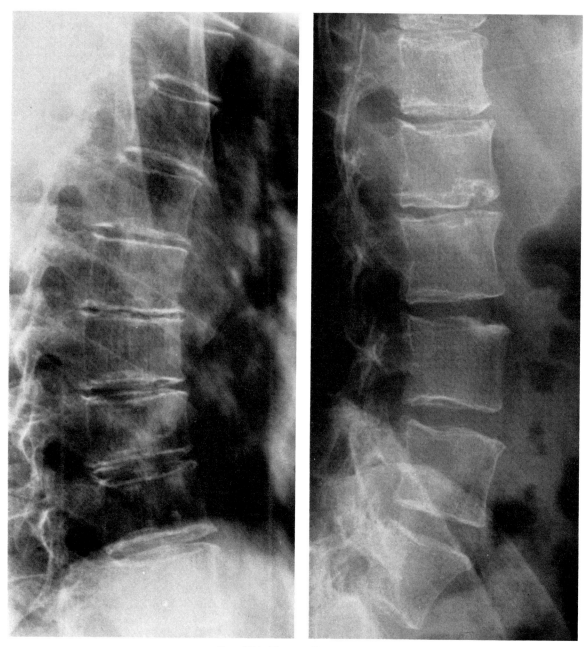

FIG. 1553. TURNER SYNDROME

This 44-year-old woman with Turner syndrome shows generalized osteoporosis, changes simulating osteochondritis, and Schmorl's nodes. (*Left*) Thoracic spine; (*right*) lumbar spine. (Courtesy of A. E. O'Hara and S. E. Abram: Radiographic clues to cytogenetic disease. *Texas Medicine,* 67: 84, 1971, © Texas Medical Association, Austin, Tex.)

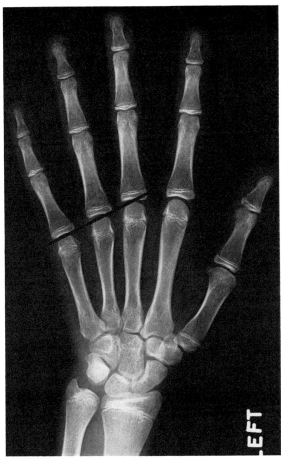

FIG. 1554. TURNER SYNDROME
Metacarpal sign is borderline. A line drawn through the distal ends of the fourth and fifth metacarpals touches the distal end of the third metacarpal. Deossification of the bones and delayed union of this 23-year-old patient.

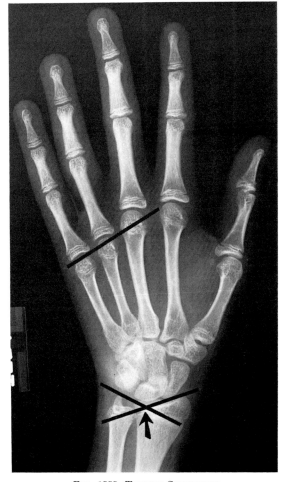

FIG. 1555. TURNER SYNDROME
Positive metacarpal sign is present. A line drawn through the distal ends of the fourth and fifth metacarpals passes through the third metacarpal. The carpal angle measures 130° (*arrow*). This is within normal limits; 117° or less is positive.

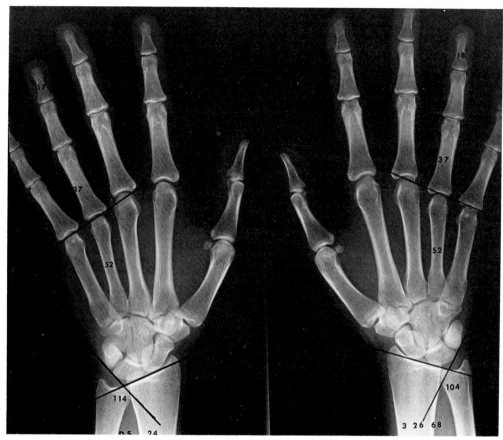

Fig. 1556. Turner Syndrome
Positive bilateral metacarpal sign but no evidence of phalangeal preponderance. Positive carpal angles of 114° and 104°. (Courtesy of A. E. O'Hara and S. E. Abram: Radiographic clues to cytogenetic disease. *Texas Medicine, 67:* 84, 1971, © Texas Medical Association, Austin, Tex.)

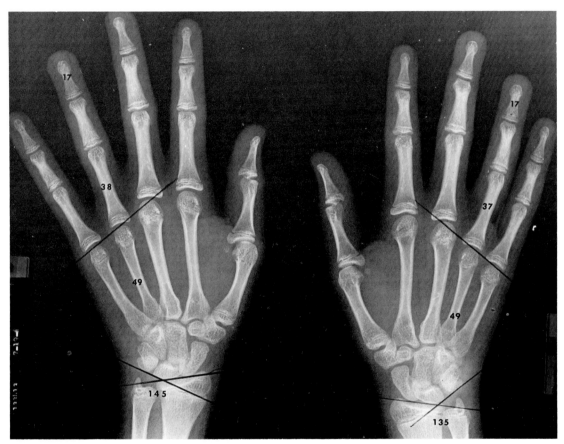

Fig. 1557. Turner Syndrome
Bilateral positive metacarpal signs, normal carpal angles of 145° and 135°, and phalangeal preponderance. (Courtesy of A. E. O'Hara and S. E. Abram: Radiographic clues to cytogenetic disease. *Texas Medicine,* 67: 84, 1971, © Texas Medical Association, Austin, Tex.)

The carpal sign[17] is due to the acute angle formed by the proximal row of carpal bones. Kosowicz[17] measured this angle by extending two tangents: (1) from the proximal borders of the scaphoid and lunate and (2) from the triquetrum and lunate (Fig. 1555). The angle at the intersection of these tangents is the carpal angle. In 466 normal subjects the mean value of the carpal angle was found to be 131.5°, whereas in patients with ovarian dysgenesis the mean carpal angle was found to be 117° or less, which he suggested should constitute the positive carpal sign (Fig. 1556). This sign is not related to dwarfism, but seems to be related to the positive metacarpal sign.

Kosowicz[17] also described "phalangeal preponderance" over the metacarpal. In normal subjects the length of the fourth metacarpal is equal to the sum of the lengths of the proximal and distal phalanges. In some cases of gonadal aplasia the total length of the fourth metacarpal by 3 mm or more, which is termed "phalangeal preponderance" (Fig. 1556 and 1557). Changes in the feet are similar to those seen in the hands (Fig. 1558). Shortening of the middle phalanx of the fifth and index fingers is reported[12]; this may also be seen in Down syndrome.

An increase in the carrying angle (cubitus valgus) of the elbow is one of Turner original triad. This is easier to demonstrate clinically than radiographically. Of greater significance is the radial tilt of the articular surface of the trochlea[16] (Fig. 1559). There may be a small bridged sella (Fig. 1560), a basal angle increased more than 140°, and basilar impression.[16]

In many women the pelvic inlet is of the male type (android). The pubic arch is narrowed and the sacrosciatic notches small. The epiphyses of the iliac crests are often late to fuse. The spine may show deossification, scoliosis, and kyphosis. Osteochondritis of the vertebral plates also occurs (Fig. 1553). The centra of the lumbar vertebrae appear squared. Anomalous development of C1 and C2 is common, especially hypoplasia of C1[18] and anomalous development of the odontoid (Fig. 1561). Rib and clavicular contour are often abnormally thin and narrow.[7] There may be pseudonotching of the ribs.

The knees are deformed much as in Blount disease[19] (Fig. 1551). The medial tibial plateau is depressed and the tibial metaphysis projects medially, becoming beaklike. The medial femoral condyle is larger than normal and extends downward below the level of the lateral condyle.

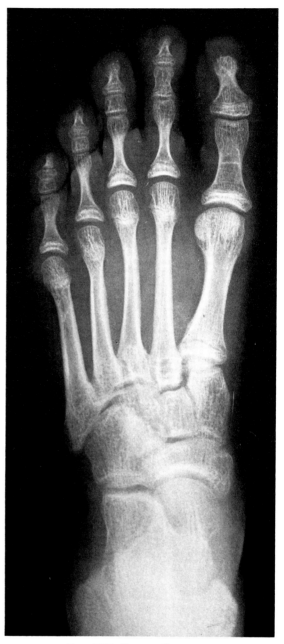

FIG. 1558. TURNER SYNDROME
Changes in the feet are similar to those in the hands. The first, fourth, and fifth metatarsals are short. Deossification and delayed union of the epiphyses.

An enlarged pituitary fossa or changes in the sellar contour show on polytomography and suggest hyperplasia or microadenoma formation of the pituitary gland.[20]

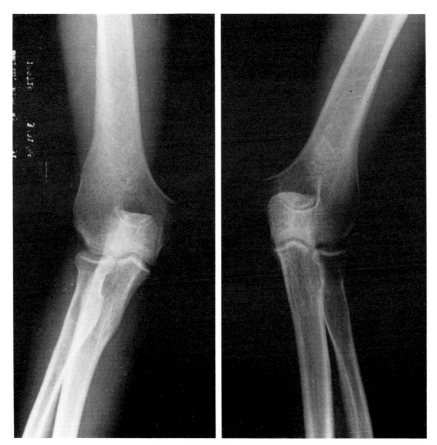

FIG. 1559. TURNER SYNDROME
Increase in the carrying angle and a radial tilt of the articular surface of the trochlea.

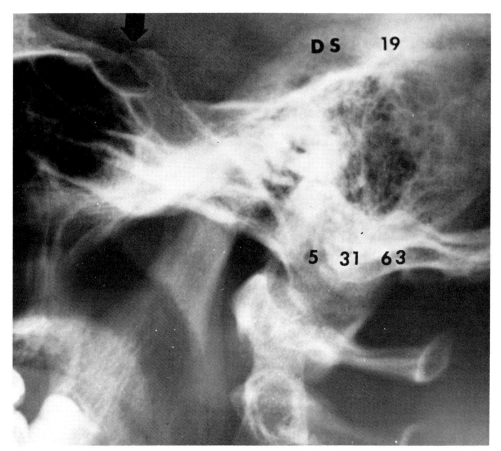

FIG. 1560. TURNER SYNDROME
Small pituitary fossa with bony bridging of the clinoid processes (*arrow*). (Courtesy of A. E. O'Hara and S. E. Abram: Radiographic clues to cytogenetic disease. *Texas Medicine, 67:* 84, 1971, © Texas Medical Association, Austin, Tex.)

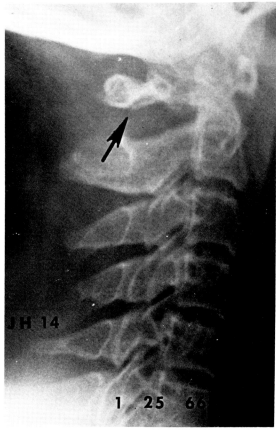

FIG. 1561. TURNER SYNDROME
Small posterior arch of C1 in this 14-year-old patient. (Courtesy of A. E. O'Hara and S. E. Abram: Radiographic clues to cytogenetic disease. *Texas Medicine, 67:* 84, 1971, © Texas Medical Association, Austin, Tex.)

NOONAN SYNDROME
(PSEUDO-TURNER OR MALE TURNER SYNDROME)

Noonan syndrome is a hereditable disease and not due to any demonstrable chromosomal abnormality. It is included in this section for distinction from Turner syndrome (Table 56).

Noonan syndrome[1-8] is a term reserved for patients who manifest somatic abnormalities similar to those of Turner syndrome, but with the karyotype normal except in a few isolated cases of mosaicism. The familial[9-12] pattern suggests either an X-linked or an autosomal transmission.

Clinical Features. Some of the clinical features of Turner syndrome are present, but with greater variability. Patients may be of normal height, and their gonadal development vary from agonadism to normal. Mental retardation and hypertelorism are common with Noonan syndrome, rare in Turner. Webbed neck and increased carrying angle of the elbow are common in both. Congenital heart disease is a feature of both conditions, but in Noonan syndrome it is usually right-sided and tends to be pulmonary stenosis or occasionally patent ductus arteriosus. Interventricular septal defect may also occur. In Turner syndrome the cardiac abnormalities are usually left-sided (coarctation). Other clinical features of Noonan syndrome include dental malocclusion, sternal deformity, and mandibular hypoplasia.[13-15] Although some patients with Noonan syndrome share features of Turner syndrome, the total phenotype allows its differentiation.[16]

Roentgenographic Features. The radiographic features are similar to Turner syndrome but less commonly observed.

Sternal abnormalities are most frequent, usually pectus cavus or pectus carinatum.[8] The sternum is short with the cephalad aspect protruding forward and the caudad portions inward. The heart may be enlarged secondary to pulmonary stenosis or interatrial septal defect. Osteoporosis, cubitus valgus, and retarded bone age are present in less than half the patients.

Dental malocclusion and mandibular hypoplasia are common. Other variable findings include biparietal foramina, dolichocephaly, microcephaly, cranial enlargement, and bitemporal bulging.[8]

Intestinal lymphangiectasia may occur, producing thick mucosal folds of the duodenum and small intestine.

Coronal clefts of the spine may be present and eventration of the diaphragm occurs infrequently.

Riggs[8] reported a high incidence of such urinary abnormalities as rotational errors, duplication, hydronephrosis, and large redundant extrarenal pelvis. Jackson and Lefrak[9] reported that the syndrome may be confused with Marfan syndrome because of the palatal shape, skeletal abnormalities, visual difficulties, and cardiac murmurs; however, clinical analysis of the physical characteristics, cardiac abnormality, and eye changes should easily identify Noonan syndrome.

TABLE 56. COMPARISON OF TURNER SYNDROME WITH NOONAN SYNDROME[a]

Finding	Turner Syndrome	Noonan Syndrome
Chromosomes	Abnormal (often XO)	Normal
Short stature	Constant	Variable
Gonads	Female gonadal dysgenesis	Varies from agonadism to normal gonad function and fertility of both sexes
Congenital heart disease	Left heart usually involved (aortic stenosis and coarctation)	More common with right heart usually involved (often pulmonic stenosis)
Mental retardation	Rare	Common
Dental malocclusion	Rare	Common
Renal anomalies	Less common	Frequent
Webbed neck	Common	Common
Radiographic bone abnormalities	Common	Less common

[a] From W. Riggs: *Radiology*, 96: 393, 1970.[8]

KLINEFELTER SYNDROME

Chromosomal analysis usually shows an XXY sex component although XXYY has been reported.[1-10]

Clinical Features. The usual clinical features are similar in the XXY and the XXYY forms. They consist of a tall, thin build, testicular atrophy with seminiferous tubular hyalinization, a normal or slightly underdeveloped penis, azoospermia, gynecomastia, paucity of hair on the face and chest and a female pubic escutcheon, mild mental retardation, and sterility. The urinary gonadotrophins are high and 17-ketosteroids are slightly low to markedly decreased. Dermatoglyphics may be useful for diagnosis and may help differentiate the XXY and XXYY forms.[11] Patients with XXYY karyotype usually present with typical Klinefelter syndrome,[12] except for an increase in the incidence and degree of mental retardation and more severe congenital hypoplasia. Patients with a chromosome complement of XXXY may be considered to be the extreme end of the Klinefelter spectrum. They generally have abnormal facies, somewhat suggestive of mongolism, with hypertelorism, malformed ears, incurving fifth fingers, muscular hypotonia, hypogonadism, and severe mental retardation.

Among the other sex chromosome abnormalities which have been reported are the XXX, XXXX, and XXXXX females. Unlike the analogous "super female," they generally have a surprisingly normal phenotype.

Roentgenographic Features. The roentgenographic features are not consistent. In more severe forms there may be delayed bone maturation (Fig. 1562), involving both the appearance and fusion of the ossification centers. Other ra-

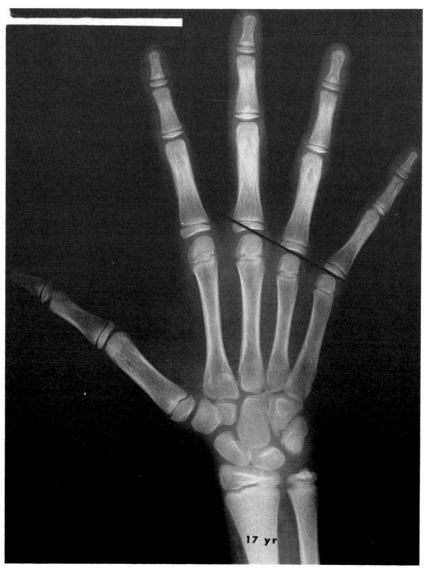

FIG. 1562. KLINEFELTER SYNDROME

A 17-year-old male with retardation of bone maturation equivalent to age 12. A positive metacarpal sign is present. (Courtesy of A. E. O'Hara and S. E. Abram: Radiographic clues to cytogenetic disease. *Texas Medicine, 67:* 84, 1971, © Texas Medical Association, Austin, Tex.)

diographic findings include the metacarpal sign, accessory epiphyses of the second metacarpal bilaterally, and failure of the frontal sinuses to develop.

In the very severe forms of the syndrome extensive bony abnormalities may involve almost any portion of the body. Among the most frequent are radial ulnar synostosis or dislocation, elongated distal ulna, coxa valga, scoliosis and kyphosis and thick or abnormally segmented sternum. Acromegalic features may also be present. In the super female group the patients are usually over 6 feet tall, with subnormal intelligence, and frequently antisocial behavior. They may have all the findings of the other forms of Klinefelter syndrome.

The sella may be small and bridged, or enlarged, or show localized enlargement on polytomography.

AUTOSOMAL ABNORMALITIES (TRISOMY DEFECTS)

The autosomal trisomy defects are usually the result of nondisjunction, but may also be due to translocation defects. The incidence is directly related to the maternal age.

Four distinctive clinical syndromes are associated with autosomal abnormalities: (1) mongolism with an extra chromosome in group 21-22 (G group) and called trisomy 21-22 or Down syndrome, (2) trisomy 18,[1,2] (3) trisomy 13-15 or D-trisomy,[3] and (4) cri-du-chat syndrome[4] due to a deletion of a short arm of the B-group.

In the future other trisomies will be described. Some trisomies not discussed below include the trisomy for the short arm of chromosome-9 (T9p).[5-7]

Chromosomal banding techniques, the site of genetic determination of structural aberrations within Group C of chromosomes (Nos. 6–12 and X), are available now in most cytogenetic laboratories.

Mongolism (Trisomy 21-22)

Before the advent of karyotype studies, the diagnosis of mongolism in early infancy was difficult. The karyotype studies, however, show an added chromosome at Group G, so that the pattern is 47 XX or 47 XY (Fig. 1563). It is believed that both trisomy 21 and 22 can show the same clinical picture.[1]

Roentgenographic Features. In 1956, Caffey and Ross[2] described the diagnostic roentgenographic changes in the pelvis of mongoloid infants: flaring of the iliac wings, flattening of the acetabular roofs, and tapering of the ischia. Caffey[2] found the acetabular index, the sum of the acetabular angle and the iliac angle, the most sensitive indication of mongolism. Astley[3] has reported that with less than 60°, mongolism is very probable, between 60° and 68° it is probable and between 68° and 78° it is improbable. Nicolis and Sacchetti[4] devised a slightly different criterion for distinction between the normal and mongol pelvis that statistically proves more discriminating. This, termed the pelvic index, is defined by the formula 0.30× (mean acetabular angle) + 0.42× (mean iliac angle). The discriminant value between normals and mongols is 25.2. The other pelvic findings are flaring of the iliac crest and tapering of the ischial rami (Fig. 1564).

Hypoplastic and triangular middle and distal phalanges of the fifth finger are other common roentgenographic signs[5] (Fig. 1565). Occasionally, however, this may occur in normal individuals, cretins, and achondroplastic dwarfs. Pseudoepiphyses of the first and second metacarpals may occur. The skull may exhibit microcrania of the brachycephalic type. The roof of the mouth is highly arched. Cranial sutures frequently show delayed closure.

There may be only 11 ribs[6,7] (Fig. 1566). The manubrium may have two or three centers of ossification instead of one[8,9] (Fig. 1567). A left to right shunt may occur and cause cardiac enlargement and prominent pulmonary vasculature.[10-13] Dislocation of C1–C2 occurs (Fig. 1568). The anterior borders of the vertebral centra may be straight or concave rather than convex as is usual in infancy (Fig. 1569). The centra are also higher and narrower. The lateral lumbar index, obtained by the ratio of the horizontal to the vertical diameters of the L2 centra, is 0.93 for mongols and 1.28 in normals under 2 years[14] (Fig. 1570).

Trisomy E (Trisomy 17-18) Syndrome

Trisomy 17-18 syndrome is a condition in which the karyotype shows 47 chromosomes.[1-3] The additional chromosomes occur in the 16-18 group, and may be either in the 17 or 18 location (Fig. 1571). The maternal age is found to be advanced, as is the case in mongolism, and the child rarely survives infancy.

Clinical Features. The ears are low-set and often deformed, and the mouth is round. The buccal cavity is unusually small. A small mandible is the most constant feature of this syndrome, but is not confined to the trisomy 17-18 syn-

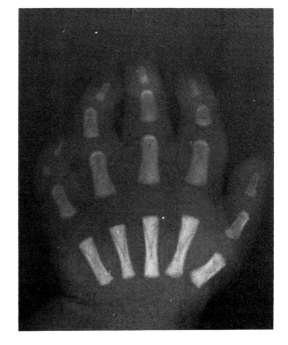

FIG. 1565. MONGOLISM
Hypoplastic and triangular middle phalanx of the fifth finger.

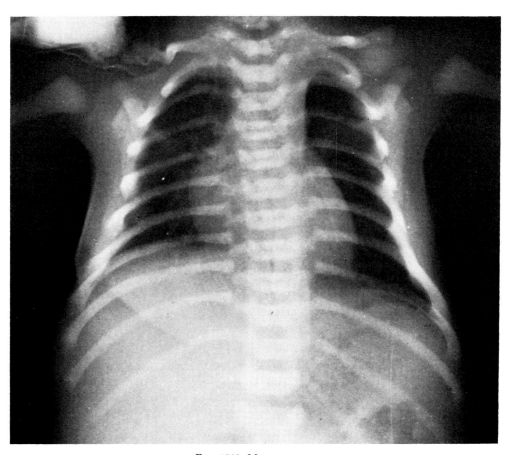

FIG. 1566. MONGOLISM
There are only 11 pairs of ribs.

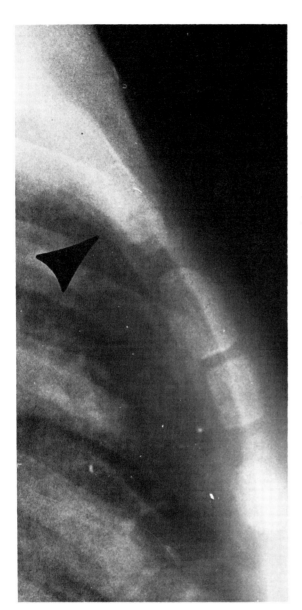

Fig. 1567. MONGOLISM
Multiple ossification centers in the manubrium (*arrow*).

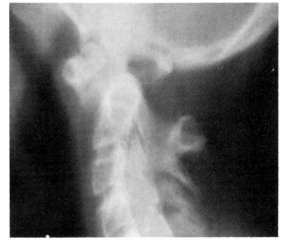

Fig. 1568. MONGOLISM
Body section films reveal dislocation of C1–C2.

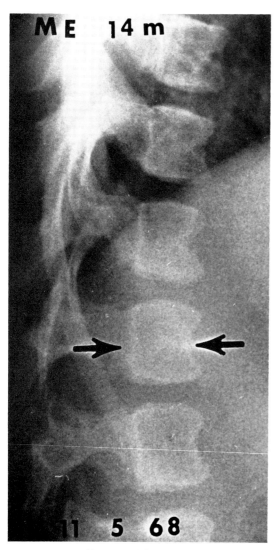

FIG. 1569. MONGOLISM
Anterior concavity of the vertebral bodies (*arrows*) and positive lumbar indices. (Courtesy of A. E. O'Hara and S. E. Abram: Radiographic clues to cytogenetic disease. *Texas Medicine, 67:* 84, 1971, © Texas Medical Association, Austin, Tex.)

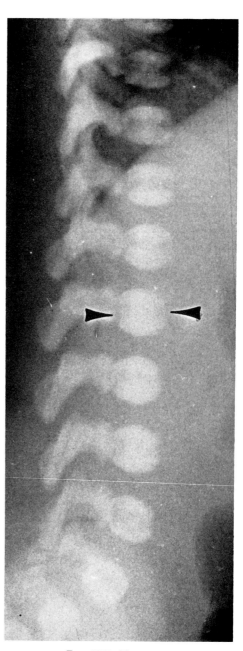

FIG. 1570. MONGOLISM
Positive lumbar indices are present.

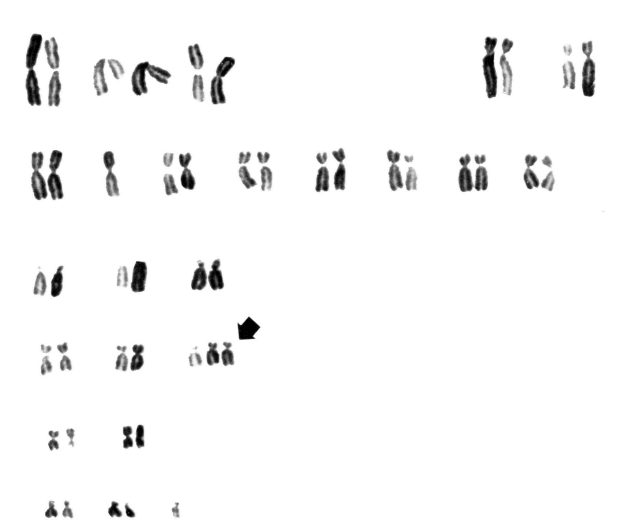

Fig. 1571. Trisomy 17–18 Syndrome
Extra chromosome in Group E (*arrow*).

drome. The hands reveal an ulnar deviation of the third, fourth, and fifth digits.

Congenital heart disease, hernias, renal anomalies, and eventration of the diaphragm may all be found in association with chromosomal abnormalities. When these changes do occur, they should suggest one of the trisomy syndromes.

Other anomalies include overlapping fingers, rocker bottom or clubfoot deformity, and a small pelvis with limited hip abduction.[1, 3, 4-10]

Roentgenographic Features. Individually, the roentgenographic features of trisomy 17-18 do not clearly indicate the syndrome but, when they occur in combination, suggest it strongly.

A hypoplastic mandible is the most constant finding (Fig. 1572). The sternum is often hypoplastic and the anteroposterior diameter of the thorax is increased (Fig. 1573). Hypoplastic clav-

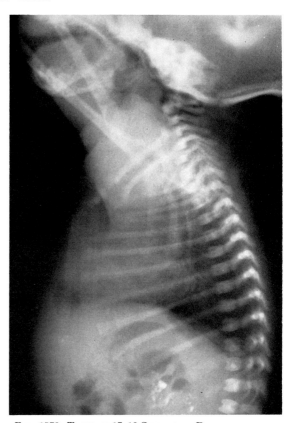

FIG. 1573. TRISOMY 17-18 SYNDROME DEMONSTRATING HYPOPLASIA OF STERNUM AND INCREASED ANTEROPOSTERIOR DIAMETER OF THORAX
(Previously published by E. B. Singleton, H. S. Rosenberg, and S. J. Yang in *The Radiologic Clinics of North America*, 2: 281, 1964[11]; reprinted by permission.)

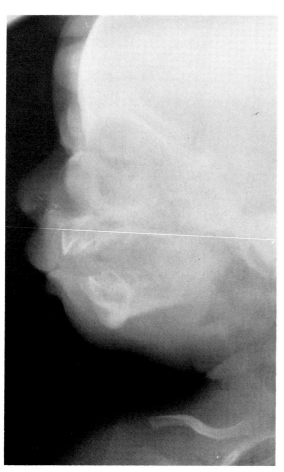

FIG. 1572. TRISOMY 17-18 DEFECT SHOWING MANDIBULAR HYPOPLASIA
(Previously published by E. B. Singleton, H. S. Rosenberg, and S. J. Yang in *The Radiologic Clinics of North America*, 2: 281, 1964[11]; reprinted by permission.)

icles may occur (Fig. 1574). Congenital heart disease is responsible for an enlarged cardiac silhouette (Fig. 1575). Patent ductus arteriosus and ventricular septal defects are the most common findings.[3] The ribs are slender and tapered (Fig. 1575). The pelvis is usually small and the forward rotation of the iliac wings result in the antimongoloid configuration (Fig. 1576).

The ulnar deviation of the third, fourth, and fifth fingers produces a rather wide gap between the second and third fingers (Fig. 1577). The metacarpals are not affected. Overlapping of the finger is frequent (Fig. 1578). Varus deformities of the forefoot and dorsiflexion of the toes and vertical talus are frequent (Fig. 1579). Other anomalies include rocker bottom or clubfoot deformity, a small pelvis with limited hip abduction,[1, 4-10] and persistent metopic suture (Fig. 1580).

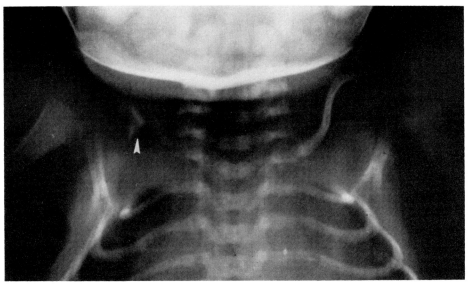

FIG. 1574. TRISOMY 17-18
Hypoplastic right clavicle (*arrow*). This may be mistaken for cleidocranial dysostosis. (Courtesy of A. E. O'Hara and S. E. Abram: Radiographic clues to cytogenetic disease. *Texas Medicine, 67:* 84, 1971, © Texas Medical Association, Austin, Tex.)

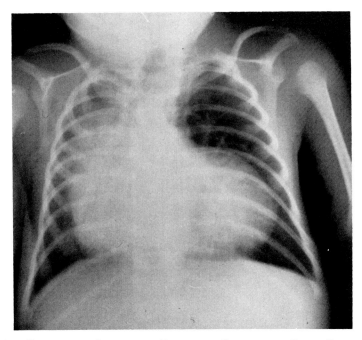

FIG. 1575. TRISOMY 17-18 SYNDROME IN INFANT WITH PULMONARY STENOSIS AND LARGE PATENT DUCTUS ARTERIOSUS
Ribs are thin and tapered with irregularity of the posterior border. (Previously published by E. B. Singleton, H. S. Rosenberg, and S. J. Yang in *The Radiologic Clinics of North America, 2:* 281, 1964[11]; reprinted by permission.)

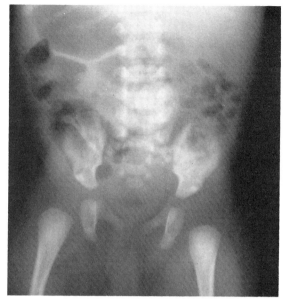

FIG. 1576. TRISOMY 17-18 SHOWING NARROW ILIAC CREST AND LARGE ACETABULAR ANGLES
Previously published by E. B. Singleton, H. S. Rosenberg, and S. J. Yang in *The Radiologic Clinics of North America*, 2: 281, 1964[11]; reprinted by permission.)

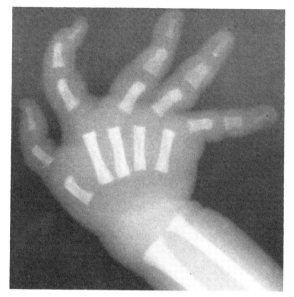

FIG. 1577. TRISOMY 17-18 SYNDROME SHOWING ULNA DEVIATION OF THIRD, FOURTH, AND FIFTH FINGERS
(Previously published by E. B. Singleton, H. S. Rosenberg, and S. J. Yang in *The Radiologic Clinics of North America*, 2: 281, 1964[11]; reprinted by permission.)

Trisomy D (Trisomy 13-15) Syndrome

In this condition a multitude of physical defects are described.[1-10] The eyes are particularly affected, showing coloboma, cataracts and microphthalmia. Many of the other defects are common to other trisomies, and include cleft lip and palate, corneal opacities, and abnormalities of the optic nerve and chiasm.

Roentgenographic Features. These are not distinctive, and may be the same as for trisomy 17-18. They include micrognathism and anomalies of the hand. An antimongoloid configuration of the pelvis has not been described in association with trisomy 13-15, and its absence may aid in distinguishing this from 17-18 trisomy.[11]

Cri-du-Chat Syndrome

The patient's karyotype contains the normal number of 46 chromosomes but the short arms are deleted in one of the Group B chromosomes.[1-10]

The peculiar high-pitched cry of the infant is the hallmark of this syndrome. Other features include low-set ears, hypertelorism, microcephaly, strabismus, and abnormal dermatoglyphics. There is marked mental and growth retardation.

Roentgenographic Features. Microcephaly, hypertelorism (which must be differentiated

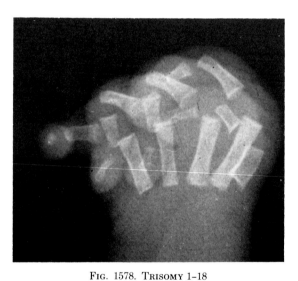

FIG. 1578. TRISOMY 1-18
Characteristic overlapping of the fingers. (Same patient as Figure 1574.) (Courtesy of A. E. O'Hara and S. E. Abram: Radiographic clues to cytogenetic disease. *Texas Medicine*, 67: 84, 1971, © Texas Medical Association, Austin, Tex.)

from trisomy 13-15),[4,11] small mandible (Fig. 1581), and faulty long bone development secondary to muscle hypotonia[1] may be seen. Other less constant features are agenesis of the corpus callosum, horseshoe kidney, and congenital heart disease.

Measurements of hand radiographs reveal that

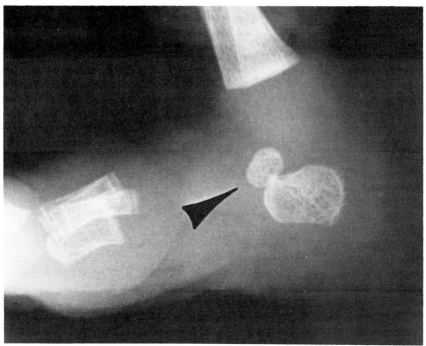

Fig. 1579. Trisomy 17–18
Lateral view of the foot reveals a vertical talus (*arrow*). (Same patient as seen in Figures 1574 and 1578.) (Courtesy of A. E. O'Hara and S. E. Abram: Radiographic clues to cytogenetic disease. *Texas Medicine,* 67: 84, 1971, © Texas Medical Association, Austin, Tex.)

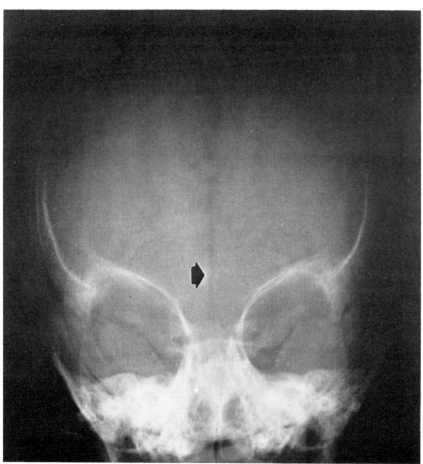

Fig. 1580. Trisomy 17–18
Persistent metopic suture (*arrow*). (Same patient as Figures 1574 and 1579.) (Courtesy of A. E. O'Hara and S. E. Abram: Radiographic clues to cytogenetic disease. *Texas Medicine,* 67: 84, 1971, © Texas Medical Association, Austin, Tex.)

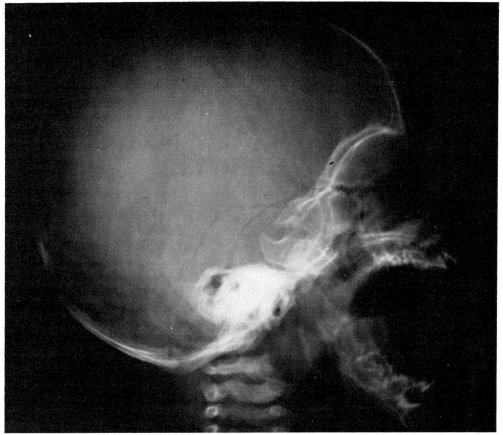

FIG. 1581. CRI-DU-CHAT SYNDROME
Microcephaly and a small mandible.

they are smaller than normal, and in most the third, fourth, and fifth metacarpals are disproportionately short and the second, third, fourth, and fifth proximal phalanges disproportionately long.[12]

The radiographic characteristics are not specific, as in trisomy 13-15, trisomy 18, and mongolism, and the clinical findings must be correlated with the phenotype and radiographic manifestations for the correct diagnosis.[1]

REFERENCES

Chromosomal Abnormalities

1. Hsu, T. C.: Mammalian chromosomes in vitro. J Hered, 43: 167, 1952.
2. Tjio, J. H., and Levan, A.: The chromosome number of man. Hereditas, 42: 1, 1956.
3. Painter, T. S.: Studies in mammalian spermatogenesis; II. The spermatogenesis of man. J Exp Zool, 37: 291, 1923.
4. Lejeune, J., Turpin, R., and Gautier, M.: Le mongolisme, premier example d'aberration autosomigue humaine. Am J Hum Genet, 11: 41, 1959.
5. Eggen R. R.: Cytogenetics; review of recent advances in a new field of clinical pathology. Am J Clin Pathol, 39: 3, 1963.
6. Conen, P. E. and Bell, A. G.: Chromosomal aberration in an infant following the use of diagnostic x-rays. Pediatrics, 31: 72, 1963.
7. Rosenfield, R. L., Briebart, S., Isaacs, H., Klevit, H. D., and Mellman, W. J.: Trisomy of chromosomes 13-15 and 17-18; association with infantile arteriosclerosis. Am J Med Sci, 244: 763, 1962.
8. Barr, M. L., and Bertram, E. G.: A morphological distinction between neurones of the male and female, and the nucleolar satellite during accelerated nucleoprotein synthesis. Nature, 163: 676, 1949.
9. Moore, K. L., and Barr, M. L.: Smears from oral mucosa in the detection of chromosomal sex. Lancet, 2: 57, 1955.
10. Davidson, W. M., and Smith, D. R.: A morphological sex difference in the polymorphonuclear neutrophil leukocytes. Br Med J, 2: 6, 1954.

Turner Syndrome

1. Turner, H. H.: A syndrome of infantilism, congenital webbed neck and cubitis valgus. Endocrinology, 23: 566, 1938.

2. Polani, P. E., Hunter, W. F., and Lennos, B.: Chromosomal sex in Turner's syndrome with coarctation of the aorta. Lancet, 2: 120, 1954.
3. Ford, C. E., Jones, K. W., Polani, P. E., Almeida, J. C. D., and Briggs, J. H.: A sex chromosome anomaly in a case of gonadal dysgenesis (Turner's syndrome). Lancet, 1: 711, 1959.
4. Noonan, J. A., and Ehmke, D. A.: Associated noncardiac malformation in children with congenital heart disease. J Pediatr, 63: 468, 1963.
5. Noonan, J. A.: Hypertelorism with Turner phenotype. Am J Dis Child, 116: 373, 1968.
6. Summitt, R. L. : Turner's syndrome and Noonan's syndrome. J. Pediatr, 74: 155, 1969.
7. Astley, R.: Chromosomal abnormalities in childhood with particular reference to Turner's syndrome and mongolism. Br J Radiol, 36: 2, 1963.
8. Avin, J.: The male Turner's syndrome. Am J Dis Child, 91: 630, 1956.
9. Elliott, G. A., Sandler, A., and Rabinowitz, D.: Gonadal dysgenesis in 3 sisters. J. Clin Endocrinol, 19: 955, 1959.
10. Federman, D. D.: Disorders of sexual development. N Engl J Med, 277: 351, 1967.
11. Kaufmann, P.: Two cases of Turner's syndrome. Schweiz Med. Wochnschr, 85: 1027, 1955.
12. Levin, J., and Kupperman, H. S.: Skeletal abnormalities in gonadal dysgenesis. Arch Intern Med, 113: 730, 1961.
13. Therman, E., Patau, K., Smith, D. W., and Demars, R. I.: The D syndrome and XO gonadal dysgenesis in 2 sisters. Am J Hum Genet, 13: 193, 1961.
14. Kosowicz, J.: Skeletal changes in Turner's syndrome and their significance in differential diagnosis. Bull Pol Med Sci Hist, 2: 23, 1959.
15. Archibald, R. M., Finby, N., and DeVita, F.: Endocrine significance of short metacarpals. J Clin Endocrinol, 19: 1312, 1959.
16. Keats, T. E., Burns, T. W.: The radiographic manifestations of gonadal dysgenesis. Radiol Clin North Am, 2: 297, 1964.
17. Kosowicz, J.: The carpal sign in gonadal dysgenesis. J. Clin Endocrinol, 22: 949, 1962.
18. Finby, N., and Archibald, R. M.: Skeletal abnormalities associated with gonadal dysgenesis. AJR, 89: 1222, 1963.
19. Kosowicz, J.: The deformity of the medial tibial condyle in 19 cases of gonadal dysgenesis. J Bone Joint Surg, 42A: 600, 1960.
20. Samaan, N. A., Stepanas, A. V., Danziger, J., and Trujillo, J.: Reactive pituitary abnormalities in patients with Klinefelter's and Turner's syndromes. Arch Intern Med, 139: 198, 1979.

Noonan Syndrome

1. Baker, D. H., Berdon, W. E., Morishima, A., et al.: Turner's syndrome and pseudo-Turner's syndrome. AJR, 100: 40, 1967.
2. Celermajer, J. M., Bowdler, J. D., and Cohen, D. H.: Pulmonary stenosis in patients with the Turner phenotype in the male. Am J Dis Child, 116: 351, 1968.
3. Chaves-Carballo, E., Hayles, A. B.: Ullrich-Turner syndrome in the male; review of the literature and report of a case with lymphocytic (Hashimoto's) thyroiditis. Mayo Clin Proc, 41: 843, 1966.
4. Kaplan, M. S., Opitz, J. M., and Gosset, F. R.: Noonan's syndrome; a case with elevated serum alkaline phosphatase levels and malignant schwannoma of the left forearm. Am J Dis Child, 116: 359, 1968.
5. Opitz, J. M., Summitt, R. L., and Sarto, G. E.: Noonan's syndrome in girls; a genocopy of the Ullrich-Turner syndrome. J Pediatr, 67: 968, 1965.
6. Summitt, R. L., Wilroy, R. S., Jr., and Camacho, A. M.: Further studies in Noonan's syndrome (abstract). South Med J, 60: 1361, 1967.
7. Wright, N. L., Summitt, R. L., and Ainger, L. E.: Noonan's syndrome and Ebstein's malformation of the tricuspid valve. Am J Dis Child, 116: 367, 1968.
8. Riggs, W.: Roentgen findings in Noonan's syndrome. Radiology, 96: 393, 1970.
9. Jackson, L. G., and Lefrak, S.: Familial occurrence of the Noonan syndrome. In *The Clinical Delineation of Birth Defects*; *V. Phenotypic Aspects of Chromosomal Aberrations*, p. 36. Williams & Wilkins, Baltimore, 1969.
10. Summitt, R. L.: Familial Goldenhar syndrome. In *The Clinical Delineation of Birth Defects*; *II. Malformation Syndromes*, p. 106. Williams & Wilkins, Baltimore, 1969.
11. Nora, J. J., and Sinha, A. K.: Inheritance of the Turner phenotype. In *The Clinical Delineation of Birth Defects*; *V. Phenotypic Aspects of Chromosomal Aberrations*, p. 29. Williams & Wilkins, Baltimore, 1969.
12. Polani, P. E.: Turner phenotype with normal sex chromosomes. In *The Clinical Delineation of Birth Defects*; *V. Phenotypic Aspects of Chromosomal Aberrations*, p. 24. Williams & Wilkins, Baltimore, 1969.
13. Noonan, J. A., and Ehmke, D. A.: Associated noncardiac malformations in children with congenital heart disease (abstract). J Pediatr, 63: 488, 1968.
14. Summitt, R. L. Opitz, J. M., and Smith, D. W.: Noonan's syndrome in the male (abstract). J Pediatr, 67: 936, 1965.
15. Opitz, J. M., Sarto, G. E. and Summitt, R. L.: Turner's syndrome and phenotype. Lancet, 2: 282, 1966.
16. Summitt, R. L.: The Noonan syndrome. In *The Clinical Delineation of Birth Defects*; *V. Phenotypic Aspects of Chromosomal Aberrations*, p. 39. Williams & Wilkins, Baltimore, 1969.

Klinefelter Syndrome

1. Alter, M.: Dermatoglyphics in birth defects. In *The Clinical Delineation of Birth Defects. III. Limb Malformations*, p. 103. Williams & Wilkins, Baltimore, 1969.
2. Ferrier, P. E., and Ferrier, S. A.: XXYY Klinefelter's syndrome; case report and a study of the Y chromosomes DNA replication pattern. Ann Genet, 11: 145, 1968.
3. Uchida, I. A., Miller, J. R., and Soltan, H. C.: Dermatoglyphics associated with the XXYY chromosome complement. Am J Hum Genet, 16: 284, 1964.
4. Alter, M., Gorlin, R., Yunis, J. J., Peagler, F., and Bruhl, H.: Dermatoglyphics in XXYY Klinefelter's syndrome. Am J Hum Genet, 18: 507, 1966.
5. Robinsin, G. C., Miller, J. R., and Kamburoff, T. D.: Klinefelter's syndrome with the XXYY sex chromosome complex. J. Pediatr, 65: 226, 1964.
6. Herbeuval, R., Gilgenkarntz, S., Guerci, O., and Thibaut, G.: Syndrome de Klinefelter a formule XXYY. Presse Med, 4: 2987, 1965.
7. Schlegel, R. J., Aspillaga, M. J., Neu, R., and Gardner, L. I.: Studies on a boy with XXYY chromosome constitution. Pediatrics, 36: 113, 1965.
8. Knorr, D., Lersch, B., and Zang, K. D.: Atypisches Klinefelter Syndrom mit dem Chromosomensatz 44 + XXYY im Kindesalter. Z Kinderheilkd, 95: 74, 1966.
9. Garcia, H. O., Borgaonkar, D. S., and Richardson, F.: XXYY syndrome in a prepubertal male. Johns Hopkins Med J, 121: 31, 1967.
10. Sachsse, W., Overzier, C., and Knolle, J.: Eine Sonderform des Klinefelter syndrome; Mosaik mit verdoppelun des Y-Chromosome (2X-2Y form). Dtsch Med Wochenschr, 92: 1213, 1967.
11. Forbes, A. P.: Fingerprints and palm prints (dermatoglyphics) and palmar flexion creases in gonadal dysgen-

esis, pseudohypoparathyroidism and Klinefelter's syndrome. New Engl J Med, *270:* 1268, 1964.
12. Townes, P. L., Zeigler, N. A., and Scheiner, A. P.: An XXXY variant of Klinefelter's syndrome in prepubertal boy. J Pediatr, *67:* 410, 1965.
13. Samaan, N. A., Stepanas, A. V., Danziger, J., and Trujillo, J.: Reactive pituitary abnormalities in patients with Klinefelter's and Turner's syndromes. Arch Intern Med, *139:* 198, 1979.

Autosomal Abnormalities

1. Edwards, J. H., Harnden, D. G., and Cameron, A. H., et al: A new trisomic syndrome, Lancet, *1:* 797, 1960.
2. James, A. E., Jr., Gelcourt, C. L., Atkins, A., and Janower, M. L.: Trisomy 18. Radiology, *92:* 37, 1969.
3. Patau, K., et al.: Multiple congenital anomalies caused by an extra autosome. Lancet, *1:* 790, 1960.
4. Lejeune, Jr., et al. Trois cas de diletion partille due bras court d'un chromosome 5. C R Acad Sci, *257:* 3098, 1963.
5. Rethore, M. O., Larget-Piet, L., Abonyi, D. et al: Sur quatre cas de trisomie pour le bras court du chromosome 9. Individualisation d'une nouvelle entite morbide. Ann Genet (Paris), *13:* 217, 1970.
6. Schinzel, A.: Autosomal chromosome aberrations. A review of the clinical syndromes caused by structural chromosome aberrations, mosaic-trisomies 8 and 9, and triploidy. Ergeb Inn Med Kinderheilkd, *38:* 37, 1976.
7. Schinzel, A.: Trisomy 9p, a chromosome aberration with distinct radiologic findings. Radiology *130:* 125, 1979.

Mongolism

1. Zellweger, H., Mikamo, K., and Abbo, G.: An unusual translocation in a case of mongolism. J Pediatr, *62:* 225, 1963.
2. Caffey, J., and Ross, S.: Mongolism (mongoloid deficiency) during early infancy—some newly recognized diagnostic changes in the pelvic bones. Pediatrics, *17:* 642, 1956.
3. Astley, R.: Chromosomal abnormalities in childhood with particular reference to Turner's syndrome and mongolism. Br J Radiol, *36:* 2, 1963.
4. Nicolis, F. B., and Sacchetti, G.: Nomogram for x-ray evaluation of some morphologic anomalies of pelvis in diagnosis of mongolism. Pediatrics, *32:* 1074, 1963.
5. Hefke, H. W.: Roentgenologic study of anomalies of the hands in 100 cases of mongolism. Am J Dis Child, *60:* 1319, 1940.
6. Beber, B. A.: A new radiographic finding in mongolism. Radiology, *86:* 332, 1966.
7. Pionner, R., and Depraz, A.: Les anomalies costales d'orginie congenitalee (etude statiszue d'apres 10,000 radiographs). Radiol Clin, *25:* 170, 1956.
8. Chu, E. H. Y., and Giles, N. H.: Human chromosome complements in normal somatic cells in culture. Am J Hum Genet, *11:* 63, 1959.
9. Horns, J. W., and O'Loughlin, B. J.: Multiple manubrial ossification centers in mongolism. AJR *93:* 395, 1965.
10. Berg, J. M., Crome, L., and France, N. E.: Congenital cardiac malformation in mongolism. Br Heart J, *22:* 331, 1960.
11. Evans, P. R.: Cardiac anomalies in mongolism. Br Heart J, *12:* 258, 1950.
12. Jue, K. L.: Anomalies origin of pulmonary arteries from pulmonary trunk ("crossed pulmonary arteries"); observations in a case of 18 trisomy syndrome. Am Heart J, *71:* 807, 1966.
13. Liu, M. C., and Corlett, K.: A study of congenital heart defects in mongolism. Arch Dis Child, *34:* 410, 1959.
14. Rabinowitz, J. G., and Moseley, J. E.: Lateral lumbar spine in Down's syndrome; a new roentgen feature. Radiology, *83:* 74, 1964.

Trisomy E (Trisomy 17-18) Syndrome

1. Edwards, J. H., Harnden, D. G., Cameron, A. H.: A new trisomic syndrome. Lancet, *1:* 787, 1960.
2. Patau, K., Smith, D. W., Therme, E., Inhorn, S. L., and Wagner, H. P.: Multiple congenital anomalies caused by an extra autosome. Lancet, *1:* 790, 1960.
3. Moseley, J. E., Wolf, B. S., and Gottlieb, M. I.: The trisomy 17-18 syndrome; roentgen features. AJR *89:* 905, 1963.
4. Butler, L. J., Snodgrass, G. J., and France, N. E.: E (16-18) trisomy syndrome; analysis of 13 cases. Arch Dis Child, *40:* 600, 1965.
5. Holman, G. H., Erkman, B., Zacharias, D. L.: The 18 trisomy syndrome, 2 new clinical variants. New Engl J Med, *268:* 928, 1963.
6. Jue, K. L.: Anomalies origin of pulmonary arteries from pulmonary trunk (" crossed pulmonary arteries"); observation in a case of 18 trisomy syndrome. Am Heart J, *71:* 807, 1966.
7. Miller, J. Q., Rostafinski, M. J., and Hyde, M. S.: A defective extra chromosome associated with clinical 17-18 trisomy syndrome. Pediatrics, *36:* 135, 1965.
8. Punnett, H. H., Pinsky, L. and DiGeorge, A. M., et al.: Familial reciprocal C/18 translocation. Am J Hum Genet, *18:* 572, 1956.
9. Smith, D. W.: The no. 18 trisomy and D trisomy syndromes. Pediatr Clin North Am, *10:* 398, 1963.
10. Windmiller, J., Marks, J. F., and Reimold, E. W., et al.: Trisomy 18 with biliary atresia. J Pediatr, *67:* 327, 1965.

Trisomy D (Trisomy 13-15) Syndrome

1. Atkins, L., and Keenan, M. E.: Probably 3/13-15 chromosome translocation with D trisomy syndrome. J Pediatr, *67:* 874, 1965.
2. Bain, A. D.: Normal trisomy 13-15 mosaicism in two infants. Arch Dis Child, *40:* 442, 1965.
3. Conen, P. D., Erkman, B., and Metaxotov, C.: The "D" syndrome, report of 4 trisomic and 1 D/D translocation cases. Am J Dis Child, *111:* 236, 1966.
4. Erkman, B., Basrur, U. R., and Conen, P. E.: D/D translocation "D" syndrome. J Pediatr, *67:* 270, 1965.
5. Hect, F., Loop, J. W., and Graham, C. B.: The radiologic phenotype of the D trisomy syndrome; abnormalities of the ribs and pelvis. J Pediatr, *67:* 870, 1965.
6. Kakulas, B. A., and Rosman, N. P.: 13-15 trisomy in 8 cases of arrhinencephaly. Lancet, *2:* 717, 1965.
7. Miller, J. Q., Picard, E. H., and Alkan, M. K.: A specific congenital brain defect (arrhinencephaly) in 13-15 trisomy. New Engl J Med, *268:* 120, 1963.
8. Snodgrass, G. J., Butler, L. J., France, N. E., et al.: The D (13-15) trisomy syndrome; an analysis of 7 examples. Arch dis Child, *41:* 250, 1966.
9. Smith, D. W.: The no. 18 trisomy and D trisomy syndromes. Pediatr Clin North Am, *10:* 398, 1963.
10. Therman, E., Patau, K., Smith, D. W., and Demars, R.: The D syndrome and XO gonadal dysgenesis in 2 sisters. Am J Hum Genet, *13:* 193, 1961.
11. Singleton, E. B., Rosenberg, H. S., and Yang, S. J.: The radiographic manifestations of chromosomal abnormalities. Radiol Clin North Am, *2:* 281, 1964.

Cri-du-Chat Syndrome

1. James, A. E., Belcourt, C. L., Atkins, L., and Janower, M. L.: Trisomy 18. Radiology, *92:* 37, 1969.

2. Edwards, J. H., Harnden, D. G., Cameron, A. H., Cross, V. M., and Wolf, O. H.: A new trisomic syndrome. Lancet, *1:* 797, 1960.
3. Patau, K.: Multiple congenital anomalies caused by an extra autosome. Lancet, *1:* 790, 1960.
4. Lejeune, J., et al: Trois cas de diletion partille du bras court d'un chromosome 5. C R Acad Sci, *257:* 3098, 1963.
5. James, A. E. Jr., Feingold, M., Atkins, L., and Janower, M. L.: The cri-du-chat syndrome. Radiology, *92:* 50, 1969.
6. Dumars, K. W., Gaskill, C., and Kitzmiller, N.: Le cri-du-chat (crying cat) syndrome. Am J Dis Child, *108:* 533, 1964.
7. Warburton, D., et al.: Distinction between chromosome 4 and chromosome 5 by replication pattern and length of long and short arms. Am J Hum Genet, *19:* 399, 1967.
8. Alvin, J.: The male Turner's syndrome. Am J Dis Child, *91:* 630, 1956.
9. Engel, E., et al.: Apparent cri-du-chat and "antimongolism" in one patient. Lancet, *1:* 1130, 1966.
10. Kajii, T., Homma, T., Dikawa, K., et al.: Cri-du-chat syndrome. Arch Dis Child, *41:* 97, 1966.
11. Poncet, E., Lafoucade, J., and Sha, J., et al.: Les troubles de la voix et les malformations, laryngees dans la maladie par aberration chromosomique de "maladie du cri-du-chat." Ann Otolaryngol, *82:* 862, 1965.
12. Fenger, K., and Niebuhr, E.: Measurements on hand radiographs from 32 cri-du-chat probands. Radiology, *129:* 137, 1978.

18

Lipidoses

The lipidoses are a group of diseases in which there is excess accumulation of lipid in the reticuloendothelial system and/or the nervous system. They are due to a disturbance of sphingolipid and/or cholesterol metabolism, probably the result of an enzymatic deficiency[1] (Table 57). Only the lipidoses producing osseous disturbances will be discussed in this chapter.

The sphingolipids are a class of ceramide (*n*-acyl-derivative of sphingosine or a related long-chain base) to which phosphate or hexose (glycosphingolipid) is linked. Sphingomyelin is the predominant glycosphingolipid in animal tissue.[2, 3] Ceramide is the generic name for this class of compounds (sphingolipids) and two well defined groups are described, depending on the long-chain fatty acid; gangliosides,[4-6] containing predominantly C-18 acids, and cerebrosides and sphingomyelin, containing predominantly C-22–24 acids.[7-10] The sphingolipids most often implicated in lipidoses are sphingomyelin (gangliosides) and ceramide hexides (cerebrosides).[11]

At the lower end of the lipid disease spectrum are diseases due to accumulation of cholesterol and which predominantly affect the reticuloendothelial system. The simplest form is hypercholesterolemia,[12] which may produce bone changes, particularly in the hands, of osteolytic lesions with sclerotic margins.[12-18] Histologically the other diseases resemble an inflammatory process in granulomatous form, with secondary accumulation of lipid depending on the disease stage. Included are eosinophilic granuloma, Hand-Schüller-Christian disease, and Letterer-Siwe disease, which probably represent different phases of the same syndrome, histiocytosis.

At the other end of the spectrum are lipidoses of complex lipids, predominantly affecting the nervous system and/or the reticuloendothelial system. These have no inflammatory characteristics, appear to be purely metabolic, and probably represent an enzyme deficiency based on chemical chromosomal abnormalities. Included in this group are Niemann-Pick disease, Gaucher disease, Tay-Sachs disease, Fabry disease, familial amaurotic idiocy, and metachromatic leukodystrophy.

Lipogranulomatosis and familial neurovisceral lipidosis are lipidoses with features from both ends of the spectrum. Neuronal tissue involvement indicates metabolic deficiency, but the granulomatous features suggest an inflammatory nature.

TABLE 57. ACCUMULATION OF LIPIDS IN LIPIDOSES

Disease	Lipid	Accumulation	
		Reticuloendothelial	Neuronal
Histiocytosis	Cholesterol	×	0
Gaucher	Cerebroside	×	0[a]
Niemann-Pick	Sphingomyelin and cholesterol	×	×
Farber	Ganglioside	×	×
Neurovisceral lipidosis	Ganglioside	×	×
Tay-Sachs	Ganglioside	0	×
Familial amaurotic idiocy	Ganglioside	0	×
Fabry	Ceramide trihexide	×	×
Metachromatic leukodystrophy	Ganglioside	×	×

[a] May be found in acute infantile cases.

HISTIOCYTOSIS X

Synonyms: Hand-Schüller-Christian disease, eosinophilic granuloma, Letterer-Siwe disease, xanthomatosis, reticuloendotheliosis, and reticulosis.

Histiocytosis X includes eosinophilic granuloma, Letterer-Siwe disease, and Hand-Schüller-Christian disease, all of which are phases of a single entity.[1]

History. Hand,[2,3] in 1893, described a syndrome in a 3-year-old boy with diabetes insipidus, exophthalmos, bronzed skin, hepatosplenomegaly, and poor development, which was thought to be an atypical form of tuberculosis. Schüller,[4] in 1915, Christian,[5,6] in 1919, and Hand,[7] in 1921, agreed that the triad of bone defects, diabetes insipidus, and exophthalmos were signs of a previously unrecognized disease entity. Schüller[8] contended that its cause was primary pituitary dysfunction, whereas Hand[7] believed this to be secondary. Thompson et al.[9] later found that the pituitary was not involved and suggested that abnormality of the hypothalamus was responsible for the diabetes insipidus. Schüller[8] (1926) restudied his own cases, those of Christian, and one of Hechstetter's, and concluded that the disease was of pituitary origin; he called it "dysostosis hypophysaria." Rowland,[10] in 1928, was the first to bring attention to the lipidization of the histiocyte. He suggested that ingestion of lipids by reticulocytes might be the cause of the disease, which he considered a xanthomatous degeneration of the skeleton. Cowie and Magee,[11] Lichty,[12] and Freud et al.[13] helped to establish the finding of high lipid content (primarily cholesterol) in the diseased tissues. This led Smith,[14] in 1935, to classify Hand-Schüller-Christian disease as a lipoid histiocytosis. In 1938, Thannhauser and Magendantz[15] considered the disease a familial disorder of the intracellular enzyme system, allowing cholesterol to accumulate within the histiocytes and transforming it into a xanthoma cell. They believed that the disease was a primary xanthomatosis.

In 1936, Abt and Denholz,[16] first suggested that a group of symptoms in children, manifested by petechia, anemia, hepatosplenomegaly, multiple osseous lesions, and lymphadenopathy, was, in fact, a disease entity. Since Letterer[17] described the first case and Siwe[18] reported a group of cases, it became known as Letterer-Siwe disease. In 1940, Wallgren[19] was the first to conclude that Hand-Schüller-Christian disease and Letterer-Siwe disease are similar and belong in the same group. Also in 1940, Lichtenstein and Jaffe[20] suggested the name "eosinophilic granuloma" to describe a bone lesion they had observed personally. Thereafter, Farber,[21] Gross and Jacox,[22] and Mallory[23] agreed with Wallgren that Letterer-Siwe disease and Hand-Schüller-Christian disease were manifestations of the same basic disease process and that eosinophilic granuloma should be added to this group. Lichtenstein[1] in 1953 reviewed all the previous histopathologic material and agreed that eosinophilic granuloma, Hand-Schüller-Christian disease, and Letterer-Siwe disease were inflammatory histiocytoses, and that the several diseases were actually different phases of the same entity.

He proposed the name "histiocytosis X." Lichtenstein divided histiocytosis X into three phases and suggested the eosinophilic granuloma was confined to bone. However, case reports show that eosinophilic granuloma may be polyostotic and affect the viscera also. Arcomano et al.[24] modified Lichtenstein's classification, which considers the phases of the disease with regard to age incidence and prognosis, as follows: (1) Letterer-Siwe disease, an acute disseminative disease of infants involving both osseous and soft tissue lesions with the soft tissue lesions predominating, and having a very poor prognosis; (2) Hand-Schüller-Christian disease, a disseminated disease of children and young adults having a more chronic course than Letterer-Siwe disease and a generally better prognosis; (3) eosinophilic granuloma, a chronic disease of children and young adults, the lesions of which are most often osseous and monostotic, and the prognosis excellent; but when the lesions are multiple or involve the viscera the prognosis is similar to that of Hand-Schüller-Christian disease. Eosinophilic granuloma can occur in older adults, but these cases often eventually prove to be lymphoma.

Pathology. The early phase, Letterer-Siwe disease, reveals a generalized involvement of reticulum cells, some containing lipids. This lipid may not be obvious, and may be confused with leukemia. In the later phase, or Hand-Schüller-Christian disease, there is proliferation of histiocytes which may be closely packed and simulate a malignant tumor, such as Ewing sarcoma. Later in this phase an influx of leukocytes, particularly eosinophils, so simulates an inflammatory reaction that the bone lesions may be mistaken for bacterial osteomyelitis. The reticulum cells then accumulate lipid material and become foam cells. Eventual tissue necrosis leads to scarification and collagenization. Finally the lesions fibrose, and the scarred tissue reveals none of the lipid containing histiocytes or the inflammatory cells. The presence of eosinophils and histiocytes, with or without lipidization, associated with areas of necrosis, is the histologic picture of eosinophilic granuloma. Hand-Schüller-Christian disease can simulate eosinophilic granuloma and actually may not be distinguishable histologically.

In summary, the histologic appearance of histiocytosis X is variable and passes through phases simulating leukemia, Ewing's sarcoma, acute inflammatory disease, lymphoma, and fibrosarcoma.

Clinical Features. Although Letterer-Siwe disease, Hand-Schüller-Christian disease, and eosinophilic granuloma may be the same entity, it seems worthwhile to maintain the clinical division because of the differences in their courses and prognoses.

CLINICAL FEATURES OF LETTERER-SIWE DISEASE. The onset occurs between the ages of several weeks and 2 years. The clinical criteria outlined by Abt and Denholz[16] are splenomegaly, hepatomegaly, lymphadenopathy, anemia, hemorrhagic tendency, and skeletal lesions. Intermittent fever and failure to grow are common. This phase is the most malignant and may terminate fatally in several weeks. At times, the course is altered by corticosteroids, but not all patients will respond.

CLINICAL FEATURES OF HAND-SCHÜLLER-CHRISTIAN DISEASE. Hand-Schüller-Christian disease is a more chronic phase. It begins in early childhood although symptoms may not appear

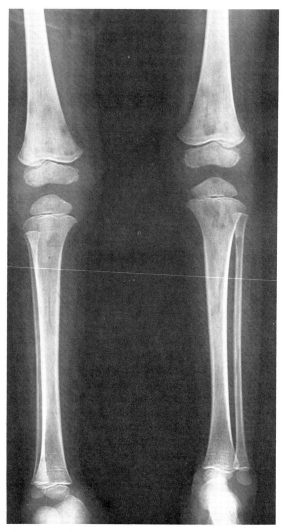

FIG. 1582. HISTIOCYTOSIS X (LETTERER-SIWE DISEASE)
Well defined osteolytic lesions occur in the metaphyses. (Courtesy of John W. Hope, Children's Hospital, Philadelphia, Pa.)

until the 3rd decade. The triad of exophthalmos, diabetes insipidus, and skeletal lesions is seen in only 10% of cases, and clinical complaints usually include only one or two of the three. The most common manifestation is osteolytic skull lesions, with overlying soft tissue nodules that may regress spontaneously even while new lesions may be forming.

Diabetes insipidus is present in approximately one-half of the patients. It appears early, and may be the presenting symptom. It persists unless treated with Pitressin, though the osseous and visceral lesions may become quiescent.

Exophthalmos, unilateral or bilateral, is present in one-third of the cases, and is occasionally the presenting complaint. Often the proptosis is associated with orbital wall destruction. Otitis media, associated with changes in the mastoid air cells and destructive lesions of the petrous portions of the temporal, is also a common presenting complaint. Many patients lose teeth due to extensive involvement of the mandible, and painful and bleeding gum lesions unrelated to mandibular disease frequently occur.

Local or generalized eczematoid skin reactions are easily mistaken for seborrheic dermatitis. The more severe cases may show the disturbances in growth and development often associated with diabetes insipidus. The course of the disease is seldom rapidly fatal; it is usually chronic, with spontaneous remissions and exacerbations.

CLINICAL FEATURES OF EOSINOPHILIC GRANULOMA. This clinical type occurs most frequently in children between the ages of 3 and 10. It is rare past the age of 15, although the authors have seen it in a 62-year-old woman. Because the incidence in older persons is so slight, the physician should suspect lymphoma when he encounters a histologic diagnosis of eosinophilic granuloma in an elderly patient.

Solitary painful bone lesions are the most common finding, although they may be disseminated and polyostotic. The skull is usually involved, but lesions have been found in all bones. The prognosis is excellent; the monostotic lesions will respond to x-ray therapy or heal spontaneously. The disseminated form will follow a course similar to that of Hand-Schüller-Christian disease.

Roentgenographic Features. The roentgenographic appearance of the osteolytic lesions of the three phases of histiocytosis is similar. Destructive punched-out areas, varying in size from a few millimeters to several centimeters in diameter, occur in both the membranous and long bones (Figs. 1582 and 1583). The borders are well

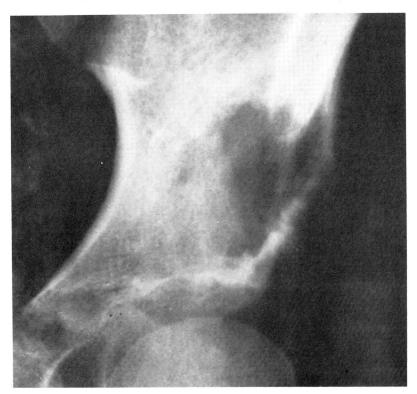

FIG. 1583. EOSINOPHILIC GRANULOMA
Large osteolytic lesion extending to the superior margin of the acetabulum. The periphery is well defined and irregular. The soft tissue mass is present.

defined, and less than half will reveal marginal reactive sclerosis.[25-29] The long bone lesions are medullary, as shown by the erosion of the endosteal layer of the cortex (Fig. 1584), and most often found in the metaphyseal or epiphyseal portions. Periosteal reaction is frequent, and its osteolytic area usually suggests sarcoma (Fig. 1585), but it may be differentiated by its appearance as a solid layer of new bone (Figs. 1586–1589), rather than an interrupted malignant reactive periosteal change. The periosteal reaction is only a few millimeters in depth. There may be interrupted periosteal reaction with some or many lamellations. This reactive change means aggressiveness but not necessarily malignancy; if it is associated with a solid reaction, the solid reaction takes precedence as indicating a benign lesion. A solid periosteal reaction indicates benignancy, an interrupted reaction represents aggressiveness but not necessarily malignancy. Long bone lesions usually do not affect the joint (Figs. 1590–1596), but sometimes involve contiguous areas of bone on either side of the joint. Sequestration may occur within the destructive lesion (Fig. 1586). Although periosteal reaction is common, true bone expansion occurs only in disease of long standing (Fig. 1597). Pathologic fracture is uncommon.[30]

The single or multiple lesions in the skull begin as well defined, small, punched-out areas, usually with no margin of reactive bone (Figs. 1598 and 1599). Although the majority of the margins are well demarcated, they may be ragged and irregular, with small peripheral lesions. Because this is more common in Letterer-Siwe disease than in the other two types, the irregular margin may point to the more malignant form.[25] The lesions tend to erode both tables of the skull; the outer table is more extensively destroyed, producing, at times, characteristic double contour. Lesions grow to considerable size and their initial ovoid or round contour becomes serpiginous (Figs. 1600–1602) ("geographic skull"). Soft tissue masses on the scalp can be palpated and demonstrated roentgenographically, to overlay the lytic processes in the calvaria. The lesions may develop rapidly and become radiologically evident over a 6-week period.[27] They can extend to the epidural space and beneath the dura into the brain substance. Large areas of destruction in the occipital bone lead to platybasia.[5]

The sella turcica is occasionally destroyed; this is not related to the incidence of diabetes insipidus, which may not be accompanied by sellar abnormality, while sellar destruction does occur without diabetes insipidus.

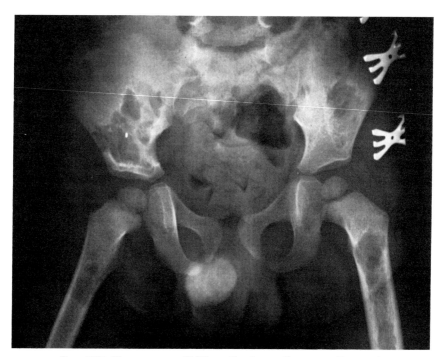

FIG. 1584. HISTIOCYTOSIS X (HAND-SCHÜLLER-CHRISTIAN DISEASE)
Long bone lesions are medullary as evidenced by erosion of the endosteal layer of the cortex. The lesions are well defined and some show a sclerotic border. (Courtesy of John W. Hope, Children's Hospital, Philadelphia, Pa.)

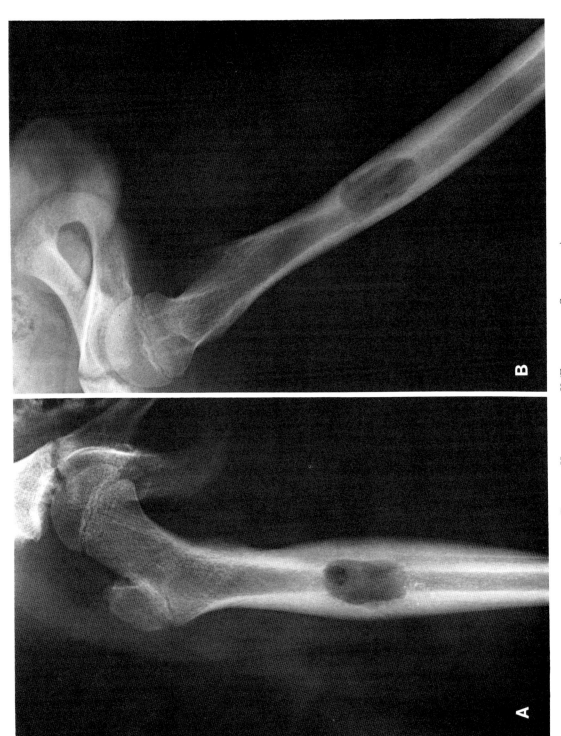

FIG. 1585. HISTIOCYTOSIS X (EOSINOPHILIC GRANULOMA)

Initial impression was Ewing sarcoma. The short transition between the tumor and host bone and the solid periosteal reaction indicates the true diagnosis. (A) Anteroposterior view; (B) lateral projection. (Courtesy of Dr. Beth Edeiken, M. D. Anderson Hospital and Tumor Institute, Houston, Tex.)

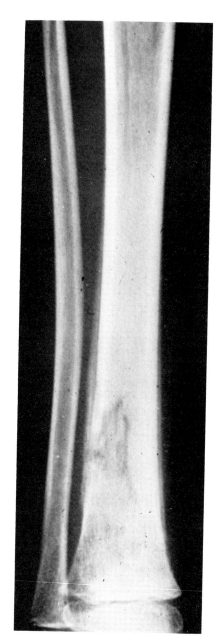

FIG. 1586. HISTIOCYTOSIS X (EOSINOPHILIC GRANULOMA)
Sequestration within the lesion and the solid periosteal reaction indicating benignancy.

FIG. 1587. HISTIOCYTOSIS X (EOSINOPHILIC GRANULOMA)
Extensive destructive lesion with periosteal reaction suggested Ewing sarcoma. The solid type periosteal reaction indicates benignancy. (Courtesy of Paul K. Berg, Doctors Hospital, Staten Island, N.Y.)

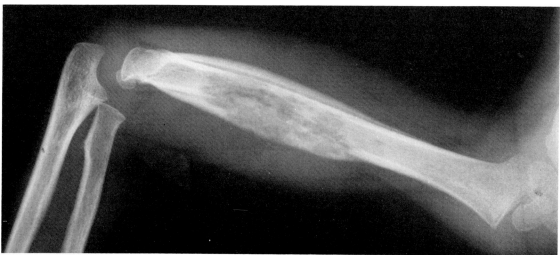

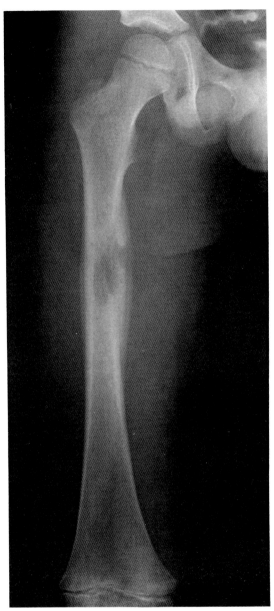

Fig. 1588. Histiocytosis X (Eosinophilic Granuloma) Ill defined destructive lesion in the femur may well be a malignant tumor. The solid periosteal reaction on the medial aspect suggests eosinophilic granuloma.

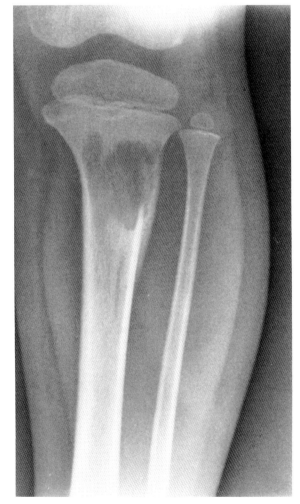

Fig. 1589. Histiocytosis X (Eosinophilic Granuloma) A well defined osteolytic area. The solid periosteal reaction on the lateral aspect of the tibia, adjacent to the destroyed bone, indicates a benign lesion. (Courtesy of Dr. Beth Edeiken, M. D. Anderson Hospital and Tumor Institute, Houston, Tex.)

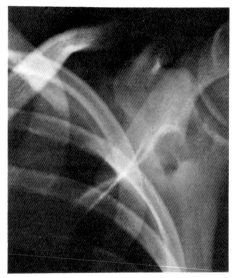

FIG. 1590. HISTIOCYTOSIS X (EOSINOPHILIC GRANULOMA) Well defined osteolytic area in the scapula.

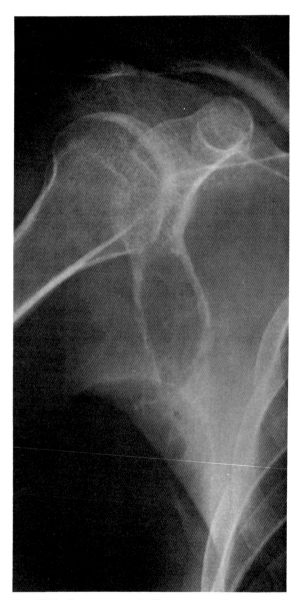

FIG. 1591. HISTIOCYTOSIS X (EOSINOPHILIC GRANULOMA) Multiple osteolytic lesions in the lateral aspect of the scapula have distinct borders.

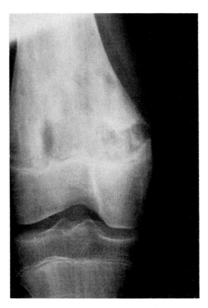

Fig. 1592. Histiocytosis X (Eosinophilic Granuloma)
Lesions extend to the epiphyseal line but do not cross it. They are well defined. (Courtesy of John W. Hope, Children's Hospital, Philadelphia, Pa.)

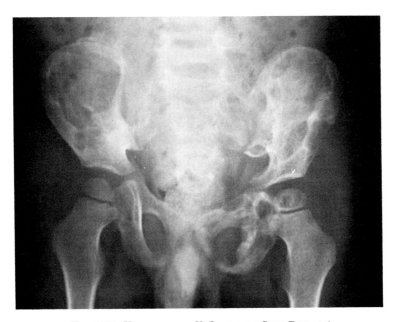

Fig. 1593. Histiocytosis X (Letterer-Siwe Disease)
Extensive lesions in the ilia, the pubic bones, and the femoral epiphyses. The lesions respect the epiphyseal lines and the joint spaces. (Courtesy of George T. Wohl, Philadelphia General Hospital, Philadelphia, Pa.)

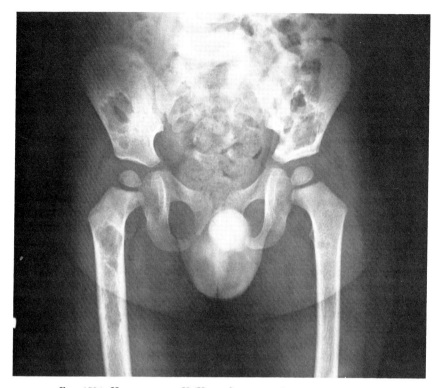

FIG. 1594. HISTIOCYTOSIS X (HAND-SCHÜLLER-CHRISTIAN DISEASE)

Well defined lesions in multiple bones may or may not have a sclerotic border. The joints are not involved. (Courtesy of John W. Hope, Children's Hospital, Philadelphia, Pa.)

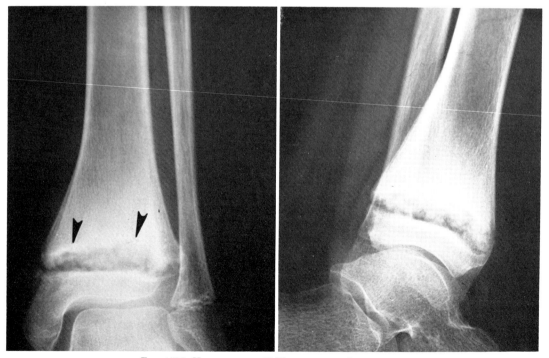

FIG. 1595. HISTIOCYTOSIS X (EOSINOPHILIC GRANULOMA)

Destructive lesions above the epiphyseal line in the tibia. The lesions do not cross the epiphyseal plate (*arrows*). There is some periosteal reaction on the medial margin of the tibia. (*Left*) Anteroposterior projecton; (*right*) lateral projection. (Courtesy of John W. Hope, Children's Hospital, Philadelphia, Pa.)

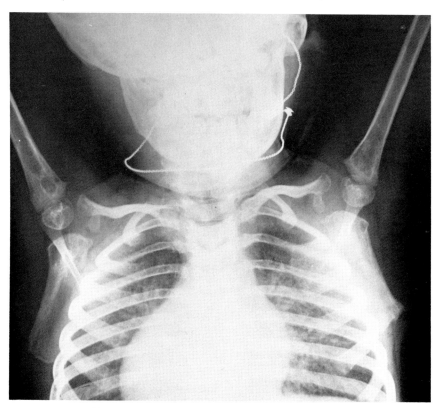

FIG. 1596. HISTIOCYTOSIS X (LETTERER-SIWE DISEASE)
Destructive lesions in the humeri and the scapula. (Courtesy of John W. Hope, Children's Hospital, Philadelphia, Pa.)

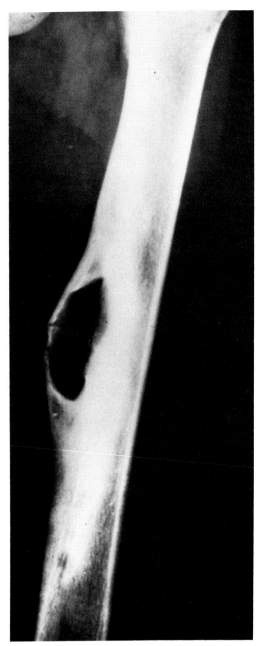

Fig. 1597. Histiocytosis X (Eosinophilic Granuloma)
Long standing lesions may cause expansion of bone.

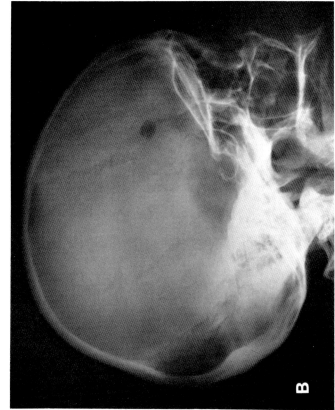
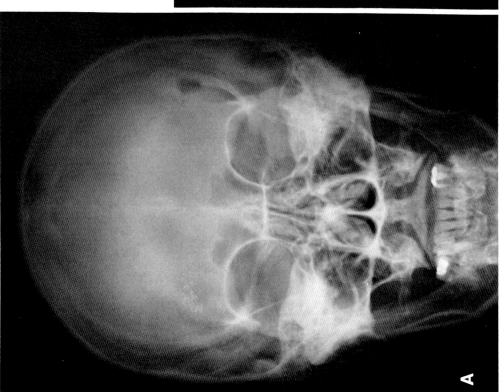

Fig. 1598. Histiocytosis X (Eosinophilic Granuloma) Single well defined small osteolytic lesion in the left frontal bone. (A) Posteroanterior view, (B) lateral projection.

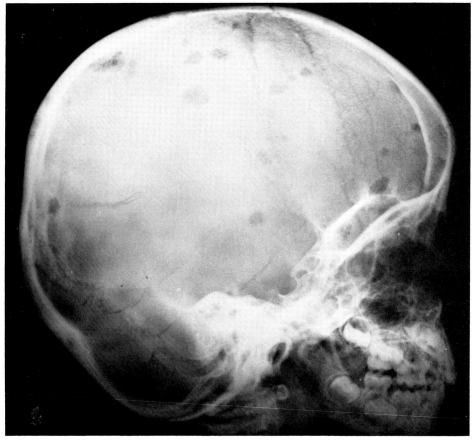

FIG. 1599. HISTIOCYTOSIS X (LETTERER-SIWE DISEASE)
Single or multiple lesions in the skull begin as well defined small punched-out areas which have no margin of reactive bone. (Courtesy of John W. Hope, Children's Hospital, Philadelphia, Pa.)

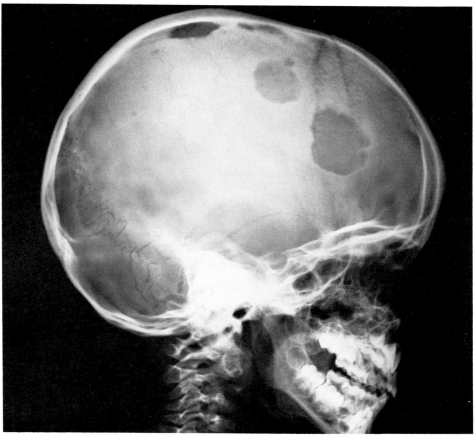

FIG. 1600. HISTIOCYTOSIS X (LETTERER-SIWE DISEASE)
As the lesions enlarge they lose their initial ovoid or round contour and become serpiginous. (Courtesy of George T. Wohl, Philadelphia General Hospital, Philadelphia, Pa.)

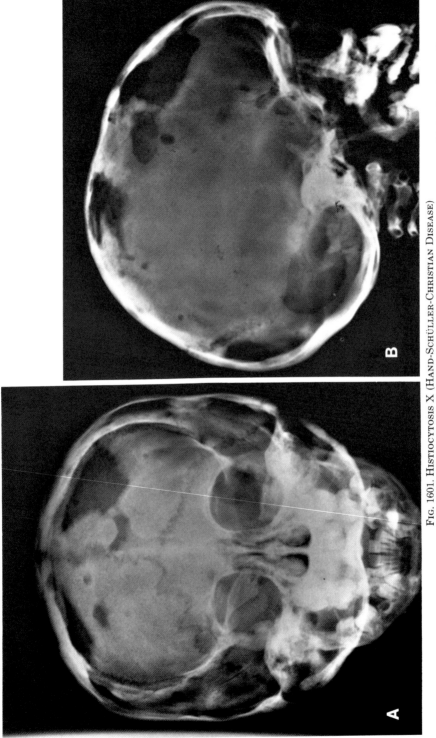

FIG. 1601. HISTIOCYTOSIS X (HAND-SCHÜLLER-CHRISTIAN DISEASE)
(A) Anteroposterior view of the skull reveals the geographic contour of the destructive lesions. The large lesions have a geographic border; the smaller ones maintain their ovoid configuration. (B) Lateral projection. (C) Pelvis. The pelvic lesions are well defined and extensive. Most of the lesions have sclerotic borders. (Courtesy of Dr. Beth Edeiken, M. D. Anderson Hospital and Tumor Institute, Houston, Tex.)

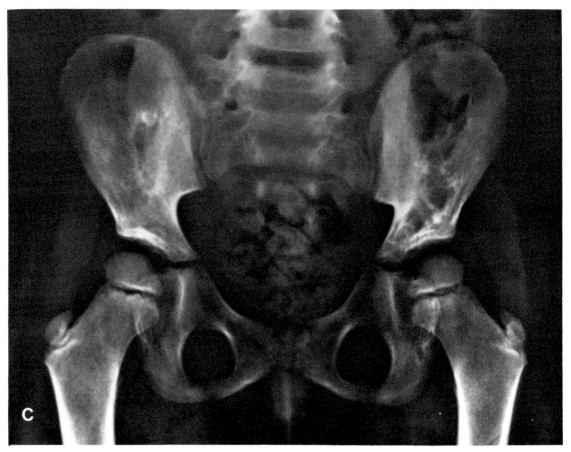

Fig. 1601 (C)

metabolic disturbance triggered the process leading to excessive production of histiocytes containing a cerebroside known as "kerasin."

The etiology of Gaucher disease is still obscure, but it probably represents a deficiency of β-glucosidase.[5-7] The cells characterizing it are formed by the accumulation of galactoside cerebrosides, glucoside cerebrosides and polycerebrosides within the cytoplasm of reticulum cells.[8] It is believed that a dysfunction of reticulum cells results from enzymatic disturbances which lead simultaneously to increased synthesis and storage of cerebrosides within the cells.[9]

The hereditary mode is by three possible routes[8]: (1) autosomal recessive in the majority, (2) autosomal dominant in a few, and (3) autosomal recessive predominantly in families in which neurologic involvement is always present. Gaucher disease probably arises as a result of sporadic mutation.[10] Once this has occurred, the disease is transmitted as an autosomal dominant.[10]

Clinical Features. Two forms of Gaucher disease are commonly recognized: the infantile, which runs a rapidly fatal course, and the more classic form, which appears somewhat later in life. The fulminating infantile form of Gaucher disease appears as early as 2 weeks of age. Almost all cases of the acute type have occurred among non-Jewish population.[11] The central nervous system is affected, and gives rise to feeding problems, hypertonicity, and even opisthotonos. (So much like Niemann-Pick disease in this form of the disease that the two may sometimes be distinguished only by chemical analysis of involved tissues.) As liver and spleen enlarge, the infant's abdomen protrudes. The skin and conjunctival pigmentation of the classic form is usually not found in this type. Death usually ensues within the first 2 years of life.

The adult form of Gaucher disease most often strikes older children and young adults; however, it may strike in middle age and even past 60; the oldest patient recorded was 79.[12-15] Although Gaucher disease, unlike the infantile type, is found most commonly among Jews, it occurs in all races.

The most frequent complaint is distress in the left upper abdominal region, caused by the marked characteristic splenomegaly. Hepatomegaly also is common. Ascites and impairment of liver function occurs.

Anemia (normochromic or hypochromic), often accompanied by leukopenia and thrombocytopenia, is responsible for the lassitude and easy fatigability of Gaucher disease. Petechia and gross bleeding from the gastrointestinal tract are not uncommon.

Yellowish-brown pigmentation of portions of the conjunctiva (pinguecula) and skin, especially in the lower extremities,[16] confirms the diagnosis. This is the result of hemochromatosis,[17] hemosiderin deposits,[18] or deposits of a melanin derivative,[15] and sometimes disappears following splenectomy.[14]

There may be dull bone pain caused by Gaucher cells packing the bone marrow. At times, crises of acute bone pain, exquisite tenderness, and elevated temperature simulate osteomyelitis. Attempts at open drainage can lead to intractable draining sinuses.[19] Bone crises spontaneously subside within 2 weeks, but, the weakened bones tend to fracture easily. Aseptic necrosis, secondary to bone infarction, is common, especially in the hips.

The diagnosis can be made upon detecting Gaucher cells in the bone marrow, liver, or spleen; they are never found in the peripheral blood. Although patients with adult Gaucher disease often die of intercurrent infection or hemorrhage before the age of 40, many live to 60 or 70.

Pathologic Features. Gaucher cells are found throughout the reticuloendothelial system, including the spleen, liver, bone marrow, and lymph nodes. Infiltrations have also been found in the kidney, thymus, tonsils, adrenals, and lungs. In later stages, the bone marrow may become fibrotic and the Gaucher cells sparse or absent. The pathogenic course consists of Gaucher cell infiltration, hemorrhage, atrophy, and, finally, fibrosis.

Roentgenographic Features. The roentgenographic features vary, depending on the number of Gaucher cells packed into the reticuloendothelial system. Bone changes are present in approximately half of the patients.[20-24] Splenomegaly, commonly present and marked, may be mild or even absent in the presence of extensive bony change. Hepatomegaly is also usual, but may be more difficult to evaluate. Occasionally, diffuse pulmonary infiltrations are found in the acute form of the disease.

The most striking roentgenographic changes occur in the osseous system. Early examination reveals bone pathology in 75% of the cases, especially the femur.[25] Where there are bone symptoms, there are usually bone lesions.[26] The symptoms can arise from aseptic necrosis, fracture, joint space narrowing with associated aseptic necrosis or bone lesions, and/or lesions of the vertebrae. Silverstein and Kelly[26] found anemia in 4% of patients with no osseous change and 78% of patients with osseous change. Also 10 of 28 patients with bone changes had splenectomy, whereas 10 patients without osseous changes did not have this procedure. Although the relationship of anemia, splenectomy, and osseous changes may be related to the severity of disease, it is possible that anemia and splenectomy contribute to the osseous change, the former by stimulating more marrow space for invasion by Gaucher cells, and the latter by eliminating a large reservoir of Gaucher cells.[26]

Deossification is often dramatic, even though there may be no definite osteolytic lesions (Fig. 1606). Pathologic fractures of the long bones and compression fractures of the vertebral centra are frequent (Figs. 1607 and 1608).

The striking cortical thinning that adds to the general deossification is caused by packed marrow encroaching on the endosteum (Figs. 1609).

Occasionally, there is bulging of localized segments of tubular bones due to focal collections of marrow foam cells (Fig. 1610). "Erlenmeyer flask" deformity of the distal femora is common, the result of lack of modeling (Fig. 1611), and indicates active disease during the formative years. It is often ascribed to Gaucher disease alone, but may occur in other conditions, and is not specific.

Numerous sharply circumscribed osteolytic areas may be present throughout the skeleton (Fig. 1612). These resemble metastatic carcinoma or multiple myeloma. There may be associated periosteal reaction (cloaking). With long standing disease the medullary portions become myelosclerotic (Figs. 1613 and 1614) due either

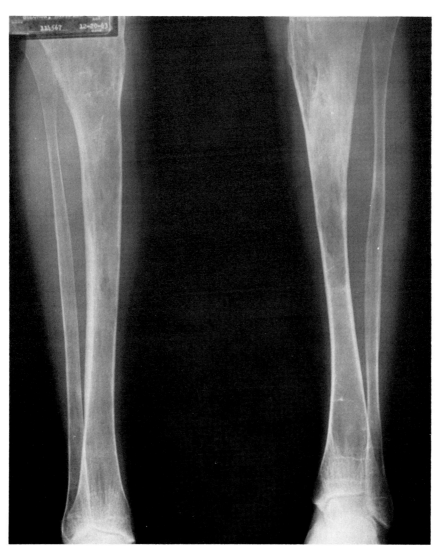

FIG. 1606. GAUCHER DISEASE
Deossification often is dramatic. Multiple osteolytic lesions in the tibia. Marked cortical thinning due to the encroachment by Gaucher cells.

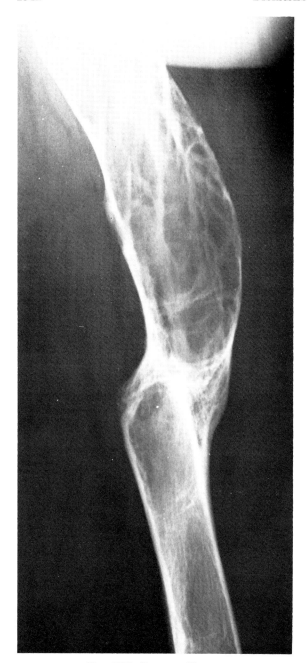

Fig. 1607. Gaucher Disease
Pathologic fractures are common. There is widening of the bone due to infiltration of Gaucher cells.

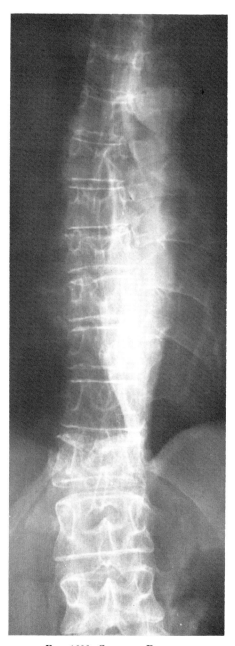

Fig. 1608. Gaucher Disease
Marked deossification of the vertebrae with several compression fractures.

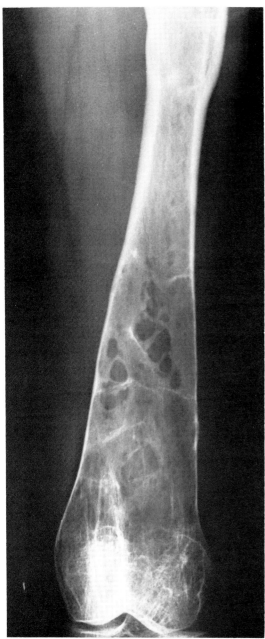

Fig. 1609. Gaucher Disease
Infiltration of the marrow has caused multiple osteolytic defects, considerable cortical thinning, and "Erlenmeyer flask" deformity.

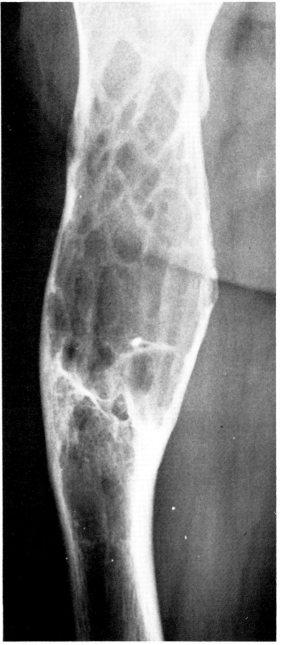

Fig. 1610. Gaucher Disease
Localized widening of the bone due to focal collections of marrow foam cells.

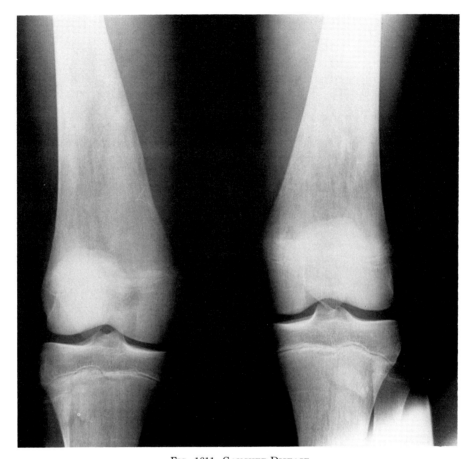

FIG. 1611. GAUCHER DISEASE
"Erlenmeyer flask" deformity of the distal femurs results from the lack of modeling. (Courtesy of John W. Hope, Children's Hospital, Philadelphia, Pa.)

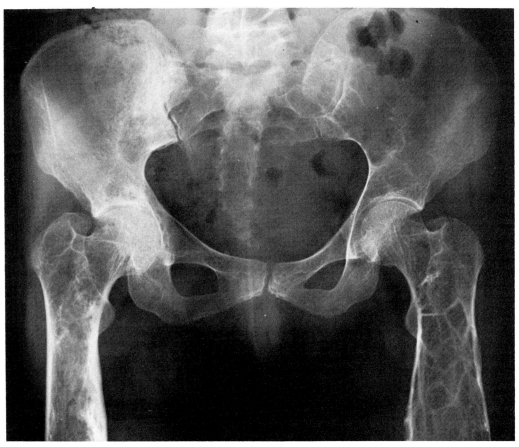

Fig. 1612. Gaucher Disease

Numerous sharply circumscribed osteolytic areas may be present throughout the skeleton and in both femurs and iliac bones.

to osseous metaplasia of connective tissue or secondary bone repair.[15, 16, 27] Aseptic necrosis is common, especially of the femoral heads (Figs. 1615–1617), commonly involving the humeral head and the bones of wrists and ankles. Bone infarctions are common (Fig. 1618), occurring in the metaphyses of the long bones. The small bones of hands and feet are usually spared. The skull rarely, if ever, presents osteolytic areas, though the diploe may be packed with foam cells. Several inconclusive reports of roentgen changes

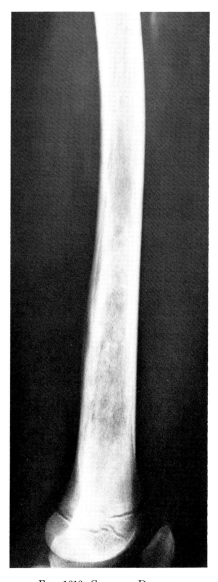

FIG. 1613. GAUCHER DISEASE
Myelosclerotic changes in the distal end of the femur. These may occur in long standing Gaucher disease. It is due to osseous metaplasia and secondary bone repair. (Courtesy of John W. Hope, Children's Hospital, Philadelphia, Pa.)

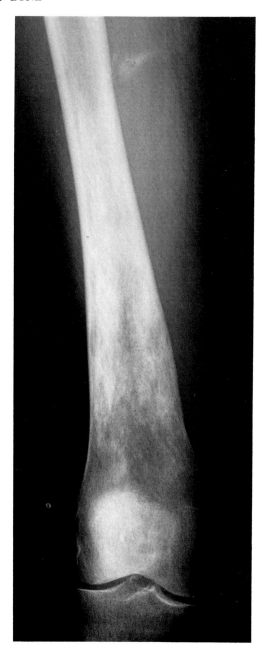

FIG. 1614. GAUCHER DISEASE
Myelosclerotic changes and "Erlenmeyer flask" deformity.

in the skull may be venous lakes or osteoporosis.[28]

Cartilage is not directly affected, and the joint spaces not narrowed. Occasionally, however, the weakening of the subchondral bone results in multiple fractures, and repeated insults produce degenerative arthritis and articular cartilage destruction with crippling deformities (Fig. 1618).

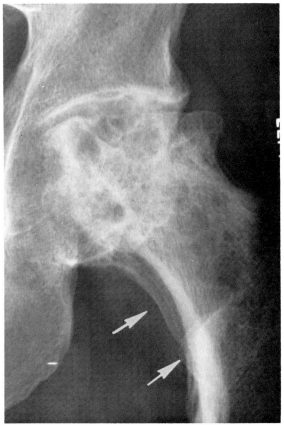

FIG. 1615. GAUCHER DISEASE
Aseptic necrosis of the head of the femur. Periosteal cloaking on the medial aspect of the neck (*arrows*).

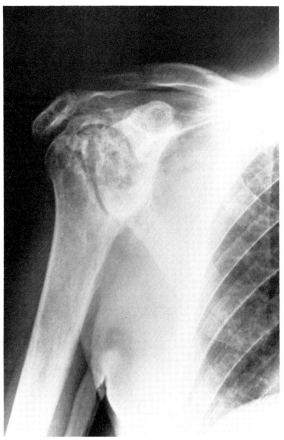

FIG. 1616. GAUCHER DISEASE
Aseptic necrosis of the humeral head.

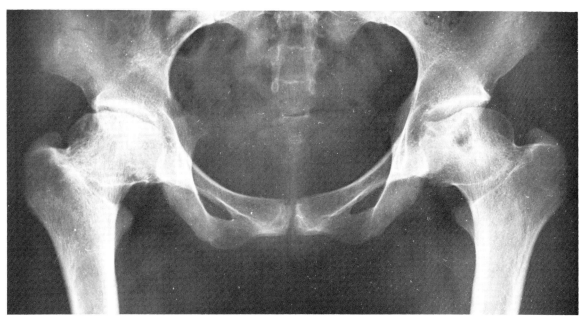

FIG. 1617. GAUCHER DISEASE
Deformity of both hips due to aseptic necrosis of the femoral heads.

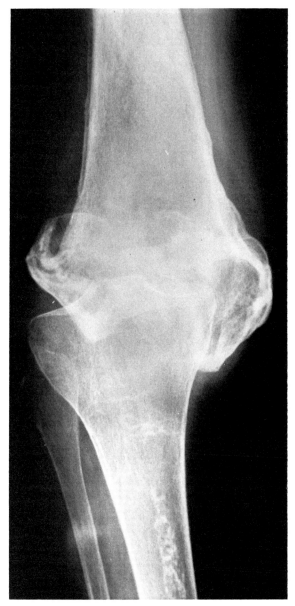

FIG. 1618. GAUCHER DISEASE
Weakening of the subchondral bone with collapse and deformity. Bone infarct in the tibia.

Over 90% of patients have orthopaedic complications at some time during their disease.[29] These include nonspecific bone pain, pseudo-osteomyelitis, pyogenic osteomyelitis, joint pain, aseptic necrosis of the femoral heads, pathologic fractures, spinal malalignment, bone deformity and generalized stunting of skeletal growth due to spinal deformity.[6, 17, 30-36]

NIEMANN-PICK DISEASE

Niemann-Pick disease is a syndrome with a congenital error of phospholipid metabolism inherited by autosomal recessive transmission.[1] It occurs in early infancy and, as a rule, causes death within a year. It resembles Gaucher disease in that there is extensive proliferation of reticulum cells in bone marrow with splenomegaly and hepatomegaly.

The disease has been classified[2,3] into four apparent phenotypes, Groups A–D, all of which exhibit sphingomyelin and cholesterol accumulation within foam cell units[4] of the pattern expected in an inborn error of metabolism syndrome (Table 58).[1]

Lachman has evaluated the roentgenographic features of the four groups, which may be found in Table 59.[1]

Marked deossification (Fig. 1619) may be the only skeletal change. Areas of rarefaction, similar to those in Gaucher disease, will occur if the patient survives.

TABLE 59. ROENTGENOGRAPHIC FEATURES OF NIEMANN-PICK DISEASE[a]

Feature	Group			
	A	B	C	D
Hepatosplenomegaly	8/8	4/4	7/7	3/3
Pulmonary infiltrates	8/8	4/4	6/9	0/7
Bone changes:				
1. Osteoporosis and coxa valga	5/8	1/4	9/9	0/7
2. Long bone marrow cavity expansion and modeling defects	1/8	2/4	0/9	0/7
3. Metacarpal widening	1/8	2/4	4/7	0/7

[a] From R. Lachman, A. Crocker, J. Schulman, and R. Strand: *Radiology, 108:* 659, 1973.[1]

TABLE 58. COMMON CLINICAL FORMS OF NIEMANN-PICK DISEASE[a]

Group	Enlargement of Liver and Spleen	Onset of CNS symptoms	Elevation of Lipids		Age at Death (yr)
			Blood	Tissue	
A. "Classical"	++++	Early infancy	++	++++	1–2
B. Heavy visceral involvement, normal CNS	++++	None seen	++	++++	
C. Moderate course	++	Late infancy	0	++	3–7
D. Nova Scotian form	+++ (usually)	Early to middle childhood	0	++	12–30

[a] From R. Lachman, A. Crocker, J. Schulman, and R. Strand: *Radiology, 108:* 659, 1973.[1]

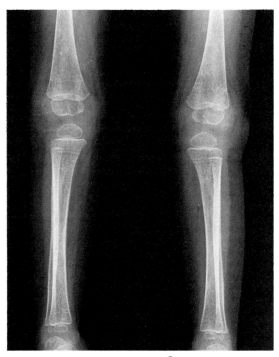 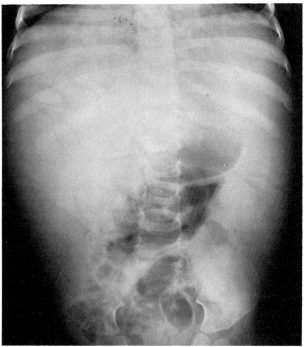

FIG. 1619. NIEMANN-PICK DISEASE
(*Left*) Deossification. Radiolucent lines in the metaphyseal areas of the tibia indicate nutritional disturbance as well. (*Right*) There is marked skeletal deossification, hepatomegaly, and splenomegaly. (Courtesy of John W. Hope, Children's Hospital, Philadelphia, Pa.)

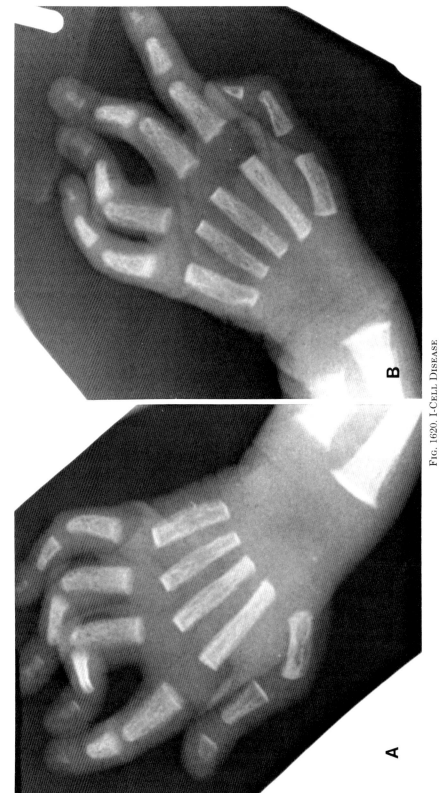

Fig. 1620. I-Cell Disease

(*A* and *B*) *Hands*. Deossification and undermodeling of the bones. The excessive periosteal new bone formation has caused diaphyseal expansion. (*C*) *Chest*. Broadening of the ribs and scoliosis. (*D*) Pelvic dysplasia with hypoplasia of the basilar portions of the ilia and flaring of the iliac wings. Dislocation of the right hip and deossification of the femurs with widening of the diaphyses. (Courtesy of Dr. Rogelio Moncada, Loyola University, Maywood, Ill.)

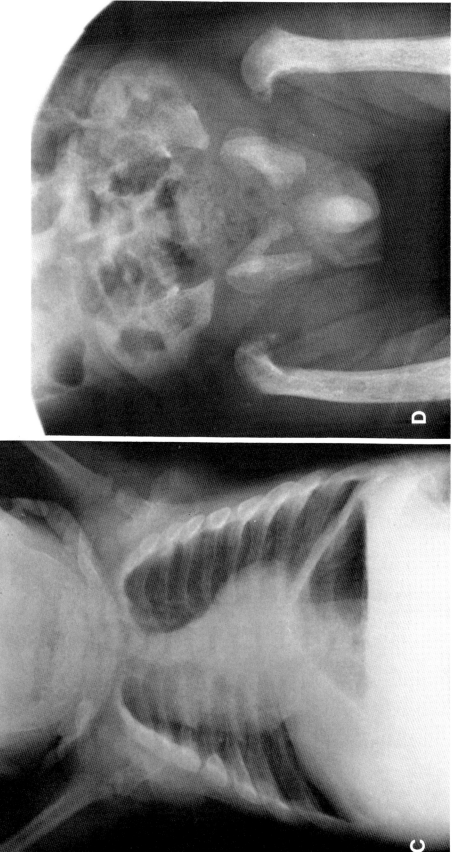

Fig. 1620 (*C* and *D*)

I-CELL DISEASE

I-cell disease is clinically similar to Hurler disease (MPS-I) except that the patients excrete a normal amount of urinary acid mucopolysaccharides. It was termed "inclusion disease" (I-cell) because fibroblasts grown from skin biopsies of these patients contained a large number of dark inclusions in the cytoplasm (Fig. 1620).[1]

Clinical Features. It is probably transmitted by an autosomal recessive mode.

The features, present at birth, consist of severe mental retardation and growth failure, coarse facial features, abdominal prominence with mild hepatomegaly, umbilical and inguinal hernia, diastasis recti, nasal discharge, respiratory and ear infections, and small hemangiomas. LeRoy et al.[1] reported the clinical differences from Hurler syndrome: (1) changes are present at birth, progress rapidly for 2 years, and then stabilize. (2) Growth impairment is severe; growth does not, as in Hurler syndrome, accelerate in the first year. (3) Mental retardation is very severe. (4) The corneas are clear. (5) Hepatomegaly and splenomegaly are slight. (6) Hirsutism is absent and hemangiomas are reported.

LeRoy et al. reported two types of laboratory studies which distinguish it from Hurler disease: (1) there is no significant excretion of MPS and (2) there are no Reilly granules in the neutrophils or Mittwoch bodies in the lymphocytes.[2]

Roentgenographic Features. The roentgenographic features are the same as with Hurler disease and are present at birth.[3, 4]

FAMILIAL NEUROVISCERAL LIPIDOSIS

Synonyms: Tay-Sachs syndrome with visceral involvement,[1] a metabolic neurovisceral disorder with accumulation of an unidentified substance,[2] pseudo-Hurler disease,[3] familial neurovisceral lipidosis,[4] systemic late infantile lipidoses,[5] generalized gangliosidoses,[6] G_{M2} gangliosidoses, and disseminated lipogranulomatosis.

Familial neurovisceral lipodosis is due to an inborn error of ganglioside metabolism, which results in an accumulation of lipids in the histiocytes of neuronal tissue and many viscera.

Pathologic Features. The most striking feature is "foam cell" histiocytosis of most of the viscera, commonly involving the bone marrow, the liver, lymph nodes, spleen, and colon. There is also involvement of the nervous system by a ballooning process resembling that of Tay-Sachs disease. There is usually a distinctive cytoplasmic swelling of glomerular epithelium[1-3] which differentiates the condition from Tay-Sachs and Niemann-Pick disease. This finding may be absent.[5, 7]

The histochemical properties of the accumulated material suggest a ganglioside[1-3] structurally different from that found in Tay-Sachs.[6, 7] Niemann-Pick storage disease contains a sphingomyelin, and Hurler syndrome a mucopolysaccharide. In Tay-Sachs disease, the ganglioside accumulates predominately in the nervous system.

Clinical Features. The clinical features described by Landing et al.[4] are severe progressive cerebral degeneration beginning neonatally or in early infancy, and usually leading to death within 2 years.

There may be facial edema at birth, producing coarse features resembling Hurler syndrome, or edema of the extremities. Hydrocephalus has been reported. The abdomen protrudes, due to hepatosplenomegaly. Growth is slow; motor retardation is noted early, perhaps with delay in sitting, and inability to grasp objects at 3 months. Macular eczematoid rashes have been reported. The cherry red spots of the macula, present in some cases, are easily confused with Tay-Sachs syndrome. Respiratory infections are common, and the most frequent cause of death.

Roentgenographic Features. The roentgenographic features of familial neurovisceral lipidosis may be indistinguishable from Hurler syndrome. However, Hurler syndrome rarely produces distinctive changes at birth, although an infantile form was described by Caffey[8] which, if it is truly Hurler syndrome renders the two conditions indistinguishable roentgenographically.

In the skull, there may be no changes, but hypertelorism and hydrocephalus commonly occur. There may be a deep anterior recess in the pituitary fossa, resembling the one found in Hurler syndrome. The calvaria may be dense and delayed developmental features may occur.

In the spine, the centra may be poorly developed, with small oval ossification centers. Hypoplastic vertebrae with beaking are common, especially in the first three lumbar and the lower dorsal segments. Hemivertebrae have also been reported in the low dorsal and upper cervical regions. The iliac wings tend to be flared and the acetabulum is flat and shallow.

The long bones often have expanded medullary cavities more marked in the ends, so that the cortices are appreciably thinner in the proximal and distal thirds than in the middle third. The trabeculations may be coarsened. Subperiosteal new bone formation may occur, especially in the humeri and femora. The ends of the long bones may be constricted; all proximal metacarpals may show this change.

Epiphyseal maturation may be delayed, with irregular ossification giving the effect of fragmentation. The epiphyseal lines may be indistinct and epiphyseal cupping is common. The epiphyseal plates of the radius and/or ulna are sometimes tilted, and face each other.

Because of constriction at the vertebral junctions and widening of the remainder of the ribs, the ribs often appear spatulous. A second constriction may occur, lateral to the usual site of constriction. The medullary cavities are expanded and the cortices thin.

Differential Diagnosis. Familial neurovisceral lipidosis is most often mistaken for infantile Hurler syndrome. The roentgenographic differentiation may be impossible, but, at times, the ends of the long bones are flared, whereas in Hurler syndrome they are usually constricted. The spatulous ribs may also closely simulate the appearance in Hurler syndrome, but when a second constriction is present differentiation may be possible. Although Hurler syndrome does not commonly produce distinctive changes in the first few weeks or months after birth, Caffey[8] described 2 cases in infants. If this is indeed Hurler syndrome, then differentiation will depend on histochemical evaluation.

Tay-Sachs disease may be ruled out in the presence of hydrocephalus or visceral involvement. Although retarded motor function may cause thinned cortices, and expanded medullary cavities may be present in Tay-Sachs disease, the other changes of familial neurovisceral lipidoses are not present. Histochemical studies will differentiate the two conditions.

Niemann-Pick disease may present expanded medullary cavities and thin cortices, but there is no cortical cloaking. Histochemical studies will disclose an accumulation of sphingomyelin rather than the ganglioside of neurovisceral lipidoses.

DISSEMINATED LIPOGRANULOMATOSIS (FARBER DISEASE)

Farber,[1] described a symptom complex occurring in infants consisting of dysphonia, laryngeal stridor, nodular joint enlargement, and infiltrative lesions of skin and subcutaneous tissues which he labeled disseminated lipogranulomatosis.

Clinical Features. The clinical features of a hoarse weak cry and swelling of the extremities are strikingly consistent and usually begin in the neonatal period. There is rapid progression characterized by generalized joint swelling, subcutaneous and periarticular granulomas, intermittent fever, dyspnea, and lymphadenopathy.[1,2] Death usually results from respiratory failure within 2 years. Survivals of 6 years[3] and 12 years[4] without nervous system involvement are recorded.

Pathologic Features. The pathologic features are disseminated periarticular and soft tissue and visceral granulomas which contain foam cells. At times, there is necrosis of collagen elements in the granuloma suggesting rheumatoid arthritis but the presence of foam cells, the absence of palisading, and the wide spread involvement distinguishes the two conditions.[2] There is also a lipid storage disease of neuronal tissue.

Biochemical analysis indicates a primary lipid accumulation of ceramide and gangliosides[5] in the granulomatous lesions and neuronormal tissues which indicates lipidosis rather than a mucopolysaccharidosis previously suggested.[6]

Roentgenographic Features. The most striking roentgenographic feature is capsular distention of multiple joints[7] especially prominent in the hand, elbow, and knee (Fig. 1621). Juxtaarticular bone erosions by soft tissue granulomas may occur (Fig. 1621). The joint granulomas may cause subluxation or dislocation. The muscle and fascial planes are normal. Deossification may occur due to disuse or corticosteroid therapy. The changes may simulate MPS-I (Hurler syndrome) but they are present in the newborn.

Although similar roentgenographic features occur in patients with juvenile rheumatoid arthritis, the pain and occurrence after infancy distinguishes the two.

Bone erosions may occur with other lipidoses, but joint capsular distention is not a feature. Arthrogryposis may be distinguished by the abnormal muscle masses.

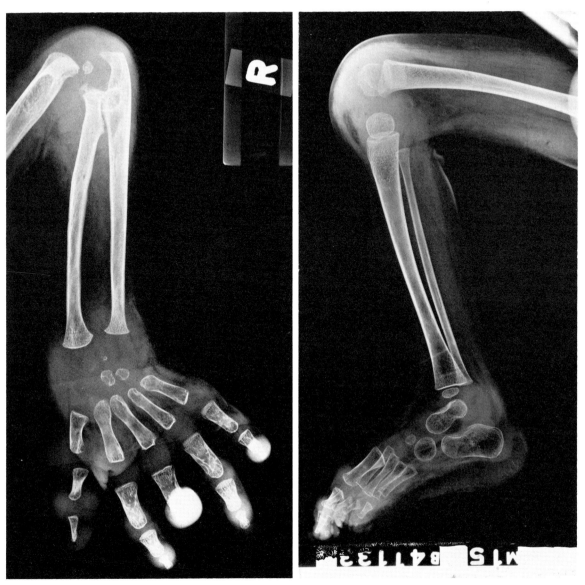

Fig. 1621. Disseminated Lipogranulomatosis (Farber Disease)
(*Left*) Roentgenogram of the right forearm reveals marked capsular distention of the elbow, wrist and interphalangeal joints. Erosion and subluxation of the elbow joint. Marked deossification and deformity of the soft tissues of the thumb. The proximal ends of the metacarpals are slightly pointed, somewhat similar to Hurler syndrome. (*Right*) Capsular distention of the knee, ankle, and interphalangeal joints. (Courtesy of G. Schultze and E. K. Lang: *Radiology, 74:* 428, 1960,[7] © The Radiological Society of North America, Syracuse, N.Y.)

WOLMAN DISEASE

Wolman disease is a familial lipidosis, its distinctive features due to an accumulation of cholesterol in visceral foam cells.[1-5]

The major characteristics appear in the neonatal period, and consist of poor development, failure to gain weight, gastrointestinal symptoms, protuberant abdomen due to hepatomegaly, and extensive bilateral adrenal calcification. Death occurs within the first few months of life.

Roentgenographic Features. The distinctive features are hepatosplenomegaly and extensive bilateral calcification of adrenals which are enlarged but maintain their normal shape. (Fig. 1622). Although no bone changes have been reported, the marrow may be packed with foam cells, and changes in the osseous system, similar to those of other lipidoses, would be likely to occur.[6]

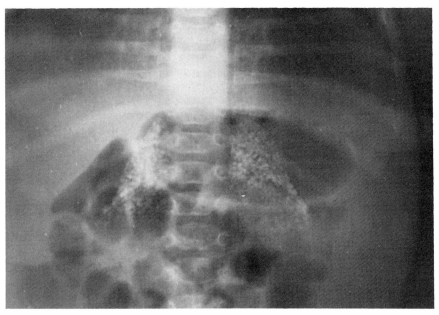

FIG. 1622. WOLMAN DISEASE
The adrenals are calcified. Deossification due to packing of the marrow with foam cells. (Courtesy of Dr. Murray Dalinka, Wilford Hall Hospital, Lackland Air Force Base, Tex.)

REFERENCES

Lipidosis

1. Gatt, S.: Enzymatic hydrolysis and synthesis of ceramides. J Biol Chem, 238: 3131, 1963.
2. Radin, N. S., and Akahori, Y.: Fatty acids of human brain and spleen cerebrosides. Fed Proc, 20: 269, 1961.
3. Trams, E. G., Guiffrida, L. E., and Karmen, A.: Gas chromatographic analysis of long chain fatty acids in gangliosides. Nature, 193: 680, 1962.
4. Klenk, E.: Neuraminsaure, das Spaltprodukt eines neuen Gehirnlipoids. Z Physiol Chem, 268: 50, 1941.
5. Klenk, E.: Uber die Ganglioside des Gehirns bei der infantilen amaurotischen Idiotie von Typus Tay-Sachs. Ber. Deutsch Chem Gesellsch, 75: 1632, 1942.
6. Klenk, E.: Uber die Ganglioside, eine neue Gruppe von zuckerhaltigen Gehirnlipoiden. Z Physiol Chem, 273: 76, 1942.
7. Carter, H. E., Glick, F. J., Norris, W. P., and Phillips, G. E.: Biochemistry of the sphingolipides; III. Structure of sphingosine. J Biol Chem, 170: 285, 1947.
8. Carter, H. E.: *Sphingolipides in Chemistry of Lipides as Related to Atherosclerosis*, p. 82, I. H. Page (ed.). Charles C Thomas, Springfield, Ill., 1958.
9. Grob, C. A., and Gadient, F.: Die synthese des sphingosins und seiner stereoisomeren. Helv Chim Acta, 40: 1145, 1957.
10. Shapiro, D., Segel, H., and Flowers, H. M.: The total synthesis of sphingosine. J Am Chem Soc, 80: 1194, 1958.
11. Rouser, G., Kritchevsky, G., Heller, D., and Lieber, E.: Lipid composition of beef brain, beef liver, and the sea anemone; two approaches to quantitative fractionation of complex lipid mixtures. J Am Oil Chem Soc, 40: 425, 1963.
12. Ansell, B. M., and Bywaters, E. G. L.: Histiocytic bone and joint disease. Ann Rheum Dis, 16: 503, 1957.
13. March, H. C., Gilbert, P. D., and Kain, T. M.: Hypercholesteremic xanthomata of the tendons. AJR, 77: 109, 1957.
14. Kovac, A., Kuo, Y-Z, and Sagar, V.: Radiographic and radioisotope evaluation of intraosseous xanthoma. Br J Radiol, 49: 281, 1976.
15. Bjersand, A. J.: Bone changes in hypercholesterolemia. Radiology, 130: 101, 1979.
16. Merril, A. S.: Case of xanthoma showing multiple bone lesions. AJR, 7: 480, 1920.
17. Lanzetta, A.: Solitary xanthoma of bone. Panminerva Med, 6: 212, 1964.
18. Levy, R. I., and Fredrickson, D. S.: Diagnosis and management of hyperlipoproteinemia. Am J Cardiol, 22: 576, 1968.

Histiocytosis X

1. Lichenstein, L.: Histiocytosis X; integration of eosinophilic granuloma of bone, "Letterer-Siwe disease" and "Schüller-Christian disease" as related manifestations of a single nosologic entity. Arch Pathol, 56: 84, 1953.
2. Hand, A., Jr.: Polyuria and tuberculosis. Arch Pediatr, 10: 673, 1893.
3. Hand, A.: General tuberculosis. Trans Pathol Soc, 16: 282, 1893.
4. Schüller, A.: Uber eingenartige Schadeldefekte im Jugendalter. Fortschr Geb Rontegenstr, 23: 12, 1915.
5. Christian, H. A.: Defect in membranous bones, exophthalmos and diabetes insipidus; unusual syndrome of dyspituitarism. *Med Contrib Biol Res*, 1: 391, 1919.
6. Christian, H. A.: Defects in membranous bones, exophthalmos and diabetes insipidus; unusual syndrome of dyspituitarism. Med Clin North Am, 3: 849, 1919–1920.
7. Hand, A.: Defects of membranous bones, exophthalmos and polyuria in childhood; is it dyspituitarism? Am J Med Sci, 162: 509, 1921.

8. Schüller, A.: Dysostosis hypophysaria. Br J Radiol, *31:* 156, 1926.
9. Thompson, C. Q., Keegan, J. J., and Dunn, A. D.: Defects of membranous bones, exophthalmos and diabetes insipidus. Arch Intern Med, *36:* 650, 1925.
10. Rowland, R. S.: Xanthomatosis and reticuloendothelial system; correlation of unidentified group of cases described as defects in membranous bones, exophthalmos and diabetes insipidus (Christian's syndrome). Arch Intern Med, *42:* 611, 1928.
11. Cowie, D. M., and Magee, M. C.: Lipoids and lipoid diseases; lipoid content of tissues in Schüller-Christian's disease (xanthomatosis) and review of literature on lipoid content of human tissues. Arch Intern Med, *53:* 391, 1934.
12. Lichty, D. E.: Lipoids and lipoid diseases; xanthomatosis (Schüller-Christian's type). Arch Intern Med, *53:* 379, 1934.
13. Freud, P., Grossman, L., and Dragutsky, D.: Acute idiopathic cholesterol granulomatosis. Am J Dis Child, *62:* 776, 1941.
14. Smith, L. A.: Xanthomatosis involving bone (lipoid histiocytosis). Radiology, *24:* 521, 1935.
15. Thannhauser, S. J., and Magendantz, H.: Different clinical groups of xanthomatous diseases; clinical physiological study of 22 cases. Ann Intern Med, *11:* 1662, 1938.
16. Abt, A. F., and Denholz, E. J.: Letterer-Siwe's disease; splenohepatomegaly, associated with widespread hyperplasia of nonlipoid storing macrophages; discussion of so-called reticuloendothelioses. Am J Dis Child, *51:* 499, 1936.
17. Letterer, E.: Aleukamische Retikulose. (Ein Beitrag zu den proliferativen Erkrankungen des Retikuloendothelialaparates). Frankfurt Z Pathol, *30:* 377, 1924.
18. Siwe, S. A.: Die Reticuloendotheliose—ein neues Krankheitsbild unter den Hepatosplenomegalien. Z Kinderheilkd, *55:* 212, 1933.
19. Wallgren, A.: Systemic reticuloendothelial granuloma; nonlipoid reticuloendotheliosis and Schüller-Christian disease. Am J Dis Child, *60:* 471, 1940.
20. Lichtenstein, L., and Jaffe, H. L.: Eosinophilic granuloma of bone, with report of a case. Am J Pathol, *16:* 595, 1940.
21. Farber, S.: Nature of some diseases ascribed to disorders of lipid metabolism. Am J Dis Child, *68:* 350, 1944.
22. Gross, P., and Jacox, H. W.: Eosinophilic granuloma and certain other reticuloendothelial hyperplasia of bones; comparison of clinical, radiologic, and pathologic features. Am J Med Sci, *203:* 673, 1942.
23. Mallory, T. B.: Pathology; diseases of bone. New Engl J Med, *277:* 955, 1942.
24. Arcomano, J. P., et al.: Histiocytosis X. Radiology, *85:* 663, 1961.
25. Takahashi, M., Martel, W., and Oberman H. A.: The variable roentgenographic appearance of idiopathic histiocytosis. Clin Radiol, *17:* 48, 1966.
26. Avery, M. E., McAfee, J. G., and Guild, H. G.: The Course and prognosis of reticuloendotheliosis (eosinophilic granuloma, Schüller-Christian disease and Letterer-Siwe disease); a study of forty cases. Am J Med, *22:* 636, 1957.
27. Hodgson, J. R., Kennedy, R. L., and Camp, J. D.: Reticuloendotheliosis. Radiology, *57:* 642, 1951.
28. Dundon, C. C., Williams, H. A., and Laipply. T. C.: Eosinophilic granuloma of bone. Radiology, *47:* 433, 1946.
29. Hamilton, J. B., Barner, J. L., Kennedy, P. C., and McCort, J. J.: The osseous manifestations of eosinophilic granuloma; report of 9 cases. Radiology, *47:* 445, 1946.
30. Hodgson, J. R., Kennedy, R. L. J., and Camp, J. D.: Reticuloendotheliosis. Radiology, *57:* 642, 1951.
31. Kayne, J., and Freiberger, R. H.: Eosinophilic granuloma of the spine without vertebra plana; a report of 2 unusual cases. Radiology, *92:* 1188, 1969.
32. Pinckney, L., and Parker, B. R.: Myelosclerosis and myelofibrosis in treated histiocytosis-X. AJR, *129:* 521, 1977.

Gaucher Disease

1. Gaucher, E.: De L'epithelioma primitif de la rate. Theses de Paris, 1882.
2. Mandelbaum, F. S., and Downey, H.: The histopathology and biology of Gaucher's disease (large-cell splenomegaly). Folia Haematol, *20:* 139, 1916.
3. Epstein, E.: Beitrag zur Chemie der Gaucherschen Krankheit. Biochem Z, *145:* 398, 1924.
4. Lieb, H.: Cerebrosidspeicherung bei splenomegalie typhus Gaucher. Z Physiol Chem. *140:* 305, 1924.
5. Losman, M. J.: Beta-glucosidase activity as a diagnostic index of Gaucher's disease. S Afr Med J, *48:* 1150, 1974.
6. Myers, H. S., Cremin, B. J., Beighton, P., and Sacks, S.: Chronic Gaucher's disease; radiological findings in 17 South African cases. Br J Radiol, *48:* 465–469, 1975.
7. Wan Ho, M., Seck, J., Schmidt, D., Veath, M. L., Johnson, W., Brady, R. O., and O'Brien, J. S.: Adult Gaucher's disease; kindred studies and demonstration of a deficiency of acid beta-glucosidase in cultured fibroblasts. Am J Hum Genet, *24:* 37, 1972.
8. Hsia, D. Y-Y, Naylor, J., and Bigler, J.: Gaucher's disease; report of 2 cases in father and son and review of the literature. New Engl J Med, *261:* 164, 1959.
9. Ottenstein, B., Schmidt, G., and Thannhauser, S. J.: Studies concerning the pathogenesis of Gaucher's disease. Blood, *2:* 1250, 1948.
10. Groen, H.: The hereditary mechanisms of Gaucher's disease. Blood, *3:* 1238, 1948.
11. van Creveld, S.: The lipidoses. Adv Pediatr, *6:* 190, 1953.
12. Petit, J. V., and Schleicher, E. M.: Atypical Gaucher's disease. Am J Clin Pathol, *13:* 260, 1943.
13. Horsley, J. S., Jr., Baker, J. P., and Apperly, F. L.: Gaucher's disease of late onset with kidney involvement and huge spleen. Am J Med Sci, *190:* 511, 1935.
14. Mandelbaum, H., Berger, L., and Lederer, M.: Gaucher's disease and hemolytic anemia and marked thrombopenia; improvement after removal of spleen, weighing 6822 grams. Ann Intern Med, *16:* 438, 1942.
15. Wechsler, H. F., and Gustafson, E.: Gaucher's disease associated with multiple telangiectases in an elderly woman. NY State J Med, *40:* 133, 1940.
16. Bloem, T. F., Groen, J., and Postma, C.: Gaucher's disease. Q J Med, *5:* 517, 1936.
17. Pick, L.: Classification of diseases of lipoid metabolism and Gaucher's disease. Am J Med Sci, *185:* 453, 1933.
18. Rusca, C. L.: Sul morbo di Gaucher; contributo allo studio delle malattie sistematische dell apparato amohimopojectica. Haematologica, I. Arch, *2:* 441, 1921.
19. Strickland, B.: Skeletal manifestations of Gaucher's disease with some unusual findings. Br J Radiol. *31:* 246, 1958.
20. Levin, B.: Gaucher's disease; clinical and roentgenologic manifestations. AJR, *85:* 686, 1961.
21. Reich, C., Seife, M., and Kessler, B. J.: Gaucher's disease; review and discussion of 20 cases. Medicine, *30:* 1, 1951.
22. Windholz, F., and Foster, S. E.: Sclerosis of bones in Gaucher's disease. AJR, *60:* 246, 1948.
23. Medoff, A. S., and Bayrd, E. D.: Gaucher's disease in 29 cases; hematologic complications and effect of splenectomy. Ann Intern Med, *40:* 481, 1954.
24. Chang-Lo, M., Yam, L. T., and Rubenstone, A. I.:

Gaucher's disease; review of the literature and report of 12 new cases. Am J Med Sci, 78: 303, 1967.
25. Thannhauser, S. J.: *Lipidoses: Diseases of the Intracellular Lipid Metabolism.* Grune & Stratton, New York, 1958.
26. Silverstein, M. N., and Kelly, P. J.: Osteoarticular manifestations of Gaucher's disease. Am J Med Sci, 5: 85, 1967.
27. Gordon, G. L.: Osseous Gaucher's disease; report of 2 cases in siblings. Am J Med, 8: 332, 1950.
28. Fairbank, T.: *An Atlas of General Affections of the Skeleton.* Williams & Wilkins, Baltimore, 1951.
29. Goldblatt, J., Sacks, S., and Beighton, P.: The orthopedic aspects of Gaucher disease. Clin Orthop, 137: 208, 1978.
30. Beighton, P., and Sacks, S.: Gaucher's disease in Southern Africa. S Afr Med J, 48: 1295, 1974.
31. Matoth, Y., and Fried, K.: Chronic Gaucher's disease. Clinical observations on 34 patients. Isr J Med Sci, 1: 521, 1965.
32. Noyes, F. R., and Smith, W. S.: Bone crises and chronic osteomyelitis in Gaucher's disease. Clin Orthop, 79: 132, 1971.
33. Sacks, S.: Arthritis in Gaucher's disease. "R." J. Int League Against Rheum, III/2: 131, 1973.
34. Sacks, S.: Osteitis in Gaucher's disease. S Afr J Surg, 4: 161, 1971.
35. Schein, A. J., and Arkin, A. M.: The classic: Hip joint involvement in Gaucher's disease. Clin Orthop, 90: 4, 1973.
36. Yossipovitch, Z. H., Herman, G., and Makin, M.: Aseptic osteomyelitis in Gaucher's disease. Isr J Med Sci, 1: 531, 1965.

Niemann-Pick Disease

1. Lachman, R., Crocker, A., Schulman, J., and Strand, R.: Radiological findings in Niemann-Pick disease. Radiology, 108: 659, 1973.
2. Crocker, A. C.: The cerebral defect in Tay-Sachs disease and Niemann-Pick disease. J Neurochem, 7: 69, 1961.
3. Crocker, A. C., and Farber, S.: Niemann-Pick disease; a review of 18 patients. Medicine, 37: 1, 1958.
4. Frederickson, D. S.: Sphingomyelin lipidosis; Niemann-Pick disease. In *Metabolic Basis of Inherited Disease,* Ed. 2, J. B. Stanbury, J. B. Wyngaarden, and D. S. Fredrickson, (eds.). McGraw-Hill, New York, 1966.

I-Cell Disease

1. LeRoy, J. G., DeMars, R. I., and Opitz, J. M.: I-Cell disease. In *The Clinical Delineation of Birth Defects; IV. Skeletal Dysplasias,* p. 174. Williams & Wilkins, Baltimore, 1969.
2. Mittwoch, U.: Inculsions of mucopolysaccharide in the lymphocytes of patients with gargoylism. Nature, 191: 315, 1961.
3. Taber, P., Gyepes, M. T., Philippart, M., and Ling, S.: Roentgenographic manifestations of LeRoy's I-cell disease. AJR, 118: 213, 1973.
4. Patriguin, H. B., Kaplan, P., Kind, H. P., and Giedion, A.: Neonatal mucolipidosis II (I-cell disease); clinical and radiologic features in three cases. AJR, 129: 37, 1977.

Familial Neurovisceral Lipidoses

1. Norman, R. M., et al.: Tay-Sachs disease with visceral involvement and its relationship to Niemann-Pick's disease. J Pathol, Bacteriol, 78: 409, 1959.
2. Craig, J. M., Clarke, J. T., and Banker, B. Q.: Metabolic neurovisceral disorder with accumulation of unidentified substances; variant of Hurler's syndrome. Am J Dis Child, 98: 577, 1959.
3. Landing, B. H., and Rubinstein, J. H.: Biopsy diagnosis of neurologic diseases in childhood with emphasis on lipidosis. In *Cerebral Sphingolipidosis,* S. M. Aronson and B. W. Volk (eds.). Academic Press, New York, 1962.
4. Landing, B. H., Silverman, F. N., Craig, J. M., Jacoby, M. D., Lahey, M. E., and Chadwich, D. L.: Familial neurovisceral lipidosis. Am J Dis Child, 108: 503, 1964.
5. Gonatas, K., and Gonatas, J.: Ultrastructural and biochemical observations on a case of systemic late infantile lipidosis and its relationship to Tay-Sachs disease and gargoylism. J Neuropathol Exp Neurol, 24: 318, 1965.
6. O'Brien, J. S., Stern, M. B., Landing, B. H., O'Brien, K. O., and Donnell, G. N.: Generalized gangliosidosis, Am J Dis Child, 109: 338, 1965.
7. Suzki, K., Suzuki, K., and Chen, G. C.: Morphological, histochemical and biochemical studies on a case of systemic late infantile lipidosis (generalized gangliosidosis).
8. Caffey, J.: Gargoylism; prenatal and early postnatal bone lesions, and their early postnatal evolution. Bull Hosp Joint Dis, 12: 38, 1951.

Disseminated Lipogranulomatosis

1. Farber, S.: A lipid metabolic disorder—disseminated lipogranulomatosis; a syndrome with similarity to and important differences from Niemann-Pick and Hans-Schüller-Christian disease (abstract). Am J Dis Child, 84: 499, 1952.
2. Farber, S., Cohen, J., and Uzman, L. Lahut: Lipogranulomatosis; a new lipo-glyco-protein "storage" disease. J Mt Sinai Hosp, 24: 816, 1957.
3. Crocker, A. C., Cohen, J., and Farber, S.: The lipogranulomatosis syndrome; review, with report of patient showing milder involvement. In *Inborn Disorders of Sphingolipid Metabolism: Proceedings of the Third International Symposium on Cerebral Sphingolipidoses,* S. M. Aronson (ed.). Pergamon Press, New York, 1967.
4. Zetterstrom, R.: Disseminated lipogranulomatosis (Farber's disease). Acta Paediatr, 47: 501, 1958.
5. Prensky, A. L., Ferreira, G., Carr, S., and Moser, H. W.: Ceramide and ganglioside accumulation in Farber's lipogranulomatosis. Proc Soc Biol Med, 126: 725, 1967.
6. Abul-Haj, S. K., Martz, D. G., Douglas, W. F., and Geppert, L. J.: Farber's disease; report of a case with observation on its histogenesis and notes on the nature of the stored material. J Pediatr, 61: 221, 1962.
7. Schultze, G., and Lang, E. K.: Disseminated lipogranulomatosis. Radiology, 74: 428, 1960.

Wolman Disease

1. Wolman, M., Sterk, V. V., Gatt, S., and Frenkel, M.: Primary familial xanthomatosis with involvement and calcification of the adrenals. Pediatrics, 28: 742, 1961.
2. Crocker, A. C., Vawter, G. F., Neuhauser, E. B. D., and Rosowsky, A.: Wolman's disease; 3 new patients with a recently described lipidosis. Pediatrics, 35: 627, 1965.
3. Sandison, A. T.: A form of lipidosis of the adrenal cortex in an infant. Arch Dis Child, 30: 538, 1955.
4. Abramov, A., Schorr, S., and Wolman, M.: Generalized xanthomatosis with calcified adrenals. J Dis Child, 91: 282, 1956.
5. Neuhauser, E. B. D., Kirkpatrick, J. A., and Weinstraub, H. A.: Wolman's disease, a new lipidosis. Ann Radiol, 8: 175, 1965.
6. Harrison, R. B., and Francke, Jr., P.: Radiographic findings in Wolman's disease. Radiology, 124: 188, 1977.

Author Index

Aase, J., 1436
Abbasy, M., 517
Abbott, M. H., 1184
Abel, M., 659
Abell, M., 457
Abrahams, O. L., 582
Abramson, N., 310
Aceto, T., Jr., 1184
Achenbach, W., 913
Acheson, E. D., 506
Acheson, R. M., 1149, 1154
Ackerman, A. J., 127, 1387
Ackerman, L. V., 83, 116, 212, 267, 277, 622, 659, 1108, 1365
Ackerman, L. W., 383
Acosta, E., 805
Adam, H., 1403
Adams, M., 1233
Adams, M. S., 1189
Adams, W. E., 642
Adb-el-Hafeez, M., 242
Adler, J. J., 642
Adler, S., 965
Aegerter, E., 33, 107, 116, 165, 182, 205, 227, 257, 277, 757, 995, 1195, 1295, 1358, 1359, 1365, 1402
Aegerter, E. E., 23, 808, 1108
Aegerter, E. R., 1377
Afshani, E., 1230
Aga, V., 208
Agarval, B., 642
Agarwal, K. N., 1073, 1365
Agnew, C. H., 1095
Agnew, J. E., 660
Aguayo, A., 319
Ahlback, S., 710, 711
Ahrens, E. H., Jr., 617, 660
Ahuja, S. C., 711
Aiello, C. L., 138
Aird, I., 909
Aitken, J. K., 1334
Akahoshi, Y., 181
Akers, W. A., 1338
Akeson, W. H., 1359
Alagille, D., 1351
Alarcon-Segovia, D., 557
Alavi, M. S., 1430
Alazraki, N. P., 705
Albert, J., 659
Albores-Saavedra, J., 256
Albrand, O. W., 517
Albrecht, C., 1181, 1351
Albright, F., 6, 830, 831, 853, 858, 867, 879, 880, 912, 913, 917, 964, 966
Albright, R., 1095
Alderman, M. H., 951
Alexander, W. J., 1239
Algan, B., 1334
Allan, C. J., 208
Alldred, A. J., 1418
Allen, A. C., 659

Allen, A. R., 757
Allen, C. E. L., 1397
Allen, D. H., 116
Allen, D. M., 1303
Allen, H. C., 227
Allen, P. W., 121
Allen, R. P., 138
Alliez, J., 1365
Allison, A. C., 722, 1181
Almagor, G., 951
Alpert, M., 277, 457, 808, 995
Alterman, S., 912, 913, 1182
Altman, J., 968
Altman, K. I., 1300
Altner, P. C., 1140
Amatruda, T., 183
Ambos, M., 977
Ames, R., 1084
Amin, P. H., 1381
Ammann, F., 1184
Amstutz, H. C., 687
Amuzo, S. J., 1214
Anderson, C. E., 380
Anderson, C. K., 909
Anderson, D. E., 1313, 1314
Anderson, D. G., 713
Anderson, H., 1341
Anderson, I. F., 1181
Anderson, L. D., 1108
Anderson, L. G., 1442
Anderson, M., 1148
Anderson, N. O., 1182
Anderson, R. L., Jr., 83
Anderson, W. A. D., 1144
Anderson, W. B., 396
Anderson, W. W., 924
Andersson, O., 517
Andreas, E., 1095
Andren, L., 1084, 1181
Andrew, J. D., 1227
Andrews, G. S., 1397
Andrews, J. R., 227
Angell, M., 730
Angle, C. R., 1207
Angst, H., 1334
Angus Muir, W., 582
Ansell, B. M., 476, 479, 495, 499, 506, 660
Ansell, I. D., 582
Anson, S. G., 158
Anspach, W. E., 873, 1416
Anton, H. C., 1172
Anton, J. I., 1243
Appel, H., 623
Aranda, D., 1184
Arango, O., 1326
Arbuckle, R. K., 995
Archibald, R., 1269
Arden, G. P., 457
Argaud, M. R., 391
Ariel, I. M., 255
Arkless, R., 1286

Arlen, M., 121, 259, 1140
Arlet, J., 544
Arnaudet, L., 1243
Arnold, W. D., 138
Arnstein, A. R., 843, 853
Asatoor, A. M., 501
Asboe-Hansen, G., 375
Ascenzi, A., 1072
Asch, M. D., 1351
Asch, T., 127, 138
Ash, J. M., 115, 1402
Ashby, D. W., 1386
Asher, J. D., 832
Ashley, L. M., 1188
Ask-Upmark, E., 1080
Asnes, R. S., 1343
Asp, D. M., 1181
Aspillaga, M. J., 1303
Asscher, A. W., 158
Aston, J. K., 208
Aterman, K., 35
Atkins, C. J., 572
Atkins, L., 1340
Atkinson, M., 617, 844
Atkinson, W. G., 1377
Au, W. Y. W., 1347
Aub, J. C., 880, 964
Auerbach, O., 768
Aufses, A. H., 642
Aufses, B. H., 642
August, P. J., 660
Auld, R. M., 923
Austin, J. H., 1182
Austin, L. T., 968
Avioli, L. V., 905
Axer, A., 696
Axhausen, G., 623
Ayala, A. G., 206
Azar, H. A., 1145
Azimi, F., 1436

Babbiani, G., 373
Bach, C., 1268
Bach, F., 457
Back, D. K., 1168
Badgley, C. E., 1017
Badgley, C. F., 1028
Badin, J., 506
Badley, B. W. D., 495
Baehner, R. L., 723, 754
Baensch, W. E., 319
Baer, R. W., 1331
Baez-Giangreco, A., 121
Baggenstoss, A. H., 457, 617
Bahr, A. L., 242
Baijens, J. K., 698
Baikie, A. G., et al., 506
Bailey, J. A., II, 1191
Bailey, R. W., 517

xi

AUTHOR INDEX

Bailey, W., 1089
Baillie, J., 1189
Baily, R. A. J., 1028
Baird, D. S. C., 1377
Baird, I. M., 935
Baker, D. H., 700, 1052
Baker, E. R., 340
Baker, H. L., Jr., 242
Baker, L. A., 530
Baker, N., 714
Baker, P. L., 380
Baker, S. L., 867, 923
Bakke, S. N., 805
Bakos, L., 545
Bakwin, H., 951, 1184, 1242
Baldursson, H., 1095
Balikian, J. P., 909
Ball, J., 506
Ball, R. E., 457
Ball, R. P., 1047
Balsan, S., 1377
Bamatter, F., 1184
Bank, S., 683, 717
Banna, M., 631
Bansal, S. C., 1146
Baradnay, G., 623
Barard, C. W., 622
Barasch, E., 642
Barba, W. P., II, 817
Barbaric, Z. L., 457
Barbe, P., 1268
Barber, C. G., 705, 1377
Barber, G. W., 1332
Barber, H. S., 457
Barcellos, J. M. P., 391
Barcia, P., 799
Barclay, N., 642
Bardeen, C. R., 1148
Barden, R. P., 277
Bardsley, J. L., 54
Barker, J. D. P., 966
Barlett, R., 124
Barlow, K. A., 572, 582, 617
Barlow, T. G., 1084
Barnard, L., 387
Barnes, C. G., 554
Barnes, R., 223, 256, 517
Barness, L. A., 1331
Barnicot, N. A., 1347
Baroutsou, E., 951
Barr, D. P., 831
Barrett, A. M., 873
Barrington, A., 1244, 1402
Barry, H. C., 976, 993
Barry, R. J., 994
Barry, W. F., Jr., 62
Barsky, A. J., 1380
Barta, C. K., 54, 700
Bartelheimer, H., 935
Bartlett, L. H., 54
Bartolomei, B., 380
Barton, C. J., 1048
Barton, D. I., 1095
Barton, E. M., 530
Bartter, F. C., 1095, 1182
Baruch, H. H., 396
Bass, M. H., 208
Bassen, F. A., 601
Basset, F., 387

Bateman, D., 1230
Batsakis, J. G., 330, 391
Batstone, J. H. F., 1181
Baty, J. M., 340, 480
Batzdorf, U., 49
Baud, C. A., 577
Bauer, G. C. H., 710, 713
Bauer, J. M., 918
Bauer, W., 417, 500, 506, 609, 867, 880, 912, 964, 994
Bauer, W. H., 1030
Baum, J., 506
Bauman, E., 584
Baumann, P. A., 1180, 1225
Baumann, R. R., 445
Bautovitch, G. J., 705
Bayrd, E. D., 1030
Bayyuk, S. I., 1145
Bayyuk, W., 1145
Beabout, J. W., 49, 60, 83, 127, 182, 206, 242, 255, 256, 385
Beachley, M. C., 387
Beals, R. K., 1180, 1219, 1286
Bean, W., 1244
Beare, M., 375
Bearn, A. G., 501, 614
Beatty, D. C., 457
Beautyman, W., 1302
Beauvais, P., 1401
Beaver, P. C., 872
Bebe, M., 1381
Becker, F. P., 255
Becker, J. E., 858
Becker, M., 1230
Becker, M. A., 530
Becker, M. H., 683
Bednar, R., 1033
Beierwaltes, W. H., 909
Beighton, P., 1230
Beiler, D. D., 1335
Bek, V., 711
Belanger, L. F., 6
Belau, P. G., 1181
Belcourt, C. L., 1340
Bell, A. L., 380
Bell, G., 340
Bell, J., 1286
Bell, N. H., 913
Bell, R. L., 517
Bell, T. K., 909
Bellamy, R., 1089
Beller, M., 965
Belot, J., 1286
Belsky, H., 319
Benedek, T. G., 532
Benedek, T. O., 311
Benedetti, G. B., 95, 256
Benesch, R., 227
Benesch, R. E., 227
Benjamin, B., 709, 1334
Bennet, G. A., 609
Bennett, B., 1033
Bennett, H. S., 817
Bennett, P. H., 553
Bennett, R. B., 722
Bennett, W. A., 1030
Bennington, J.L., 1033
Benton, C., 1362
Berard, C. W., 242, 257

Beraud, C., 1192, 1207
Berdon, W. E., 700, 1052
Berenberg, W., 1073
Berendes, H., 754
Berens, D. L., 445
Berg, G., 1387
Berg, J. W., 378
Berg, N. O., 1033
Bergeim, O., 964
Bergenstal, D. M., 227
Berger, P. E., 32
Bergmann, E. W., 517
Bergsma, D., 1184
Bergstrand, C. G., 913
Berk, M. E., 554
Berk, R. N., 572, 808
Berke, I., 1073
Berkley, H. J., 1423
Berkman, Y. M., 242
Berlin, C., 375
Berlin, R., 1181
Berlung, G., 845
Bernard, J., 1269
Bernardino, M. E., 195
Bernett, C. H., 1286
Bernstein, A., 313, 557
Bernstein, D., 255
Bernstein, D. S., 1144
Bernstein, S. A., 1107
Bernstein, S. S., 554
Berres, H. H., 375
Berry, G. A., 1313
Berstein, J., 208
Bertellotti, L., 375
Bertheau, J., M., 1430
Bertolotti, M., 1243
Besemann, E. F., 378, 380
Besse, B. E., Jr., 165, 480
Bessent, R. G., 340
Bessler, V., 62
Bessone, J. E., 225
Beumer, H., 912
Beutler, E., 582
Beveridge, B., 935
Bhakkaviziam, A., 165
Bhardwaj, O. P., 1365
Bhardwaj, P., 1073
Bianchine, J. W., 1182
Bianco, A. J., Jr., 722
Bianco, C. J., 830
Bickel, W. H., 35, 138, 149, 457, 557, 683, 687, 1377
Bidder, U., 1180
Biedmond, A., 1286
Biehl, G., 114
Biehler, R., 805
Bienefeld, C. H., 1244
Biesecker, J. L., 165
Bigelow, D. R., 722
Bigelow, E. L., 375
Bignami, A., 659
Bigongiari, L. R., 601
Bigorgne, J. C., 552
Bigozzi, U., 1322
Biller, H. F., 242
Binenfeld, C., 1224
Binet, E. F., 1362
Bingold, A. C., 1410
Binkley, G. W., 1313

Birch-Jensen, A., 1303
Birkett, A. N., 517
Birtwell, W. M., 867, 872
Biskis, B. O., 181
Bitnum, S., 476, 506
Bittel, D., 1268
Bivings, L., 149
Black, B. M., 909
Black, J., 1184
Black, M., 504
Blackburn, C. R. B., 660
Blackfan, K. D., 1300, 1303
Blades, B., 1359
Blades, J. F., 1334
Blaisdell, J. L., 993
Blank, C. E., 1340
Blank, E., 1190
Blank, H., 659
Blank, N., 62
Blatt, E. S., 242
Blattner, R. J., 800
Bledsoe, R., 1107
Bleehan, S. S., 532
Bleshman, M. H., 72
Blizzard, R. M., 1184
Bloch, K. J., 311, 504
Bloch, S., 808
Block, J. B., 1314
Blockey, N. J., 1223
Bloom, D., 1184
Bloom, G., 375
Bloomberg, E., 853, 867
Bloomberg, T. J., 700
Bloomer, H. A., 923
Bloomfield, J. A., 1207
Blount, W. P., 722
Bluestone, R., 506
Blumberg, B. S., 722, 1181
Blumgart, L. H., 340
Bobechko, W. P., 679, 1289
Bocher, J., 572
Bocketl, R., 313
Bockus, H. L., 660
Bodenhoff, J., 1227
Boeck, C., 808
Boguist, L., 383
Bohatirchuk, F., 6
Bohatirchuk, F. A., 6
Bohm, A., 913
Bohme, A., 722, 1181
Bohne, W. H., 710, 711
Bohr, H., 680
Bohrer, S. P., 1059, 1070
Bohrer, S. P. S., 1059
Boinet, M. E., 1262
Bolam, R. M., 1334
Boldero, J. L., 696
Boldrey, E. B., 54
Bole, G. G., 475
Bollack, C., 722
Bonakdarpour, A., 165, 711, 808
Bonduelle, M., 994
Bonfiglio, M., 54, 1181, 1397
Bonnet, P., 1334
Bonse, G., 149
Book, J. A., 1219
Booth, J. R., 330
Boothroy, B., 6
Boots, R. H., 478

Borchard, 631
Borenstein, D. B., 1181
Boreux, G., 1184
Borghi, A., 1322
Borglin, N. E., 1084
Borm, D., 909
Born, P. F., 1159
Borrone, C., 1182
Bortz, A. I., 659
Bosma, J. F., 1442
Bossu, A., 1334
Bost, F. C., 1436
Boswell, W. et al., 601
Bosworth, D. M., 768
Bothwell, T. H., 582
Boudin, G., 501
Boujor, A., 495
Boulle, J., 1213
Bourneville, D., 1386
Bouroncle, B. A., 1074
Boussina, L., 577
Bouvier, M., 843
Bove, K. E., 1028
Bovill, E. G., Jr., 195
Bowen, P., 1266
Bowerman, J. W., 966
Bowness, J. M., 1095
Boxer, L. A., 1073
Boycott, H. E., 1080
Boyd, H. B., 844
Boyer, C. W., Jr., 208
Bozsoky, S., 545
Braband, H., 554
Brace, K.C., 227
Brackett, N. C., Jr., 312
Brady, L. P., 54
Brady, L. W., 340, 391
Bragg, D. G., 1125
Braidwood, A. S., 380
Brailsford, J. F., 637, 846, 968, 1181, 1267, 1283, 1377
Braimbridge, C. V., 651
Brain, W. R., 1386
Brannan, D., 1080
Brash, J. C., 456
Brashear, H. R., 659
Bray, C., 1332
Bregeon, C., 552
Breimer, C. W., 651
Brennan, B. B., 659
Brennan, R. E., 383
Brenner, R. J., 115
Brenton, D. P., 845, 1331, 1332
Brett, E. M., 1331
Brewer, E., 601
Brewer, E. J., Jr., 478
Brewerton, D., 506
Brialsford, J. F., 1269
Brian, A. T., 457
Bridges, R. A., 754
Brighton, C. T., 383, 499, 572
Brill, J. M., 572
Brill, P. W., 727, 1331
Brinkhous, K., 591
Brinkmann, E., 375
Brocher, J. E., 1184, 1243
Broder, H. M., 105
Broderick, T. W., 552
Broders, A. C., 138, 149, 330, 651

Brodey, P. A., 808
Brodie, A. G., 495
Brodsky, I., 1030
Brody, G. L., 138, 499
Brogdon, B. G., 1100, 1230
Brogren, N., 375
Brombart, M., 1283, 1334
Bronsky, D., 313, 557, 913
Brook, C. G. D., 709
Brook, J., 687
Brotherton, B. J., 699
Brower, A. C., 127, 544, 642
Brower, T. D., 808
Brown, A., 995, 1071
Brown, C. E., 645, 1310
Brown, D. C., 699
Brown, E. M., Jr., 428
Brown, G. A., 1377
Brown, J. C., 1107
Brown, P., 506
Brown, R. H., 116
Brown, W., 183
Browne, S. G., 1432
Browning, C. H., 1080
Brubaker, D. A., 754
Bruce, R. S., 1233
Bruce, W., 659
Bruch, H., 995
Brunat, M., 122
Brunner, S., 1196
Brunschwig, A., 22, 257, 697, 714
Brunton, F. J., 1310
Bryant, J. S., 1189
Bryk, D., 35
Bubanj, R., 330
Buchanan, R. C., 1340
Buchanan, W. W., 504
Buchignani, J. S., 1442
Buchman, J. J., 754, 1057
Buckley, J. J., 872
Buckman, J., 684
Buckwalter, J. A., 912
Bucy, P. C., 127, 1403
Budd, J. W., 257
Budenz, G. C., 1387
Budzinska, A., 1268
Bufkin, W. J., 116
Buhl, L., 1387
Buhring, H., 909
Bull, D. C., 387
Bullough, P., 691, 696, 713, 714, 977, 1108
Bullough, P. G., 256, 373, 679, 711, 1289
Bunch, W. H., 696
Bundens, W. D., 383
Bundens, W. D., Jr., 572
Bunim, J. J., 504
Buraczewski, J., 90, 165, 273
Burch, F. E., 1330
Burch, J. E., 1059
Burd, R., 642
Burgan, D. W., 623
Burgener, F. A., 965
Burgert, E. O., Jr., 1300
Buri, J., 1438
Burke, E., 1235
Burker, E. C., 722
Burnett, C. H., 853, 913, 1182
Burnham, J., 557
Burns, J. E., 645

AUTHOR INDEX

Burr, C. W., 1423
Burrows, A., 1334
Burton, C. C., 714
Burus, R. R., 846
Busch, K. F. B., 1392
Buschke, A., 1392
Bussey, H. J. R., 1433
Butler, A. M., 853, 867, 912, 994, 1181
Butler, H. G., 1266
Butler, R. W., 701
Butt, H. R., 617
Buttner, A., 1017, 1028
Byers, P. D., 35, 651
Bywater, E. G., 506, 552
Bywater, E. G. L., 499, 532
Bywaters, E. G. L., 445, 479, 500, 506, 572, 660, 905

Cabello-Campos, J. M., 380
Cade, S., 181, 1125
Cadili, D. G., 313
Caesar, R., 1185
Caffey, J., 62, 111, 116, 354, 705, 774, 775, 817, 822, 858, 917, 951, 1033, 1069, 1073, 1084, 1120, 1125, 1181, 1192, 1225, 1227, 1242, 1286, 1343, 1351, 1359, 1401, 1410, 1418, 1421
Caffey, J. D., 483
Caffey, J. P., 1181
Caffey, M., 506
Cahan, W. C., 165
Cahan, W. G., 1131
Caille, B., 501
Caldicott, W. J. H., 54
Caldwell, J. A., 844
Calenoff, L., 1146
Cameron, E. C., 935
Cameron, J. A. P., 1235
Cameron, J. M., 696
Cameron, J. R., 830
Camitta, B. M., 1073
Camp, J. D., 650, 964, 1148
Campanacci, M., 1017, 1028
Campbell, C. J., 1397
Campbell, C. P., 1181
Campbell, J. A. H., 387
Campbell, J. C., 83
Campbell, R. E., 62
Campeti, L., 138
Canary, J. J., 923
Canellos, G. P., 330
Caner, J. E., 557
Caner, J. E. Z., 572
Canet, L., 495
Cangir, A., 642
Canivet, J., 909
Canizares, O., 532
Cannon, J. F., 1403
Cantagrel, U., 1243
Capitano, M. A., 1230
Caplan, A., 457
Caplan, H. I., 445
Caplan, R. M., 149, 375
Capp, C. S., 127
Carbonara, A. D., 311
Carbone, P. P., 325, 1144
Cardenas-Ramirez, L., 256
Carey, E. J., 687
Carey, L. S., 1269

Carey, M., 1377
Carey, M. C., 1332
Carezzano, P., 1423
Carlson, D. H., 165
Carmel, P., 442, 506
Carmel, P. W., 1403
Caro, M. R., 1313
Carpenter, E. B., 277
Carpenter, G. G., 1343
Carrington, H. T., 1059
Carroll, D. S., 684, 808, 1047
Carron, R., 1192, 1207
Carsen, G. M., 1033
Carson, J. J., 754
Carson, N. A. J., 1331, 1332
Carter, A. R., 453
Carter, C., 1084
Carter, C. O., 1334
Carter, M., 1310
Carter, M. E., 479
Carter, T. M., 1148
Carter, W. J., 1387
Cary, R., 66
Case records of Massachusetts General Hospital, 54, 90, 208, 257
Casey, B., 805
Cash, J. R., 1080
Cassady, J. R., 700, 1052
Casselman, E. S., 1365
Cassidy, J., 479
Cassidy, J. T., 479, 499, 506, 1030
Castillo, B. A., 445
Castle, W. B., 181
Castleman, B., 149, 879, 909, 912
Castro, E. B., 330
Casuccio, C., 6
Catelier, C., 1381
Cathie, I. A. B., 1300
Catterall, A., 696
Catto, M., 256, 387
Caughey, D. E., 532
Cavallino, R., 501
Cavanagh, R. C., 121
Cavanaugh, J. J., 642
Cavina, C., 1322
Cawley, E. P., 1095
Cayotte, L., 1430
Cederquist, E., 879
Cervansky, J., 585
Cervenka, J., 1184
Cetina, J. A., 557
Chabner, B. A., 330
Chadwick, D. L., 754
Chamberlain, W. E., 805
Chambers, G., 257, 622
Chan, R. N. W., 517
Chandler, F. A., 35, 1436
Chang, C. H., 1433
Changus, G. W., 378
Chase, N. E., 1021
Chavez, E., 256
Chaykin, L. B., 530
Cheatum, D. E., 506
Chebat, J., 1377
Cheesman, A. D., 330
Chen, J. C., 853
Chen, S. H., 181
Cheney, B. A., 935
Cheney, W. D., 1385

Chevalier, J., 552
Chigot, P. L., 1401
Child, D. I., 617
Child, D. L., 660
Chiroff, R. T., 885
Choremis, C., 951
Chormley, R., K., 273
Chou, S. J., 853
Chrenka, P., 380
Christensen, W. R., 1095
Christie, A., 1326
Christie, F. H., 872
Christmas, R. O., 832
Christopherson, W. M., 378, 995
Chu, F. C. H., 1125
Chu, H. L., 853
Chuinard, E. G., 1397
Chung, S. M. K., 499, 709
Chytil, M., 1033
Cilley, E. I. L., 342, 1148
Citrin, D. L., 340
Ciuliano, V. J., 623
Civatte, J., 375
Claflin, D., 867
Clagett, O. T., 642
Claisse R., 994
Clarisse, P. T., 319
Clark, D. H., 1377
Clark, D. M., 1161
Clark, G. M., 808
Clark, J. A., 475
Clark, J. D., 1326
Clark, M., 808
Clark, R. L., 506
Clark, W. S., 506
Clarke, A. K., 617
Clarke, T. E., 696
Clawson, D. K., 1359, 1381
Clay, A., 457
Clayton, M. L., 456, 517
Clearfield, R., 1359
Cleland, R. A., 754
Clement, R., 1397
Clemett, A. R., 817
Clemmens, R. L., 1343
Clemmesen, S., 445
Clendenning, W. E., 1314
Clennar, G., 572
Cleveland, M., 768
Cleveland, R. H., 1377
Cleveland, D. A., 1080
Cliff, M. M., 711
Clifton, W. M., 873
Close, A. S., 1080
Clough, P. W. L., 966, 968
Cloup, M., 122
Cobb, N., 1377
Coccaro, P. J., 1442
Cocchi, U., 1191, 1230
Cockshott, W. P., 1048
Codman, E. A., 83, 263, 1105
Coffin, G. S., 1312
Cohen, H., 1302
Cohen, J., 105, 138, 149, 256
Cohen, J. A., 295
Cohen, L., 582, 705
Cohen, M., 1233
Cohen, M. L., 916
Cohen, M. M., 1189

AUTHOR INDEX

Cohen, M. M., Jr., 1181
Cohen, N. H., 1334
Cohen, P., 225, 325, 1144
Cohen, R., 1357
Cohn, W. L., 124
Colaert, J., 1059
Cole, G. J., 1432
Cole, H. N., 1338
Cole, T. B., 330
Cole, W. R., 1052
Colella, A. C., 909
Coleman, H. M., 995
Coleman, S. S., 90, 273
Coley, B. L., 263, 273, 277, 993, 995
Collins, D. H., 608, 966
Collins, J. J., 1144
Collins, L. C., 642
Combes-Hamelle, A., 1397
Committee of the American Rheumatism Association, 506
Compere, C. L., 642
Compere, E. L., 642
Compere, S. L., 1028
Comroe, B. I., 417
Condon, V. R., 951
Conerly, E. B., 1331
Congdon, C. C., 391
Conkling, W. S., 1326
Conley, C. L., 1047
Conlon, P., 181
Connor, T. B., 909, 923
Conrad, M. B., 1085
Conrad, M. E., 617
Conradi, E., 1203, 1230
Conte, N., 909
Cook, A. J., 1442
Cook, P. L., 325, 1144, 1262
Cook, W. A., 1313
Coombes, F. S., 554
Cooney, J. P., 805
Cooper, A. P., 263
Cooper, G., Jr., 1120
Cooper, J. P., 591
Cooper, J. R., 1386
Cooper, R. R., 206
Cope, C. L., 995
Cope, O., 909
Copeland, M. M., 83, 208, 223, 265, 273, 297
Copeland, N. N., 206
Copeland, W. S. C., 417
Copp, D. H., 6, 935
Corcos, J., 457
Cordier, J., 1334
Corley, C. C., Jr., 1073
Corley, C. L., 1397
Corlin, R. J., 1189
Corn, O., 517
Coss, J., 478
Coste, F., 387
Costello, J. l., 913
Costin, G., 1377
Cotler, J. M., 591
Cotter, W. B., 1181
Cotton, R. E., 651
Coupatez, G., 1334
Courtecuisse, W., 1351
Courts, A., 6
Coury, C., 642

Cousin, J., 722
Coutts, R. D., 445
Coventry, M. A., 277
Coventry, M. B., 83, 181, 380, 659, 1334
Covey, G. W., 918, 1080
Cowell, H. R., 1286
Cox, D. W., 1084
Crabbe, P. A., 311
Cracco, J. B., 1343
Craig, J. M., 138, 149
Craig, M. D., 1227
Cramer, L. M., 1380
Crammer, F. J., 1403
Crain, D. C., 614
Crane, J., 881
Cranley, R. E., 1201
Crasselt, C., 1381
Craver, L. F., 297, 354
Crawford, T., 391, 905
Cremin, B. J., 759, 1206, 1230
Crevasse, L., 1227
Creveld, S., 1227
Crile, G., Jr., 313
Crittenden, J. J., 623
Croll, M. N., 340
Crome, L., 1340
Cronberg, N. E., 1206
Crooks, F., 517
Crosby, E. H., 805
Crosby, W. H., 1302
Crosett, A. D., Jr., 1387
Cross, H. E., 1181, 1239
Crow, N. E., 1230
Crow, R. S., 313
Cruess, R. L., 499, 700
Cruickshank, B., 107, 457, 506, 507
Csakany, G., 545
Csonka, G. W., 530, 532, 536
Cudkowicz, L., 457
Cuevillas, A. R., 396
Culberson, J. D., 391
Cullen, J. C., 722
Cunningingham, M. P., 259
Cupps, R. E., 267
Currarino, G., 873
Curratino, G., 1188
Currey, H. F. L., 572
Currey, H. L. F., 572
Curtis, A. C., 1095
Curtis, B. H., 1436
Curtis, L. E., 1163
Cusworth, D. C., 1331, 1332
Cutler, S. J., 330
Cutter, E., 1416
Cuttner, J., 312
Czernizk, P., 138
Czipott, Z., 623

Dabska, M., 165
Dachi, S. F., 1181
DaCosta, J. C., 964
Daescgner, C. W., 1326
Daeschner, W., 1283
Dahlin, D., 242, 380
Dahlin, D. C., 35, 54, 60, 83, 90, 95, 101, 127, 149, 158, 165, 181, 206, 207, 225, 227, 236, 242, 255, 256, 257, 267, 273, 277, 279, 280, 300, 383, 385, 387, 391, 622, 687, 1030, 1403

Dahlin, D. S., 54
Dalinka, M. D., 383
Dalinka, M. K., 208, 604, 771, 1131
Dalsgaard-Nielsen, T., 1387
Daly, D., 1377
Dameshek, W., 1303
Danarj, T. J., 872
Danau, M., 986
Dandy, W. E., 1191
Danelius, G., 1289
D'Angelo, W. A., 475
D'Angio, G. J., 1286
Daniele, R. P., 309
Danigelis, J. A., 699, 771
Danks, D. M., 1230
Danlos, M., 1334
Dannenberg, M., 1243
Danon, F., 311
Danowski, T. S., 918
Danziger, J., 205
Dargeon, H. W., 277
Darling, R. C., 1227
Dartois, A. M., 1377
Das Guptz, T. K., 208
Dash, J., 340, 681
Dass, R., 554
d'Aubique, R. M., 687
Daudet, M., 122
Daves, M. L., 994
Davidson, A. G. F., 935
Davidson, H. B., 313
Davidson, K. C., 1365
Davidson, L. M., 1377
Davidson, P. T., 771
Davidson, R. T., 1189, 1233, 1313
Davies, B., 1332
Davies, D. R., 923
Davis, J. A., 1239
Davis, J. G., 923
Dawson, E. G., 49
Dawson, I. M. P., 1433
Dawson, W. B., 319
Dayer, L., 1227
Deacon, O. W., 651
DeBoer, W. G. R. M., 208
Debre, R., 1334
DeBuen, S., 256
DeBusk, F. L., 1438
DeBussere, A., 683
Deceker, J. P., 659
Decker, F. H., 342
Decker, J. L., 532, 557, 572
Decker, R. D., 1403
Deffrenne, P., 1214
Degos, R., 375
Deiss, W. P., Jr., 1181
Delage, J. M., 1303
Deland, F. H., 1365
Delannoy, E., 995
DeLima, C. S. P., 696
Della Cella, G., 1182
DeLorimier, A. A., 844
Deluca, H. F., 913
DeMarchi, E., 843
DeMarinis, F., 1188
Demars, R. I., 1185
DeMyer, W., 1362
Denko, C. W., 532
Denko, J. V., 90, 935

AUTHOR INDEX

Dennison, W. M., 727
Denny-Brown, D., 1423
Dent, C. E., 845, 867, 873, 905, 923, 924, 935, 1181, 1182, 1331, 1377
DePalma, A., 507
DePalma, A. F., 35, 591, 757, 966
DePalma, D., 417
deProspero, J. D., 380
Derot, M., 1381
DeSanto, D. A., 651
deSantos, L. A., 181, 195, 206
DeSeze, S., 572, 582, 993
Desmarais, M. H. L., 445
Desmond, M. M., 800
Dessecker, C., 722
Detrick, J. E., 831
de Veber, L. L., 1189
DeVilliers, D. R., 387
Devine, K. D., 242
deVries, A., 501
Dewar, J. P., 330
Dewbury, K. C., 966
DeWet, I. S., 35
Dewing, S. B., 517
Dhar, N., 1073
Diamond, H. S., 539
Diamond, L. K., 1300, 1303
Diamond, P. E., 183
Diaz-Jouanen, E., 557
Dichiro, G., 396
Dickerson, W. W., 1386, 1387
Dickson, D. D., 964
Dickson, E. C., 793
Dickson, F. D., 417
Diepeveen, W. P., 380
DiGeorge, A. M., 1182
Diggs, L. W., 1052
Dihlmann, W., 553
Dikland, R., 380
Dillehunt, R. B., 1397
Dillon, R., 1331
Dimas, M. D., 380
Dine, M. S., 1362
Diner, W. C., 642
DiTata, D., 442, 506, 545
Dixon, A. S., 445, 554
Dixon, A. St. J., 453, 457, 532, 552, 572, 905
Dixon, T. F., 1100
Doan, C. A., 1074
Dobbs, W. J., 909
Dockeray, G. C., 651
Dockerty, M. B., 273, 380, 994
Doctor, V. M., 83
Dodd, G. D., 208, 1313, 1314
Dodds, W. J., 557, 572, 683, 905
Dodge, H. W., Jr., 1403
Dodge, J. A., 1181
Dodge, O. G., 993
Dodge, W. F., 1095
Dodson, J. H., 1397
Doe, R. P., 879
Doege, T. C., 1189
Dogaeva, M. A., 1403
Dohn, D. F., 965
Dolan, P. A., 208
Dolin, S., 330
Donaldson, J. S., 1286
Donhauser, J. L., 1080

Donnell, G. N., 916
Donnelly, B., 645
Donnelly, M., 1300
Donner, M. W., 62, 107, 1325
Donner, R., 380
Doppman, J. L., 457, 909
Dorfman, H. D., 32, 35, 57, 158, 165, 206, 383, 622, 1108
Dorfman, R. F., 330
Dorst, J., 1201
Dorst, J. P., 1181, 1201, 1332, 1386
Doty, S. B., 1201
Douady, D., 1357
Doub, H. P., 701
Dow, C. J., 1331
Dowling, E. A., 256
Doyle, F. H., 830, 935, 951, 1172
Drake, T. G., 912
Drapeau, P., 1213
Drash, A. L., 183
Drescher, E., 124
Dreskin, E. A., 935
Drew, C. E., 642
Dreyfuss, J. R., 722
Drezner, M. K., 913
Drinker, C. K., 181
Drinker, K. R., 181
Driver, J. R., 1338
Dubin, A., 913
DuBois, 1334
DuBois, J. J., 805
Dubrisay, J., 1381
Ducroquet, R., 1377
Dudding, B. A., 1189
Dudley, H. R., Jr., 994
Duff, I. F., 659
Duffy, J., 387
Duggan, C. A., 1343
Dujarrier, L., 993
Dukes, E., 1325
Dunbar, J. S., 1239
Duncan, H., 554, 843
Duncan, J. W., 499
Duncan, T. R., 1172
Dunea, G., 557
Duner, H., 375
Dunn, E. J., 242, 255, 257, 622
Dunn, F. H., 1343
Dunn, H. G., 1033
Dunn, J. S., 1080
Dunn, V., 951
Durand, P., 1182
Durenne, J. M., 1095
Duret, M. H., 1095
Durham, D. G., 1334
Dusenberry, J. F., Jr., 1355
Dutta, P., 1188
Duvall, A. J., III, 330
Duwell, H. J., 909
Dwosh, I. L., 530
Dyckman, J., 623
Dyke, C. G., 968, 1314
Dykes, R. G., 1190
Dykstra, O. H., 1072
Dymling, J. F., 1181
Dymock, I. W., 582
Dziadiw, R., 1080
Dziewiatkowski, D. D., 227

Earle, K. V., 1432
Eaton, L. M., 912
Eaton, W. L., 1326
Edeiken, J., 11, 35, 101, 158, 212, 417, 504, 507, 617, 682, 711, 757, 857, 938, 966, 1131, 1146, 1282, 1359, 1365, 1381, 1436
Edgerton, M. T., 1184
Edmonds, J., 479, 506
Edmonson, A. S., 1028
Edstrom, G., 479, 506, 517
Edwards, J. H., 1084, 1189
Eek, S., 817, 858
Eells, R. W., 1140
Egeland, J. A., 1180, 1203, 1224, 1227, 1239
Egeland, R., 1239
Eggers, G. W. N., 1381
Eggstein, M., 552
Ehrenfried, A., 1402
Ehrlich, G. E., 659
Eickhoff, U., 1365
Eiger, M. S., 951
Einstein, A., 499
Einstein, A. B., 312
Eisenberg, E., 881, 909, 951
Eisenberg, S. H., 242, 257, 622
Eisenstadt, H. B., 1381
Eistein, R., 659
Ejsmont, K., 256
Ekengren, K., 913
Eldridge, R., 1180, 1182, 1224, 1227, 1239, 1365
Eldrige, R., 1199, 1203
El-Gindi, S., 242
Elguezabel, A., 35
Elissalde, B., 1381
El-Khoury, G. Y., 700
Elkin, M., 501, 572, 614
Ell, P. J., 340, 681
Ellefsen, F., 479, 506
Ellenbogen, N. E., 1059
Elliot, G. B., 380
Ellis, R. W. B., 1227
Ellman, M. H., 572
Ellman, P., 457, 504
Elmore, S. M., 312, 1357
Eloesser, L., 631
elSallab, R. A., 445
Elsbach, L., 1181
Elwood, J. S., 457
Emami-Ahari, Z., 1357
Emmerson, B. T., 557
Enachesco, L., 722
Engel, M. B., 495
Engelmann, G., 1410
Engels, E. P., 319
Engeset, A., 774
Engh, C. A., 116
Engleman, E. P., 500, 530, 532
English, P. B., 552
Enloe, L. T., 909
Enneking, W., 32
Enneking, W. F., 54, 181, 373
Ennis, J. T., 1052
Enriquez, P., 373
Ensinck, J., 1144
Enterline, H. T., 391
Entin, M. A., 1303
Enzinger, F. M., 121, 122, 256

Enzinger, R. M., 116
Epps, C. H., Jr., 709
Epstein, B. S., 340, 391, 768
Epstein, D. A., 1402
Erdheim, J., 554
Erich, J. B., 181
Erikson, U., 723
Ernsting, G., 1397
Escamilla, R. F., 1163, 1266
Esguerra-Gomez, G., 805
Esteban, J., 227
Estrada, R. L., 1377
Estren, S., 311, 1303
Etter, L., 1330
Evans, A. N. W., 1381
Evans, C. D., 659
Evans, D. K., 623
Evans, D. L., 699
Evans, D. M. D., 993
Evans, E. B., 1095
Evans, F. G., 975
Evans, G. A., 114
Evans, H. W., 517
Evans, J. A., 387
Evans, J. W., 684, 1047
Evans, K. T., 319
Evans, P. G., 1262
Evens, R. G., 909
Everall, J. D., 532
Evison, G., 319
Ewing, J., 83, 149, 263, 273, 297, 342
Ewing, M. R., 330
Exeter, N. H., 705
Exton-Smith, A. N., 845
Eyler, P. W., 330
Eyre-Brook, A. L., 259, 1028
Eyre-Brooke, A. L., 280
Eyring, E. J., 951
Eysholdt, K. G., 1017, 1028

Faber, H. K., 793
Faget, G. H., 805
Fahey, J. L., 312, 582
Fahlgren, H., 1107
Fairbank, H. A. T., 1192, 1230, 1277, 1289, 1326, 1396, 1397
Fairbank, T., 601, 1181, 1230, 1239, 1245, 1277
Fairbank, T. J., 1286
Fairbanks, V. F., 582
Fairchild, R. D., 699
Faires, J. S., 571, 573
Fairly, G. H., 319
Fairweather, D. B. I., 873
Falconer, M. A., 995
Falek, A., 1322
Falk, S., 277
Falkenheim, C., 912
Fallet, G. H., 577
Falls, H. F., 1181, 1313
Falvo, K. A., 1289
Fanconi, G., 923, 951, 1302
Fanning, J. P., 1073
Fansleau, H. A., 121
Farber, S. J., 475
Farhat, S. M., 517
Farman, J., 138, 387
Farr, G. H., 206
Farrell, C., 212

Farrell, F. J., 1181
Farrell, J. J., 924
Farreras-Valenti, P., 375
Farth, W. F., 572
Fateh, H., 1074
Fathie, K., 387
Faure, C., 387, 1401
Favara, B. E., 1184
Fawcett, J., 1080
Fechner, R. E., 165, 206, 257, 273, 383, 622, 659
Feher, M., 445
Feigenbaum, J., 267
Feingold, M., 1189
Feist, J. H., 623
Feldman, E. B., 570
Feldman, F., 49, 90, 373, 1073
Feldman, L., 1331
Feldman, S. M., 642
Feller, E. R., 501, 614
Fellows, K. E., 642
Felman, A. H., 1180
Fels, S. L., 1188
Felson, B., 1004
Fenn, P. A., 867
Fenton, S. S. A., 909
Ferguson, A. B., 181, 696, 1377
Ferguson, A. D., 1059
Ferguson-Smith, M. A., 1266
Fernandez, P. C., 631
Ferraro, E. F., 330
Ferreira, F. S., 380
Ferriman, D., 1341
Fetterman, L. E., 1432
Fetterman, R. A., 1377
Fevre, M., 1028
Fialho, F., 391
Fidler, M. W., 709
Field, C., 1332
Field, C. M. B., 1331
Field, M. H., 935
Figi, F. A., 330
Filipsson, R., 913
Finby, N., 501, 614, 1269
Finch, C. A., 582
Finch, S. C., 582
Findelstein, J. B., 181
Fine, G., 1235
Finkel, A. J., 1125
Finkel, M. P., 181
Finkelstein, J. B., 1131
Finklestein, J. B., 206
Finnegan, T. L., 696
Finnerud, C. W., 375
Finnestrom, O., 1159
Fiore, J. M., 149
Firooznia, H., 396, 642
Fischer, D. S., 642
Fischer, H., 62
Fischer, W., 1365
Fish, A., 659
Fishbein, M., 1144
Fisher, B., 378
Fisher, E. R., 378
Fisher, H. P., Jr., 642
Fisher, J. H., 387
Fisher, M. T., 951
Fisher, O. D., 873
Fisher, R. I., 699

Fisher, R. L., 696, 1436
Fitch, K. D., 1184
Fitz, C. R., 1359
Fiveash, A., 208
Flanagan, B., 832
Flandin, C., 1243
Flavell, G., 642
Fleisch, H., 858, 1296
Fleishmajer, R., 124
Fleming, J. L., 1235
Fletcher, A. G., Jr., 659
Fletcher, B. D., 1181, 1332
Fletcher, D. E., 440
Fletcher, E., 659, 1326
Fletcher, G. H., 1125
Fletcher, R. F., 923
Florkiewicz, L., 722
Flory, C. D., 1148
Flynn, M. J., 993, 1351
Follis, R. H., 846
Follis, R. H., Jr., 1289
Fong, E. E., 1421
Font, R. L., 256
Fontaine, G., 1335
Fontaine, R., 722
Foote, F. W., 255
Foote, F. W., Jr., 208, 330, 378, 383, 993
Forbes, A. P., 1286
Forbes, C. D., 604
Forbus, W. D., 793
Ford, D. K., 532
Ford, N., 1210
Fordham, E. W., 645
Forestier, F., 507
Forestier, J., 517, 530, 552, 553
Forgacs, S., 554
Forkner, C. E., Jr., 333
Forkner, C. E., r., 480
Forland, M., 1180
Fornasier, V. L., 149
Forrest, J., 623
Forrester, D. M., 1107
Forrester, R. M., 1268
Forrestier, M., 808
Forster, W. G., 1101
Forsyth, C. C., 480
Fos, S., 1184
Foster, J. B., 631
Foster, J. B. T., 1080
Foster, L. S., 993
Foster, S. E., 1074
Foukas, M., 1377
Foulk, W. T., 617
Fourman, P., 660
Fournier, A., 722
Fowler, F. D., 1365
Fowler, G. B., 1154
Fox, J. E., 391
Fox, J. M., 659
Fox, L., 727
Fox, T. A., 935
Foxworthy, D. T., 530
Fraccaro, M., 1225
Fraenkel, G. J., 601
Fraker, K., 557
Frame, B., 530, 843, 853, 1235
Franceschetti, A., 1184
Francis, C. C., 1149
Francis, K., 183, 227

Francis, K. C., 255, 993
Francisco, C. B., 107
Francois, R., 1214
Francone, M. V., 138
Franghenheim, P., 1028
Frangione, B., 311
Frank, G. R., 257, 622
Frank, J. L., 709
Frank, L., 532
Frankel, R. S., 295
Franken, E. A., Jr., 1403
Franklin, E. C., 311
Franklin, E. L., 1397
Franklyn, P. P., 1438
Frantz, A. G., 913
Franzen, S., 375
Fraser, D., 872, 935
Fraser, E., 909
Fraser, F. C., 1184, 1227, 1230, 1239
Fraser, G. R., 1268
Fraser, J., 727
Fraser, R., 6, 1377
Frazell, E. L., 993
Freehafer, A. A., 517, 698
Freeman, C. D., Jr., 1338
Freeman, J. M., 1386
Frei, E., III, 333, 480
Freiberger, R. H., 35, 54, 508, 651, 698, 1189, 1190
Freidman, B., 637
Freiman, D. G., 808
Freire-Maia, N., 1225
Frejka, B., 1326
French, N. Y., 1149
Frendel, E. P., 457
Frenkel, E. P., 313
Freriks, D. J., 817
Freudlich, I. M., 504
Freund, E., 183, 965
Freund, U., 553
Freundenthal, W., 659
Freundlich, I. M., 905
Frey, K. W., 909
Frich, J. C., Jr., 373
Fried, A., 1266
Fried, B. M., 1310
Friedell, H. L., 354
Frieder, M., 121
Friedl, E., 319
Friedlander, J., 1331
Friedman, B., 256, 267
Friedman, E., 951
Friedman, J., 1347
Friedman, M., 651, 845, 867, 993, 1181, 1182
Friedmann, A. J., 1268
Fries, G. T., 330
Fries, J. F., 475
Fries, J. W., 1021
Frigo, M., 1028
Frimpter, G., 951
Front, D., 951
Frost, H., 843
Frost, H. M., 5, 227, 852, 853
Frost, J. F., 149
Frost, T. T., 768
Fruchter, Z., 722
Fry, E. I., 1154
Fu, Y., 373

Fu, Y. S., 256
Fukuma, H., 256
Fulginiti, V. A., 1239
Fulkerson, J. P., 858, 1296
Fuller, D. E., 1334
Funderburk, E. E., Jr., 808
Funk, E. H., 964
Furey, J. G., 517
Furnival, C. M., 340

Gabrielsen, L. H., 858
Gage, H. C., 696
Galbraith, R. M., 617
Gall, J. C., Jr., 1182, 1189, 1233
Gallagher, T., 805
Gallardo, H., 83
Galloway, J. D. B., 651
Gama, G., 1381
Ganda, O. P., 445
Garavaglia, C., 1085
Garceau, G. J., 107
Garcia, M., 95
Garcia Capurro, F., 799
Garcia-Moral, Carlos A., 383
Gardner, E. J., 1433
Gardner, F. H., 325
Gardner, H., 1030
Gardner, L. I., 1303
Gardner, M. J., 966
Gareiso, A., 1243
Garland, L. H., 805
Garn, S. M., 830, 1154, 1188
Garriga, S., 1302
Garrod, A. B., 554
Garrow, J. S., 909
Gartner, R. J. W., 552
Gastou, P., 375
Gates, D., 700
Gates, D. J., 499
Gatter, R. A., 572, 905
Gattereau, A., 1332
Gatti, R. A., 1183, 1239
Gaucher, A., 1429, 1430
Gaucher, P., 1429
Gaull, G. E., 951, 1331, 1332
Gauthier, J., 571
Gavras, M., 698
Gawlik, Z., 256
Gay, B. B., 1343
Gay, B. B., Jr., 1201
Gayler, B. W., 62, 107, 242, 1100
Gazale, W. J., 659
Gehweiler, J. A., 651
Geiger, D. W., 935
Geller, W., 1331
Gellis, S. S., 1189
Gelman, M., 242
Genant, H. K., 195, 445, 475, 705
Genieser, N. B., 642
Genner, B. A., 116
Gerald, B., 798, 1189, 1322
Gerald, B. E., 1387
Gerald, P. S., 1189
Gerber, F. H., 340
Gerdon-Cohen, J., 830
Gerrard, J. W., 817, 1351
Gerritsen, T., 1322, 1331
Gershon-Cohen, J., 1239
Gershuni, D. H., 696

Gerster, J. C., 577
Gerundo, M., 107
Geschickter, C. F., 83, 206, 208, 223, 257, 265, 273, 280, 1108
Gewurz, H., 754
Ghormley, R. K., 165, 651, 699, 964, 994, 1095
Ghrist, D., 417
Giaccai, L., 1385
Giannini, R. E., 138
Gibbons, T. G., 623
Gibbs, D. D., 642
Gibson, H. J., 445, 457
Gibson, J. B., 1332
Gibson, T., 642
Giedion, A., 951, 1180, 1181, 1207, 1286
Giffen, H. K., 1338
Gikas, P. W., 380
Gilday, D. L., 49, 115, 1402
Gilford, H., 1438
Gill, A. B., 696
Gill, A. J., 532
Gill, J. A., 798
Gillespie, J. B., 1397
Gilman, S., 1073
Gilmer, W. S., 1028, 1108
Gilmer, W. S., Jr., 808
Gilsanz, V., 1377
Gimlette, T. M. D., 1166, 1168
Ginsberg, M., 475
Ginsberg, M. H., 445
Ginsburg, L. D., 1338
Giordano, A., 1289
Girdany, B. R., 1181, 1190, 1230, 1286
Gitkis, F. L., 1227
Giuliano, V. J., 617
Giunti, A., 1017, 1028
Giustra, P. E., 54
Glasberg, S., 312
Glasser, G. H., 1181
Glay, A., 711
Glen, A. C. A., 951
Glenner, G. G., 311
Glicksman, A., 183
Glidewell, O., 325
Glimcher, M. J., 4, 722
Globus, J. H., 1387
Glover, L. P., 1330
Glynn, J. L., 35
Gmyrek, D., 1302
Go, E. B., 808
Goergen, T. G., 49, 571, 705
Goergen, T. G., et al., 530
Goethals, P. L., 242
Goff, C. W., 696
Gofton, J. P., 619
Gofton, J. P., et al., 619
Gokay, H., 1403
Gold, A. P., 1386
Gold, R. H., 545, 660, 1052, 1386
Goldberg, A., 1030
Goldberg, M. B., 1163
Goldblatt, M., 759
Golden, A., 1184
Golden, R., 1030
Goldenberg, A., 478
Goldenberg, D. B., 1403
Goldenberg, R. R., 35, 182, 225
Goldfisher, S., 1310

Goldhaber, P., 832
Golding, D. N., 501, 572, 614
Golding, F. C., 507, 867
Golding, J. S., 1048
Golding, J. S. R., 54, 225, 709
Golding, M., 723
Golding, P., 617
Golding, P. L., 660
Goldman, A., 698
Goldman, A. B., 698, 700, 977, 1108
Goldman, A. S., 757
Goldman, L., 881
Goldman, R. L., 121, 255, 256, 257, 391
Goldman, R. T., 330
Goldmany, A. B., 1380
Goldschmidt, B., 1302
Goldstein, G. S., 49
Gole, L., 1310
Goltz, R. W., 1313, 1338
Gomes, S. P., 1341
Goncharova, R. F., 1403
Gonnella, J. S., 375
Gonzales-Crussi, F., 1207
Goobar, J. E., 808
Good, A. E., 457, 517, 536
Good, R. A., 727, 754, 1183
Goodier, T. E. W., 504
Goodkin, R., 1021
Goodman, J., 375
Goodman, M. A., 183
Goodman, P. C., 1377
Goodreau, J. J., 340
Goodrich, E. B., 872
Gordon, D. A., 313, 471
Gordon, G. S., 881
Gordon, H., 313
Gordon, I. R. S., 1362
Gore, D. R., 49
Gore, I., 1331
Gorlin, F. J., 1338
Gorlin, R., 1227
Gorlin, R. J., 517, 1180, 1181, 1182, 1184, 1225, 1286, 1313, 1338, 1340
Gorlin, R. L., 1313
Gorman, A. A., 817
Gossling, H. R., 696
Gottlieb, J. S., 1387
Gottschalk, A., 909
Gottschalk, R. G., 227
Goudie, R. B., 660
Gould, E. P., 1365
Gould, G. M., 1334
Goust, J. M., 501
Grabias, S., 107
Grabies, S., 255
Graham, C. B., 1159, 1286, 1436
Graham, D. C., 506, 517
Graham, E. A., 1359
Graham, G., 659
Graham, J. B., 1233
Graham, J. K., 1313
Graham, R. C., 582
Grainger, R. G., 536, 976
Gram, P. B., 1235
Gramiak, R., 138
Granelli, U., 659
Grant, B. P., 1125
Grant, I. W., 457
Grant, M., 604

Grant, S., 256
Grasset, E., 609
Gratzek, F. R., 1131
Graucob, E., 1185
Grauthoff, H. J., 1365
Gray, J. M., 517
Graziansky, W., 1283
Graziansky, W. P., 722
Grebe, H., 1225
Green, B., 642
Green, G. J., 1387
Green, L. R., Jr., 1227
Green, W. T., 727
Greenberg, L. A., 800
Greenblatt, R. B., 1163
Greene, M. L., 554
Greenfield, G. B., 642
Greenfield, M. M., 651
Greenspan, A., 35
Greenspan, R. H., 1181
Greep, R. O., 5
Greer, R., B., III, 255, 257
Greer, R. B., 242
Greer, R. B., III, 622
Gregan, J. C. F., 457
Gregg, N. M., 800
Gregory, C. F., 107
Greig, D. M., 1326
Greig, W. R., 604
Greinacher, I., 1180, 1206, 1227, 1239
Grekin, R. H., 1095
Grepl, J., 1357
Greulich, W. W., 1148, 1159
Grewar, D., 845
Greyberg, R. H., 918
Grieg, W., R., 340
Griffith, G., 1326
Griffith, G. C., 832
Griffiths, G. J., 295
Grillo, L. A., 642
Grinvalsky, H. T., 1181
Grisolia, A., 517
Griswold, M. H., 330
Groh, C., 1306
Grokoest, A. W., 591
Gromer, R. C., 330
Gros, C. M., 313
Gross, B. G., 375
Gross, C. W., 1021
Gross, H., 1306
Gross, R. E., 373
Gross, S. W., 517
Grossberger, S., 327
Grossman, H., 501
Grossman, L., 1059
Grossman, M., 905
Grossman, M. S., 1185
Grossman, R. E., 1030
Gruber, G. B., 1326
Gruber, W. F., 832
Gruca, A., 1028
Grude, H. E., 1334
Gruenwald, P., 457
Grunberg, L. A., 1377
Gruneberg, H., 1347
Gruner, O. C., 993
Gruskay, F. L., 754
Guccion, J. G., 256
Guilleminet, M., 1028

Guiot, G., 993
Guise, E. R., 227
Guivarch, J., 1377
Gump, H., 227
Guncaga, J., 1095
Gunderson, C. H., 1181
Gunther, H., 1340
Guntheroth, W. G., 1381
Gupta, D. N., 208
Gupta, R. C., 843
Gurdjian, J. E., 965
Guri, D., 1144
Gurlt, E., 1289
Gutman, A. B., 557, 964
Gutstein, W. H., 1387
Guttman, S., 1184
Guttmann, L., 517
Guy, E., 1243
Guyer, P. B., 966, 968, 1310
Guzzo, F., 1359, 1381
Gwinn, J. L., 333, 1184, 1302

Haas, L. L., 1362
Haber, E., 311
Haber, S. L., 1033
Hackenbroch, M., 1235
Hacket, L. J., Jr., 995
Hacuhanefioglu, U., 1095
Hadders, H. N., 951, 958, 1181, 1335
Hadfield, G., 387, 396
Hadley, L. A., 54
Hagelstam, L., 1365
Haggars, M. E., 683, 1057
Haggart, G. E., 544
Haggert, A., 913
Hahne, H. H., 771
Haines, S. F., 912
Haining, R. G., 1174
Hajdu, S. L., 208
Hajek, J. V., 181
Halbertsma, T., 1072
Hall, A. P., 457
Hall, B. E., et al., 557
Hall, F., 417
Hall, G. S., 1387
Hall, J. E., 54, 83, 700
Hall, J. G., 1181, 1184, 1219
Hall, T., 183
Hallel, T., 698
Halliday, D. R., 149
Halliday, J., 1181
Halpenny, G. W., 923
Halpern, M., et al., 54
Halpern, S. E., 705
Halvorsen, S., 858
Ham, A. W., 1
Hambach, R., 138
Hambrick, G. W., Jr., 1310
Hamburg, A. E., 1052
Hamby, W. B., 1403
Hamilton, E., 572, 582, 617
Hamilton, E. B. D., 572, 582, 617
Hamilton, G. W., 539
Hamilton, R. D., 517
Hammarsten, J. F., 642
Hammond, W. G., 909
Hampers, C. L., 909, 935
Hampton, A. O., 994, 1181
Hamrin, B., 375

AUTHOR INDEX

Han, S. Y., 642
Hanaoka, H., 267
Hancock, J. A. H., 532
Hancox, N. M., 6
Hand, A. M., 1227
Handa, J., 1365
Handel, S. R., 205
Handmaker, H., 730
Hanelin, J., 1277
Hanelin, L. G., 54
Hanissian, A. S., 1203
Hannay, M. G., 375
Hannon, R. R., 853
Hanoaka, H., 256
Hanse, J. F., 1181
Hansen, G. C., 1052
Hansen, H. G., 1300
Hansen, J. F., 951, 958
Hansen, J. L., 642
Hansen, O. H., 1182
Hansen, S. T., 517
Hanson, D. J., 909
Happle, R., 1230
Harbin, M., 696
Hardin, J. C., 1313
Hardoff, R., 951
Hardon, B., 1180
Hardy, J. D., 909
Hardy, R., 116, 1432
Hardy, S. B., 208
Hare, H. F., 544, 1330
Hargrave, R. L., 1182
Harkess, J. W., 1095
Harkins, T. D., 1388
Harkis, G. K., 1347
Harkness, A. H., 532
Harle, T. S., 330
Harley, J. B., 325
Harmer, C. L., 645
Harmon, P. H., 22
Harms, L., 1385
Harnasch, H., 1385
Harper, F. R., 642
Harper, P. V., 909
Harper, R. A., 1144
Harpey, J. P., 501
Harrell, G. T., 808
Harris, H., 935
Harris, H. A., 1230
Harris, R., 138
Harris, V. J., 817
Harris, W. H., 853, 994
Harris, W. O., Jr., 312
Harris, W. R., 679
Harrison, H. C., 1184
Harrison, H. E., 936, 1182, 1184
Harrison, M. H. M., 5, 698
Harrison, R. B., 637
Harrison, W. J., 506
Hart, F., 506
Hart, F. D., 506
Hart, M. S., 127
Hart, V. L., 1084
Hartley, J., 1107
Hartman, J. T., 499, 700, 965
Hartmann, W. H., 383, 879
Harvey, J. P., Jr., 457
Harview, J. N., 107
Harville, W. E., 457

Harwick, R. D., 656
Harwood-Nash, D. C., 1359, 1365
Haselwood, D. H., 571
Hashimoto, K., 375
Haskell, C. M., 330
Haskin, M. E., 90, 572, 993, 1351
Hass, G. M., 1100
Hass, J., 1436
Hass, R., 1436
Hasselmann, C. M., 375
Hassler, E., 1231
Hasterlik, R. J., 1125
Hastings, A. B., 881
Hatcher, C. H., 83, 111, 417
Hathaway, B. E., 557
Hathaway, W. E., 1239
Hattner, R. S., 115
Haubold, V., 909
Hauser, F. B., 504
Haussler, M., 913
Havard, C. W. H., 375
Havens, F. Z., 330
Hawes, L. E., 35
Hawk, P. B., 964
Hawkins, C. F., 1421
Hawkins, M. R., 1182
Hay, M. C., 165
Hayem, F., 1213
Hayes, J. T., 138, 1214
Hayes, R. L., 729
Haynes, C. D., 208
Haynie, T. P., 909
Hayreh, S. S., 554
Head, G. L., 993, 1351
Headington, J. T., 380
Healy, M., Jr., 1151
Heaney, R. P., 853
Heard, G. E., 1359
Heatly, C. A., 330
Hecht, H. L., 90
Heckmann, J. D., 642
Hedensio, B., 1181
Hedhammar, A., 22
Hefke, H. W., 1189
Hefner, R. A., 1188
Heilbrun, N., 1111
Heiner, J. W., 539
Heinonen, O. P., 552
Helferich, H., 1326
Hellenschmied, R., 375
Hellstrom, J., 880
Hellwig, C. A., 330
Hellyer, D. T., 817
Hemming, V. G., 771
Hench, P. K., 457
Hendel, D., 696
Henderson, E. D., 227, 242, 256, 1397
Henderson, J. F., 554
Henkin, W. A., 1048
Henneman, P. H., 1286
Hennequet, A., 122
Henrard, J. C., 553
Henrichsen, E., 1095
Henry, M. J., 757
Hensinger, R. N., 1286
Hensley, G. T., 1030
Henson, P. S., 417
Henze, K. G., 935
Heppenstall, R. B., 499, 709

Herbert, F. K., 935
Herdman, R. C., 1180, 1206, 1227, 1239
Herdt, J. R., 1314
Heremans, J. F., 310, 311
Hermel, M. B., 54, 1239
Herrmann, L. G., 844
Hersh, A. H., 1188
Herz, D. A., 1386
Herzenberg, J., 1080
Heschl, R., 335
Heselson, N. G., 1230
Hess, A. F., 861
Hesse, J. C., 1438
Hetter, H. J., 805
Heublein, G. W., 1365
Heully, F., 1429
Heurtaux, A., 651
Hewett, B. V., 1047
Hewitt, M., 659
Heys, F. M., 800
Higinbotham, N. L., 181, 255, 1131, 1140
Hilal, S. K., 49
Hilbish, T. F., 333, 1095, 1144
Hill, M. C., 1365
Hillenius, L., 1359
Hilliard, C., 1232
Hills, A. G., 912, 913, 924, 1182
Himaldi, G. M., 391
Hinck, V. C., 517
Hinkel, C. L., 1335
Hirsch, F., 1326
Hirschberg, M., 805
Hitselberger, W. E., 1365
Hitzig, W. H., 1180
Hjort, G. H., 380
Hock, W. S., et al., 1377
Hodes, P. J., 35, 101, 158, 428, 507, 682, 711, 938, 966
Hodge, A. J., 4
Hodges, F. J., 1365
Hodges, P. C., 257, 697, 714, 1148
Hodgkinson, A., 909, 1033
Hodson, C. J., 905, 924
Hoefnagel, D., 1189
Hoerr, N. L., 1149
Hoffbauer, F. W., 617, 660
Hoffman, F. L., 181
Hoffman, H. C., 340
Hoffman, R. R., Jr., 62
Hogan, J. M., 905
Hogeman, K. E., 1181
Holbrook, W. P., 417
Holden, J. D., 1184, 1338
Holder, A. K., 832
Holdsworth, C. D., 660
Holdsworth, D. E., 572, 591
Hollan, P., 227
Hollander, J., 557
Hollander, J. L., 428, 440, 483, 571
Hollin, S. A., 517
Holling, H. E., 557
Hollister, D. W., 1201
Holly, L. E., 1392
Holman, G. H., 817
Holman, G. J., 642
Holmann, G., 1235

AUTHOR INDEX

Holmes, B., 727, 754
Holmes, L. B., 1233
Holmes, R. A., 642
Holmstrom, E. G., 1131
Holt, J., 479
Holt, J. F., 808, 1180, 1181, 1214, 1242, 1357, 1358, 1359, 1365, 1386
Holt, J. R., 1334
Holtermueller, K., 1207
Holubar, K., 659
Honet, J. C., 517
Hong, R., 1183
Horan, F. T., 354
Horler, A. R., 457
Horn, J. S., 35
Horn, R. C., Jr., 330, 659
Horne, C., 183, 259
Horner, R. H., 1181
Horowitz, I., 517, 771
Horowitz, S. J., 951
Horton, B. T., 149
Horwith, M., 951
Horwitz, H., 601
Horwitz, T., 391, 631
Hosain, P., 340
Hosking, G. E., 572
Hosoi, K., 1357
Hostetler, J. A., 1180, 1199, 1239
Houang, M. T. W., 845
Hough, G., 1084
Houkom, S. S., 417
Houser, W., 1386
Houston, C. S., 1189, 1351
Howard, J. E., 880, 923, 1100
Howel, J. B., 1313
Howell, J. B., 1313, 1338
Howell, R. R., 557
Howland, J., 860
Howorth, B., 1084
Howorth, M. B., 696, 798, 1105
Hoyningen-Huene, C. B., 557
Hu, C. H., 1080
Huang, T., 1331
Hubault, A., 572, 582, 993
Hubble, D., 1300
Huepel, R. C., 1416
Huffer, W. E., 843
Hug, I., 1095
Hughes, E. S., 703
Hughes, G. N. F., 1201
Hughes, G. R., 554
Hughes, J. L., 968
Hughes, R. E., 571
Hughes, R. L., 1365
Huguley, C. M., Jr., 1073
Hulu, N., 312
Humphrey, J. G., 319
Hunder, G. G., 557, 683
Hunermann, C., 1230
Hunt, A. D., Jr., 829
Hunt, D. D., 417, 757
Hunt, E. E., Jr., 1149
Hunt, J., 1235
Hunt, J. C., 1365
Hunter, D., 965, 994, 1245
Hunter, T., 517
Hured, D., 311
Hurt, R. L., 1396
Husson, G. S., 1227

Hustu, H. O., 391
Hutcheson, J., 604
Hutchinson, J., 1326, 1438
Hutchinson, L. A., 1230
Hutter, R. V. P., 181, 225, 255, 383, 993
Huvos, A. G., 165, 206, 225, 255, 373, 1140

Iancu, T. C., 951
Ibsen, K. J., 1095
Ichiseki, H., 1436
Imerslund, O., 1303
Immelman, E. J., 683, 717
Inclan, A., 1095
Ingelmans, P., 1286
Ingelrans, P., 995
Inglis, K., 1380
Ingram, V. M., 1065
Inman, V. I., 1397
Innis, J. J., 700
Inoue, T., 722
Iob, L. V., 1101
Iribe, A., 256
Irvin, W. S., 475
Irvine-Jones, E., 1331
Isdale, I. C., 687
Isersky, C., 311
Isitman, A. T., 642
Isobe, T., 309
Israel, E., 1095
Israel, H. L., 808
Israel-Asselain, R., 1377
Itano, H. A., 1047
Ivanissevich, O., 799
Ives, E. J., 1351
Ivins, J. C., 127, 273, 300, 385
Ivinsk, J. C., 54
Izatt, M. M., 696

Jabbour, J. T., 1343
Jackson, D., 1144
Jackson, D. C., 319
Jackson, D. M., 319, 1174
Jackson, H., 456
Jackson, H., Jr., 297
Jackson, J. B. S., 149
Jackson, M. A., 1355
Jackson, R. P., 57
Jackson, W. P. U., 1277
Jackson, W. T., 1095
Jacobasch, G., 1302
Jacobs, B., 709
Jacobs, J. E., 138
Jacobs, P., 225, 479, 506, 966, 1084
Jacobs, P. A., 255
Jacobs, P. H., 375
Jacobsohn, L., 242
Jacobson, B. M., 554
Jacobson, D., 846
Jacobson, H. G., 327, 1074, 1095, 1331, 1334, 1359
Jacobson, S. A., 255
Jacobson, W., 768
Jacqueline, F., 495, 507, 517, 544
Jaffe, H., 554
Jaffe, H. J., 1095
Jaffe, H. L., 35, 54, 62, 83, 90, 106, 107, 111, 113, 116, 127, 149, 181, 206, 210, 223, 255, 257, 259, 263, 267, 273, 300, 313, 354, 380, 383, 387, 622, 651, 965, 994, 995, 1028, 1030, 1256
Jaffe, J. H., 35
Jaffe, M., 181
Jaffee, I. S., 642
Jaffres, R., 582
Jahn, R., 935
Jajic, I., 545
Jakovcic, S., 554
Jamane, B., 242
James, A. E., Jr., 805, 1340
Jameson, R. M., 517
Jamsen, M., 1203
Janda, W. E., 517
Jankulov, D., 722
Janower, M. L., 993, 1340, 1351
Jansen, M., 1235, 1277
Janssens, P. G., 771
Jarvi, O. H., 208
Jarvis, J. L., 1416
Javid, B., 1357
Jayson, M. I., 445, 506
Jayson, M. I. V., 453
Jedden, H., 182
Jeffrey, M. R., 506, 554
Jensen, F., 538
Jensen, O. A., 538
Jensen, P. A., 582
Jensen, P. S., 572, 584
Jervis, G. A., 1322
Jessar, R. A., 428
Jessner, M., 1338
Jeune, M., 1192, 1207
Jimenea, C. B., 530
Jing, B., 1314
Jirasek, A., 1033
Joanny, J., 1397
Job, C. K., 805
Joffe, N., 1365
Johnson, A. C., 805
Johnson, A. D., 90
Johnson, E. W., 267
Johnson, E. W., Jr., 54
Johnson, H. H., Jr., 1313
Johnson, H. M., 659
Johnson, J. G., 1433
Johnson, J. R., 909
Johnson, K. A., 845
Johnson, K. W., 295
Johnson, L., 107, 149, 263
Johnson, L. C., 116, 387, 609, 1108
Johnson, P. M., 49, 645
Johnson, R. E., 280
Johnson, S., 805
Johnson, T. H., 539
Johnston, C. C., 913
Johnston, C. C., Jr., 1181
Johnston, G. S., 295
Johnston, J. A., 1163
Johnston, R. B., Jr., 754
Joiner, C. L., 923
Jolly, S. S., 1146
Jones, A. E., 295
Jones, B. S., 499, 709
Jones, D. M., 623
Jones, D. T., 517, 1239
Jones, E. E., 1331
Jones, F. E., 659
Jones, J. D., 872, 1181

Liebaldt, G., 1207
Lieber, A., 62, 1089
Lieberg, O. U., 517
Lieberman, P. H., 378
Liebman, C., 1095
Liebow, A. A., 872
Lie-Injo L. E., 1072
Liess, G., 1283
Lievre, J. A., 1381
Light, W., 923
Likhoded, V. I., 723
Lilia, B., 1181
Lilien, D., 115
Lin, J. P., 1021
Lin, R., 445
Lin, S. R., 504, 808
Lindbom, A., 54, 1392
Lindell, T. D., 723
Lindemann, K., 1235
Lindenberg, Richard, 1184
Linder, L., 1181
Lindquist, B., 845
Lindsay, S., 1030
Linduall, N., 54
Ling, S. R., 1172
Lingg, C., 545
Linossier, M., 1199, 1219
Linovitz, R. J., 49
Linsman, J. F., 1146
Lipscomb, P. R., 35, 158
Lipsey, A. I., 375
Lisbona, R., 49, 729
Lisser, H., 1163, 1266
Lissia, G., 375
Lister, J., 457
Litchman, H. M., 623
Little, E. G., 375
Little, H. A., 313, 471
Little, K., 6
Littler, W. H., 1380
Littman, A., 1189
Litwin, S. D., 457
Liu, E., 935
Liu, S. H., 853
Liyanage, S. P., 453
Ljungberg, O., 879
Lloyd, P. C., 879
Lloyd-Roberts, G. C., 1028
Loaec, Y., 1192, 1207
Lobstein, J. G. C. F. M., 1289
Locke, E. A., 965
Locke, G. B., 457
Lockie, L. M., 445
Lodmell, E. A., 1359
Lodwick, G. S., 9, 107
Loewenthal, M., 295, 375
Loewi, G., 608
Logan, W., 1080
Lohr, B., 909
Lohr, W., 722
Loitman, B. S., 35, 54
Long, R. E., 771
Longmore, J. B., 1268
Longscope, W. T., 808
Looney, W. B., 993, 1351
Loop, J. W., 1359
Looser, E., 857
LoPresti, J. M., et al., 846
Lorber, A., 457

Lord, T., 1181
Lorentzon, R., 383
Lospalluto, J., 506
Lotom, M., 1266
Loubry, P., 122
Love, F., 1332
Low-Beer, A., 1326
Lowe, H. G., 499, 700
Lowell, D. M., 1030
Lowell, F., 1148
Lowney, E. D., 445
Lowrey, C. W., 700
Lowry, K., Jr., 208
Lowy, M., 651
Lozner, E. L., 312
Lubs, H. A., 1181
Lubtezki, D., 711
Lucas, G. L., 1181
Lucas, J. F., 1140
Lucas, P., 905
Luck, V. J., 881
Lumb, G. A., 923, 1095
Lumberton, N. C., 1377
Lumsden, K., 506
Lund, B. A., 387
Lunn, H. F., 771
Luntz, M. H., 506
Luppi, A., 457
Lutwak, L., 1326
Lutz, W., 375
Lyall, J., 993
Lynch, H. T., 1181
Lyon, G., 1182, 1381
Lyon, J. A., Jr., 1343
Lysakowska, J., 90, 273

Macaluso, M. P., 1033
MacCarthy, J. M. T., 1332
MacCarty, C. S., 391
MacCarty, W. C., Jr., 1387
MacDonald, L., 257
MacDonald, R. A., 582
MacDonald, W. B., 1184
MacEwen, G. D., 698, 1436
Mach, K., 659
Macher, E., 1230
MacIver, J. E., 1048
Mack, R. P., 267
MacKenzie, D. H., 121, 380, 387
Mackey, E. A., 800
MacLachian, E. A., 912
MacLean, A. D., 849
MacLellan, D. L., 54
Macleod, W., 1184
Madigan, R., 1403
Madsen, E. T., 1377
Maeyama, L., 256
Maffucci, A., 1244, 1262
Magill, F. B., 1303
Magsamen, B. F., 867
Magyar, E., 445
Maher, F., 1235
Mahloudji, M., 1184
Mahoney, C. P., 1387
Mahoubi, S., 1230
Mahoudeau, D., 1381
Maiello, M., 1322
Mainzer, F., 445
Mair, W. F., 1326

Maisel, B. W., 127
Maisel, W., 138
Maisels, D. O., 1181
Maitland, D. G., 1230
Maitrepierre, J., 571, 843
Mak, E-B., 722
Malawista, S. E., 557, 754
Maldague, B. E., 54, 703
Malers, E., 1310
Malghem, J., 703
Malghem, J. J., 54
Mall, J. C., 623
Mallory, G. K., 582
Mallory, T. B., 879
Maly, V., 138
Mamelle, J. C., 122
Manchon, F., 554, 1146
Manciaux, M., 1214
Mande, R., 122
Mandel, L., 554
Mandell, A. J., 1423
Mandell, G. A., 1365
Maner, G. D., 829
Mangalik, V. S., 380
Mankin, H. J., 107, 255, 457, 499, 572
Mann, J. B., 912, 913, 924, 1182
Marble, A., 867
Marche, J., 536
Marchetta, F. C., 330
Marcial-Rogas, R. A., 387
Marcove, R., 183, 227
Marcove, R. C., 165, 181, 225, 227, 255, 256, 373, 383
Marcus, H., 1387
Mardock, M. G., 54
Marek, F. M., 116
Margulies, M. E., 456
Margulies, S. I., 506
Marie, J., 122, 1381, 1438
Marie, P., 1191, 1416
Marinozzi, V., 1072, 1080
Mariteaux, P., 1192
Markiewicz, C., 124
Marks, I. N., 683, 717
Marks, S. C., 1347
Maroteaux, P., 1180, 1181, 1182, 1199, 1201, 1203, 1206, 1210, 1219, 1239, 1244, 1267, 1269, 1335, 1351, 1354, 1397, 1401
Marsas, B. E., 913
Marsh, B. W., 54
Marshall, G., 637
Marshall, W. C., 951
Martel, W., 456, 457, 479, 499, 506, 517, 659, 700
Martin, C., 1322
Martin, C. I., 642
Martin, E. D., 709
Martin, J. H., 867
Martin, M. M., 631
Martin, N. H., 905
Martin, S. H., 1362
Martineau, J., 1334
Martinez, A., 723
Martin-Scott, I., 1338
Martinson, F. D., 330
Martland, H. S., 181, 1125
Martoni, L., 1300
Martsolf, J. T., 1343
Marty, R., 340

Marvel, J. P., 499
Marvel, J. P., Jr., 709
Marwah, V., 417, 757
Marx, L. H., 1030
Marx, W. J., 617, 660
Maseritz, I. A., 280
Maseritz, I. H., 1108
Masi, A. T., 475
Mason, A. S., 660
Mason, G., 1188
Mason, R. M., 532, 554, 572
Mason, R M., 554
Massa, E. V., 208
Massaro, A. F., 702
Massaro, D., 1377
Masse, G., 1149
Masse, J. L., 1365
Masson, P., 1357
Mastragostino, S., 391
Mastrogostino, S., 396
Masuda, M., 256
Mata, J., 604
Matheson, I., 1397
Matthews, J. A., 660
Matthews, M. J., 1377
Matthews, T., 517
Matles, A. L., 158
Matson, D. D., 1181, 1365
Matsoukas, J., 698
Matsuno, T., 182
Matthews, B. F., 609
Matthews, J. A., 660
Matthews, M. J., et al., 1377
Matthews, T., 517
Matthias, H-H., 1230
Mattioli, M., 475
Matxunaga, R., 1080
Maudsley, R. H., 114, 457, 1277
Maugey, F., 1213
Maurer, H. J., 319
Maurer, H. S., 1073
Maurer, R. C., 709
Maurer, R. M., 1325
Maurer, W. J., 909
Maxfield, W., 354
Maximow, A., 1
May, C. D., et al., 845
Mayer, L., 57
Mayne, J. G., 532
Mayne, V., 1230
Mayoral, A., 805
Mazabraud, A., 256
Mazur, A., 845
McAfee, J. G., 1325
McAlister, W., 1201
McAllister, W. H., 1073, 1201
McArthur, R. G., 1189
McBride, J. A., 506
McCabe, B. F., 1365
McCallus, R. I., 683
McCance, R. A., 872, 873, 1182
McCann, J. S., 1331
McCann, P., 846
McCarthy, D. J., Jr., 572
McCarty, D., 1030
McCarty, D. J., 445, 572, 905
McCarty, D. J., Jr., 557, 571, 573
McCauley, R. G. K., 705
McClain, E. I., 1403

McClendon, J. L., 1313
McCluskie, R. A., 642
McCollum, D. E., 557, 570, 659
McCollum, E. B., 858
McCombs, R. K., 49
McCormack, L. J., 273, 659
McCracken, J. P., 554
McCrary, W. E., 1095
McCredie, J., 1397
McCrorie, W. D. C., 1131
McCune, D. J., 912, 995
McDonald, J. E., 905
McDonald, J. R., 642
McDonald, L., 1332
McDonald, P., 319
McDonald, R., 1302
McDougall, A., 380
McDowell, F. W., 121
McEwen, C., 440, 506, 545, 617
McFarland, B., 1028
McFarland, B. L., 1436
McFarland, G., Jr., 35
McFarland, W., 1334
McFarlane, P. S., 754
McGarland, G. B., Jr., 255
McGavran, M. H., 242, 255, 257, 622
McGee, B. J., 1357
McGeown, M. G., 909
McGibbon, K. C., 698
McHenry, L. C., 387
McIntosh, R., 846, 1227
McIntyre, M. S., 1207
McIvor, J., 572
McIvor, J. M., 572
McJimsey, B. A., 1313
McKay, J. W., 54
McKeever, C., 843
McKenna, R. J., 181, 182
McKenzie, I., 1080
McKibbin, B., 699
McKuscik, V., 1203
McKusick, B. A., 1331, 1334
McKusick, V., 1180, 1235
McKusick, V. A., 966, 1180, 1181, 1182, 1184, 1190, 1199, 1201, 1219, 1224, 1227, 1233, 1239, 1289, 1326, 1330, 1340, 1386
McLaughlin, C. L., 309
McLaughlin, C. W., 879
McLean, F. C., 6, 881
McLeod, R. A., 49, 60, 83, 101, 182, 1386
McMaster, J. H., 183
McMurray, A., 1146
McMurray, G. A., 1423
McMurray, J. S., 754
McNew, J., 1365
McPeak, C. N., 1346
Meary, R., 256
Medley, B. E., 1386
Meema, H. E., 830, 881, 1166
Meema, S., 830
Megyesi, K., 183
Mehrotra, L., 380
Meister, L., 687
Meljer, R., 1184
Mellanby, E., 858, 917
Melnick, J. C., 1181, 1307
Melnick, J. L., 800
Melsom, R. S., 805
Melson, G. L., 909

Melton, J., 319
Mendelzun, R., 501
Mendenhall, J. T., 909
Mendl, K., 340
Mendlowitz, M., 645
Meredith, S. C., 1355
Merkow, L. P., 373
Merrill, J. P., 909, 935
Merritt, A. D., 1181
Merz, E. H., 1325
Meschan, L., 62
Meschen, I., 1230
Meskin, L. H., 1182, 1189, 1338
Messina, E. J., 683
Meszaros, W. T., 1359, 1381
Methur, M., 1146
Metrakos, J. D., 1227
Metzger, A. L., 479, 506, 660
Meurman, K. O. A., 457
Meyer, I., 993
Meyer, L., 1334
Meyer-Borstel, H., 968
Meyerding, H. W., 1089
Michaelis, L. L., 1206
Michaux, J. L., 310
Michell, I. C., 614
Michell, R. C., 501
Michelson, M. R., 499
Michener, W. M., 557
Micigovsky, B. B., 6
Mickelson, M. R., 700, 709
Middlemiss, H., 1144
Middlemiss, J. H., 554, 591, 659
Mielants, H., 557, 683
Miettinen, O. S., 993, 1351
Mike, V., 181, 225
Mikkelsen, W., 457
Milgram, J. E., 633
Milgram, J. W., 623, 710, 1033
Milkman, L. A., 857
Millard, D. R., Jr., 1181
Millard, P. H., 845
Milledge, R. D., 1387
Miller, A. S., 165
Miller, G., 1300
Miller, H. G., 935
Miller, J. L., 631
Miller, P. R., 709
Miller, R. J., 1189
Miller, S., 1235
Miller, T. R., 165, 225, 255
Miller, W. C., 1432
Miller, W. T., 656
Mills, J. A., 311
Milne, M. D., 501
Minagi, H., 1300, 1302
Mindelzun, R., 572, 614
Minear, W. L., 844
Minkowitz, F., 1380
Minkowitz, S., 1380
Miric, V., 722
Mirra, J. M., 138, 227, 255, 256, 373, 660
Mistilis, S. P., 660
Mital, N., 539
Mitchell, A. G., 923
Mitchell, C. L., 227
Mitchell, N., 700
Mitrovic, D., 582
Mitsudo, S. M., 951

Mitty, H. A., 1331
Mitus, A., 279
Mladick, R. A., 1380
Moe, J. H., 35, 1331
Moehlig, R. C., 965
Mohanta, K. D., 1365
Molle, W. E., 313
Moller-Christensen, V., 805
Mollon, R. A. B., 1377
Moloshik, R. E., 1189, 1190
Monckeberg, 105
Monnet, P., 122, 1214
Monroe, J., 1184
Montagne, J-P., 1401
Montgomery, H., 659
Montgomery, M. D., 530
Montgomery, W. W., 1021
Moolten, S. E., 1386
Moon, T. E., 279
Moore, B. H., 1365, 1380
Moore, C. V., 845
Moore, J. E., 1377
Moore, R. C., 1207
Moore, R. D., 700
Moorehead, W. G., Jr., 709
Moran, F. T., 1101
Moran, T. J., 591
Mordin, B. E., 617
Morehead, C. D., 1343
Moreira, G., 951
Morettin, L. B., 121
Morgan, D. B., 923
Morgan, H. J., 831
Morgan, J. D., 729
Mori, P. A., 1242
Morin, M. E., 54
Morin, P., 1403
Morisette, R. A., 917
Morreels, C. L., 1332
Morresan, G. M., 35
Morris, H., 706
Morris, J. M., 1397
Morris, J. W., 1144
Morris, M. L., 698
Morris, R. I., 479, 506
Morrison, A. B., 873
Morrow, C. S., 457
Morrow, G., 1331
Morrow, I. H., 817
Morse, D., Jr., 208
Morson, B. C., 1433
Mortan, A. D., 380
Morton, K. S., 54
Mortureaux, Y., 1322
Moseley, J. E., 138, 208, 354, 480, 591, 684, 1047, 1048, 1057, 1071, 1074, 1189, 1190
Moser, E., 651
Moshang, T., 1159
Mosher, J. F., 1365, 1381
Mosier, D., 1266
Moskowitz, H., 417, 757
Moskowitz, R. W., 572, 637
Moss, M. S., 1326
Motta, C., 1214
Mottet, N. K., 1381
Motulsky, A. G., 1233
Mouchet, A., 1286, 1381
Moule, B., 604
Mouledous, P., 1214
Moulias, R., 501
Mounier-Kuhn, P., 1334
Movson, I. J., 808
Mudd, S. H., 1331
Mueller, C. E., 506
Mueller, E., 121
Mueller, H. A., 49
Muheim, G., 711
Muhletaler, C. A., 506
Muir, A. R., 582
Muir, C. S., 256
Mujahed, Z., 387
Mulchahy, F., 1074
Mulder, D. W., 35
Muller, G. M., 1236
Muller, S. A., 1397
Mullick, P., 1214
Mulligan, L., 1377
Mullins, B. P., 517
Mullins, F., 242, 257, 622
Multz, C. V., 445
Munn, J. D., 935
Munro, M., 1423
Munson, P. L., 881
Murdoch, J. L., 1219
Murdoch, L., 1219
Murdock, J. R., 805
Murphy, F. P., 242, 622
Murphy, J. A., 380
Murphy, J. M., 965
Murphy, K. J., 1314
Murphy, W. R., 267
Murray, J. A., 181, 195, 206
Murray, J. E., 1334
Murray, M. R., 373, 387, 1357
Murray, R. C., 54
Murray, R. O., 620, 1087, 1331, 1334, 1359, 1397
Murray, R. S., 532
Musser, A. W., 659
Musser, J. H., 714
Muthukrishnan, N., 1355
Myers, B. W., 506
Myers, M., 457
Myerson, R. M., 808

Nachlas, I. W., 1108
Nadler, H., 1322
Nager, G. T., 1365
Nagura, S., 710
Naji, A. F., 380
Nanca, C. N., 1365
Nance, W. E., 1357
Nanda, B. K., 1365
Nasca, R., 499
Nasca, R. J., 499
Nash, H. H., 456, 517
Nassim, J. R., 1377
Nathan, D. G., 727, 754
Nathan, M. H., 885
Nathanson, L., 183, 532
National Conference Medical Nomenclature, 506
Nauhauser, E. B. D., et al, 1410
Navin, J. J., 208
Naylor, B., 1359
Nazzaro, P., 659
Neal, E. G., 585
Nedwich, A., 124
Needles, C. F., 1181, 1307
Neel, J. V., 1047
Neelon, F. A., 913
Neill, D. W., 873, 1331, 1332
Nelms, D. K., 757
Nelson, B., 729
Nelson, D. A., 312
Nelson, P., 242, 255, 257, 622
Netherlands Committee on Bone Tumours, 107
Neuhauser, E. B. D., 873, 1195
Neville, W. E., 380
Nevin, N. C., 1181
Newcombe, D. S., 1283
Newman, H., 805
Newman, P. H., 651
Newmark, H., 1107
Newton, T. H., 909
Newton, W., Jr., 279
Nezelof, C., 256
Ng, A. B. P., 582
Ngan, H., 62
Nice, C. M., 354
Nice, C. M., Jr., 1047
Nichols, G., Jr., 832
Nicholson, D. C., 572
Nieburgs, H. W., 1163
Nielson, J. L., Jr., 1245
Niemann, I., 1347
Niemann, N., 1214
Niemeyer, T., 1089
Nievergelt, K., 1224
Nilsson, L. R., 1302
Niordson, A. M., 375
Nishimoto, T., 181
Niwayama, G., 49, 445, 553, 571
Nixon, G. W., 138, 333
Nixon, J. E., 257, 622
Noack, M., 1340
Nobuhara, K., 722
Noel, H. N., 703
Noetzli, M., 923
Nomaland, R., 1313
Nora, J. J., 1189
Norby, D. E., 1189
Norcross, B. M., 445
Norden, B. E., 844
Nordenskjold, A., 1387
Nordin, B. E. C., 6, 1377
Norgaard, F., 445
Norman, A., 54, 633, 691
Norman, D., 373
Norman, M. E., 1362
Normandale, P. A., 1181
Norris, C. B., 1338
Norum, R. A., 1181
Nosheny, S. Z., 659
Notman, D. D., 475
Noto, P., 1334
Novello, A., 1028
Nowlan, F. B., 1334
Noyes, W. D., 1080
Nukko, C. K., 1145
Numaguchi, Y., 544
Nussey, A. M., 1346

AUTHOR INDEX

Nuttall, J., 700
Nyhan, W. L., 557

Oates, J. K., 532, 536
Oberling, C., 273
O'Brien, J. E., 373
O'Brien, R. M., 994
Ochs, H. D., 506
Ochsner, E. H., 1403
Ochsner, S., 457
Ockerman, P. A., 1185
O'Connell, D. J., 617, 660
O'Connor, S. J., 1017, 1028
O'Dell, C. W., 49
O'Donnell, A. A., 757
O'Duffy, J. D., 572
Oechlecker, F., 631
Oechler, H. W., 1365
Ogbeide, M., 642
Ogden, W. S., 709
Ogryzlo, M. A., 313
Oh, S. K., 1365
O'Hara, A. E., 121, 1343
Ohnsorge, J., 642
Olcott, C. T., 1330
Oldhoff, J., 208
O'Leary, J., 642
Oliveira, de S., :623
Oliver, I., 501
Oliver, R. A. M., 1306
Olken, S. M., 1107
Ollendorff, H., 1392
Ollier, L., 1244
Olpp, J. L., 1108
Olson, W. H., 49
Omenn, G. S., 479, 506
O'Neil, M. T., 570
Onitsuka, H., 62
Ophulus, W., 793
Opitz, J. M., 1180, 1181, 1185, 1322
Oppewheimer, E. H., 1331
Oreopoulos, D., 909
Ormerod, E. L., 1289
Orofino, C., 700
Ortega, L., 631
Ortiz, A. C., 585
Ortolani, M., 1084, 1085
Osborn, J. F., 701
Osgood, C. P., 517
Osserman, E. F., 309
Ossipowski, B., 375
Oster, J., 1189
Otis, J., 383
Otis, R. D., 124
Otte, W. K., 909
Ottenheimer, E. J., 1184
Owano, L. R., 391
Owen, B. J., 659
Owen, P. S., 554
Owen, R. H., 1335
Owens, W. I., 808
Ozonoff, M. B., 699, 858, 1296
Ozzello, L., 373

Pabst, H. W., 909
Pachter, M. R., 909
Packard, F. A., 965
Packer, B. D., 1334
Pagani, P. A., 1017

Page, A. R., 727, 754
Page, J., 517
Page, J. W., 456
Paget, J., 254, 964
Paige, M., 1004
Paik, C., 557
Painter, C. F., 1326
Pajewski, M., 817, 822
Palacios, E., 54, 554
Palesi, S., 1300
Pallis, C., 1365
Paloyan, D., 879, 909
Paloyan, E., 879, 909
Pantazopoulis, T. H., 698
Papadatos, C., 951
Papademetroiu, T., 1181, 1397
Papaioannou, A. C., 913
Papo, Y., 295
Papvasiliou, C. G., 642
Pardo, M., 373
Parenti, G. C., 1225
Parienty, R., 387
Park, E. A., 861, 873
Park, J. K., 312
Park, W. M., 114
Parker, A. S., 1330
Parker, B. R., 1073
Parker, F., Jr., 297
Parker, M. D., 475
Parkman, P., 1227
Parks, E. A., 846
Parks, H., 1423
Parks, J. S., 709
Parlee, D. E., 905
Paronen, I., 532
Parrot, J., 774
Parrot, M. J., 1191
Parson, W., 853, 913, 1286
Parsons, F. M., 1033
Partington, M. W., 1207
Pascher, F., 1334
Paschlau, G., 727
Pascoe, H. R., 330
Pashayan, H., 1184
Patchefsky, A. S., 383, 1377
Patel, M. R., 383
Paterson, D., 165
Paterson, D. C., 1108
Paterson, D. E., 805
Path, F. C., 35
Paton, W. D. M., 683
Patrassi, G., 1030
Patterson, C. D., 457
Patterson, J. T., 642
Patterson, R. J., 687
Patton, J. T., 536, 582
Paul, L., 1235
Paul, R. G., 935
Paul, S. S., 1214
Paul, W. D., 918
Pauli, A., 722
Pauling, L., 1047
Pautrier, M., 1334
Pavon, S. J., 256
Payne, M. A., 617, 660
Payne, T. R., 845
Pear, B. L., 354, 1030
Pearlman, A. W., 993
Pearlman, D. S., 1239

Pearlman, H. S., 383
Pearson, C. M., 457
Pearson, K. D., 312
Pearson, T. A., 313
Pease, G. L., 1300
Pedroso, R. B., 380
Peison, B., 267
Peitzman, S. J., 631
Pellarano, J. C., 1243
Peltier, L. F., 517
Pendergrass, E. P., 1365
Penning, L., 1089
Penrose, L. S., 1219, 1340, 1386
Pepin, B., 501
Perez, M. A., 378, 380
Perez-Modrego, S., 227
Perin, F., 1059
Perkins, H. R., 1100
Perlow, S., 659
Perman, G., 582
Pernow, B., 375
Perold, S. M., 582
Perri, G., 601
Perri, J. A., 457
Perry, H., 1334
Perry, H. O., 532
Perry, P. E., 965
Pesek, J., 557
Peter, J. B., 614
Peters, H. J., 1095
Peters, J. P., 918
Petersen, J., 604
Peterson, H. O., 1362
Peterson, L. F. A., 872
Peterson, O. S., 817
Peterson, O. S., Jr., 817
Peterson, R., 935
Peterson, W. C., 1338
Peterson, W. C., Jr., 1338
Petry, R., 909
Pettigrew, J. D., 1074
Pfeiffer, R. A., 1343
Phalen, G: S., 659
Phemister, D. B., 181, 223, 257, 417, 679, 680, 697, 710, 714, 1191
Phil, D., 325
Phillips, C. A., 800
Piatt, E. D., 1433
Pickard, N. S., 478, 479
Pickering, R., 754
Pickett, P. T., 570
Pickleman, J. R., 879
Picot, C., 1214
Pierce, D. S., 935
Pierce, J. A., 457
Pierson, M., 1214
Pigorini, F., 909
Pilling, G. P., 1303
Pinckney, L., 1073
Pindborg, J. J., 1184, 1340
Pinkus, H., 532
Pinto, R. S., 396
Piper, R. K., 1331
Pirnar, T., 1195
Pisani, G., 1268
Pitt, M. J., 1365, 1381
Pittman, M. R., 256
Pizzetti, M., 1214
Platt, N., 1183

AUTHOR INDEX

Plenk, H. P., 1433
Plotz, C. M., 506, 532
Pochaczevsky, R., 1207
Pock-Steen, O. C., 380
Poker, N., 1269
Pol, P., 1189, 1326
Polga, J. P., 909
Poli, E., 138
Poligard, A., 6
Pollack, A. D., 1331
Pollack, E. S., 330
Polle, C. A., 800
Polley, H. F., 557, 642, 659
Pomerance, H. H., 1183
Pomeroy, T. C., 280, 295
Pommer, G., 105, 867
Pommer, J., 6
Ponseti, I., 700
Ponseti, I. V., 499, 709, 923
Ponseti, L., 54
Ponten, B., 723
Pool, J. L., 391
Poole, A. G., 330
Popert, A. J., 532
Popovtzer, M. M., 843
Popowitz, L., 83
Poppel, M. H., 327, 832, 1074, 1095
Poppen, J. L., 391
Porter, D. C., 1331
Porter, J. M., 723
Portis, R. B., 90, 479
Posch, J. L., 1380
Poske, R. M., 530
Potanos, J. N., 391
Potchen, E. J., 909
Potts, W. J., 631
Poumeau-Delille, G., 1334
Powell, R. C., 557
Poznanski, A. K., 1182, 1188
Prader, A., 1180
Prager, R. J., 623
Praharaj, K. C., 1365
Prandoni, A. G., 138
Prat, D., 799
Pratt, A. D., 1004
Pratt, G. B., 1188
Pratt, J. H., 554
Pratt, P. W., 311
Pratt, T. L. C., 456, 517
Preger, L., 1386
Prentice, C. R., 604
Preston, J. M., 1365
Prevot, J. H., 909
Price, C. H. G., 182, 208, 259, 659, 1028
Price, L. W., 457
Prichard, J. E., 54
Priest, J. H., 1189
Pritchard, J. E., 923, 994
Pruzanski, A., 471
Pruzanski, A. W., 313
Pruzanski, W., 1233
Pryor, J. W., 1148
Psaume, J., 1189, 1286
Ptacek, L. J., 1322
Pugh, D. C., 54
Pugh, D. G., 90, 149, 165, 544, 557, 651, 881, 883, 923, 994, 1090, 1101, 1108, 1124, 1365
Pugh, D. J., 659

Pujman, J., 138
Pulvertaft, R. J. V., 330
Purisch, M., 1184
Purser, D. W., 456, 517
Pusitz, M. E., 107
Putnam, C. E., 572
Putschar, W. G., 107
Putschar, W. J. G., 387
Putti, V., 1397
Pyle, E., 1239
Pyle, S. I., 111, 1148, 1149, 1159
Pyle, W. L., 1334
Pyorala, K., 552
Pyrah, L. N., 909

Quelce-Salgado, A., 1180, 1225
Quie, P., 727
Quie, P. G., 727, 754

Raap, G., 1230
Rabhan, W. N., 116
Rachman, R., 208
Radkowski, M. A., 1033
Raim, J., 1059
Raine, D. N., 1331
Raine, G. E. T., 517
Raisz, L. G., 1347
Rajic, D. S., 1189
Rallison, M. L., 951
Ramage, D., 1386
Ramilo, J., 817
Ramot, B., 312
Ramsey, P. L., 1286
Rancier, L. F., 604
Randelli, G., 116
Rannie, I., 1262
Ranninger, K., 909
Rao, P. L., 1214
Rao, P. S., 1403
Rapp, G. R., 517
Rapp, I. H., 116
Rappaport, H., 312
Raskind, R., 256
Rasmussen, K., 501
Rassin, D. K., 1332
Ratcliff, R. G., 683
Rater, C. J., 1181
Rathbun, J. C., 872
Rathery, M., 1381
Ratliff, H. C., 698
Ratner, H., 1207
Ravault, P. P., 571
Raven, R. W., 457
Ravenna, F., 1199
Ravits, H. G., 1338
Rawson, A. J., 330
Rawson, R. J., 183
Ray, E. S., 642
Raymond, J., 340, 681
Razelli, A., 1283
Reames, P. M., 457
Reback, S., 965
Rech, M. J., 256
Record, R. G., 1084
Reddy, R., 805
Redisch, W., 683
Reed, J. O., 208
Reed, R. J., 255

Reed, W. B., 532
Reeder, W. H., 872
Reeves, J. R., 124
Reeves, R. J., 319, 1095
Regato, J. A., 277
Reggs, B. L., 830
Reichlin, M., 475
Reid, M. R., 149
Reid, R. T., 457
Reifenstein, E. C., 830, 831, 858, 879, 912, 917, 994, 995, 1027, 1095, 1181, 1277, 1286
Reifenstein, E. C., Jr., 830, 853, 913, 995, 1095, 1100
Reilly, E. B., 375
Reilly, W. A., 1233, 1266
Reinberg, S. A., 722
Reineke, H. G., 844
Reinhoff, W. F., III, 1403
Reinke, R. T., 554
Reisman, M., 1286
Reisner, D., 808
Reiter, H., 532
Remigio, P. A., 1181
Rene, R. M., 457
Rengachary, S. S., 256
Renier, J. C., 843
Rennell, C., 445
Renner, R. R., 312
Rennie, W., 700
Renton, P., 845
Resnick, D., 49, 445, 517, 530, 552, 553, 554, 571, 612
Resnick, D. L., 882
Retz, L. D., 396
Retz, L. D., Jr., 387
Reuben, M. S., 1365
Revenna, F., 1219
Reyersbach, G. C., 873
Reynolds, D. F., 532
Reynolds, G. G., 517
Reynolds, J., 1047, 1052, 1070
Reynolds, W. E., 500
Rhangos, W. C., 659
Rhinehart, B. A., 1416
Rhoton, A. L., 517
Rhyne, R. R., 623
Ribbing, S., 1396, 1410
Rich, C., 1144
Richard, R. C., 1433
Richards, W. C. D., 601
Richardson, G. O., 935
Richardson, R. E., 935
Richin, P. F., 1377
Richmond, J., 495
Ricker, W., 808
Riddervold, H. O., 1206
Riddoch, G., 964
Ridings, G. R., 277
Riggs, B. L., 872
Riggs, W., 1438
Riggs, W., Jr., 1070
Riggs, W. W., Jr., 1203
Rigler, L. G., 1131
Riley, H. D., Jr., 1326
Riley, J. D., 557
Riley, M. J., 479, 506
Rimoin, D. L., 1181, 1184, 1191, 1201, 1351

Ringrose, E. J., 1334
Rinsky, L. A., 517
Rinvik, R., 1300
Riordan, D. C., 1303
Rios-Dalenz, J. L., 121
Ritchie, G. W., 722
Rittenberg, G. M., 330, 557
Ritvo, M., 1286
Rivelis, M., 506, 508
Roaf, R., 1268
Robbins, H., 633
Robbins, S. L., 730
Robb-Smith, A. H. T., 995
Robert, F., 808
Robert, J. M., 1203, 1206, 1214
Robertson, D. E., 727
Robertson, W., 845
Robichon, J., 6
Robins, J., 552
Robins, P. R., 1331
Robinson, A., 1107
Robinson, G. C., 1189
Robinson, R., 4
Roche, A. F., 1149
Rochlin, D. B., 391
Rockett, J.F., 1070
Rodan, G. P., 532
Rodman, T., 808
Rodnan, G. P., 457, 539
Rodriquez, H. A., 122
Roeckrath, W., 1421
Rogers, C. I., 330
Rogers, J. V., Jr., 1073
Rogers, W. A., 517
Rohlfing,, B. H., 1052
Rohman, C. G., 1188
Rohmann, C. G., 830, 1154
Rohoim, K., 1144
Rohwedder, H. J., 1181, 1351
Rolleston, H. D., 1326
Rollhauser, H., 1334
Romanus, R., 517
Rona, G., 457
Roos, A., 853
Root, L., 1289
Rosai, J., 116, 380, 659
Rosen, P. A., 517
Rosen, R., 601
Rosen, R. A., 843
Rosenbaum, H. D., 965
Rosenberg, D., 122
Rosenberg, E. F., 457
Rosenberg, H. S., 208
Rosenberg, M., 456
Rosenberg, M. A., 517, 642, 645
Rosenberg, R. N., 1021
Rosenbloom, A. H., 1030
Rosenbloom, A. L., 1180
Rosenbloom, F. M., 554
Rosenblum, C., 1033
Rosengren, J. E., 181
Rosenoer, V. M., 501, 614
Rosenquist, C. J., 138
Rosenstirn, J., 1326
Rosenthal, A., 642
Rosenthal, I. M., 1073
Rosenthal, R. E., 1201
Rosenthall, L., 49, 645, 729
Rosevear, J., 1235

Roske, G., 817
Rosman, C., 375
Rosner, F., 312
Ross, A. T., 1386
Ross, C. E., 582
Ross, C. F., 387, 396
Ross, P., 1080
Rosselin, G., 1381
Rossi, E., 1334
Rostenberg, A., 659
Rotch, T. M., 1148
Rotes-Querol, J., 507, 517, 552
Roth, L. J., 313
Rothfield, N., 478
Rothman, P. E., 917
Rothman, S., 504
Rothney, E. W., 1227
Rothschild, S., 499
Rothstein, J., 457
Rothstein, T., 1403
Rotzler, A., 122
Roussakoff, A. V., 6
Rovin, S., 1181
Rowe, C. N., 683
Rowe, C. W., 1057
Rowlands, D. T., Jr., 309
Rowley, K. A., 440
Roy, C., 1351
Royer, P., 1377
Ruangivit, V., 1239
Ruangwit, U., 1180
Rubens-Duval, A., 711
Rubenstein, D., 445
Rubenstein, L., 1387
Rubenstein, M., 1334
Ruberman, W., 312
Rubin, P., 227, 1181, 1196, 1199, 1210, 1219, 1340
Rubinstein, H. M., 722
Rubinstein, J. H., 1184, 1312
Ruderman, M., 478
Rudhe, U. L. F., 916
Rudnicki, R. D., 478
Rudolph, A. J., 800
Rudowski, W., 90, 273
Ruiz, G., 138
Rundles, R. W., 319
Rushton, J. G., 35, 1403
Russell, D. G., 1387
Russell, D. S., 993
Russell, W., 1172
Rutten, F. J., 771
Rutishauser, E., 609, 1381
Ruvalcaba, R. H. A., 1387
Ruzicka, F. F., 1359
Ryan, S. F., 1397
Ryder, C. T., 1084
Ryffel, H., 722
Rynes, R. I., 572, 591
Rypins, E. L., 319

Saacebra, J. A., 380
Saba, M. M., 1365
Sabinas, A. O., 35
Sacco, J. J., 35
Sachatello, C. R., 1433
Sack, H., 909
Sage, H. H., 387
Sagher, F., 375

Sahlin, D. C., 256
Saigal, S., 1214
Sainton, P., 1416
Sairanen, E., 479
Salama, M., 242
Salama, R., 295
Saleeby, E. R., 1080
Salem, E. P., 35
Salerno, N., 1359
Salle, B., 122, 1214
Salm, R., 208
Salman, I., 387
Salmon, P. R., 506
Salmon, S. E., 308
Salter, R. B., 1084
Salvador, A. H., 256
Salvadore, A. H., 236
Salvati, E. M., 698
Salyer, W. R., 1033
Samilson, R. L., 1397
Samppinato, F., 604
Sams, A., 645
Samuelson, S. M., 1073
Sandell, H. J., 631
Sandford, J. P., 617
Sankey, H. H., 1286
Santacroce, A., 843
Santarelli, A. G., 623
Sante, L. R., 994
Santoro, A. J., 273
Santos, J. V., 696
Sapelier, J., 1214
Sareen, C. K., 1387
Sarosi, G., 879
Sarwar, M., 1365
Sashin, D., 793
Sassin, J. F., 1021
Saunders, J. B. C., 380, 1397
Sauvegrain, J., 1213
Savart, P., 1192
Saville, P. D., 831, 1377
Sawyer, P. N., et al., 723
Sawyer, W. R., 60
Saxena, K. M., 1377
Saxl, O., 340
Say, B., 1073
Scalettar, R., 532
Scandellari, C., 909
Scanion, R. L., 650
Scanlon, P. W., 277
Scann, A., 879
Schabel, S. I., 330, 557
Schaefer, H. G., 623
Schajowicz, F., 54, 83, 138, 225, 280, 396, 698, 993
Schaller, J., 476, 506
Schaller, J. G., 479, 506
Schallock, G., 1231
Scham, S. M., 1436
Schaper, G., 1334
Scharfman, W., 1335
Schatz, D. L., 881, 1166
Schatzki, R., 723
Schauerte, E. W., 1181, 1340
Schaumann, J., 808
Schecter, L., 709
Scheer, G. E., 935
Scheinberg, I. H., 501
Scheinberg, J. H., 614

AUTHOR INDEX

Scheinberg, L. H., 572
Schen, R. J., 375
Schenk, R. K., 951
Schere, S., 1243
Scherer, E., 62
Scheuer, P. J., 582, 617, 660
Scheuermann, H., 700
Schilder, J. H., 1227
Schiller, A. L., 993
Schiller, K. F. R., 642
Schiller, W., 1080
Schilling, A., 325
Schimke, R. N., 879, 1331
Schinz, H. R., 319
Schippers, J. C., 845
Schlachter, L., 227
Schlaeger, R., 591
Schlagenhaufer, F., 642
Schlosser, W., 122
Schlosstein, L., 506
Schlumberger, H. G., 994
Schmid, F., 1203, 1236
Schmidt, E. R., 428
Schmidt, M. C., 572
Schmidt, R., 1322
Schmike, R. N., 879
Schmitt, F. A., 4
Schmitt-Rohde, J. M., 935
Schmorl, G., 554, 858, 965
Schneeberg, N. G., 857
Schneider, B., 313
Schneider, G., 623
Schneider, R., 483, 506, 700
Schneider, R. C., 517
Schnitka, T. K., 1181
Schoen, D., 552
Scholder-Oehmichen, C., 375
Scholy, D. A., 830
Schorr, S., 138
Schorsch, H., 1359, 1381
Schorsch, H. A., 642
Schottstaedt, E. R., 1436
Schrantz, J., 499
Schrantz, J. L., 499
Schreiber, M., 121
Schroeder, F., 265
Schrurmacher, E., 1268
Schryver, H. F., 22
Schull, E., 478
Schull, W. J., 1181
Schuller, A., 966
Schulman, A., 1433
Schulman, L., 35, 57
Schulte, W. C., 623
Schultz, M. D., 278
Schultze, W. H., 1080
Schumacher, H. R., 501, 582, 614, 631
Schwartz, A., 637, 1377
Schwartz, D. T., 995
Schwartz, E. E., 651
Schwartz, F. D., 617
Schwarz, E., 659
Schwarz, G. A., 965
Schwarzweller, F., 1181
Schwinn, C. P., 181, 182
Scott, A. J., 572
Scott, C. I., 1184
Scott, C. I., Jr., 1185
Scott, C. L., 1180

Scott, J. T., 445, 457, 905
Scott, R. B., 312, 375, 1059
Scott, S. G., 506
Scott, W. F., 378
Scotvold, M. J., 1387
Scranton, P. E., Jr., 183
Scrimger, F. A., 993
Scriver, J. B., 1230
Scrurer, T. B., 1239
Seah, C. S., 256
Seakins, J., 951
Seaman, A. J., 1300
Seaman, W. B., 457, 968
Sear, H. R., 1351, 1410
Sears, K. A., 885
Sears, W. P., 256
Seawright, A. A., 552
Sebes, J. L., 1052
Sechel, H. P., 1313
Sedano, H. O., 1184, 1338
Sedlezky, I., 483
Seedat, Y. K., 808
Seedorff, K. S., 1289
Seeger, J. F., 506
Seegmiller, J. E., 557, 584
Segal, A. T., 1322
Seibert, J., 1438
Selby, S., 114
Seldinger, S. I., 909
Seliger, G., 642
Seligmann, M., 311
Selinsky, H., 1387
Selke, A. C., Jr., 722
Selke, W. J., 1181
Selye, H., 935, 1095, 1100
Semalaigne, G., 1334
Semian, D. W., 1028
Semple, T., 642
Sengpiel, G. W., 1359
Serre, H., 986
Seth, H. N., 256
Sevel, D., 256
Sevitt, S., 35
Shafer, S. J., 659
Shaff, M. I., 1206
Shaffer, H. A., 642
Shah, M. M., 1073
Shalit, I. E., 532
Shanahan, J. R., 428
Shannon, J. C., 727
Shanoff, L. B., 208
Shantharaj, S., 1073
Shapiro, J. H., 327, 1074
Shapiro, M., 642
Shapiro, R., 829
Shapiro, R. F., 571, 612
Shapiro, S. K., 1334
Sharma, O. M. P., 642
Sharnoff, J. G., 319
Sharp, G. C., 475
Sharp, G. S., 993
Sharp, J., 456, 517
Shatin, H., 532
Shattie, S. J., 310
Shaub, M., 601
Shaul, S., 552
Shaver, J. A., et al., 532
Shaw, D. G., 845
Shaw, E. W., 722

Shaw, N. E., 1028
Shaw, S., 1306
Sheffield, L. J., 1230
Sheikh, M. A., 208
Shelburne, S. A., 879
Sheldon, J. H., 582
Shephard, E., 1277
Sherlock, S., 617, 660, 844
Sherman, M. S., 35, 54, 83, 817
Sherman, R., 149
Sherman, R. C., 255
Sherman, R. S., 83, 165, 300, 313, 383
Shertzer, J. H., 1377
Shetty, M. V. K., 1355
Shidnia, H., 1125
Shillito, J., Jr., 1181
Shimizu, Y., 1365
Shimkin, M. O., 256
Shintani, J., 375
Shipley, F. H., 378
Shipley, P. G., 858
Shipp, F. L., 544
Shiraki, M., 116
Shirazi, P. H., 645
Shirkey, H. S., 754
Shkolnik, A., 642
Shmerling, D. H., 1180
Shoji, H., 225
Sholkoff, S. D., 545
Shopfner, C. E., 138
Shorhe, H. B., 846
Shorr, E., 831
Short, C. I., 500
Shuler, S. E., 1354
Shulman, H. S., 107
Shumacher, H. R., 642
Shurfleff, D. B., 1381
Sicher, H., 5
Siegel, B. A., 1073
Siegel, S., 227
Siegelman, S. S., 310, 965, 1095
Siegling, J. A., 1397
Siemsen, J. K., 687
Siggers, D., 1201
Siggers, D. C., 1201
Sigler, J. W., 530
Sigurdsson, G., 951
Silber, R., 832
Silberberg, R., 1201
Silberman, F. S., 396
Silberman, W. W., 35
Silfverskiold, N., 1267, 1283
Silver, C. M., 623
Silver, H. J., 1189
Silver, H. K., 1189, 1286
Silver, J. W., 383
Silver, T. M., 475
Silverman, F. N., 335, 354, 480, 1154, 1196, 1203, 1210, 1233, 1282, 1338, 1362
Silverman, J. L., 1182
Silverman, S., 881
Silverman, W. A., 817, 1084
Silverstein, M. N., 1073
Sim, F. H., 242
Simmonds, N., 858
Simmons, C. R., 330
Simmons, D. J., 1140
Simon, B., 642
Simon, G., 651

Simon, H., 116, 313
Simon, L., 986
Simon, M. A., 1355
Simon, N., 601
Simon, R., 181
Simon, S. D., 623
Simons, H. M., 445
Simpson, S. L., 964
Simril, W. A., 696
Sinclair, A. M., 35
Sinclair, R. J. G., 457
Singer, D. E., 642
Singer, J. M., 506
Singer, R. A., 905
Singer, S. J., 1047
Singh, A., 554, 1146
Singh, M., 256, 705
Singleton, E. B., 208, 330, 1283, 1410
Sinha, G. P., 1380
Sinha, K., 1189
Sinha, T. K., 913
Sipple, J. H., 879
Sire, D. J., 1397
Siren, M., 375
Sirsat, M. V., 83
Sisson, J. C., 909
Sissons, H. A., 22, 54, 380, 651, 1235, 1236
Sitaj, S., 571
Sjogren, H. S., 504
Sjolin, K. E., 1357
Sjolin, S., 1303
Skinsnes, O. K., 585
Sklar, G., 993
Sklaroff, D. M., 340
Sklaroff, R. B., 340
Skouteris, A., 1377
Slater, P. E., 935
Slaughter, D. P., 1131
Slaughter, W. H., 885
Sledge, C. B., 499
Slepian, A., 1403
Slifkin, M., 373
Slocumb, C. H., 417
Slullitel, I., 993
Smetana, H. F., 378
Smigh, M., 660
Smillie, I. S., 698
Smith, B. F., 500
Smith, C. E., 793
Smith, C. H., 1065, 1072
Smith, C. K., 1423
Smith, D. W., 1179, 1180, 1306, 1322, 1436
Smith, E. E., 557
Smith, E. M., 713
Smith, E. W., 1047
Smith, F., 277
Smith, F. W., 909, 935
Smith, H. L., 1227
Smith, H. P., 586
Smith, H. P., Jr., 586
Smith, J. H., 651
Smith, J. L., 1181
Smith, J. V., 1033
Smith, J. W., 617, 1380
Smith, L. W., 1302, 1358
Smith, M., 617
Smith, O. E., 1421
Smith, P., 994, 1181
Smith, P. H., 913, 1286

Smith, P. M., 572, 582
Smith, R., 909, 1357
Smith, R. C., 319
Smith, R. J., 1188
Smith, S. D., 1089
Smith, S. P., 723
Smith, S. W., 1332
Smith, T. T., 1189
Smukler, N. M., 428, 617
Smyth, C. J., 483, 843
Smyth, F. S., 1233
Smyth, W. T., 149
Smythe, H. A., 506
Smythe, V., 417, 757
Snapper, I., 913, 994
Snapper, L., 313
Snavely, J. R., 909
Snell, D., 1148
Snelling, C. C., 1071
Snure, H., 829
Snyder, A. J., 591
Snyder, C. H., 696
Snyder, R. E., 300
Sobel, E. H., 873
Soddard, S. E., 1188
Sodee, D. B., 909
Soderberg, G., 54
Soeters, J. M., 1326
Soffa, D. J., 1397
Sokoloff, L., 325, 557
Solarino, G. B., 843
Solente, G., 1310
Solnica, J., 582
Solomon, A., 309
Solomon, L., 1181
Solomons, C. C., 1296
Solonen, K. A., 1180, 1224
Som, M. L., 242
Somerville, J., 557
Sommers, S. C., 1074
Sommerville, R. G., 754
Sones, M., 808
Sonnenschein, A., 554
Sontag, L. W., 111, 1148
Soong, K. Y., 165, 181, 182
Sooy, F. A., 54
Sorensern, L. B., 554
Sorenson, J., 830
Soriano, M., 554, 1146
Sorrel, E., 1436
Sorrel-Dejerine, J., 1381
Sors, C., 1377
Sorsdahl, O. S., 1080
Sosman, J. L., 572, 591
Sosman, M. C., 966, 1095, 1144
Soule, A. B., Jr., 1416
Soule, E. H., 208, 255, 279, 373, 659
Soulie, P., 1334
Soyannwo, M. A. O., 909
Spaet, T. H., 1074
Spaeth, G. L., 1332
Spahr, A., 1235
Spahr-Hartmann, I., 1235
Spanier, S. S., 373
Sparagana, M., 1140
Sparkes, R. S., 1381
Speed, J. S., 378
Speer, D. P., 1095
Speirs, A. L., 754

Speiser, F., 993
Spencer, H., 872
Spender, R. P., 340
Spengler, D. M., 517
Spengos, M., 572, 614
Spiegel, M. B., 1243
Spillane, J. D., 631
Spinzig, E. W., 1432
Spira, M., 208
Spjut, H., 54
Spjut, H. J., 83, 149, 165, 206, 212, 257, 385, 622
Spodheim, M., 722
Sprackman, T. J., 1230
Sprague, B. L., 1377
Spranger, H. W., 1185
Spranger, J., 1181, 1201, 1239, 1267, 1268, 1269, 1351
Spranger, J. W., 1180, 1185, 1206, 1219, 1227, 1230, 1351
Spritzer, H. W., 539
Sprunt, K., 727
Sraer, C., 1381
Srinivasan, C. K., 383
St. Aubin, P. M., 1181, 1340
Stabenau, J. R., 480
Stack, B. H., 457
Stafne, E. C., 968
Stalmann, A., 1357
Stam, H., 208
Stambler, A. A., 1182
Stamp, W. G., 935
Stanbury, S. W., 923, 1095
Stanger, J. K., 683
Stanley, B., 1403
Stansfeld, A. G., 114, 659
Staple, T. W., 319, 623
Staples, O. S., 1423
Starobin, S. G., 138
Starshak, R. J., 1365
Stasney, R. J., 380
Stauffer, H. M., 995
Stearns, G., 923
Stecher, R. M., 1030, 1188
Steel, H. H., 62, 1182, 1381
Steele, J. D., 965
Stehr, L., 1351
Steida, A., 62
Stein, G. L., 808
Stein, I., 965
Stein, J. M., 319
Stein, R., 965
Steinbach, H., 557, 1300
Steinbach, H. L., 54, 572, 584, 683, 832, 881, 905, 913, 916, 917, 923, 1172, 1286, 1302, 1386
Steinberg, A. G., 1181
Steinberg, I., 1331
Steinberg, V. L., 506, 554
Steindler, A., 774
Steiner, G. C., 256, 383
Steiner, G. M., 138
Steiner, K., 1334
Steiner, R. E., 909
Steiner, R. M., 843
Steinitz, H., 1101
Steinlauf, P., 225
Stejskal, J., 1033
Stemmerman, M. G., 768

AUTHOR INDEX

Stemmermann, G. N., 951
Stenkvist, B., 1073
Stephani, S., 1140
Stern, A. M., 1182, 1189, 1233
Stern, C., 585
Stern, M. H., 1030
Stern, P. H., 913
Sternlieb, I., 501, 614
Sternlieb, L., 572
Stevanovic, D. V., 642
Stevens, J., 387
Stevenson, A. W., 757
Stevenson, C. A., 1144
Stevenson, F. H., 1377
Stewart, F., 378
Stewart, F. W., 378, 383, 387, 995
Stewart, J. R., 158
Stewart, J. S., 1286
Stewart, M. J., 330, 1028
St. Geme, J. W., 1182, 1189
Stickler, G., 1235
Stickler, G. B., 1181
Stieda, A., 1392
Stiff, R. H., 330
Still, F. G., 478
Still, W. J. S., 312
Stillman, B. C., 698
Stock, F. E., 149
Stocks, P., 1244, 1402
Stoddard, A., 701
Stoker, D. J., 700
Stone, H. B., III, 330
Stone, M. J., 313, 457
Stones, D. A., 539
Stonham, C., 1326
Storey, G. L., 506
Stout, A. P., 121, 122, 208, 225, 255, 256, 257, 273, 330, 373, 387, 391, 1357
Stover, C. N., 1214
Stovin, J. J., 1343
Stovin, P. G. I., 642
Stowe, F. R., 1181
Straith, F. E., 1313
Strand, G., 1338
Strang, C., 1262
Stransky, E., 1300
Strasburger, A. F., 1182
Strates, B. S., 1095
Straus, F. H., 879
Strauss, I., 1387
Streda, A., 711
Streeto, J. M., 913
Stretton, A. O. W., 1065
Strong, E. W., 330
Stroud, C. E., 375
Stubbins, S. G., 1377
Stuber, J. L., 554
Stuhl, L., 1429
Sturman, J. A., 1332
Styner, J., 1296
Sudeck, P., 844
Suess, J. F., 1303
Suffecool, S. L., 1377
Suh, S. M., 951
Sulamaa, M., 1180, 1224
Sullivan, C. R., 242
Sullivan, R. C., 622
Summerfeldt, P., 995
Summitt, R. L., 1184

Sundaram, M., 536
Sussman, M. L., 1074
Sutherland, B. S., 1184
Sutherland, C. G., 342
Sutphin, A., 912
Sutro, C. J., 651, 995, 1107, 1334
Svab, V., 1392
Swain, R. W., 227
Swanson, W. W., 1101
Swanton, M. C., 591
Swarm, R. L., 227
Swartzendruber, D. C., 729
Swee, R. G., 49
Sweetman, L., 557
Sweetman, R., 1334
Sweetnam, R., 182
Swenson, P. C., 277
Swerdlow, R. S., 242
Swettenham, K. V., 572
Swezey, R., 506
Swezey, R. L., 843
Swischuk, L. E., 1365
Swoboda, J. W., 951
Sylim-Rapoport, I., 1302
Symchych, P., 727
Symmers, W., 1030
Szymanski, F. J., 659

Taddei, L., 1322
Tahal, N., 504
Taira, Y., 1080
Tait, G. B. S., 457
Taitz, L. S., 917
Taketa, R. M., 554
Takeuchi, S., 181
Talbot, N. B., 912
Talbott, J. H., 554, 557
Talerman, A., 225, 445
Talmage, R. V., 5
Tampas, J. P., 817
Tamyakopoulos, S., 1144
Tan, E. M., 475
Tanaka, T., 1080
Tanner, F. H., 1095
Tanner, J. J., 1151
Tanner, M., 442, 506
Tapia, J., 923
Tashjian, A. H., 913
Tausk, K., 1325
Taussig, H. B., 1331
Taveras, J. M., 993
Taves, D. R., 325
Taybi, H., 913, 1199, 1214, 1310, 1312, 1343
Taylor, A. L., 330, 659
Taylor, A. R., 457, 623
Taylor, B., 1206
Taylor, H. K., 717
Taylor, K. B., 601
Taylor, P. E., 1080
Taylor, R. G., 517, 793
Taylor, T. K. F., 165, 517
Taylor, W. B., 1313
Teates, C. D., 642
Tebbi, K., 1073
Tefft, M., 279
Tegi, S., 124
Telfer, N., 49
Temtamy, S., 1182, 1340

Temtamy, S. A., 1181, 1188, 1340, 1341
Teng, C. T., 885, 1283
Teng, P., 124
Tennenouse, A., 1347
Teplick, J. G., 993, 1351, 1386
Tepperman, H. M., 881
Terasaki, P., 506
Terry, D. W., 642
Terry, W. D., 311
Tetamy, S., 1286
Teutschlaender, O., 1095
Thannhauser, S. J., 994, 1325
Theander, G., 375
Thelander, H. E., 722
Thevenard, A., 1381
Thieffry, S., 1182, 1381
Thiemann, H., 722, 1267
Thier, C. J., 1181
Thomas, A., 383
Thomas, A. E., 500
Thomas, C., 1334
Thomas, C. C., 808
Thomas, C. P., 642
Thomas, D. A., 1203
Thomas, E. D., 506
Thomas, H. M., Jr., 1166
Thomas, K. E., 1433
Thomas, L. B., 333, 480
Thomas, W. C., 1100
Thompson, D. W., 319
Thompson, E. A., 1421
Thompson, F. R., 768
Thompson, M., 457
Thompson, N. M., 1397
Thompson, R. C., Jr., 951
Thompson, R. P. H., 617, 660
Thompson, T. C., 35
Thompson, W. L., 116
Thomson, A. D., 267, 651
Thomson, J. E. M., 1095
Thomson, J. G., 557
Thorne, F. L., 1380
Thorne, M. G., 923
Thuline, H. C., 1189
Thune, S., 479, 506
Tighe, J. R., 532
Tikva, P., 501
Tilden, I. L., 659
Tillema, D. A., 709
Tillman, B. P., 158
Tinney, W. S., 557
Tips, R. L., 1181
Tishler, J. M., 722
Todd, T. W., 1162
Tofft, M., 256
Tokura, N., 1326
Tolksdorf, M., 1185
Tomesevic, M., 722
Tondreau, R. L., 428
Toomey, J. A., 917
Topazian, R. G., 387
Topfer, D., 127
Torg, J., 867
Torg, J. S., 1182, 1381
Tornblom, M., 1310
Torok, L., 375
Torun, B., 951
Touraine, A., 1310
Touraine, R., 375

AUTHOR INDEX

Tourtellotte, C. D., 867
Travers, B., 263
Treacher-Collins, E., 1313
Treadwell, B. L., 554
Treger, A., 1189
Trentani, C., 1017
Treta, J., 729
Treves, S., 340
Trevor, D., 1286
Trippel, J. G., 722
Trueta, J., 5, 609, 696, 698
Trujillo, M. M., 1182
Truscott, D. E., 208
Trzcinska-Dabrowska, Z., 256
Tsachalos, P., 557
Tubbs, F. E., 1227
Tucker, A. S., 138
Tucker, H. A., 1432
Tuli, S. M., 1380
Turcotte, B., 90
Turkington, R. W., 1334
Turnbull, H. M., 867, 994, 1334
Turner, J. W., 354
Turner, J. W. A., 965
Turner, M., 183, 259
Turner-Warnick, M., 660
Turunen, M. J., 457
Twigg, H. L., 500
Tyars, M. D., 1334

Ubbens, R., 1335
Udoff, E. J., 475
Uehlinger, E., 206, 951, 1302, 1326
Uehra, H., 255
Uhlendorf, B. W., 1331
Uhlig, H., 1268
Ulin, R., 1286
Umansky, R., 1189, 1322
Unger, L. J., 861
Unni, K., 127
Unni, K. K., 182, 206, 242, 385
Upton, G., 183
Urbanek, T., 585
Urbaniak, J. R., 709
Urist, M. J., 1095
Urist, M. R., 6
Utsinger, D., 571
Utisinger, P. D., 612
Utz, V. W., 1351
Uzel, A. R., 83

Vahvanen, V., 457
Vainio, P. V., 208
Valentin, B., 1436
Valle, A. R., 1359
Van, B., 845
van Berkum, K. A. P., 1184
Van Buchem, F. S. P., 951, 958, 1181, 1335
VanBuskirk, F. N., 817
VanBuskirk, F. W., 817
van Creveld, S., 1326
Vandepitte, J., 1059
Van Der Heul, R. O., 206
Van der Hoeven, L. H., 771
Vanderpool, D. W., 257
van Der Sar, A., 771
Van Epps, E. F., 1181

van Ere Jeught, J., 683
Van Herpe, L., 1377
Vasilas, A., 387
Vassar, P. S., 255
Vastine, J. H., II, 1326
Vastine, M. F., 1326
Vaughan, B. F., 935
Vaughan, J., 1125
Vautrin, D., 1429
Vauzelle, J. L., 1214
Vawter, G. F., 279
Veller, K., 817
Vellios, F., 1181
Vendt, E. R., 872
Verger, P., 1322
Vermess, M., 312
Verner, E. W., 225, 256
Verocay, J., 1357
Verspyck, R., 387
Vertes, V., 637
Vetter, H., 107, 387
Veveque, B., 1438
Veys, E. M., 683
Vicinus, J. H., 1154
Vickers, C. W., 54
Videback, A., 319
Vieta, J. O., 354
Vignon, G., 571
Villaumey, J., 711
Vincent, J., 6
Vincent, M., 659
Vincenti, N. H., 330
Vining, C. W., 557
Vinogradova, T. P., 6
Vinson, H. A., 723
Vinstein, A. L., 1403
Vockers, R. A., 1313
Vogl, A., 1310
Vogt, E. C., 340, 480
Vogt, W., 552
Vohra, V. G., 281
Volberg, F. M., Jr., 22
Voluter, G., 456, 517
von Albertini, A., 1095
Von Haam, E., 532
von Recklinghausen, F., 1357
von Rosen, S., 1084
von Studnitz, W., 879
Voorhess, M. L., 1303
Voorhoeve, N., 1244, 1396
Vrolik, W., 1289
Vulliamy, D. G., 1181
Vure, E., 817, 822

Waaler, E., 805
Wachtel, H., 1392
Wadia, R. S., 1438
Wagers, L. T., 1397
Wagner, B., 830
Wagner, M. L., 1303
Wagoner, G. W., 829
Waine, H., 609
Waisman, H. A., 1322, 1331
Wakem, C. J., 532
Waldemar, Ogryzlo, M. A., 471
Waldenstrom, H., 499, 696, 700
Waldenstrom, J., 312, 375
Walder, D. N., 683

Waldman, I., 1188
Walker, B. A., 1219
Walker, C. H. M., 1423
Walker, D. G., 1347
Walker, E. A., Jr., 387
Walker, E. T., 1421
Walker, G. D., 1347
Walker, J. C., Jr., 1184
Walker, R. J., 582
Walker, W. F., 723
Wallace, K. M., 651
Wallace, S., 205
Wallace, W. M., 935
Wallis, L. A., 127, 138
Wallyn, R., 1331
Walpin, L., 965
Walsh, F. B., 880
Walshe, J. M., 501, 614
Walsje, J. M., 572
Walters, M. N. L., 935
Walton, J. E., 912
Wang, C. C., 278, 330
Wang, S. H., 853
Wanke, R., 909
Warburg, M., 1338
Ward, H. P., 1074
Ward, L. E., 457, 475, 1181
Warden, M. J., 124
Wardle, E. N., 582
Warin, R. P., 659
Warkany, J., 1184, 1266
Warner, N. E., 1033
Warnock, C. G., 501
Warrick, C. K., 995
Wasserman, D., 1421
Wasserman, L. R., 1074
Watchi, J. M., 1381
Waterhouse, A. M., 1159
Watkinson, G., 506
Watne, A. L., 1433
Watson, J. D., 457
Watson, L., 867, 1182
Watt, T. L., 445
Waugh, W., 1277
Wayson, N. E., 805
Weatherall, J. A., 1144
Weatherall, S. M., 1144
Weaver, A. L., 539
Weaver, T. S., 1266
Webb, W. R., 1377
Weber, F. P., 375, 457, 504, 659, 1334
Weber, H. M., 530
Weber, K., 1365
Webster, G. D., Jr., 923
Webster, M. D., 965
Wechsler, L., 375
Wedgewood, R. J., 476, 506
Weed, L. A., 759, 771
Weens, H. S., 645, 1310, 1421
Wegner, G., 775
Wei, W. C., 1403
Weichert, K. A., 1362
Weidmann, S. M., 1144
Weil, S., 1436
Weilbaecher, R. G., 1332
Weill, J., 1243, 1334
Weinberger, H. J., 532, 660
Weiner, H., 1334
Weinfeld, A., 1030

AUTHOR INDEX

Weingarten, R. J., 872
Weinmann, J. P., 5
Weippl, G., 1306
Weisman, S. L., 295
Weismann-Netter, R., 1429
Weisner, K. B., 571, 612
Weiss, A., 881
Weiss, L., 1188
Weitzman, G., 572
Welch, J. P., 1188
Weldon, W. V., 532
Welfling, J., 572, 582
Weller, M. H., 682, 711
Weller, M. P., 938
Wells, C. E. C., 631
Wells, I. C., 1047
Wells, J., 457
Wendeberg, B., 1181
Wener, H., 909
Wenger, D. R., 499, 709
Wenley, W. G., 554
Went, L. N., 1048
Wenzel, H. G., 1334
Wenzl, J. E., 722
Wepler, W., 532
Werne, S., 456
Werner, A., 456, 517
Werner, I., 1073
Wescott, R. J., 1365
Wesenberg, R. L., 1184, 1302
Wessel, A. B., 1084
Wexler, I. B., 1207
Weyer, G. W., 537
Weyers, H., 1181
Whalen, J. P., 22, 951
Whaley, K., 660
Whedon, G. D., 831
Wheeler, C. E., 1095
Wheeler, E., 845
Whelan, D., 1184
Whelton, J. J., 660
Whelton, M., 617
Whitaker, P. H., 1388
Whitby, L. G., 905
White, J. G., 754
Whitehouse, G. H., 295
Whitehouse, R. H., 1151
Whitehouse, W., 457
Whitesides, T. E., 935
Whitesides, T. E., Jr., 116
Whitlock, H. M., 918
Whitridge, J., Jr., 1125
Whitten, C. F., 771
Wichtl, O., 1206
Widmann, B., 1365
Wiedemann, H., 1201
Wiedemann, H. R., 1181, 1185, 1207, 1230, 1239, 1267, 1268, 1269, 1351
Wierman, W. H., 642
Wiernik, P. H., 313, 457
Wigley, R. A. D., 506
Wilde, M. D., 273
Wilder, C. S., 330
Wiles, P., 1245
Wilhelmi, A. E., 881
Wilk, L. H., 696
Wilkinson, J., 1084, 1334
Wilkinson, J. A., 1084
Wilkinson, M., 506

Wilkinson, R. H., 54, 165, 1377
Willcox, A., 923
Williams, B. R., 1201
Williams, E. D., 879
Williams, G. A., 1347
Williams, J. H., 817
Williams, J. L., 711
Williams, R., 572, 582, 617, 660
Williams, R. H., 831
Williams, T. F., 1182
Williamson, J., 660
Williamson, J. J., 1313
Willis, J. B., 1028
Willis, R. A., 277, 380, 396, 1244
Wills, M. R., 935
Wilner, D., 149, 383
Wilske, K. R., 532
Wilson, C. L., 1106
Wilson, D. W., 1214
Wilson, F. C., Jr., 54
Wilson, H., 255
Wilson, J. K. V., 333, 480
Wilson, J. N., 1190
Wilson, J. S., 195
Wilson, J. W., 651
Wilson, P. D., 651, 700, 709
Wilson, R. J., 539
Wilson, S. R., 107
Wiltshaw, E., 330
Wimberger, H., 775
Winchester, P., 22, 727
Windholz, F., 1074, 1392
Windhorst, D. B., 727, 754
Winkelmann, R. D., 475
Winkelmann, R. K., 375
Winkler, A. W., 918
Winter, P. F., 49
Winter, R. B., 1331
Winters, R. W., 1182
Wintrobe, M. M., 557, 1080
Wiot, J. F., 1004
Wirth, J. E., 256
Witkowski, R., 1302
Wittbom-Cigen, G., 479
Wittenborg, M. H., 1084
Witthom-Cigen, G., 506
Witwicki, T., 256
Wohl, M. J., 504
Wolbach, S. B., 373, 917
Woldring, J. W., 958
Woldring, M. G., 951, 1181
Wolf, B. S., 1387
Wolf, E. L., 700
Wolf, H. L., 935
Wolf, M. D., 683
Wolfe, D. C., 138
Wolfe, R. R., 623
Wolff, S., 1436
Wolfson, J. J., 727, 754
Wolkow, M., 584
Wollenius, G., 1310
Wolter, F. D., 1365
Wond, W. K., 808
Wood, B. J., 1189
Wood, E. H., 391, 993
Wood, P. H., 476
Woodard, H. Q., 1131
Woodhouse, N. J. Y., 951
Woodrow, H., 1148

Woodruff, F. P., 517
Wooten, W. B., 181
Workman, J. B., 909
Worrall, T., 1403
Worthington, J. W., 557, 683
Wouters, H., 445
Woyke, S., 124
Wranne, L., 1303
Wren, M. W. G., 1310
Wright, A. D., 1172
Wright, B., 552
Wright, C. J. E., 659
Wright, E. M., 1357
Wright, J. C., 909
Wright, J. L., 83
Wright, R., 506
Wright, V., 506, 532
Wu, F., 22
Wu, K. K., 227
Wu, W. Q., 256
Wyatt, J. P., 1074
Wyllie, J. C., 659
Wyllie, W. G., 965
Wynne-Davies, R., 1084

Yacoub, M. H., 642
Yaghmai, I., 387
Yakovac, W. C., et al., 771
Yamamuro, T., 181
Yannakos, D., 951
Yardley, K. H., 994
Yates, B. W., 1181
Yates, J. W., 1189
Yau, A. C. M., 517
Yden, S., 517
Yendt, E. R., 923
Yentis, L., 325
Yeoman, P., 445, 453
Yoneda, S., 1365
Yonezawa, H., 181
Young, A. A., 617
Young, A. C., 532
Young, A. W., Jr., 1397
Young, D. A., 808, 916, 917, 1286
Young, D. F., 1365
Young, H. H., 90, 149, 591
Young, I., 659
Young, I. S., 1172
Young, J. K., 1080
Young, L. W., 457
Young, R. S., 1207
Young, R. W., 6
Young, W. B., 1235
Yow, M. D., 800
Yu, T. F., 445, 557

Zabriskie, J., 1286
Zadek, I., 727
Zadek, R. E., 968
Zannoni, U. G., 584
Zarabi, M., 1357
Zarembski, P. M., 1033
Zarkowsky, H. S., 1073
Zausner, J., 396
Zawadzki, Z. A., 311
Zawisch, C., 1346
Zeitel, B. E., 1095

AUTHOR INDEX

Zellweger, H., 1206
Zener, J. C., 808
Zervas, N. T., 1359
Zetterqvist, P., 1233
Zheutlin, B., 1095
Zielinska, K., 256
Ziff, M., 442, 506
Ziliotto, D., 909
Zimmer, E. A., 554
Zimmer, L. E., 256
Zimmerman, C., 1421
Zimmerman, H. B., 935
Zipkin, I., 325
Zitnan, D., 571
Zonntag, A., 909
Zornoza, J., 642
Zubrow, A. B., 709
Zucker, L., 506
Zvaifer, N. J., 506
Zvi Even-Paz, 375
Zychowicz, C., 1199

Subject Index

Achilles tendon calcification, 1117
Achondrogenesis, 1203, 1225
 Brazilian type, 1125, 1127
 Grebe type, 1125, 1127
 lethal type, 1125
 Patenti-Fraccarro type, 1125, 1126
Achondroplasia, 1191
 femur, 1192, 1198
 fibula, 1198
 hand, 1193
 humerus, 1198
 mesomelic dwarfism and, 1224
 pelvis, 1197, 1198
 spine, 1195
 tibia, 1194, 1198
 vertebra, 1196, 1199
Acrocephalosyndactyly, 1340
 type I, 1340-1342
 type II, 1343
 type III, 1343
 type IV, 1343
 type V, 1342-1346
Acrocephopolysyndactyly, 1340
Acrodinia multilente, 1383
Acromegaly, skeletal maturation and, 1168, 1173-1175
Acro-osteolysis, 1385, 1386
Actinomyces bovis, osteomyelitis, 793
Actinomycosis, 792
Adamantinoma, 378, 383
 femur, 382, 384
 jaw, 380
 tibia, 381
Adrenocortical steroids, metabolic bone disease, 830
Agammaglobulinemia, lymphopenic, 1239
Agnogenic myeloid metaplasia, 1073
 femur, 1075, 1078
 foot, 1079
 humerus, 1078
 iliac bones, 1075, 1080
 pelvis, 1074, 1075
 rib, 1079
 sclerosis of thoracic bones, 1079
 spine, 1077
 tibia, 1078
 vertebra, 1075-1077
Ainhum, 1432
Albers-Schönberg disease, 1346
Albright syndrome, 994
Amsterdam dwarfism, 1322
Amyloidosis, 1030
 classification, 1031
 disorders associated with, 1031
 humerus, 1032
 pelvis, 1032
 radius, 1033
Anemia
 acquired, 1072
 congenital hemolytic, 1071
 hereditary, 1047
 iron deficiency, 1072
 skull, 1073
 sickle cell, 1047
 femur, 1057, 1059, 1060, 1062
 growth disturbances in distal ilium, 1061
 hand, 1050, 1053, 1054, 1063
 hip, 1051
 humerus, 1049, 1054, 1055, 1063
 pelvis, 1048, 1051
 skull, 1052, 1058
 spine, 1051
 tibia, 1056
 vertebra, 1048, 1049
Aneurysm, malignant, 383
Angioblastoma, 383
 malignant, 378
Angioendothelioma, 383
Angiofibroma, humerus, 385
Angioma, malignant, 383
 metastasizing, 383
Angiomatosis, cystic, 138
 carpal, 142
 femur, 140, 141, 144, 147, 151, 154
 fibula, 142, 145
 foot, 148
 humerus, 154
 pelvic, 146, 147, 151
 phalanges, 155
 radius, 142
 rib, 150, 153
 skull, 150, 153
 spine, 145-147, 153, 154
 tibia, 143, 144
 ulna, 142
 vertebra, 146
Angiomyosarcoma, 383
Angiosarcoma, 383
 pelvic, 389
Ankylosing hyperostosis, *see* Hyperostosis
Ankylosing spondylitis, *see* Spondylitis
Apert syndrome, 1340
Apophysitis, calcaneus, 708
Arachnodactyly, 1330, 1331
Arteriovenous fistula, 149
 tibia and fibula, 155
Arteriovenous malformation, 127
 fibula, 135
 hand, 132-134
 pelvic, 136, 137
Arthritis, 414
 bacterial, 417
 brucellar, 439
 hip, 440
 childhood onset, 476
 classification, 415
 clinical signs compared with roentgen diagnosis, 415
 gonorrheal, 438, 439
 ankle, 439
 hand, 438
 idiochondrolysis, 499
 Jaccoud, 500
 juvenile rheumatoid, 476
 classification, 476
 femoral head, 487
 foot, 482
 hand, 477, 478, 480, 481, 484, 485, 487-489, 493, 496, 497
 hip, 483, 491, 493, 498
 insidious systemic type, 479
 knee, 480, 486, 490, 493
 micrognathia, 495
 nonsystemic type, 479
 pauciarticular type, 479
 spine, 489, 492
 systemic type, 479
 tibial shortening, 494
 vertebra, 492
 wrist soft tissue swelling, 485
 mixed connective tissue disease, 475
 primary biliary cirrhosis, 664
 psoriatic, 544
 ankle, 549
 elbow, 546
 foot, 547
 hand, 545, 546, 548-551
 knee, 549, 552
 sacroiliac, 540
 spine, 551
 syndesmophyte formation in vertebrae, 551
 pyogenic, 428
 corticosteroid complications, 437
 foot, 434
 hip, 437
 hip joint subluxation, 433
 humeral head, 436
 soft tissue swelling, 432
 spine, 435
 rheumatoid, 440
 ankle, 468
 atlantoaxial joint, 461
 bilateral navicular subluxation in, 463
 cervical spine degeneration, 459, 460
 cyst formation in ulna, 445
 elbow, 452
 foot, 442
 giant synovial cysts, 457, 462
 hand, 444, 455, 463, 466, 467, 469, 473-475
 hip, 451, 456, 468, 470, 471
 knee, 443, 450, 452, 462, 472
 manubrial dislocation, 464
 metacarpal, 446-450
 metatarsal, 449
 odontoid, 450, 460
 pseudocyst formation, 452, 453

Arthritis—*continued*
  rheumatoid—*continued*
    radioulnar subluxation, 454
    robust type, 456, 458
    scleroderma and, 441
    sclerotic healing, 453
    shoulder, 470
    soft tissue contractures of hand, 464
    spine, 459–461
    tibiotalar joint, 457
    ulnar styloid, 454
    wrist, 472
  tuberculous, 417
    carpal bone deossification, 418
    corticosteroid complications, 428
    cystlike destruction, 421, 425
    elbow, 419
    femoral head, 424
    foot, 427
    healed hip, 426
    healing, 417
    interphalangeal joint, 423
    invasive stage, 417
    kissing sequestra, 425
    knee, 420–422
    quiescent stage, 417
    sacroiliac, 424
    soft tissue swelling, 419
    spine, 428–430
    tissue destruction, 417
Arthritis urethritica, 530
Arthropathy, neurotrophic, 631
  diabetic, 632–634
  foot, 639
  hip, 641
  knee, 638
  spine, 641
  spur formation, 642
  syphilis, 634, 635, 638, 640
  tabes dorsalis, 636
Arthrosis, degenerative, knee, 622
Avulsive cortical irregularity, 116

Barlow disease, 845
Bismuth poisoning, 1125
Blastomycosis, 796
  calcaneus, 797
  fibula, 798
  hand, 797, 798
  knee, 797
Blennorrheal arthritis, idiopathic, 530
Blount disease, 705, 706
Bone
  cyst, *see* Cyst
  calcification mechanism, 2
  derivation, 1
  development in embryo, 4
  endochondral growth, 5
  epiphyseal, growth, 5
  geographic destruction, 9
  histopathology, 1
  hyaline, histology, 3
  intramembranous, formation, 4
  ischemia, *see* Ischemia
  island
    benign tumor, 62
    iliac region, 63, 64

  lesion, radiologic approach to, general, 8
  mineral, 3
  modeling, 6
  moth-eaten destruction, 9
  necrosis, 679, 681, 682
    appositional new bone formation, 683
    avascular, 686
    bilateral ischemic, 689
    femoral head, 694
    normal, 680
    nutrition, 6
    periosteal cloaking, due to Gaucher disease, 21
  periosteal reaction, 11
    amorphous, 23, 28
    cause, 13
    dense elliptical, 14, 18
    dense undulating, 14
    interrupted, 14
    lamellar nodular, 24, 25
    lamellated, 22, 23
    perpendicular, 23, 26, 27
    solid, 11, 18–20
    solid thin type, 17
    solid undulating, 15, 16
    sunburst, 23, 26, 27
    thin undulating, 14
    type, 13
    types, 13
  permeative destruction, 9
  physiology, 1
  resorption, 5, 680
  revascularization, 679, 682
  tumor, 4
    age incidence, 31
    benign, 33
    classification, 33
    malignant, 31, 181
    malignant metastasis, 340
  tumorlike conditions, 4
Bow legs, 707
  anterior tibial, 1429–1431
Brown tumor
  hyperparathyroidism, 886, 896, 898–911
  renal osteodystrophy, 941
Bursitis, 1105

Caffey syndrome, 817
Caisson disease
  bone infarction, 714–718, 720, 721
  ischemic necrosis of femur, 721
Calcification
  Achilles tendon, 1117
  prepatellar bursa tendons, 1107
  soft tissue
    leg, 1106
    systemic disease, 1100
  tendons of shoulder, 1105, 1106
Calcified island of bone, 62
Calcified medullary defect, 62
Calcinosis
  acrosclerosis with, 1101
  circumscripta, 1101
  dermatomyositis with, 1104
  hand, 1102

  interstitial, 1101
  tumoral, 1095
    elbow, 1098
    forearm, 1096
    pelvis, 1098
    radionuclide scan, 1099
    shoulder, 1097
  universalis, 1101
Calcium, metabolic bone disease, 829
Calcium oxalosis, 1033
Calcium pyrophosphate dihydrate deposition disease, 571
Calvé disease, 702
Calvé-Legg-Perthes disease, 696
Camptodactyly, 1188
Cartilage
  calcification mechanism, 2
  calcified, resorption, 5
  development, 2
  formation in embryo, 1
  growth, 1
  histopathology, 1
  mature, 1
  mineral availability for, 3
  nutrition, 2
  physiology, 1
  relationship to blood vessels, 2
  resorption, 2
  tumor, differentiation of, 255
Cartilage-hair hypoplasia, 1199, 1236
Chalk bones, 1346
Chondroblastoma, 83
  differentiation from ischemic necrosis, 92
  epiphysis, 84, 89
  femur, 85, 86, 93
  greater trochanter, 86, 90, 94
  humerus, 87, 92
  metacarpal, 90
  metatarsal, 88
  sphenoid sinus, 91
Chondrocalcinosis, 571
  elbow, 579
  hip, 579
  knee, 572–575, 578
  secondary to hyperparathyroidism, 576
  wrist, 576, 581
Chondrodysplasia, Conradi-Hunermann type, differentiated from rhizomelic, 1233
Chondrodystrophia, 1191
  calcificans congenita, 1230–1233
  rhizomelic type, 1231
Chondrodystrophia fetalis, 1191
  calcificans, 1230
  hypoplastica, 1230
Chondrodystrophic dwarfism, 1191, 1199
Chondroectodermal dysplasia, 1227–1229
Chondrolysis, idiopathic, hip, 499
Chondroma, 62
  femur, 67
  hand, 65, 66
  juxtacortical, 70
  tibia, 67, 69
Chondromatosis
  external, 1402

internal, 1244
synovial, 622
  elbow, 623
  hip, 625, 628
  knee, 623, 624, 626, 627, 630
  trabeculation in, 629
Chondromyxoid fibroma, 90
  calcaneus, 98, 100
  femur, 96
    transition to chondrosarcoma, 102
  fibula, 95
  fourth proximal phalanx, 99
  humerus, 97
  rib, 96, 99
  thumb, 100
  tibia, 101
Chondroplasia, hereditary deforming, 1402
Chondrosarcoma, 223
  acetabulum, 235, 238
  central, 225, 229, 232, 234, 237, 248
  clear cell, 242, 252
    femur, 250
  extraskeletal, 254
  femur, 230, 231, 234, 239–243
  fibula, 232
  humerus, 249
  iliac bone, 232
  mandible, 247
  mesenchymal, 255
  new bone formation, 13
  peripheral, 246, 251–254, 256, 257
  rib, 255
  sacrum, 231, 233, 236, 239
  scapula, 251, 253
  skull, incidence of, 246
  snowflake pattern of calcified chondroid matrix, 32
  sphenoid bone, 244
  sternum, 228, 245, 251
  synovial, 257
Chordoma
  pelvic, 395
  vertebra, 396, 397
Chromosome
  abnormalities, 1487
    autosomal, 1505
  karyotype
    D1 trisomy, 1490
    female, 1488
    Klinefelter syndrome, 1493
    male, 1489
    mongolism, 1506
    trisomy 17-18, 1511
    Turner syndrome, 1492
  meiosis, 1487
  mitosis, 1487
  morphology
    alterations, 1488
    normal, 1488
    normal, 1488
  sex, possible constitutions of, 1491
Cirrhosis, primary biliary, arthritis of, 664
Clear cell chondrosarcoma, 242, 250, 252
Cleidocranial dysostosis, 1416–1420
Clinodactyly, 1188
Coccidioides immitis, osteomyelitis, 793

Coccidioidomycosis, 793
  hand, 794
  sacroiliac, 796
  vertebra, 795, 796
Cockayne syndrome, compared with homocystinuria, progeria and bird-headed dwarfs, 1438
Codman triangle, 14
  due to hemorrhage, 22
  due to osteosarcoma, 22
Compact island, 62
Conjunctivourethrosynovial syndrome, 530
Cooley anemia, 1064
Cornelia de Lange syndrome, 1322–1324
Cortical abrasion, 116
Coxa plana, 696
Cretinism, skeletal maturation in, 1163–1166
Cri-du-chat syndrome, 1514, 1516
Crooked fingers, 1188
  association with other abnormalities, 1189
  hereditary bowing of terminal phalanges, 1190, 1191
Cushing syndrome, osteoporosis, 840–842
Cyst
  aneurysmal, 149
    callus formation, 177
    clavicle, 163
    extraosseous, 164
    femur, 160, 175, 176
    fibula, 173
    foot, 159, 165
    hand, 161
    iliac bone, 178
    intraosseous, 167, 169
    ischium, 175
    maxillary sinus, 180
    metacarpal, 171
    mixed intra- and extraosseous, 170
    patella, 170
    periosteal reaction, 179
    rapid growth and diagnosis, 174
    rib, 171
    sacrum, 180
    skull, 162
    spine, 166, 168
    talus, 163
    vertebra, 172
  hemorrhagic
    humerus, 109
    metacarpal, 111
    middle phalanx of hand, 110
    radius, 109
  inclusion, 111
    thumb, 112
  solitary, 105–108
    humerus, 108
  unicameral, 106

Dactylitis, tuberculous, 768, 769
Dactylolysis spontanea, 1432
Dermatoarthritis, lipoid, 659
Desmoid
  cortical, 116

medial condyle, 117–119
periosteal, 116
subperiosteal, 116
Diaphyseal aclasis, 1402
Diastrophic dwarfism, 1199, 1214–1218
Disappearing bone, 149
Dwarfism
  bird-headed, compared with Cockayne syndrome, homocystinuria and progeria, 1438
  chondrodystrophic, 1191, 1199
  diastrophic, 1199, 1214–1218
  mesomelic, 1223–1225
  metatrophic, 1201, 1202
  pseudoachondroplastic, 1200
  thanatophoric, 1203
Dyschondroplasia, 1244
Dyschondrosteosis, 1223, 1243, 1244
Dyscrasias, plasma cell, 305, 308
Dysosteosclerosis, 1351–1354
Dysostosis
  cleidocranial, 1416–1420
    compared with pycnodysostosis and osteopetrosis, 1357
  metaphyseal, 1235
  multiplex, 1465
  peripheral, 1283–1286
Dysplasia, 1179
  anomalies of hand, foot, and arm, 1187
  asphyxiating thoracic, 1203
  chondroectodermal, 1227–1229
  classification, 1180
    International Nomenclature (1977), 1180
    known pathogenesis, 1184
    McKusick and Scott, 1183
    Rubin, 1185
    unknown pathogenesis, 1183
  combined mesoectodermal, 1338
  congenital ectodermal, 1338
  ectodermal and mesodermal, 1338
  epiphyseal, 1267
  epiphysialis hemimelica, 1286–1288
  epiphysialis punctata, 1203, 1230
  familial metaphyseal, 1239
  fibrous, 994
    compared with osteofibrous dysplasia, 1030
    femur, 997–1003, 1006, 1011, 1016–1020
    hand, 996, 997, 999, 1007, 1012
    hip, 1006
    humerus, 997, 1000, 1001, 1004, 1010
    leg, 1009
    malignant degeneration in, 1026
    pelvis, 996, 1002, 1006, 1017
    pubis, 1015
    radius, 1014
    rib, 1006, 1008
    skull, 999, 1016, 1021–1025
    spine, 1013
    technetium uptake in, 1004
    tibia, 997, 998, 1005, 1015
    vertebra, 1013
  focal dermatophalangeal, 1338
  hereditary, 1179
    International Nomenclature (1977), 1180

Dysplasia—*continued*
  hereditary—*continued*
    McKusick and Scott classification, 1183
    Melnick-Needles, 1307–1309
    metaphyseal, 1239
    multiplex, epiphyseal, 1277–1282
    osteofibrous, 1028
      compared with fibrous dysplasia, 1030
      pseudoarthrosis and, 1027
      tibia, 1028, 1029
    pattern of malformation, 1179
    progressive diaphyseal, 1410
      femur, 1411, 1412
      fibula, 1410
      humerus, 1413
      rib, 1414
      skull, 1415
      tibia, 1410
    puncticularis, epiphyseal, 1230
    spondyloepiphyseal, 1267, 1275
      congenita, 1268, 1270–1274
      differentiation from Morquio disease, 1268
      metaphyseal, 1267
      tarda, 1269, 1276
      types, 1267
Dystelphalangy, 1190
Dystrophic bone disease, 829
Dystrophy
  familial asphyxiating thoracic, 1210–1213
  infantile thoracic, 1210
  thoracic asphyxiating, 1210
  thoraco-pelvic-phalangeal, 1210

Echinococcus, osteomyelitis, 798, 799
Ehlers-Danlos syndrome, 1334
Ellis-van Creveld syndrome, 1203, 1224, 1227–1229
Enchondroma, 62
  femur, 68, 75
  fifth finger, 65
  multiple, 1244
Enchondromatosis, 1244
  chondrosarcoma with, 1258
  femur, 1249, 1250, 1253, 1254, 1256
  fibula, 1252
  foot, 1257, 1261
  forearm, 1260
  hand, 1257, 1260
  humerus, 1248, 1258
  leg, 1245–1247, 1255
  osteosarcoma with, 1259
  pelvis, 1258, 1261
  rib, 1251
  scapula, 1259
  tibia, 1252, 1261
Endothelioma, 383
  diffuse, 273
  intravascular, 383
  multiple, 383
Engelmann-Camurati disease, 1410
Engelmann disease, 1410
Eosinophilic granuloma, 1521
  solid periosteal reaction in, 19, 20
Epiphyseal dysostosis, 1214

Epiphyseal dysplasia multiplex, 1277–1282
Epiphyseal plate, anatomy, 5
Erythrogenesis imperfecta, 1300–1302
  congenital anomalies associated with, 1300
Eunuchoidism, skeletal maturation and, 1174
Ewing tumor, 273, 289, 290, 292, 293
  acetabulum, 281
  amorphous new bone formation, 28
  femur, 282, 286–288, 294, 298
  fibula, 299
  humerus, 285, 288
  ilium, 285
  mandible, 280
  metacarpal, 284, 302
  pubis, 295, 296
  radiation therapy results, 299
  rib, 280, 283, 284
    operative specimen, 283
  scapula, 294, 295
  skull, 291, 294
  spine, 298
  tibia, 301
  vertebra, 297, 298, 300
Exostosis
  cartilaginous, 1402
  hereditary multiple, 1402
    ankle, 1405
    hand, 1406
    humerus, 1402
    knee, 1403, 1404, 1408
    pelvis, 1407, 1409
    wrist, 1404

Familial onycho-osteodysplasia, 1421–1423
Familial osteodysplasia, 1442
Fanconi syndrome, 936, 1302–1305
  congenital anomalies associated with, 1303
  hand, 924
  knee, 924
  skeletal deformities associated with, 1303
Farber disease, 1553
  lipid accumulation in, 1521
Fibrocellulitis, 1326
Fibrogenesis imperfecta, 867
  foot, 871
  hand, 870
  knee, 872
Fibroma
  desmoplastic, 116
    humerus, 120
    ulna, 121
  juvenile aponeurotic, 121
    metacarpal, 123
    radius and ulna, 122
    wrist, 124
  nonossifying, 113
  nonosteogenic, 113
    femur, 115
    fibula, 114
Fibromatosis, 121
  age of onset, 125
  aggressive infantile, 122

  cite of lesion, 125
  classification, 121
  congenital generalized, 122
    femur, 126
    humerus, 126
    metacarpal, 126
  diffuse infantile, 122
  histology, 125
Fibrosarcoma, 257, 265–267
  femur, 260, 262, 263
  fibula, 259
  iliac bone, 258
  metacarpal, 262
  skull, 261
  wrist, 264
Fibrositis ossificans progressiva, 1326
Fibrous cortical defect, 111
  femur, 113
Fibrous dysplasia, *see under* Dysplasia
Fistula
  arteriovenous, 149
    tibia and fibula, 155
Fluorine poisoning, 1144
  pelvis, 1145
  spine, 1145
Focal sclerosis, 62
Fong disease, 1421–1423
Forestier disease, 552
Fragilitas ossium, 1289
Freiberg infarction, 708
Frostbite injury
  foot, 723
  hand, 722

Ganglion, intraosseous, *see* Intraosseous ganglion
Gangliosidosis
  generalized, 1552
  G_{M2}, 1552
Gargoylism, 1465–1470
Garner syndrome, 1433–1435
Garré, sclerosing osteomyelits of, 753, 754
Gaucher disease, 1539
  Erlenmyer flask deformity, 1543, 1544, 1546
  femur, 1544, 1546, 1547
  humerus, 1547
  lipid accumulation in, 1521
  marrow foam cells, 1543
  pelvis, 1545, 1547
  periosteal cloaking due to, 21
  tibia, 1541, 1548
  vertebra, 1542
Giant cell tumor, 263, 272, 274–278
  benign, 268
  acetabulum, 270
  femur, 269
  hand, 271
  spine, 270
  foot, 273
  malignant, 279
  tibia, 277
Goltz syndrome, 1338–1340
Gonadal hormone, metabolic bone disease, 830
Gout, 557
  carpal bone erosion, 562

SUBJECT INDEX

elbow tophus, 561
femoral head necrosis, 571
hand, 563, 564, 566
metatarsal erosion, 560, 561
odontoid fracture, 570
sacroiliac, 567
spine, 568, 569
tophus destruction of phalanx, 565
Granuloma, reticulohistiocytic, 659
Granulomatous disease of childhood, chronic, 754–756
Graves disease, 1169
Greulich-Pyle *Atlas*, determination of skeletal maturation, 1148, 1152–1155
Growth hormone, metabolic bone disease, 830

Halisteresis, 6
Hamartoma, 101
  mesenchymatous, 104
  rib, 105
Hand-Schüller-Christian disease, 1521
Haversian system, 6
Heavy chain disease, 311
Hemangioblastoma, 383
Hemangioendothelial sarcoma, 383
Hemangioendothelioma, 273, 383
Hemangioma, 127
  epiphysis, 130
  foot, 139
  malignant, 383
  pelvic, 131, 156, 157
  soft tissue, 138
  sunburst periosteal reaction, 129
  vertebra, 127, 128
Hemangiomatosis, 138
Hemangiopericytoma, 383, 387
  femur, 391
  sacrum, 390
Hemangiosarcoma, 383
Hematopoiesis
  extramedullary, 1080
  sickle cell anemia and, 1081
Hemochromomatosis, 582
  hand, 583
  hip, 583
Hemophilia, 587
  bilateral coxa valga, 601
  elbow, 591
  femoral neck, 591, 594
  femur, 603
  foot, 589, 597, 600, 604
  hand, 605–608
  hip, 593, 595
  knee, 588, 590, 596–599, 602, 604
  shoulder, 592
  thumb hemorrhage, 602
Hepatolenticular degeneration, 501
Hereditary multiple exostosis, 1402
  ankle, 1405
  hand, 1406
  humerus, 1402
  knee, 1403, 1404, 1408
  pelvis, 1407, 1409
  wrist, 1404
Hip
  congenital dislocation, 1084, 1086, 1087

normal adduction, 1085
Histiocytes, giant cell, 659
Histiocytoma
  malignant fibrous, 373
    clavicle, 376
    femur, 376
Histiocytomatosis, giant cell, 659
Histiocytosis X, 1521
  acetabulum, 1523
  femur, 1525
  hip, 1524
  humerus, 1526, 1531
  leg, 1522, 1526, 1527
  lipid accumulation in, 1521
  mandible, 1539
  pelvis, 1529, 1530, 1537
  scapula, 1528, 1531
  skull, 1533–1536, 1538
  tibia, 1530
  vertebra, 1539
Hodgkin disease, 363, 365, 367
Holt-Oram syndrome, 1233–1235
Homocystinuria, 1331
  compared with Cockayne syndrome, progeria and bird-headed dwarfs, 1438
  cystathionine synthase deficiency and, 1332–1334
Hunter syndrome, 1471–1473
  compared with other mucopolysaccharidoses, 1466
  elbow, 1472
  hand, 1473
  hip, 1472
  shoulder, 1472
  vertebra, 1473
Hurler syndrome, 1465
  compared with other mucopolysaccharidoses, 1466
  hand, 1466, 1467
  rib, 1468
  skull, 1469
  vertebra, 1470
Hypercalcemia, idiopathic, 918, 921, 922
Hypergonadism, skeletal maturation in, 1176
Hyperostosis
  ankylosing, 552
    hip, 558
    spine, 553–556, 558, 559
  corticalis deformans juvenilis, 951
  diffuse idiopathic skeletal, 552
  flowing, 1397
  generalized cortical, 1335–1338
  infantile cortical, 817–821
    chronic, 822
Hyperparathyroidism, 879
  acetabulum, 900
  adenoma as cause, 892, 911
  brown tumor, 886, 896, 898–911
  clavicle, 910
  contrasted with pseudohypoparathyroidism, 880
  femur, 899, 905, 910
  hand, 882, 883, 885, 887, 907, 910
  humerus, 898
  knee, 911
  newborn, 890, 891

adenoma as cause, 892
pelvis, 899, 903, 908
phalangeal cortical striation, 884
rib, 898, 902
skull, 881, 893–895, 906
spine, 897
sternum, 906
teeth, 896
tibia, 888, 889, 901, 908
vertebra, 901, 909
Hyperphosphatasemia, chronic familial, 951
Hyperphosphatasia
  arm, 954
  femur, 957, 961
  foot, 958
  hand, 955, 962
  hereditary, 951
  leg, 958
  pelvis, 956, 963, 964
  rib, 960
  skull, 952, 953, 959
  spine, 963
Hyperpituitarism, skeletal maturation and, 1168
Hyperthyroidism
  renal osteodystrophy and, 939, 940, 942, 946
  skeletal maturation in, 1166
Hypertrophicans, 1310
Hypervitaminosis A, 917
Hypervitaminosis D, 918
  gout and, 919, 920
  leg, 920
  rheumatoid arthritis and, 919
  skull, 922
  spine, 921, 922
  tophaceous gout and, 919
Hypochondroplasia, 1199, 1219–1222
Hypoparathyroidism, 912
  knee, 916
  sacroiliac, 915
  skull, 913, 914
  spine, 915
Hypophosphatasia, 872
  arm, 876, 908
  clinical groups, 872
  femur, 878, 879
  group I, 872, 873
  group II, 872, 873
  group III, 872, 873
  group IV, 873
  humerus, 877
  legs, 874
  skull, 875
  tibia, 878, 879
Hypopituitarism, skeletal maturation and, 1168–1172
Hypoplasia, focal dermal, 1338–1340
Hypothyroidism, skeletal maturation in, 1163–1166

I-cell disease, 1552
  hand, 1550
  pelvis, 1551
  rib, 1551
Idiopathic chondrolysis of hip, 499
Idiopathic juvenile osteoporosis, 845

Iliac horns, 1421–1423
Infantile cortical hyperostosis, 817–821
　chronic, 822
Infarcts, 715, 716, 719, 720
　Caisson disease and, 718, 720, 721
　pancreatitis and, 717
Intraosseous ganglion, 111
　scapula, 112
Ischemia, 679
　diaphysometaphyseal, 683–685
　epiphysometaphyseal, 684, 686–693
　necrotic
　　common sites of, 696
　　etiology, 695
　　femoral head after fracture, 714
　　proximal fragment of carpal scaphoid, 714
　　thrombosis, 695
　　trauma, 695
　　vascular wall disease, 695
　　vertebra, 703

Jaccoud arthritis, 500
Jansen metaphyseal dysostosis, 1235, 1237–1239
Jaundice, osteoporosis, 844
Jeune disease, 1210
Juvenile polyarthritis, 476
Juxtacortical osteosarcoma, 209

Karyotype
　D1 trisomy, 1490
　female, 1488
　Klinefelter syndrome, 1493
　male, 1489
　mongolism, 1506
　trisomy 17-18, 1511
　Turner syndrome, 1492
Kayser-Fleisher ring, 501
Keratoconjunctivitis sicca, 504
Kienböck disease, 709
Kirner deformity, 1190, 1191
Kissing sequestra, tuberculous arthritis, 422, 425
Klinefelter syndrome, 1503
　hand, 1504
　karyotype, 1493
Kniest syndrome, 1201, 1204, 1205
Köhler disease, 704
Kompakten Knochenkerne, 62
Kyphosis dorsalis juvenilis, 700

Larsen syndrome, 1436, 1437
Laurence-Moon-Biedl-Bardet syndrome, 1266
　associated anomalies with, 1266
Lead poisoning, 1120
　femur, 1121, 1122
　fibula, 1121
　hand, 1121
　spine, 1123
　tibia, 1121
Leprosy, 804
　acute phase, 805
　chronic burns, 807
　healing phase, 805
　leg, 806
　phalanx, 804, 805

Leptocytosis, 1064
Letter-Siwe disease, 1521
Leukanemia, 1073
Leukemia, 330
　acute, 336
　bone lesion incidence, 335
　femur, 337, 338, 343
　knees, 341
　metacarpal, 339
　newborn, 334
　pubic, 340
　radius, 339, 342
　spine, 341
　tibia, 343
　ulna, 339
　osteoporosis, 335
Lipidosis, 1520
　familial neurovisceral, 1552
　Farber disease, 1553, 1554
　Gaucher disease, 1539–1548
　histiocytosis X, 1521–1539
　I-cell disease, 1550–1552
　Niemann-Pick disease, 1548, 1549
　systemic late infantile, 1552
　Wolman disease, 1554, 1555
Lipochondrodystrophy, 1465
Lipogranulomatosis, disseminated, 1552–1554
Lipoid dermatoarthritis, 659
Lipoma, 394
　calcaneus, 105
　ossifying, 392
Liposarcoma, 387
　soft tissue, with no bone involvement, 393
Lupus erythematosus, finger deviation and flexion contractures, 465
Lymphangioma, tibia, 143
Lymphangiomatosis, 138
　foot, 148
Lymphangiosarcoma, 383
Lymphoma, vertebra, 365, 366
Lymphopenic agammaglobulinemia, 1239
Lymphosarcoma
　mandible, 369
　skull, 364
　tibia, 368

Macrodystrophia lipomatosa, 1380
Madelung deformity, 1243, 1244
Maffucci syndrome, 1262–1265
Marble bones, 1346
Marfan syndrome, 1330
Maroteaux-Lamy syndrome, 1481
　compared with other mucopolysaccharidoses, 1466
　femur, 1484
　hand, 1482
　hip, 1482, 1484
　rib, 1482
　skull, 1482–1484
　vertebra, 1483
Mastocytosis, 373
　classification of, 376
　femur, 379, 380
　iliac bone, 378
　skull, 377

urticaria pigmentosa, 373
vertebra, 379
Medial distal metaphyseal femoral irregularity, 116
Mediterranean anemia, 1064
Megakaryocytic myelosis, 1073
Melnick-Needles dysplasia, 1307–1309
Melorheostosis, 1397
　foot, 1400
　hand, 1397, 1399
　leg, 1399, 1400
　rib, 1401
　sclerotome sites, 1398
Mesenchymoma
　benign, 101
　scapula, 103
　malignant, 396
Mesomelic dwarfism, 1223–1225
Metabolic bone disease, 829
Metal poisoning, osseous manifestations of, 1120
Metatrophic dwarfism, 1201, 1202
Micrognathia, juvenile rheumatoid arthritis, 495
Micromelia, 1191
Mollities ossium, 1289
Mongolism, 1505
　hand, 1508
　karyotype, 1506
　manubrium, 1509
　pelvis, 1507
　rib, 1508
　spine, 1509, 1510
Monostotic fibrous dysplasia, 994
Morquio disease, differentiation from spondyloepiphyseal dysplasia congenita, 1268
Morquio syndrome, 1474
　compared with other mucopolysaccharidoses, 1466
　femur, 1477, 1478
　hand, 1480
　humerus, 1479
　leg, 1480
　rib, 1479
　vertebra, 1475, 1476
Mucopolysaccharidosis, 1465
　elbow, 1472
　femur, 1477, 1478, 1484
　genetic types, comparison of, 1466
　hand, 1466, 1467, 1473, 1480, 1482
　hip, 1468, 1472, 1482
　humerus, 1479
　Hunter syndrome, 1471–1473
　Hurler syndrome, 1465–1470
　leg, 1480
　Maroteaux-Lamy syndrome, 1481–1484
　Morquio syndrome, 1474–1480
　rib, 1468, 1479, 1482
　Sanfilippo syndrome, 1474
　Scheie syndrome, 1481
　shoulder, 1472
　skull, 1469, 1482–1484
　vertebra, 1470, 1473, 1475, 1476, 1483
Mycobacterium leprae, osteomyelitis, 804
Myeloid metaplasia, 1073
Myeloid sclerosis, 1073

SUBJECT INDEX

Myeloma
  extramedullary, 330
  multiple, 312, 330
    abnormality of serum proteins in, 313
    clavicle, 316
    femur, 317, 329
    hand, 325
    humerus, 316, 319
    iliac bone, 324
    pathologic fracture in, 329
    pelvic, 318, 323
    radius, 328
    sclerosis and, 326
    skull, 314, 315, 318, 320–322
    vertebra, 327, 328
  solitary
    acetabulum, 332
    iliac bone, 332
    vertebra, 331
Myelomatosis, 327
Myelosclerosis, 1073
Myositis ossificans, 1108
  elbow, 1115
  femur, 1109
  forearm, 1114
  hand, 1115
  humerus, 1110
  localized, secondary to trauma, 1108
  neurologic disorders and, 1111
  paraplegia and, 1111
  post-traumatic, 1109
  radius, 1112
  rib, 1110
Myositis ossificans progressiva, 1326–1330

Necrosis
  aseptic, femoral head, 699
  radium, 1125, 1129–1139
Neuroblastoma, metastatic, 362
  skull, 363
Neurofibromatosis, 1357
  arachnoid cysts and, 1362
  auditory canal, 1363
  elbow, 1372
  facial, 1366, 1367
  femur, 1369, 1376
  foot, 1373, 1374
  hand, 1374
  kyphosis and, 1368
  leg, 1370, 1371, 1377
  neurofibrosarcoma and, 1379
  osteomalacia and, 1378
  rib, 1375
  scoliosis in, 1358
    bone grafts for stability, 1359
  sphenoid bone absent, 1362, 1364
  spine, 1360, 1361
  tibia, 1373, 1376
Neurotrophic arthropathy, see Arthropathy
Nevoid basal cell carcinoma syndrome, 1313
  dentigerous cysts, 1316
  hand, 1314, 1321
  mandible, 1313
  radius, 1314
  rib, 1317
  sacrotuberous ligament, calcification, 1321
  scapula, 1315
  sella turcica, 1320
  skull, 1319
  soft tissue calcification, 1320
  spine, 1318
Niemann-Pick disease, 1548
  lipid accumulation in, 1521
  spine, 1549
  tibia, 1549
Nievergelt syndrome, 1224
Noonan syndrome, 1503
  compared with Turner syndrome, 1503

Ochronosis, 584
  intervertebral disks, 585
  sacroiliac, 587
  spine, 586
Ollier disease, 1244
Omoblastoma, 273
Onycho-osteodysplasia, familial, 1421–1423
Osgood-Schlatter disease, 703
Ossificans progessiva, 1326
Ossification
  post-traumatic, 1108
  soft tissue, 1100
Osteitis
  condensing, clavicle, 540, 544
  radium, 1125
Osteitis condensans ilii, 539, 544
Osteitis deformans, 964, 968
Osteitis fibrosa, renal osteodystrophy, 924
Osteitis fibrosa disseminata, 994
Osteoarthritis, 608
  erosive
    hand, 615, 616
    knee, 616
  femoral neck, 612
  hand, 613, 614, 617, 618
  hip, 618, 619, 621
    classification, 620
    incidence of types, 620
  knee, 611
  shoulder, 610
Osteoarthropathy
  chronic idiopathic hypertrophic, 650
    foot, 650
    hand, 650
  dense undulating periosteal reaction in, 16
  hypertrophic, 642
  pulmonary, 642
    ankle, 645, 647
    femur, 644
    foot, 648
    forearm, 643, 647
    hand, 643, 649
    osteosarcoma and, 646
  solid thin periosteal reaction in, 17
Osteoblastoma, 54
  femur, 61
  lytic, 58
  patella, 60
  spinous process, 58
  talus, 60
  transverse process, 57
  vertebra, 59
Osteochondritis deformans juvenilis, 696
Osteochondritis dissecans, 686, 709–712
  bilateral, 711
  simulation in normal variation, 712
Osteochondrodystrophy, 1465
Osteochondroma
  femur, 75, 76, 80
  intraarticular, pelvis, 82, 83
  malignancy subsequent to, 84
  multiple, 1402
  radius, 76
  scapula, 72
  solitary, 70
    femur, 78, 79
    humerus, 73, 74, 77
    pelvis, 71, 81
    tibia, 77
Osteochondromatosis, 622
  hereditary, 1402
Osteochondrosis, 700
  calcaneus, 708
    bilateral, 708
  common sites of, 696
  femoral head, 696, 697
  femur, 698
  ischemic necrosis and, 694
  lunate bone, 709, 710
  medial tibial condyle, 705
  metatarsal, 708, 709
  spinal, 702
  tarsal scaphoid, 704, 705
  tibial tuberosity, 703
  vertebral, 702
Osteodermatopathia, 1310
Osteodysplasia, familial, 1442
Osteodystrophy, renal, 923
  antivitamin D factor, 923
  arm, 926
  arterial calcification, 945
  brown tumor, 941
  chronic acidosis, 923, 925
  clavicle, 933
  Fanconi syndrome, 924, 936
  femur, 927, 928, 930, 944
  hand, 936, 939, 945–949
  humerus, 929, 930, 937, 940, 946
  hyperparathyroidism and, 939, 940, 942, 946
  hypoplasia of renal artery, 931
  knee, 934, 939
  lacy subperiosteal bone resorption, 932
  osteitis fibrosa, 924
  osteomalacia and, 923
  osteosclerosis, 935, 950
  pelvis, 927, 928, 930
  rachitic rosary deossification, 930
  rickets and, 924
  rugger jersey appearance of spine, 950
  sacroiliac, 928, 929
  skull, 941, 942, 951
  soft tissue calcification, 946–949
  spine, 950
  teeth, 943
  tibia, 932
  tubular acidosis, 937
  tubular syndrome, 935
  vitamin D insensitivity, 923
  vitamin D-resistant rickets, 935

SUBJECT INDEX

Osteoectasia, familial, 951
Osteogenesis imperfecta, 1286, 1289, 1290
  congenita, 1289
  cystic type, 1294, 1295
  femur, 1298
  fractures in, 1298, 1299
  in utero, 1291
  slender bone type, 1293, 1295
  tarda, 1289
  thick bone type, 1292, 1295
  vertebra, 1297
  wormian bones of skull, 1296
Osteolysis
  essential, 1381, 1382, 1384, 1385
  massive, 149
    pelvic, 156, 157
    rib, 158
Osteoma, 33
  osteoid, 35
    acetabulum, 38
    cancellous, 49
    carponavicular, 45
    cortical, 35, 40, 41, 47
    dense elliptical periosteal reaction in, 18
    femur, 37, 39, 42, 46, 52, 54
    finger, 44
    humerus, 49
    metacarpal, 51
    patella, 35
    phalanx, 51
    radius, 48
    rib, 36
    scapula, 37
    solid periosteal reaction in, 19
    subperiosteal, 49, 55
    talus, 53
    tibia, 43, 50
    ulna, 49
    vertebra, 56, 57
  parietal bone, 34
Osteomalacia, 852
  adult, 853
  classification, 853
  femur, 855, 857-859
  fibula, 855
  metatarsal pseudofractures, 859
  pseudofractures
    femur, 859
    metatarsal, 859
  pubic ramus, 857
  renal osteodystrophy, 923
  rib, 857
  scapula, 857
  sprue, 854, 856, 859
  vertebra, 854
  osteomyelitis, 793
Osteomyelitis, 727
  acute cellulitis and, 733
  antibiotic therapy, 757
  brucellar, 752, 753
  lamellar nodular periosteal reaction in, 25
  mycobacterial, atypical, 771, 773
  pyogenic, 727
    ankle, 730, 731
    calcaneus, 737
    clavicle, 740

eburnation, 743
femur, 742, 745-747, 749, 751
fibula, 741, 744
hip, 732, 737
knee, 730, 734-736, 738
localized bone abscess, 729
necrosis, 728
new bone formation, 728
radius, 741, 742, 748
serpiginous tract, 750-752
shoulder, 739
tibia, 750, 751
sclerosing, of Garré, 753, 754
solid periosteal reaction in, 18
sporotrichous, 756
tuberculous, lamellar nodular periosteal reaction in, 24
Osteonecrosis
  knee, 711, 713
  primary
    with osteoarthritis, 713
    with rheumatoid arthritis, 713
Osteopathia condensans dessiminata, 1392
Osteopathia hyperostotica scleroticans multiplex infantilis, 1410
Osteopathia striata, 1396
Osteopetrosis, 1346-1350
  compared with pycnodysostosis and cleidocranial dysostosis, 1357
  generalisata, 1346
  Léri type, 1397
Osteopoikilosis, 1392
  femur, 1396
  foot, 1392
  hand, 1395
  humerus, 1395
  pelvis, 1393, 1394
  tibia, 1396
  vertebra, 1396
Osteoporosis, 830
  acute, 837, 843
  disuse, 831, 836
  femoral head, 835
  foot, 844
  heparin, 832
  hypogonadism, 837
  idiopathic juvenile, 845
  jaundice, 844
  pelvis, 838
  physiologic classification, 830
  postmenopausal, 831
  protein deficiency, 832
  regional migratory, 834, 843
  senile, 831
  transitory, 843
    femoral head, 843
  vertebra, 838, 840, 841
Osteoporosis circumscripta, Paget disease, 966, 967
Osteopsathyrosis idiopathica, 1289
Osteosarcoma, 181
  age incidence of, 181
  central, 181
  classification, 182
  clavicle, 195
  Codman triangle secondary to, 22
  extraosseous, 208

femur, 184, 187, 190, 192-194, 199-202, 204
fibula, 188, 205
humerus, 183, 187, 188, 190, 204
iliac bone, 197
increased density in, 185
juxtacortical, 209
kidney primary, 209
lamellated periosteal reaction secondary to, 23
maxillary sinus, 189
metastasis to lungs and lymph nodes, 206, 207
mortality, 182
  site of lesion and, 182
new bone formation, 12
parosteal, 209
  femur, 212, 216, 224
  pelvic, 224
  radius, 225
pelvic, 192
periosteal, 206
  operative specimen, 203
perpendicular periosteal reaction in, 26, 27
radium-induced, 1129, 1140, 1142
radius, 199
rib, 32, 196
sacrum, 197
scapula, 184
spine, 191
subtrochanter, 200
sunburst periosteal reaction in, 26, 27
telangiectatic type, 184, 199-201
tibia, 186, 198, 203
treated as giant cell tumor, 273
Osteosarcomatosis, 208
  eburnated increase in density, 210
  hand, 210
  pelvic, 211
Osteosclerosis, 1346
  generalisata, 1346
  renal osteodystrophy, 935, 950
Osteosclerotic anemia, 1073
Osteosi eburnizzani monomelica, 1397
Ovarian agenesis, skeletal maturation and, 1174, 1176
Oxalosis, 1033
  elbow, 1034
  femur, 1036
  foot, 1036
  forearm, 1034
  hand, 1034
  humerus, 1034
  leg, 1036
  pelvis, 1035
  spine, 1035

Pachydermoperiostosis, 1310, 1311
Paget disease, 964
  ankle, 971, 972
  calcaneus, 992
  clavicle, 973, 982
  femur, 977, 983, 993
  juvenile, 951
  knee, 969-972, 983
  leg, 975, 976
  long bone effects, 968

SUBJECT INDEX

pelvis, 977, 978, 980–982, 987, 992
pubis, 974, 979, 991
rib, 982, 993
sarcomatous change in, 993
skull, 966–968, 982
spine, 983, 986
vertebra, 977, 978, 984, 985, 988–990
Pain, congenital insensitivity to, 1423
  calcaneus, 1429
  femur, 1424, 1428
  hip, 1426, 1427
  knee, 1429
  thumb, 1427
  tibia, 1425
Parathormone, metabolic bone disease, 829
Parathyroid hormone, metabolic bone disease, 829
Parosteal osteosarcoma, 209
Pellegrini-Stieda disease, 1108
Periosteal reaction
  amorphous, 23, 28
  cause, 13
  dense
    elliptical, 14, 18
    undulating, 14
  in osteosarcoma, 26, 27
  interrupted, 14
    cause, 13
    type, 13
  lamellar nodular, 24, 25
  lamellated, 22, 23
  perpendicular, 23, 26, 27
  solid, 18–20
    thin type, 17
    undulating, 15, 16
  sunburst, 23, 26, 27
  thin undulating, 14
  types, 13
Peripheral dysostosis, 1283–1286
Perithelioma, 383
Perthes disease, bilateral, 699
Pfaundler-Hurler disease, 1465
Pfeiffer syndrome, 1342–1346
Phantom bone, 149
Phosphorus
  metabolic bone disease, 829
  poisoning, 1124
    knee, 1124
    pelvis, 1124
Plasma cell dyscrasias, 305, 308
Plasmacytoma
  rib, 331
  sacrum, 333
  solitary, 330
Plasmacytosis, 310
Polychondritis, relapsing, 538
  ear cauliflower configuration, 542
  foot, 543
  hand, 541
  sacroiliac, 543
  spine, 543
Polydysenteric arthritis, 530
Polyhistioma, 255
Polyostotic fibrous dysplasia, 994
Progeria, 1438–1442
  comparison with Cockayne syndrome, homocystinuria and bird-headed dwarfs, 1438
Progressive diaphyseal dysplasia, see Dysplasia
Progressive myositis ossificans, 1326–1330
Pseudoachondroplastic dwarfism, 1200
Pseudoarthrosis, tibia, 1371
Pseudogout, 571
  cervical spine arthrosis, 577
  compared with acute gouty arthritis, 581
  elbow, 580, 581
  hand, 580
Pseudo-Hurler syndrome, 1552
Pseudohypoparathyroidism, 880, 913, 916, 917
Pseudoleukemia, 1073
Pseudo-pseudohypoparathyroidism, 917
Pulmonary osteoarthropathy, see Osteoarthropathy
Pycnodysostosis, 1354–1356
  compared with osteopetrosis and cleidocranial dysostosis, 1357
Pyle disease, 1239, 1241, 1242
Pyrophosphate arthropathy, 571

Radial aplasia-thrombocytopenia, 1306
Radium
  necrosis, 1125
    clavicle, 1138
    humerus, 1138
    knee, 1143
    mandible, 1129
    pelvis, 1130–1137, 1142
    rib, 1138, 1139
    scapula, 1138
  poisoning, 1125
    femur, 1127, 1128
    fibula, 1127
    hip, 1144
    humerus, 1143
    mandible, 1128
    skull, 1126
    tibia, 1126
Radium-induced neoplasia, 1125, 1131
Reiter syndrome, 530
  calcaneus, 534
    spur formation, 535
  chronic stages, 535
  foot, 531, 533, 535
  hip, 537
  knee, 532
  periarticular deossification, 535
  peripheral joint effects, 535
  sacroiliac effects, 536, 538
  symphysis pubis, 536
Relapsing polychondritis, see Polychondritis
Renal osteodystrophy, see Osteodystrophy
Reticuloendotheliosis, 1521
Reticulohistiocytoma, 659
Reticulohistiocytosis
  giant cell, 659
  medullary, 661–664
  multicentric, 659
Reticulosis, 1521
Reticulum cell sarcoma, see under Sarcoma
Rheumatoid arthritis, see under Arthritis
Rickets, 858
  chemical pathology, 860
  concavity of metaphyses, 860, 862, 863
  congenital biliary obstruction as cause, 866
  development of bone in, 858
  femur, 868
  hand, 861, 868
  healing, 860, 865
  pelvis, 868
  radius, 861, 868
  renal osteodystrophy, 924
  scapula, 868
  temporary zone of calcification, 860, 862–864
  ulna, 861, 868
Rothmund syndrome, 1325
Rubella, osteomyelitis, 800
Rubinstein-Taybi syndrome, 1312

Sanfilippo syndrome, 1474
  compared with other mucopolysaccharidoses, 1466
Sarcoidosis, 808
  femur, 811
  foot, 815
  hand, 809, 810, 812–814
  humerus, 811
  rib, 816
  toe, 812, 813
  vertebra, 816
Sarcoma
  alveolar soft-part, 378
  endothelia, 273
  granulation tissue, 383
  hemangioendothelial, 383, 385
    radius, 386
    tibia, 387
    vertebra, 388
  parosteal, 209, 220, 222, 223, 226
    femur, 213, 219
    fibula, 215
    mastoid, 217
    maxilla, 218
    tibia, 214, 227
  periosteal osteogenic, 206
  primitive multipotential primary, 255
  radiation-induced, 1141
  reticulum cell, 297, 305, 306
    clavicle, 310
    femur, 304, 308, 312
    humerus, 304
    iliac bone, 311
    pelvic, 303
    scapula, 307, 309
Scheie syndrome, 1481
  compared with other mucopolysaccharidoses, 1466
Scheuermann disease, 700
Schmid metaphyseal dysostosis, 1236, 1240
Scleroderma, 1102
  rheumatoid arthritis and, 441
  tibia, 1103

SUBJECT INDEX

Sclerosis, tuberous, 1386
  foot, 1388
  hand, 1390
  leg, 1389
  skull, 1387
  spine, 1391
Sclerotic bone island, 62
Sclerotome, in melorheostosis, 1398
Scurvy
  atrophic line, 849
  corner sign, 846, 848
  cortical thinning, 846
  dense metaphyseal line, 846, 847
  ground glass osteoporosis, 846
  halo ossification center, 846
  healing phase, 850–852
  infantile, 845
  lateral spurs, 846
  metaphyseal fractures, 849, 857
  soft tissue edema, 849
  spinal changes, 849
  subperiosteal hematomas, 846
  white line of, 848, 849, 851
Sjögren syndrome, 504
  costophrenic sulcus, 506
  hand, 505
  sialectasis in parotid gland, 504
Skeletal maturation, 1148
  acromegaly effect on, 1168, 1173–1175
  centers of ossification, time of appearance of, 1150, 1151
  differences, hands, 1161
  endocrine disturbances and, 1163
  eunuchoidism effect on, 1174
  Greulich-Pyle *Atlas*
    boys, 1152, 1153
    girls, 1154, 1155
  hypergonadism and, 1176
  hyperpituitarism effect on, 1168
  hyperthyroidism effect on, 1166
  hypopituitarism effect on, 1168–1172
  hypothyroidism effect on, 1163
  mental retardation in, 1162
  methods for determining, 1148, 1160
    Camp and Cilley, 1148
    Garn, 1157–1159
    Greulich-Pyle, 1148
    Hodges, 1148
    sampling method, 1159
    Tanner-Whitehouse, 1151
  ovarian agenesis effect on, 1174, 1176
  pattern of, 1162
  precocious puberty and, 1176
  rate of, 1162
  thyroid acropachy effect on, 1166
  thyrotoxicosis effect on, 1167
  variation, degree, maximal, 1161
Slipped femoral epiphysis, 700, 701
Smallpox, osteomyelitis, 801–803
Soft tissue calcification, 1100
Spahr metaphyseal dysostosis, 1236
Spherocytosis, hereditary, 1071
Spina ventosa, 768
Spondylitis
  ankylosing, 506
    bamboo spine, 520
    dislocation of cervical vertebra, 525
    fixed amphiarthrodial joints, 507
    fracture of spine, 527–529
    freely movable diarthroidal joints, 507
    juvenile rheumatoid arthritis, 476
    osteitis, 519
    plantar fascia calcification, 524
    pseudoarthrosis and, 526
    sacroiliacs, 507–513
    soft tissue calcification, 519–521
    spine, 510, 514–519, 521
    symphysis pubis, 510, 511, 522, 523
    vertebral body squaring, 508
    whiskering type periostitis, 517
  tuberculous, healed, 431
Spondyloepiphyseal dysplasia, pseudoachondroplastic, 1199
Spondylolisthesis, 1089–1093
  Meyerding method for measuring degree of, 1092
Sporotrichosis osteomyelitis, 756
Staphylococcus aureus
  cause of pyogenic arthritis, 428
  pyogenic osteomyelitis, 727
Steatorrhea, osteomalacia, 867
Still disease, 478, 480
Stippled epiphyses, 1230
Subperiosteal abrasion, 116
Subperiosteal cortical defect, 116
Sudeck atrophy, 833, 839, 844
Syndesmophyte
  ankylosing hyperostosis, 553–556, 560
  psoriatic arthritis, 551
Synovial chondromatosis, *see* Chondromatosis
Synovioma, benign, 659
Synovitis
  extra-articular localized nodular, 659
  intra-articular localized nodular, 659
  pigmented villonodular, 651
    acetabulum, 655
    chondrolysis and, 656
    distal phalanx, 657, 658
    femoral neck, 652, 655
    great trochanter, 657
    hip, 656
    knee, 653, 654
Syphilis, 774
  acquired, 785
    clavicle, 786, 788
    femur, 786, 787
    foot, 789
    gumma of soft tissue, 789
    hands, 791
    leg, 785
    radius, 788
    skull, 790
    ulna, 787
  congenital, 774
    femur, 781
    forearm, 775–777, 779, 780
    humerus, 782
    late, 784
    leg, 780, 782, 783
    tibia, 778, 779

Tanner-Whitehouse, determination of skeletal maturation, 1151, 1156
Tarsoepiphyseal aclasis, 1286
Taxopachyosteose diaphysaire tibio-peroniere, 1429
Tay-Sachs disease, 1552
  lipid accumulation in, 1521
Tendinitis, retropharyngeal, 1107
Thalassemia, 1064
  femur, 1064, 1065
  foot, 1066
  hand, 1069, 1070
  rib, 1066
  skull, 1071, 1072
  vertebra, 1067, 1068
Thanatophoric dwarfism, 1203, 1206–1209
Thiemann disease, 722
Thyrocalcitonin, metabolic bone disease, 830
Thyroid acropachy, skeletal maturation in, 1166
Thyroid hormone, metabolic bone disease, 830
Tibia, anterior bowing, 1429–1431
Touraine-Solente-Golé syndrome, 1310
Trisomy defects, 1505
Trisomy D syndrome, 1514
Trisomy E syndrome, 1505
Trisomy 13-15 syndrome, 1514
Trisomy 17-18 syndrome, 1505
  clavicle, 1513
  hand, 1514
  hip, 1514
  karyotype, 1511
  mandibular hypoplasia, 1512
  pulmonary stenosis, 1513
  sternum, 1512
Trisomy 21-22 syndrome, 1505
Tuberculosis, 757
  abscess and, 767
  acetabulum, 759
  acromion, 770
  calcaneus, 761
  dactylitis, 768, 769
  disk space, 530
  femur, 758, 765
  foot, 762, 766
  great trochanter, 772
  humerus, 760
  metatarsal, 765
  sacroiliac, 772
  spine, 770
  sternum, 764
  tibia, 760
  vertebra, 763, 771
Tuberculous spondylitis, healed, 431
Tuberous sclerosis, *see* Sclerosis
Tumor
  benign, histiocytic origin, 375
  malignant vascular, 383
  metastatic malignancy, 340, 348
    bone scanning in, 340
    clavicle, 349, 352
    femur, 352, 359, 372
    fibula, 374
    fingers, 344, 346
    foot, 345, 347
    histiocytic origin, 375
    humerus, 356, 360
    ischial ramus, 370
    knee, 358, 373
    neuroblastoma, 362
    pelvic, 353, 355–357

radius, 350, 371
sacrum, 351
spine, 351, 361
vertebra, 354
vascular, 385
Turner syndrome, 1494
compared with Noonan syndrome, 1503
foot, 1500
hand, 1497–1499
karyotype, 1492
knee, 1494
pelvis, 1495
spine, 1496
Typus degenerativus amstelodamensis, 1322

Urethritis, nongonococcal, 530
Uroarthritis, infectious, 530
Urticaria pigmentosa, 373

Van Buchem disease, 1335–1338
Varicosity, long standing, cause of solid undulating periosteal reaction, 15
Venereal arthritis, 530
Vertebra, developmental notching of, 702
Vertebra plana, 702
Vertebral epiphysitis, 700
Vitamin D
deficiency, 858
metabolic bone disease, 829
Von Recklinghausen disease, 1357

Waldenström macroglobulinemia, 312
Weismann-Netter syndrome, 1429
Wilson disease, 501
hand, 502
humerus, 501
knee, 502
spine, 503
Wolman disease, 1554, 1555

Xanthofibroma, 659
Xanthogranuloma, 113
Xanthoma, 113
Xanthomatosis, 1521
nondiabetic cutaneous, 659
normocholesterolemic, 659
Xerostomia, 504